Regularity and Complexity in Dynamical Systems

Albert C. J. Luo

Regularity and Complexity in Dynamical Systems

Springer

Albert C. J. Luo
School of Engineering
Mechanical and Industrial Engineering
Southern Illinois University
Edwardsville, IL 62026-1805
USA
e-mail: aluo@siue.edu

ISBN 978-1-4614-6168-5 (Softcover)
ISBN 978-1-4614-1523-7 (Hardcover) e-ISBN 978-1-4614-1524-4
DOI 10.1007/978-1-4614-1524-4
Springer New York Dordrecht Heidelberg London

Library of Congress Control Number: 2011942215

Cover design: eStudio Calamar S.L.

Printed on acid-free paper

(Corrected at 2nd printing 2012)

Springer is part of Springer Science+Business Media (www.springer.com)

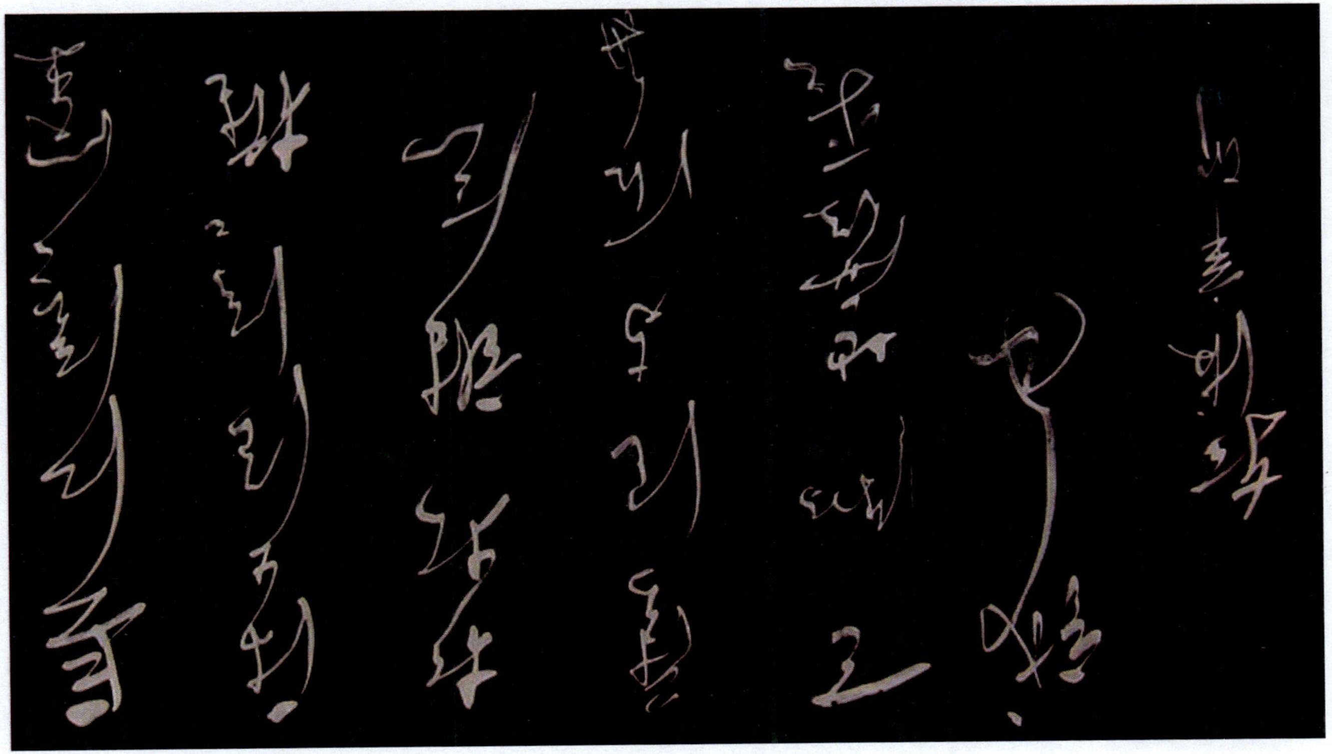

Preface

This book provides a different view to look at complex dynamics of dynamical systems from the author's research and teaching experience. The author hopes this book can provide a better understanding of complexity and chaos caused by nonlinearity, discontinuity, switching and impulses. The materials in this book are scattered into six chapters plus two appendices.

The stability, stability switching and bifurcation of equilibriums and fixed points for nonlinear continuous and discrete dynamical systems are systemically presented in Chaps. 1 and 2, which is different from the traditional presentation. In Chap. 3, the fractal theory based on single and multiple generators with single and multiple measures are discussed, and the fractality of chaos in nonlinear discrete dynamical systems are presented from self-similar geometric structures. In Chap. 4, the Ying–Yang theory of nonlinear discrete dynamical systems is presented for the complete dynamics of discrete dynamical systems. In addition, the companion and synchronization of discrete dynamical systems are discussed to describe dynamical relations between different discrete dynamical systems. In Chap. 5, nonlinear dynamics of switching systems with impulses will be discussed. In Chap. 6, mapping dynamics is presented as an extension of symbolic dynamics to describe periodic flows and chaos in discontinuous dynamical systems. The grazing phenomenon is a key to investigate the strange attractor fragmentation in discontinuous dynamics, which can be easily extended to chaos caused by global transversality in nonlinear continuous dynamical systems. To help one easily read the main body, linear continuous and discrete dynamical systems are discussed in Appendices A and B. The author believes that the presentation arrangement may not always be reasonable. The author sincerely hopes readers point out and criticizes for improvement.

Finally, I would like to appreciate my students (Jianzhe Huang, Yu Guo, Yang Wang, Bing Xue) for applying the theories to practical systems and completing numerical computations. Herein, I would like to thank my wife (Sherry X. Huang) and my children (Yanyi Luo, Robin Ruo-Bing Luo and Robert Zong-Yuan Luo) again for tolerance, patience, understanding and support.

Edwardsville, Illinois Albert C. J. Luo

Contents

Chapter 1
Nonlinear Continuous Dynamical Systems

In this Chapter, basic concepts of nonlinear dynamical systems will be presented as a review material. Local theory, global theory and bifurcation theory of nonlinear dynamical systems will be discussed. The stability switching and bifurcation on specific eigenvectors of the linearized system at equilibrium will be presented. The higher singularity and stability for nonlinear systems on the specific eigenvectors will be developed. In addition, a periodically excited Duffing oscillator with cubic damping and constant force will be discussed as an application. The stability of approximate periodic solutions of such a Duffing oscillator will be discussed.

1.1 Continuous Dynamical Systems

Definition 1.1 For $I \subseteq \mathscr{R}$, $\Omega \subseteq \mathscr{R}^n$ and $\Lambda \subseteq \mathscr{R}^m$, consider a vector function $\mathbf{f} : \Omega \times I \times \Lambda \to \mathscr{R}^n$ which is $C^r(r \geq 1)$-continuous, and there is an ordinary differential equation in a form of

$$\dot{\mathbf{x}} = \mathbf{f}(\mathbf{x}, t, \mathbf{p}) \text{ for } t \in I, \mathbf{x} \in \Omega \text{ and } \mathbf{p} \in \Lambda \tag{1.1}$$

where $\dot{\mathbf{x}} = d\mathbf{x}/dt$ is differentiation with respect to time t, which is simply called the velocity vector of the state variables $\mathbf{x}$. With an initial condition of $\mathbf{x}(t_0) = \mathbf{x}_0$, the solution of Eq. (1.1) is given by

$$\mathbf{x}(t) = \mathbf{\Phi}(\mathbf{x}_0, t - t_0, \mathbf{p}). \tag{1.2}$$

(i) The ordinary differential equation with the initial condition is called a *dynamical system*.
(ii) The vector function $\mathbf{f}(\mathbf{x}, t, \mathbf{p})$ is called a *vector field* on domain Ω.
(iii) The solution $\mathbf{\Phi}(\mathbf{x}_0, t - t_0, \mathbf{p})$ is called the *flow* of dynamical systems.
(iv) The projection of the solution $\mathbf{\Phi}(\mathbf{x}_0, t - t_0, \mathbf{p})$ on domain Ω is called the trajectory, phase curve or orbit of dynamical system, which is defined as

$$\Gamma = \{\mathbf{x}(t) \in \Omega | \mathbf{x}(t) = \mathbf{\Phi}(\mathbf{x}_0, t - t_0, \mathbf{p}) \text{ for } t \in I\} \subset \Omega. \tag{1.3}$$

A. C. J. Luo, *Regularity and Complexity in Dynamical Systems*,
DOI: 10.1007/978-1-4614-1524-4_1, © Springer Science+Business Media, LLC 2012

Definition 1.2 If the vector field of the dynamical system in Eq. (1.1) is independent of time, such a system is called an autonomous dynamical system. Thus, Eq. (1.1) becomes

$$\dot{\mathbf{x}} = \mathbf{f}(\mathbf{x}, \mathbf{p}) \text{ for } t \in I \subseteq \mathscr{R}, \mathbf{x} \in \Omega \subseteq \mathscr{R}^n \text{ and } \mathbf{p} \in \Lambda \subseteq \mathscr{R}^m. \tag{1.4}$$

Otherwise, such a system is called non-autonomous dynamical systems if the vector field of the dynamical system in Eq. (1.1) is dependent on time and state variables.

Definition 1.3 For a vector function $\mathbf{f} \in \mathscr{R}^n$, $\mathbf{f} : \mathscr{R}^n \to \mathscr{R}^n$. The operator norm of $\mathbf{f}$ is defined by

$$||\mathbf{f}|| = \sum_{k=1}^{n} \max_{||\mathbf{x}|| \leq 1, t \in I} |f_i(\mathbf{x}, t)|. \tag{1.5}$$

For an $n \times n$ matrix $\mathbf{f}(\mathbf{x}, \mathbf{p}) = \mathbf{A}\mathbf{x}$ with $\mathbf{A} = (a_{ij})_{n \times n}$, the corresponding norm is defined by

$$||\mathbf{A}|| = \sum_{i,j=1}^{n} |a_{ij}|. \tag{1.6}$$

Definition 1.4 For a vector function $\mathbf{x}(t) = (x_1, x_2, \cdots, x_n)^{\mathrm{T}} \in \mathscr{R}^n$, the derivative and integral of $\mathbf{x}(t)$ are defined by

$$\begin{aligned} \frac{d\mathbf{x}(t)}{dt} &= (\frac{dx_1(t)}{dt}, \frac{dx_2(t)}{dt}, \cdots, \frac{dx_n(t)}{dt})^{\mathrm{T}}, \\ \int \mathbf{x}(t)dt &= (\int x_1(t)dt, \int x_2(t)dt, \cdots, \int x_n(t)dt)^{\mathrm{T}}. \end{aligned} \tag{1.7}$$

For an $n \times n$ matrix $\mathbf{A} = (a_{ij})_{n \times n}$, the corresponding derivative and integral are defined by

$$\frac{d\mathbf{A}(t)}{dt} = (\frac{da_{ij}(t)}{dt})_{n \times n} \text{ and } \int \mathbf{A}(t)dt = (\int a_{ij}(t)dt)_{n \times n}. \tag{1.8}$$

Definition 1.5 For $I \subseteq \mathscr{R}$, $\Omega \subseteq \mathscr{R}^n$ and $\Lambda \subseteq \mathscr{R}^m$, the vector function $\mathbf{f}(\mathbf{x}, t, \mathbf{p})$ with $\mathbf{f} : \Omega \times I \times \Lambda \to \mathscr{R}^n$ is differentiable at $\mathbf{x}_0 \in \Omega$ if

$$\left.\frac{\partial \mathbf{f}(\mathbf{x}, t, \mathbf{p})}{\partial \mathbf{x}}\right|_{(\mathbf{x}_0, t, \mathbf{p})} = \lim_{\Delta \mathbf{x} \to \mathbf{0}} \frac{\mathbf{f}(\mathbf{x}_0 + \Delta \mathbf{x}, t, \mathbf{p}) - \mathbf{f}(\mathbf{x}_0, t, \mathbf{p})}{\Delta \mathbf{x}}. \tag{1.9}$$

$\partial \mathbf{f}/\partial \mathbf{x}$ is called the spatial derivative of $\mathbf{f}(\mathbf{x}, t, \mathbf{p})$ at $\mathbf{x}_0$, and the derivative is given by the Jacobian matrix

$$\frac{\partial \mathbf{f}(\mathbf{x}, t, \mathbf{p})}{\partial \mathbf{x}} = \left[\frac{\partial f_i}{\partial x_j}\right]_{n \times n}. \tag{1.10}$$

Definition 1.6 For $I \subseteq \mathscr{R}$, $\Omega \subseteq \mathscr{R}^n$ and $\Lambda \subseteq \mathscr{R}^m$, consider a vector function $\mathbf{f}(\mathbf{x}, t, \mathbf{p})$ with $\mathbf{f} : \Omega \times I \times \Lambda \to \mathscr{R}^n$, $t \in I$ and $\mathbf{x} \in \Omega$ and $\mathbf{p} \in \Lambda$. The vector function $\mathbf{f}(\mathbf{x}, t, \mathbf{p})$ satisfies the Lipschitz condition with respect to $\mathbf{x}$ for $\Omega \times I \times \Lambda$,

$$||\mathbf{f}(\mathbf{x}_2, t, \mathbf{p}) - \mathbf{f}(\mathbf{x}_1, t, \mathbf{p})|| \le L||\mathbf{x}_2 - \mathbf{x}_1|| \tag{1.11}$$

with $\mathbf{x}_1, \mathbf{x}_2 \in \Omega$ and L a constant. The constant L is called the Lipschitz constant.

Theorem 1.1 *Consider a dynamical system as*

$$\dot{\mathbf{x}} = \mathbf{f}(\mathbf{x}, t, \mathbf{p}) \;\text{ with }\; \mathbf{x}(t_0) = \mathbf{x}_0 \tag{1.12}$$

with $t_0, t \in I = [t_1, t_2], \mathbf{x} \in \Omega = \{\mathbf{x} | \, ||\mathbf{x} - \mathbf{x}_0|| \le d\}$ *and* $\mathbf{p} \in \Lambda$. *If the vector function* $\mathbf{f}(\mathbf{x}, t, \mathbf{p})$ *is* C^r*-continuous* ($r \ge 1$) *in* $G = \Omega \times I \times \Lambda$, *then the dynamical system in Eq.* (1.12) *has one and only one solution* $\mathbf{\Phi}(\mathbf{x}_0, t - t_0, \mathbf{p})$ *for*

$$|t - t_0| \le \min(t_2 - t_1, d/M) \;\text{ with }\; M = \max_G ||\mathbf{f}||. \tag{1.13}$$

Proof The proof of this theorem can be referred to the book by Coddington and Levinson (1955). ■

Theorem 1.2 (Gronwall) *Suppose there is a continuous real valued function* $g(t) \ge 0$ *to satisfy*

$$g(t) \le \delta_1 \int_{t_0}^{t} g(\tau) d\tau + \delta_2 \tag{1.14}$$

for all $t \in [t_0, t_1]$ *and* δ_1 *and* δ_2 *are positive constants. For* $t \in [t_0, t_1]$, *one obtains*

$$g(t) \le \delta_2 e^{\delta_1 (t - t_0)}. \tag{1.15}$$

Proof For $t \in [t_0, t_1]$, consider

$$G(t) = \delta_1 \int_{t_0}^{t} g(\tau) d\tau + \delta_2.$$

Since $\delta_1 > 0$ and $\delta_2 > 0$ are constants, one obtains $G(t) > 0$ and $G(t) \ge g(t)$ for $t \in [t_0, t_1]$. The derivative of the foregoing equation gives

$$\dot{G}(t) = \delta_1 g(t).$$

Further,

$$\frac{\dot{G}(t)}{G(t)} = \frac{\delta_1 g(t)}{G(t)} \le \frac{\delta_1 G(t)}{G(t)} = \delta_1,$$

so

$$\frac{d}{dt}(\log G(t)) \le \delta_1 \Rightarrow \log G(t) \le \delta_1(t - t_0) - \log G(t_0).$$

In other words, for all $t \in [t_0, t_1]$

$$G(t) \le G(t_0)e^{\delta_1(t-t_0)} = \delta_2 e^{\delta_1(t-t_0)}.$$

Therefore, for all $t \in [t_0, t_1]$

$$g(t) \le \delta_2 e^{\delta_1(t-t_0)}.$$

■

Theorem 1.3 *Consider a dynamical system as* $\dot{\mathbf{x}} = \mathbf{f}(\mathbf{x}, t, \mathbf{p})$ *with* $\mathbf{x}(t_0) = \mathbf{x}_0$ *in Eq.* (1.12) *with* $t_0, t \in I = [t_1, t_2], \mathbf{x} \in \Omega = \{\mathbf{x} | \, ||\mathbf{x} - \mathbf{x}_0|| \le d\}$ *and* $\mathbf{p} \in \Lambda$. *The vector function* $\mathbf{f}(\mathbf{x}, t, \mathbf{p})$ *is* C^r*-continuous* ($r \ge 1$) *in* $G = \Omega \times I \times \Lambda$. *If the solution of* $\dot{\mathbf{x}} = \mathbf{f}(\mathbf{x}, t, \mathbf{p})$*with* $\mathbf{x}(t_0) = \mathbf{x}_0$ *is* $\mathbf{x}(t)$ *on* G *and the solution of* $\dot{\mathbf{y}} = \mathbf{f}(\mathbf{y}, t, \mathbf{p})$ *with* $\mathbf{y}(t_0) = \mathbf{y}_0$ *is* $\mathbf{y}(t)$ *on* G. *For a given* $\varepsilon > 0$, *if* $||\mathbf{x}_0 - \mathbf{y}_0|| \le \varepsilon$, *then*

$$||\mathbf{x}(t) - \mathbf{y}(t)|| \le \varepsilon e^{L(t-t_0)} \text{ on } I \times \Lambda. \tag{1.16}$$

Proof From the method of successive approximations, with the local Lipschitz condition, the two initial value problems become

$$\mathbf{x}(t) = \mathbf{x}_0 + \int_{t_0}^{t} \mathbf{f}(\mathbf{x}, \tau, \mathbf{p}) d\tau \;\text{ and }\; \mathbf{y}(t) = \mathbf{y}_0 + \int_{t_0}^{t} \mathbf{f}(\mathbf{y}, \tau, \mathbf{p}) d\tau.$$

Thus,

$$\mathbf{x}(t) - \mathbf{y}(t) = \mathbf{x}_0 - \mathbf{y}_0 + \int_{t_0}^{t} (\mathbf{f}(\mathbf{x}, \tau, \mathbf{p}) - \mathbf{f}(\mathbf{y}, \tau, \mathbf{p})) d\tau,$$

$$||\mathbf{x}(t) - \mathbf{y}(t)|| \le ||\mathbf{x}_0 - \mathbf{y}_0|| + \int_{t_0}^{t} ||\mathbf{f}(\mathbf{x}, \tau, \mathbf{p}) - \mathbf{f}(\mathbf{y}, \tau, \mathbf{p})|| d\tau.$$

Using the local Lipschitz condition of $||\mathbf{f}(\mathbf{x}, \tau, \mathbf{p}) - \mathbf{f}(\mathbf{y}, \tau, \mathbf{p})|| \le L||\mathbf{x} - \mathbf{y}||$ gives

$$||\mathbf{x}(t) - \mathbf{y}(t)|| \le \varepsilon + \int_{t_0}^{t} L||\mathbf{x}(\tau) - \mathbf{y}(\tau)|| d\tau.$$

So, the Gronwall's inequality gives

$$||\mathbf{x}(t) - \mathbf{y}(t)|| \le \varepsilon e^{L(t-t_0)}.$$

This theorem is proved. ■

1.2 Equilibriums and Stability

Definition 1.7 Consider a metric space Ω and $\Omega_\alpha \subseteq \Omega\,(\alpha = 1, 2, \cdots)$.

(i) A map $\mathbf{h}$ is called a homeomorphism of Ω_α onto Ω_β $(\alpha, \beta = 1, 2, \cdots)$ if the map $\mathbf{h} : \Omega_\alpha \to \Omega_\beta$ is continuous and one-to-one, and $\mathbf{h}^{-1} : \Omega_\beta \to \Omega_\alpha$ is continuous.

(ii) Two sets Ω_α and Ω_β are homeomorphic or topologically equivalent if there is a homeomorphism of Ω_α onto Ω_β.

Definition 1.8 A connected, metric space Ω with an open cover $\{\Omega_\alpha\}$ (i.e., $\Omega = \bigcup_\alpha \Omega_\alpha$) is called an n-dimensional, $C^r\,(r \geq 1)$ differentiable manifold if the following properties exist.

(i) There is an open unit ball $B = \{\mathbf{x} \in \mathscr{R}^n \mid ||\mathbf{x}|| < 1\}$.
(ii) For all α, there is a homeomorphism $\mathbf{h}_\alpha : \Omega_\alpha \to B$.
(iii) If $\mathbf{h}_\alpha : \Omega_\alpha \to B$ and $\mathbf{h}_\beta : \Omega_\beta \to B$ are homeomorphisms for $\Omega_\alpha \cap \Omega_\beta \neq \varnothing$, then there is a C^r-differentiable map $\mathbf{h} = \mathbf{h}_\alpha \circ \mathbf{h}_\beta^{-1}$ for $\mathbf{h}_\alpha(\Omega_\alpha \cap \Omega_\beta) \subset \mathscr{R}^n$ and $\mathbf{h}_\beta(\Omega_\alpha \cap \Omega_\beta) \subset \mathscr{R}^n$ with

$$\mathbf{h} : \mathbf{h}_\beta(\Omega_\alpha \cap \Omega_\beta) \to \mathbf{h}_\alpha(\Omega_\alpha \cap \Omega_\beta), \tag{1.17}$$

and for all $\mathbf{x} \in \mathbf{h}_\beta(\Omega_\alpha \cap \Omega_\beta)$, the Jacobian determinant $\det D\mathbf{h}(\mathbf{x}) \neq 0$.

The manifold Ω is called to be analytic if the maps $\mathbf{h} = \mathbf{h}_\alpha \circ \mathbf{h}_\beta^{-1}$ are analytic.

Definition 1.9 Consider an autonomous, nonlinear dynamical system $\dot{\mathbf{x}} = \mathbf{f}(\mathbf{x}, \mathbf{p})$ in Eq. (1.4). A point $\mathbf{x}^* \in \Omega$ is called an equilibrium point or critical point of a nonlinear system $\dot{\mathbf{x}} = \mathbf{f}(\mathbf{x}, \mathbf{p})$ if

$$\mathbf{f}(\mathbf{x}^*, \mathbf{p}) = \mathbf{0}. \tag{1.18}$$

The linearized system of the nonlinear system $\dot{\mathbf{x}} = \mathbf{f}(\mathbf{x}, \mathbf{p})$ in Eq. (1.4) at the equilibrium point $\mathbf{x}^*$ is given by

$$\dot{\mathbf{y}} = D\mathbf{f}(\mathbf{x}^*, \mathbf{p})\mathbf{y} \text{ where } \mathbf{y} = \mathbf{x} - \mathbf{x}^*. \tag{1.19}$$

Definition 1.10 Consider an n-dimensional, autonomous, nonlinear dynamical system $\dot{\mathbf{x}} = \mathbf{f}(\mathbf{x}, \mathbf{p})$ in Eq. (1.4) with an equilibrium point $\mathbf{x}^*$. The linearized system of the nonlinear system at the equilibrium point $\mathbf{x}^*$ is $\dot{\mathbf{y}} = D\mathbf{f}(\mathbf{x}^*, \mathbf{p})\mathbf{y}$ $(\mathbf{y} = \mathbf{x} - \mathbf{x}^*)$ in Eq. (1.19). The matrix $D\mathbf{f}(\mathbf{x}^*, \mathbf{p})$ possesses n eigenvalues λ_k $(k = 1, 2, \cdots n)$. Set $N = \{1, 2, \cdots, n\}$, $N_i = \{l_1, l_2, \cdots, l_{n_i}\} \cup \varnothing$ with $l_{j_i} \in N$ $(j_i = 1, 2, \cdots, n_i; i = 1, 2, \cdots, 6)$ and $\Sigma_{i=1}^3 n_i = n$. $\cup_{i=1}^3 N_i = N$ and $N_i \cap N_l = \varnothing\,(l \neq i)$. $N_i = \varnothing$ if $n_i = 0$. The corresponding vectors for the negative, positive and zero eigenvalues of $D\mathbf{f}(\mathbf{x}^*, \mathbf{p})$ are $\{\mathbf{u}_{k_i}\}$ $(k_i \in N_i,\ i = 1, 2, 3)$, respectively. The stable, unstable and invariant subspaces of the linearized nonlinear system in Eq. (1.19) are defined as

$$\begin{aligned}
\mathscr{E}^{\mathrm{s}} &= span\left\{\mathbf{u}_k|(D\mathbf{f}(\mathbf{x}^*,\mathbf{p})-\lambda_k\mathbf{I})\mathbf{u}_k=\mathbf{0},\lambda_k<0,k\in N_1\subseteq N\cup\varnothing\right\};\\
\mathscr{E}^{\mathrm{u}} &= span\left\{\mathbf{u}_k|(D\mathbf{f}(\mathbf{x}^*,\mathbf{p})-\lambda_k\mathbf{I})\mathbf{u}_k=\mathbf{0},\lambda_k>0,k\in N_2\subseteq N\cup\varnothing\right\};\\
\mathscr{E}^{\mathrm{i}} &= span\left\{\mathbf{u}_k|(D\mathbf{f}(\mathbf{x}^*,\mathbf{p})-\lambda_k\mathbf{I})\mathbf{u}_k=\mathbf{0},\lambda_k=0,k\in N_3\subseteq N\cup\varnothing\right\}.
\end{aligned}\tag{1.20}$$

Definition 1.11 Consider an $2n$-dimensional, autonomous dynamical system $\dot{\mathbf{x}}=\mathbf{f}(\mathbf{x},\mathbf{p})$ in Eq. (1.4) with an equilibrium point $\mathbf{x}^*$. The linearized system of the nonlinear system at the equilibrium point $\mathbf{x}^*$ is $\dot{\mathbf{y}}=D\mathbf{f}(\mathbf{x}^*,\mathbf{p})\mathbf{y}\ (\mathbf{y}=\mathbf{x}-\mathbf{x}^*)$ in Eq. (1.19). The matrix $D\mathbf{f}(\mathbf{x}^*,\mathbf{p})$ has complex eigenvalues $\alpha_k\pm\mathbf{i}\beta_k$ with eigenvectors $\mathbf{u}_k\pm\mathbf{i}\mathbf{v}_k$ $(k\in\{1,2,\cdots,n\})$ and the base of vector is

$$\mathbf{B}=\{\mathbf{u}_1,\mathbf{v}_1,\cdots,\mathbf{u}_k,\mathbf{v}_k,\cdots,\mathbf{u}_n,\mathbf{v}_n\}.\tag{1.21}$$

The stable, unstable, center subspaces of Eq. (1.19) are linear subspaces spanned by $\{\mathbf{u}_k,\mathbf{v}_k\}$ $(k\in N_i,\ i=1,2,3)$, respectively. Set $N_i=\{i_1,i_2,\cdots,i_{n_i}\}\cup\varnothing\subseteq N\cup\varnothing$ and $N=\{1,2,\cdots,n\}$ with $l_j\in N\ (j=1,2,\cdots,n_i,i=1,2,3)$ and $\Sigma_{i=1}^3 n_i=n$. $\cup_{i=1}^3 N_i=N$ and $N_i\cap N_l=\varnothing(l\neq i)$. $N_i=\varnothing$ if $n_i=0$. The stable, unstable, center subspaces of the linearized nonlinear system in Eq. (1.19) are defined as

$$\begin{aligned}
\mathscr{E}^{\mathrm{s}} &= span\left\{(\mathbf{u}_k,\mathbf{v}_k)\left|\begin{array}{l}\alpha_k<0,\beta_k\neq 0,\\ (D\mathbf{f}(\mathbf{x}^*,\mathbf{p})-(\alpha_k\pm\mathbf{i}\beta_k)\mathbf{I})\,(\mathbf{u}_k\pm\mathbf{i}\mathbf{v}_k)=\mathbf{0},\\ k\in N_1\subseteq\{1,2,\cdots,n\}\cup\varnothing\end{array}\right.\right\};\\
\mathscr{E}^{\mathrm{u}} &= span\left\{(\mathbf{u}_k,\mathbf{v}_k)\left|\begin{array}{l}\alpha_k>0,\beta_k\neq 0,\\ (D\mathbf{f}(\mathbf{x}^*,\mathbf{p})-(\alpha_k\pm\mathbf{i}\beta_k)\mathbf{I})\,(\mathbf{u}_k\pm\mathbf{i}\mathbf{v}_k)=\mathbf{0},\\ k\in N_2\subseteq\{1,2,\cdots,n\}\cup\varnothing\end{array}\right.\right\};\\
\mathscr{E}^{\mathrm{c}} &= span\left\{(\mathbf{u}_k,\mathbf{v}_k)\left|\begin{array}{l}\alpha_k=0,\beta_k\neq 0,\\ (D\mathbf{f}(\mathbf{x}^*,\mathbf{p})-(\alpha_k\pm\mathbf{i}\beta_k)\mathbf{I})\,(\mathbf{u}_k\pm\mathbf{i}\mathbf{v}_k)=\mathbf{0},\\ k\in N_3\subseteq\{1,2,\cdots,n\}\cup\varnothing\end{array}\right.\right\}.
\end{aligned}\tag{1.22}$$

Theorem 1.4 *Consider an n-dimensional, autonomous, nonlinear dynamical system* $\dot{\mathbf{x}}=\mathbf{f}(\mathbf{x},\mathbf{p})$ *in Eq.* (1.4) *with an equilibrium point* $\mathbf{x}^*$. *The linearized system of the nonlinear system at the equilibrium point* $\mathbf{x}^*$ *is* $\dot{\mathbf{y}}=D\mathbf{f}(\mathbf{x}^*,\mathbf{p})\mathbf{y}\ (\mathbf{y}=\mathbf{x}-\mathbf{x}^*)$ *in Eq.*(1.19). *The eigenspace of* $D\mathbf{f}(\mathbf{x}^*,\mathbf{p})$ *(i.e.,* $\mathscr{E}\subseteq\mathscr{R}^n$*) in the linearized dynamical system is expressed by direct sum of three subspaces*

$$\mathscr{E}=\mathscr{E}^s\oplus\mathscr{E}^u\oplus\mathscr{E}^c\tag{1.23}$$

where $\mathscr{E}^s$, $\mathscr{E}^u$ and $\mathscr{E}^c$ *are the stable, unstable and center spaces* $\mathscr{E}^s$, $\mathscr{E}^u$ *and* $\mathscr{E}^c$, *respectively.*

Proof This proof is the same as the linear system in Appendix A. ∎

Definition 1.12 Consider an n-dimensional, autonomous, nonlinear dynamical system $\dot{\mathbf{x}}=\mathbf{f}(\mathbf{x},\mathbf{p})$ in Eq. (1.4) with an equilibrium point $\mathbf{x}^*$ and $\mathbf{f}(\mathbf{x},\mathbf{p})$ is C^r $(r\geq 1)$-continuous in a neighborhood of the equilibrium $\mathbf{x}^*$. The corresponding solution is $\mathbf{x}(t)=\mathbf{\Phi}(\mathbf{x}_0,t-t_0,\mathbf{p})=\mathbf{\Phi}_t(\mathbf{x}_0)$. The linearized system of the nonlinear system at the equilibrium point $\mathbf{x}^*$ is $\dot{\mathbf{y}} = D\mathbf{f}(\mathbf{x}^*, \mathbf{p})\mathbf{y}$ ($\mathbf{y} = \mathbf{x} - \mathbf{x}^*$) in Eq. (1.19). Suppose there is a neighborhood of the equilibrium $\mathbf{x}^*$ as $U(\mathbf{x}^*) \subset \Omega$, and in the neighborhood

$$\lim_{||\mathbf{y}|| \to 0} \frac{||\mathbf{f}(\mathbf{x}^* + \mathbf{y}, \mathbf{p}) - D\mathbf{f}(\mathbf{x}^*, \mathbf{p})\mathbf{y}||}{||\mathbf{y}||} = 0. \tag{1.24}$$

(i) A C^r invariant manifold

$$\mathscr{S}_{loc}(\mathbf{x}, \mathbf{x}^*) = \{\mathbf{x} \in U(\mathbf{x}^*) | \lim_{t \to \infty} \mathbf{x}(t) = \mathbf{x}^*, \mathbf{x}(t) \in U(\mathbf{x}^*) \text{ for all } t \geq 0\} \tag{1.25}$$

is called the local stable manifold of $\mathbf{x}^*$, and the corresponding global stable manifold is defined as

$$\mathscr{S}(\mathbf{x}, \mathbf{x}^*) = \cup_{t \leq 0} \mathbf{\Phi}_t(\mathscr{S}_{loc}(\mathbf{x}, \mathbf{x}^*)). \tag{1.26}$$

(ii) A C^r invariant manifold

$$\mathscr{U}_{loc}(\mathbf{x}, \mathbf{x}^*) = \{\mathbf{x} \in U(\mathbf{x}^*) | \lim_{t \to -\infty} \mathbf{x}(t) = \mathbf{x}^*, \mathbf{x}(t) \in U(\mathbf{x}^*) \text{ for all } t \leq 0\} \tag{1.27}$$

is called the unstable manifold of $\mathbf{x}^*$, and the corresponding global unstable manifold is defined as

$$\mathscr{U}(\mathbf{x}, \mathbf{x}^*) = \cup_{t \geq 0} \mathbf{\Phi}_t(\mathscr{U}_{loc}(\mathbf{x}, \mathbf{x}^*)). \tag{1.28}$$

(iii) A C^{r-1} invariant manifold $\mathscr{C}_{loc}(\mathbf{x}, \mathbf{x}^*)$ is called the center manifold of $\mathbf{x}^*$ if $\mathscr{C}_{loc}(\mathbf{x}, \mathbf{x}^*)$ possesses the same dimension of $\mathscr{E}^c$ for $\mathbf{x}^* \in \mathscr{S}(\mathbf{x}, \mathbf{x}^*)$, and the tangential space of $\mathscr{C}_{loc}(\mathbf{x}, \mathbf{x}^*)$ is identical to $\mathscr{E}^c$.

The stable and unstable manifolds are unique, but the center manifold is not unique. If the nonlinear vector field $\mathbf{f}$ is C^∞-continuous, then a C^r center manifold can be found for any $r < \infty$.

Theorem 1.5 *Consider an n-dimensional, autonomous, nonlinear dynamical system* $\dot{\mathbf{x}} = \mathbf{f}(\mathbf{x}, \mathbf{p})$ *in Eq.* (1.4) *with a hyperbolic equilibrium point* $\mathbf{x}^*$ *and* $\mathbf{f}(\mathbf{x}, \mathbf{p})$ *is* C^r ($r \geq 1$)*-continuous in a neighborhood of the equilibrium* $\mathbf{x}^*$. *The corresponding solution is* $\mathbf{x}(t) = \mathbf{\Phi}(\mathbf{x}_0, t - t_0, \mathbf{p}) = \mathbf{\Phi}_t(\mathbf{x}_0)$. *The linearized system of the nonlinear system at the equilibrium point* $\mathbf{x}^*$ *is* $\dot{\mathbf{y}} = D\mathbf{f}(\mathbf{x}^*, \mathbf{p})\mathbf{y}$ ($\mathbf{y} = \mathbf{x} - \mathbf{x}^*$) *in Eq.* (1.19). *Suppose there is a neighborhood of the hyperbolic equilibrium* $\mathbf{x}^*$ *as* $U(\mathbf{x}^*) \subset \Omega$. *If the homeomorphism between the local invariant subspace* $E(\mathbf{x}, \mathbf{x}^*) \subset U(\mathbf{x}^*)$ *under the flow* $\mathbf{\Phi}(\mathbf{x}_0, t - t_0, \mathbf{p})$ *of* $\dot{\mathbf{x}} = \mathbf{f}(\mathbf{x}, \mathbf{p})$ *in Eq.* (1.4) *and the eigenspace* $\mathscr{E}$ *of the linearized system exists with the condition in Eq.* (1.24), *the local invariant subspace is decomposed by*

$$E(\mathbf{x}, \mathbf{x}^*) = \mathscr{S}_{loc}(\mathbf{x}, \mathbf{x}^*) \oplus \mathscr{U}_{loc}(\mathbf{x}, \mathbf{x}^*). \tag{1.29}$$

(a) *The local stable invariant manifold* $\mathscr{S}_{loc}(\mathbf{x}, \mathbf{x}^*)$ *possesses the following properties:*

(i) *for* $\mathbf{x}^* \in \mathscr{S}_{loc}(\mathbf{x}, \mathbf{x}^*)$, $\mathscr{S}_{loc}(\mathbf{x}, \mathbf{x}^*)$ *possesses the same dimension of* $\mathscr{E}^s$ *and the tangential space of* $\mathscr{S}_{loc}(\mathbf{x}, \mathbf{x}^*)$ *is identical to* $\mathscr{E}^s$;

(ii) *for* $\mathbf{x}_0 \in \mathscr{S}_{loc}(\mathbf{x}, \mathbf{x}^*)$, $\mathbf{x}(t) \in \mathscr{S}_{loc}(\mathbf{x}, \mathbf{x}^*)$ *for all time* $t \geq t_0$ *and* $\lim_{t\to\infty} \mathbf{x}(t) = \mathbf{x}^*$;

(iii) *for* $\mathbf{x}_0 \notin \mathscr{S}_{loc}(\mathbf{x}, \mathbf{x}^*)$, $||\mathbf{x} - \mathbf{x}^*|| \geq \delta$ *for* $\delta > 0$ *with* $t \geq t_1 \geq t_0$.

(b) *The local unstable invariant manifold* $\mathscr{U}_{loc}(\mathbf{x}, \mathbf{x}^*)$ *possesses the following properties:*

(i) *for* $\mathbf{x}^* \in \mathscr{U}_{loc}(\mathbf{x}, \mathbf{x}^*)$, $\mathscr{U}_{loc}(\mathbf{x}, \mathbf{x}^*)$ *possesses the same dimension of* $\mathscr{E}^u$ *and the tangential space of* $\mathscr{U}_{loc}(\mathbf{x}, \mathbf{x}^*)$ *is identical to* $\mathscr{E}^u$;

(ii) *for* $\mathbf{x}_0 \in \mathscr{U}_{loc}(\mathbf{x}, \mathbf{x}^*)$, $\mathbf{x}(t) \in \mathscr{U}_{loc}(\mathbf{x}, \mathbf{x}^*)$ *for all time* $t \leq t_0$ *and* $\lim_{t\to-\infty} \mathbf{x}(t) = \mathbf{x}^*$;

(iii) *for* $\mathbf{x}_0 \notin \mathscr{U}_{loc}(\mathbf{x}, \mathbf{x}^*)$, $||\mathbf{x} - \mathbf{x}^*|| \geq \delta$ *for* $\delta > 0$ *with* $t \leq t_1 \leq t_0$.

Proof The proof for stable and unstable manifolds can be referred to Hartman (1964). The proof for center manifold can be referenced to Marsden and McCracken (1976) or Carr (1981). ■

Theorem 1.6 *Consider an n-dimensional, autonomous, nonlinear dynamical system* $\dot{\mathbf{x}} = \mathbf{f}(\mathbf{x}, \mathbf{p})$ *in Eq. (1.4) with an equilibrium point* $\mathbf{x}^*$. *Suppose there is a neighborhood of the equilibrium* $\mathbf{x}^*$ *as* $U(\mathbf{x}^*) \subset \Omega$, *then* $\mathbf{f}(\mathbf{x}, \mathbf{p})$ *is* C^r $(r \geq 1)$*-continuous in a neighborhood of the equilibrium* $\mathbf{x}^*$. *The corresponding solution is* $\mathbf{x}(t) = \mathbf{\Phi}(\mathbf{x}_0, t - t_0, \mathbf{p})$. *The linearized system of the nonlinear system at the equilibrium point* $\mathbf{x}^*$ *is* $\dot{\mathbf{y}} = D\mathbf{f}(\mathbf{x}^*, \mathbf{p})\mathbf{y}$ $(\mathbf{y} = \mathbf{x} - \mathbf{x}^*)$ *in Eq. (1.19). If the homeomorphism between the local invariant subspace* $E(\mathbf{x}, \mathbf{x}^*) \subset U(\mathbf{x}^*)$ *under the flow* $\mathbf{\Phi}(\mathbf{x}_0, t - t_0, \mathbf{p})$ *of* $\dot{\mathbf{x}} = \mathbf{f}(\mathbf{x}, \mathbf{p})$ *in Eq. (1.4) and the eigenspace* $\mathscr{E}$ *of the linearized system exists with the condition in Eq. (1.24), in addition to the local stable and unstable invariant manifolds, there is a* C^{r-1} *center manifold* $\mathscr{C}_{loc}(\mathbf{x}, \mathbf{x}^*)$. *The center manifold possesses the same dimension of* $\mathscr{E}^c$ *for* $\mathbf{x}^* \in \mathscr{C}_{loc}(\mathbf{x}, \mathbf{x}^*)$, *and the tangential space of* $\mathscr{C}_{loc}(\mathbf{x}, \mathbf{x}^*)$ *is identical to* $\mathscr{E}^c$. *Thus, the local invariant subspace is decomposed by*

$$E(\mathbf{x}, \mathbf{x}^*) = \mathscr{S}_{loc}(\mathbf{x}, \mathbf{x}^*) \oplus \mathscr{U}_{loc}(\mathbf{x}, \mathbf{x}^*) \oplus \mathscr{C}_{loc}(\mathbf{x}, \mathbf{x}^*). \tag{1.30}$$

Proof The proof for stable and unstable manifolds can be referred to Hartman (1964). The proof for center manifold can be referenced to Marsden and McCracken (1976) or Carr (1981). ■

Definition 1.13 Consider an n-dimensional, autonomous, nonlinear dynamical system $\dot{\mathbf{x}} = \mathbf{f}(\mathbf{x}, \mathbf{p})$ in Eq. (1.4) with an equilibrium point $\mathbf{x}^*$ and $\mathbf{f}(\mathbf{x}, \mathbf{p})$ is C^r $(r \geq 1)$-continuous in a neighborhood of the equilibrium $\mathbf{x}^*$.

(i) The equilibrium $\mathbf{x}^*$ is stable if all $\varepsilon > 0$, there is a $\delta > 0$ such that for all $\mathbf{x}_0 \in U_\delta(\mathbf{x}^*)$ where $U_\delta(\mathbf{x}^*) = \{\mathbf{x} |\, ||\mathbf{x} - \mathbf{x}^*|| < \delta\}$ and $t \geq 0$,

$$\mathbf{\Phi}(\mathbf{x}_0, t - t_0, \mathbf{p}) \in U_\varepsilon(\mathbf{x}^*). \tag{1.31}$$

(ii) The equilibrium $\mathbf{x}^*$ is unstable if it is not stable or if all $\varepsilon > 0$, there is a $\delta > 0$ such that for all $\mathbf{x}_0 \in U_\delta(\mathbf{x}^*)$ where $U_\delta(\mathbf{x}^*) = \{\mathbf{x}|\,||\mathbf{x} - \mathbf{x}^*|| < \delta\}$ and $t \geq t_1 > 0$,

$$\mathbf{\Phi}(\mathbf{x}_0, t - t_0, \mathbf{p}) \notin U_\varepsilon(\mathbf{x}^*). \tag{1.32}$$

(iii) The equilibrium $\mathbf{x}^*$ is asymptotically stable if all $\varepsilon > 0$, there is a $\delta > 0$ such that for all $\mathbf{x}_0 \in U_\delta(\mathbf{x}^*)$ where $U_\delta(\mathbf{x}^*) = \{\mathbf{x}|\,||\mathbf{x} - \mathbf{x}^*|| < \delta\}$ and $t \geq 0$,

$$\lim_{t\to\infty} \mathbf{\Phi}(\mathbf{x}_0, t - t_0, \mathbf{p}) = \mathbf{x}^*. \tag{1.33}$$

(iv) The equilibrium $\mathbf{x}^*$ is asymptotically unstable stable if all $\varepsilon > 0$, there is a $\delta > 0$ such that for all $\mathbf{x}_0 \in U_\delta(\mathbf{x}^*)$ where $U_\delta(\mathbf{x}^*) = \{\mathbf{x}|\,||\mathbf{x} - \mathbf{x}^*|| < \delta\}$ and $t \leq 0$,

$$\lim_{t\to-\infty} \mathbf{\Phi}(\mathbf{x}_0, t - t_0, \mathbf{p}) = \mathbf{x}^*. \tag{1.34}$$

Definition 1.14 Consider an n-dimensional, autonomous, nonlinear dynamical system $\dot{\mathbf{x}} = \mathbf{f}(\mathbf{x}, \mathbf{p})$ in Eq. (1.4) with an equilibrium point $\mathbf{x}^*$. Suppose there is a neighborhood of the equilibrium $\mathbf{x}^*$ as $U(\mathbf{x}^*) \subset \Omega$, then $\mathbf{f}(\mathbf{x}, \mathbf{p})$ is C^r $(r \geq 1)$-continuous and Eq. (1.24) holds in $U(\mathbf{x}^*) \subset \Omega$. The corresponding solution is $\mathbf{x}(t) = \mathbf{\Phi}(\mathbf{x}_0, t - t_0, \mathbf{p})$. For a linearized dynamical system in Eq. (1.19), consider a real eigenvalue λ_k of matrix $D\mathbf{f}(\mathbf{x}^*, \mathbf{p})$ $(k \in N = \{1, 2, \cdots, n\})$ with an eigenvector $\mathbf{v}_k$. For $\mathbf{y}^{(k)} = c^{(k)}\mathbf{v}_k$, $\dot{\mathbf{y}}^{(k)} = \dot{c}^{(k)}\mathbf{v}_k = \lambda_k c^{(k)}\mathbf{v}_k$, thus $\dot{c}^{(k)} = \lambda_k c^{(k)}$.

(i) $\mathbf{x}^{(k)}$ at the equilibrium $\mathbf{x}^*$ on the direction $\mathbf{v}_k$ is stable if

$$\lim_{t\to\infty} c^{(k)} = \lim_{t\to\infty} c_0^{(k)} e^{\lambda_k t} = 0 \text{ for } \lambda_k < 0. \tag{1.35}$$

(ii) $\mathbf{x}^{(k)}$ at the equilibrium $\mathbf{x}^*$ on the direction $\mathbf{v}_k$ is unstable if

$$\lim_{t\to\infty} |c^{(k)}| = \lim_{t\to\infty} |c_0^{(k)} e^{\lambda_k t}| = \infty \text{ for } \lambda_k > 0. \tag{1.36}$$

(iii) $\mathbf{x}^{(i)}$ at the equilibrium $\mathbf{x}^*$ on the direction $\mathbf{v}_k$ is uncertain (critical) if

$$\lim_{t\to\infty} c^{(k)} = \lim_{t\to\infty} e^{\lambda_k t} c_0^{(k)} = c_0^{(k)} \text{ for } \lambda_k = 0. \tag{1.37}$$

Definition 1.15 Consider a $2n$-dimensional, autonomous, nonlinear dynamical system $\dot{\mathbf{x}} = \mathbf{f}(\mathbf{x}, \mathbf{p})$ in Eq. (1.4) with an equilibrium point $\mathbf{x}^*$. Suppose there is a neighborhood of the equilibrium $\mathbf{x}^*$ as $U(\mathbf{x}^*) \subset \Omega$, then $\mathbf{f}(\mathbf{x}, \mathbf{p})$ is C^r $(r \geq 1)$-continuous and Eq. (1.24) holds in $U(\mathbf{x}^*) \subset \Omega$. The corresponding solution is $\mathbf{x}(t) = \mathbf{\Phi}(\mathbf{x}_0, t - t_0, \mathbf{p})$. For a linearized dynamical system in Eq. (1.19), consider a pair of complex eigenvalue $\alpha_k \pm \mathbf{i}\beta_k$ $(k \in N = \{1, 2, \cdots, n\}$, $\mathbf{i} = \sqrt{-1})$ of matrix $D\mathbf{f}(\mathbf{x}^*, \mathbf{p})$ with a pair of eigenvectors $\mathbf{u}_k \pm \mathbf{i}\mathbf{v}_k$. On the invariant plane of $(\mathbf{u}_k, \mathbf{v}_k)$, consider $\mathbf{y}^{(k)} = \mathbf{y}_+^{(k)} + \mathbf{y}_-^{(k)}$ with

$$\mathbf{y}^{(k)} = c^{(k)}\mathbf{u}_k + d^{(k)}\mathbf{v}_k, \dot{\mathbf{y}}^{(k)} = \dot{c}^{(k)}\mathbf{u}_k + \dot{d}^{(k)}\mathbf{v}_k. \tag{1.38}$$

Thus, $\mathbf{c}^{(k)} = (c^{(k)}, d^{(k)})^{\mathrm{T}}$ with

$$\dot{\mathbf{c}}^{(k)} = \mathbf{E}_k\mathbf{c}^{(k)} \Rightarrow \mathbf{c}^{(k)} = e^{\alpha_k t}\mathbf{B}_k\mathbf{c}_0^{(k)} \tag{1.39}$$

where

$$\mathbf{E}_k = \begin{bmatrix} \alpha_k & \beta_k \\ -\beta_k & \alpha_k \end{bmatrix} \text{ and } \mathbf{B}_k = \begin{bmatrix} \cos\beta_k t & \sin\beta_k t \\ -\sin\beta_k t & \cos\beta_k t \end{bmatrix}. \tag{1.40}$$

(i) $\mathbf{x}^{(k)}$ at the equilibrium $\mathbf{x}^*$ on the plane of $(\mathbf{u}_k, \mathbf{v}_k)$ is spirally stable if

$$\lim_{t\to\infty} ||\mathbf{c}^{(k)}|| = \lim_{t\to\infty} e^{\alpha_k t}||\mathbf{B}_k|| \times ||\mathbf{c}_0^{(k)}|| = 0 \text{ for } \mathrm{Re}\lambda_k = \alpha_k < 0. \tag{1.41}$$

(ii) $\mathbf{x}^{(k)}$ at the equilibrium $\mathbf{x}^*$ on the plane of $(\mathbf{u}_k, \mathbf{v}_k)$ is spirally unstable if

$$\lim_{t\to\infty} ||\mathbf{c}^{(k)}|| = \lim_{t\to\infty} e^{\alpha_k t}||\mathbf{B}_k|| \times ||\mathbf{c}_0^{(k)}|| = \infty \text{ for } \mathrm{Re}\lambda_k = \alpha_k > 0. \tag{1.42}$$

(iii) $\mathbf{x}^{(k)}$ at the equilibrium $\mathbf{x}^*$ on the plane of $(\mathbf{u}_k, \mathbf{v}_k)$ is on the invariant circle if

$$\lim_{t\to\infty} ||\mathbf{c}^{(k)}|| = \lim_{t\to\infty} e^{\alpha_k t}||\mathbf{B}_k|| \times ||\mathbf{c}_0^{(k)}|| = ||\mathbf{c}_0^{(k)}|| \text{ for } \mathrm{Re}\lambda_k = \alpha_k = 0. \tag{1.43}$$

(iv) $\mathbf{x}^{(k)}$ at the equilibrium $\mathbf{x}^*$ on the plane of $(\mathbf{u}_k, \mathbf{v}_k)$ is degenerate in the direction of $\mathbf{u}_k$ if $\mathrm{Im}\lambda_k = 0$.

Definition 1.16 Consider an n-dimensional, autonomous, nonlinear dynamical system $\dot{\mathbf{x}} = \mathbf{f}(\mathbf{x}, \mathbf{p})$ in Eq. (1.4) with an equilibrium point $\mathbf{x}^*$. Suppose there is a neighborhood of the equilibrium $\mathbf{x}^*$ as $U(\mathbf{x}^*) \subset \Omega$, then $\mathbf{f}(\mathbf{x}, \mathbf{p})$ is C^r ($r \geq 1$)-continuous in the neighborhood, and Eq. (1.24) holds. The corresponding solution is $\mathbf{x}(t) = \mathbf{\Phi}(\mathbf{x}_0, t - t_0, \mathbf{p})$. The linearized system of the nonlinear system at the equilibrium point $\mathbf{x}^*$ is $\dot{\mathbf{y}} = D\mathbf{f}(\mathbf{x}^*, \mathbf{p})\mathbf{y}$ ($\mathbf{y} = \mathbf{x} - \mathbf{x}^*$) in Eq. (1.19).

(i) The equilibrium $\mathbf{x}^*$ is said *a hyperbolic equilibrium* if none of eigenvalues of $D\mathbf{f}(\mathbf{x}^*, \mathbf{p})$ is zero real part (i.e., $\mathrm{Re}\lambda_k \neq 0$ $(k = 1, 2, \cdots, n)$).
(ii) The equilibrium $\mathbf{x}^*$ is said *a sink* if all of eigenvalues of $D\mathbf{f}(\mathbf{x}^*, \mathbf{p})$ have negative real parts (i.e., $\mathrm{Re}\lambda_k < 0$ $(k = 1, 2, \cdots, n)$).
(iii) The equilibrium $\mathbf{x}^*$ is said *a source* if all of eigenvalues of $D\mathbf{f}(\mathbf{x}^*, \mathbf{p})$ have positive real parts (i.e., $\mathrm{Re}\lambda_k > 0$ $(k = 1, 2, \cdots, n)$).
(iv) The equilibrium $\mathbf{x}^*$ is said *a saddle* if it is a hyperbolic equilibrium and $D\mathbf{f}(\mathbf{x}^*, \mathbf{p})$ have at least one eigenvalue with a positive real part (i.e., $\mathrm{Re}\lambda_j > 0$ $(j \in \{1, 2, \cdots, n\})$ and one with a negative real part (i.e., $\mathrm{Re}\lambda_k < 0$ $(k \in \{1, 2, \cdots, n\})$).
(v) The equilibrium $\mathbf{x}^*$ is called a center if all of eigenvalues of $D\mathbf{f}(\mathbf{x}^*, \mathbf{p})$ have zero real parts (i.e., $Re\lambda_j = 0$ $(j = 1, 2, \cdots, n)$) with distinct eigenvalues.

(vi) The equilibrium $\mathbf{x}^*$ is called a stable node if all of eigenvalues of $D\mathbf{f}(\mathbf{x}^*, \mathbf{p})$ are real $\lambda_k < 0$ $(k = 1, 2, \cdots n)$.
(vii) The equilibrium $\mathbf{x}^*$ is called an unstable node if all of eigenvalues of $D\mathbf{f}(\mathbf{x}^*, \mathbf{p})$ are real $\lambda_k > 0$ $(k = 1, 2, \cdots n)$.
(viii) The equilibrium $\mathbf{x}^*$ is called a degenerate case if all of eigenvalues of $D\mathbf{f}(\mathbf{x}^*, \mathbf{p})$ are zero $\lambda_k = 0$ $(k = 1, 2, \cdots n)$.

As in Appendix A, the refined classification of the linearized nonlinear system at equilibrium points should be discussed. The generalized stability and bifurcation of flows in linearized, nonlinear dynamical systems in Eq. (1.4) will be discussed as follows.

Definition 1.17 Consider an n-dimensional, autonomous, nonlinear dynamical system $\dot{\mathbf{x}} = \mathbf{f}(\mathbf{x}, \mathbf{p})$ in Eq. (1.4) with an equilibrium point $\mathbf{x}^*$. Suppose there is a neighborhood of the equilibrium $\mathbf{x}^*$ as $U(\mathbf{x}^*) \subset \Omega$, then $\mathbf{f}(\mathbf{x}, \mathbf{p})$ is C^r $(r \geq 1)$-continuous in the neighborhood, and Eq. (1.24) holds. The corresponding solution is $\mathbf{x}(t) = \mathbf{\Phi}(\mathbf{x}_0, t - t_0, \mathbf{p})$. The matrix $D\mathbf{f}(\mathbf{x}^*, \mathbf{p})$ in Eq. (1.19) possesses n eigenvalues λ_k $(k = 1, 2, \cdots, n)$. Set $N = \{1, 2, \cdots, m, m+1, (n-m)/2\}$, $N_i = \{i_1, i_2, \cdots, i_{n_i}\} \cup \varnothing$ with $i_j \in N$ $(j = 1, 2, \cdots, n_i,\ i = 1, 2, \cdots, 6)$, $\Sigma_{i=1}^3 n_i = m$ and $2\Sigma_{i=4}^6 n_i = n - m$. $\cup_{i=1}^6 N_i = N$ and $N_i \cap N_l = \varnothing (l \neq i)$. $N_i = \varnothing$ if $n_i = 0$. The matrix $D\mathbf{f}(\mathbf{x}^*, \mathbf{p})$ possesses n_1-stable, n_2-unstable and n_3-invariant real eigenvectors plus n_4-stable, n_5-unstable and n_6-center pairs of complex eigenvectors. Without repeated complex eigenvalues of $\mathrm{Re}\lambda_k = 0$ $(k \in N_3 \cup N_6)$, the flow $\mathbf{\Phi}(t)$ of the nonlinear system $\dot{\mathbf{x}} = \mathbf{f}(\mathbf{x}, \mathbf{p})$ is an $(n_1 : n_2 : [n_3; m_3] \| n_4 : n_5 : n_6)$ flow in the neighborhood of $\mathbf{x}^*$. However, with repeated complex eigenvalues of $\mathrm{Re}\lambda_k = 0$ $(k \in N_3 \cup N_6)$, the flow $\mathbf{\Phi}(t)$ of the nonlinear system $\dot{\mathbf{x}} = \mathbf{f}(\mathbf{x}, \mathbf{p})$ is an $(n_1 : n_2 : [n_3; \kappa_3] \| n_4 : n_5 : [n_6, l; \boldsymbol{\kappa}_6])$ flow in the neighborhood of x^*. $\kappa_3 \in \{\varnothing, m_3\}$, $\kappa_6 = (\boldsymbol{\kappa}_{61}, \boldsymbol{\kappa}_{62}, \cdots, \boldsymbol{\kappa}_{6l})^{\mathrm{T}}$ with $\kappa_{6i} \in \{\varnothing, m_{6i}\} (i = 1, 2, \cdots, l)$. $\mathbf{m}_6 = (m_{61}, m_{62}, \cdots, m_{6l})^{\mathrm{T}}$. The meanings of notations in the aforementioned structures are defined as follows:

(i) n_1 represents exponential sinks on n_1-directions of $\mathbf{v}_k$ $(k \in N_1)$ if $\lambda_k < 0$ $(k \in N_1$ and $1 \leq n_1 \leq m)$ with distinct or repeated eigenvalues.
(ii) n_2 represents exponential sources on n_2-directions of $\mathbf{v}_k$ $(k \in N_2)$ if $\lambda_k > 0$ $(k \in N_2$ and $1 \leq n_2 \leq m)$ with distinct or repeated eigenvalues.
(iii) $n_3 = 1$ represents an invariant center on 1-direction of $\mathbf{v}_k$ $(k \in N_3)$ if $\lambda_k = 0$ $(k \in N_3$ and $n_3 = 1)$.
(iv) n_4 represents spiral sinks on n_4-pairs of $(\mathbf{u}_k, \mathbf{v}_k) (k \in N_4)$ if $\mathrm{Re}\lambda_k < 0$ and $\mathrm{Im}\lambda_k \neq 0$ $(k \in N_4$ and $1 \leq n_4 \leq (n - m)/2)$ with distinct or repeated eigenvalues.
(v) n_5 represents spiral sources on n_5-pairs of $(\mathbf{u}_k, \mathbf{v}_k)$ $(k \in N_5)$ if $\mathrm{Re}\lambda_k > 0$ and $\mathrm{Im}\lambda_k \neq 0$ $(k \in N_5$ and $1 \leq n_5 \leq (n - m)/2)$ with distinct or repeated eigenvalues.
(vi) n_6 represents invariant centers on n_6-pairs of $(\mathbf{u}_k, \mathbf{v}_k)$ $(k \in N_6)$ if $\mathrm{Re}\lambda_k = 0$ and $\mathrm{Im}\lambda_k \neq 0$ $(k \in N_6$ and $1 \leq n_6 \leq (n - m)/2)$ with distinct eigenvalues
(vii) $\varnothing$ represents none if $n_i = 0$ $(i \in \{1, 2, \cdots, 6\})$.

(viii) $[n_3; m_3]$ represents invariant centers on $(n_3 - m_3)$-directions of $\mathbf{v}_{k_3}$ $(k_3 \in N_3)$ and sources in m_3-directions of $\mathbf{v}_{j_3}(j_3 \in N_3$ and $j_3 \neq k_3)$ if $\lambda_k = 0\,(k \in N_3$ and $n_3 \leq m)$ with the $(m_3 + 1)$th-order nilpotent matrix $\mathbf{N}_3^{m_3+1} = \mathbf{0}$ $(0 < m_3 \leq n_3 - 1)$.

(ix) $[n_3; \varnothing]$ represents invariant centers on n_3-directions of $\mathbf{v}_k$ $(k \in N_3)$ if $\lambda_k = 0\,(k \in N_3$ and $1 < n_3 \leq m)$ with a nilpotent matrix $\mathbf{N}_3 = \mathbf{0}$.

(x) $[n_6, l; \mathbf{m}_6]$ represents invariant centers on $(n_6 - \Sigma_{s=1}^{l} m_{6s})$-pairs of $(\mathbf{u}_{k_6}, \mathbf{v}_{k_6})$ $(k_6 \in N_6)$, and sources in $\Sigma_{s=1}^{l} m_{6s}$-pairs of $(\mathbf{u}_{j_6}, \mathbf{v}_{j_6})$ $(j_6 \in N_6$ and $j_6 \neq k_6)$ if $\mathrm{Re}\lambda_k = 0$ and $\mathrm{Im}\lambda_k \neq 0\,(k \in N_6$ and $n_6 \leq (n - m)/2)$ for $\Sigma_{s=1}^{l} m_{6s}$-pairs of repeated eigenvalues with the $(m_{6s} + 1)$th-order nilpotent matrix $\mathbf{N}_6^{m_{6s}+1} = \mathbf{0}\,(0 < m_{6s} \leq l, s = 1, 2, \cdots, l)$.

(xi) $[n_6, l; \varnothing]$ represents invariant centers on n_6-pairs of $(\mathbf{u}_k, \mathbf{v}_k)\,(k \in N_6)$ if $\mathrm{Re}\lambda_k = 0$ and $\mathrm{Im}\lambda_k \neq 0\,(k \in N_6$ and $1 \leq n_6 \leq (n - m)/2)$ for $(l + 1)$ pairs of repeated eigenvalues with a nilpotent matrix $\mathbf{N}_6 = \mathbf{0}$.

Definition 1.18 Consider an n-dimensional, autonomous, nonlinear dynamical system $\dot{\mathbf{x}} = \mathbf{f}(\mathbf{x}, \mathbf{p})$ in Eq. (1.4) with an equilibrium point $\mathbf{x}^*$. Suppose there is a neighborhood of the equilibrium $\mathbf{x}^*$ as $U(\mathbf{x}^*) \subset \Omega$, then $\mathbf{f}(\mathbf{x}, \mathbf{p})$ is C^r $(r \geq 1)$-continuous in the neighborhood, and Eq. (1.24) holds. The corresponding solution is $\mathbf{x}(t) = \mathbf{\Phi}(\mathbf{x}_0, t - t_0, \mathbf{p})$. The matrix $D\mathbf{f}(\mathbf{x}^*, \mathbf{p})$ in Eq. (1.19) possesses n eigenvalues λ_k $(k = 1, 2, \cdots, n)$. Set $N = \{1, 2, \cdots, m, m + 1, \cdots, (n - m)/2\}$, $N_i = \{i_1, i_2, \cdots, i_{n_i}\} \cup \varnothing$ with $i_j \in N\,(j = 1, 2, \cdots, n_i,\ i = 1, 2, \cdots, 6)$, $\sum_{i=1}^{3} n_i = m$ and $2\Sigma_{i=4}^{6} n_i = n - m$. $\cup_{i=1}^{6} N_i = N$ and $N_i \cap N_l = \varnothing\ (l \neq i)$. $N_i = \varnothing$ if $n_i = 0$. The matrix $D\mathbf{f}(\mathbf{x}^*, \mathbf{p})$ possesses n_1-stable, n_2-unstable and n_3-invariant real eigenvectors plus n_4-stable, n_5-unstable and n_6-center pairs of complex eigenvectors. $k_3 \in \{\varnothing, m_3\}$. $\mathbf{k} = (k_{61}, k_{62}, \cdots, k_{6l})^T$ with $k_{6i} \in \{\varnothing, m_{6i}\}(i = 1, 2, \cdots, l)$. The flow $\mathbf{\Phi}(t)$ of the nonlinear system $\dot{\mathbf{x}} = \mathbf{f}(\mathbf{x}, \mathbf{p})$ is an $(n_1 : n_2 : [n_3 : \kappa_3] \| n_4 : n_5 : [n_6, l; \boldsymbol{\kappa}_6])$ flow in the neighborhood of $\mathbf{x}^*$.$\kappa_3 \in \{\varnothing, m_3\}$, $\boldsymbol{\kappa}_6 = (\kappa_{61}, \kappa_{62}, \cdots, \kappa_{6l})^{\mathrm{T}}$ with $\kappa_{6i} \in \{\varnothing, m_{6i}\}\ (i = 1, 2, \cdots, l)$. $\mathbf{m}_6 = (m_{61}, m_{62}, \cdots, m_{6l})^{\mathrm{T}}$.

I. *Non-degenerate cases*

(i) The equilibrium point $\mathbf{x}^*$ is an $(n_1 : n_2 : \varnothing | n_4 : n_5 : \varnothing)$ hyperbolic point (or saddle) for the nonlinear system.

(ii) The equilibrium point $\mathbf{x}^*$ is an $(n_1 : \varnothing : \varnothing | n_4 : \varnothing : \varnothing)$ sink for the nonlinear system.

(iii) The equilibrium point $\mathbf{x}^*$ is an $(\varnothing : n_2 : \varnothing | \varnothing : n_5 : \varnothing)$ source for the nonlinear system.

(iv) The equilibrium point $\mathbf{x}^*$ is an $(\varnothing : \varnothing : \varnothing | \varnothing : \varnothing : n/2)$ center for the nonlinear system.

(v) The equilibrium point $\mathbf{x}^*$ is an $(\varnothing : \varnothing : \varnothing | \varnothing : \varnothing : [n/2, l; \varnothing])$ center for the nonlinear system.

(vi) The equilibrium point $\mathbf{x}^*$ is an $(\varnothing : \varnothing : \varnothing | \varnothing : \varnothing : [n/2, l; \mathbf{m}_6])$ point for the nonlinear system.

(vii) The equilibrium point $\mathbf{x}^*$ is an $(n_1 : \varnothing : \varnothing | n_4 : \varnothing : n_6)$ point for the nonlinear system.
(viii) The equilibrium point $\mathbf{x}^*$ is an $(\varnothing : n_2 : \varnothing | \varnothing : n_5 : n_6)$ point for the nonlinear system.
(ix) The equilibrium point $\mathbf{x}^*$ is an $(n_1 : n_2 : \varnothing | n_4 : n_5 : n_6)$ point for the nonlinear system.

II. *Simple degenerate cases*

(i) The equilibrium point $\mathbf{x}^*$ is an $(\varnothing : \varnothing : [n; \varnothing] | \varnothing : \varnothing : \varnothing)$-invariant (or static) center for the nonlinear system.
(ii) The equilibrium point $\mathbf{x}^*$ is an $(\varnothing : \varnothing : [n; m_3] | \varnothing : \varnothing : \varnothing)$ point for the nonlinear system.
(iii) The equilibrium point $\mathbf{x}^*$ is an $(\varnothing : \varnothing : [n_3; \varnothing] | \varnothing : \varnothing : n_6)$ point for the nonlinear system.
(iv) The equilibrium point $\mathbf{x}^*$ is an $(\varnothing : \varnothing : [n_3; m_3] | \varnothing : \varnothing : n_6)$ point for the nonlinear system
(v) The equilibrium point $\mathbf{x}^*$ is an $(\varnothing : \varnothing : [n_3; \varnothing] | \varnothing : \varnothing : [n_6, l; \varnothing])$ point for the nonlinear system.
(vi) The equilibrium point $\mathbf{x}^*$ is an $(\varnothing : \varnothing : [n_3; m_3] | \varnothing : \varnothing : [n_6, l; \varnothing])$ point for the nonlinear system.
(vii) The equilibrium point $\mathbf{x}^*$ is an $(\varnothing : \varnothing : [n_3; \varnothing] | \varnothing : \varnothing : [n_6, l; \mathbf{m}_6])$ point for the nonlinear system.
(viii) The equilibrium point $\mathbf{x}^*$ is an $(\varnothing : \varnothing : [n_3; m_3] | \varnothing : \varnothing : [n_6, l; \mathbf{m}_6])$ point for the nonlinear system.

III. *Complex degenerate cases*

(i) The equilibrium point $\mathbf{x}^*$ is an $(n_1 : \varnothing : [n_3; \varnothing] | n_4 : \varnothing : \varnothing)$ point for the nonlinear system.
(ii) The equilibrium point $\mathbf{x}^*$ is an $(n_1 : \varnothing : [n_3; m_3] | n_4 : \varnothing : \varnothing)$ point for the nonlinear system.
(iii) The equilibrium point $\mathbf{x}^*$ is an $(\varnothing : n_2 : [n_3; \varnothing] | \varnothing : n_5 : \varnothing)$ point for the nonlinear system.
(iv) The equilibrium point $\mathbf{x}^*$ is an $(\varnothing : n_2 : [n_3; m_3] | \varnothing : n_5 : \varnothing)$ point for the nonlinear system.
(v) The equilibrium point $\mathbf{x}^*$ is an $(n_1 : \varnothing : [n_3; \varnothing] | n_4 : \varnothing : n_6)$ point for the nonlinear system.
(vi) The equilibrium point $\mathbf{x}^*$ is an $(n_1 : \varnothing : [n_3; m_3] | n_4 : \varnothing : n_6)$ point for the nonlinear system.
(vii) The equilibrium point $\mathbf{x}^*$ is an $(\varnothing : n_2 : [n_3; \varnothing] | \varnothing : n_5 : n_6)$ point for the nonlinear system.
(viii) The equilibrium point $\mathbf{x}^*$ is an $(\varnothing : n_2 : [n_3; m_3] | \varnothing : n_5 : n_6)$ point for the nonlinear system.
(ix) The equilibrium point $\mathbf{x}^*$ is an $(n_1 : \varnothing : [n_3; \varnothing] | n_4 : \varnothing : [n_6, l; \boldsymbol{\kappa}_6])$ point for the nonlinear system.

(x) The equilibrium point $\mathbf{x}^*$ is an $(n_1 : \varnothing : [n_3; m_3]|n_4 : \varnothing : [n_6, l; \kappa_6])$ point for the nonlinear system.
(xi) The equilibrium point $\mathbf{x}^*$ is an $(\varnothing : n_2 : [n_3; \varnothing]|\varnothing : n_5 : [n_6, l; \kappa_6])$ point for the nonlinear system.
(xii) The equilibrium point $\mathbf{x}^*$ is an $(\varnothing : n_2 : [n_3; m_3]|\varnothing : n_5 : [n_6, l; \kappa_6])$ point for the nonlinear system.

Definition 1.19 Consider an n-dimensional, autonomous, nonlinear dynamical system $\dot{\mathbf{x}} = \mathbf{f}(\mathbf{x}, \mathbf{p})$ in Eq. (1.4) with an equilibrium point $\mathbf{x}^*$. Suppose there is a neighborhood of the equilibrium $\mathbf{x}^*$ as $U(\mathbf{x}^*) \subset \Omega$, then $\mathbf{f}(\mathbf{x}, \mathbf{p})$ is C^r $(r \geq 1)$-continuous in the neighborhood, and Eq. (1.24) holds. The corresponding solution is $\mathbf{x}(t) = \mathbf{\Phi}(\mathbf{x}_0, t - t_0, \mathbf{p})$. The matrix $D\mathbf{f}(\mathbf{x}^*, \mathbf{p})$ in Eq. (1.19) possesses n eigenvalues λ_k $(k = 1, 2, \cdots n)$. Set $N = \{1, 2, \cdots, n\}$, $N_i = \{i_1, i_2, \cdots, i_{n_i}\} \cup \varnothing$ with $i_j \in N$ $(j = 1, 2, \cdots, n_i,\ i = 1, 2, 3)$ and $\Sigma_{i=1}^{3} n_i = n$. $\cup_{i=1}^{3} N_i = N$ and $N_i \cap N_l = \varnothing (l \neq i)$. $N_i = \varnothing$ if $n_i = 0$. The matrix $D\mathbf{f}(\mathbf{x}^*, \mathbf{p})$ possesses n_1-stable, n_2-unstable and n_3-invariant real eigenvectors. Without repeated eigenvalues of $\lambda_k = 0$ $(k \in N_3)$, the flow $\mathbf{\Phi}(t)$ of the nonlinear system $\dot{\mathbf{x}} = \mathbf{f}(\mathbf{x}, \mathbf{p})$ in Eq. (1.4) is an $(n_1 : n_2 : \varnothing|$ or $(n_1 : n_2 : 1|$ local flow in the neighborhood of equilibrium point $\mathbf{x}^*$. However, with repeated eigenvalues of $\lambda_\kappa = 0$ $(\kappa \in N_3)$, the flow $\mathbf{\Phi}(t)$ of the nonlinear system $\dot{\mathbf{x}} = \mathbf{f}(\mathbf{x}, \mathbf{p})$ in Eq. (1.4) is an $(n_1 : n_2 : [n_3; m_3]|$ local flow in the neighborhood of equilibrium point $\mathbf{x}^*$.

I. *Non-degenerate cases*

(i) The equilibrium point $\mathbf{x}^*$ is an $(n : \varnothing : \varnothing|$-stable node for the nonlinear system.
(ii) The equilibrium point $\mathbf{x}^*$ is an $(\varnothing : n : \varnothing|$-unstable node for the nonlinear system.
(iii) The equilibrium point $\mathbf{x}^*$ is an $(n_1 : n_2 : \varnothing|$-saddle for the nonlinear system.

II. *Degenerate cases*

(i) The equilibrium point $\mathbf{x}^*$ is an $(n_1 : n_2 : 1|$-critical state for the nonlinear system.
(ii) The equilibrium point $\mathbf{x}^*$ is an $(n_1 : n_2 : [n_3; \varnothing]|$ point for the nonlinear system.
(iii) The equilibrium point $\mathbf{x}^*$ is an $(n_1 : n_2 : [n_3; m_3]|$ point for the nonlinear system.

Definition 1.20 Consider a $2n$-dimensional, autonomous, nonlinear dynamical system $\dot{\mathbf{x}} = \mathbf{f}(\mathbf{x}, \mathbf{p})$ in Eq. (1.4) with an equilibrium point $\mathbf{x}^*$. Suppose there is a neighborhood of the equilibrium $\mathbf{x}^*$ as $U(\mathbf{x}^*) \subset \Omega$, then $\mathbf{f}(\mathbf{x}, \mathbf{p})$ is C^r $(r \geq 1)$-continuous in the neighborhood, and Eq. (1.24) holds. The corresponding solution is $\mathbf{x}(t) = \mathbf{\Phi}(\mathbf{x}_0, t - t_0, \mathbf{p})$. The matrix $D\mathbf{f}(\mathbf{x}^*, \mathbf{p})$ in Eq. (1.19) possesses n-pairs of complex eigenvalues $(k = 1, 2, \cdots, n)$. Set $N = \{1, 2, \cdots, n\}$, $N_i = \{i_1, i_2, \cdots, i_{n_i}\} \cup \varnothing$ with $l_j \in N$ $(j = 1, 2, \cdots, n_i; i = 4, 5, 6)$ and $\Sigma_{i=4}^{6} n_i = n$. $\cup_{i=4}^{6} N_i = N$ and $N_i \cap N_l = \varnothing (l \neq i)$. $N_i = \varnothing$ if $n_i = 0$. The matrix $D\mathbf{f}(\mathbf{x}^*, \mathbf{p})$ possesses n_4-stable,

n_5-unstable and n_6-center pairs of complex eigenvectors. Without repeated eigenvalues of $\mathrm{Re}\lambda_k = 0\,(k \in N_6)$, the flow $\mathbf{\Phi}(t)$ of the nonlinear system $\dot{\mathbf{x}} = \mathbf{f}(\mathbf{x}, \mathbf{p})$ in Eq. (1.4) is an $|n_4 : n_5 : n_6)$ local flow in the neighborhood of equilibrium point $\mathbf{x}^*$. However, with repeated eigenvalues of $\mathrm{Re}\lambda_k = 0\,(k \in N_6)$, the flow $\mathbf{\Phi}(t)$ of the nonlinear system $\dot{\mathbf{x}} = \mathbf{f}(\mathbf{x}, \mathbf{p})$ in Eq. (1.4) is an $|n_4 : n_5 : [n_6, l; \mathbf{k}_6])$ local flow in the neighborhood of equilibrium point $\mathbf{x}^*$. $\boldsymbol{\kappa}_6 = (\kappa_{61}, \kappa_{62}, \cdots, \kappa_{6l})$ with $\kappa_{6i} \in \{\varnothing, m_{6i}\}\,(i = 1, 2, \cdots, l)$. $\mathbf{m}_6 = (m_{61}, m_{62}, \cdots, m_{6l})^T$

I. *Non-degenerate cases*

(i) The equilibrium point $\mathbf{x}^*$ is an $|n : \varnothing : \varnothing)$-spiral sink for the nonlinear system.
(ii) The equilibrium point $\mathbf{x}^*$ is an $|\varnothing : n : \varnothing)$-spiral source for the nonlinear system.
(iii) The equilibrium point $\mathbf{x}^*$ is an $|\varnothing : \varnothing : n)$-center for the nonlinear system.
(iv) The equilibrium point $\mathbf{x}^*$ is an $|n_4 : n_5 : \varnothing)$-spiral saddle for the nonlinear system.

II. *Quasi-degenerate cases*

(i) The equilibrium point $\mathbf{x}^*$ is an $|n_4 : \varnothing : n_6)$-point for the nonlinear system.
(ii) The equilibrium point $\mathbf{x}^*$ is an $|\varnothing : n_5 : n_6)$-point for the nonlinear system.
(iii) The equilibrium point $\mathbf{x}^*$ is an $|n_4 : \varnothing : [n_6, l; \varnothing])$-point for the nonlinear system.
(iv) The equilibrium point $\mathbf{x}^*$ is an $|n_4 : \varnothing : [n_6, l; \boldsymbol{\kappa}_6])$-point for the nonlinear system.
(v) The equilibrium point $\mathbf{x}^*$ is an $|\varnothing : n_5 : [n_6; \varnothing])$-point for the nonlinear system.
(vi) The equilibrium point $\mathbf{x}^*$ is an $|\varnothing : n_5 : [n_6, l; \boldsymbol{\kappa}_6])$-point for the nonlinear system.

1.3 Bifurcation and Stability Switching

As before, the dynamical characteristics of equilibriums in nonlinear dynamical systems in Eq. (1.4) are based on the given parameters. With varying parameters in dynamical systems, the corresponding dynamical behaviors will change qualitatively. The qualitative switching of dynamical behaviors in dynamical systems is called *bifurcation* and the corresponding parameter values are called *bifurcation values*. To understand the qualitative changes of dynamical behaviors of nonlinear systems with parameters in the neighborhood of equilibriums, the bifurcation theory for equilibrium of nonlinear dynamical system in Eq. (1.4) will be presented. $D_{\mathbf{x}}() = \partial()/\partial\mathbf{x}$ and $D_{\mathbf{p}}() = \partial()/\partial\mathbf{p}$ will be adopted from now on. For no specific notice, $D \equiv D_{\mathbf{x}}$.

Definition 1.21 Consider an n-dimensional, autonomous, nonlinear dynamical system $\dot{\mathbf{x}} = \mathbf{f}(\mathbf{x}, \mathbf{p})$ in Eq. (1.4) with an equilibrium point $(\mathbf{x}^*, \mathbf{p})$. Suppose there is a neighborhood of the equilibrium $\mathbf{x}^*$ as $U(\mathbf{x}^*) \subset \Omega$, and in the neighborhood, equation (1.24) holds. The linearized system of the nonlinear system at the equilibrium point $(\mathbf{x}^*, \mathbf{p})$ is $\dot{\mathbf{y}} = D_{\mathbf{x}}\mathbf{f}(\mathbf{x}^*, \mathbf{p})\mathbf{y}$ $(\mathbf{y} = \mathbf{x} - \mathbf{x}^*)$ in Eq. (1.19).

(i) The equilibrium point $(\mathbf{x}_0^*, \mathbf{p}_0)$ is called *the switching point* of equilibrium solutions if $D_{\mathbf{x}}\mathbf{f}(\mathbf{x}^*, \mathbf{p})$ at $(\mathbf{x}_0^*, \mathbf{p}_0)$ possesses at least one more real eigenvalue (or one more pair of complex eigenvalues) with zero real part.
(ii) The value $\mathbf{p}_0$ in Eq. (1.4) is called *a switching value* of $\mathbf{p}$ if the dynamical characteristics at point $(\mathbf{x}_0^*, \mathbf{p}_0)$ change from one state to another state.
(iii) The equilibrium point $(\mathbf{x}_0^*, \mathbf{p}_0)$ is called the *bifurcation point* of equilibrium solutions if $D_{\mathbf{x}}\mathbf{f}(\mathbf{x}^*, \mathbf{p})$ at $(\mathbf{x}_0^*, \mathbf{p}_0)$ possesses at least one more real eigenvalue (or one more pair of complex eigenvalues) with zero real part, and more than one branches of equilibrium solutions appear or disappear.
(iv) The value $\mathbf{p}_0$ in Eq. (1.4) is called a *bifurcation value* of $\mathbf{p}$ if the dynamical characteristics at point $(\mathbf{x}_0^*, \mathbf{p}_0)$ change from one stable state into another unstable state.

Definition 1.22 Consider an n-dimensional, autonomous, nonlinear dynamical system $\dot{\mathbf{x}} = \mathbf{f}(\mathbf{x}, \mathbf{p})$ in Eq. (1.4) with an equilibrium point $\mathbf{x}^*$. Suppose there is a neighborhood of the equilibrium $\mathbf{x}^*$ as $U(\mathbf{x}^*) \subset \Omega$, then $\mathbf{f}(\mathbf{x}, \mathbf{p})$ is C^r $(r \geq 1)$-continuous in the neighborhood, and Eq. (1.24) holds. The corresponding solution is $\mathbf{x}(t) = \mathbf{\Phi}(\mathbf{x}_0, t - t_0, \mathbf{p})$. The matrix $D\mathbf{f}(\mathbf{x}^*, \mathbf{p})$ in Eq. (1.19) possesses n eigenvalues λ_k $(k = 1, 2, \cdots, n)$. Set $N = \{1, 2, \cdots, m, m+1, \cdots, (n-m)/2\}$, $N_i = \{i_1, i_2, \cdots, i_{n_i}\} \cup \varnothing$ with $i_j \in N$ $(j = 1, 2, \cdots, n_i,\ i = 1, 2, \cdots, 6)$, $\Sigma_{i=1}^{3} n_i = m$ and $2\Sigma_{i=4}^{6} n_i = n - m$. $\cup_{i=1}^{6} N_i = N$ and $N_i \cap N_l = \varnothing (l \neq i)$. $N_i = \varnothing$ if $n_i = 0$. The matrix $D\mathbf{f}(\mathbf{x}^*, \mathbf{p})$ possesses n_1-stable, n_2-unstable and n_3-invariant real eigenvectors plus n_4-stable, n_5-unstable and n_6-center pairs of complex eigenvectors $\kappa_3 \in \{\varnothing, m_3\}$. $\boldsymbol{\kappa}_6 = (\kappa_{61}, \kappa_{62}, \cdots, \kappa_{6l})^{\mathrm{T}}$ with $\kappa_{6i} \in \{\varnothing, m_{6i}\} (i = 1, 2, \cdots, l)$.

I. *Simple switching and bifurcation*

(i) An $(n_1 : n_2 : 1|n_4 : n_5 : \varnothing)$ state of equilibrium point $(\mathbf{x}_0^*, \mathbf{p}_0)$ for the nonlinear system is a *switching* of the $(n_1 : n_2 + 1 : \varnothing|n_4 : n_5 : \varnothing)$-spiral saddle and $(n_1 + 1 : n_2 : \varnothing|n_4 : n_5 : \varnothing)$-spiral saddle of equilibrium point $(\mathbf{x}^*, \mathbf{p})$.
(ii) An $(n_1 - 1 : \varnothing : 1|n_4 : \varnothing : \varnothing)$ state of equilibrium point $(\mathbf{x}_0^*, \mathbf{p}_0)$ for the nonlinear system is a *stable bifurcation* of its $(n_1 - 1 : 1 : \varnothing|n_4 : \varnothing : \varnothing)$-spiral saddle and $(n_1 : \varnothing : \varnothing|n_4 : \varnothing : \varnothing)$-spiral sink of equilibrium point $(\mathbf{x}^*, \mathbf{p})$.
(iii) An $(\varnothing : n_2 - 1 : 1|\varnothing : n_5 : \varnothing)$ state of equilibrium point $(\mathbf{x}_0^*, \mathbf{p}_0)$ for the nonlinear system is an unstable *bifurcation* of its$(1 : n_2 - 1 : \varnothing|\varnothing : n_5 : \varnothing)$-spiral saddle and $(\varnothing : n_2 : \varnothing|\varnothing : n_5 : \varnothing)$-spiral source of equilibrium point $(\mathbf{x}^*, \mathbf{p})$.

(iv) An $(n_1:n_2:\varnothing|n_4:n_5:1)$ state of equilibrium point $(\mathbf{x}_0^*, \mathbf{p}_0)$ for the nonlinear system is a *Hopf bifurcation* of its $(n_1:n_2:\varnothing|n_4+1:n_5:\varnothing)$-spiral saddle and $(n_1:n_2:\varnothing|n_4:n_5+1:\varnothing)$-spiral saddle of equilibrium point $(\mathbf{x}^*, \mathbf{p})$.

(v) An $(n_1:\varnothing:\varnothing|n_4-1:\varnothing:1)$ state of equilibrium point $(\mathbf{x}_0^*, \mathbf{p}_0)$ for the nonlinear system is a *stable Hopf bifurcation* of its $(n_1:\varnothing:\varnothing|n_4-1:1:\varnothing)$-spiral saddle and $(n_1:\varnothing:\varnothing|n_4:\varnothing:\varnothing)$-spiral sink of equilibrium point $(\mathbf{x}^*, \mathbf{p})$.

(vi) An $(\varnothing:n_2:\varnothing|\varnothing:n_5-1:1)$ state of equilibrium point $(\mathbf{x}_0^*, \mathbf{p}_0)$ for the nonlinear system is an *unstable Hopf bifurcation* of its $(\varnothing:n_2:\varnothing|1:n_5-1:\varnothing)$-spiral saddle and $(\varnothing:n_2:\varnothing|\varnothing:n_5:\varnothing)$-spiral source of equilibrium point $(\mathbf{x}^*, \mathbf{p})$.

(vii) An $(n_1:n_2:1|n_4:n_5:n_6)$ state of equilibrium point $(\mathbf{x}_0^*, \mathbf{p}_0)$ for the nonlinear system is a switching of the $(n_1+1:n_2:\varnothing|n_4:n_5:n_6)$ state and $(n_1:n_2+1:\varnothing|n_4:n_5:n_6)$ state of equilibrium point $(\mathbf{x}^*, \mathbf{p})$.

(viii) An $(n_1:n_2:1|n_4:n_5:[n_6;l;\kappa_6])$ state of equilibrium point $(\mathbf{x}_0^*, \mathbf{p}_0)$ for the nonlinear system is a switching of its $(n_1+1:n_2:\varnothing|n_4:n_5:[n_6;l;\kappa_6])$ state and $(n_1:n_2+1:\varnothing|n_4:n_5:[n_6;l;\kappa_6])$ state of equilibrium point $(\mathbf{x}^*, \mathbf{p})$.

(ix) An $(n_1:n_2:\varnothing|n_4:n_5:n_6+1)$ state of equilibrium point $(\mathbf{x}_0^*, \mathbf{p}_0)$ for the nonlinear system is a *Hopf* switching of its $(n_1:n_2:\varnothing|n_4+1:n_5:n_6)$ state and $(n_1:n_2:\varnothing|n_4:n_5+1:n_6)$ state of equilibrium point $(\mathbf{x}^*, \mathbf{p})$.

(x) An $(n_1:n_2:[n_3;\varnothing]|n_4:n_5:n_6+1)$ state of equilibrium point $(\mathbf{x}_0^*, \mathbf{p}_0)$ for the nonlinear system is a *Hopf* switching of its $(n_1:n_2:[n_3;\varnothing]|n_4+1:n_5:n_6)$ state and $(n_1:n_2:[n_3;\varnothing]|n_4:n_5+1:n_6)$ state of equilibrium point $(\mathbf{x}^*, \mathbf{p})$.

II. *Complex switching*

(i) An $(n_1:n_2:[n_3;\kappa_3]|n_4:n_5:n_6)$-state of equilibrium point $(\mathbf{x}_0^*, \mathbf{p}_0)$ for the nonlinear system is a *degenerate switching* of its $(n_1+n_3:n_2:\varnothing|n_4:n_5:n_6)$ state and $(n_1:n_2+n_3:\varnothing|n_4:n_5:n_6)$ state of equilibrium point $(\mathbf{x}^*, \mathbf{p})$.

(ii) An $(n_1-m:n_2:[n_3+m;\kappa_3]|n_4:n_5:n_6)$-state of equilibrium point $(\mathbf{x}_0^*, \mathbf{p}_0)$ for the nonlinear system is a degenerate *switching* of its $(n_1:n_2:[n_3;\kappa_3']|n_4:n_5:n_6)$ state and $(n_1-m:n_2+m:[n_3;\kappa_3']|n_4:n_5:n_6)$ state of equilibrium point $(\mathbf{x}^*, \mathbf{p})$.

(iii) An $(n_1:n_2-m:[n_3+m;\kappa_3]|n_4:n_5:n_6)$-state of equilibrium point $(\mathbf{x}_0^*, \mathbf{p}_0)$ for the nonlinear system is a switching of its $(n_1:n_2:[n_3;\kappa_3']|n_4:n_5:n_6)$-state and $(n_1+m:n_2-m:[n_3;\kappa_3']|n_4:n_5:n_6)$ state of equilibrium point $(\mathbf{x}^*, \mathbf{p})$.

(iv) An $(n_1-m_1:n_2:[n_3+m_1;\kappa_3]|n_4:n_5:n_6)$-state of equilibrium point $(\mathbf{x}_0^*, \mathbf{p}_0)$ for the nonlinear system is a switching of $(n_1:n_2:[n_3;\kappa_3']|n_4:n_5:n_6)$ state and $(n_1-m_1:n_2+m_2:[n_3+m_1-m_2;\kappa_3'']|n_4:n_5:n_6)$ state of equilibrium point $(\mathbf{x}^*, \mathbf{p})$.

(v) An $(n_1 : n_2 - m_2 : [n_3 + m_2; \kappa_3] | n_4 : n_5 : n_6)$-state of equilibrium point $(\mathbf{x}_0^*, \mathbf{p}_0)$ for the nonlinear system is a degenerate switching of its $(n_1 : n_2 : [n_3; \kappa_3'] | n_4 : n_5 : n_6)$ state and $(n_1 + m_1 : n_2 - m_2 : [n_3 + m_2 - m_1; \kappa_3''] | n_4 : n_5 : n_6)$ state of equilibrium point $(\mathbf{x}^*, \mathbf{p})$.

(vi) An $(n_1 : n_2 : [n_3; \kappa_3] | n_4 : n_5 : [n_6, l; \boldsymbol{\kappa}_6])$ state of equilibrium point $(\mathbf{x}_0^*, \mathbf{p}_0)$ for the nonlinear system is a degenerate Hopf switching of its $(n_1 : n_2 : [n_3; \kappa_3] | n_6 + n_4 : n_5 : \varnothing)$ state and $(n_1 : n_2 : [n_3; \kappa_3] | n_4 : n_5 + n_6 : \varnothing)$ state of equilibrium point $(\mathbf{x}^*, \mathbf{p})$.

(vii) An $(n_1 : n_2 : [n_3; \kappa_3] | n_4 - m : n_5 : [n_6 + m, l_1; \boldsymbol{\kappa}_6])$ state of equilibrium point $(\mathbf{x}_0^*, \mathbf{p}_0)$ for the nonlinear system is a degenerate Hopf switching of its $(n_1 : n_2 : [n_3; \kappa_3] | n_4 : n_5 : [n_6, l_2; \boldsymbol{\kappa}_6])$ state and $(n_1 : n_2 : [n_3; \kappa_3] | n_4 - m : n_5 + m : [n_6, l_2; \boldsymbol{\kappa}_6])$ state of equilibrium point $(\mathbf{x}^*, \mathbf{p})$.

(viii) An $(n_1 : n_2 : [n_3; \kappa_3] | n_4 : n_5 - m : [n_6 + m, l + m; \boldsymbol{\kappa}_6])$ of equilibrium point $(\mathbf{x}_0^*, \mathbf{p}_0)$ for the nonlinear system is a degenerate Hopf switching of its $(n_1 : n_2 : [n_3; \kappa_3] | n_4 : n_5 : [n_6, l; \boldsymbol{\kappa}_6])$ state and $(n_1 : n_2 : [n_3; \kappa_3] | n_4 + m : n_5 - m : [n_6, l; \boldsymbol{\kappa}_6])$ state of equilibrium point $(\mathbf{x}^*, \mathbf{p})$.

(ix) An $(n_1 : n_2 : [n_3; \kappa_3] | n_4 : n_5 - m : [n_6 + m, l_1; \boldsymbol{\kappa}_6])$ state of equilibrium point $(\mathbf{x}_0^*, \mathbf{p}_0)$ for the nonlinear system is a degenerate Hopf boundary of its $(n_1 : n_2 : [n_3; \kappa_3] | n_4 : n_5 : [n_6, l_2; \boldsymbol{\kappa}_6])$ state and $(n_1 : n_2 : [n_3; \kappa_3] | n_4 + m; n_5 - m : [n_6, l_2; \boldsymbol{\kappa}_6'])$ state of equilibrium point $(\mathbf{x}^*, \mathbf{p})$.

(x) An $(n_1 : n_2 : [n_3; \kappa_3] | n_4 - m_4 : n_5 : [n_6 + m_4, l_1; \boldsymbol{\kappa}_6])$ state of equilibrium point $(\mathbf{x}_0^*, \mathbf{p}_0)$ for the nonlinear system is a degenerate Hopf switching of $(n_1 : n_2 : [n_3; \kappa_3] | n_4 : n_5 : [n_6, l_2; \boldsymbol{\kappa}_6'])$ state and $(n_1 : n_2 : [n_3; \kappa_3] | n_4 - m_4 : n_5 + m_5 : [n_6 + m_4 - m_5, l_3; \boldsymbol{\kappa}_6''])$ state of equilibrium point $(\mathbf{x}^*, \mathbf{p})$.

Definition 1.23 Consider an n-dimensional, autonomous, nonlinear dynamical system $\dot{\mathbf{x}} = \mathbf{f}(\mathbf{x}, \mathbf{p})$ in Eq. (1.4) with an equilibrium point $\mathbf{x}^*$. $\kappa_3 \in \{\varnothing, m_3\}$. Suppose there is a neighborhood of the equilibrium $\mathbf{x}^*$ as $U(\mathbf{x}^*) \subset \Omega$, then $\mathbf{f}(\mathbf{x}, \mathbf{p})$ is C^r $(r \geq 1)$-continuous in the neighborhood, and Eq. (1.24) holds. The corresponding solution is $\mathbf{x}(t) = \boldsymbol{\Phi}(\mathbf{x}_0, t - t_0, \mathbf{p})$. The matrix $D\mathbf{f}(\mathbf{x}^*, \mathbf{p})$ in Eq. (1.19) possesses n eigenvalues λ_k $(k = 1, 2, \cdots n)$. Set $N = \{1, 2, \cdots, n\}$, $N_i = \{i_1, i_2, \cdots, i_{n_i}\} \cup \varnothing$ with $i_j \in N$ $(j = 1, 2, \cdots, n_i,\ i = 1, 2, 3)$ and $\Sigma_{i=1}^3 n_i = n$. $\cup_{i=1}^3 N_i = N$ and $N_i \cap N_l = \varnothing$ $(l \neq i)$. $N_i = \varnothing$ if $n_i = 0$. The matrix $D\mathbf{f}(\mathbf{x}^*, \mathbf{p})$ possesses n_1-stable, n_2-unstable and n_3-invariant real eigenvectors. Without repeated eigenvalues of $\lambda_k = 0$ $(k \in N_3)$, the flow $\boldsymbol{\Phi}(t)$ of the nonlinear system $\dot{\mathbf{x}} = \mathbf{f}(\mathbf{x}, \mathbf{p})$ in Eq. (1.4) is an $(n_1 : n_2 : \varnothing|$ or $(n_1 : n_2 : 1|$ local flow in the neighborhood of equilibrium point $\mathbf{x}^*$. However, with repeated eigenvalues of $\lambda_k = 0$ $(k \in N_3)$, the flow $\boldsymbol{\Phi}(t)$ of the nonlinear system $\dot{\mathbf{x}} = \mathbf{f}(\mathbf{x}, \mathbf{p})$ in Eq. (1.4) is an $(n_1 : n_2 : [n_3; m_3]|$ local flow in the neighborhood of equilibrium point $\mathbf{x}^*$. $\kappa_3 \in \{\varnothing, m_3\}$.

I. *Simple switching and bifurcation*

(i) An $(n_1 : n_2 : 1|$ state of equilibrium point $(\mathbf{x}_0^*, \mathbf{p}_0)$ for the nonlinear system is a *saddle-saddle switching* of its $(n_1 + 1 : n_2 : \varnothing|$-saddle and $(n_1 : n_2 + 1 : \varnothing|$-saddle of equilibrium point $(\mathbf{x}^*, \mathbf{p})$.

(ii) An $(n-1:\varnothing:1|$ state of equilibrium point $(\mathbf{x}_0^*, \mathbf{p}_0)$ for the nonlinear system is a *stable saddle-node bifurcation* of its $(n-1:1:\varnothing|$-saddle and $(n:\varnothing:\varnothing|$-stable node of equilibrium point $(\mathbf{x}^*, \mathbf{p})$.

(iii) An $(\varnothing:n-1:1|$ state of equilibrium point $(\mathbf{x}_0^*, \mathbf{p}_0)$ for the nonlinear system is an *unstable saddle-node bifurcation* of $(1:n-1:\varnothing|$-saddle and $(\varnothing:n:\varnothing|$-unstable node of equilibrium point $(\mathbf{x}^*, \mathbf{p})$.

(iv) An $(n_1-1:n_2:[n_3+1;\kappa_3]|$ state of equilibrium point $(\mathbf{x}_0^*, \mathbf{p}_0)$ for the nonlinear system is a *degenerate saddle-saddle switching* of $(n_1:n_2:[n_3;\kappa_3]|$-degenerate saddle and $(n_1-1:n_2+1:[n_3;\kappa_3]|$-degenerate saddle of equilibrium point $(\mathbf{x}^*, \mathbf{p})$.

II. *Complex switching*

(i) An $(n_1:n_2:[n_3;\kappa_3]|$ state of equilibrium point $(\mathbf{x}_0^*, \mathbf{p}_0)$ for the nonlinear system is a *degenerate switching* of $(n_1+n_3:n_2:\varnothing|$-saddle and $(n_1:n_2+n_3:\varnothing|$-saddle of equilibrium point $(\mathbf{x}^*, \mathbf{p})$.

(ii) An $(n_1-m:n_2:[n_3+m;\kappa_3]|$ state of equilibrium point $(\mathbf{x}_0^*, \mathbf{p}_0)$ for the nonlinear system is a *degenerate switching* of the $(n_1:n_2:[n_3;\kappa_3']|$ state and $(n_1-m:n_2+m:[n_3;\kappa_3']|$ state of equilibrium point $(\mathbf{x}^*, \mathbf{p})$.

(iii) An $(n_1:n_2-m:[n_3+m;\kappa_3]|$ state of equilibrium point $(\mathbf{x}_0^*, \mathbf{p}_0)$ for the nonlinear system is a *degenerate switching* of its $(n_1:n_2:[n_3;\kappa_3']|$-state and $(n_1+m:n_2-m:[n_3;\kappa_3']|$ state of equilibrium point $(\mathbf{x}^*, \mathbf{p})$.

(iv) An $(n_1-m_1:n_2:[n_3+m_1;\kappa_3]|$-state of equilibrium point $(\mathbf{x}_0^*, \mathbf{p}_0)$ for the nonlinear system is a *degenerate switching* of $(n_1:n_2:[n_3;\kappa_3']|$ state and $(n_1-m_1:n_2+m_2:[n_3+m_1-m_2;\kappa_3'']|$ state of equilibrium point $(\mathbf{x}^*, \mathbf{p})$.

(v) An $(n_1:n_2-m_2:[n_3+m_2;\kappa_3]|$-state of equilibrium point $(\mathbf{x}_0^*, \mathbf{p}_0)$ for the nonlinear system is a *degenerate switching* of $(n_1:n_2:[n_3;\kappa_3']|$-state and $(n_1+m_1:n_2-m_2:[n_3+m_2-m_1;\kappa_3'']|$-state of equilibrium point $(\mathbf{x}^*, \mathbf{p})$.

Definition 1.24 Consider a $2n$-dimensional, autonomous, nonlinear dynamical system $\dot{\mathbf{x}} = \mathbf{f}(\mathbf{x}, \mathbf{p})$ in Eq. (1.4) with an equilibrium point $\mathbf{x}^*$. Suppose there is a neighborhood of the equilibrium $\mathbf{x}^*$ as $U(\mathbf{x}^*) \subset \Omega$, then $\mathbf{f}(\mathbf{x}, \mathbf{p})$ is C^r $(r \geq 1)$-continuous in the neighborhood, and Eq. (1.24) holds. The corresponding solution is $\mathbf{x}(t) = \mathbf{\Phi}(\mathbf{x}_0, t-t_0, \mathbf{p})$. The matrix $D\mathbf{f}(\mathbf{x}^*, \mathbf{p})$ in Eq. (1.19) possesses n-pairs of complex eigenvalues $(k = 1, 2, \cdots, n)$. Set $N = \{1, 2, \cdots, n\}$, $N_i = \{i_1, i_2, \cdots, i_{n_i}\} \cup \varnothing$ with $l_j \in N$ $(j = 1, 2, \cdots, n_i; i = 4, 5, 6)$ and $\Sigma_{i=4}^{6} n_i = n$. $\cup_{i=4}^{6} N_i = N$ and $N_i \cap N_l = \varnothing (l \neq i)$. $N_i = \varnothing$ if $n_j = 0$. The matrix $D\mathbf{f}(\mathbf{x}^*, \mathbf{p})$ possesses n_4-stable, n_5-unstable and n_6-center pairs of complex eigenvectors. Without repeated eigenvalues of $Re\lambda_k = 0$ $(k \in N_6)$, the flow $\mathbf{\Phi}(t)$ of the nonlinear system $\dot{\mathbf{x}} = \mathbf{f}(\mathbf{x}, \mathbf{p})$ in Eq. (1.4) is an $|n_4 : n_5 : n_6)$ local flow in the neighborhood of equilibrium point $\mathbf{x}^*$. However, with repeated eigenvalues of $\text{Re}\lambda_k = 0\,(k \in N_6)$, the flow $\mathbf{\Phi}(t)$ of the nonlinear system $\dot{\mathbf{x}} = \mathbf{f}(\mathbf{x}, \mathbf{p})$ in Eq. (1.4) is an $|n_4 : n_5 : [n_6, l; \boldsymbol{\kappa}_6])$ local flow in the neighborhood of equilibrium point $\mathbf{x}^*$. $\boldsymbol{\kappa}_6 = (\kappa_{61}, \kappa_{62}, \cdots, \kappa_{6l})^{\mathrm{T}}$ with $\kappa_{6i} \in \{\varnothing, m_{6i}\} (i = 1, 2, \cdots, l)$.

I. *Simple switching and bifurcation*

(i) An $|n_4 : n_5 : 1)$-state of equilibrium point $(\mathbf{x}_0^*, \mathbf{p}_0)$ for the nonlinear system is a Hopf switching of its $|n_4 + 1 : n_5 : \varnothing)$-spiral saddle and $|n_4 : n_5 + 1 : \varnothing)$-spiral saddle of equilibrium point $(\mathbf{x}^*, \mathbf{p})$.
(ii) An $|n_4 : \varnothing : 1)$-state equilibrium point $(\mathbf{x}_0^*, \mathbf{p}_0)$ for the nonlinear system is a stable Hopf bifurcation of its $|n_4 + 1 : \varnothing : \varnothing)$-spiral sink and $|n_4 : 1 : \varnothing)$-spiral saddle of equilibrium point $(\mathbf{x}^*, \mathbf{p})$.
(iii) An $|\varnothing : n_5 : 1)$-state equilibrium point $(\mathbf{x}_0^*, \mathbf{p}_0)$ for the nonlinear system is an unstable Hopf bifurcation of $|\varnothing : n_5 + 1 : \varnothing)$-spiral source and $|1 : n_5 : \varnothing)$-spiral saddle of equilibrium point $(\mathbf{x}^*, \mathbf{p})$.
(iv) An $|n_4 : n_5 : n_6 + 1)$-state equilibrium point $(\mathbf{x}_0^*, \mathbf{p}_0)$ for the nonlinear system is a Hopf switching of its $|n_4 + 1 : n_5 : n_6)$-state and $|n_4 : n_5 + 1 : n_6)$-state of equilibrium point $(\mathbf{x}^*, \mathbf{p})$.
(v) An $|n_4 : \varnothing : n_6 + 1)$-state equilibrium point $(\mathbf{x}_0^*, \mathbf{p}_0)$ for the nonlinear system is a Hopf switching of its $|n_4 + 1 : \varnothing : n_6)$-state and $|n_4 : 1 : n_6)$-state of equilibrium point $(\mathbf{x}^*, \mathbf{p})$.
(vi) A $|\varnothing : n_5 : n_6 + 1)$-state equilibrium point $(\mathbf{x}_0^*, \mathbf{p}_0)$ for the nonlinear system is a Hopf switching of $|\varnothing : n_5 + 1 : n_6)$-state source and $|1 : n_5 : n_6)$-state of equilibrium point $(\mathbf{x}^*, \mathbf{p})$.

II. *Complex Hopf switching*

(i) An $|n_4 : n_5 : [n_6, l; \boldsymbol{\kappa}_6])$-state of equilibrium point $(\mathbf{x}_0^*, \mathbf{p}_0)$ for the nonlinear system is a degenerate Hopf switching of its $|n_6 + n_4 : n_5 : \varnothing)$-spiral saddle and $|n_4 : n_5 + n_6 : \varnothing)$-spiral saddle of equilibrium point $(\mathbf{x}^*, \mathbf{p})$.
(ii) An $|n_4 - m : n_5 : [n_6 + m, l_2; \boldsymbol{\kappa}_6])$-state of equilibrium point $(\mathbf{x}_0^*, \mathbf{p}_0)$ for the nonlinear system is a degenerate Hopf switching of its $|n_4 : n_5 : [n_6, l_1; \boldsymbol{\kappa}_6'])$-state and $|n_4 - m : n_5 + m : [n_6, l_3; \boldsymbol{\kappa}_6''])$-state of equilibrium point $(\mathbf{x}^*, \mathbf{p})$.
(iii) An $|n_4 : n_5 - m : [n_6 + m, l_2; \boldsymbol{\kappa}_6])$-state of equilibrium point $(\mathbf{x}_0^*, \mathbf{p}_0)$ for the nonlinear system is a degenerate Hopf switching of its $|n_4 : n_5 : [n_6, l_1; \boldsymbol{\kappa}_6'])$-state and $|n_4 + m : n_5 - m : [n_6, l_3; \boldsymbol{\kappa}_6''])$-state of equilibrium point $(\mathbf{x}^*, \mathbf{p})$.
(iv) An $|n_4 - m_4 : n_5 : [n_6 + m_4, l_2; \boldsymbol{\kappa}_6])$ state of equilibrium point $(\mathbf{x}_0^*, \mathbf{p}_0)$ for the nonlinear system is a degenerate Hopf switching of its $|n_4 : n_5 : [n_6; l_1; \boldsymbol{\kappa}_6'])$-state and $|n_4 - m_4 : n_5 + m_5 : [n_6 + m_4 - m_5, l_3; \boldsymbol{\kappa}_6''])$-state of equilibrium point $(\mathbf{x}^*, \mathbf{p})$.
(v) An $|n_4 : n_5 - m_5 : [n_6 + m_5; l_2; \boldsymbol{\kappa}_6])$-state of equilibrium point $(\mathbf{x}_0^*, \mathbf{p}_0)$ for the nonlinear system is a degenerate Hopf switching of $|n_4 : n_5 : [n_6, l_1; \boldsymbol{\kappa}_6])$-state and $|n_4 + m_4 : n_5 - m_5 : [n_6 + m_5 - m_4, l_3; \boldsymbol{\kappa}_6''])$-state of equilibrium point $(\mathbf{x}^*, \mathbf{p})$.

1.3.1 Stability and Switching

To extend the idea of Definitions 1.14 and 1.15, a new function will be defined to determine the stability and the stability state switching.

Definition 1.25 Consider an n-dimensional, autonomous, nonlinear dynamical system $\dot{\mathbf{x}} = \mathbf{f}(\mathbf{x}, \mathbf{p})$ in Eq. (1.4) with an equilibrium point $\mathbf{x}^*$ and $\mathbf{f}(\mathbf{x}, \mathbf{p})$ is C^r $(r \geq 1)$-continuous in a neighborhood of the equilibrium $\mathbf{x}^*$. The corresponding solution is $\mathbf{x}(t) = \mathbf{\Phi}(\mathbf{x}_0, t - t_0, \mathbf{p})$. Suppose $U(\mathbf{x}^*) \subset \Omega$ is a neighborhood of equilibrium $\mathbf{x}^*$, and there are n linearly independent vectors $\mathbf{v}_k$ $(k = 1, 2, \cdots, n)$. For a perturbation of equilibrium $\mathbf{y} = \mathbf{x} - \mathbf{x}^*$, let $\mathbf{y}^{(k)} = c_k \mathbf{v}_k$ and $\dot{\mathbf{y}}^{(k)} = \dot{c}_k \mathbf{v}_k$,

$$s_k = \mathbf{v}_k^{\mathrm{T}} \cdot \mathbf{y} = \mathbf{v}_k^{\mathrm{T}} \cdot (\mathbf{x} - \mathbf{x}^*) \tag{1.44}$$

where $s_k = c_k ||\mathbf{v}_k||^2$. Define the following functions

$$G_k(\mathbf{x}, \mathbf{p}) = \mathbf{v}_k^{\mathrm{T}} \cdot \mathbf{f}(\mathbf{x}, \mathbf{p}) \tag{1.45}$$

and

$$\begin{aligned} G_{s_k}^{(1)}(\mathbf{x}, \mathbf{p}) &= \mathbf{v}_k^{\mathrm{T}} \cdot D_{s_k} \mathbf{f}(\mathbf{x}(s_k), \mathbf{p}) \\ &= \mathbf{v}_k^{\mathrm{T}} \cdot D_{\mathbf{x}} \mathbf{f}(\mathbf{x}(s_k), \mathbf{p}) \partial_{c_k} \mathbf{x} \partial_{s_k} c_k \\ &= \mathbf{v}_k^{\mathrm{T}} \cdot D_{\mathbf{x}} \mathbf{f}(\mathbf{x}(s_k), \mathbf{p}) \mathbf{v}_k ||\mathbf{v}_k||^{-2}, \end{aligned} \tag{1.46}$$

$$\begin{aligned} G_{s_k}^{(m)}(\mathbf{x}, \mathbf{p}) &= \mathbf{v}_k^{\mathrm{T}} \cdot D_{s_k}^{(m)} \mathbf{f}(\mathbf{x}(s_k), \mathbf{p}) \\ &= \mathbf{v}_k^{\mathrm{T}} \cdot D_{s_k}(D_{s_k}^{(m-1)} \mathbf{f}(\mathbf{x}(s_k), \mathbf{p})) \end{aligned} \tag{1.47}$$

where $D_{s_k}(\cdot) = \partial(\cdot)/\partial s_k$ and $D_{s_k}^{(m)}(\cdot) = D_{s_k}(D_{s_k}^{(m-1)}(\cdot))$. $G_{s_k}^{(0)}(\mathbf{x}, \mathbf{p}) = G_k(\mathbf{x}, \mathbf{p})$ if $m = 0$.

Definition 1.26 Consider an n-dimensional, autonomous, nonlinear dynamical system $\dot{\mathbf{x}} = \mathbf{f}(\mathbf{x}, \mathbf{p})$ in Eq. (1.4) with an equilibrium point $\mathbf{x}^*$ and $\mathbf{f}(\mathbf{x}, \mathbf{p})$ is C^r $(r \geq 1)$-continuous in a neighborhood of the equilibrium $\mathbf{x}^*$. The corresponding solution is $\mathbf{x}(t) = \mathbf{\Phi}(\mathbf{x}_0, t - t_0, \mathbf{p})$. Suppose $U(\mathbf{x}^*) \subset \Omega$ is a neighborhood of equilibrium $\mathbf{x}^*$, and there are n linearly independent vectors $\mathbf{v}_k$ $(k = 1, 2, \cdots, n)$. For a perturbation of equilibrium $\mathbf{y} = \mathbf{x} - \mathbf{x}^*$, let $\mathbf{y}^{(k)} = c_k \mathbf{v}_k$ and $\dot{\mathbf{y}}^{(k)} = \dot{c}_k \mathbf{v}_k$.

(i) $\mathbf{x}^{(k)}$ at the equilibrium $\mathbf{x}^*$ on the direction $\mathbf{v}_k$ is stable if

$$\begin{aligned} &\mathbf{v}_k^{\mathrm{T}} \cdot (\mathbf{x}(t + \varepsilon) - \mathbf{x}(t)) < 0 \ \text{ for } \ \mathbf{v}_k^{\mathrm{T}} \cdot (\mathbf{x}(t) - \mathbf{x}^*) > 0; \\ &\mathbf{v}_k^{\mathrm{T}} \cdot (\mathbf{x}(t + \varepsilon) - \mathbf{x}(t)) > 0 \ \text{ for } \ \mathbf{v}_k^{\mathrm{T}} \cdot (\mathbf{x}(t) - \mathbf{x}^*) < 0; \end{aligned} \tag{1.48}$$

for all $\mathbf{x} \in U(\mathbf{x}^*) \subset \Omega$ and all $t \in [t_0, \infty)$. The equilibrium $\mathbf{x}^*$ is called the sink (or stable node) on the direction $\mathbf{v}_k$.

(ii) $\mathbf{x}^{(k)}$ at the equilibrium $\mathbf{x}^*$ on the direction $\mathbf{v}_k$ is unstable if

$$\begin{aligned}
&\mathbf{v}_k^{\mathrm{T}} \cdot (\mathbf{x}(t+\varepsilon) - \mathbf{x}(t)) > 0 \text{ for } \mathbf{v}_k^{\mathrm{T}} \cdot (\mathbf{x}(t) - \mathbf{x}^*) > 0; \\
&\mathbf{v}_k^{\mathrm{T}} \cdot (\mathbf{x}(t+\varepsilon) - \mathbf{x}(t)) < 0 \text{ for } \mathbf{v}_k^{\mathrm{T}} \cdot (\mathbf{x}(t) - \mathbf{x}^*) < 0;
\end{aligned} \tag{1.49}$$

for all $\mathbf{x} \in U(\mathbf{x}^*) \subset \Omega$ and all $t \in [t_0, \infty)$. The equilibrium $\mathbf{x}^*$ is called the source (or unstable node) on the direction $\mathbf{v}_k$.

(iii) $\mathbf{x}^{(k)}$ at the equilibrium $\mathbf{x}^*$ on the direction $\mathbf{v}_k$ is increasingly unstable if

$$\begin{aligned}
&\mathbf{v}_k^{\mathrm{T}} \cdot (\mathbf{x}(t+\varepsilon) - \mathbf{x}(t)) > 0 \text{ for } \mathbf{v}_k^{\mathrm{T}} \cdot (\mathbf{x}(t) - \mathbf{x}^*) > 0; \\
&\mathbf{v}_k^{\mathrm{T}} \cdot (\mathbf{x}(t+\varepsilon) - \mathbf{x}(t)) > 0 \text{ for } \mathbf{v}_k^{\mathrm{T}} \cdot (\mathbf{x}(t) - \mathbf{x}^*) < 0;
\end{aligned} \tag{1.50}$$

for all $\mathbf{x} \in U(\mathbf{x}^*) \subset \Omega$ and all $t \in [t_0, \infty)$. The equilibrium $\mathbf{x}^*$ is called the increasing saddle-node on the direction $\mathbf{v}_k$.

(iv) $\mathbf{x}^{(k)}$ at the equilibrium $\mathbf{x}^*$ on the direction $\mathbf{v}_k$ is decreasingly unstable if

$$\begin{aligned}
&\mathbf{v}_k^{\mathrm{T}} \cdot (\mathbf{x}(t+\varepsilon) - \mathbf{x}(t)) < 0 \text{ for } \mathbf{v}_k^{\mathrm{T}} \cdot (\mathbf{x}(t) - \mathbf{x}^*) > 0; \\
&\mathbf{v}_k^{\mathrm{T}} \cdot (\mathbf{x}(t+\varepsilon) - \mathbf{x}(t)) < 0 \text{ for } \mathbf{v}_k^{\mathrm{T}} \cdot (\mathbf{x}(t) - \mathbf{x}^*) < 0;
\end{aligned} \tag{1.51}$$

for all $\mathbf{x} \in U(\mathbf{x}^*) \subset \Omega$ and all $t \in [t_0, \infty)$. The equilibrium $\mathbf{x}^*$ is called the decreasing saddle-node on the direction $\mathbf{v}_k$.

(v) $\mathbf{x}^{(i)}$ at the equilibrium $\mathbf{x}^*$ on the direction $\mathbf{v}_k$ is invariant if

$$\begin{aligned}
&\mathbf{v}_k^{\mathrm{T}} \cdot (\mathbf{x}(t+\varepsilon) - \mathbf{x}(t)) = 0 \\
&\text{for } \mathbf{v}_k^{\mathrm{T}} \cdot (\mathbf{x}(t) - \mathbf{x}^*) \neq 0;
\end{aligned} \tag{1.52}$$

for all $\mathbf{x} \in U(\mathbf{x}^*) \subset \Omega$ and all $t \in [t_0, \infty)$. The equilibrium $\mathbf{x}^*$ is called to be degenerate on the direction $\mathbf{v}_k$.

Theorem 1.7 *Consider an n-dimensional, autonomous, nonlinear dynamical system* $\dot{\mathbf{x}} = \mathbf{f}(\mathbf{x}, \mathbf{p})$ *in Eq. (1.4) with an equilibrium point* $\mathbf{x}^*$ *and* $\mathbf{f}(\mathbf{x}, \mathbf{p})$ *is* C^r $(r \geq 1)$*-continuous in a neighborhood of the equilibrium* $\mathbf{x}^*$ *(i.e.,* $U(\mathbf{x}^*) \subset \Omega$*). The corresponding solution is* $\mathbf{x}(t) = \boldsymbol{\Phi}(\mathbf{x}_0, t-t_0, \mathbf{p})$*. Suppose Eq. (1.24) holds in* $U(\mathbf{x}^*) \subset \Omega$*. For a linearized dynamical system in Eq. (1.19), consider a real eigenvalue* λ_k *of matrix* $D\mathbf{f}(\mathbf{x}^*, \mathbf{p})$ $(k \in N = \{1, 2, \cdots, n\})$ *with an eigenvector* $\mathbf{v}_k$*. Let* $\mathbf{y}^{(k)} = c_k \mathbf{v}_k$ *and* $\dot{\mathbf{y}}^{(k)} = \dot{c}_k \mathbf{v}_k$, $s_k = \mathbf{v}_k^{\mathrm{T}} \cdot \mathbf{y} = \mathbf{v}_k^{\mathrm{T}} \cdot (\mathbf{x} - \mathbf{x}^*)$ *in Eq. (1.44) with* $s_k = c_k ||\mathbf{v}_k||^2$*. Define*

$$\dot{s}_k = \mathbf{v}_k^{\mathrm{T}} \cdot \dot{\mathbf{y}} = \mathbf{v}_k^{\mathrm{T}} \cdot \dot{\mathbf{x}} = \mathbf{v}_k^{\mathrm{T}} \cdot \mathbf{f}(\mathbf{x}, \mathbf{p}). \tag{1.53}$$

(i) $\mathbf{x}^{(k)}$ *at the equilibrium* $\mathbf{x}^*$ *on the direction* $\mathbf{v}_k$ *is stable if and only if*

$$\begin{aligned}
&G_k(\mathbf{x}, \mathbf{p}) = \mathbf{v}_k^{\mathrm{T}} \cdot \mathbf{f}(\mathbf{x}, \mathbf{p}) < 0 \text{ for } s_k = \mathbf{v}_k^{\mathrm{T}} \cdot (\mathbf{x}(t) - \mathbf{x}^*) > 0; \\
&G_k(\mathbf{x}, \mathbf{p}) = \mathbf{v}_k^{\mathrm{T}} \cdot \mathbf{f}(\mathbf{x}, \mathbf{p}) > 0 \text{ for } s_k = \mathbf{v}_k^{\mathrm{T}} \cdot (\mathbf{x}(t) - \mathbf{x}^*) < 0
\end{aligned} \tag{1.54}$$

for all $\mathbf{x} \in U(\mathbf{x}^*) \subset \Omega$ *and all* $t \in [t_0, \infty)$.

(ii) $\mathbf{x}^{(k)}$ *at the equilibrium* $\mathbf{x}^*$ *on the direction* $\mathbf{v}_k$ *is unstable if and only if*

$$\begin{aligned} G_k(\mathbf{x}, \mathbf{p}) &= \mathbf{v}_k^{\mathrm{T}} \cdot \mathbf{f}(\mathbf{x}, \mathbf{p}) > 0 \ \text{ for } \ s_k = \mathbf{v}_k^{\mathrm{T}} \cdot (\mathbf{x}(t) - \mathbf{x}^*) > 0; \\ G_k(\mathbf{x}, \mathbf{p}) &= \mathbf{v}_k^{\mathrm{T}} \cdot \mathbf{f}(\mathbf{x}, \mathbf{p}) < 0 \ \text{ for } \ s_k = \mathbf{v}_k^{\mathrm{T}} \cdot (\mathbf{x}(t) - \mathbf{x}^*) < 0 \end{aligned} \tag{1.55}$$

for all $\mathbf{x} \in U(\mathbf{x}^*) \subset \Omega$ *and all* $t \in [t_0, \infty)$.

(iii) $\mathbf{x}^{(k)}$ *at the equilibrium* $\mathbf{x}^*$ *on the direction* $\mathbf{v}_k$ *is increasingly unstable if and only if*

$$\begin{aligned} G_k(\mathbf{x}, \mathbf{p}) &= \mathbf{v}_k^{\mathrm{T}} \cdot \mathbf{f}(\mathbf{x}, \mathbf{p}) > 0 \ \text{ for } \ s_k = \mathbf{v}_k^{\mathrm{T}} \cdot (\mathbf{x}(t) - \mathbf{x}^*) > 0; \\ G_k(\mathbf{x}, \mathbf{p}) &= \mathbf{v}_k^{\mathrm{T}} \cdot \mathbf{f}(\mathbf{x}, \mathbf{p}) > 0 \ \text{ for } \ s_k = \mathbf{v}_k^{\mathrm{T}} \cdot (\mathbf{x}(t) - \mathbf{x}^*) < 0 \end{aligned} \tag{1.56}$$

for all $\mathbf{x} \in U(\mathbf{x}^*) \subset \Omega$ *and all* $t \in [t_0, \infty)$.

(iv) $\mathbf{x}^{(k)}$ *at the equilibrium* $\mathbf{x}^*$ *on the direction* $\mathbf{v}_k$ *is decreasingly unstable if and only if*

$$\begin{aligned} G_k(\mathbf{x}, \mathbf{p}) &= \mathbf{v}_k^{\mathrm{T}} \cdot \mathbf{f}(\mathbf{x}, \mathbf{p}) < 0 \ \text{ for } \ s_k = \mathbf{v}_k^{\mathrm{T}} \cdot (\mathbf{x}(t) - \mathbf{x}^*) > 0; \\ G_k(\mathbf{x}, \mathbf{p}) &= \mathbf{v}_k^{\mathrm{T}} \cdot \mathbf{f}(\mathbf{x}, \mathbf{p}) < 0 \ \text{ for } \ s_k = \mathbf{v}_k^{\mathrm{T}} \cdot (\mathbf{x}(t) - \mathbf{x}^*) < 0 \end{aligned} \tag{1.57}$$

for all $\mathbf{x} \in U(\mathbf{x}^*) \subset \Omega$ *and all* $t \in [t_0, \infty)$.

(v) $\mathbf{x}^{(i)}$ *at the equilibrium* $\mathbf{x}^*$ *on the direction* $\mathbf{v}_k$ *is invariant if*

$$G_k(\mathbf{x}, \mathbf{p}) = \mathbf{v}_k^{\mathrm{T}} \cdot \mathbf{f}(\mathbf{x}, \mathbf{p}) = 0 \tag{1.58}$$

for all $\mathbf{x} \in U(\mathbf{x}^*) \subset \Omega$ *and all* $t \in [t_0, \infty)$.

Proof Because

$$\begin{aligned} \mathbf{v}_k^{\mathrm{T}} \cdot (\mathbf{x}(t+\varepsilon) - \mathbf{x}(t)) &= \mathbf{v}_k^{\mathrm{T}} \cdot (\mathbf{x}(t) + \dot{\mathbf{x}}(t)\varepsilon + o(\varepsilon) - \mathbf{x}(t)) \\ &= \mathbf{v}_k^{\mathrm{T}} \cdot \dot{\mathbf{x}}(t)\varepsilon + o(\varepsilon) \end{aligned}$$

and $\dot{\mathbf{x}} = \mathbf{f}(\mathbf{x}, \mathbf{p})$, we have

$$\begin{aligned} \mathbf{v}_k^{\mathrm{T}} \cdot (\mathbf{x}(t+\varepsilon) - \mathbf{x}(t)) &= \mathbf{v}_k^{\mathrm{T}} \cdot \mathbf{f}(\mathbf{x}, \mathbf{p})\varepsilon + o(\varepsilon) \\ &= G_k(\mathbf{x}, \mathbf{p})\varepsilon + o(\varepsilon). \end{aligned}$$

(i) Due to any selection of $\varepsilon > 0$, for $s_k = \mathbf{v}_k^{\mathrm{T}} \cdot (\mathbf{x}(t) - \mathbf{x}^*) < 0$

$$\mathbf{v}_k^{\mathrm{T}} \cdot (\mathbf{x}(t+\varepsilon) - \mathbf{x}(t)) < 0 \ \text{ if } \ G_k(\mathbf{x}, \mathbf{p}) < 0$$

vice versa; and for $s_k = \mathbf{v}_k^{\mathrm{T}} \cdot (\mathbf{x}(t) - \mathbf{x}^*) > 0$

$$\mathbf{v}_k^{\mathrm{T}} \cdot (\mathbf{x}(t+\varepsilon) - \mathbf{x}(t)) > 0 \ \text{ if } \ G_k(\mathbf{x}, \mathbf{p}) > 0$$

vice versa.

(ii) For $s_k = \mathbf{v}_k^{\mathrm{T}} \cdot (\mathbf{x}(t) - \mathbf{x}^*) > 0$

$$\mathbf{v}_k^{\mathrm{T}} \cdot (\mathbf{x}(t+\varepsilon) - \mathbf{x}(t)) > 0 \text{ if } G_k(\mathbf{x}, \mathbf{p}) > 0$$

vice versa; and for $s_k = \mathbf{v}_k^{\mathrm{T}} \cdot (\mathbf{x}(t) - \mathbf{x}^*) < 0$

$$\mathbf{v}_k^{\mathrm{T}} \cdot (\mathbf{x}(t+\varepsilon) - \mathbf{x}(t)) < 0 \text{ if } G_k(\mathbf{x}, \mathbf{p}) < 0$$

vice versa.

(iii) For $s_k = \mathbf{v}_k^{\mathrm{T}} \cdot (\mathbf{x}(t) - \mathbf{x}^*) > 0$

$$\mathbf{v}_k^{\mathrm{T}} \cdot (\mathbf{x}(t+\varepsilon) - \mathbf{x}(t)) > 0 \text{ if } G_k(\mathbf{x}, \mathbf{p}) > 0$$

vice versa; and for $s_k = \mathbf{v}_k^{\mathrm{T}} \cdot (\mathbf{x}(t) - \mathbf{x}^*) < 0$

$$\mathbf{v}_k^{\mathrm{T}} \cdot (\mathbf{x}(t+\varepsilon) - \mathbf{x}(t)) > 0 \text{ if } G_k(\mathbf{x}, \mathbf{p}) > 0$$

vice versa.

(iv) For $s_k = \mathbf{v}_k^{\mathrm{T}} \cdot (\mathbf{x}(t) - \mathbf{x}^*) > 0$

$$\mathbf{v}_k^{\mathrm{T}} \cdot (\mathbf{x}(t+\varepsilon) - \mathbf{x}(t)) < 0 \text{ if } G_k(\mathbf{x}, \mathbf{p}) < 0$$

vice versa; and for $s_k = \mathbf{v}_k^{\mathrm{T}} \cdot (\mathbf{x}(t) - \mathbf{x}^*) < 0$

$$\mathbf{v}_k^{\mathrm{T}} \cdot (\mathbf{x}(t+\varepsilon) - \mathbf{x}(t)) < 0 \text{ if } G_k(\mathbf{x}, \mathbf{p}) < 0$$

vice versa.

(v) For $s_k = \mathbf{v}_k^{\mathrm{T}} \cdot (\mathbf{x}(t) - \mathbf{x}^*) > 0$

$$\mathbf{v}_k^{\mathrm{T}} \cdot (\mathbf{x}(t+\varepsilon) - \mathbf{x}(t)) = 0 \text{ if } G_k(\mathbf{x}, \mathbf{p}) = 0$$

vice versa. Similarly, for $s_k = \mathbf{v}_k^{\mathrm{T}} \cdot (\mathbf{x}(t) - \mathbf{x}^*) < 0$

$$\mathbf{v}_k^{\mathrm{T}} \cdot (\mathbf{x}(t+\varepsilon) - \mathbf{x}(t)) = 0 \text{ if } G_k(\mathbf{x}, \mathbf{p}) = 0$$

vice versa. The theorem is proved. ■

Theorem 1.8 *Consider an n-dimensional, autonomous, nonlinear dynamical system* $\dot{\mathbf{x}} = \mathbf{f}(\mathbf{x}, \mathbf{p})$ *in Eq.* (1.4) *with an equilibrium point* $\mathbf{x}^*$ *and* $\mathbf{f}(\mathbf{x}, \mathbf{p})$ *is* C^r $(r \geq 1)$-*continuous in a neighborhood of the equilibrium* $\mathbf{x}^*$ *(i.e.,*$U(\mathbf{x}^*) \subset \Omega$*). The corresponding solution is* $\mathbf{x}(t) = \boldsymbol{\Phi}(\mathbf{x}_0, t-t_0, \mathbf{p})$. *Suppose Eq.* (1.24) *holds in* $U(\mathbf{x}^*) \subset \Omega$. *For a linearized dynamical system in Eq.* (1.19), *consider a real eigenvalue* λ_k *of matrix* $D\mathbf{f}(\mathbf{x}^*, \mathbf{p})$ $(k \in N = \{1, 2, \cdots, n\})$ *with an eigenvector* $\mathbf{v}_k$. *Let* $\mathbf{y}^{(k)} = c_k \mathbf{v}_k$ *and* $\dot{\mathbf{y}}^{(k)} = \dot{c}_k \mathbf{v}_k$, $s_k = \mathbf{v}_k^{\mathrm{T}} \cdot \mathbf{y} = \mathbf{v}_k^{\mathrm{T}} \cdot (\mathbf{x} - \mathbf{x}^*)$ *in Eq.* (1.44) *with* $s_k = c_k ||\mathbf{v}_k||^2$. *Define* $\dot{s}_k = \mathbf{v}_k^{\mathrm{T}} \cdot \mathbf{f}(\mathbf{x}, \mathbf{p})$ *in Eq.* (1.53). *Suppose* $||G_k^{(2)}(\mathbf{x}^*, \mathbf{p})|| < \infty$.

(i) $\mathbf{x}^{(k)}$ *at the equilibrium* $\mathbf{x}^*$ *on the direction* $\mathbf{v}_k$ *is stable if and only if*

$$G_{s_k}^{(1)}(\mathbf{x}^*, \mathbf{p}) = \lambda_k < 0 \tag{1.59}$$

for all $\mathbf{x} \in U(\mathbf{x}^*) \subset \Omega$ *and all* $t \in [t_0, \infty)$.

(ii) $\mathbf{x}^{(k)}$ *at the equilibrium* $\mathbf{x}^*$ *on the direction* $\mathbf{v}_k$ *is unstable if and only if*

$$G_{s_k}^{(1)}(\mathbf{x}^*, \mathbf{p}) = \lambda_k > 0 \tag{1.60}$$

for all $\mathbf{x} \in U(\mathbf{x}^*) \subset \Omega$ *and all* $t \in [t_0, \infty)$.

(iii) $\mathbf{x}^{(k)}$ *at the equilibrium* $\mathbf{x}^*$ *on the direction* $\mathbf{v}_k$ *is increasingly unstable if and only if*

$$G_{s_k}^{(1)}(\mathbf{x}^*, \mathbf{p}) = \lambda_k = 0, \quad \text{and} \quad G_{s_k}^{(2)}(\mathbf{x}^*, \mathbf{p}) > 0 \tag{1.61}$$

for all $\mathbf{x} \in U(\mathbf{x}^*) \subset \Omega$ *and all* $t \in [t_0, \infty)$.

(iv) $\mathbf{x}^{(k)}$ *at the equilibrium* $\mathbf{x}^*$ *on the direction* $\mathbf{v}_k$ *is decreasingly unstable if and only if*

$$G_{s_k}^{(1)}(\mathbf{x}^*, \mathbf{p}) = \lambda_k = 0, \quad \text{and} \quad G_{s_k}^{(2)}(\mathbf{x}^*, \mathbf{p}) < 0 \tag{1.62}$$

for all $\mathbf{x} \in U(\mathbf{x}^*) \subset \Omega$ *and all* $t \in [t_0, \infty)$.

(v) $\mathbf{x}^{(i)}$ *at the equilibrium* $\mathbf{x}^*$ *on the direction* $\mathbf{v}_k$ *is invariant if and only if*

$$G_{s_k}^{(m)}(\mathbf{x}^*, \mathbf{p}) = 0 \quad (m = 0, 1, 2, \cdots) \tag{1.63}$$

for all $\mathbf{x} \in U(\mathbf{x}^*) \subset \Omega$ *and all* $t \in [t_0, \infty)$.

Proof For $\mathbf{x} = \mathbf{x}^*$, $s_k = 0$. Using Taylor series expansion gives

$$\begin{aligned}
\dot{s}_k &= \mathbf{v}_k^{\mathrm{T}} \cdot \mathbf{f}(\mathbf{x}, \mathbf{p}) \\
&= \mathbf{v}_k^{\mathrm{T}} \cdot (\mathbf{f}(\mathbf{x}^*, \mathbf{p}) + D_{s_k}\mathbf{f}(\mathbf{x}^*, \mathbf{p}) s_k) + o(s_k) \\
&= [\mathbf{v}_k^{\mathrm{T}} \cdot D_{s_k}\mathbf{f}(\mathbf{x}^*, \mathbf{p})] s_k + o(s_k) \\
&= G_{s_k}^{(1)}(\mathbf{x}^*, \mathbf{p}) s_k + o(s_k)
\end{aligned}$$

and

$$\begin{aligned}
G_{s_k}^{(1)}(\mathbf{x}^*, \mathbf{p}) &= \mathbf{v}_k^{\mathrm{T}} \cdot D_{\mathbf{x}}\mathbf{f}(\mathbf{x}(s_k), \mathbf{p}) \partial_{c_k}\mathbf{x} \partial_{s_k} c_k \\
&= \mathbf{v}_k^{\mathrm{T}} \cdot D_{\mathbf{x}}\mathbf{f}(\mathbf{x}(s_k), \mathbf{p}) \mathbf{v}_k ||\mathbf{v}_k||^{-2} \\
&= \lambda_k.
\end{aligned}$$

Thus,

$$\dot{s}_k = G_{s_k}^{(1)}(\mathbf{x}^*, \mathbf{p}) s_k + o(s_k) = \lambda_k s_k + o(s_k).$$

(i) For $s_k > 0$

$$G_k(\mathbf{x}, \mathbf{p}) = \dot{s}_k = \lambda_k s_k < 0.$$

and for $s_k < 0$

$$G_k(\mathbf{x}, \mathbf{p}) = \dot{s}_k = \lambda_k s_k > 0.$$

Thus, $G_{s_k}^{(1)}(\mathbf{x}^*, \mathbf{p}) = \lambda_k < 0$.

(ii) For $s_k > 0$

$$G_k(\mathbf{x}, \mathbf{p}) = \dot{s}_k = \lambda_k s_k > 0.$$

and for $s_k < 0$

$$G_k(\mathbf{x}, \mathbf{p}) = \dot{s}_k = \lambda_k s_k < 0.$$

Thus, $G_{s_k}^{(1)}(\mathbf{x}^*, \mathbf{p}) = \lambda_k > 0$.

(iii) For $s_k > 0$

$$G_k(\mathbf{x}, \mathbf{p}) = \dot{s}_k = \lambda_k s_k > 0.$$

and for $s_k < 0$

$$G_k(\mathbf{x}, \mathbf{p}) = \dot{s}_k = \lambda_k s_k > 0.$$

Thus, $G_{s_k}^{(1)}(\mathbf{x}^*, \mathbf{p}) = \lambda_k = 0$ and the higher order should be considered. The higher-order Taylor series expansion gives

$$\begin{aligned}
\dot{s}_k &= \mathbf{v}_k^{\mathrm{T}} \cdot \mathbf{f}(\mathbf{x}, \mathbf{p}) \\
&= \mathbf{v}_k^{\mathrm{T}} \cdot (\mathbf{f}(\mathbf{x}^*, \mathbf{p}) + D_{s_k}\mathbf{f}(\mathbf{x}^*, \mathbf{p})s_k + \frac{1}{2!}D_{s_k}^2\mathbf{f}(\mathbf{x}^*, \mathbf{p})s_k^2) + o(s_k^2) \\
&= \frac{1}{2!}[\mathbf{v}_k^{\mathrm{T}} \cdot D_{s_k}^2\mathbf{f}(\mathbf{x}^*, \mathbf{p})]s_k^2 + o(s_k^2) = \frac{1}{2!}G_{s_k}^{(2)}(\mathbf{x}^*, \mathbf{p})s_k^2 + o(s_k^2).
\end{aligned}$$

For $s_k > 0$

$$G_k(\mathbf{x}, \mathbf{p}) = \dot{s}_k = \frac{1}{2!}G_{s_k}^{(2)}(\mathbf{x}^*, \mathbf{p})s_k^2 > 0.$$

and for $s_k < 0$

$$G_k(\mathbf{x}, \mathbf{p}) = \dot{s}_k = \frac{1}{2!}G_{s_k}^{(2)}(\mathbf{x}^*, \mathbf{p})s_k^2 > 0.$$

So we have

$$G_{s_k}^{(2)}(\mathbf{x}^*, \mathbf{p}) > 0.$$

(iv) Similar to (iii), we have $G_k^{(1)}(\mathbf{x}^*, \mathbf{p}) = \lambda_k = 0$. In addition, for $s_k > 0$

$$G_k(\mathbf{x}, \mathbf{p}) = \dot{s}_k = \frac{1}{2!} G_{s_k}^{(2)}(\mathbf{x}^*, \mathbf{p}) s_k^2 < 0.$$

and for $s_k < 0$

$$G_k(\mathbf{x}, \mathbf{p}) = \dot{s}_k = \frac{1}{2!} G_{s_k}^{(2)}(\mathbf{x}^*, \mathbf{p}) s_k^2 < 0.$$

So

$$G_{s_k}^{(2)}(\mathbf{x}^*, \mathbf{p}) < 0.$$

(v) Using Taylor series expansion yields

$$\dot{s}_k = \mathbf{v}_k^{\mathrm{T}} \cdot \mathbf{f}(\mathbf{x}, \mathbf{p}) = \sum_{m=1}^{N} \frac{1}{m!} G_{s_k}^{(m)}(\mathbf{x}^*, \mathbf{p}) s_k^m + o(s_k^N) = 0$$

$$(N = 1, 2, \cdots)$$

for any selected values of s_k. Thus only if

$$G_{s_k}^{(m)}(\mathbf{x}^*, \mathbf{p}) = 0 \ (m = 1, 2, \cdots)$$

the above equation holds, vice versa. The theorem is proved. ∎

Definition 1.27 Consider an n-dimensional, autonomous, nonlinear dynamical system $\dot{\mathbf{x}} = \mathbf{f}(\mathbf{x}, \mathbf{p})$ in Eq. (1.4) with an equilibrium point $\mathbf{x}^*$ and $\mathbf{f}(\mathbf{x}, \mathbf{p})$ is C^r ($r \geq 1$)-continuous in a neighborhood of the equilibrium $\mathbf{x}^*$ (i.e.,$U(\mathbf{x}^*) \subset \Omega$). The corresponding solution is $\mathbf{x}(t) = \mathbf{\Phi}(\mathbf{x}_0, t - t_0, \mathbf{p})$. Suppose Eq. (1.24) holds in $U(\mathbf{x}^*) \subset \Omega$. For a linearized dynamical system in Eq. (1.19), consider a real eigenvalue λ_k of matrix $D\mathbf{f}(\mathbf{x}^*, \mathbf{p})$ ($k \in N = \{1, 2, \cdots, n\}$) with an eigenvector $\mathbf{v}_k$ and let $\mathbf{y}^{(k)} = c_k \mathbf{v}_k$.

(i) $\mathbf{x}^{(k)}$ at the equilibrium $\mathbf{x}^*$ on the direction $\mathbf{v}_k$ is stable of the $(2m_k + 1)$th order if

$$\begin{aligned} &G_{s_k}^{(r_k)}(\mathbf{x}^*, \mathbf{p}) = 0,\ r_k = 0, 1, 2, \cdots, 2m_k; \\ &\mathbf{v}_k^{\mathrm{T}} \cdot (\mathbf{x}(t+\varepsilon) - \mathbf{x}(t)) < 0 \text{ for } \mathbf{v}_k^{\mathrm{T}} \cdot (\mathbf{x}(t) - \mathbf{x}^*) > 0; \\ &\mathbf{v}_k^{\mathrm{T}} \cdot (\mathbf{x}(t+\varepsilon) - \mathbf{x}(t)) > 0 \text{ for } \mathbf{v}_k^{\mathrm{T}} \cdot (\mathbf{x}(t) - \mathbf{x}^*) < 0 \end{aligned} \tag{1.64}$$

for all $\mathbf{x} \in U(\mathbf{x}^*) \subset \Omega$ and all $t \in [t_0, \infty)$. The equilibrium $\mathbf{x}^*$ is called the sink (or stable node) of the $(2m_k + 1)$th order on the direction $\mathbf{v}_k$.

(ii) $\mathbf{x}^{(k)}$ at the equilibrium $\mathbf{x}^*$ on the direction $\mathbf{v}_k$ is unstable of the $(2m_k + 1)$th order if

$$\begin{aligned}
&G_{s_k}^{(r_k)}(\mathbf{x}^*, \mathbf{p}) = 0, r_k = 0, 1, 2, \cdots, 2m_k;\\
&\mathbf{v}_k^{\mathrm{T}} \cdot (\mathbf{x}(t+\varepsilon) - \mathbf{x}(t)) > 0 \text{ for } \mathbf{v}_k^{\mathrm{T}} \cdot (\mathbf{x}(t) - \mathbf{x}^*) > 0;\\
&\mathbf{v}_k^{\mathrm{T}} \cdot (\mathbf{x}(t+\varepsilon) - \mathbf{x}(t)) < 0 \text{ for } \mathbf{v}_k^{\mathrm{T}} \cdot (\mathbf{x}(t) - \mathbf{x}^*) < 0
\end{aligned} \tag{1.65}$$

for all $\mathbf{x} \in U(\mathbf{x}^*) \subset \Omega$ and all $t \in [t_0, \infty)$. The equilibrium $\mathbf{x}^*$ is called the source (or unstable node) of the $(2m_k + 1)$th order on the direction $\mathbf{v}_k$.

(iii) $\mathbf{x}^{(k)}$ at the equilibrium $\mathbf{x}^*$ on the direction $\mathbf{v}_k$ is increasingly unstable of the $(2m_k)$th order if

$$\begin{aligned}
&G_{s_k}^{(r_k)}(\mathbf{x}^*, \mathbf{p}) = 0, r_k = 0, 1, 2, \cdots, 2m_k - 1;\\
&\mathbf{v}_k^{\mathrm{T}} \cdot (\mathbf{x}(t+\varepsilon) - \mathbf{x}(t)) > 0 \text{ for } \mathbf{v}_k^{\mathrm{T}} \cdot (\mathbf{x}(t) - \mathbf{x}^*) > 0;\\
&\mathbf{v}_k^{\mathrm{T}} \cdot (\mathbf{x}(t+\varepsilon) - \mathbf{x}(t)) > 0 \text{ for } \mathbf{v}_k^{\mathrm{T}} \cdot (\mathbf{x}(t) - \mathbf{x}^*) < 0
\end{aligned} \tag{1.66}$$

for all $\mathbf{x} \in U(\mathbf{x}^*) \subset \Omega$ and all $t \in [t_0, \infty)$. The equilibrium $\mathbf{x}^*$ is called the increasing saddle of the $(2m_k)$th order on the direction $\mathbf{v}_k$.

(iv) $\mathbf{x}^{(k)}$ at the equilibrium $\mathbf{x}^*$ on the direction $\mathbf{v}_k$ is decreasingly unstable of the $(2m_k)$th order if

$$\begin{aligned}
&G_{s_k}^{(r_k)}(\mathbf{x}^*, \mathbf{p}) = 0, r_k = 0, 1, 2, \cdots, 2m_k - 1;\\
&\mathbf{v}_k^{\mathrm{T}} \cdot (\mathbf{x}(t+\varepsilon) - \mathbf{x}(t)) < 0 \text{ for } \mathbf{v}_k^{\mathrm{T}} \cdot (\mathbf{x}(t) - \mathbf{x}^*) > 0;\\
&\mathbf{v}_k^{\mathrm{T}} \cdot (\mathbf{x}(t+\varepsilon) - \mathbf{x}(t)) < 0 \text{ for } \mathbf{v}_k^{\mathrm{T}} \cdot (\mathbf{x}(t) - \mathbf{x}^*) < 0
\end{aligned} \tag{1.67}$$

for all $\mathbf{x} \in U(\mathbf{x}^*) \subset \Omega$ and all $t \in [t_0, \infty)$. The equilibrium $\mathbf{x}^*$ is called the decreasing saddle of the $(2m_k)$th order on the direction $\mathbf{v}_k$.

Theorem 1.9 *Consider an n-dimensional, autonomous, nonlinear dynamical system* $\dot{\mathbf{x}} = \mathbf{f}(\mathbf{x}, \mathbf{p})$ *in Eq.* (1.4) *with an equilibrium point* $\mathbf{x}^*$ *and* $\mathbf{f}(\mathbf{x}, \mathbf{p})$ *is* C^r $(r \geq 1)$*-continuous in a neighborhood of the equilibrium* $\mathbf{x}^*$ *(i.e.,*$U(\mathbf{x}^*) \subset \Omega$*). The corresponding solution is* $\mathbf{x}(t) = \mathbf{\Phi}(\mathbf{x}_0, t - t_0, \mathbf{p})$*. Suppose Eq.* (1.24) *holds in* $U(\mathbf{x}^*) \subset \Omega$*. For a linearized dynamical system in Eq.* (1.19)*, consider a real eigenvalue* λ_k *of matrix* $D\mathbf{f}(\mathbf{x}^*, \mathbf{p})$ $(k \in N = \{1, 2, \cdots, n\})$ *with an eigenvector* $\mathbf{v}_k$ *and let* $\mathbf{y}^{(k)} = c_k \mathbf{v}_k$.

(i) $\mathbf{x}^{(k)}$ *at the equilibrium* $\mathbf{x}^*$ *on the direction* $\mathbf{v}_k$ *is stable of the* $(2m_k + 1)$th *order if and only if*

$$\begin{aligned}
&G_{s_k}^{(r_k)}(\mathbf{x}^*, \mathbf{p}) = 0, r_k = 0, 1, 2, \cdots, 2m_k;\\
&G_{s_k}^{(2m_k+1)}(\mathbf{x}^*, \mathbf{p}) < 0
\end{aligned} \tag{1.68}$$

for all $\mathbf{x} \in U(\mathbf{x}^*) \subset \Omega$ *and all* $t \in [t_0, \infty)$.

(ii) $\mathbf{x}^{(k)}$ *at the equilibrium* $\mathbf{x}^*$ *on the direction* $\mathbf{v}_k$ *is unstable of the* $(2m_k + 1)$th *order if and only if*

$$\begin{aligned}
&G_{s_k}^{(r_k)}(\mathbf{x}^*, \mathbf{p}) = 0, r_k = 0, 1, 2, \cdots, 2m_k;\\
&G_{s_k}^{(2m_k+1)}(\mathbf{x}^*, \mathbf{p}) > 0
\end{aligned} \tag{1.69}$$

for all $\mathbf{x} \in U(\mathbf{x}^*) \subset \Omega$ *and all* $t \in [t_0, \infty)$.

(iii) $\mathbf{x}^{(k)}$ *at the equilibrium* $\mathbf{x}^*$ *on the direction* $\mathbf{v}_k$ *is increasingly unstable of the* $(2m_k)$th *order if and only if*

$$G_{s_k}^{(r_k)}(\mathbf{x}^*, \mathbf{p}) = 0, r_k = 0, 1, 2, \cdots, 2m_k - 1; \\ G_{s_k}^{(2m_k)}(\mathbf{x}^*, \mathbf{p}) > 0 \tag{1.70}$$

for all $\mathbf{x} \in U(\mathbf{x}^*) \subset \Omega$ *and all* $t \in [t_0, \infty)$.

(iv) $\mathbf{x}^{(k)}$ *at the equilibrium* $\mathbf{x}^*$ *on the direction* $\mathbf{v}_k$ *is decreasingly unstable of the* $(2m_k)$th *order if and only if*

$$G_{s_k}^{(r_k)}(\mathbf{x}^*, \mathbf{p}) = 0, r_k = 0, 1, 2, \cdots, 2m_k - 1; \\ G_{s_k}^{(2m_k)}(\mathbf{x}^*, \mathbf{p}) < 0 \tag{1.71}$$

for all $\mathbf{x} \in U(\mathbf{x}^*) \subset \Omega$ *and all* $t \in [t_0, \infty)$.

Proof For $\mathbf{x} = \mathbf{x}^*$, $s_k = 0$. Using the Taylor series expansion gives

$$\begin{aligned} \dot{s}_k &= \mathbf{v}_k^{\mathrm{T}} \cdot \mathbf{f}(\mathbf{x}, \mathbf{p}) \\ &= \sum\nolimits_{r_k=1}^{2m_k} \frac{1}{r_k!} G_{s_k}^{(r_k)}(\mathbf{x}^*, \mathbf{p}) s_k^{r_k} + \frac{1}{(2m_k+1)!} G_{s_k}^{(2m_k+1)}(\mathbf{x}^*, \mathbf{p}) s_k^{2m_k+1} + o(s_k^{2m_k+1}) \end{aligned}$$

and

$$G_{s_k}^{(r_k)}(\mathbf{x}^*, \mathbf{p}) = 0 \text{ for } r_k = 0, 1, 2, \cdots, 2m_k;$$

$$\dot{s}_k = \mathbf{v}_k^{\mathrm{T}} \cdot \mathbf{f}(\mathbf{x}, \mathbf{p}) = \frac{1}{(2m_k+1)!} G_{s_k}^{(2m_k+1)}(\mathbf{x}^*, \mathbf{p}) s_k^{2m_k+1}.$$

(i) For $s_k > 0$

$$G_k(\mathbf{x}, \mathbf{p}) = \dot{s}_k = \frac{1}{(2m_k+1)!} G_{s_k}^{(2m_k+1)}(\mathbf{x}^*, \mathbf{p}) s_k^{2m_k+1} < 0,$$

and for $s_k < 0$

$$G_k(\mathbf{x}, \mathbf{p}) = \dot{s}_k = \frac{1}{(2m_k+1)!} G_{s_k}^{(2m_k+1)}(\mathbf{x}^*, \mathbf{p}) s_k^{2m_k+1} > 0.$$

Thus, $G_{s_k}^{(2m_k+1)}(\mathbf{x}^*, \mathbf{p}) < 0$.

(ii) For $s_k > 0$

$$G_k(\mathbf{x}, \mathbf{p}) = \dot{s}_k = \frac{1}{(2m_k+1)!} G_{s_k}^{(2m_k+1)}(\mathbf{x}^*, \mathbf{p}) s_k^{2m_k+1} > 0,$$

and for $s_k < 0$

$$G_k(\mathbf{x}, \mathbf{p}) = \dot{s}_k = \frac{1}{(2m_k+1)!} G_{s_k}^{(2m_k+1)}(\mathbf{x}^*, \mathbf{p}) s_k^{2m_k+1} < 0.$$

Thus, $G_{s_k}^{(2m_k+1)}(\mathbf{x}^*, \mathbf{p}) > 0$.

(iii) For $\mathbf{x} = \mathbf{x}^*$, $s_k = 0$. Using the Taylor series expansion gives

$$\dot{s}_k = \mathbf{v}_k^{\mathrm{T}} \cdot \mathbf{f}(\mathbf{x}, \mathbf{p}) = \sum_{r_k=1}^{2m_k-1} \frac{1}{r_k!} G_{s_k}^{(r_k)}(\mathbf{x}^*, \mathbf{p}) s_k^{r_k} + \frac{1}{(2m_k)!} G_{s_k}^{(2m_k)}(\mathbf{x}^*, \mathbf{p}) s_k^{2m_k} + o(s_k^{2m_k})$$

and

$$G_{s_k}^{(r_k)}(\mathbf{x}^*, \mathbf{p}) = 0 \text{ for } r_k = 0, 1, \cdots, 2m_k - 1.$$

Thus,

$$\dot{s}_k = \mathbf{v}_k^{\mathrm{T}} \cdot \mathbf{f}(\mathbf{x}, \mathbf{p}) = \frac{1}{(2m_k)!} G_{s_k}^{(2m_k)}(\mathbf{x}^*, \mathbf{p}) s_k^{2m_k}.$$

For $s_k > 0$

$$G_k(\mathbf{x}, \mathbf{p}) = \dot{s}_k = \frac{1}{(2m_k)!} G_{s_k}^{(2m_k)}(\mathbf{x}^*, \mathbf{p}) s_k^{2m_k} > 0,$$

and for $s_k < 0$

$$G_k(\mathbf{x}, \mathbf{p}) = \dot{s}_k = \frac{1}{(2m_k)!} G_{s_k}^{(2m_k)}(\mathbf{x}^*, \mathbf{p}) s_k^{2m_k} > 0.$$

So we have

$$G_{s_k}^{(2m_k)}(\mathbf{x}^*, \mathbf{p}) > 0.$$

(iv) Similar to (iii), for $s_k > 0$

$$G_k(\mathbf{x}, \mathbf{p}) = \dot{s}_k = \frac{1}{(2m_k)!} G_{s_k}^{(2m_k)}(\mathbf{x}^*, \mathbf{p}) s_k^{2m_k} < 0,$$

and for $s_k < 0$

$$G_k(\mathbf{x}, \mathbf{p}) = \dot{s}_k = \frac{1}{(2m_k)!} G_{s_k}^{(2m_k)}(\mathbf{x}^*, \mathbf{p}) s_k^{2m_k} < 0.$$

So

$$G_{s_k}^{(2m_k)}(\mathbf{x}^*, \mathbf{p}) < 0.$$

The theorem is proved.

■

Definition 1.28 Consider an n-dimensional, autonomous, nonlinear dynamical system $\dot{\mathbf{x}} = \mathbf{f}(\mathbf{x}, \mathbf{p})$ in Eq. (1.4) with an equilibrium point $\mathbf{x}^*$ and $\mathbf{f}(\mathbf{x}, \mathbf{p})$ is C^r $(r \geq 1)$-continuous in a neighborhood of the equilibrium $\mathbf{x}^*$ (i.e.,$U(\mathbf{x}^*) \subset \Omega$). The corresponding solution is $\mathbf{x}(t) = \mathbf{\Phi}(\mathbf{x}_0, t - t_0, \mathbf{p})$. Suppose Eq. (1.24) holds in $U(\mathbf{x}^*) \subset \Omega$.

For a linearized dynamical system in Eq. (1.19), consider a pair of complex eigenvalue $\alpha_k \pm \mathbf{i}\beta_k$ ($k \in N = \{1, 2, \cdots, n\}$, $\mathbf{i} = \sqrt{-1}$) of matrix $D\mathbf{f}(\mathbf{x}^*, \mathbf{p})$ with a pair of eigenvectors $\mathbf{u}_k \pm \mathbf{i}\mathbf{v}_k$. On the invariant plane of $(\mathbf{u}_k, \mathbf{v}_k)$, consider $\mathbf{r}_k = \mathbf{y} = \mathbf{y}_+^{(k)} + \mathbf{y}_-^{(k)}$ with

$$\begin{aligned} \mathbf{r}_k &= c_k \mathbf{u}_k + d_k \mathbf{v}_k = r_k \mathbf{e}_{r_k}, \\ \dot{\mathbf{r}}_k &= \dot{c}_k \mathbf{u}_k + \dot{d}_k \mathbf{v}_k = \dot{r}_k \mathbf{e}_{r_k} + r_k \dot{\mathbf{e}}_{r_k} \end{aligned} \tag{1.72}$$

and

$$\begin{aligned} &c_k = \frac{1}{\Delta}[\Delta_2(\mathbf{u}_k^{\mathrm{T}} \cdot \mathbf{y}) - \Delta_{12}(\mathbf{v}_k^{\mathrm{T}} \cdot \mathbf{y})] \text{ and } d_k = \frac{1}{\Delta}[\Delta_1(\mathbf{v}_k^{\mathrm{T}} \cdot \mathbf{y}) - \Delta_{12}(\mathbf{u}_k^{\mathrm{T}} \cdot \mathbf{y})] \\ &\Delta_1 = ||\mathbf{u}_k||^2, \Delta_2 = ||\mathbf{v}_k||^2, \Delta_{12} = \mathbf{u}_k^{\mathrm{T}} \cdot \mathbf{v}_k \text{ and } \Delta = \Delta_1 \Delta_2 - \Delta_{12}^2. \end{aligned} \tag{1.73}$$

Consider a polar coordinate of (r_k, θ_k) defined by

$$\begin{aligned} c_k &= r_k \cos\theta_k, \text{ and } d_k = r_k \sin\theta_k; \\ r_k &= \sqrt{c_k^2 + d_k^2}, \text{ and } \theta_k = \arctan d_k / c_k; \end{aligned} \tag{1.74}$$

$$\begin{aligned} &\mathbf{e}_{r_k} = \cos\theta_k \mathbf{u}_k + \sin\theta_k \mathbf{v}_k \text{ and } \mathbf{e}_{\theta_k} = -\cos\theta_k \mathbf{u}_k^{\perp}\Delta_3 + \sin\theta_k \mathbf{v}_k^{\perp}\Delta_4 \\ &\Delta_3 = \mathbf{v}_k^{\mathrm{T}} \cdot \mathbf{u}_k^{\perp} \text{ and } \Delta_4 = \mathbf{u}_k^{\mathrm{T}} \cdot \mathbf{v}_k^{\perp} \end{aligned} \tag{1.75}$$

where $\mathbf{u}_k^{\perp}$ and $\mathbf{v}_k^{\perp}$ are the normal vectors of $\mathbf{u}_k$ and $\mathbf{v}_k$, respectively.

$$\begin{aligned} \dot{c}_k &= \frac{1}{\Delta}[\Delta_2 G_{c_k}(\mathbf{x}, \mathbf{p}) - \Delta_{12} G_{d_k}(\mathbf{x}, \mathbf{p})] \\ \dot{d}_k &= \frac{1}{\Delta}[\Delta_1 G_{d_k}(\mathbf{x}, \mathbf{p}) - \Delta_{12} G_{d_k}(\mathbf{x}, \mathbf{p})] \end{aligned} \tag{1.76}$$

where

$$\begin{aligned} G_{c_k}(\mathbf{x}, \mathbf{p}) &= \mathbf{u}_k^{\mathrm{T}} \cdot \mathbf{f}(\mathbf{x}, \mathbf{p}) = \sum_{m=1}^{\infty} \frac{1}{m!} G_{c_k}^{(m)}(\mathbf{x}^*, \mathbf{p}) r_k^m, \\ G_{d_k}(\mathbf{x}, \mathbf{p}) &= \mathbf{v}_k^{\mathrm{T}} \cdot \mathbf{f}(\mathbf{x}, \mathbf{p}) = \sum_{m=1}^{\infty} \frac{1}{m!} G_{d_k}^{(m)}(\mathbf{x}^*, \mathbf{p}) r_k^m; \end{aligned} \tag{1.77}$$

$$\begin{aligned} G_{c_k}^{(m)}(\mathbf{x}^*, \mathbf{p}) &= \mathbf{u}_k^{\mathrm{T}} \cdot \partial_{\mathbf{x}}^{(m)} \mathbf{f}(\mathbf{x}, \mathbf{p})[\mathbf{u}_k \cos\theta_k + \mathbf{v}_k \sin\theta_k]^m \Big|_{(\mathbf{x}^*, \mathbf{p})}, \\ G_{d_k}^{(m)}(\mathbf{x}^*, \mathbf{p}) &= \mathbf{v}_k^{\mathrm{T}} \cdot \partial_{\mathbf{x}}^{(m)} \mathbf{f}(\mathbf{x}, \mathbf{p})[\mathbf{u}_k \cos\theta_k + \mathbf{v}_k \sin\theta_k]^m \Big|_{(\mathbf{x}^*, \mathbf{p})}. \end{aligned} \tag{1.78}$$

Thus

$$\begin{aligned}
\dot{r}_k &= \dot{c}_k \cos\theta_k + \dot{d}_k \sin\theta_k = \sum_{m=1}^{\infty} \frac{1}{m!} G_{r_k}^{(m)}(\theta_k) r_k^m \\
\dot{\theta}_k &= r_k^{-1}(\dot{d}_k \cos\theta_k - \dot{c}_k \sin\theta_k) = r_k^{-1} \sum_{m=1}^{\infty} \frac{1}{m!} G_{\theta_k}^{(m)}(\theta_k) r_k^{m-1}
\end{aligned} \tag{1.79}$$

where

$$\begin{aligned}
G_{r_k}^{(m)}(\theta_k) &= \frac{1}{\Delta}[(\Delta_2 \cos\theta_k - \Delta_{12} \sin\theta_k)\mathbf{u}_k^{\mathrm{T}} + (\Delta_2 \sin\theta_k - \Delta_{12} \cos\theta_k)\mathbf{v}_k^{\mathrm{T}}] \\
&\quad \cdot \partial_{\mathbf{x}}^{(m)} \mathbf{f}(\mathbf{x}, \mathbf{p})(\mathbf{u}_k \cos\theta_k + \mathbf{v}_k \sin\theta_k)^m \Big|_{(\mathbf{x}^*, \mathbf{p})}, \\
G_{\theta_k}^{(m)}(\theta_k) &= -\frac{1}{\Delta}[(\Delta_2 \sin\theta_k + \Delta_{12} \cos\theta_k)\mathbf{u}_k^{\mathrm{T}} - (\Delta_1 \cos\theta_k - \Delta_{12} \sin\theta_k)\mathbf{v}_k^{\mathrm{T}}] \\
&\quad \cdot \partial_{\mathbf{x}}^{(m)} \mathbf{f}(\mathbf{x}, \mathbf{p})(\mathbf{u}_k \cos\theta_k + \mathbf{v}_k \sin\theta_k)^m \Big|_{(\mathbf{x}^*, \mathbf{p})}.
\end{aligned} \tag{1.80}$$

From the foregoing definition, consider the first order terms of G-function

$$\begin{aligned}
G_{c_k}^{(1)}(\mathbf{x}, \mathbf{p}) &= G_{c_k 1}^{(1)}(\mathbf{x}, \mathbf{p}) + G_{c_k 2}^{(1)}(\mathbf{x}, \mathbf{p}) \\
G_{d_k}^{(1)}(\mathbf{x}, \mathbf{p}) &= G_{d_k 1}^{(1)}(\mathbf{x}, \mathbf{p}) + G_{d_k 2}^{(1)}(\mathbf{x}, \mathbf{p})
\end{aligned} \tag{1.81}$$

where

$$\begin{aligned}
G_{c_k 1}^{(1)}(\mathbf{x}, \mathbf{p}) &= \mathbf{u}_k^{\mathrm{T}} \cdot D_{\mathbf{x}} \mathbf{f}(\mathbf{x}, \mathbf{p}) \partial_{c_k} \mathbf{x} = \mathbf{u}_k^{\mathrm{T}} \cdot D_{\mathbf{x}} \mathbf{f}(\mathbf{x}, \mathbf{p}) \mathbf{u}_k \\
&= \mathbf{u}_k^{\mathrm{T}} \cdot (-\beta_k \mathbf{v}_k + \alpha_k \mathbf{u}_k) = \alpha_k \Delta_1 - \beta_k \Delta_{12}, \\
G_{c_k 2}^{(1)}(\mathbf{x}, \mathbf{p}) &= \mathbf{u}_k^{\mathrm{T}} \cdot D_{\mathbf{x}} \mathbf{f}(\mathbf{x}, \mathbf{p}) \partial_{d_k} \mathbf{x} = \mathbf{u}_k^{\mathrm{T}} \cdot D_{\mathbf{x}} \mathbf{f}(\mathbf{x}, \mathbf{p}) \mathbf{v}_k \\
&= \mathbf{u}_k^{\mathrm{T}} \cdot (\beta_k \mathbf{u}_k + \alpha_k \mathbf{v}_k) = \alpha_k \Delta_{12} + \beta_k \Delta_1;
\end{aligned} \tag{1.82}$$

and

$$\begin{aligned}
G_{d_k 1}^{(1)}(\mathbf{x}, \mathbf{p}) &= \mathbf{v}_k^{\mathrm{T}} \cdot D_{\mathbf{x}} \mathbf{f}(\mathbf{x}, \mathbf{p}) \partial_{c_k} \mathbf{x} = \mathbf{v}_k^{\mathrm{T}} \cdot D_{\mathbf{x}} \mathbf{f}(\mathbf{x}, \mathbf{p}) \mathbf{u}_k \\
&= \mathbf{v}_k^{\mathrm{T}} \cdot (-\beta_k \mathbf{v}_k + \alpha_k \mathbf{u}_k) = -\beta_k \Delta_2 + \alpha_k \Delta_{12}, \\
G_{d_k 2}^{(1)}(\mathbf{x}, \mathbf{p}) &= \mathbf{v}_k^{\mathrm{T}} \cdot D_{\mathbf{x}} \mathbf{f}(\mathbf{x}, \mathbf{p}) \partial_{d_k} \mathbf{x} = \mathbf{v}_k^{\mathrm{T}} \cdot D_{\mathbf{x}} \mathbf{f}(\mathbf{x}, \mathbf{p}) \mathbf{v}_k \\
&= \mathbf{v}_k^{\mathrm{T}} \cdot (\beta_k \mathbf{u}_k + \alpha_k \mathbf{v}_k) = \alpha_k \Delta_2 + \beta_k \Delta_{12}.
\end{aligned} \tag{1.83}$$

Substitution of Eqs. (1.81)–(1.83) into Eq. (1.78) gives

$$\begin{aligned}
G_{c_k}^{(1)}(\mathbf{x}, \mathbf{p}) &= G_{c_k 1}^{(1)}(\mathbf{x}, \mathbf{p}) \cos\theta_k + G_{c_k 2}^{(1)}(\mathbf{x}, \mathbf{p}) \sin\theta_k \\
&= (\alpha_k \Delta_1 - \beta_k \Delta_{12}) \cos\theta_k + (\alpha_k \Delta_{12} + \beta_k \Delta_1) \sin\theta_k, \\
G_{d_k}^{(1)}(\mathbf{x}, \mathbf{p}) &= G_{d_k 1}^{(1)}(\mathbf{x}, \mathbf{p}) \cos\theta_k + G_{d_k 2}^{(1)}(\mathbf{x}, \mathbf{p}) \sin\theta_k \\
&= (-\beta_k \Delta_2 + \alpha_k \Delta_{12}) \cos\theta_k + (\alpha_k \Delta_2 + \beta_k \Delta_{12}) \sin\theta_k.
\end{aligned} \tag{1.84}$$

From Eq. (1.80), we have

$$G_{r_k}^{(1)}(\theta_k) = \frac{1}{\Delta}[(G_{c_k}^{(1)}\Delta_2 - G_{d_k}^{(1)}\Delta_{12})\cos\theta_k + (G_{d_k}^{(1)}\Delta_1 - G_{c_k}^{(1)}\Delta_{12})\sin\theta_k] = \alpha_k;$$
$$G_{\theta_k}^{(1)}(\theta_k) = \frac{1}{\Delta}[(G_{d_k}^{(1)}\Delta_1 - G_{c_k}^{(1)}\Delta_{12})\cos\theta_k - (G_{c_k}^{(1)}\Delta_2 - G_{d_k}^{(1)}\Delta_{12})\sin\theta_k] = -\beta_k. \tag{1.85}$$

Furthermore, Eq. (1.79) gives

$$\dot{r}_k = \alpha_k r_k + o(r_k) \text{ and } \dot{\theta}_k r_k = -\beta_k r_k + o(r_k). \tag{1.86}$$

As $r_k << 1$ and $r_k \to 0$, we have

$$\dot{r}_k = \alpha_k r_k \text{ and } \dot{\theta}_k = -\beta_k. \tag{1.87}$$

With an initial condition of $r_k = r_k^0$ and $\theta_k = \theta_k^0$, the corresponding solution of Eq. (1.87) is

$$r_k = r_k^0 e^{\alpha_k t} \text{ and } \theta_k = -\beta_k t + \theta_k^0 \tag{1.88}$$

and

$$\begin{aligned} c_k &= r_k^0 e^{\alpha_k t}\cos(-\beta_k t + \theta_k^0) = e^{\alpha_k t}[\cos(\beta_k t)c_k^0 + \sin(\beta_k t)d_k^0]; \\ d_k &= r_k^0 e^{\alpha_k t}\sin(-\beta_k t + \theta_k^0) = e^{\alpha_k t}[-\sin(\beta_k t)c_k^0 + \cos(\beta_k t)d_k^0]. \end{aligned} \tag{1.89}$$

Letting $\mathbf{c}^{(k)} = (c^{(k)}, d^{(k)})^{\mathrm{T}}$, we have

$$\dot{\mathbf{c}}^{(k)} = \mathbf{E}_k \mathbf{c}^{(k)} \Rightarrow \mathbf{c}^{(k)} = e^{\alpha_k t}\mathbf{B}_k \mathbf{c}_0^{(k)} \tag{1.90}$$

where

$$\mathbf{E}_i = \begin{bmatrix} \alpha_k & \beta_k \\ -\beta_k & \alpha_k \end{bmatrix} \text{ and } \mathbf{B}_k = \begin{bmatrix} \cos\beta_k t & \sin\beta_k t \\ -\sin\beta_k t & \cos\beta_k t \end{bmatrix}. \tag{1.91}$$

If $G_{r_k}^{(m)}(\theta_k)$ and $G_{\theta_k}^{(m)}(\theta_k)$ are dependent on θ_k, Eq. (1.79) gives the dynamical systems based on the polar coordinates on the invariant plane of $(\mathbf{u}_k, \mathbf{v}_k)$ of matrix $D\mathbf{f}(\mathbf{x}^*, \mathbf{p})$ with a pair of eigenvectors $\mathbf{u}_k \pm \mathbf{i}\mathbf{v}_k$. If $G_{r_k}^{(m)}(\theta_k)$ and $G_{\theta_k}^{(m)}(\theta_k)$ are independent of θ_k, the deformed dynamical system on the plane of $(\mathbf{u}_k, \mathbf{v}_k)$ is dependent on r_k, then the G-functions can be used to determine the stability of $\mathbf{x}^{(k)}$ at the equilibrium $\mathbf{x}^*$ on the plane of $(\mathbf{u}_k, \mathbf{v}_k)$.

Definition 1.29 Consider an n-dimensional, autonomous, nonlinear dynamical system $\dot{\mathbf{x}} = \mathbf{f}(\mathbf{x}, \mathbf{p})$ in Eq. (1.4) with an equilibrium point $\mathbf{x}^*$ and $\mathbf{f}(\mathbf{x}, \mathbf{p})$ is C^r $(r \geq 1)$-continuous in a neighborhood of the equilibrium $\mathbf{x}^*$. The corresponding solution is $\mathbf{x}(t) = \mathbf{\Phi}(\mathbf{x}_0, t - t_0, \mathbf{p})$. Suppose $U(\mathbf{x}^*) \subset \Omega$ is a neighborhood of equilibrium $\mathbf{x}^*$. For a linearized dynamical system in Eq. (1.19), consider a pair of complex eigenvalue $\alpha_k \pm \mathbf{i}\beta_k$ $(k \in N = \{1, 2, \cdots, n\}$, $\mathbf{i} = \sqrt{-1})$ of matrix $D\mathbf{f}(\mathbf{x}^*, \mathbf{p})$ with a pair of eigenvectors $\mathbf{u}_k \pm \mathbf{i}\mathbf{v}_k$. On the invariant plane of $(\mathbf{u}_k, \mathbf{v}_k)$, consider $\mathbf{y}^{(k)} = \mathbf{y}_+^{(k)} + \mathbf{y}_-^{(k)}$ with Eqs. (1.72) and (1.74). For any arbitrarily small $\varepsilon > 0$, the stability of the equilibrium $\mathbf{x}^*$ on the invariant plane of$(\mathbf{u}_k, \mathbf{v}_k)$ can be determined.

(i) $\mathbf{x}^{(k)}$ at the equilibrium $\mathbf{x}^*$ on the plane of $(\mathbf{u}_k, \mathbf{v}_k)$ is spirally stable if

$$r_k(t+\varepsilon) - r_k(t) < 0. \tag{1.92}$$

(ii) $\mathbf{x}^{(k)}$ at the equilibrium $\mathbf{x}^*$ on the plane of $(\mathbf{u}_k, \mathbf{v}_k)$ is spirally unstable if

$$r_k(t+\varepsilon) - r_k(t) > 0. \tag{1.93}$$

(iii) $\mathbf{x}^{(k)}$ at the equilibrium $\mathbf{x}^*$ on the plane of $(\mathbf{u}_k, \mathbf{v}_k)$ is stable with the m_kth-order singularity if for $\theta_k \in [0, 2\pi]$

$$\begin{aligned} &G_{r_k}^{(s_k)}(\theta_k) = 0 \;\text{ for } s_k = 0, 1, 2, \cdots, m_k - 1 \\ &r_k(t+\varepsilon) - r_k(t) < 0. \end{aligned} \tag{1.94}$$

(iv) $\mathbf{x}^{(k)}$ at the equilibrium $\mathbf{x}^*$ on the plane of $(\mathbf{u}_k, \mathbf{v}_k)$ is spirally unstable with the m_kth-order singularity if for $\theta_k \in [0, 2\pi]$

$$\begin{aligned} &G_{r_k}^{(s_k)}(\theta_k) = 0 \;\text{ for } s_k = 0, 1, 2, \cdots, m_k - 1 \\ &r_k(t+\varepsilon) - r_k(t) > 0. \end{aligned} \tag{1.95}$$

(v) $\mathbf{x}^{(k)}$ at the equilibrium $\mathbf{x}^*$ on the plane of $(\mathbf{u}_k, \mathbf{v}_k)$ is circular if for $\theta_k \in [0, 2\pi]$

$$r_k(t+\varepsilon) - r_k(t) = 0. \tag{1.96}$$

(vi) $\mathbf{x}^{(k)}$ at the equilibrium $\mathbf{x}^*$ on the plane of $(\mathbf{u}_k, \mathbf{v}_k)$ is degenerate in the direction of $\mathbf{u}_k$ if

$$\beta_k = 0 \;\text{ and }\; \theta_k(t+\varepsilon) - \theta_k(t) = 0. \tag{1.97}$$

Theorem 1.10 *Consider an n-dimensional, autonomous, nonlinear dynamical system $\dot{\mathbf{x}} = \mathbf{f}(\mathbf{x}, \mathbf{p})$ in Eq. (1.4) with an equilibrium point $\mathbf{x}^*$ and $\mathbf{f}(\mathbf{x}, \mathbf{p})$ is C^r $(r \geq 1)$-continuous in a neighborhood of the equilibrium $\mathbf{x}^*$. The corresponding solution is $\mathbf{x}(t) = \mathbf{\Phi}(\mathbf{x}_0, t - t_0, \mathbf{p})$. Suppose $U(\mathbf{x}^*) \subset \Omega$ is a neighborhood of equilibrium $\mathbf{x}^*$. For a linearized dynamical system in Eq. (1.19), consider a pair of complex eigenvalue $\alpha_k \pm \mathbf{i}\beta_k$ $(k \in N = \{1, 2, \cdots, n\}$, $\mathbf{i} = \sqrt{-1})$ of matrix $D\mathbf{f}(\mathbf{x}^*, \mathbf{p})$ with a pair of eigenvectors $\mathbf{u}_k \pm \mathbf{i}\mathbf{v}_k$. On the invariant plane of $(\mathbf{u}_k, \mathbf{v}_k)$, consider $\mathbf{y}^{(k)} = \mathbf{y}_+^{(k)} + \mathbf{y}_-^{(k)}$ with Eqs. (1.72) and (1.74) with $G_{r_k}^{(s_k)}(\theta_k) = const$. For any arbitrarily small $\varepsilon > 0$, the stability of the equilibrium $\mathbf{x}^*$ on the invariant plane of $(\mathbf{u}_k, \mathbf{v}_k)$ can be determined.*

(i) *$\mathbf{x}^{(k)}$ at the equilibrium $\mathbf{x}^*$ on the plane of $(\mathbf{u}_k, \mathbf{v}_k)$ is spirally stable if and only if*

$$G_{r_k}^{(1)}(\theta_k) = \alpha_k < 0. \tag{1.98}$$

(ii) $\mathbf{x}^{(k)}$ *at the equilibrium* $\mathbf{x}^*$ *on the plane of* $(\mathbf{u}_k, \mathbf{v}_k)$ *is spirally unstable if and only if*

$$G_{r_k}^{(1)}(\theta_k) = \alpha_k > 0. \tag{1.99}$$

(iii) $\mathbf{x}^{(k)}$ *at the equilibrium* $\mathbf{x}^*$ *on the plane of* $(\mathbf{u}_k, \mathbf{v}_k)$ *is stable with the* m_k*th-order singularity if and only if for* $\theta_k \in [0, 2\pi]$

$$\begin{aligned} &G_{r_k}^{(s_k)}(\theta_k) = 0 \text{ for } s_k = 1, 2, \cdots, m_k - 1 \\ &\text{and } G_{r_k}^{(m_k)}(\theta_k) < 0. \end{aligned} \tag{1.100}$$

(iv) $\mathbf{x}^{(k)}$ *at the equilibrium* $\mathbf{x}^*$ *on the plane of* $(\mathbf{u}_k, \mathbf{v}_k)$ *is spirally unstable with the* m_k*th-order singularity if and only if for* $\theta_k \in [0, 2\pi]$

$$\begin{aligned} &G_{r_k}^{(s_k)}(\theta_k) = 0 \text{ for } s_k = 1, 2, \cdots, m_k - 1 \\ &\text{and } G_{r_k}^{(m_k)}(\theta_k) > 0. \end{aligned} \tag{1.101}$$

(v) $\mathbf{x}^{(k)}$ *at the equilibrium* $\mathbf{x}^*$ *on the plane of* $(\mathbf{u}_k, \mathbf{v}_k)$ *is circular if and only if for* $\theta_k \in [0, 2\pi]$

$$G_{r_k}^{(s_k)}(\theta_k) = 0 \text{ for } s_k = 1, 2, \cdots. \tag{1.102}$$

(vi) $\mathbf{x}^{(k)}$ *at the equilibrium* $\mathbf{x}^*$ *on the plane of* $(\mathbf{u}_k, \mathbf{v}_k)$ *is degenerate in the direction of* $\mathbf{u}_k$ *if and only if*

$$\mathrm{Im}\lambda_k = \beta_k = 0 \text{ and } G_{\theta_k}^{(s_k)}(\theta_k) = 0 \text{ for } s_k = 2, 3, \cdots. \tag{1.103}$$

Proof For $\mathbf{x} = \mathbf{x}^*$, $s_k = 0$. The first order approximation of $\dot{c}_k$ and $\dot{d}_k$ in the Taylor series expansion gives

$$\begin{aligned} \dot{c}_k &= \frac{1}{\Delta}[\Delta_2 G_{c_k}^{(1)}(\mathbf{x}, \mathbf{p}) - \Delta_{12} G_{d_k}^{(1)}(\mathbf{x}, \mathbf{p})] \\ \dot{d}_k &= \frac{1}{\Delta}[\Delta_1 G_{d_k}^{(1)}(\mathbf{x}, \mathbf{p}) - \Delta_{12} G_{c_k}^{(1)}(\mathbf{x}, \mathbf{p})] \end{aligned}$$

where $r_k = \sqrt{c_k^2 + d_k^2}$ and

$$\begin{aligned} G_{c_k 1}^{(1)}(\mathbf{x}, \mathbf{p}) &= \mathbf{u}_k^{\mathrm{T}} \cdot D_{\mathbf{x}}\mathbf{f}(\mathbf{x}, \mathbf{p})\partial_{c_k}\mathbf{x} = \mathbf{u}_k^{\mathrm{T}} \cdot D_{\mathbf{x}}\mathbf{f}(\mathbf{x}, \mathbf{p})\mathbf{u}_k \\ &= \mathbf{u}_k^{\mathrm{T}} \cdot (-\beta_k \mathbf{v}_k + \alpha_k \mathbf{u}_k) = \alpha_k \Delta_1 - \beta_k \Delta_{12}, \\ G_{c_k 2}^{(1)}(\mathbf{x}, \mathbf{p}) &= \mathbf{u}_k^{\mathrm{T}} \cdot D_{\mathbf{x}}\mathbf{f}(\mathbf{x}, \mathbf{p})\partial_{d_k}\mathbf{x} = \mathbf{u}_k^{\mathrm{T}} \cdot D_{\mathbf{x}}\mathbf{f}(\mathbf{x}, \mathbf{p})\mathbf{v}_k \\ &= \mathbf{u}_k^{\mathrm{T}} \cdot (\beta_k \mathbf{u}_k + \alpha_k \mathbf{v}_k) = \alpha_k \Delta_{12} + \beta_k \Delta_1; \end{aligned}$$

and

$$
\begin{aligned}
G^{(1)}_{d_k1}(\mathbf{x},\mathbf{p}) &= \mathbf{v}_k^{\mathrm{T}}\cdot D_{\mathbf{x}}\mathbf{f}(\mathbf{x},\mathbf{p})\partial_{c_k}\mathbf{x} = \mathbf{v}_k^{\mathrm{T}}\cdot D_{\mathbf{x}}\mathbf{f}(\mathbf{x},\mathbf{p})\mathbf{u}_k\\
&= \mathbf{v}_k^{\mathrm{T}}\cdot(-\beta_k\mathbf{v}_k+\alpha_k\mathbf{u}_k) = -\beta_k\Delta_2+\alpha_k\Delta_{12},\\
G^{(1)}_{d_k2}(\mathbf{x},\mathbf{p}) &= \mathbf{v}_k^{\mathrm{T}}\cdot D_{\mathbf{x}}\mathbf{f}(\mathbf{x},\mathbf{p})\partial_{d_k}\mathbf{x} = \mathbf{v}_k^{\mathrm{T}}\cdot D_{\mathbf{x}}\mathbf{f}(\mathbf{x},\mathbf{p})\mathbf{v}_k\\
&= \mathbf{v}_k^{\mathrm{T}}\cdot(\beta_k\mathbf{u}_k+\alpha_k\mathbf{v}_k) = \alpha_k\Delta_2+\beta_k\Delta_{12}.
\end{aligned}
$$

Therefore, using

$$
\begin{aligned}
G^{(1)}_{c_k}(\mathbf{x},\mathbf{p}) &= G^{(1)}_{c_k1}(\mathbf{x},\mathbf{p})c_k + G^{(1)}_{c_k2}(\mathbf{x},\mathbf{p})d_k\\
G^{(1)}_{d_k}(\mathbf{x},\mathbf{p}) &= G^{(1)}_{d_k1}(\mathbf{x},\mathbf{p})c_k + G^{(1)}_{d_k2}(\mathbf{x},\mathbf{p})d_k
\end{aligned}
$$

to the first order approximation of $\dot{c}_k$ and $\dot{d}_k$ yields

$$
\dot{c}_k = \alpha_k c_k + \beta_k d_k \quad \text{and} \quad \dot{d}_k = -\beta_k c_k + \alpha_k d_k,
$$

or

$$
\begin{bmatrix}\dot{c}_k\\ \dot{d}_k\end{bmatrix} = \begin{bmatrix}\alpha_k & \beta_k\\ -\beta_k & \alpha_k\end{bmatrix}\begin{bmatrix}c_k\\ d_k\end{bmatrix}.
$$

Introduce the rotation coordinates $(\mathbf{e}_{r_k},\mathbf{e}_{\theta_k})$

$$
\mathbf{r}_k = c_k\mathbf{u}_k + d_k\mathbf{v}_k = r_k\mathbf{e}_{r_k},
$$

where

$$
\begin{aligned}
c_k &= r_k\cos\theta_k,\ d_k = r_k\sin\theta_k;\\
\mathbf{e}_{r_k} &= \cos\theta_k\mathbf{u}_k + \sin\theta_k\mathbf{v}_k,\\
\mathbf{e}_{\theta_k} &= -\cos\theta_k\mathbf{u}_k^{\perp}\Delta_3 + \sin\theta_k\mathbf{v}_k^{\perp}\Delta_4
\end{aligned}
$$

and

$$
\begin{aligned}
\dot{\mathbf{r}}_k &= \dot{c}_k\mathbf{u}_k + \dot{d}_k\mathbf{v}_k = \dot{r}_k\mathbf{e}_{r_k} + r_k\dot{\mathbf{e}}_{r_k},\\
\dot{\mathbf{e}}_{r_k} &= -\dot{\theta}_k\mathbf{u}_k\sin\theta_k + \dot{\theta}_k\mathbf{v}_k\cos\theta_k.
\end{aligned}
$$

Thus

$$
\begin{aligned}
\dot{r}_k &= \dot{c}_k\cos\theta_k + \dot{d}_k\sin\theta_k,\\
\dot{\theta}_k &= r_k^{-1}(\dot{d}_k\cos\theta_k - \dot{c}_k\sin\theta_k).
\end{aligned}
$$

For the first approximation of the relative change rate in the $\mathbf{e}_{r_k}$ direction, we obtain

$$
\begin{aligned}
\dot{r}_k &= (\alpha_k c_k + \beta_k d_k)\cos\theta_k + (-\beta_k c_k + \alpha_k d_k)\sin\theta_k\\
&= \alpha_k r_k.
\end{aligned}
$$

Further

$$\dot{r}_k = \alpha_k r_k.$$

Similarly, the first approximation of rotation speed in the hoop direction is

$$\begin{aligned}\dot{\theta}_k r_k &= (-\beta_k c_k + \alpha_k d_k)\cos\theta_k + (\alpha_k c_k + \beta_k d_k)\sin\theta_k \\ &= -\beta_k r_k\end{aligned}$$

so

$$\dot{\theta}_k r_k = -\beta_k r_k \Rightarrow \dot{\theta}_k = -\beta_k.$$

Therefore,

$$G^{(1)}_{r_k}(\theta_k) = \alpha_k \text{ and } G^{(1)}_{\theta_k}(\theta_k) = -\beta_k.$$

In fact, the relative change rate in the $\mathbf{e}_{r_k}$ direction is of interest. The corresponding higher-order expression is given by

$$\dot{r}_k = \sum\nolimits_{s_k=1}^{m_k-1} \frac{1}{s_k!} G^{(s_k)}_{r_k}(\theta_k) r_k^{s_k} + \frac{1}{m_k!} G^{(m_k)}_{r_k}(\theta_k) r_k^{m_k} + o(r_k^{m_k}).$$

Because for $\varepsilon > 0$ and $\varepsilon \to 0$,

$$\begin{aligned}r_k(t+\varepsilon) - r_k(t) &= \dot{r}_k \varepsilon \\ &= \varepsilon \sum\nolimits_{s_k=1}^{m_k-1} \frac{1}{s_k!} G^{(s_k)}_{r_k}(\theta_k) r_k^{s_k} + \varepsilon \frac{1}{m_k!} G^{(m_k)}_{r_k}(\theta_k) r_k^{m_k} + o(\varepsilon r_k^{m_k}).\end{aligned}$$

(i) For equilibrium stability, $r_k > 0$ and $r_k \to 0$. If $G^{(1)}_{r_k}(\theta_k) = \alpha_k \neq 0$, we have

$$\dot{r}_k = G^{(1)}_{r_k}(\theta_k) r_k = \alpha_k r_k.$$

Due to $r_k > 0$, if $\alpha_k < 0$, then $\dot{r}_k < 0$. Therefore,

$$r_k(t+\varepsilon) - r_k(t) = \dot{r}_k \varepsilon < 0$$

which implies $\mathbf{x}^{(k)}$ at the equilibrium $\mathbf{x}^*$ on the plane of $(\mathbf{u}_k, \mathbf{v}_k)$ is spirally stable, vice versa.

(ii) Due to $r_k > 0$, if $\alpha_k > 0$, then $\dot{r}_k > 0$. Thus,

$$r_k(t+\varepsilon) - r_k(t) = \dot{r}_k \varepsilon > 0,$$

which implies $\mathbf{x}^{(k)}$ at the equilibrium $\mathbf{x}^*$ on the plane of $(\mathbf{u}_k, \mathbf{v}_k)$ is spirally unstable, vice versa.

(iii) If for $\theta_k \in [0, 2\pi]$ the following conditions exist

$$G_{r_k}^{(s_k)}(\theta_k) = 0 \text{ for } s_k = 1, 2, \cdots, m_k - 1;$$
$$G_{r_k}^{(m_k)}(\theta_k) \neq 0, \text{ and } |G_{r_k}^{(s_k)}(\theta_k)| < \infty \text{ for } s_k = m_k + 1, m_k + 2, \cdots,$$

then the higher-order terms can be ignored, i.e.,

$$\dot{r}_k = \frac{1}{m_k!} G_{r_k}^{(m_k)}(\theta_k) r_k^{m_k}.$$

If $G_{r_k}^{(m_k)}(\theta_k)$ is independent of θ_k (i.e., $G_{r_k}^{(m_k)}(\theta_k) = \text{const}$), it can be used to determine the equilibrium stability. Due to $r_k > 0$, if $G_{r_k}^{(m_k)}(\theta_k) < 0$, then $\dot{r}_k < 0$. Therefore,

$$r_k(t + \varepsilon) - r_k(t) = \dot{r}_k \varepsilon < 0.$$

In other words, $\mathbf{x}^{(k)}$ at the equilibrium $\mathbf{x}^*$ on the plane of $(\mathbf{u}_k, \mathbf{v}_k)$ is spirally stable with the m_kth-order singularity, vice versa.

(iv) Due to $r_k > 0$, if $G_{r_k}^{(m_k)}(\theta_k) > 0$, then $\dot{r}_k > 0$. Therefore,

$$r_k(t + \varepsilon) - r_k(t) = \dot{r}_k \varepsilon > 0.$$

In other words, $\mathbf{x}^{(k)}$ at the equilibrium $\mathbf{x}^*$ on the plane of $(\mathbf{u}_k, \mathbf{v}_k)$ is spirally unstable with the m_kth-order singularity, vice versa.

(v) If for $\theta_k \in [0, 2\pi]$ the following conditions exist

$$G_{r_k}^{(s_k)}(\theta_k) = 0 \text{ for } s_k = 1, 2, \cdots,$$

then

$$r_k(t + \varepsilon) - r_k(t) = \dot{r}_k \varepsilon = 0$$

vice versa. Therefore $r_k(t)$ is constant. $\mathbf{x}^{(k)}$ at the equilibrium $\mathbf{x}^*$ on the plane of $(\mathbf{u}_k, \mathbf{v}_k)$ is circular.

(vi) Consider

$$\begin{aligned}\dot{\theta}_k \varepsilon &= \theta_k(t + \varepsilon) - \theta_k(t) \\ &= \varepsilon\Big[-\beta_k + \sum\nolimits_{s_k=2}^{m_k-1} \frac{1}{S_k!} G_{\theta_k}^{(s_k)}(\theta_k) r_k^{s_k-1} + \frac{1}{m_k!} G_{\theta_k}^{(m_k)}(\theta_k) r_k^{m_k-1} + o(r_k^{m_k-1})\Big].\end{aligned}$$

If for $\theta_k \in [0, 2\pi]$ the following conditions exist

$$\beta_k = 0 \text{ and } G_{\theta_k}^{(s_k)}(\theta_k) = 0 \text{ for } s_k = 2, 3, \cdots.$$

Then

$$\theta_k(t+\varepsilon) - \theta_k(t) = \dot{\theta}_k \varepsilon = 0.$$

Therefore, $\mathbf{x}^{(k)}$ at the equilibrium $\mathbf{x}^*$ on the plane of $(\mathbf{u}_k, \mathbf{v}_k)$ is degenerate in the direction of $\mathbf{u}_k$. This theorem is proved. ■

Note that $G_{r_k}^{(s_k)}(\theta_k) = const$ requires $s_k = 2m_k - 1$ and one obtains $G_{r_k}^{(s_k)}(\theta_k) = 0$ for $s_k = 2m_k$.

1.3.2 Bifurcations

Definition 1.30 Consider an n-dimensional, autonomous, nonlinear dynamical system $\dot{\mathbf{x}} = \mathbf{f}(\mathbf{x}, \mathbf{p})$ in Eq. (1.4) with an equilibrium point $\mathbf{x}^*$ and $\mathbf{f}(\mathbf{x}, \mathbf{p})$ is C^r $(r \geq 1)$-continuous in a neighborhood of the equilibrium $\mathbf{x}^*$ (i.e., $U(\mathbf{x}^*) \subset \Omega$). The corresponding solution is $\mathbf{x}(t) = \mathbf{\Phi}(\mathbf{x}_0, t - t_0, \mathbf{p})$. Suppose Eq. (1.24) holds in $U(\mathbf{x}^*) \subset \Omega$. For a linearized dynamical system in Eq. (1.19), consider a real eigenvalue λ_k of matrix $D\mathbf{f}(\mathbf{x}^*, \mathbf{p}^*)$ $(k \in N = \{1, 2, \cdots, n\})$ with an eigenvector $\mathbf{v}_k$. Suppose one of n independent solutions $\mathbf{y} = c_k \mathbf{v}_k$ and $\dot{\mathbf{y}} = \dot{c}_k \mathbf{v}_k$,

$$s_k = \mathbf{v}_k^{\mathrm{T}} \cdot \mathbf{y} = \mathbf{v}_k^{\mathrm{T}} \cdot (\mathbf{x} - \mathbf{x}^*) \tag{1.104}$$

where $s_k = c_k \|\mathbf{v}_k\|^2$.

$$\dot{s}_k = \mathbf{v}_k^{\mathrm{T}} \cdot \dot{\mathbf{y}} = \mathbf{v}_k^{\mathrm{T}} \cdot \dot{\mathbf{x}} = \mathbf{v}_k^{\mathrm{T}} \cdot \mathbf{f}(\mathbf{x}, \mathbf{p}). \tag{1.105}$$

In the vicinity of point $(\mathbf{x}_0^*, \mathbf{p}_0)$, $\mathbf{v}_k^{\mathrm{T}} \cdot \mathbf{f}(\mathbf{x}, \mathbf{p})$ can be expended for $(0 < \theta < 1)$ as

$$\begin{aligned}
&\mathbf{v}_k^{\mathrm{T}} \cdot \mathbf{f}(\mathbf{x}, \mathbf{p}) \\
&= a_k (s_k - s_{k0}^*) + \mathbf{b}_k^{\mathrm{T}} \cdot (\mathbf{p} - \mathbf{p}_0) + \sum_{q=2}^{m} \sum_{r=0}^{q} \frac{1}{q!} C_q^r \mathbf{a}_k^{(q-r,r)} (s_k - s_{k0}^*)^{q-r} (\mathbf{p} - \mathbf{p}_0)^r \\
&+ \frac{1}{(m+1)!} [(s_k - s_{k0}^*) \partial_{s_k} + (\mathbf{p} - \mathbf{p}_0) \partial_{\mathbf{p}}]^{m+1} (\mathbf{v}_k^{\mathrm{T}} \cdot \mathbf{f}(\mathbf{x}_0^* + \theta \Delta \mathbf{x}, \mathbf{p}_0 + \theta \Delta \mathbf{p}))
\end{aligned} \tag{1.106}$$

where

$$\begin{aligned}
a_k &= \mathbf{v}_k^{\mathrm{T}} \cdot \partial_{s_k} \mathbf{f}(\mathbf{x}, \mathbf{p}) \Big|_{(\mathbf{x}_0^*, \mathbf{p}_0)}, \\
\mathbf{b}_k^{\mathrm{T}} &= \mathbf{v}_k^{\mathrm{T}} \cdot \partial_{\mathbf{p}} \mathbf{f}(\mathbf{x}, \mathbf{p}) \Big|_{(\mathbf{x}_0^*, \mathbf{p}_0)}, \\
\mathbf{a}_k^{(r,s)} &= \mathbf{v}_k^{\mathrm{T}} \cdot \partial_{s_k}^{(r)} \partial_{\mathbf{p}}^{(s)} \mathbf{f}(\mathbf{x}, \mathbf{p}) \Big|_{(\mathbf{x}_0^*, \mathbf{p}_0)}.
\end{aligned} \tag{1.107}$$

If $a_k = 0$ and $\mathbf{p} = \mathbf{p}_0$, the stability of current equilibrium $\mathbf{x}^*$ on an eigenvector $\mathbf{v}_k$ changes from stable to unstable state (or from unstable to stable state). The bifurcation manifold in the direction of $\mathbf{v}_k$ is determined by

$$\mathbf{b}_k^{\mathrm{T}} \cdot (\mathbf{p}-\mathbf{p}_0) + \sum_{q=2}^{m}\sum_{r=0}^{q} \frac{1}{q!} C_q^r \mathbf{a}_k^{(q-r,r)} (s_k - s_{k0}^*)^{q-r} (\mathbf{p}-\mathbf{p}_0)^r = 0. \tag{1.108}$$

In the neighborhood of $(\mathbf{x}_0^*, \mathbf{p}_0)$, when other components of equilibrium $\mathbf{x}^*$ on the eigenvector of $\mathbf{v}_j$ for all $j \neq k$, $(j, k \in N)$ do not change their stability states, equation (1.108) possesses l-branch solutions of equilibrium s_k^* $(0 < l \leq m)$ with l_1-stable and l_2-unstable solutions $(l_1, l_2 \in \{0, 1, 2, \cdots, l\})$. Such l-branch solutions are called the bifurcation solutions of equilibrium $\mathbf{x}^*$ on the eigenvector of $\mathbf{v}_k$ in the neighborhood of $(\mathbf{x}_0^*, \mathbf{p}_0)$. Such a bifurcation at point $(\mathbf{x}_0^*, \mathbf{p}_0)$ is called the hyperbolic bifurcation of mth-order on the eigenvector of $\mathbf{v}_k$.

Three special cases are defined as

(i) If

$$\mathbf{a}_k^{(1,1)} = \mathbf{0} \text{ and } \mathbf{b}_k^{\mathrm{T}} \cdot (\mathbf{p}-\mathbf{p}_0) + \frac{1}{2!} a_k^{(2,0)} (s_k^* - s_{k0}^*)^2 = 0 \tag{1.109}$$

where

$$\begin{aligned} a_k^{(2,0)} &= \mathbf{v}_k^{\mathrm{T}} \cdot \partial_{s_k}^{(2)} \partial_{\mathbf{p}}^{(0)} \mathbf{f}(\mathbf{x},\mathbf{p})\Big|_{(\mathbf{x}_0^*,\mathbf{p}_0)} = \mathbf{v}_k^{\mathrm{T}} \cdot \partial_{s_k}^{(2)} \mathbf{f}(\mathbf{x},\mathbf{p})\Big|_{(\mathbf{x}_0^*,\mathbf{p}_0)} \\ &= \mathbf{v}_k^{\mathrm{T}} \cdot \partial_{\mathbf{x}}^{(2)} \mathbf{f}(\mathbf{x},\mathbf{p})(\mathbf{v}_k\mathbf{v}_k)\Big|_{(\mathbf{x}_0^*,\mathbf{p}_0)} = G_k^{(2)}(\mathbf{x}_0^*, \mathbf{p}_0) \neq 0, \\ \mathbf{b}_k^{\mathrm{T}} &= \mathbf{v}_k^{\mathrm{T}} \cdot \partial_{\mathbf{p}} \mathbf{f}(\mathbf{x},\mathbf{p})\Big|_{(\mathbf{x}_0^*,\mathbf{p}_0)} \neq \mathbf{0}, \end{aligned} \tag{1.110}$$

$$a_k^{(2,0)} \times [\mathbf{b}_k^{\mathrm{T}} \cdot (\mathbf{p}-\mathbf{p}_0)] < 0, \tag{1.111}$$

such a bifurcation at point $(\mathbf{x}_0^*, \mathbf{p}_0)$ is called the *saddle-node* bifurcation on the eigenvector of $\mathbf{v}_k$.

(ii) If

$$\begin{aligned} &\mathbf{b}_\alpha^{\mathrm{T}} \cdot (\mathbf{p}-\mathbf{p}_0) = 0 \\ &\mathbf{a}_k^{(1,1)} \cdot (\mathbf{p}-\mathbf{p}_0)(s_k^* - s_{k0}^*) + \frac{1}{2!} a_k^{(2,0)} (s_k^* - s_{k0}^*)^2 = 0 \end{aligned} \tag{1.112}$$

where

$$\begin{aligned} a_k^{(2,0)} &= \mathbf{v}_k^{\mathrm{T}} \cdot \partial_{s_k}^{(2)} \partial_{\mathbf{p}}^{(0)} \mathbf{f}(\mathbf{x},\mathbf{p})\Big|_{(\mathbf{x}_0^*,\mathbf{p}_0)} = \mathbf{v}_k^{\mathrm{T}} \cdot \partial_{s_k}^{(2)} \mathbf{f}(\mathbf{x},\mathbf{p})\Big|_{(\mathbf{x}_0^*,\mathbf{p}_0)} \\ &= \mathbf{v}_k^{\mathrm{T}} \cdot \partial_{\mathbf{x}}^{(2)} \mathbf{f}(\mathbf{x},\mathbf{p})(\mathbf{v}_k\mathbf{v}_k)\Big|_{(\mathbf{x}_0^*,\mathbf{p}_0)} = G_k^{(2)}(\mathbf{x}_0^*, \mathbf{p}_0) \neq 0, \\ \mathbf{a}_k^{(1,1)} &= \mathbf{v}_k^{\mathrm{T}} \cdot \partial_{s_k}^{(1)} \partial_{\mathbf{p}}^{(1)} \mathbf{f}(\mathbf{x},\mathbf{p})\Big|_{(\mathbf{x}_0^*,\mathbf{p}_0)} = \mathbf{v}_k^{\mathrm{T}} \cdot \partial_{s_k} \partial_{\mathbf{p}} \mathbf{f}(\mathbf{x},\mathbf{p})\Big|_{(\mathbf{x}_0^*,\mathbf{p}_0)} \\ &= \mathbf{v}_k^{\mathrm{T}} \cdot \partial_{\mathbf{x}} \partial_{\mathbf{p}} \mathbf{f}(\mathbf{x},\mathbf{p})\mathbf{v}_k\Big|_{(\mathbf{x}_0^*,\mathbf{p}_0)} \neq \mathbf{0}, \end{aligned} \tag{1.113}$$

$$a_k^{(2,0)} \times [\mathbf{a}_k^{(1,1)} \cdot (\mathbf{p} - \mathbf{p}_0)] < 0, \tag{1.114}$$

such a bifurcation at point $(\mathbf{x}_0^*, \mathbf{p}_0)$ is called the *transcritical* bifurcation on the eigenvector of $\mathbf{v}_k$.

(iii) If

$$\begin{aligned} &\mathbf{b}_\alpha^{\mathrm{T}} \cdot (\mathbf{p} - \mathbf{p}_0) = 0, a_k^{(2,0)} = 0, \mathbf{a}_k^{(2,1)} = 0, \mathbf{a}_k^{(1,2)} = 0, \\ &\mathbf{a}_k^{(1,1)} \cdot (\mathbf{p} - \mathbf{p}_0)(s_k^* - s_{k0}^*) + \frac{1}{3!} a_k^{(3,0)} (s_k^* - s_{k0}^*)^3 = 0 \end{aligned} \tag{1.115}$$

where

$$\begin{aligned} a_k^{(3,0)} &= \mathbf{v}_k^{\mathrm{T}} \cdot \partial_{s_k}^{(3)} \partial_{\mathbf{p}}^{(0)} \mathbf{f}(\mathbf{x}, \mathbf{p}) \Big|_{(\mathbf{x}_0^*, \mathbf{p}_0)} = \mathbf{v}_k^{\mathrm{T}} \cdot \partial_{s_k}^{(3)} \mathbf{f}(\mathbf{x}, \mathbf{p}) \Big|_{(\mathbf{x}_0^*, \mathbf{p}_0)} \\ &= \mathbf{v}_k^{\mathrm{T}} \cdot \partial_{\mathbf{x}}^{(3)} \mathbf{f}(\mathbf{x}, \mathbf{p})(\mathbf{v}_k \mathbf{v}_k \mathbf{v}_k) \Big|_{(\mathbf{x}_0^*, \mathbf{p}_0)} = G_k^{(3)}(\mathbf{x}_0^*, \mathbf{p}_0) \neq 0, \\ \mathbf{a}_k^{(1,1)} &= \mathbf{v}_k^{\mathrm{T}} \cdot \partial_{s_k}^{(1)} \partial_{\mathbf{p}}^{(1)} \mathbf{f}(\mathbf{x}, \mathbf{p}) \Big|_{(\mathbf{x}_0^*, \mathbf{p}_0)} = \mathbf{v}_k^{\mathrm{T}} \cdot \partial_{s_k} \partial_{\mathbf{p}} \mathbf{f}(\mathbf{x}, \mathbf{p}) \Big|_{(\mathbf{x}_0^*, \mathbf{p}_0)} \\ &= \mathbf{v}_k^{\mathrm{T}} \cdot \partial_{\mathbf{x}} \partial_{\mathbf{p}} \mathbf{f}(\mathbf{x}, \mathbf{p}) \mathbf{v}_k \Big|_{(\mathbf{x}_0^*, \mathbf{p}_0)} \neq \mathbf{0}, \end{aligned} \tag{1.116}$$

$$a_k^{(3,0)} \times [\mathbf{a}_k^{(1,1)} \cdot (\mathbf{p} - \mathbf{p}_0)] < 0, \tag{1.117}$$

such a bifurcation at point $(\mathbf{x}_0^*, \mathbf{p}_0)$ is called the *pitchfork* bifurcation on the eigenvector of $\mathbf{v}_k$.

The above three special case can be discussed through 1-D systems and intuitive illustrations are presented in Fig. 1.1 for a better understanding of bifurcation. Similarly, other cases on the eigenvector of $\mathbf{v}_k$ can be discussed from Eq. (1.108). In Fig. 1.1, the bifurcation point is also represented by a solid circular symbol. The stable and unstable equilibrium branches are given by solid and dashed curves, respectively. The vector fields are represented by lines with arrows. If no equilibriums exist, such a region is shaded.

Consider a saddle-node bifurcation in 1-D system

$$\dot{x} = f(x, p) \equiv p - x^2. \tag{1.118}$$

The corresponding equilibriums of the foregoing equation are $x^* = \pm\sqrt{p}\,(p > 0)$ and no any equilibriums exist for $p < 0$. From Eq. (1.118), the linearized equation in the vicinity of the equilibrium with $y = x - x^*$ is

$$\dot{y} = Df(x^*, p)y = -2x^* y. \tag{1.119}$$

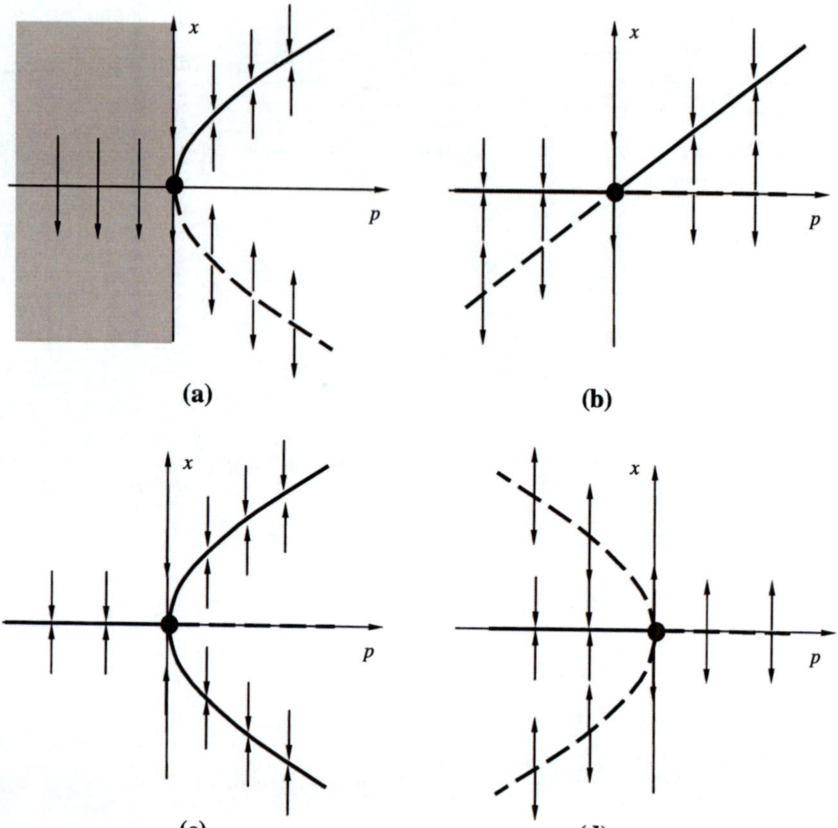

Fig. 1.1 Bifurcation diagrams: **a** saddle-node bifurcation, **b** transcritical bifurcation, **c** pitchfork bifurcation for stable-symmetry and **d** pitchfork bifurcation for unstable-symmetry

For the branch of $x^* = +\sqrt{p}\ (p > 0)$, the equilibrium is stable due to $Df(x^*, p) < 0$. However, for the branch of $x^* = -\sqrt{p}\ (p > 0)$, such an equilibrium is unstable due to $Df(x^*, p) > 0$. For $p = p_0 = 0$, we have $x^* = x_0^* = 0$ and $Df(x_0^*, p_0) = 0$. $D^2 f(x_0^*, p_0) = -2$ is needed. Thus

$$\dot{y} = D^2 f(x^*, p) y^2 = -2y^2. \tag{1.120}$$

At $(x_0^*, p_0) = (0, 0)$, the flow vector field is always less than zero. The equilibrium point $(x_0^*, p_0) = (0, 0)$ is bifurcation point, which is a decreasing saddle of the second order. For $p < 0$, the vector field of Eq. (1.118) is always less than zero without any equilibriums. The equilibrium varying with parameter p is sketched in Fig. 1.1a. On the left side of x-axes, no equilibrium exist, so only the vector field is presented.

Consider a transcritical bifurcation as

$$\dot{x} = f(x, p) \equiv px - x^2. \tag{1.121}$$

The equilibriums of the foregoing equation are $x^* = 0, p$. From Eq. (1.121), the linearized equation in the vicinity of the equilibrium with $y = x - x^*$ is

$$\dot{y} = Df(x^*, p)y = (p - 2x^*)y. \tag{1.122}$$

For the branch of $x^* = 0\,(p > 0)$, the equilibrium is unstable due to $Df(x^*, p) > 0$. For the branch of $x^* = p\,(p > 0)$, such an equilibrium is stable due to $Df(x^*, p) < 0$. However, for the branch of $x^* = 0\,(p < 0)$, the equilibrium is stable due to $Df(x^*, p) < 0$. For the branch of $x^* = p\,(p < 0)$, such an equilibrium is unstable due to $Df(x^*, p) > 0$. For $p = p_0 = 0$, $x^* = x_0^* = 0$and $Df(x_0^*, p_0) = 0$ are obtained. $D^2 f(x_0^*, p_0) = -2$ is needed. Thus the variational equation at the equilibrium is in Eq. (1.120). At $(x_0^*, p_0) = (0, 0)$, the flow vector field is always less than zero. The equilibrium point $(x_0^*, p_0) = (0, 0)$is bifurcation point, which is a decreasing saddle of the second order. The equilibrium varying with parameter p is sketched in Fig. 1.1b.

Consider the pitchfork bifurcation with stable-symmetry

$$\dot{x} = px - x^3 \tag{1.123}$$

from which its equilibriums are $x^* = 0, \pm\sqrt{p}\,(p > 0)$ and $x^* = 0\,(p \le 0)$. From Eq. (1.123), the linearized equation in the vicinity of the equilibrium with $y = x - x^*$ is

$$\dot{y} = Df(x^*, p)y = [p - 3(x^*)^2]y. \tag{1.124}$$

For the branch of $x^* = 0\ (p > 0)$, the equilibrium is unstable due to $Df(x^*, p) > 0$. For the branches of $x^* = \pm\sqrt{p}\ (p > 0)$, such equilibriums are stable due to $Df(x^*, p) < 0$. However, for the branch of $x^* = 0\ (p < 0)$, the equilibrium is stable due to $Df(x^*, p) < 0$. For $p = p_0 = 0$, $x^* = x_0^* = 0$ and $Df(x_0^*, p_0) = 0$ are obtained. Because of $D^2 f(x_0^*, p_0) = -6x_0^* = 0$, we have $D^3 f(x_0^*, p_0) = -6 < 0$. Thus the variational equation at the equilibrium is

$$\dot{y} = D^3 f(x^*, p)y^3 = -6y^3. \tag{1.125}$$

At $(x_0^*, p_0) = (0, 0)$, the flow vector field is less than zero for $y > 0$ and the flow vector field is greater than zero for $y < 0$. The equilibrium point $(x_0^*, p_0) = (0, 0)$ is bifurcation point, which is a sink of the third order. The equilibrium varying with parameter p is sketched in Fig. 1.1c.

Consider the pitchfork bifurcation for unstable-symmetry as

$$\dot{x} = px + x^3 \tag{1.126}$$

from which its equilibriums are $x^* = 0, \pm\sqrt{-p}\ (p < 0)$ and $x^* = 0\,(p \ge 0)$. From Eq. (1.126), the linearized equation in the vicinity of the equilibrium with $y = x - x^*$ is

$$\dot{y} = Df(x^*, p)y = [p + 3(x^*)^2]y. \tag{1.127}$$

For the branch of $x^* = 0\,(p < 0)$, the equilibrium is stable due to $Df(x^*, p) < 0$. For the branches of $x^* = \pm\sqrt{-p}\,(p < 0)$, such equilibriums are unstable due to $Df(x^*, p) > 0$. However, for the branch of $x^* = 0\,(p > 0)$, the equilibrium is unstable due to $Df(x^*, p) > 0$. For $p = p_0 = 0$, $x^* = x_0^* = 0$ and $Df(x_0^*, p_0) = 0$ are obtained. Since $D^2 f(x_0^*, p_0) = 6x_0^* = 0$, we have $Df(x^*, p) < 0$. Thus the variational equation at the equilibrium is

$$\dot{y} = D^3 f(x^*, p)y^3 = +6y^3. \tag{1.128}$$

At $(x_0^*, p_0) = (0, 0)$, the flow vector field is bigger than zero for $y > 0$ and the flow vector field is less than zero for $y < 0$. The equilibrium point $(x_0^*, p_0) = (0, 0)$ is bifurcation point, which is a source of the third order. The equilibrium varying with parameter p is sketched in Fig. 1.1d.

From the analysis, the bifurcation points possess the higher-order singularity of the flow in dynamical system. For the saddle-node bifurcation, the $(2m)$th order singularity of the flow at the bifurcation point exists as a saddle of the $(2m)$th order. For the transcritical bifurcation, the $(2m)$th order singularity of the flow at the bifurcation point exists as a saddle of the $(2m)$th order. However, for the stable pitchfork bifurcation, the $(2m+1)$th order singularity of the flow at the bifurcation point exists as a sink of the $(2m+1)$th order. For the unstable pitchfork bifurcation, the $(2m+1)$th order singularity of the flow at the bifurcation point exists as a source of the $(2m+1)$th order.

Definition 1.31 Consider an n-dimensional, autonomous, nonlinear dynamical system $\dot{\mathbf{x}} = \mathbf{f}(\mathbf{x}, \mathbf{p})$ in Eq. (1.4) with an equilibrium point $\mathbf{x}^*$ and $\mathbf{f}(\mathbf{x}, \mathbf{p})$ is C^r $(r \geq 1)$-continuous in a neighborhood of the equilibrium $\mathbf{x}^*$. The corresponding solution is $\mathbf{x}(t) = \mathbf{\Phi}(\mathbf{x}_0, t - t_0, \mathbf{p})$. Suppose $U(\mathbf{x}^*) \subset \Omega$ is a neighborhood of equilibrium $\mathbf{x}^*$. For a linearized dynamical system in Eq. (1.19), consider a pair of complex eigenvalue $\alpha_k \pm \mathbf{i}\beta_k$ $(k \in N = \{1, 2, \cdots, n\}$, $\mathbf{i} = \sqrt{-1})$ of matrix $D\mathbf{f}(\mathbf{x}^*, \mathbf{p})$ with a pair of eigenvectors $\mathbf{u}_k \pm \mathbf{i}v_k$. On the invariant plane of $(\mathbf{u}_k, \mathbf{v}_k)$, consider $\mathbf{r}_k = \mathbf{y} = \mathbf{y}_+^{(k)} + \mathbf{y}_-^{(k)}$ with

$$\begin{aligned} \mathbf{r}_k &= c_k \mathbf{u}_k + d_k \mathbf{v}_k = r_k \mathbf{e}_{r_k}, \\ \dot{\mathbf{r}}_k &= \dot{c}_k \mathbf{u}_k + \dot{d}_k \mathbf{v}_k = \dot{r}_k \mathbf{e}_{r_k} + r_k \dot{\mathbf{e}}_{r_k} \end{aligned} \tag{1.129}$$

and

$$\begin{aligned} c_k &= \frac{1}{\Delta}[\Delta_2(\mathbf{u}_k^{\mathrm{T}} \cdot \mathbf{y}) - \Delta_{12}(\mathbf{v}_k^{\mathrm{T}} \cdot \mathbf{y})] \text{ and } d_k = \frac{1}{\Delta}[\Delta_1(\mathbf{v}_k^{\mathrm{T}} \cdot \mathbf{y}) - \Delta_{12}(\mathbf{u}_k^{\mathrm{T}} \cdot \mathbf{y})] \\ \Delta_1 &= ||\mathbf{u}_k||^2, \Delta_2 = ||\mathbf{v}_k||^2, \Delta_{12} = \mathbf{u}_k^{\mathrm{T}} \cdot \mathbf{v}_k \text{ and } \Delta = \Delta_1\Delta_2 - \Delta_{12}^2. \end{aligned} \tag{1.130}$$

Consider a polar coordinate of (r_k, θ_k) defined by

$$\begin{aligned}
c_k &= r_k \cos\theta_k, \quad \text{and} \quad d_k = r_k \sin\theta_k;\\
r_k &= \sqrt{c_k^2 + d_k^2}, \quad \text{and} \quad \theta_k = \arctan d_k/c_k;\\
\mathbf{e}_{r_k} &= \cos\theta_k \mathbf{u}_k + \sin\theta_k \mathbf{v}_k \quad \text{and}\\
\mathbf{e}_{\theta_k} &= -\cos\theta_k \mathbf{u}_k^{\perp}\Delta_3 + \sin\theta_k \mathbf{v}_k^{\perp}\Delta_4\\
\Delta_3 &= \mathbf{v}_k^{\mathrm{T}} \cdot \mathbf{u}_k^{\perp} \quad \text{and} \quad \Delta_4 = \mathbf{u}_k^{\mathrm{T}} \cdot \mathbf{v}_k^{\perp}.
\end{aligned} \tag{1.131}$$

Thus

$$\begin{aligned}
\dot{c}_k &= \frac{1}{\Delta}[\Delta_2 G_{c_k}(\mathbf{x},\mathbf{p}) - \Delta_{12} G_{d_k}(\mathbf{x},\mathbf{p})]\\
\dot{d}_k &= \frac{1}{\Delta}[\Delta_1 G_{d_k}(\mathbf{x},\mathbf{p}) - \Delta_{12} G_{d_k}(\mathbf{x},\mathbf{p})]
\end{aligned} \tag{1.132}$$

where

$$\begin{aligned}
G_{c_k}(\mathbf{x},\mathbf{p}) =& \mathbf{u}_k^{\mathrm{T}} \cdot \mathbf{f}(\mathbf{x},\mathbf{p}) = \mathbf{a}_k^{\mathrm{T}} \cdot (\mathbf{p}-\mathbf{p}_0) + a_{k11}(c_k - c_{k0}^*) + a_{k12}(d_k - d_{k0}^*)\\
&+ \sum\nolimits_{q=2}^{m} \sum\nolimits_{r=0}^{q} \frac{1}{q!} C_q^r \mathbf{G}_{c_k}^{(q-r,r)}(\mathbf{x}^*,\mathbf{p}_0)(\mathbf{p}-\mathbf{p}_0)^r r_k^{q-r}\\
&+ \frac{1}{(m+1)!}[(c_k - c_{k0}^*)\partial_{c_k} + (d_k - d_{k0}^*)\partial_{d_k} + (\mathbf{p}-\mathbf{p}_0)\partial_{\mathbf{p}}]^{m+1}\\
&\times (\mathbf{u}_k^{\mathrm{T}} \cdot \mathbf{f}(\mathbf{x}_0^* + \theta\Delta\mathbf{x}, \mathbf{p}_0 + \theta\Delta\mathbf{p})),
\end{aligned} \tag{1.133a}$$

$$\begin{aligned}
G_{d_k}(\mathbf{x},\mathbf{p}) =& \mathbf{v}_k^{\mathrm{T}} \cdot \mathbf{f}(\mathbf{x},\mathbf{p}) = \mathbf{b}_k^{\mathrm{T}} \cdot (\mathbf{p}-\mathbf{p}_0) + a_{k21}(c_k - c_{k0}^*) + a_{k22}(d_k - d_{k0}^*)\\
&+ \sum\nolimits_{q=2}^{m} \sum\nolimits_{r=0}^{q} \frac{1}{q!} C_q^r \mathbf{G}_{c_k}^{(q-r,r)}(\mathbf{x}^*,\mathbf{p}_0)(\mathbf{p}-\mathbf{p}_0)^r r_k^{q-r};\\
&\frac{1}{(m+1)!} + [(c_k - c_{k0}^*)\partial_{c_k} + (d_k - d_{k0}^*)\partial_{d_k} + (\mathbf{p}-\mathbf{p}_0)\partial_{\mathbf{p}}]^{m+1}\\
&\times (\mathbf{v}_k^{\mathrm{T}} \cdot \mathbf{f}(\mathbf{x}_0^* + \theta\Delta\mathbf{x}, \mathbf{p}_0 + \theta\Delta\mathbf{p}))
\end{aligned} \tag{1.133b}$$

and

$$\begin{aligned}
\mathbf{G}_{c_k}^{(s,r)}(\mathbf{x}^*,\mathbf{p}) &= \mathbf{u}_k^{\mathrm{T}} \cdot [\partial_{\mathbf{x}}()\mathbf{u}_k \cos\theta_k + \partial_{\mathbf{x}}()\mathbf{v}_k \sin\theta_k]^s \partial_{\mathbf{p}}^{(r)} \mathbf{f}(\mathbf{x},\mathbf{p})\Big|_{(\mathbf{x}^*,\mathbf{p})},\\
\mathbf{G}_{d_k}^{(s,r)}(\mathbf{x}^*,\mathbf{p}) &= \mathbf{v}_k^{\mathrm{T}} \cdot [\partial_{\mathbf{x}}()\mathbf{u}_k \cos\theta_k + \partial_{\mathbf{x}}()\mathbf{v}_k \sin\theta_k]^s \partial_{\mathbf{p}}^{(r)} \mathbf{f}(\mathbf{x},\mathbf{p})\Big|_{(\mathbf{x}^*,\mathbf{p})};
\end{aligned} \tag{1.134}$$

$$\begin{aligned}
\mathbf{a}_k^{\mathrm{T}} &= \mathbf{u}_k^{\mathrm{T}} \cdot \partial_{\mathbf{p}} \mathbf{f}(\mathbf{x},\mathbf{p}), \mathbf{b}_k^{\mathrm{T}} = \mathbf{v}_k^{\mathrm{T}} \cdot \partial_{\mathbf{p}} \mathbf{f}(\mathbf{x},\mathbf{p});\\
a_{k11} &= \mathbf{u}_k^{\mathrm{T}} \cdot \partial_{\mathbf{x}} \mathbf{f}(\mathbf{x},\mathbf{p})\mathbf{u}_k, a_{k12} = \mathbf{u}_k^{\mathrm{T}} \cdot \partial_{\mathbf{x}} \mathbf{f}(\mathbf{x},\mathbf{p})\mathbf{v}_k;\\
a_{k21} &= \mathbf{v}_k^{\mathrm{T}} \cdot \partial_{\mathbf{x}} \mathbf{f}(\mathbf{x},\mathbf{p})\mathbf{u}_k, a_{k22} = \mathbf{v}_k^{\mathrm{T}} \cdot \partial_{\mathbf{x}} \mathbf{f}(\mathbf{x},\mathbf{p})\mathbf{v}_k.
\end{aligned} \tag{1.135}$$

Thus

$$
\begin{aligned}
\dot{r}_k &= \dot{c}_k \cos\theta_k + \dot{d}_k \sin\theta_k \\
&= \sum_{q=1}^{m} \sum_{r=0}^{q} \frac{1}{q!} C_q^r \mathbf{G}_{r_k}^{(q-r,r)}(\theta_k, \mathbf{p}_0)(\mathbf{p}-\mathbf{p}_0)^{q-r} r_k^r \\
\dot{\theta}_k &= r_k^{-1}(\dot{d}_k \cos\theta_k - \dot{c}_k \sin\theta_k) \\
&= \sum_{q=1}^{m} \sum_{r=0}^{q} \frac{1}{q!} C_q^r \mathbf{G}_{\theta_k}^{(q-r,r)}(\theta_k, \mathbf{p}_0)(\mathbf{p}-\mathbf{p}_0)^{q-r} r_k^r
\end{aligned} \tag{1.136}
$$

where

$$
\begin{aligned}
\mathbf{G}_{r_k}^{(m-r,r)}(\theta_k, \mathbf{p}_0) &= \frac{1}{\Delta}[(\Delta_2\cos\theta_k - \Delta_{12}\sin\theta_k)\mathbf{G}_{c_k}^{(m-r,r)}(\mathbf{x}^*, \mathbf{p}_0) \\
&\quad + (\Delta_2\sin\theta_k - \Delta_{12}\cos\theta_k)\mathbf{G}_{d_k}^{(m-r,r)}(\mathbf{x}^*, \mathbf{p}_0)], \\
\mathbf{G}_{\theta_k}^{(m-r,r)}(\theta_k, \mathbf{p}_0) &= -\frac{1}{\Delta}[(\Delta_2\sin\theta_k + \Delta_{12}\cos\theta_k)\mathbf{G}_{c_k}^{(m-r,r)}(\mathbf{x}^*, \mathbf{p}_0) \\
&\quad - (\Delta_1\cos\theta_k - \Delta_{12}\sin\theta_k)\mathbf{G}_{d_k}^{(m-r,r)}(\mathbf{x}^*, \mathbf{p}_0)].
\end{aligned} \tag{1.137}
$$

Suppose

$$
\mathbf{a}_k^{\mathrm{T}} \cdot (\mathbf{p}-\mathbf{p}_0) = 0 \ \text{ and } \ \mathbf{b}_k^{\mathrm{T}} \cdot (\mathbf{p}-\mathbf{p}_0) = 0 \tag{1.138}
$$

then

$$
\begin{aligned}
\dot{r}_k &= (\alpha_k + \mathbf{G}_{r_k}^{(1,1)}(\theta_k, \mathbf{p}_0)\cdot(\mathbf{p}-\mathbf{p}_0))r_k + \frac{1}{3!}G_{r_k}^{(3,0)}(\theta_k, \mathbf{p}_0)r_k^3 + o(r_k^3) \\
\dot{\theta}_k &= \beta_k + \mathbf{G}_{\theta_k}^{(1,1)}(\theta_k, \mathbf{p}_0)\cdot(\mathbf{p}-\mathbf{p}_0) + \frac{1}{3!}\mathbf{G}_{\theta_k}^{(3,0)}(\theta_k, \mathbf{p}_0)r_k^2 + o(r_k^2)
\end{aligned} \tag{1.139}
$$

where

$$
\begin{aligned}
&\mathbf{G}_{r_k}^{(1,1)}(\theta_k, \mathbf{p}_0) = \mathbf{G}_{r_k}^{(1,1)}(\mathbf{p}_0) \ \text{ and } \ G_{r_k}^{(3,0)}(\theta_k, \mathbf{p}_0) = G_{r_k}^{(3,0)}(\mathbf{p}_0) \\
&\mathbf{G}_{\theta_k}^{(1,1)}(\theta_k, \mathbf{p}_0) = \mathbf{G}_{\theta_k}^{(1,1)}(\mathbf{p}_0) \ \text{ and } \ G_{\theta_k}^{(3,0)}(\theta_k, \mathbf{p}_0) = G_{\theta_k}^{(3,0)}(\mathbf{p}_0).
\end{aligned} \tag{1.140}
$$

If $\alpha_k = 0$ and $\mathbf{p} = \mathbf{p}_0$, the stability of current equilibrium $\mathbf{x}^*$ on an eigenvector plane of $(\mathbf{u}_k, \mathbf{v}_k)$ changes from stable to unstable state (or from unstable to stable state). The bifurcation manifold in the direction of $\mathbf{v}_k$ is determined by

$$
\begin{aligned}
(\alpha_{k0} + \mathbf{G}_{r_k}^{(1,1)}(\theta_k, \mathbf{p}_0)\cdot(\mathbf{p}-\mathbf{p}_0))r_k + \frac{1}{3!}G_{r_k}^{(3,0)}(\theta_k, \mathbf{p}_0)r_k^3 &= 0 \\
\beta_{k0} + \mathbf{G}_{\theta_k}^{(1,1)}(\theta_k, \mathbf{p}_0)\cdot(\mathbf{p}-\mathbf{p}_0) + \frac{1}{3!}\mathbf{G}_{\theta_k}^{(3,0)}(\theta_k, \mathbf{p}_0)r_k^2 &= 0
\end{aligned} \tag{1.141}
$$

where

$$\begin{aligned} &\mathbf{G}_{r_k}^{(1,1)}(\theta_k, \mathbf{p}_0) = \partial_{\mathbf{p}} \alpha_k|_{(\mathbf{x}_0^*, \mathbf{p}_0)} \neq \mathbf{0} \\ &[\mathbf{G}_{r_k}^{(1,1)}(\theta_k, \mathbf{p}_0) \cdot (\mathbf{p} - \mathbf{p}_0)] \times G_{r_k}^{(3,0)}(\theta_k, \mathbf{p}_0) < 0. \end{aligned} \tag{1.142}$$

Such a bifurcation at point $(\mathbf{x}_0^*, \mathbf{p}_0)$ is called the Hopf bifurcation on the eigenvector plane of $(\mathbf{u}_k, \mathbf{v}_k)$.

For the repeating eigenvalues of $D\mathbf{f}(\mathbf{x}^*, \mathbf{p})$, the bifurcation of equilibrium can be similarly discussed in the foregoing two Theorems 1.9 and 1.10. Herein, such a procedure will not be repeated.

Consider a dynamical system

$$\begin{aligned} \dot{x} &= [\alpha d + a(x^2 + y^2)]x + [\beta + \alpha c + b(x^2 + y^2)]y, \\ \dot{y} &= -[\beta + \alpha c + b(x^2 + y^2)]y + [\alpha d + a(x^2 + y^2)]y. \end{aligned} \tag{1.143}$$

Setting

$$r^2 = x^2 + y^2 \text{ with } x = r\cos\theta \text{ and } y = r\sin\theta, \tag{1.144}$$

we have

$$\begin{aligned} \dot{r} &= [\alpha d + ar^2]r, \\ \dot{\theta} &= \beta_0 + \alpha c + br^2. \end{aligned} \tag{1.145}$$

The equilibrium is

$$\begin{aligned} &r_1^* = 0 \text{ for } \alpha \in (-\infty, +\infty) \\ &r_2^* = (-\alpha d/a)^{1/2} \text{ for } (\alpha d) \times a < 0. \end{aligned} \tag{1.146}$$

If $d \neq 0$, we have

$$Df_r(r^*, \alpha) = \alpha d + 3ar^{*^2}. \tag{1.147}$$

For $r_1^* = 0$, $Df_r = \alpha d$. For $d > 0$, this equilibrium is stable as $\alpha < 0$ or unstable as $\alpha > 0$. This equilibrium is critical point for $\alpha = 0$. However, for $d < 0$, this equilibrium is stable as $\alpha > 0$ or unstable as $\alpha < 0$. The equilibrium of $r_2^* = (-\alpha d/a)^{1/2}$ requires $(\alpha d) \times a < 0$. For $a \times d > 0$, such equilibrium solution exists for $\alpha < 0$ and for $a \times d < 0$, the equilibrium existence condition is $\alpha > 0$. From $Df_r = -2\alpha d$, for $a \times d > 0$, the equilibrium is stable for $(d > 0, a > 0)$ and unstable for $(d < 0, a < 0)$. For $a \times d < 0$, the equilibrium is stable for $(d < 0,\ a > 0)$ and unstable for $(d > 0,\ a < 0)$. For $\alpha = 0$, we have $r^* = 0$ and

$$Df_r(r^*, \alpha) = \alpha d = 0 \text{ and } D_\alpha Df_r(r^*, \alpha) = d. \tag{1.148}$$

Therefore, for $\alpha = 0$, $r^* = 0$ is stable for $d > 0$ and $r^* = 0$ is unstable for $d < 0$. The bifurcation of equilibrium at point $(r^*, \alpha) = (0, 0)$ is the Hopf bifurcation. The

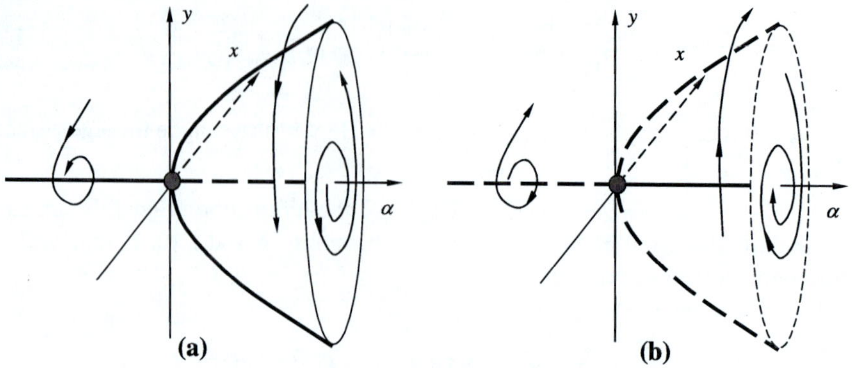

Fig. 1.2 Hopf bifurcations: **a** supercritical ($d > 0, a > 0$) and **b** subcritical ($d < 0, a > 0$)

Hopf bifurcation with stable focus ($d > 0$) is supercritical. The Hopf bifurcation with unstable focus ($d < 0$) is subcritical. The supercritical and subcritical Hopf bifurcation is shown in Fig. 1.2a and b. The solid lines and curves represent stable equilibrium. The dashed lines and curves represent unstable equilibrium. Since $\dot{\theta} = \beta$ is constant and $r^* \neq 0$, one gets a periodic motion on the circle.

On the other hand, $D^2 f_r(r^*, \alpha) = 6ar^* = 0$ for $r^* = 0. D^3 f_r(r^*, \alpha) = 6a$ is obtained. Further, $\dot{y} = D^3 f_r(r^*, \alpha)y^3 = 6ay^3$ ($y = r - r^*$). If $a > 0$, the vector field is greater than zero if $y > 0$ and less than zero if $y < 0$. For this case, the bifurcation point possesses a source flow of the third-order. The bifurcation branch is unstable. From Eq. (1.146), we have $\alpha d < 0$ for such a unstable bifurcation because of $(\alpha d) \times a < 0$. If $a < 0$, the vector field is less than zero if $y > 0$ and greater than zero if $y < 0$. For such a case, the bifurcation point possesses a sink flow of the third-order. The bifurcation branch is stable. From Eq. (1.146), we have $\alpha d > 0$ for such a stable bifurcation due to $(\alpha d) \times a < 0$.

From the analysis, the Hopf bifurcation points possess the higher-order singularity of the flow in dynamical system in the corresponding radial direction. For the stable Hopf bifurcation, the *m*th order singularity of the flow at the bifurcation point exists as a sink of the *m*th order in the radial direction. For the unstable Hopf bifurcation, the *m*th order singularity of the flow at the bifurcation point exists as a source of the *m*th order in the radial direction.

The stability and bifurcation for 2-D dynamic system are summarized in Fig. 1.3 with $\det(D\mathbf{f}) = \det(D\mathbf{f}(\mathbf{x}_0^*, \mathbf{p}_0))$ and $\mathrm{tr}(D\mathbf{f}) = \mathrm{tr}(D\mathbf{f}(\mathbf{x}_0^*, \mathbf{p}_0))$. The thick dashed lines are bifurcation lines. The stability of equilibriums is given by the eigenvalues in complex plane. The stability of equilibriums for higher dimensional systems can be identified by using a naming of stability for linear dynamical systems in Appendix A. The saddle-node bifurcation possesses stable saddle-node bifurcation (critical) and unstable saddle-node bifurcation (degenerate).

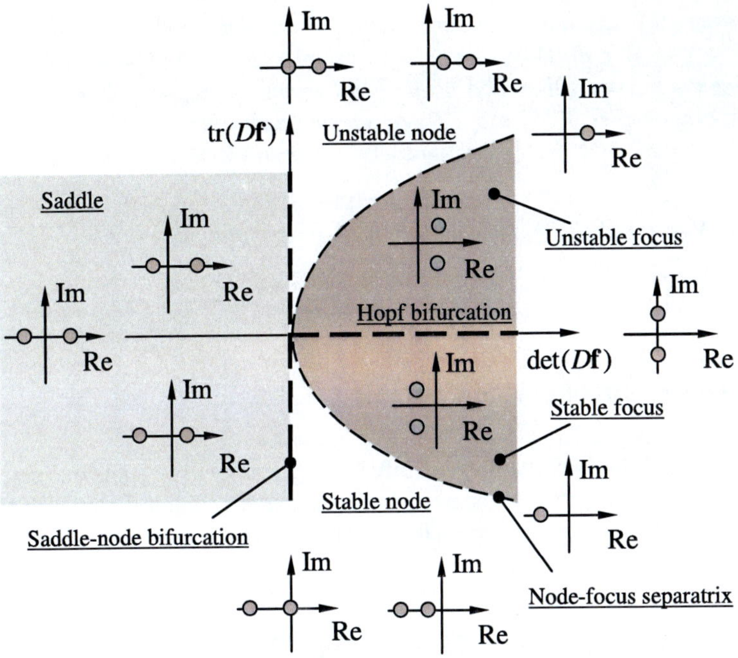

Fig. 1.3 Stability and bifurcation diagrams through the complex plane of eigenvalues for 2D-dynamical systems

1.3.3 Lyapunov Functions and Stability

Consider an n-dimensional, autonomous, nonlinear dynamical system $\dot{\mathbf{x}} = \mathbf{f}(\mathbf{x}, \mathbf{p})$ in Eq. (1.4). Let $V : U \to \mathscr{R}$ be a differentiable function defined in a neighborhood of equilibrium $\mathbf{x}^*$ on $U/\{\mathbf{x}^*\}$. A function $\dot{V} : U \to \mathscr{R}$ defined by

$$\dot{V}(\mathbf{x}) = DV(\mathbf{x}) = \mathbf{n}^{\mathrm{T}} \cdot \mathbf{f}(\mathbf{x}, t) \tag{1.149}$$

where $\mathbf{n} = (\partial_{x_1} V, \cdots, \partial_{x_n} V)^{\mathrm{T}}$ and $\mathbf{f}(\mathbf{x}, t) = (f_1, \cdots, f_n)^{\mathrm{T}}$. The generalized case will be discussed in Chap. 4.

Definition 1.32 Consider an n-dimensional, autonomous, nonlinear dynamical system $\dot{\mathbf{x}} = \mathbf{f}(\mathbf{x}, \mathbf{p})$ in Eq. (1.4) with an equilibrium point $\mathbf{x}^*$ and $\mathbf{f}(\mathbf{x}, \mathbf{p})$ is C^r $(r \geq 1)$-continuous in a neighborhood of the equilibrium $\mathbf{x}^*$. Suppose $U(\mathbf{x}^*) \subset \Omega$ is a neighborhood of equilibrium $\mathbf{x}^*$. There is a continuous function $V : U \to \mathscr{R}$ which is differentiable on $U/\{\mathbf{x}^*\}$, such that $V(\mathbf{x}^*) = 0$ and $V(\mathbf{x}) > 0$ if $\mathbf{x} \neq \mathbf{x}^*$.

(i) If $\dot{V} \leq 0$ in $U/\{\mathbf{x}^*\}$, the function V is called *a Lyapunov function* for equilibrium $\mathbf{x}^*$.
(ii) If $\dot{V} < 0$ in $U/\{\mathbf{x}^*\}$, the function V is called a *strict Lyapunov function* for equilibrium $\mathbf{x}^*$.

Theorem 1.11 *Consider an n-dimensional, autonomous, nonlinear dynamical system* $\dot{\mathbf{x}} = \mathbf{f}(\mathbf{x}, \mathbf{p})$ *in Eq.* (1.4) *with an equilibrium point* $\mathbf{x}^*$ *and* $\mathbf{f}(\mathbf{x}, \mathbf{p})$ *is* C^r $(r \geq 1)$*-continuous in a neighborhood of the equilibrium* $\mathbf{x}^*$. *Suppose* $U(\mathbf{x}^*) \subset \Omega$ *is a neighborhood of equilibrium* $\mathbf{x}^*$. *There is a continuous function* $V : U \to \mathscr{R}$ *which is differentiable on* $U/\{\mathbf{x}^*\}$, *such that* $V(\mathbf{x}^*) = 0$ *and* $V(\mathbf{x}) > 0$ *if* $\mathbf{x} \neq \mathbf{x}^*$.

(i) *If* $\dot{V} \leq 0$ *in* $U/\{\mathbf{x}^*\}$, *the equilibrium* $\mathbf{x}^*$ *is stable.*
(ii) *If* $\dot{V} < 0$ *in* $U/\{\mathbf{x}^*\}$, *the equilibrium* $\mathbf{x}^*$ *is asymptotically stable.*

Proof The proof can be referred to Hirsch and Smale (1974). ■

1.4 Approximate Periodic Motions

In this section, the local stability and bifurcation theory will be applied to determine the stability and bifurcation of approximate solutions of a nonlinear oscillator. Herein, a generalized harmonic balance method is presented as in Luo and Huang (2011).

1.4.1 A Generalized Harmonic Balance Method

Consider a nonlinear dynamical system as

$$\ddot{\mathbf{x}} + \mathbf{f}(\dot{\mathbf{x}}, \mathbf{x}, t) = \mathbf{0} \tag{1.150}$$

where $\mathbf{f}(\dot{\mathbf{x}}, \mathbf{x}, t)$ is a nonlinear function vector and is periodic for time with $T = 2\pi/\Omega$. Assume an approximate generalized periodic solution for the steady-state motion of Eq. (1.150) in the form of

$$\mathbf{x}^*(t) = \mathbf{a}_0(t) + \sum_{k=1}^{N} \mathbf{b}_k(t)\cos(k\Omega t) + \mathbf{c}_k(t)\sin(k\Omega t). \tag{1.151}$$

Then the first and second order derivatives of $\mathbf{x}^*(t)$ are

$$\dot{\mathbf{x}}^*(t) = \dot{\mathbf{a}}_0 + \sum_{k=1}^{N} [\dot{\mathbf{b}}_k + k\Omega\mathbf{c}_k]\cos(k\Omega t) + [\dot{\mathbf{c}}_k - k\Omega\mathbf{b}_k]\sin(k\Omega t), \tag{1.152}$$

$$\begin{aligned}\ddot{\mathbf{x}}^*(t) = \ddot{\mathbf{a}}_0 + \sum_{k=1}^{N} \{&[\ddot{\mathbf{b}}_k + 2k\Omega\dot{\mathbf{c}}_k - (k\Omega)^2\mathbf{b}_k]\cos(k\Omega t)\\ &+ [\ddot{\mathbf{c}}_k - 2k\Omega\dot{\mathbf{b}}_k - (k\Omega)^2\mathbf{c}_k]\sin(k\Omega t)\}.\end{aligned} \tag{1.153}$$

Suppose $\mathbf{a}_0(t)$, $\mathbf{b}_k(t)$ and $\mathbf{c}_k(t)$ vary slowly with time. Substitution of Eqs. (1.151)–(1.153) into Eq. (1.150) and averaging for each harmonic terms of $\cos(k\Omega t)$ and $\sin(k\Omega t)(k = 1, 2, \cdots)$ gives

$$
\begin{aligned}
&\ddot{\mathbf{a}}_0 + \mathbf{F}_0(\dot{\mathbf{a}}_0, \dot{\mathbf{b}}, \dot{\mathbf{c}}, \mathbf{a}_0, \mathbf{b}, \mathbf{c}) = \mathbf{0}, \\
&\ddot{\mathbf{b}}_k + 2\Omega\mathbf{k}_1\dot{\mathbf{c}}_k - \Omega^2\mathbf{k}_2\mathbf{b}_k + \mathbf{F}_{1k}(\dot{\mathbf{a}}_0, \dot{\mathbf{b}}, \dot{\mathbf{c}}, \mathbf{a}_0, \mathbf{b}, \mathbf{c}) = \mathbf{0}, \\
&\ddot{\mathbf{c}}_k - 2\Omega\mathbf{k}_1\dot{\mathbf{b}}_k - \Omega^2\mathbf{k}_2\mathbf{c}_k + \mathbf{F}_{2k}(\dot{\mathbf{a}}_0, \dot{\mathbf{b}}, \dot{\mathbf{c}}, \mathbf{a}_0, \mathbf{b}, \mathbf{c}) = \mathbf{0}; \\
&k = 1, 2, \cdots, N
\end{aligned} \tag{1.154}
$$

where

$$
\begin{aligned}
\mathbf{k}_1 &= diag(\mathbf{I}_{n\times n}, 2\mathbf{I}_{n\times n}, \cdots, N\mathbf{I}_{n\times n}), \\
\mathbf{k}_2 &= diag(\mathbf{I}_{n\times n}, 2^2\mathbf{I}_{n\times n}, \cdots, N^2\mathbf{I}_{n\times n}), \\
\mathbf{b} &= (\mathbf{b}_1, \mathbf{b}_2, \cdots, \mathbf{b}_N)^{\mathrm{T}} \text{ and } \mathbf{c} = (\mathbf{c}_1, \mathbf{c}_2, \cdots, \mathbf{c}_N)^{\mathrm{T}}, \\
\mathbf{F}_1 &= (\mathbf{F}_{11}, \mathbf{F}_{12}, \cdots, \mathbf{F}_{1N})^{\mathrm{T}} \text{ and } \mathbf{F}_2 = (\mathbf{F}_{21}, \mathbf{F}_{22}, \cdots, \mathbf{F}_{2N})^{\mathrm{T}} \\
&\text{for } N = 1, 2, \cdots, \infty
\end{aligned} \tag{1.155}
$$

and

$$
\begin{aligned}
\mathbf{F}_0(\mathbf{a}_0, \mathbf{b}, \mathbf{c}, \dot{\mathbf{a}}_0, \dot{\mathbf{b}}, \dot{\mathbf{c}}) &= \frac{1}{T}\int_0^T \mathbf{f}(\dot{\mathbf{x}}^*, \mathbf{x}^*, t)dt; \\
\mathbf{F}_{1k}(\mathbf{a}_0, \mathbf{b}, \mathbf{c}, \dot{\mathbf{a}}_0, \dot{\mathbf{b}}, \dot{\mathbf{c}}) &= \frac{2}{T}\int_0^T \mathbf{f}(\dot{\mathbf{x}}^*, \mathbf{x}^*, t)\cos(k\Omega t)dt, \\
\mathbf{F}_{2k}(\mathbf{a}_0, \mathbf{b}, \mathbf{c}, \dot{\mathbf{a}}_0, \dot{\mathbf{b}}, \dot{\mathbf{c}}) &= \frac{2}{T}\int_0^T \mathbf{f}(\dot{\mathbf{x}}^*, \mathbf{x}^*, t)\sin(k\Omega t)dt; \\
&\text{for } k = 1, 2, \cdots, N.
\end{aligned} \tag{1.156}
$$

Without the assumption of slowly varying with time, the averaging cannot be done for the dynamical system in Eq. (1.150) with the approximate solutions. The approximate solution in Eq. (1.151) is treated as a transformation, which can be applied to obtain any transient solution also. However, once the assumption of slow varying with time is used, the form in Eq. (1.151) is an approximate solution for steady-state motion in Eq. (1.150).

Setting $\mathbf{z} = (\mathbf{a}_0, \mathbf{b}, \mathbf{c})^{\mathrm{T}}$ and $\dot{\mathbf{z}} = \mathbf{z}_1$, one obtains

$$
\mathbf{g} = (-\mathbf{F}_0, -\mathbf{F}_1 - 2\Omega\mathbf{k}_1\dot{\mathbf{c}} + \Omega^2\mathbf{k}_2\mathbf{b}, -\mathbf{F}_2 + 2\Omega\mathbf{k}_1\dot{\mathbf{b}} + \Omega^2\mathbf{k}_2\mathbf{c})^{\mathrm{T}}. \tag{1.157}
$$

Equation (1.154) becomes

$$
\dot{\mathbf{z}} = \mathbf{z}_1 \text{ and } \dot{\mathbf{z}}_1 = \mathbf{g}(\mathbf{z}, \dot{\mathbf{z}}_1). \tag{1.158}
$$

If $\mathbf{z}_1 = \mathbf{0}$, the equilibrium points are given by $\mathbf{g}(\mathbf{0}, \mathbf{z}^*) = \mathbf{0}_{1\times n(2N+1)}$, i.e.,

$$
\begin{aligned}
&\mathbf{F}_0(\mathbf{a}_0^*, \mathbf{b}^*, \mathbf{c}^*, \mathbf{0}, \mathbf{0}, \mathbf{0}) = \mathbf{0}, \\
&\mathbf{F}_1(\mathbf{a}_0^*, \mathbf{b}^*, \mathbf{c}^*, \mathbf{0}, \mathbf{0}, \mathbf{0}) - \Omega^2\mathbf{k}_2\mathbf{b} = \mathbf{0}, \\
&\mathbf{F}_2(\mathbf{a}_0^*, \mathbf{b}^*, \mathbf{c}^*, \mathbf{0}, \mathbf{0}, \mathbf{0}) - \Omega^2\mathbf{k}_2\mathbf{c} = \mathbf{0}.
\end{aligned} \tag{1.159}
$$

The foregoing equation is given by the traditional harmonic balance method. Once the equilibrium point of $\mathbf{z}^* = (\mathbf{a}_0^*, \mathbf{b}^*, \mathbf{c}^*)^{\mathrm{T}}$ is obtained, the approximate solution in Eq. (1.151) is obtained, which gives the steady-state solution of dynamical systems in Eq. (1.150). The stability of approximate solution can be determined from Eq. (1.159). Let $\mathbf{y} = (\mathbf{z}, \mathbf{z}_1)^{\mathrm{T}}$ and $\mathbf{f} = (\mathbf{z}_1, \mathbf{g})^{\mathrm{T}}$. Equation (1.158) becomes $\dot{\mathbf{y}} = \mathbf{f}(\mathbf{y})$. The linearized equation at the equilibrium point $\mathbf{y}^* = (\mathbf{z}^*, \mathbf{0})^{\mathrm{T}}$ is given by

$$\Delta\dot{\mathbf{y}} = D\mathbf{f}(\mathbf{y}^*)\Delta\mathbf{y} \text{ and } D\mathbf{f}(\mathbf{y}^*) = \partial\mathbf{f}(\mathbf{y})/\partial\mathbf{y}\big|_{\mathbf{y}^*}. \tag{1.160}$$

The corresponding eigenvalue analysis requires

$$\left|D\mathbf{f}(\mathbf{y}^*) - \lambda\mathbf{I}_{2n(2N+1)\times 2n(2N+1)}\right| = 0. \tag{1.161}$$

The eigenvalues of $D\mathbf{f}(\mathbf{y}^*)$ are classified as

$$(n_1, n_2, n_3 | n_4, n_5, n_6) \tag{1.162}$$

where n_1 is the total number of negative real eigenvalues, n_2 is the total number of positive real eigenvalues, n_2 is the total number of zero real eigenvalues; n_4 is the total pair number of complex eigenvalues with negative real parts, n_5 is the total pair number of complex eigenvalues with positive real parts, n_6 is the total pair number of complex eigenvalues with zero real parts. If all eigenvalues possess negative real parts, the approximate steady-state solution is stable. If one of eigenvalues possesses positive real parts, the approximate steady-state solution is unstable. The corresponding boundary is the bifurcation condition, including saddle-node bifurcation, Hopf bifurcation and so on.

1.4.2 A Nonlinear Duffing Oscillator

Consider a periodically forced Duffing oscillator as

$$\ddot{x} + d_1\dot{x} + d_2\dot{x}^3 + a_1 x + a_2 x^3 = Q + Q_0\cos\Omega t. \tag{1.163}$$

From the foregoing equation, the standard form is

$$\ddot{x} + f(x, \dot{x}, t) = 0, \tag{1.164}$$

where

$$f(x, \dot{x}, t) = d_1\dot{x} + d_2\dot{x}^3 + a_1 x + a_2 x^3 - Q - Q_0\cos\Omega t. \tag{1.165}$$

The periodic solution for one degree of freedom system is assumed as

$$x^*(t) = a_0(t) + \sum_{k=1}^{N} b_k(t)\cos(k\Omega t) + c_k(t)\sin(k\Omega t). \tag{1.166}$$

The first and second order derivatives of $x^*(t)$ are

$$\dot{x}^*(t) = \dot{a}_0 + \sum_{k=1}^{N} [\dot{b}_k + k\Omega c_k]\cos(k\Omega t) + [\dot{c}_k - k\Omega b_k]\sin(k\Omega t), \quad (1.167)$$

$$\begin{aligned}\ddot{x}^*(t) = \ddot{a}_0 + \sum_{k=1}^{N} \{[\ddot{b}_k + 2k\Omega\dot{c}_k - (k\Omega)^2 b_k]\cos(k\Omega t) \\ + [\ddot{c}_k - 2k\Omega\dot{b}_k - (k\Omega)^2 c_k]\sin(k\Omega t)\}.\end{aligned} \quad (1.168)$$

Application of Eqs. (1.166)–(1.168) into Eq. (1.163) and averaging all terms of $\cos(n\Omega t)$ and $\sin(n\Omega t)$ term gives for $k = 1, 2, \cdots, N$

$$\begin{aligned}&\ddot{a}_0 + F_0(a_0, \mathbf{b}, \mathbf{c}, \dot{a}_0, \dot{\mathbf{b}}, \dot{\mathbf{c}}) = 0;\\ &\ddot{b}_k + (2k\Omega)\dot{c}_k - (k\Omega)^2 b_k + F_{1k}(a_0, \mathbf{b}, \mathbf{c}, \dot{a}_0, \dot{\mathbf{b}}, \dot{\mathbf{c}}) = 0,\\ &\ddot{c}_k - (2k\Omega)\dot{b}_k - (k\Omega)^2 c_k + F_{2k}(a_0, \mathbf{b}, \mathbf{c}, \dot{a}_0, \dot{\mathbf{b}}, \dot{\mathbf{c}}) = 0.\end{aligned} \quad (1.169)$$

The coefficients of constant, $\cos k\Omega t$ and $\sin k\Omega t$ for the function of $f(x, \dot{x}, t)$ are

$$\begin{aligned}&F_0(a_0, \mathbf{b}, \mathbf{c}, \dot{a}_0, \dot{\mathbf{b}}, \dot{\mathbf{c}}) = d_1\dot{a} + d_2 f_1^{(0)} + a_1 a + a_2 f_2^{(0)} - Q;\\ &F_{1k}(a_0, \mathbf{b}, \mathbf{c}, \dot{a}_0, \dot{\mathbf{b}}, \dot{\mathbf{c}}) = d_1(\dot{b}_k + k\Omega c_k) + d_2 f_{1k}^{(c)} + a_1 b_k + a_2 f_{2k}^{(c)} - Q_0\,\delta_k^1;\\ &F_{2k}(a_0, \mathbf{b}, \mathbf{c}, \dot{a}_0, \dot{\mathbf{b}}, \dot{\mathbf{c}}) = d_1(\dot{c}_k - k\Omega b_k) + d_2 f_{1k}^{(s)} + a_1 c_k + a_2 f_{2k}^{(s)}\end{aligned} \quad (1.170)$$

where

$$\begin{aligned}f_1^{(0)} = \dot{a}_0^3 + \sum_{l=1}^{N}\sum_{j=1}^{N}\sum_{i=1}^{N}\Bigg[&\frac{3\dot{a}_0}{2N}(\dot{b}_i + i\Omega c_i)(\dot{b}_j + j\Omega c_j)\,\delta_{i-j}^0\\ &+ \frac{3\dot{a}_0}{2N}(\dot{c}_i - i\Omega b_i)(\dot{c}_j - j\Omega b_j)\,\delta_{i-j}^0\\ &+ \frac{1}{4}(\dot{b}_i + i\Omega c_i)(\dot{b}_j + j\Omega c_j)(\dot{b}_l + k\Omega c_l)(\delta_{i-j-l}^0 + \delta_{i-j+l}^0 + \delta_{i+j-l}^0)\\ &+ \frac{3}{4}(\dot{b}_i + i\Omega c_i)(\dot{c}_j - j\Omega b_j)(\dot{c}_l - l\Omega b_l)(\delta_{l-j-i}^0 + \delta_{l-j+i}^0 - \delta_{l+j-i}^0)\Bigg],\end{aligned} \quad (1.171)$$

$$\begin{aligned}f_2^{(0)} = a_0^3 + \sum_{l=1}^{N}\sum_{j=1}^{N}\sum_{i=1}^{N}\Bigg[&\frac{3a_0}{2N} b_i b_j\,\delta_{i-j}^0 + \frac{3a_0}{2N} c_i c_j\,\delta_{i-j}^0\\ &+ \frac{1}{4} b_i b_j b_l(\delta_{i-j-l}^0 + \delta_{i-j+l}^0 + \delta_{i+j-l}^0)\\ &+ \frac{3}{4} b_i c_j c_l(\delta_{i-j+l}^0 + \delta_{i+j-l}^0 - \delta_{i-j-l}^0)\Bigg],\end{aligned} \quad (1.172)$$

$$
\begin{aligned}
f_{1k}^{(c)} = \sum_{l=1}^{N}\sum_{j=1}^{N}\sum_{i=1}^{N}\Big[& 3(\frac{\dot{a}_0}{N})^2(\dot{b}_i + i\Omega c_i)\,\delta_i^k \\
& + \frac{3\dot{a}_0}{2N}(\dot{b}_i + i\Omega c_i)(\dot{b}_j + j\Omega c_j)(\delta_{|i-j|}^k + \delta_{i+j}^k) \\
& + \frac{3\dot{a}_0}{2N}(\dot{c}_i - i\Omega b_i)(\dot{c}_j - j\Omega b_j)(\delta_{|i-j|}^k - \delta_{i+j}^k) \\
& + \frac{1}{4}(\dot{b}_i + i\Omega c_i)(\dot{b}_j + j\Omega c_j)(\dot{b}_l + l\Omega c_l) \\
& \times (\delta_{|i-j-l|}^k + \delta_{i+j+l}^k + \delta_{|i-j+l|}^k + \delta_{|i+j-l|}^k) \\
& + \frac{3}{4}(\dot{b}_i + i\Omega c_i)(\dot{c}_j - j\Omega b_j)(\dot{c}_l - l\Omega b_l) \\
& \times(\delta_{|i+j-l|}^k + \delta_{|i-j+l|}^k - \delta_{i+j+l}^k - \delta_{|i-j-l|}^k)\Big],
\end{aligned} \tag{1.173}
$$

$$
\begin{aligned}
f_{2k}^{(c)} = \sum_{l=1}^{N}\sum_{j=1}^{N}\sum_{i=1}^{N}\Big[& 3(\frac{a_0}{N})^2 b_i\,\delta_i^k + \frac{3a_0}{2N} b_i b_j (\delta_{i+j}^k + \delta_{|i-j|}^k) \\
& + \frac{3a_0}{2N} c_i c_j (\delta_{|i-j|}^k - \delta_{i+j}^k) \\
& \frac{1}{4} b_i b_j b_l (\delta_{|i-j-l|}^k + \delta_{i+j+l}^k + \delta_{|i-j+l|}^k + \delta_{|i+j-l|}^k) \\
& + \frac{3}{4} b_i c_j c_l (\delta_{|i+j-l|}^k + \delta_{|i-j+l|}^k - \delta_{i+j+l}^k - \delta_{|i-j-l|}^k)\Big],
\end{aligned} \tag{1.174}
$$

$$
\begin{aligned}
f_{1k}^{(s)} = \sum_{l=1}^{N}\sum_{j=1}^{N}\sum_{i=1}^{N}\Bigg\{ & 3\left(\frac{\dot{a}_0}{N}\right)^2 (\dot{c}_i - i\Omega b_i)\,\delta_i^k \\
& + \frac{3\dot{a}_0}{N}(\dot{c}_i - i\Omega b_i)(\dot{b}_j + j\Omega c_j)[\operatorname{sgn}(i-j)\,\delta_{|i-j|}^k + \delta_{i+j}^k] \\
& + \frac{3}{4}(\dot{b}_i + i\Omega c_i)(\dot{b}_j + j\Omega c_j)(\dot{c}_l - l\Omega b_l) \\
& \times [\delta_{i+j+l}^k + \operatorname{sgn}(i-j+l)\,\delta_{|i-j+l|}^k - \operatorname{sgn}(i+j-l)\,\delta_{|i+j-l|}^k \\
& - \operatorname{sgn}(i-j-l)\,\delta_{|i-j-l|}^k] + \frac{1}{4}(\dot{c}_i - i\Omega b_i)(\dot{c}_j - j\Omega b_j)(\dot{c}_l - l\Omega b_l) \\
& \times [\operatorname{sgn}(i-j+l)\,\delta_{|i-j+l|}^k - \delta_{i+j+l}^k + \operatorname{sgn}(i+j-l)\,\delta_{|i+j-l|}^k \\
& - \operatorname{sgn}(i-j-l)\,\delta_{|i-j-l|}^k]\Bigg\},
\end{aligned} \tag{1.175}
$$

$$f_{2k}^{(s)} = \sum_{l=1}^{N}\sum_{j=1}^{N}\sum_{i=1}^{N}\Big\{3(\frac{a_0}{N})^2 c_i\,\delta_i^k + \frac{3a_0}{N} b_i c_j[\delta_{i+j}^k - \text{sgn}(i-j)\,\delta_{|i-j|}^k]$$
$$+\frac{3}{4} b_i b_j c_l[\delta_{i+j+l}^k + \text{sgn}(i-j+l)\,\delta_{|i-j+l|}^k - \text{sgn}(i+j-l)\,\delta_{|i+j-l|}^k$$
$$-\text{sgn}(i-j-l)\,\delta_{|i-j-l|}^k] + \frac{1}{4} c_i c_j c_l[\text{sgn}(i-j+l)\,\delta_{|i-j+l|}^k$$
$$+\text{sgn}(i+j-l)\,\delta_{|i+j-l|}^k - \delta_{i+j+l}^k - \text{sgn}(i-j-k)\,\delta_{|i-j-l|}^k]\Big\}. \tag{1.176}$$

Define

$$\begin{aligned}
\mathbf{z} &= (a_0, \mathbf{b}^{\mathrm{T}}, \mathbf{c}^{\mathrm{T}},)^{\mathrm{T}} = (a_0, b_1, \cdots, b_N, c_1, \cdots, c_N)^{\mathrm{T}} \\
&\equiv (z_0, z_1, \cdots, z_{2N})^{\mathrm{T}}, \\
\mathbf{z}_1 &= \dot{\mathbf{z}} = (\dot{a}_0, \dot{\mathbf{b}}^{\mathrm{T}}, \dot{\mathbf{c}}^{\mathrm{T}},)^{\mathrm{T}} = (\dot{a}_0, \dot{b}_1, \cdots, \dot{b}_N, \dot{c}_1, \cdots, \dot{c}_N)^{\mathrm{T}} \\
&\equiv (\dot{z}_0, \dot{z}_1, \cdots, \dot{z}_{2N})^{\mathrm{T}}
\end{aligned} \tag{1.177}$$

where

$$\mathbf{b} = (b_1, b_2, \cdots, b_N)^{\mathrm{T}} \text{ and } \mathbf{c} = (c_1, c_2, \cdots, c_N)^{\mathrm{T}}. \tag{1.178}$$

Equation (1.169) can be expressed in the form of vector field as

$$\dot{\mathbf{z}} = \mathbf{z}_1 \text{ and } \dot{\mathbf{z}}_1 = \mathbf{g}(\mathbf{z}, \mathbf{z}_1) \tag{1.179}$$

where

$$\mathbf{g}(\mathbf{z}, \mathbf{z}_1) = \begin{pmatrix} F^{(0)}(\mathbf{z}, \mathbf{z}_1) \\ \mathbf{F}_1(\mathbf{z}, \mathbf{z}_1) - 2\Omega\mathbf{k}_1\dot{\mathbf{b}} + \Omega^2\mathbf{k}_2\mathbf{b} \\ \mathbf{F}_2(\mathbf{z}, \mathbf{z}_1) + 2\Omega\mathbf{k}_1\dot{\mathbf{c}} + \Omega^2\mathbf{k}_2\mathbf{c} \end{pmatrix} \tag{1.180}$$

and

$$\begin{aligned}
&\mathbf{k}_1 = diag(1, 2, \cdots, N) \text{ and } \mathbf{k}_2 = diag(1, 2^2, \cdots, N^2) \\
&\mathbf{F}_1 = (F_{11}, F_{12}, \cdots, F_{1N})^{\mathrm{T}} \text{ and } \mathbf{F}_2 = (F_{21}, F_{22}, \cdots, F_{2N})^{\mathrm{T}} \\
&\text{for } N = 1, 2, \cdots, \infty.
\end{aligned} \tag{1.181}$$

Letting

$$\mathbf{y} = (\mathbf{z}, \mathbf{z}_1) \text{ and } \mathbf{f} = (\mathbf{z}_1, \mathbf{g})^{\mathrm{T}}. \tag{1.182}$$

Equation (1.179) becomes

$$\dot{\mathbf{y}} = \mathbf{f}(\mathbf{y}). \tag{1.183}$$

The steady-state solutions for periodic motion can be obtained by setting $\dot{\mathbf{y}} = \mathbf{0}$, i.e.,

$$\begin{aligned} &F_0(a_0^*, \mathbf{b}^*, \mathbf{c}^*, 0, \mathbf{0}, \mathbf{0}) = 0, \\ &-\Omega^2 \mathbf{k}_2 \mathbf{b}^* + \mathbf{F}_1(a_0^*, \mathbf{b}^*, \mathbf{c}^*, 0, \mathbf{0}, \mathbf{0}) = 0, \\ &-\Omega^2 \mathbf{k}_2 \mathbf{c}^* + \mathbf{F}_2(a_0^*, \mathbf{b}^*, \mathbf{c}^*, 0, \mathbf{0}, \mathbf{0}) = 0. \end{aligned} \tag{1.184}$$

The $(2N+1)$nonlinear equations in Eq. (1.184) are solved by the Newton-Raphson method. As in the previous section, the linearized equation at the equilibrium point $\mathbf{y}^* = (\mathbf{z}^*, \mathbf{0})^{\mathrm{T}}$ is given by

$$\Delta\dot{\mathbf{y}} = D\mathbf{f}(\mathbf{y}^*)\Delta\mathbf{y} \quad \text{and} \quad D\mathbf{f}(\mathbf{y}^*) = \partial\mathbf{f}(\mathbf{y})/\partial\mathbf{y}|_{\mathbf{y}^*}. \tag{1.185}$$

The corresponding eigenvalues are determined by

$$\left|D\mathbf{f}(\mathbf{y}^*) - \lambda\mathbf{I}_{2(2N+1)\times 2(2N+1)}\right| = 0. \tag{1.186}$$

If $\mathrm{Re}(\lambda_k) < 0$ $(k = 1, 2, \cdots, 2(2N+1))$, the approximate steady-state solution $\mathbf{y}^*$ with truncation of $\cos(N\Omega t)$ and $\sin(N\Omega t)$ is stable. If $\mathrm{Re}(\lambda_k) < 0\,(k \in \{1, 2, \cdots, 2(2N+1)\})$, the truncated approximate steady-state solution is unstable. The corresponding boundary between the stable and unstable solution is given by the bifurcation condition, including saddle-node bifurcation, Hopf bifurcation and so on.

1.4.3 Approximate Solutions

From such approximate, analytical solutions, the stability and bifurcation analysis can be done through eigenvalue analysis. The equilibrium solution can be obtained from Eq. (1.184) by using the Newton-Raphson method as an example, and the stability analysis will be presented. The system parameters are

$$d_1 = 0.5, d_2 = 0.05, a_1 = -1.0, a_2 = 10, Q = 1.0, Q_0 = 10.0. \tag{1.187}$$

The backbone curves of response amplitude varying with excitation frequency Ω are illustrated in Fig. 1.3, where harmonic amplitude and phase are defined by

$$A_k \equiv \sqrt{b_k^2 + c_k^2}, \varphi_k = \arctan\frac{c_k}{b_k} \tag{1.188}$$

and the corresponding solution in Eq. (1.166) in

$$x^*(t) = a_0 + \sum_{k=1}^{N} A_k \cos(k\Omega t - \varphi_k). \tag{1.189}$$

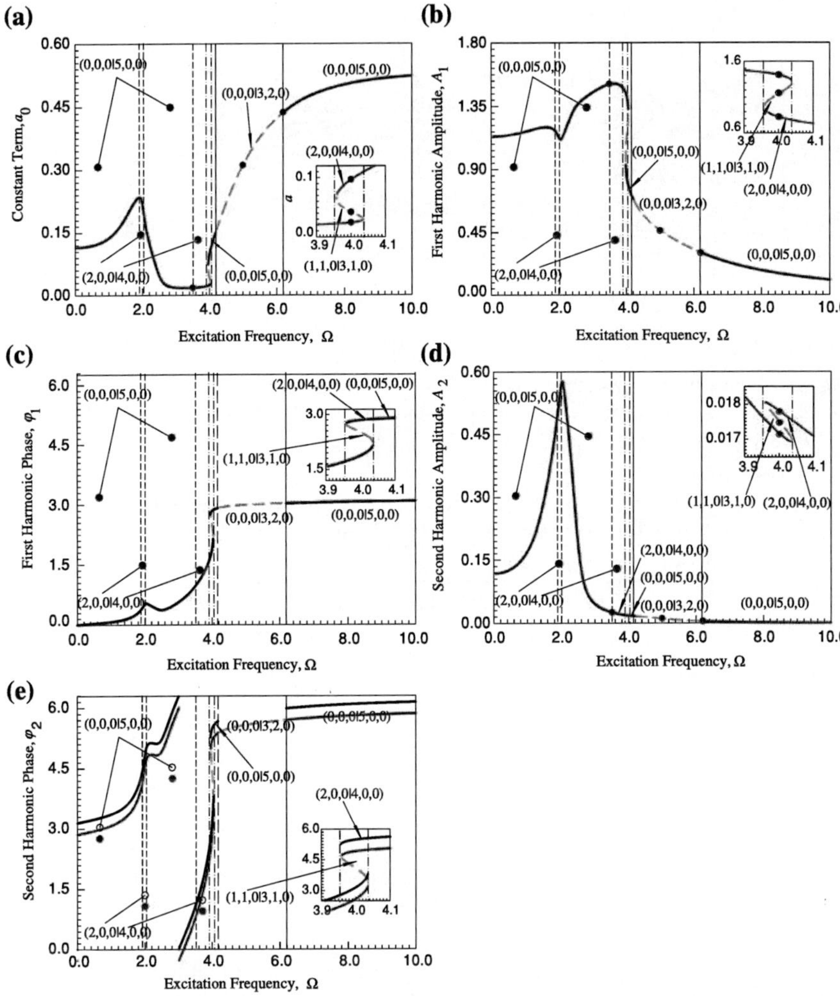

Fig. 1.4 Analytical prediction of periodic solutions based on two harmonic terms (HB2): **a** constant term (a_0), **b** first harmonic amplitude (A_1), **c** first harmonic phase (φ_1), **d** second harmonic amplitude (A_2), **e** second harmonic phase (φ_2). ($d_1 = 0.5$, $d_2 = 0.05$, $a_1 = -1.0$, $a_2 = 10$, $Q = 1.0$, $Q_0 = 10.0$)

The analytical approximate solutions for periodic motion are based on two harmonic terms (HB2) as presented in Fig. 1.4. The constant term (a_0), the first harmonic term amplitude (A_1) and phase (φ_1), the second harmonic term amplitude (A_2) and phase (φ_2) versus excitation frequency are presented in Fig. 1.4a–e, respectively. The corresponding eigenvalue analysis is given by the real and imaginary parts of eigenvalues, as shown in Fig. 1.5. The real and imaginary parts of eigenvalues

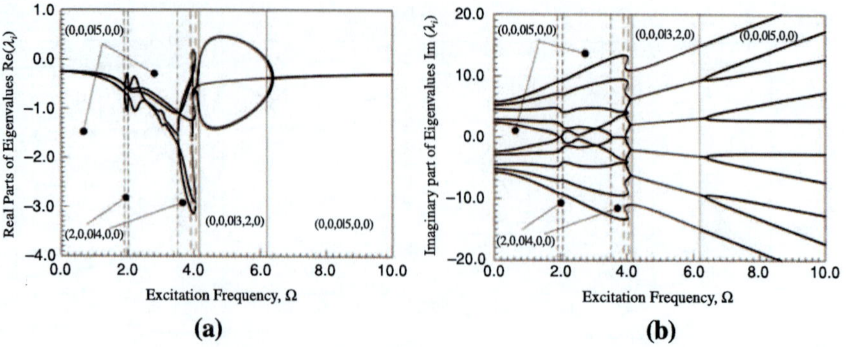

Fig. 1.5 Eigenvalue analysis of periodic solutions based on two harmonic terms (HB2): **a** real part of eigenvalues and **b** imaginary parts of eigenvalues. ($d_1 = 0.5,\ d_2 = 0.05, a_1 = -1.0,\ a_2 = 10,\ Q = 1.0,\ Q_0 = 10.0$)

Table 1.1 Stability classification of periodic motions based on the two harmonic terms (BH2)($d_1 = 0.5, d_2 = 0.05, a_1 = -1.0, a_2 = 10, Q = 1.0, Q_0 = 10.0$)

Type of eigenvalues	Excitation Frequency Ω	Stability
(0, 0, 0|5, 0, 0)	(0, 1.929), (1.988, 3.533), (3.97, 4.178), (6.195, 10]	Stable
(2, 0, 0|4, 0, 0)	(3.533, 4.035), (1.929, 1.988), (3.957, 3.97)	Stable
(1, 1, 0|3, 1, 0)	(3.956, 4.035)	Unstable
(0, 0, 0|3, 2, 0)	(4.178, 6.153)	Unstable
(0, 0, 0|4, 1, 0)	(6.153, 6.195)	Unstable

for approximate solutions with such two harmonic terms are presented in Fig. 1.5a and b, respectively. The eigenvalue analysis provides the possible stability and bifurcation analysis of the periodic motions based on such approximate solutions. The solid and dashed curves represent the stable and unstable periodic solutions based on the harmonic terms (HB2), respectively. The dot-dash vertical lines represent the "jump" phenomena, and three periodic motions exist. The bifurcations are given by solid vertical lines. The Hopf bifurcations (HB) occur at $\Omega = 4.178$ and 6.195. In addition, the eigenvalue types of periodic motions are labeled through Eq. (1.162). From such classification of eigenvalues, the stability and bifurcation characteristics are presented. The corresponding stability classification is tabulated in Table 1.1. For $\Omega = (3.956, 4.035)$, there are three periodic motions. Two stable periodic motions and one unstable periodic motion exist. As excitation frequency Ω increases, the upper branch of periodic motion for A_1 will disappear at $\Omega = 4.035$, and jumps to the lower branch. As excitation frequency Ω decreases, the bottom branch of periodic motion for A_1 will disappear and jumps to the upper branch. Similarly, the jumping phenomena can be observed in the plots of a_0, A_1, A_2, φ_1 and φ_2. The middle branch of solutions is unstable.

For a global view of stability of the approximate solutions (HB2), the corresponding parameter map is presented in Fig. 1.6. The regions of single stable periodic

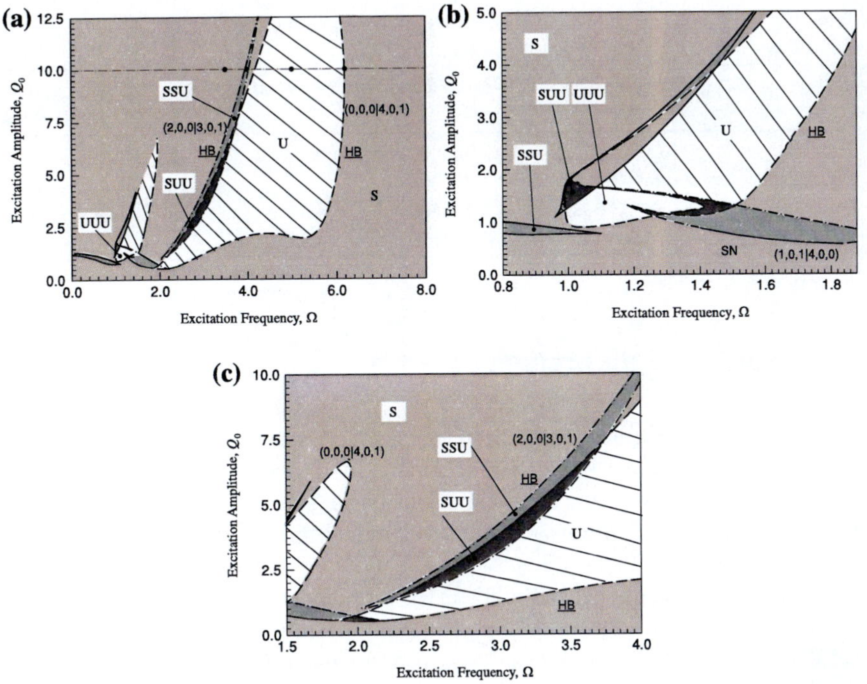

Fig. 1.6 A parameter map (Ω, Q_0) for periodic motion based on two harmonic terms (HB2): **a** Global view ($\Omega = (0.0, 10.0)$), **b** zoomed view ($\Omega = (0.8, 1.9)$), and **c** zoomed view ($\Omega = (1.5, 4.0)$) ($d_1 = 0.5,\ d_2 = 0.05,\ a_1 = -1.0,\ a_2 = 10,\ Q = 1.0$)

motion are shaded by gray color, labeled by "S". The regions of single unstable periodic motion are hatched, labeled by "U". The regions of three periodic motions are divided into three portions. The portions with two stable solutions and one unstable solution are labeled by "SSU" (red). The portions with one stable solution and two unstable solutions are labeled by "SUU" (blue), and the portions with three unstable solutions are labeled by "UUU" (yellow). This parameter map gives a better understanding of dynamical behaviors of periodic solution based on the two harmonic terms. Some region in parameter map in Fig. 1.6a is not clear. Thus, the zoomed parameter map should be presented in order to show the dynamical behavior, as shown in Fig. 1.6b and c. Similarly, for approximate solution with other higher-order harmonic terms, the parameter maps can be developed. Herein, the parameter map based on HB2 is as an example to show how to develop the parameter maps.

Table 1.2 Input data for numerical simulations of stable periodic motions ($d_1 = 0.5, d_2 = 0.05, a_1 = -1.0, a_2 = 10, Q = 1.0, Q_0 = 10.0$)

	$\Omega = 3.5$	Initial conditions ($t_0 = 0.0$)	Stable
	x_0	$\dot{x}_0$	Harmonic terms
Fig. 1.7a	0.903000	4.305000	HB1
Fig. 1.7b	0.898300	4.465300	HB2
Fig. 1.7c	0.958100	4.709600	HB3
Fig. 1.7d	0.955913	4.746170	HB4

1.4.4 Numerical Illustrations

In this section, numerical illustrations are given that are based on the analytical solutions and numerical integration schemes. The initial conditions in numerical simulation are obtained from approximate analytical solutions of periodic solutions. Input data comprising system parameters and initial conditions for numerical simulations are tabulated in Tables 1.2 and 1.3. In all plots for illustration, circular symbols give approximate solutions, and solid curves give numerical simulation results.

Four approximate solutions of a stable periodic motion are illustrated in Fig. 1.7 for $\Omega = 3.5$. From the stability analysis of approximate solution based on HB2, the solutions is a *stable* focus of $(0, 0, 0|5, 0, 0)$. In Fig. 1.7a, the trajectory of periodic motions based on one harmonic term expression (HB1) is plotted. The numerical solution has a transient motion to reach the steady-state solution. The numerical and analytical results are quite different. Thus more harmonic terms should be considered. In Fig. 1.7b, the approximate solution of periodic motion with two harmonic terms expression is adopted. Indeed, the transient motion in numerical result becomes better. However, the approximate, analytical solutions are still different from the numerical steady-state solution of periodic motions. In Fig. 1.7c, the approximate solution with three harmonic terms is used. Compared to the analytical predictions based on HB1 and HB2, the transient motion is very small before the steady-state motion is obtained, which implies that the approximate solution gives a good approximation of the steady-state periodic motion. Consider the higher-order harmonic term solutions of periodic motion. The analytical prediction based on four harmonic terms (HB4) is presented in Fig. 1.7d. Compared to the prediction of HB3, the analytical prediction of HB4 gives a solution closed to HB3. Indeed, small improvements here and where of trajectory in phase plane are presented. More harmonic terms included will improve the analytical prediction of periodic motions. However, the computation workload will increase dramatically. It is very important to find the appropriate harmonic terms to give a good approximation of periodic motion.

From the analytical prediction of HB2, during the range of $\Omega \in (3.956, 4.035)$, there are three periodic motions. One is unstable periodic motion of $(1, 1, 0|3, 1, 0)$ and two stable periodic motions are of $(2, 0, 0|4, 0, 0)$. The stable solutions based on HB1–HB4 can be illustrated. Herein, the unstable periodic motion is a spiral sad-

Table 1.3 Input data for numerical simulations of unstable periodic motions ($d_1 = 0.5, d_2 = 0.05, a_1 = -1.0, a_2 = 10, Q = 1.0, Q_0 = 10.0$)

$\Omega = 4.0(middle)$	Initial conditions ($t_0 = 0.0$)	Unstable	
	x_0	$\dot{x}_0$	Harmonic terms
Fig. 1.8a	−0.871000	2.755200	HB1
Fig. 1.8b	−0.881407	2.657200	HB2
Fig. 1.8c	−0.879248	2.194800	HB3
Fig. 1.8d	−0.879302	2.178870	HB4

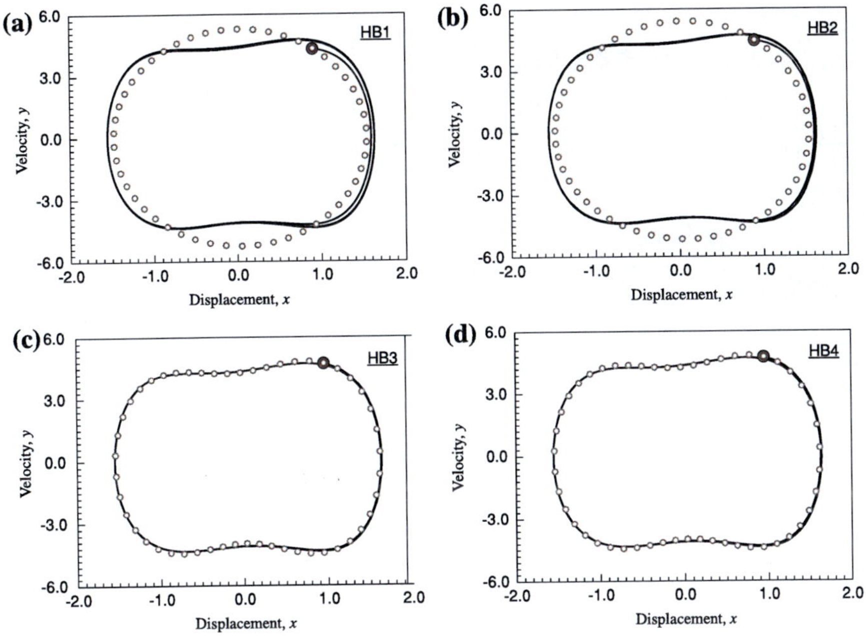

Fig. 1.7 Analytical and numerical solutions of *stable* periodic motion in phase plane ($\Omega = 3.5$): **a** HB1, **b** HB2, **c** HB3 and **d** HB4. ($d_1 = 0.5$, $d_2 = 0.05, a_1 = -1.0$, $a_2 = 10$, $Q = 1.0$, $Q_0 = 10.0$)

dle of (1, 1, 0|3, 1, 0). Consider the unstable periodic motion of (1, 1, 0|3, 1, 0) at $\Omega = 4.0$. From the approximate, analytical solutions of BH1–HB4, the numerical and analytical predictions of unstable periodic motions will be presented in Fig. 1.8a–d, respectively. From the unstable periodic motion given by the approximate, analytical solution, the numerical result of trajectory in phase space show the numerical solution reaches the stable periodic motion at $\Omega = 4.0$. In Fig. 1.8a, the unstable periodic motion is based on HB1. The unstable periodic motion moves to the outer stable periodic motions (upper branch). The numerical result of trajectory in phase plane does not stick with the analytical prediction of HB1. Therefore, the analytical approximate solution is not accurate. The triangle symbols give the outer stable periodic motion

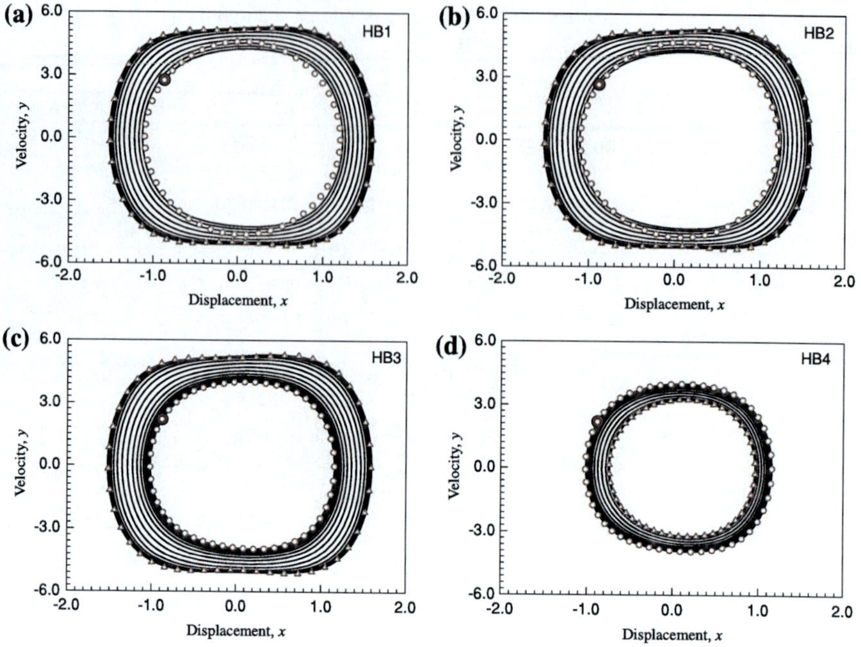

Fig. 1.8 Analytical and numerical solutions of *unstable* periodic motion in phase plane ($\Omega = 4.0$, middle branch): **a** HB1, **b** HB2, **c** HB3, **d** HB4. ($d_1 = 0.5,\ d_2 = 0.05,\ a_1 = -1.0,\ a_2 = 10,\ Q = 1.0,\ Q_0 = 10.0$)

based on the HB4 with the initial conditions of $(x_0, y_0) = (0.072388, 5.156110)$, which matches well with the numerical results. It implies that the analytical approximate solution is on the outside of unstable periodic motion. In Fig. 1.8b, the unstable periodic motion is based on HB2. The numerical result of trajectory in phase plane still moves to the outer stable periodic motion, and still does not stick with the analytical prediction of HB2, but such numerical result is better than the result given by HB1. Therefore, the analytical approximate solution is still not accurate. In Fig. 1.8c, the unstable periodic motion based on HB3 is considered for illustration. The numerical result of trajectory in phase plane still moves to the outer stable periodic motion. However, such numerical result does stick with the analytical prediction of HB3, which implies that the analytical, approximate solution of unstable periodic motion given by HB3 is close to the exact one. Since the unstable periodic motion moves to the outer stable periodic motion, the approximate solution is still on the outside of the exact motion. In Fig. 1.8d, the unstable periodic motion based on HB4 is considered for illustration. The numerical result of trajectory in phase plane still moves to the inner stable periodic motion. The analytical solution of inner stable periodic motion is based on the HB4 with $(x_0, y_0) = (-0.704000, 1.010270)$, represented by triangle symbols. The numerical result of trajectory in phase plane indeed sticks with the unstable periodic motion. Since the final state of the numerical results moves to the inner stable periodic motion, it means that the approximate, analytical solution of the unstable solution may be inside of the exact one.

For the constant term $Q=0$ without the cubic term of damping, the Duffing oscillator will be reduced to the cases given in Luo and Han (1997). For those cases, the approximate solutions with only one harmonic term can give the appropriate approximation, and the corresponding analytical conditions for stability and bifurcation were presented. However, if more nonlinear terms and constant forces are involved, the approximate solution needs more harmonic terms to get appropriate approximations. Furthermore, the parameter maps for stability and bifurcation will be more complicated.

References

Carr, J. (1981), *Applications of Center Manifold Theory*, Applied Mathematical Science **35**, Springer-Verlag, New York

Hartman, P., *Ordinary Differential Equations*, Wiley, New York, 1964 (2nd ed. Birkhauser, Boston Basel Stuttgart, 1982)

Hirsch, A.W. and Smale, S. (1974), *Differential Equations, Dynamical Systems and Linear Algebra*, Academic Press.

Luo, A.C.J. and Han, R.P.S. (1997), "A quantitative stability and bifurcation analyses of a generalized Duffing oscillator with strong nonlinearity", *Journal of Franklin Institute*, **334B**, 447–459

Luo, A.C.J. and Huang, J.Z. (2012), Approximate solutions of periodic motions in nonlinear systems via a generalized harmonic balance, *Journal of Vibration and Control*, **18**, 1661–1674

Marsden, J.E. and McCracken, M.F., *The Hopf Bifurcation and Its Applications*, Applied Mathematical Science **19**, Springer-Verlag, New York, 1976

Chapter 2
Nonlinear Discrete Dynamical Systems

In this chapter, the basic concepts of nonlinear discrete systems will be presented. The local and global theory of stability and bifurcation for nonlinear discrete systems will be discussed. The stability switching and bifurcation on specific eigenvectors of the linearized system at fixed points under specific period will be presented. The higher singularity and stability for nonlinear discrete systems on the specific eigenvectors will be developed. A few special cases in the lower dimensional maps will be presented for a better understanding of the generalized theory. The route to chaos will be discussed briefly, and the intermittency phenomena relative to specific bifurcations will be presented. The normalization group theory for 2-D discrete systems will be presented via Duffing discrete systems.

2.1 Discrete Dynamical Systems

Definition 2.1 For $\Omega_\alpha \subseteq \mathscr{R}^n$ and $\Lambda \subseteq \mathscr{R}^m$ with $\alpha \in \mathbb{Z}$, consider a vector function $\mathbf{f}_\alpha : \Omega_\alpha \times \Lambda \to \Omega_\alpha$ which is C^r $(r \geq 1)$-continuous, and there is a discrete (or difference) equation in the form of

$$\mathbf{x}_{k+1} = \mathbf{f}_\alpha(\mathbf{x}_k, \mathbf{p}_\alpha) \text{ for } \mathbf{x}_k, \mathbf{x}_{k+1} \in \Omega_\alpha, k \in \mathbb{Z} \text{ and } \mathbf{p}_\alpha \in \Lambda. \tag{2.1}$$

With an initial condition of $\mathbf{x}_k = \mathbf{x}_0$, the solution of Eq. (2.1) is given by

$$\begin{gathered}\mathbf{x}_k = \underbrace{\mathbf{f}_\alpha(\mathbf{f}_\alpha(\cdots(\mathbf{f}_\alpha}_{k}(\mathbf{x}_0, \mathbf{p}_\alpha)))) \\ \text{for } \mathbf{x}_k \in \Omega_\alpha, k \in \mathbb{Z} \text{ and } \mathbf{p} \in \Lambda.\end{gathered} \tag{2.2}$$

(i) The difference equation with the initial condition is called a *discrete dynamical system.*
(ii) The vector function $\mathbf{f}_\alpha(\mathbf{x}_k, \mathbf{p}_\alpha)$ is called a *discrete vector field* on domain Ω_α.
(iii) The solution $\mathbf{x}_k$ for each $k \in \mathbb{Z}$ is called a *flow* of discrete dynamical system.

A. C. J. Luo, *Regularity and Complexity in Dynamical Systems*,
DOI: 10.1007/978-1-4614-1524-4_2,

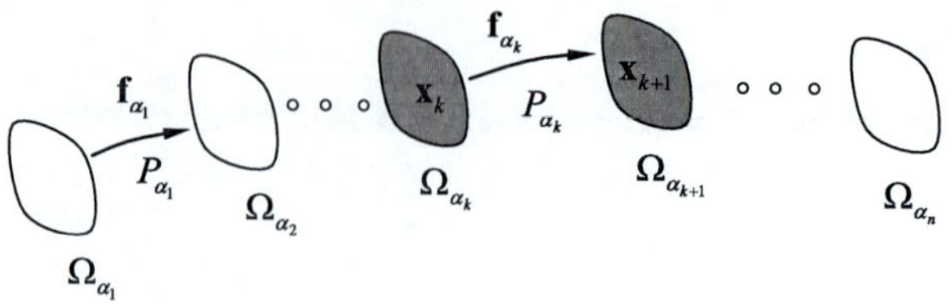

Fig. 2.1 Maps and vector functions on each sub-domain for discrete dynamical system

(iv) The solution $\mathbf{x}_k$ for all $k \in \mathbb{Z}$ on domain Ω_α is called the trajectory, phase curve or orbit of discrete dynamical system, which is defined as

$$\Gamma = \left\{\mathbf{x}_k | \mathbf{x}_{k+1} = \mathbf{f}_\alpha(\mathbf{x}_k, \mathbf{p}_\alpha) \text{ for } k \in \mathbb{Z} \text{ and } \mathbf{p}_\alpha \in \Lambda\right\} \subseteq \cup_\alpha \Omega_\alpha. \tag{2.3}$$

(v) The discrete dynamical system is called *a uniform discrete system* if

$$\mathbf{x}_{k+1} = \mathbf{f}_\alpha(\mathbf{x}_k, \mathbf{p}_\alpha) = \mathbf{f}(\mathbf{x}_k, \mathbf{p}) \text{ for } k \in \mathbb{Z} \text{ and } \mathbf{x}_k \in \Omega_\alpha. \tag{2.4}$$

Otherwise, this discrete dynamical system is called a *non-uniform discrete system.*

Definition 2.2 For the discrete dynamical system in Eq. (2.1), the relation between state $\mathbf{x}_k$ and state $\mathbf{x}_{k+1}$ $(k \in \mathbb{Z})$ is called a discrete map if

$$P_\alpha : \mathbf{x}_k \xrightarrow{\mathbf{f}_\alpha} \mathbf{x}_{k+1} \text{ and } \mathbf{x}_{k+1} = P_\alpha \mathbf{x}_k \tag{2.5}$$

with the following properties:

$$P_{(k,n)} : \mathbf{x}_k \xrightarrow{\mathbf{f}_{\alpha_1}, \mathbf{f}_{\alpha_2}, \cdots, \mathbf{f}_{\alpha_n}} \mathbf{x}_{k+n} \text{ and } \mathbf{x}_{k+n} = P_{\alpha_n} \circ P_{\alpha_{n-1}} \circ \cdots \circ P_{\alpha_1} \mathbf{x}_k \tag{2.6}$$

where

$$P_{(k;n)} = P_{\alpha_n} \circ P_{\alpha_{n-1}} \circ \cdots \circ P_{\alpha_1}. \tag{2.7}$$

If $P_{\alpha_n} = P_{\alpha_{n-1}} = \cdots = P_{\alpha_1} = P_\alpha$, then

$$P_{(\alpha;n)} \equiv P_\alpha^{(n)} = P_\alpha \circ P_\alpha \circ \cdots \circ P_\alpha \tag{2.8}$$

with

$$P_\alpha^{(n)} = P_\alpha \circ P_\alpha^{(n-1)} \text{ and } P_\alpha^{(0)} = \mathbf{I}. \tag{2.9}$$

The total map with n-different submaps is shown in Fig. 2.1. The map P_{α_k} with the relation function $\mathbf{f}_{\alpha_k}$ $(\alpha_k \in \mathbb{Z})$ is given by Eq. (2.5). The total map $P_{(k,n)}$ is given in Eq. (2.7). The domains $\Omega_{\alpha_k} (\alpha_k \in \mathbb{Z})$ can fully overlap each other or can be completely separated without any intersection.

Definition 2.3 For a vector function in $\mathbf{f}_\alpha \in \mathscr{R}^n, \mathbf{f}_\alpha : \mathscr{R}^n \to \mathscr{R}^n$. The operator norm of $\mathbf{f}_\alpha$ is defined by

$$||\mathbf{f}_\alpha|| = \sum\nolimits_{i=1}^{n} \max_{||\mathbf{x}_k|| \le 1} |f_{\alpha(i)}(\mathbf{x}_k, \mathbf{p}_\alpha)|. \tag{2.10}$$

For an $n \times n$ matrix $\mathbf{f}_\alpha(\mathbf{x}_k, \mathbf{p}_\alpha) = \mathbf{A}_\alpha \mathbf{x}_k$ and $\mathbf{A}_\alpha = (a_{ij})_{n \times n}$, the corresponding norm is defined by

$$||\mathbf{A}_\alpha|| = \sum_{i,j=1}^{n} |a_{ij}|. \tag{2.11}$$

Definition 2.4 For $\Omega_\alpha \subseteq \mathscr{R}^n$ and $\Lambda \subseteq \mathscr{R}^m$ with $\alpha \in \mathbb{Z}$, the vector function $\mathbf{f}_\alpha(\mathbf{x}_k, \mathbf{p}_\alpha)$ with $\mathbf{f}_\alpha : \Omega_\alpha \times \Lambda \to \mathscr{R}^n$ is differentiable at $\mathbf{x}_k \in \Omega_\alpha$ if

$$\left.\frac{\partial \mathbf{f}_\alpha(\mathbf{x}_k, \mathbf{p}_\alpha)}{\partial \mathbf{x}_k}\right|_{(\mathbf{x}_k, \mathbf{p})} = \lim_{\Delta \mathbf{x}_k \to \mathbf{0}} \frac{\mathbf{f}_\alpha(\mathbf{x}_k + \Delta \mathbf{x}_k, \mathbf{p}_\alpha) - \mathbf{f}_\alpha(\mathbf{x}_k, \mathbf{p}_\alpha)}{\Delta \mathbf{x}_k}. \tag{2.12}$$

$\partial \mathbf{f}_\alpha / \partial \mathbf{x}_k$ is called the spatial derivative of $\mathbf{f}_\alpha(\mathbf{x}_k, \mathbf{p}_\alpha)$ at $\mathbf{x}_k$, and the derivative is given by the Jacobian matrix

$$\frac{\partial \mathbf{f}_\alpha(\mathbf{x}_k, \mathbf{p}_\alpha)}{\partial \mathbf{x}_k} = \left[\frac{\partial f_{\alpha(i)}}{\partial x_{k(j)}}\right]_{n \times n}. \tag{2.13}$$

Definition 2.5 For $\Omega_\alpha \subseteq \mathscr{R}^n$ and $\Lambda \subseteq \mathscr{R}^m$, consider a vector function $\mathbf{f}(\mathbf{x}_k, \mathbf{p})$ with $\mathbf{f} : \Omega_\alpha \times \Lambda \to \mathscr{R}^n$, where $\mathbf{x}_k \in \Omega_\alpha$ and $\mathbf{p} \in \Lambda$ with $k \in \mathbb{Z}$. The vector function $\mathbf{f}(\mathbf{x}_k, \mathbf{p})$ satisfies the Lipschitz condition

$$||\mathbf{f}(\mathbf{y}_k, \mathbf{p}) - \mathbf{f}(\mathbf{x}_k, \mathbf{p})|| \le L||\mathbf{y}_k - \mathbf{x}_k|| \tag{2.14}$$

with $\mathbf{x}_k, \mathbf{y}_k \in \Omega_\alpha$ and L a constant. The constant L is called the Lipschitz constant.

2.2 Fixed Points and Stability

Definition 2.6 Consider a discrete, nonlinear dynamical system $\mathbf{x}_{k+1} = \mathbf{f}(\mathbf{x}_k, \mathbf{p})$ in Eq. (2.4).

(i) A point $\mathbf{x}_k^* \in \Omega_\alpha$ is called a fixed point or a period-1 solution of a discrete nonlinear system $\mathbf{x}_{k+1} = \mathbf{f}(\mathbf{x}_k, \mathbf{p})$ under map P if for $\mathbf{x}_{k+1} = \mathbf{x}_k = \mathbf{x}_k^*$

$$\mathbf{x}_k^* = \mathbf{f}(\mathbf{x}_k^*, \mathbf{p}). \tag{2.15}$$

The linearized system of the nonlinear discrete system $\mathbf{x}_{k+1} = \mathbf{f}(\mathbf{x}_k, \mathbf{p})$ in Eq. (2.4) at the fixed point $\mathbf{x}_k^*$ is given by

$$\mathbf{y}_{k+1} = DP(\mathbf{x}_k^*, \mathbf{p})\mathbf{y}_k = D\mathbf{f}(\mathbf{x}_k^*, \mathbf{p})\mathbf{y}_k \tag{2.16}$$

where

$$\mathbf{y}_k = \mathbf{x}_k - \mathbf{x}_k^* \text{ and } \mathbf{y}_{k+1} = \mathbf{x}_{k+1} - \mathbf{x}_{k+1}^*. \tag{2.17}$$

(ii) A set of points $\mathbf{x}_j^* \in \Omega_{\alpha_j}(\alpha_j \in \mathbb{Z})$ is called the fixed point set or period-1 point set of the total map $P_{(k;n)}$ with n-different submaps in nonlinear discrete system of Eq. (2.2) if

$$\begin{aligned} &\mathbf{x}_{k+j+1}^* = \mathbf{f}_{\alpha_{j'}}(\mathbf{x}_{k+j}^*, \mathbf{p}_{\alpha_{j'}}) \text{ for } j \in \mathbb{Z}_+ \text{ and } j' = \operatorname{mod}(j, n) + 1;\\ &\mathbf{x}_{k+\operatorname{mod}(j,n)}^* = \mathbf{x}_k^*. \end{aligned} \tag{2.18}$$

The linearized equation of each map in the total map $P_{(k;n)}$ gives

$$\begin{aligned} &\mathbf{y}_{k+j+1} = DP_{\alpha_{j'}}(\mathbf{x}_{k+j}^*, \mathbf{p}_{\alpha_{j'}})\mathbf{y}_{k+j} = D\mathbf{f}_{\alpha_{j'}}(\mathbf{x}_{k+j}^*, \mathbf{p}_{\alpha_{j'}})\mathbf{y}_{k+j}\\ &\text{with } \mathbf{y}_{k+j+1} = \mathbf{x}_{k+j+1} - \mathbf{x}_{k+j+1}^* \text{ and } \mathbf{y}_{k+j} = \mathbf{x}_{k+j} - \mathbf{x}_{k+j}^*\\ &\text{for } j \in \mathbb{Z}_+ \text{ and } j' = \operatorname{mod}(j, n) + 1. \end{aligned} \tag{2.19}$$

The resultant equation for the total map is

$$\mathbf{y}_{k+j+1} = DP_{(k,n)}(\mathbf{x}_k^*, \mathbf{p})\mathbf{y}_{k+j} \text{ for } j \in \mathbb{Z}_+ \tag{2.20}$$

where

$$\begin{aligned} DP_{(k,n)}(\mathbf{x}_k^*, \mathbf{p}) &= \prod\nolimits_{j=n}^{1} DP_{\alpha_j}(\mathbf{x}_{k+j-1}^*, \mathbf{p})\\ &= DP_{\alpha_n}(\mathbf{x}_{k+n-1}^*, \mathbf{p}_{\alpha_n}) \cdot \ldots \cdot DP_{\alpha_2}(\mathbf{x}_{k+1}^*, \mathbf{p}_{\alpha_2}) \cdot DP_{\alpha_1}(\mathbf{x}_k^*, \mathbf{p}_{\alpha_1})\\ &= D\mathbf{f}_{\alpha_n}(\mathbf{x}_{k+n-1}^*, \mathbf{p}_{\alpha_n}) \cdot \ldots \cdot D\mathbf{f}_{\alpha_2}(\mathbf{x}_{k+1}^*, \mathbf{p}_{\alpha_2}) \cdot D\mathbf{f}_{\alpha_1}(\mathbf{x}_k^*, \mathbf{p}_{\alpha_1}). \end{aligned} \tag{2.21}$$

The fixed point $\mathbf{x}_k^*$ lies in the intersected set of two domains Ω_k and Ω_{k+1}, as shown in Fig. 2.2. In the vicinity of the fixed point $\mathbf{x}_k^*$, the incremental relations in the two domains Ω_k and Ω_{k+1} are different. In other words, setting $\mathbf{y}_k = \mathbf{x}_k - \mathbf{x}_k^*$ and $\mathbf{y}_{k+1} = \mathbf{x}_{k+1} - \mathbf{x}_{k+1}^*$, the corresponding linearization is generated as in Eq. (2.16). Similarly, the fixed point of the total map with n-different submaps requires the intersection set of two domains Ω_k and Ω_{k+n}, which are a set of equations to obtain the fixed points from Eq. (2.18). The other values of fixed points lie in different domains, i.e., $\mathbf{x}_j^* \in \Omega_j$ $(j = k+1, k+2, \cdots, k+n-1)$, as shown in Fig. 2.3.

The corresponding linearized equations are given in Eq. (2.19). From Eq. (2.20), the local characteristics of the total map can be discussed as a single map. Thus, the dynamical characteristics for the fixed point of the single map will be discussed comprehensively, and the fixed points for the resultant map are applicable. The results can be extended to any period-m flows with $P^{(m)}$.

Definition 2.7 Consider a discrete, nonlinear dynamical system $\mathbf{x}_{k+1} = \mathbf{f}(\mathbf{x}_k, \mathbf{p})$ in Eq. (2.4) with a fixed point $\mathbf{x}_k^*$. The linearized system of the discrete nonlinear

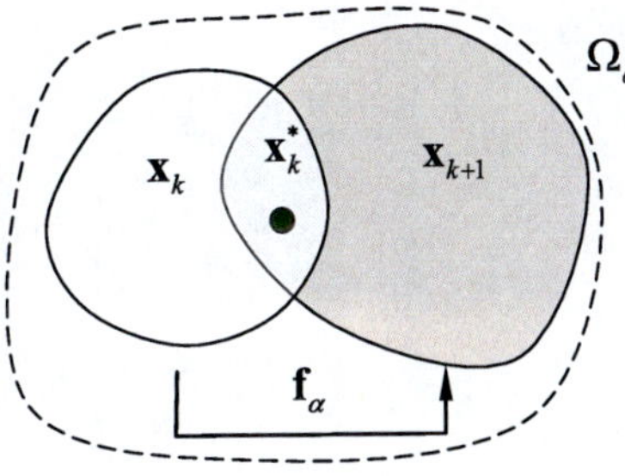

Fig. 2.2 A fixed point between domains Ω_k and Ω_{k+1} for a discrete dynamical system

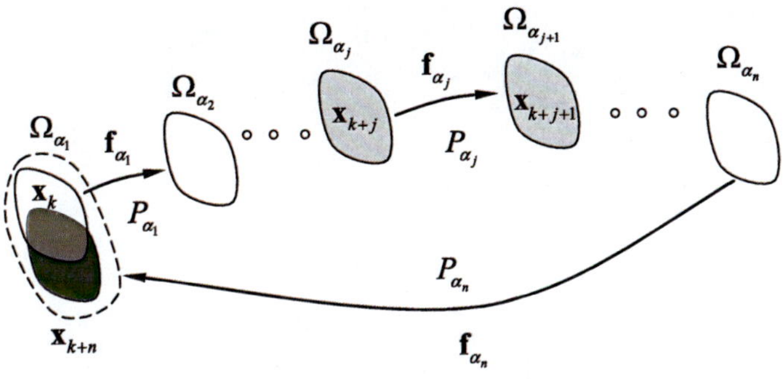

Fig. 2.3 Fixed points with n-maps for a discrete dynamical system

system in the neighborhood of $\mathbf{x}_k^*$ is $\mathbf{y}_{k+1} = D\mathbf{f}(\mathbf{x}_k^*, \mathbf{p})\mathbf{y}_k$ ($\mathbf{y}_l = \mathbf{x}_l - \mathbf{x}_k^*$ and $l = k, k+1$) in Eq. (2.16). The matrix $D\mathbf{f}(\mathbf{x}_k^*, \mathbf{p})$ possesses n_1 real eigenvalues $|\lambda_j| < 1$ ($j \in N_1$), n_2 real eigenvalues $|\lambda_j| > 1$ ($j \in N_2$), n_3 real eigenvalues $\lambda_j = 1$ ($j \in N_3$), and n_4 real eigenvalues $\lambda_j = -1$ ($j \in N_4$). Set $N = \{1, 2, \cdots, n\}$ and $N_i = \{i_1, i_2, \cdots, i_{n_i}\} \cup \varnothing$ ($i = 1, 2, 3, 4$) with $i_m \in N$ ($m = 1, 2, \cdots, n_i$). $N_i \subseteq N \cup \varnothing$, $\cup_{i=1}^{3} N_i = N$, $N_i \cap N_p = \varnothing (p \neq i)$ and $\Sigma_{i=1}^{3} n_i = n$. $N_i = \varnothing$ if $n_i = 0$. The corresponding eigenvectors for contraction, expansion, invariance and flip oscillation are $\{\mathbf{v}_j\}$ ($j \in N_i$) ($i = 1, 2, 3, 4$), respectively. The stable, unstable, invariant and flip subspaces of $\mathbf{y}_{k+1} = D\mathbf{f}(\mathbf{x}_k^*, \mathbf{p})\mathbf{y}_k$ in Eq. (2.16) are linear subspace spanned by $\{\mathbf{v}_j\}$ ($j \in N_i$) ($i = 1, 2, 3, 4$),respectively, i.e.,

$$
\begin{aligned}
\mathscr{E}^{\mathrm{s}} &= span\left\{ \mathbf{v}_j \left| \begin{array}{l} (D\mathbf{f}(\mathbf{x}_k^*, \mathbf{p}) - \lambda_j \mathbf{I})\mathbf{v}_j = \mathbf{0}, \\ |\lambda_j| < 1, j \in N_1 \subseteq N \cup \varnothing \end{array} \right. \right\}; \\
\mathscr{E}^{\mathrm{u}} &= span\left\{ \mathbf{v}_j \left| \begin{array}{l} (D\mathbf{f}(\mathbf{x}_k^*, \mathbf{p}) - \lambda_j \mathbf{I})\mathbf{v}_j = \mathbf{0}, \\ |\lambda_j| > 1, j \in N_2 \subseteq N \cup \varnothing \end{array} \right. \right\}; \\
\mathscr{E}^{\mathrm{i}} &= span\left\{ \mathbf{v}_j \left| \begin{array}{l} (D\mathbf{f}(\mathbf{x}_k^*, \mathbf{p}) - \lambda_j \mathbf{I})\mathbf{v}_j = \mathbf{0}, \\ \lambda_j = 1, j \in N_3 \subseteq N \cup \varnothing \end{array} \right. \right\}; \\
\mathscr{E}^{\mathrm{f}} &= span\left\{ \mathbf{v}_j \left| \begin{array}{l} (D\mathbf{f}(\mathbf{x}_k^*, \mathbf{p}) - \lambda_j \mathbf{I})\mathbf{v}_j = \mathbf{0}, \\ \lambda_j = -1, j \in N_4 \subseteq N \cup \varnothing \end{array} \right. \right\}.
\end{aligned}
\tag{2.22}
$$

where

$$\mathscr{E}^{\mathrm{s}} = \mathscr{E}^{\mathrm{s}}_{\mathrm{m}} \cup \mathscr{E}^{\mathrm{s}}_{\mathrm{o}} \cup \mathscr{E}^{\mathrm{s}}_{\mathrm{z}} \text{ with}$$

$$\begin{aligned}
\mathscr{E}^{\mathrm{s}}_{\mathrm{m}} &= span\left\{ \mathbf{v}_j \left| \begin{array}{l} (D\mathbf{f}(\mathbf{x}^*_k, \mathbf{p}) - \lambda_j \mathbf{I})\mathbf{v}_j = \mathbf{0}, \\ 0 < \lambda_j < 1, j \in N^{\mathrm{m}}_1 \subseteq N \cup \varnothing \end{array} \right. \right\}; \\
\mathscr{E}^{\mathrm{s}}_{\mathrm{o}} &= span\left\{ \mathbf{v}_j \left| \begin{array}{l} (D\mathbf{f}(\mathbf{x}^*_k, \mathbf{p}) - \lambda_j \mathbf{I})\mathbf{v}_j = \mathbf{0}, \\ -1 < \lambda_j < 0, j \in N^{\mathrm{o}}_1 \subseteq N \cup \varnothing \end{array} \right. \right\}; \\
\mathscr{E}^{\mathrm{s}}_{\mathrm{z}} &= span\left\{ \mathbf{v}_j \left| \begin{array}{l} (D\mathbf{f}(\mathbf{x}^*_k, \mathbf{p}) - \lambda_j \mathbf{I})\mathbf{v}_j = \mathbf{0}, \\ \lambda_j = 0, j \in N^{\mathrm{z}}_1 \subseteq N \cup \varnothing \end{array} \right. \right\};
\end{aligned} \tag{2.23}$$

$$\mathscr{E}^{\mathrm{u}} = \mathscr{E}^{\mathrm{u}}_{\mathrm{m}} \cup \mathscr{E}^{\mathrm{u}}_{\mathrm{o}} \text{ with}$$

$$\begin{aligned}
\mathscr{E}^{\mathrm{u}}_{\mathrm{m}} &= span\left\{ \mathbf{v}_j \left| \begin{array}{l} (D\mathbf{f}(\mathbf{x}^*_k, \mathbf{p}) - \lambda_j \mathbf{I})\mathbf{v}_j = \mathbf{0}, \\ \lambda_j > 1, j \in N^{\mathrm{m}}_2 \subseteq N \cup \varnothing \end{array} \right. \right\}; \\
\mathscr{E}^{\mathrm{u}}_{\mathrm{o}} &= span\left\{ \mathbf{v}_j \left| \begin{array}{l} (D\mathbf{f}(\mathbf{x}^*_k, \mathbf{p}) - \lambda_j \mathbf{I})\mathbf{v}_j = \mathbf{0}, \\ \lambda_j < -1, j \in N^{\mathrm{o}}_2 \subseteq N \cup \varnothing \end{array} \right. \right\};
\end{aligned} \tag{2.24}$$

where subscripts "m" and "o" represent the monotonic and oscillatory evolutions.

Definition 2.8 Consider a $2n$-dimensional, discrete, nonlinear dynamical system $\mathbf{x}_{k+1} = \mathbf{f}(\mathbf{x}_k, \mathbf{p})$ in Eq. (2.4) with a fixed point $\mathbf{x}^*_k$. The linearized system of the discrete nonlinear system in the neighborhood of $\mathbf{x}^*_k$ is $\mathbf{y}_{k+1} = D\mathbf{f}(\mathbf{x}^*_k, \mathbf{p})\mathbf{y}_k$ ($\mathbf{y}_l = \mathbf{x}_l - \mathbf{x}^*_k$ and $l = k, k+1$) in Eq. (2.16). The matrix $D\mathbf{f}(\mathbf{x}^*_k, \mathbf{p})$ has complex eigenvalues $\alpha_j \pm \mathbf{i}\beta_j$ with eigenvectors $\mathbf{u}_j \pm \mathbf{i}\mathbf{v}_j$ ($j \in \{1, 2, \cdots, n\}$) and the base of vector is

$$\mathbf{B} = \{\mathbf{u}_1, \mathbf{v}_1, \cdots, \mathbf{u}_j, \mathbf{v}_j, \cdots, \mathbf{u}_n, \mathbf{v}_n\}. \tag{2.25}$$

The stable, unstable and center subspaces of $\mathbf{y}_{k+1} = D\mathbf{f}_k(\mathbf{x}^*_k, \mathbf{p})\mathbf{y}_k$ in Eq. (2.16) are linear subspaces spanned by $\{\mathbf{u}_j, \mathbf{v}_j\}$ ($j \in N_i$, $i = 1, 2, 3$), respectively. $N = \{1, 2, \cdots, n\}$ plus $N_i = \{i_1, i_2, \cdots, i_{n_i}\} \cup \varnothing \subseteq N \cup \varnothing$ with $i_m \in N$ ($m = 1, 2, \cdots, n_i, i = 1, 2, 3$). $\cup_{i=1}^3 N_i = N$ with $N_i \cap N_p = \varnothing (p \neq i)$ and $\Sigma_{i=1}^3 n_i = n$. $N_i = \varnothing$ if $n_i = 0$. The stable, unstable and center subspaces of $\mathbf{y}_{k+1} = D\mathbf{f}(\mathbf{x}^*_k, \mathbf{p})\mathbf{y}_k$ in Eq. (2.16) are defined by

$$\mathscr{E}^{\mathrm{s}} = span\left\{ (\mathbf{u}_j, \mathbf{v}_j) \left| \begin{array}{l} r_j = \sqrt{\alpha_j^2 + \beta_j^2} < 1, \\ \left(D\mathbf{f}(\mathbf{x}^*_k, \mathbf{p}) - (\alpha_j \pm \mathbf{i}\beta_j)\mathbf{I}\right)(\mathbf{u}_j \pm \mathbf{i}\mathbf{v}_j) = \mathbf{0}, \\ j \in N_1 \subseteq \{1, 2, \cdots, n\} \cup \varnothing \end{array} \right. \right\};$$

$$\mathscr{E}^{\mathrm{u}} = span\left\{ (\mathbf{u}_j, \mathbf{v}_j) \left| \begin{array}{l} r_j = \sqrt{\alpha_j^2 + \beta_j^2} > 1, \\ \left(D\mathbf{f}(\mathbf{x}^*_k, \mathbf{p}) - (\alpha_j \pm \mathbf{i}\beta_j)\mathbf{I}\right)(\mathbf{u}_j \pm \mathbf{i}\mathbf{v}_j) = \mathbf{0}, \\ j \in N_2 \subseteq \{1, 2, \cdots, n\} \cup \varnothing \end{array} \right. \right\};$$

$$\mathscr{E}^c = span\left\{ (\mathbf{u}_j, \mathbf{v}_j) \left| \begin{array}{l} r_j = \sqrt{\alpha_j^2 + \beta_j^2} = 1, \\ \left(D\mathbf{f}(\mathbf{x}_k^*, \mathbf{p}) - (\alpha_j \pm \mathbf{i}\beta_j)\mathbf{I} \right) (\mathbf{u}_j \pm \mathbf{i}\mathbf{v}_j) = \mathbf{0}, \\ j \in N_3 \subseteq \{1, 2, \cdots, n\} \cup \varnothing \end{array} \right. \right\}. \tag{2.26}$$

Definition 2.9 Consider a discrete, nonlinear dynamical system $\mathbf{x}_{k+1} = \mathbf{f}(\mathbf{x}_k, \mathbf{p})$ in Eq. (2.4) with a fixed point $\mathbf{x}_k^*$. The linearized system of the discrete nonlinear system in the neighborhood of $\mathbf{x}_k^*$ is $\mathbf{y}_{k+1} = D\mathbf{f}(\mathbf{x}_k^*, \mathbf{p})\mathbf{y}_k$ ($\mathbf{y}_l = \mathbf{x}_l - \mathbf{x}_k^*$ and $l = k, k+1$) in Eq. (2.16). The fixed point or period-1point is *hyperbolic* if no eigenvalues of $D\mathbf{f}(\mathbf{x}_k^*, \mathbf{p})$ are on the unit circle (i.e., $|\lambda_j| \neq 1$ for $j = 1, 2, \cdots, n$).

Theorem 2.1 *Consider a discrete, nonlinear dynamical system* $\mathbf{x}_{k+1} = \mathbf{f}(\mathbf{x}_k, \mathbf{p})$ *in Eq. (2.4) with a fixed point* $\mathbf{x}_k^*$. *The linearized system of the discrete nonlinear system in the neighborhood of* $\mathbf{x}_k^*$ *is* $\mathbf{y}_{k+1} = D\mathbf{f}(\mathbf{x}_k^*, \mathbf{p})\mathbf{y}_k$ ($\mathbf{y}_l = \mathbf{x}_l - \mathbf{x}_k^*$ *and* $l = k, k+1$) *in Eq. (2.16). The eigenspace of* $D\mathbf{f}(\mathbf{x}_k^*, \mathbf{p})$ *(i.e.,* $\mathscr{E} \subseteq \mathscr{R}^n$*) in the linearized dynamical system is expressed by direct sum of three subspaces*

$$\mathscr{E} = \mathscr{E}^s \oplus \mathscr{E}^u \oplus \mathscr{E}^c \tag{2.27}$$

where $\mathscr{E}^s$, $\mathscr{E}^u$ and $\mathscr{E}^c$ *are the stable, unstable and center subspaces, respectively.*

Proof This proof is the same as the linear system in Appendix B. ∎

Definition 2.10 Consider a discrete, nonlinear dynamical system $\mathbf{x}_{k+1} = \mathbf{f}(\mathbf{x}_k, \mathbf{p})$ in Eq. (2.4) with a fixed point $\mathbf{x}_k^*$. Suppose there is a neighborhood of the fixed point $\mathbf{x}_k^*$ as $U_k(\mathbf{x}_k^*) \subset \Omega_k$, and in the neighborhood,

$$\lim_{||\mathbf{y}_k|| \to 0} \frac{||\mathbf{f}(\mathbf{x}_k^* + \mathbf{y}_k, \mathbf{p}) - D\mathbf{f}(\mathbf{x}_k^*, \mathbf{p})\mathbf{y}_k||}{||\mathbf{y}_k||} = 0 \tag{2.28}$$

and

$$\mathbf{y}_{k+1} = D\mathbf{f}(\mathbf{x}_k^*, \mathbf{p})\mathbf{y}_k. \tag{2.29}$$

(i) A C^r invariant manifold

$$\begin{aligned} \mathscr{S}_{loc}(\mathbf{x}_k, \mathbf{x}_k^*) = \{\mathbf{x}_k \in U(\mathbf{x}_k^*) | \lim_{j \to +\infty} \mathbf{x}_{k+j} = \mathbf{x}_k^* \text{ and} \\ \mathbf{x}_{k+j} \in U(\mathbf{x}_k^*) \text{ with } j \in \mathbb{Z}_+\} \end{aligned} \tag{2.30}$$

is called the local stable manifold of $\mathbf{x}_k^*$, and the corresponding global stable manifold is defined as

$$\begin{aligned} \mathscr{S}(\mathbf{x}_k, \mathbf{x}_k^*) &= \cup_{j \in \mathbb{Z}_-} \mathbf{f}(\mathscr{U}_{loc}(\mathbf{x}_{k+j}, \mathbf{x}_{k+j}^*)) \\ &= \cup_{j \in \mathbb{Z}_-} \mathbf{f}^{(j)}(\mathscr{U}_{loc}(\mathbf{x}_k, \mathbf{x}_k^*)). \end{aligned} \tag{2.31}$$

(ii) A C^r invariant manifold $\mathscr{U}_{loc}(\mathbf{x}_k, \mathbf{x}_k^*)$

$$\begin{aligned}\mathscr{U}_{loc}(\mathbf{x}_k, \mathbf{x}_k^*) = \{\mathbf{x}_k \in U(\mathbf{x}_k^*) | \lim_{j\to-\infty} \mathbf{x}_{k+j} = \mathbf{x}_k^* \\ \text{and } \mathbf{x}_{k+j} \in U(\mathbf{x}_k^*) \text{ with } j \in \mathbb{Z}_-\}\end{aligned} \tag{2.32}$$

is called the unstable manifold of $\mathbf{x}^*$, and the corresponding global unstable manifold is defined as

$$\begin{aligned}\mathscr{U}(\mathbf{x}_k, \mathbf{x}_k^*) &= \cup_{j\in\mathbb{Z}_+} \mathbf{f}(\mathscr{U}_{loc}(\mathbf{x}_{k+j}, \mathbf{x}_{k+j}^*)) \\ &= \cup_{j\in\mathbb{Z}_+} \mathbf{f}^{(j)}(\mathscr{U}_{loc}(\mathbf{x}_k, \mathbf{x}_k^*)).\end{aligned} \tag{2.33}$$

(iii) A C^{r-1} invariant manifold $\mathscr{C}_{loc}(\mathbf{x}, \mathbf{x}^*)$ is called the center manifold of $\mathbf{x}^*$ if $\mathscr{C}_{loc}(\mathbf{x}, \mathbf{x}^*)$ possesses the same dimension of $\mathscr{E}^c$ for $\mathbf{x}^* \in \mathscr{S}(\mathbf{x}, \mathbf{x}^*)$, and the tangential space of $\mathscr{C}_{loc}(\mathbf{x}, \mathbf{x}^*)$ is identical to $\mathscr{E}^c$.

As in continuous dynamical systems, the stable and unstable manifolds are unique, but the center manifold is not unique. If the nonlinear vector field $\mathbf{f}$ is C^∞-continuous, then a C^r center manifold can be found for any $r < \infty$.

Theorem 2.2 *Consider a discrete, nonlinear dynamical system* $\mathbf{x}_{k+1} = \mathbf{f}(\mathbf{x}_k, \mathbf{p})$ *in Eq. (2.4) with a hyperbolic fixed point* $\mathbf{x}_k^*$. *The corresponding solution is* $\mathbf{x}_{k+j} = \mathbf{f}(\mathbf{x}_{k+j-1}, \mathbf{p})$ *with* $j \in \mathbb{Z}$. *Suppose there is a neighborhood of the hyperbolic fixed point* $\mathbf{x}_k^*$ *(i.e.,* $U_k(\mathbf{x}_k^*) \subset \Omega_\alpha$*), and* $\mathbf{f}(\mathbf{x}_k, \mathbf{p})$ *is* C^r $(r \geq 1)$*-continuous in* $U_k(\mathbf{x}_k^*)$. *The linearized system is* $\mathbf{y}_{k+j+1} = D\mathbf{f}(\mathbf{x}_k^*, \mathbf{p})\mathbf{y}_{k+j}$ ($y_l = x_l - x_k^*$ and $l = k+j, k+j+1$) *in* $U_k(\mathbf{x}_k^*)$. *If the homeomorphism between the local invariant subspace* $E(\mathbf{x}_k^*) \subset U(\mathbf{x}_k^*)$ *and the eigenspace* $\mathscr{E}$ *of the linearized system exists with the condition in Eq. (2.28), the local invariant subspace is decomposed by*

$$E(\mathbf{x}_k, \mathbf{x}_k^*) = \mathscr{S}_{loc}(\mathbf{x}_k, \mathbf{x}_k^*) \oplus \mathscr{U}_{loc}(\mathbf{x}_k, \mathbf{x}_k^*). \tag{2.34}$$

(a) *The local stable invariant manifold* $\mathscr{S}_{loc}(\mathbf{x}, \mathbf{x}^*)$ *possesses the following properties:*

(i) *for* $\mathbf{x}_k^* \in \mathscr{S}_{loc}(\mathbf{x}_k, \mathbf{x}_k^*)$, $\mathscr{S}_{loc}(\mathbf{x}_k, \mathbf{x}_k^*)$ *possesses the same dimension of* $\mathscr{E}^s$ *and the tangential space of* $\mathscr{S}_{loc}(\mathbf{x}_k, \mathbf{x}_k^*)$ *is identical to* $\mathscr{E}^s$;

(ii) *for* $\mathbf{x}_k \in \mathscr{S}_{loc}(\mathbf{x}_k, \mathbf{x}_k^*)$, $\mathbf{x}_{k+j} \in \mathscr{S}_{loc}(\mathbf{x}_k, \mathbf{x}_k^*)$ *and* $\lim_{j\to\infty} \mathbf{x}_{k+j} = \mathbf{x}_k^*$ *for all* $j \in \mathbb{Z}_+$;

(iii) *for* $\mathbf{x}_k \notin \mathscr{S}_{loc}(\mathbf{x}_k, \mathbf{x}_k^*)$, $||\mathbf{x}_{k+j} - \mathbf{x}_k^*|| \geq \delta$ *for* $\delta > 0$ *with* $j, j_1 \in \mathbb{Z}_+$ *and* $j \geq j_1 \geq 0$.

(b) *The local unstable invariant manifold* $\mathscr{U}_{loc}(\mathbf{x}_k, \mathbf{x}_k^*)$ *possesses the following properties:*

(i) *for* $\mathbf{x}_k^* \in \mathscr{U}_{loc}(\mathbf{x}_k, \mathbf{x}_k^*)$, $\mathscr{U}_{loc}(\mathbf{x}_k, \mathbf{x}_k^*)$ *possesses the same dimension of* $\mathscr{E}^u$ *and the tangential space of* $\mathscr{U}_{loc}(\mathbf{x}_k, \mathbf{x}_k^*)$ *is identical to* $\mathscr{E}^u$;

(ii) *for* $\mathbf{x}_k \in \mathscr{U}_{loc}(\mathbf{x}_k, \mathbf{x}_k^*)$, $\mathbf{x}_{k+j} \in \mathscr{U}_{loc}(\mathbf{x}_k, \mathbf{x}_k^*)$ *and* $\lim_{j\to-\infty} \mathbf{x}_{k+j} = \mathbf{x}^*$ *for all* $j \in \mathbb{Z}_-$

(iii) *for* $\mathbf{x}_k \notin \mathscr{U}_{loc}(\mathbf{x}, \mathbf{x}^*)$, $||\mathbf{x}_{k+j} - \mathbf{x}_k^*|| \geq \delta$ *for* $\delta > 0$ *with* $j_1, j \in \mathbb{Z}_-$ *and* $j \leq j_1 \leq 0$.

Proof See Hirtecki (1971). ∎

Theorem 2.3 *Consider a discrete, nonlinear dynamical system* $\mathbf{x}_{k+1} = \mathbf{f}(\mathbf{x}_k, \mathbf{p})$ *in Eq.* (2.4) *with a fixed point* $\mathbf{x}_k^*$. *The corresponding solution is* $\mathbf{x}_{k+j} = \mathbf{f}(\mathbf{x}_{k+j-1}, \mathbf{p})$ *with* $j \in \mathbb{Z}$. *Suppose there is a neighborhood of the fixed point* $\mathbf{x}_k^*$ *(i.e.,* $U_k(\mathbf{x}_k^*) \subset \Omega_\alpha$*), and* $\mathbf{f}(\mathbf{x}_k, \mathbf{p})$ *is* C^r $(r \geq 1)$*−continuous in* $U_k(\mathbf{x}_k^*)$. *The linearized system is* $\mathbf{y}_{k+j+1} = D\mathbf{f}(\mathbf{x}_k^*, \mathbf{p})\mathbf{y}_{k+j}$ $(\mathbf{y}_{k+j} = \mathbf{x}_{k+j} - \mathbf{x}_k^*)$ in $U_k(\mathbf{x}_k^*)$. *If the homeomorphism between the local invariant subspace* $E(\mathbf{x}_k^*) \subset U(\mathbf{x}_k^*)$ *and the eigenspace* $\mathscr{E}$ *of the linearized system exists with the condition in Eq.* (2.28), *in addition to the local stable and unstable invariant manifolds, there is a* C^{r-1} *center manifold* $\mathscr{C}_{loc}(\mathbf{x}_k, \mathbf{x}_k^*)$. *The center manifold possesses the same dimension of* $\mathscr{E}^c$ *for* $\mathbf{x}^* \in \mathscr{C}_{loc}(\mathbf{x}_k, \mathbf{x}_k^*)$, *and the tangential space of* $\mathscr{C}_{loc}(\mathbf{x}, \mathbf{x}^*)$ *is identical to* $\mathscr{E}^c$. *Thus, the local invariant subspace is decomposed by*

$$E(\mathbf{x}_k, \mathbf{x}_k^*) = \mathscr{S}_{loc}(\mathbf{x}_k, \mathbf{x}_k^*) \oplus \mathscr{U}_{loc}(\mathbf{x}_k, \mathbf{x}_k^*) \oplus \mathscr{C}_{loc}(\mathbf{x}_k, \mathbf{x}_k^*). \tag{2.35}$$

Proof See Guckenhiemer and Holmes (1990). ∎

Definition 2.11 Consider a discrete, nonlinear dynamical system $\mathbf{x}_{k+1} = \mathbf{f}(\mathbf{x}_k, \mathbf{p})$ in Eq. (2.4) on domain $\Omega_\alpha \subseteq \mathscr{R}^n$. Suppose there is a metric space (Ω_α, ρ), then the map P under the vector function $\mathbf{f}(\mathbf{x}_k, \mathbf{p})$ is called a contraction map if

$$\rho(\mathbf{x}_{k+1}^{(1)}, \mathbf{x}_{k+1}^{(2)}) = \rho(\mathbf{f}(\mathbf{x}_k^{(1)}, \mathbf{p}), \mathbf{f}(\mathbf{x}_k^{(2)}, \mathbf{p})) \leq \lambda \rho(\mathbf{x}_k^{(1)}, \mathbf{x}_k^{(2)}) \tag{2.36}$$

for $\lambda \in (0, 1)$ and $\mathbf{x}_k^{(1)}, \mathbf{x}_k^{(2)} \in \Omega_\alpha$ with $\rho(\mathbf{x}_k^{(1)}, \mathbf{x}_k^{(2)}) = ||\mathbf{x}_k^{(1)} - \mathbf{x}_k^{(2)}||$.

Theorem 2.4 *Consider a discrete, nonlinear dynamical system* $\mathbf{x}_{k+1} = \mathbf{f}(\mathbf{x}_k, \mathbf{p})$ *in Eq.* (2.4) on domain $\Omega_\alpha \subseteq \mathscr{R}^n$. *Suppose there is a metric space* (Ω_α, ρ), *if the map* P *under the vector function* $\mathbf{f}(\mathbf{x}_k, \mathbf{p})$ *is a contraction map, then there is a unique fixed point* $\mathbf{x}_k^*$ *which is globally stable.*

Proof Consider

$$\begin{aligned}
\rho(\mathbf{x}_{k+j+1}^{(1)}, \mathbf{x}_{k+j+1}^{(2)}) &= \rho(\mathbf{f}(\mathbf{x}_{k+j}^{(1)}, \mathbf{p}), \mathbf{f}(\mathbf{x}_{k+j}^{(2)}, \mathbf{p})) \leq \lambda \rho(\mathbf{x}_{k+j}^{(1)}, \mathbf{x}_{k+j}^{(2)}) \\
&= \lambda \rho(\mathbf{f}(\mathbf{x}_{k+j-1}^{(1)}, \mathbf{p}), \mathbf{f}(\mathbf{x}_{k+j-1}^{(2)}, \mathbf{p})) \leq \lambda^2 \rho(\mathbf{x}_{k+j-1}^{(1)}, \mathbf{x}_{k+j-1}^{(2)}) \\
&\vdots \\
&= \lambda^j \rho(\mathbf{f}(\mathbf{x}_{k+1}^{(1)}, \mathbf{p}), \mathbf{f}(\mathbf{x}_{k+1}^{(2)}, \mathbf{p})) \leq \lambda^{j+1} \rho(\mathbf{x}_k^{(1)}, \mathbf{x}_k^{(2)}).
\end{aligned}$$

As $j \to \infty$ and $0 < \lambda < 1$, thus, we have

$$\lim_{j\to\infty} \rho(\mathbf{x}_{k+j+1}^{(1)}, \mathbf{x}_{k+j+1}^{(2)}) = \lim_{j\to\infty} \lambda^{j+1} \rho(\mathbf{x}_k^{(1)}, \mathbf{x}_k^{(2)}) = 0.$$

If $\mathbf{x}_{k+j+1}^{(2)} = \mathbf{x}_k^{(2)} = \mathbf{x}_k^*$, in domain $\Omega_\alpha \in \mathscr{R}^n$, we have

$$\lim_{j\to\infty} \rho(\mathbf{x}_{k+j+1}^{(1)}, \mathbf{x}_{k+j+1}^{(2)}) = \lim_{j\to\infty} ||\mathbf{x}_{k+j+1}^{(1)} - \mathbf{x}_k^*|| = 0.$$

Consider two fixed points $\mathbf{x}_{k1}^*$ and $\mathbf{x}_{k2}^*$. The above equation gives

$$\begin{aligned} ||\mathbf{x}_{k1}^* - \mathbf{x}_{k2}^*|| &= \lim_{j\to\infty} ||\mathbf{x}_{k1}^* - \mathbf{x}_{k+j+1} + \mathbf{x}_{k+j+1} - \mathbf{x}_{k2}^*|| \\ &\le \lim_{j\to\infty} ||\mathbf{x}_{k1}^* - \mathbf{x}_{k+j+1}|| + \lim_{j\to\infty} ||\mathbf{x}_{k+j+1} - \mathbf{x}_{k2}^*|| = 0. \end{aligned}$$

Therefore, the fixed point is unique and globally stable. This theorem is proved. ■

Definition 2.12 Consider a discrete, nonlinear dynamical system $\mathbf{x}_{k+1} = \mathbf{f}(\mathbf{x}_k, \mathbf{p})$ in Eq. (2.4) with a fixed point $\mathbf{x}_k^*$. The corresponding solution is given by $\mathbf{x}_{k+j} = \mathbf{f}(\mathbf{x}_{k+j-1}, \mathbf{p})$ with $j \in \mathbb{Z}$. Suppose there is a neighborhood of the fixed point $\mathbf{x}_k^*$ (i.e., $U_k(\mathbf{x}_k^*) \subset \Omega_\alpha$), and $\mathbf{f}(\mathbf{x}_k, \mathbf{p})$ is C^r $(r \ge 1)$-continuous in $U_k(\mathbf{x}_k^*)$. The linearized system is $\mathbf{y}_{k+j+1} = D\mathbf{f}(\mathbf{x}_k^*, \mathbf{p})\mathbf{y}_{k+j}$ $(\mathbf{y}_{k+j} = \mathbf{x}_{k+j} - \mathbf{x}_k^*)$ in $U_k(\mathbf{x}_k^*)$. Consider a real eigenvalue λ_i of matrix $D\mathbf{f}(\mathbf{x}_k^*, \mathbf{p})$ $(i \in N = \{1, 2, \cdots, n\})$ and there is a corresponding eigenvector $\mathbf{v}_i$. On the invariant eigenvector $\mathbf{v}_k^{(i)} = \mathbf{v}_i$, consider $\mathbf{y}_k^{(i)} = c_k^{(i)}\mathbf{v}_i$ and $\mathbf{y}_{k+1}^{(i)} = c_{k+1}^{(i)}\mathbf{v}_i = \lambda_i c_k^{(i)}\mathbf{v}_i$, thus, $c_{k+1}^{(i)} = \lambda_i c_k^{(i)}$.

(i) $\mathbf{x}_k^{(i)}$ on the direction $\mathbf{v}_i$ is stable if

$$\lim_{k\to\infty} |c_k^{(i)}| = \lim_{k\to\infty} |(\lambda_i)^k| \times |c_0^{(i)}| = 0 \text{ for } |\lambda_i| < 1. \tag{2.37}$$

(ii) $\mathbf{x}_k^{(i)}$ on the direction $\mathbf{v}_i$ is unstable if

$$\lim_{k\to\infty} |c_k^{(i)}| = \lim_{k\to\infty} |(\lambda_i)^k| \times |c_0^{(i)}| = \infty \text{ for } |\lambda_i| > 1. \tag{2.38}$$

(iii) $\mathbf{x}_k^{(i)}$ on the direction $\mathbf{v}_i$ is invariant if

$$\lim_{k\to\infty} c_k^{(i)} = \lim_{k\to\infty} (\lambda_i)^k c_0^{(i)} = c_0^{(i)} \text{ for } \lambda_i = 1. \tag{2.39}$$

(iv) $\mathbf{x}_k^{(i)}$ on the direction $\mathbf{v}_i$ is flipped if

$$\left.\begin{aligned} \lim_{2k\to\infty} c_k^{(i)} &= \lim_{2k\to\infty} (\lambda_i)^{2k} \times c_0^{(i)} = c_0^{(i)} \\ \lim_{2k+1\to\infty} c_k^{(i)} &= \lim_{2k+1\to\infty} (\lambda_i)^{2k+1} \times c_0^{(i)} = -c_0^{(i)} \end{aligned}\right\} \text{for } \lambda_i = -1. \tag{2.40}$$

(v) $\mathbf{x}_k^{(i)}$ on the direction $\mathbf{v}_i$ is degenerate if

$$c_k^{(i)} = (\lambda_i)^k c_0^{(i)} = 0 \text{ for } \lambda_i = 0. \tag{2.41}$$

Definition 2.13 Consider a discrete, nonlinear dynamical system $\mathbf{x}_{k+1} = \mathbf{f}(\mathbf{x}_k, \mathbf{p})$ in Eq. (2.4) with a fixed point $\mathbf{x}_k^*$. The corresponding solution is given by $\mathbf{x}_{k+j} = \mathbf{f}(\mathbf{x}_{k+j-1}, \mathbf{p})$ with $j \in \mathbb{Z}$. Suppose there is a neighborhood of the fixed point $\mathbf{x}_k^*$ (i.e., $U_k(\mathbf{x}_k^*) \subset \Omega_\alpha$), and $\mathbf{f}(\mathbf{x}_k, \mathbf{p})$ is C^r ($r \geq 1$)-continuous in $U_k(\mathbf{x}_k^*)$. Consider a pair of complex eigenvalues $\alpha_i \pm \mathbf{i}\beta_i$ of matrix $D\mathbf{f}(\mathbf{x}_k^*, \mathbf{p})$ ($i \in N = \{1, 2, \cdots, n\}, \mathbf{i} = \sqrt{-1}$) and there is a corresponding eigenvector $\mathbf{u}_i \pm \mathbf{i}\mathbf{v}_i$. On the invariant plane of $(\mathbf{u}_k^{(i)}, \mathbf{v}_k^{(i)}) = (\mathbf{u}_i, \mathbf{v}_i)$, consider $\mathbf{x}_k^{(i)} = \mathbf{x}_{k+}^{(i)} + \mathbf{x}_{k-}^{(i)}$ with

$$\mathbf{x}_k^{(i)} = c_k^{(i)}\mathbf{u}_i + d_k^{(i)}\mathbf{v}_i, \mathbf{x}_{k+1}^{(i)} = c_{k+1}^{(i)}\mathbf{u}_i + d_{k+1}^{(i)}\mathbf{v}_i. \tag{2.42}$$

Thus, $\mathbf{c}_k^{(i)} = (c_k^{(i)}, d_k^{(i)})^{\mathrm{T}}$ with

$$\mathbf{c}_{k+1}^{(i)} = \mathbf{E}_i\mathbf{c}_k^{(i)} = r_i\mathbf{R}_i\mathbf{c}_k^{(i)} \tag{2.43}$$

where

$$\begin{aligned} &\mathbf{E}_i = \begin{bmatrix} \alpha_i & \beta_i \\ -\beta_i & \alpha_i \end{bmatrix} \text{ and } \mathbf{R}_i = \begin{bmatrix} \cos\theta_i & \sin\theta_i \\ -\sin\theta_i & \cos\theta_i \end{bmatrix}, \\ &r_i = \sqrt{\alpha_i^2 + \beta_i^2}, \cos\theta_i = \alpha_i/r_i \text{ and } \sin\theta_i = \beta_i/r_i; \end{aligned} \tag{2.44}$$

and

$$\mathbf{E}_i^k = \begin{bmatrix} \alpha_i & \beta_i \\ -\beta_i & \alpha_i \end{bmatrix}^{\mathrm{k}} \text{ and } \mathbf{R}_i^k = \begin{bmatrix} \cos k\theta_i & \sin k\theta_i \\ -\sin k\theta_i & \cos k\theta_i \end{bmatrix}. \tag{2.45}$$

(i) $\mathbf{x}_k^{(i)}$ on the plane of $(\mathbf{u}_i, \mathbf{v}_i)$ is spirally stable if

$$\lim_{k\to\infty} ||\mathbf{c}_k^{(i)}|| = \lim_{k\to\infty} r_i^k||\mathbf{R}_i^k|| \times ||\mathbf{c}_0^{(i)}|| = 0 \text{ for } r_i = |\lambda_i| < 1. \tag{2.46}$$

(ii) $\mathbf{x}_k^{(i)}$ on the plane of $(\mathbf{u}_i, \mathbf{v}_i)$ is spirally unstable if

$$\lim_{k\to\infty} ||\mathbf{c}_k^{(i)}|| = \lim_{k\to\infty} r_i^k||\mathbf{R}_i^k|| \times ||\mathbf{c}_0^{(i)}|| = \infty \text{ for } r_i = |\lambda_i| > 1. \tag{2.47}$$

(iii) $\mathbf{x}_k^{(i)}$ on the plane of $(\mathbf{u}_i, \mathbf{v}_i)$ is on the invariant circles if,

$$||\mathbf{c}_k^{(i)}|| = r_i^k||\mathbf{R}_i^k|| \times ||\mathbf{c}_0^{(i)}|| = ||\mathbf{c}_0^{(i)}|| \text{ for } r_i = |\lambda_i| = 1. \tag{2.48}$$

(iv) $\mathbf{x}_k^{(i)}$ on the plane of $(\mathbf{u}_i, \mathbf{v}_i)$ is degenerate in the direction of $\mathbf{u}_i$ if $\beta_i = 0$.

Definition 2.14 Consider a discrete, nonlinear dynamical system $\mathbf{x}_{k+1} = \mathbf{f}(\mathbf{x}_k, \mathbf{p})$ in Eq. (2.4) with a fixed point $\mathbf{x}_k^*$. The corresponding solution is given by $\mathbf{x}_{k+j} = \mathbf{f}(\mathbf{x}_{k+j-1}, \mathbf{p})$ with $j \in \mathbb{Z}$. Suppose there is a neighborhood of the fixed point $\mathbf{x}_k^*$ (i.e., $U_k(\mathbf{x}_k^*) \subset \Omega_\alpha$), and $\mathbf{f}(\mathbf{x}_k, \mathbf{p})$ is C^r ($r \geq 1$)-continuous in $U_k(\mathbf{x}_k^*)$ with Eq. (2.28). The linearized system is $\mathbf{y}_{k+j+1} = D\mathbf{f}(\mathbf{x}_k^*, \mathbf{p})\mathbf{y}_{k+j}$ ($\mathbf{y}_{k+j} = \mathbf{x}_{k+j} - \mathbf{x}_k^*$) in $U_k(\mathbf{x}_k^*)$. The matrix $D\mathbf{f}(\mathbf{x}_k^*, \mathbf{p})$ possesses n eigenvalues λ_k ($k = 1, 2, \cdots n$).

(i) The fixed point $\mathbf{x}_k^*$ is called a hyperbolic point if $|\lambda_i| \neq 1$ $(i = 1, 2, \cdots, n)$.
(ii) The fixed point $\mathbf{x}_k^*$ is called a sink if $|\lambda_i| < 1$ $(i = 1, 2, \cdots, n)$.
(iii) The fixed point $\mathbf{x}_k^*$ is called a source if $|\lambda_i| > 1(i = \{1, 2, \cdots, n\})$.
(iv) The fixed point $\mathbf{x}_k^*$ is called a center if $|\lambda_i| = 1$ $(i = 1, 2, \cdots, n)$ with distinct eigenvalues.

Definition 2.15 Consider a discrete, nonlinear dynamical system $\mathbf{x}_{k+1} = \mathbf{f}(\mathbf{x}_k, \mathbf{p})$ in Eq. (2.4) with a fixed point $\mathbf{x}_k^*$. The corresponding solution is given by $\mathbf{x}_{k+j} = \mathbf{f}(\mathbf{x}_{k+j-1}, \mathbf{p})$ with $j \in \mathbb{Z}$. Suppose there is a neighborhood of the fixed point $\mathbf{x}_k^*$ (i.e., $U_k(\mathbf{x}_k^*) \subset \Omega_\alpha$), and $\mathbf{f}(\mathbf{x}_k, \mathbf{p})$ is C^r $(r \geq 1)$-continuous in $U_k(\mathbf{x}_k^*)$ with Eq. (2.28). The linearized system is $\mathbf{y}_{k+j+1} = D\mathbf{f}(\mathbf{x}_k^*, \mathbf{p})\mathbf{y}_{k+j}$ $(\mathbf{y}_{k+j} = \mathbf{x}_{k+j} - \mathbf{x}_k^*)$ in $U_k(\mathbf{x}_k^*)$. The matrix $D\mathbf{f}(\mathbf{x}_k^*, \mathbf{p})$ possesses n eigenvalues λ_i $(i = 1, 2, \cdots n)$.

(i) The fixed point $\mathbf{x}_k^*$ is called a stable node if $|\lambda_i| < 1(i = 1, 2, \cdots, n)$.
(ii) The fixed point $\mathbf{x}_k^*$ is called an unstable node if $|\lambda_i| > 1$ $(i = 1, 2, \cdots, n)$.
(iii) The fixed point $\mathbf{x}_k^*$ is called an $(l_1 : l_2)$-saddle if at least one $|\lambda_i| > 1(i \in L_1 \subset \{1, 2, \cdots n\})$ and the other $|\lambda_j| < 1$ $(j \in L_2 \subset \{1, 2, \cdots n\})$ with $L_1 \cup L_2 = \{1, 2, \cdots, n\}$ and $L_1 \cap L_2 = \varnothing$.
(iv) The fixed point $\mathbf{x}_k^*$ is called an lth-order degenerate case if $\lambda_i = 0$ $(i \in L \subseteq \{1, 2, \cdots n\})$.

Definition 2.16 Consider a discrete, nonlinear dynamical system $\mathbf{x}_{k+1} = \mathbf{f}(\mathbf{x}_k, \mathbf{p})$ in Eq. (2.4) with a fixed point $\mathbf{x}_k^*$. The corresponding solution is given by $\mathbf{x}_{k+j} = \mathbf{f}(\mathbf{x}_{k+j-1}, \mathbf{p})$ with $j \in \mathbb{Z}$. Suppose there is a neighborhood of the fixed point $\mathbf{x}_k^*$ (i.e., $U_k(\mathbf{x}_k^*) \subset \Omega_\alpha$), and $\mathbf{f}(\mathbf{x}_k, \mathbf{p})$ is C^r $(r \geq 1)$-continuous in $U_k(\mathbf{x}_k^*)$ with Eq. (2.28). The linearized system is $\mathbf{y}_{k+j+1} = D\mathbf{f}(\mathbf{x}_k^*, \mathbf{p})\mathbf{y}_{k+j}$ $(\mathbf{y}_{k+j} = \mathbf{x}_{k+j} - \mathbf{x}_k^*)$ in $U_k(\mathbf{x}_k^*)$. The matrix $D\mathbf{f}(\mathbf{x}_k^*, \mathbf{p})$ possesses n-pairs of complex eigenvalues λ_i $(i = 1, 2, \cdots, n)$.

(i) The fixed point $\mathbf{x}_k^*$ is called a spiral sink if $|\lambda_i| < 1$ $(i = 1, 2, \cdots, n)$ and $\operatorname{Im}\lambda_j \neq 0$ $(j \in \{1, 2, \cdots, n\})$.
(ii) fixed point $\mathbf{x}_k^*$ is called a spiral source if $|\lambda_i| > 1$ $(i = 1, 2, \cdots, n)$ with $\operatorname{Im}\lambda_j \neq 0$ $(j \in \{1, 2, \cdots, n\})$.
(iii) fixed point $\mathbf{x}_k^*$ is called a center if $|\lambda_i| = 1$ with distinct $\operatorname{Im}\lambda_i \neq 0$ $(i \in \{1, 2, \cdots, n\})$.

As in Appendix B, the refined classification of the linearized, discrete, nonlinear system at fixed points should be discussed. The generalized stability and bifurcation of flows in linearized, nonlinear dynamical systems in Eq. (2.4) will be discussed as follows.

Definition 2.17 Consider a discrete, nonlinear dynamical system $\mathbf{x}_{k+1} = \mathbf{f}(\mathbf{x}_k, \mathbf{p})$ in Eq. (2.4) with a fixed point $\mathbf{x}_k^*$. The corresponding solution is given by $\mathbf{x}_{k+j} = \mathbf{f}(\mathbf{x}_{k+j-1}, \mathbf{p})$ with $j \in \mathbb{Z}$. Suppose there is a neighborhood of the fixed point $\mathbf{x}_k^*$ (i.e., $U_k(\mathbf{x}_k^*) \subset \Omega_\alpha$), and $\mathbf{f}(\mathbf{x}_k, \mathbf{p})$ is C^r $(r \geq 1)$-continuous in $U_k(\mathbf{x}_k^*)$ with Eq. (2.28). The linearized system is $\mathbf{y}_{k+j+1} = D\mathbf{f}(\mathbf{x}_k^*, \mathbf{p})\mathbf{y}_{k+j}$ $(\mathbf{y}_{k+j} = \mathbf{x}_{k+j} - \mathbf{x}_k^*)$ in $U_k(\mathbf{x}_k^*)$. The matrix $D\mathbf{f}(\mathbf{x}_k^*, \mathbf{p})$ possesses n eigenvalues λ_i $(i = 1, 2, \cdots, n)$. Set $N = \{1, 2, \cdots, m, m+1, \cdots, (n-m)/2\}$, $N_p = \{p_1, p_2, \cdots, p_{n_p}\} \cup \varnothing$ with $p_q \in N$ $(q = 1, 2, \cdots, n_p,$

$p = 1, 2, \cdots, 7$) and $\Sigma_{p=1}^{4} n_p = m$ and $2\Sigma_{p=5}^{7} n_p = n - m$. $\cup_{p=1}^{7} N_p = N$ and $N_p \cap N_l = \varnothing (l \neq p)$.$N_p = \varnothing$ if $n_p = 0$. $N_\alpha = N_\alpha^{\mathrm{m}} \cup N_\alpha^{\mathrm{o}}$ ($\alpha = 1, 2$) and $N_\alpha^{\mathrm{m}} \cap N_\alpha^{\mathrm{o}} = \varnothing$ with $n_\alpha^{\mathrm{m}} + n_\alpha^{\mathrm{o}} = n_\alpha$, where superscripts "m" and "o" represent monotonic and oscillatory evolutions. The matrix $D\mathbf{f}(\mathbf{x}_k^*, \mathbf{p})$ possesses n_1-stable, n_2-unstable, n_3-invariant and n_4-flip real eigenvectors plus n_5-stable, n_6-unstable and n_7-center pairs of complex eigenvectors. Without repeated complex eigenvalues of $|\lambda_i| = 1 (i \in N_3 \cup N_4 \cup N_7)$, an iterative response of $\mathbf{x}_{k+1} = \mathbf{f}(\mathbf{x}_k, \mathbf{p})$ is an $([n_1^{\mathrm{m}}, n_1^{\mathrm{o}}] : [n_2^{\mathrm{m}}, n_2^{\mathrm{o}}] : [n_3; \kappa_3] : [n_4; \kappa_4] \| n_5 : n_6 : n_7)$ flow in the neighborhood of the fixed point $\mathbf{x}_k^*$. With repeated complex eigenvalues of $|\lambda_i| = 1$ $(i \in N_3 \cup N_4 \cup N_7)$, an iterative response of $\mathbf{x}_{k+1} = \mathbf{f}(\mathbf{x}_k, \mathbf{p})$ is an $([n_1^{\mathrm{m}}, n_1^{\mathrm{o}}] : [n_2^{\mathrm{m}}, n_2^{\mathrm{o}}] : [n_3; \kappa_3] : [n_4; \kappa_4] \| n_5 : n_6 : [n_7, l; \boldsymbol{\kappa}_7])$ flow in the neighborhood of the fixed point $\mathbf{x}_k^*$, where $\kappa_p \in \{\varnothing, m_p\}$ $(p = 3, 4)$, $\boldsymbol{\kappa}_7 = (\kappa_{71}, \kappa_{72}, \cdots, \kappa_{7l})^{\mathrm{T}}$ with $\kappa_{7s} \in \{\varnothing, m_{7s}\} (s = 1, 2, \cdots, l)$. The meanings of notations in the aforementioned structures are defined as follows:

(i) $[n_1^{\mathrm{m}}, n_1^{\mathrm{o}}]$ represents that there are n_1-sinks with n_1^{m}-monotonic convergence and n_1^{o}-oscillatory convergence among n_1-directions of $\mathbf{v}_i$ $(i \in N_1)$ if $|\lambda_i| < 1$ $(k \in N_1$ and $1 \leq n_1 \leq m)$ with distinct or repeated eigenvalues.

(ii) $[n_2^{\mathrm{m}}, n_2^{\mathrm{o}}]$ represents that there are n_2-sources with n_2^{m}-monotonic divergence and n_2^{o}-oscillatory divergence among n_2-directions of $\mathbf{v}_i$ $(i \in N_2)$ if $|\lambda_i| > 1$ $(k \in N_2$ and $1 \leq n_2 \leq m)$ with distinct or repeated eigenvalues.

(iii) $n_3 = 1$ represents an invariant center on 1-direction of $\mathbf{v}_i$ $(i \in N_3)$ if $\lambda_k = 1$ $(i \in N_3$ and $n_3 = 1)$.

(iv) $n_4 = 1$ represents a flip center on 1-direction of $\mathbf{v}_i$ $(i \in N_4)$ if $\lambda_i = -1$ $(i \in N_4$ and $n_4 = 1)$.

(v) n_5 represents n_5-spiral sinks on n_5-pairs of $(\mathbf{u}_i, \mathbf{v}_i)$ $(i \in N_5)$ if $|\lambda_i| < 1$ and $\mathrm{Im}\ \lambda_i \neq 0$ $(i \in N_5$ and $1 \leq n_5 \leq (n - m)/2)$ with distinct or repeated eigenvalues.

(vi) n_6 represents n_6-spiral sources on n_6-directions of $(\mathbf{u}_i, \mathbf{v}_i)$ $(i \in N_6)$ if $|\lambda_i| > 1$ and $\mathrm{Im}\ \lambda_i \neq 0$ $(i \in N_6$ and $1 \leq n_6 \leq (n - m)/2)$ with distinct or repeated eigenvalues.

(vii) n_7 represents n_7-invariant centers on n_7-pairs of $(\mathbf{u}_i, \mathbf{v}_i)$ $(i \in N_7)$ if $|\lambda_i| = 1$ and $\mathrm{Im}\ \lambda_i \neq 0$ $(i \in N_7$ and $1 \leq n_7 \leq (n - m)/2)$ with distinct eigenvalues.

(viii) $\varnothing$ represents none if $n_j = 0$ $(j \in \{1, 2, \cdots, 7\})$.

(ix) $[n_3; \kappa_3]$ represents $(n_3 - \kappa_3)$ invariant centers on $(n_3 - \kappa_3)$ directions of $\mathbf{v}_{i_3}$ $(i_3 \in N_3)$ and κ_3-sources in κ_3-directions of $\mathbf{v}_{j_3}$ $(j_3 \in N_3$ and $j_3 \neq i_3)$ if $\lambda_i = 1$ $(i \in N_3$ and $n_3 \leq m)$ with the $(\kappa_3 + 1)$th-order nilpotent matrix $\mathbf{N}_3^{\kappa_3+1} = \mathbf{0}$ $(0 < \kappa_3 \leq n_3 - 1)$.

(x) $[n_3; \varnothing]$ represents n_3 invariant centers on n_3-directions of $\mathbf{v}_i (i \in N_3)$ if $\lambda_i = 1$ $(i \in N_3$ and $1 < n_3 \leq m)$ with a nilpotent matrix $\mathbf{N}_3 = \mathbf{0}$.

(xi) $[n_4; \kappa_4]$ represents $(n_4 - \kappa_4)$ flip oscillatory centers on $(n_4 - \kappa_4)$ directions of $\mathbf{v}_{i_4}$ $(i_4 \in N_4)$ and κ_4-sources in κ_4-directions of $\mathbf{v}_{j_4}$ $(j_4 \in N_4$ and $j_4 \neq i_4)$ if $\lambda_i = -1$ $(i \in N_4$ and $n_4 \leq m)$ with the $(\kappa_4 + 1)$th-order nilpotent matrix $\mathbf{N}_4^{\kappa_4+1} = \mathbf{0}$ $(0 < \kappa_4 \leq n_4 - 1)$.

(xii) $[n_4; \varnothing]$ represents n_4 flip oscillatory centers on n_4-directions of $\mathbf{v}_i$ $(i \in N_3)$ if $\lambda_i = -1$ $(i \in N_4$ and $1 < n_4 \leq m)$ with a nilpotent matrix $\mathbf{N}_4 = \mathbf{0}$.

(xiii) $[n_7, l; \boldsymbol{\kappa}_7]$ represents $(n_7 - \sum_{s=1}^{l} \kappa_{7s})$ invariant centers on $(n_7 - \sum_{s=1}^{l} \kappa_{7s})$ pairs of $(\mathbf{u}_{i_7}, \mathbf{v}_{i_7})(i_7 \in N_7)$ and $\sum_{s=1}^{l} \kappa_{7s}$ sources on $\sum_{s=1}^{l} \kappa_{7s}$ pairs of $(\mathbf{u}_{j_7}, \mathbf{v}_{j_7})$ $(j_7 \in N_7$ and $j_7 \neq i_7)$ if $|\lambda_i| = 1$ and Im $\lambda_i \neq 0$ $(i \in N_7$ and $n_7 \leq (n-m)/2)$ for $\sum_{s=1}^{l} \kappa_{7s}$-pairs of repeated eigenvalues with the $(\kappa_{7s}+1)$th-order nilpotent matrix $\mathbf{N}_7^{\kappa_{7s}+1} = \mathbf{0}$ $(0 < \kappa_{7s} \leq l, s = 1, 2, \cdots, l)$.

(xiv) $[n_7, l; \varnothing]$ represents n_7-invariant centers on n_7-pairs of $(\mathbf{u}_i, \mathbf{v}_i)(i \in N_6)$ if $|\lambda_i| = 1$ and Im $\lambda_i \neq 0$ $(i \in N_7$ and $1 \leq n_7 \leq (n-m)/2)$ for $\sum_{s=1}^{l} \kappa_{7s}$-pairs of repeated eigenvalues with a nilpotent matrix $\mathbf{N}_7 = \mathbf{0}$.

Definition 2.18 Consider a discrete, nonlinear dynamical system $\mathbf{x}_{k+1} = \mathbf{f}(\mathbf{x}_k, \mathbf{p})$ in Eq. (2.4) with a fixed point $\mathbf{x}_k^*$. The corresponding solution is given by $\mathbf{x}_{k+j} = \mathbf{f}(\mathbf{x}_{k+j-1}, \mathbf{p})$ with $j \in \mathbb{Z}$. Suppose there is a neighborhood of the fixed point $\mathbf{x}_k^*$ (i.e., $U_k(\mathbf{x}_k^*) \subset \Omega_\alpha$), and $\mathbf{f}(\mathbf{x}_k, \mathbf{p})$ is C^r $(r \geq 1)$-continuous in $U_k(\mathbf{x}_k^*)$ with Eq. (2.28). The linearized system is $\mathbf{y}_{k+j+1} = D\mathbf{f}(\mathbf{x}_k^*, \mathbf{p})\mathbf{y}_{k+j}$ $(\mathbf{y}_{k+j} = \mathbf{x}_{k+j} - \mathbf{x}_k^*)$ in $U_k(\mathbf{x}_k^*)$. The matrix $D\mathbf{f}(\mathbf{x}_k^*, \mathbf{p})$ possesses n eigenvalues λ_i $(i = 1, 2, \cdots, n)$. Set $N = \{1, 2, \cdots, m, m+1, \cdots, (n-m)/2\}$, $N_p = \{p_1, p_2, \cdots, p_{n_p}\} \cup \varnothing$ with $p_q \in N$ $(q = 1, 2, \cdots, n_p$, $p = 1, 2, \cdots, 7)$, $\Sigma_{p=1}^{4} n_p = m$ and $2\Sigma_{p=5}^{7} n_p = n - m$. $\cup_{p=1}^{7} N_p = N$ and $N_p \cap N_l = \varnothing (l \neq p)$. $N_p = \varnothing$ if $n_p = 0$. $N_\alpha = N_\alpha^{\mathrm{m}} \cup N_\alpha^{\mathrm{o}}$ $(\alpha = 1, 2)$ and $N_\alpha^{\mathrm{m}} \cap N_\alpha^{\mathrm{o}} = \varnothing$ with $n_\alpha^{\mathrm{m}} + n_\alpha^{\mathrm{o}} = n_\alpha$, where superscripts "m" and "o" represent monotonic and oscillatory evolutions. The matrix $D\mathbf{f}(\mathbf{x}_k^*, \mathbf{p})$ possesses n_1-stable, n_2-unstable, n_3-invariant, and n_4-flip real eigenvectors plus n_5-stable, n_6-unstable and n_7-center pairs of complex eigenvectors. Without repeated complex eigenvalues of $|\lambda_k| = 1 (k \in N_3 \cup N_4 \cup N_7)$, an iterative response of $\mathbf{x}_{k+1} = \mathbf{f}(\mathbf{x}_k, \mathbf{p})$ is an $([n_1^{\mathrm{m}}, n_1^{\mathrm{o}}] : [n_2^{\mathrm{m}}, n_2^{\mathrm{o}}] : [n_3; \kappa_3] : [n_4; \kappa_4] | n_5 : n_6 : n_7)$ flow in the neighborhood of the fixed point $\mathbf{x}_k^*$. With repeated complex eigenvalues of $|\lambda_i| = 1$ $(i \in N_3 \cup N_4 \cup N_7)$, an iterative response of $\mathbf{x}_{k+1} = \mathbf{f}(\mathbf{x}_k, \mathbf{p})$ is an $([n_1^{\mathrm{m}}, n_1^{\mathrm{o}}] : [n_2^{\mathrm{m}}, n_2^{\mathrm{o}}] : [n_3; \kappa_3] : [n_4; \kappa_4] | n_5 : n_6 : [n_7, l; \boldsymbol{\kappa}_7])$ flow in the neighborhood of the fixed point $\mathbf{x}_k^*$, where $\kappa_p \in \{\varnothing, m_p\} (p = 3, 4)$, $\boldsymbol{\kappa}_7 = (\kappa_{71}, \kappa_{72}, \cdots, \kappa_{7l})^{\mathrm{T}}$ with $\kappa_{7s} = \{\varnothing, m_{7s}\} (s = 1, 2, \cdots, l)$.

I. *Non-degenerate cases*

(i) The fixed point $\mathbf{x}_k^*$ is an $([n_1^{\mathrm{m}}, n_1^{\mathrm{o}}] : [n_2^{\mathrm{m}}, n_2^{\mathrm{o}}] : \varnothing : \varnothing | n_5 : n_6 : \varnothing)$ hyperbolic point.

(ii) The fixed point $\mathbf{x}_k^*$ is an $([n_1^{\mathrm{m}}, n_1^{\mathrm{o}}] : [\varnothing, \varnothing] : \varnothing : \varnothing | n_5 : \varnothing : \varnothing)$-sink.

(iii) The fixed point $\mathbf{x}_k^*$ is an $([\varnothing, \varnothing] : [n_2^{\mathrm{m}}, n_2^{\mathrm{o}}] : \varnothing : \varnothing | \varnothing : n_6 : \varnothing)$-source.

(iv) The fixed point $\mathbf{x}_k^*$ is an $([\varnothing, \varnothing] : [\varnothing, \varnothing] : \varnothing : \varnothing | \varnothing : \varnothing : n/2)$-circular center.

(v) The fixed point $\mathbf{x}_k^*$ is an $([\varnothing, \varnothing] : [\varnothing, \varnothing] : \varnothing : \varnothing | \varnothing : \varnothing : [n/2, l; \varnothing])$-circular center.

(vi) The fixed point $\mathbf{x}_k^*$ is an $([\varnothing, \varnothing] : [\varnothing, \varnothing] : \varnothing : \varnothing | \varnothing : \varnothing : [n/2, l; \boldsymbol{\kappa}_7])$-point.

(vii) The fixed point $\mathbf{x}_k^*$ is an $([n_1^{\mathrm{m}}, n_1^{\mathrm{o}}] : [\varnothing, \varnothing] : \varnothing : \varnothing | n_5 : \varnothing : n_7)$-point.

(viii) The fixed point $\mathbf{x}_k^*$ is an $([\varnothing, \varnothing] : [n_2^{\mathrm{m}}, n_2^{\mathrm{o}}] : \varnothing : \varnothing | \varnothing : n_6 : n_7)$-point.

(ix) The fixed point $\mathbf{x}_k^*$ is an $([n_1^{\mathrm{m}}, n_1^{\mathrm{o}}] : [n_2^{\mathrm{m}}, n_2^{\mathrm{o}}] : \varnothing : \varnothing | n_5 : n_6 : n_7)$-point.

II. *Simple special cases*

(i) The fixed point $\mathbf{x}_k^*$ is an $([\varnothing,\varnothing]:[\varnothing,\varnothing]:[n;\varnothing]:\varnothing|\varnothing:\varnothing:\varnothing)$-invariant center (or static center).
(ii) The fixed point $\mathbf{x}_k^*$ is an $([\varnothing,\varnothing]:[\varnothing,\varnothing]:[n;m]:\varnothing|\varnothing:\varnothing:\varnothing)$-point.
(iii) The fixed point $\mathbf{x}_k^*$ is an $([\varnothing,\varnothing]:[\varnothing,\varnothing]:\varnothing:[n;\varnothing]|\varnothing:\varnothing:\varnothing)$-flip center.
(iv) The fixed point $\mathbf{x}_k^*$ is an $([\varnothing,\varnothing]:[\varnothing,\varnothing]:\varnothing:[n;m]|\varnothing:\varnothing:\varnothing)$-point.
(v) The fixed point $\mathbf{x}_k^*$ is an $([\varnothing,\varnothing]:[\varnothing,\varnothing]:[n_3;\kappa_3]:[n_4;\kappa_4]|\varnothing:\varnothing:\varnothing)$-point.
(vi) The fixed point $\mathbf{x}_k^*$ is an $([\varnothing,\varnothing]:[\varnothing,\varnothing]:[1;\varnothing]:[n_4;\kappa_4]|\varnothing:\varnothing:\varnothing)$-point.
(vii) The fixed point $\mathbf{x}_k^*$ is an $([\varnothing,\varnothing]:[\varnothing,\varnothing]:[n_3;\kappa_3]:[1;\varnothing]|\varnothing:\varnothing:\varnothing)$-point.
(viii) The fixed point $\mathbf{x}_k^*$ is an $([\varnothing,\varnothing]:[\varnothing,\varnothing]:[n_3;\kappa_3]:[\varnothing;\varnothing]|\varnothing:\varnothing:n_7)$-point.
(ix) The fixed point $\mathbf{x}_k^*$ is an $([\varnothing,\varnothing]:[\varnothing,\varnothing]:[1;\varnothing]:[\varnothing;\varnothing]|\varnothing:\varnothing:n_7)$-point.
(x) The fixed point $\mathbf{x}_k^*$ is an $([\varnothing,\varnothing]:[\varnothing,\varnothing]:[n_3;\kappa_3]:[\varnothing;\varnothing]|\varnothing:\varnothing:[n_7,l;\boldsymbol{\kappa}_7])$-point.
(xi) The fixed point $\mathbf{x}_k^*$ is an $([\varnothing,\varnothing]:[\varnothing,\varnothing]:[\varnothing;\varnothing]:[n_4;\kappa_4]|\varnothing:\varnothing:n_7)$-point.
(xii) The fixed point $\mathbf{x}_k^*$ is an $([\varnothing,\varnothing]:[\varnothing,\varnothing]:[\varnothing;\varnothing]:[n_4;\kappa_4]|\varnothing:\varnothing:[n_7,l;\boldsymbol{\kappa}_7])$-point.
(xiii) The fixed point $\mathbf{x}_k^*$ is an $([\varnothing,\varnothing]:[\varnothing,\varnothing]:[n_3;\kappa_3]:[n_4;\kappa_4]|\varnothing:\varnothing:n_7)$-point.
(xiv) The fixed point $\mathbf{x}_k^*$ is an $([\varnothing,\varnothing]:[\varnothing,\varnothing]:[n_3;\kappa_3]:[n_4;\kappa_4]|\varnothing:\varnothing:[n_7,l;\boldsymbol{\kappa}_7])$-point.

III. *Complex special cases*

(i) The fixed point $\mathbf{x}_k^*$ is an $([n_1^{\mathrm{m}},n_1^{\mathrm{o}}]:[n_2^{\mathrm{m}},n_2^{\mathrm{o}}]:[1;\varnothing]:[\varnothing;\varnothing]|n_5:n_6:n_7)$-point.
(ii) The fixed point $\mathbf{x}_k^*$ is an $([n_1^{\mathrm{m}},n_1^{\mathrm{o}}]:[n_2^{\mathrm{m}},n_2^{\mathrm{o}}]:[1;\varnothing]:[\varnothing;\varnothing]|n_5:n_6:[n_7,l;\boldsymbol{\kappa}_7])$-point.
(iii) The fixed point $\mathbf{x}_k^*$ is an $([n_1^{\mathrm{m}},n_1^{\mathrm{o}}]:[n_2^{\mathrm{m}},n_2^{\mathrm{o}}]:[\varnothing;\varnothing]:[1;\varnothing]|n_5:n_6:n_7)$-point.
(iv) The fixed point $\mathbf{x}_k^*$ is an $([n_1^{\mathrm{m}},n_1^{\mathrm{o}}]:[n_2^{\mathrm{m}},n_2^{\mathrm{o}}]:[\varnothing;\varnothing]:[1;\varnothing]|n_5:n_6:[n_7,l;\boldsymbol{\kappa}_7])$-point.
(v) The fixed point $\mathbf{x}_k^*$ is an $([n_1^{\mathrm{m}},n_1^{\mathrm{o}}]:[n_2^{\mathrm{m}},n_2^{\mathrm{o}}]:[n_3;\kappa_3]:[n_4;\kappa_4]|n_5:n_6:n_7)$-point.
(vi) The fixed point $\mathbf{x}_k^*$ is an $([n_1^{\mathrm{m}},n_1^{\mathrm{o}}]:[n_2^{\mathrm{m}},n_2^{\mathrm{o}}]:[n_3;\kappa_3]:[n_4;\kappa_4]|n_5:n_6:[n_7,l;\boldsymbol{\kappa}_7])$-point.

Definition 2.19 Consider a discrete, nonlinear dynamical system $\mathbf{x}_{k+1}=\mathbf{f}(\mathbf{x}_k,\mathbf{p})$ in Eq. (2.4) with a fixed point $\mathbf{x}_k^*$. The corresponding solution is given by $\mathbf{x}_{k+j}=\mathbf{f}(\mathbf{x}_{k+j-1},\mathbf{p})$ with $j\in\mathbb{Z}$. Suppose there is a neighborhood of the fixed point $\mathbf{x}_k^*$ (i.e., $U_k(\mathbf{x}_k^*)\subset\Omega$), and $\mathbf{f}(\mathbf{x}_k,\mathbf{p})$ is C^r $(r\geq 1)$-continuous in $U_k(\mathbf{x}_k^*)$ with Eq. (2.28). The linearized system is $\mathbf{y}_{k+j+1}=D\mathbf{f}(\mathbf{x}_k^*,\mathbf{p})\mathbf{y}_{k+j}$ $(\mathbf{y}_{k+j}=\mathbf{x}_{k+j}-\mathbf{x}_k^*)$ in $U_k(\mathbf{x}_k^*)$. The matrix $D\mathbf{f}(\mathbf{x}_k^*,\mathbf{p})$ possesses n eigenvalues λ_i $(i=1,2,\cdots,n)$. Set $N=\{1,2,\cdots,n\}$, $N_p=\{p_1,p_2,\cdots,p_{n_p}\}\cup\varnothing$ with $p_q\in N(q=1,2,\cdots,n_p,\ p=1,2,3,4)$ and $\Sigma_{p=1}^4 n_p=n$. $\cup_{p=1}^4 N_p=N$ and $N_p\cap N_l=\varnothing(1\neq p)$. $N_p=\varnothing$ if $n_p=0$. N_α

$= N_\alpha^{\rm m} \cup N_\alpha^{\rm o}$ ($\alpha = 1, 2$) and $N_\alpha^{\rm m} \cap N_\alpha^{\rm o} = \varnothing$ with $n_\alpha^{\rm m} + n_\alpha^{\rm o} = n_\alpha$ where superscripts "m" and "o" represent monotonic and oscillatory evolutions. The matrix $D\mathbf{f}(\mathbf{x}_k^*, \mathbf{p})$ possesses n_1-stable, n_2-unstable, n_3-invariant and n_4-flip real eigenvectors. An iterative response of $\mathbf{x}_{k+1} = \mathbf{f}(\mathbf{x}_k, \mathbf{p})$ is an $([n_1^{\rm m}, n_1^{\rm o}] : [n_2^{\rm m}, n_2^{\rm o}] : [n_3; \kappa_3] : [n_4; \kappa_4]|$ flow in the neighborhood of the fixed point $\mathbf{x}_k^*$. $\kappa_p \in \{\varnothing, m_p\}$ ($p = 3, 4$).

I. *Non-degenerate cases*

(i) The fixed point $\mathbf{x}_k^*$ is an $([n_1^{\rm m}, n_1^{\rm o}] : [n_2^{\rm m}, n_2^{\rm o}] : \varnothing : \varnothing|$ saddle.
(ii) The fixed point $\mathbf{x}_k^*$ is an $([n_1^{\rm m}, n_1^{\rm o}] : [\varnothing, \varnothing] : \varnothing : \varnothing|$-sink.
(iii) The fixed point $\mathbf{x}_k^*$ is an $([\varnothing, \varnothing] : [n_2^{\rm m}, n_2^{\rm o}] : \varnothing : \varnothing|$-source.

II. *Simple special cases*

(i) The fixed point $\mathbf{x}_k^*$ is an $([\varnothing, \varnothing] : [\varnothing, \varnothing] : [n; \varnothing] : \varnothing|$-invariant center (or static center).
(ii) The fixed point $\mathbf{x}_k^*$ is an $([\varnothing, \varnothing] : [\varnothing, \varnothing] : [n; m] : \varnothing|$-point.
(iii) The fixed point $\mathbf{x}_k^*$ is an $([\varnothing, \varnothing] : [\varnothing, \varnothing] : \varnothing : [n; \varnothing]|$-flip center.
(iv) The fixed point $\mathbf{x}_k^*$ is an $([\varnothing, \varnothing] : [\varnothing, \varnothing] : \varnothing : [n; m]|$-point.
(v) The fixed point $\mathbf{x}_k^*$ is an $([\varnothing, \varnothing] : [\varnothing, \varnothing] : [n_3; \kappa_3] : [n_4; \kappa_4]|$-point.
(vi) The fixed point $\mathbf{x}_k^*$ is an $([\varnothing, \varnothing] : [\varnothing, \varnothing] : [1; \varnothing] : [n_4; \kappa_4]|$-point.
(vii) The fixed point $\mathbf{x}_k^*$ is an $([\varnothing, \varnothing] : [\varnothing, \varnothing] : [n_3; \kappa_3] : [1; \varnothing]|$-point.
(viii) The fixed point $\mathbf{x}_k^*$ is an $([\varnothing, \varnothing] : [\varnothing, \varnothing] : [n_3; \kappa_3] : [\varnothing; \varnothing]|$-point.
(ix) The fixed point $\mathbf{x}_k^*$ is an $([\varnothing, \varnothing] : [\varnothing, \varnothing] : [\varnothing; \varnothing] : [n_4; \kappa_4]|$-point.

III. *Complex special cases*

(i) The fixed point $\mathbf{x}_k^*$ is an $([n_1^{\rm m}, n_1^{\rm o}] : [n_2^{\rm m}, n_2^{\rm o}] : [1; \varnothing] : [\varnothing; \varnothing]|$-point.
(ii) The fixed point $\mathbf{x}_k^*$ is an $([n_1^{\rm m}, n_1^{\rm o}] : [n_2^{\rm m}, n_2^{\rm o}] : [\varnothing; \varnothing] : [1; \varnothing]|$-point.
(iii) The fixed point $\mathbf{x}_k^*$ is an $([n_1^{\rm m}, n_1^{\rm o}] : [n_2^{\rm m}, n_2^{\rm o}] : [n_3; \kappa_3] : [n_4; \kappa_4]|$-point.

Definition 2.20 Consider a discrete, nonlinear dynamical system $\mathbf{x}_{k+1} = \mathbf{f}(\mathbf{x}_k, \mathbf{p}) \in \mathscr{R}^{2n}$ in Eq. (2.4) with a fixed point $\mathbf{x}_k^*$. The corresponding solution is given by $\mathbf{x}_{k+j} = \mathbf{f}(\mathbf{x}_{k+j-1}, \mathbf{p})$ with $j \in \mathbb{Z}$. Suppose there is a neighborhood of the fixed point $\mathbf{x}_k^*$ (i.e., $U_k(\mathbf{x}_k^*) \subset \Omega_\alpha$), and $\mathbf{f}(\mathbf{x}_k, \mathbf{p})$ is C^r ($r \geq 1$)-continuous in $U_k(\mathbf{x}_k^*)$ with Eq. (2.28). The linearized system is $\mathbf{y}_{k+j+1} = D\mathbf{f}(\mathbf{x}_k^*, \mathbf{p})\mathbf{y}_{k+j}$ ($\mathbf{y}_{k+j} = \mathbf{x}_{k+j} - \mathbf{x}_k^*$) in $U_k(\mathbf{x}_k^*)$. The matrix $D\mathbf{f}(\mathbf{x}_k^*, \mathbf{p})$ possesses n-pairs of eigenvalues λ_i ($i = 1, 2, \cdots, n$). Set $N = \{1, 2, \cdots, n\}$, $N_p = \{p_1, p_2, \cdots, p_{n_p}\} \cup \varnothing$ with $p_q \in N$ ($q = 1, 2, \cdots, n_p$, $p = 5, 6, 7$) and $\Sigma_{p=5}^7 n_p = n$. $\cup_{p=5}^7 N_p = N$ and $N_p \cap N_l = \varnothing (l \neq p)$. $N_p = \varnothing$ if $n_p = 0$. The matrix $D\mathbf{f}(\mathbf{x}_k^*, \mathbf{p})$ possesses n_5-stable, n_6-unstable and n_7-center pairs of complex eigenvectors. Without repeated complex eigenvalues of $|\lambda_k| = 1 (k \in N_7)$, an iterative response of $\mathbf{x}_{k+1} = \mathbf{f}(\mathbf{x}_k, \mathbf{p})$ is an $|n_5 : n_6 : n_7)$ flow. With repeated complex eigenvalues of $|\lambda_k| = 1$ ($k \in N_7$), an iterative response of $\mathbf{x}_{k+1} = \mathbf{f}(\mathbf{x}_k, \mathbf{p})$ is an $|n_5 : n_6 : [n_7, l; \kappa_7])$ flow in the neighborhood of the fixed point $\mathbf{x}_k^*$, where $\kappa_7 = (\kappa_{71}, \kappa_{72}, \cdots, \kappa_{7l})^{\rm T}$ with $\kappa_{7s} \in \{\varnothing, m_{7s}\} (s = 1, 2, \cdots, l)$.

I. *Non-degenerate cases*

(i) The fixed point $\mathbf{x}_k^*$ is an $|n_5 : n_6 : \varnothing)$ spiral hyperbolic point.

(ii) The fixed point $\mathbf{x}_k^*$ is an $|n:\varnothing:\varnothing)$ spiral sink.
(iii) The fixed point $\mathbf{x}_k^*$ is an $|\varnothing:n:\varnothing)$ spiral source.
(iv) The fixed point $\mathbf{x}_k^*$ is an $|\varnothing:\varnothing:n)$-circular center.
(v) The fixed point $\mathbf{x}_k^*$ is an $|n_5:\varnothing:n_7)$-point.
(vi) The fixed point $\mathbf{x}_k^*$ is an $|\varnothing:n_6:n_7)$-point.
(vii) The fixed point $\mathbf{x}_k^*$ is an $|n_5:n_6:n_7)$-point.

II. *Special cases*

(i) The fixed point $\mathbf{x}_k^*$ is an $|\varnothing:\varnothing:[n,l;\varnothing])$-circular center.
(ii) The fixed point $\mathbf{x}_k^*$ is an $|\varnothing:\varnothing:[n,l;\boldsymbol{\kappa}_7])$-point.
(iii) The fixed point $\mathbf{x}_k^*$ is an $|n_5:\varnothing:[n_7,l;\boldsymbol{\kappa}_7])$-point.
(iv) The fixed point $\mathbf{x}_k^*$ is an $|\varnothing:n_6:[n_7,l;\boldsymbol{\kappa}_7])$-point.
(v) The fixed point $\mathbf{x}_k^*$ is an $|n_5:n_6:[n_7,l;\boldsymbol{\kappa}_7])$-point.

2.3 Bifurcation and Stability Switching

To understand the qualitative changes of dynamical behaviors of discrete systems with parameters in the neighborhood of fixed points, the bifurcation theory for fixed points of nonlinear dynamical system in Eq. (2.4) will be investigated.

Definition 2.21 Consider a discrete, nonlinear dynamical system $\mathbf{x}_{k+1}=\mathbf{f}(\mathbf{x}_k,\mathbf{p})$ in Eq. (2.4) with a fixed point $\mathbf{x}_k^*$. The corresponding solution is given by $\mathbf{x}_{k+j}=\mathbf{f}(\mathbf{x}_{k+j-1},\mathbf{p})$ with $j\in\mathbb{Z}$. Suppose there is a neighborhood of the fixed point $\mathbf{x}_k^*$ (i.e., $U_k(\mathbf{x}_k^*)\subset\Omega$), and $\mathbf{f}(\mathbf{x}_k,\mathbf{p})$ is C^r $(r\geq 1)$-continuous in $U_k(\mathbf{x}_k^*)$ with Eq. (2.28). The linearized system is $\mathbf{y}_{k+j+1}=D\mathbf{f}(\mathbf{x}_k^*,\mathbf{p})\mathbf{y}_{k+j}$ $(\mathbf{y}_{k+j}=\mathbf{x}_{k+j}-\mathbf{x}_k^*)$ in $U_k(\mathbf{x}_k^*)$. The matrix $D\mathbf{f}(\mathbf{x}_k^*,\mathbf{p})$ possesses eigenvalues λ_i $(i=1,2,\cdots,n)$. Set $N=\{1,2,\cdots,m,m+1,\cdots,(n-m)/2\}$, $N_p=\{p_1,p_2,\cdots,p_{n_p}\}\cup\varnothing$ with $p_q\in N$ $(q=1,2,\cdots,n_p,p=1,2,\cdots,7)$, $\Sigma_{p=1}^4 n_p=m$ and $2\Sigma_{p=5}^7 n_p=n-m$. $\cup_{p=1}^7 N_p=N$ and $N_p\cap N_l=\varnothing(l\neq p)$.$N_p=\varnothing$ if $n_p=0$. $N_\alpha=N_\alpha^{\mathrm{m}}\cup N_\alpha^{\mathrm{o}}$ $(\alpha=1,2)$ and $N_\alpha^{\mathrm{m}}\cap N_\alpha^{\mathrm{o}}=\varnothing$ with $n_\alpha^{\mathrm{m}}+n_\alpha^{\mathrm{o}}=n_\alpha$, where superscripts "m" and "o" represent monotonic and oscillatory evolutions. The matrix $D\mathbf{f}(\mathbf{x}_k^*,\mathbf{p})$ possesses n_1-stable, n_2-unstable, n_3-invariant and n_4-flip real eigenvectors plus n_5-stable, n_6-unstable and n_7-center pairs of complex eigenvectors. Without repeated complex eigenvalues of $|\lambda_k|=1(k\in N_3\cup N_4\cup N_7)$, an iterative response of $\mathbf{x}_{k+1}=\mathbf{f}(\mathbf{x}_k,\mathbf{p})$ is an $([n_1^{\mathrm{m}},n_1^{\mathrm{o}}]:[n_2^{\mathrm{m}},n_2^{\mathrm{o}}]:[n_3;\kappa_3]:[n_4;\kappa_4]||\ n_5:n_6:n_7)$ flow in the neighborhood of the fixed point x_k^*. With repeated complex eigenvalues of $|\lambda_k|=1$ $(k\in N_3\cup N_4\cup N_7)$, an iterative response of $\mathbf{x}_{k+1}=\mathbf{f}(\mathbf{x}_k,\mathbf{p})$ is an $([n_1^{\mathrm{m}},n_1^{\mathrm{o}}]:[n_2^{\mathrm{m}},n_2^{\mathrm{o}}]:[n_3;\kappa_3]:[n_4;\kappa_4]||$ $n_5:n_6:[n_7,l;\boldsymbol{\kappa}_7])$ flow in the neighborhood of the fixed point $\mathbf{x}_k^*$, where $\kappa_p\in\{\varnothing,m_p\}(p=3,4)$, $\boldsymbol{\kappa}_7=(\kappa_{71},\kappa_{72},\cdot,\kappa_{7l})^{\mathrm{T}}$ with $\kappa_{7s}\in\{\varnothing,m_{7s}\}(s=1,2,\cdots,l)$.

I. *Simple switching and bifurcation*

(i) An $([n_1^{\mathrm{m}},n_1^{\mathrm{o}}]:[n_2^{\mathrm{m}},n_2^{\mathrm{o}}]:1:\varnothing|n_5:n_6:\varnothing)$ state of the fixed point $(\mathbf{x}_{k0}^*,\mathbf{p}_0)$ is a switching of the $([n_1^{\mathrm{m}},n_1^{\mathrm{o}}+1]:[n_2^{\mathrm{m}},n_2^{\mathrm{o}}]:\varnothing:\varnothing|n_5:n_6:\varnothing)$ spiral saddle and $([n_1^{\rm m}, n_1^{\rm o}] : [n_2^{\rm m}+1, n_2^{\rm o}] : \varnothing : \varnothing | n_5 : n_6 : \varnothing)$ spiral saddle for the fixed point $(\mathbf{x}_k^*, \mathbf{p})$.

(ii) An $([n_1^{\rm m}, n_1^{\rm o}] : [n_2^{\rm m}, n_2^{\rm o}] : \varnothing : 1 | n_5 : n_6 : \varnothing)$ state of the fixed point $(\mathbf{x}_{k0}^*, \mathbf{p}_0)$ is a switching of the $([n_1^{\rm m}, n_1^{\rm o}+1] : [n_2^{\rm m}, n_2^{\rm o}] : \varnothing : \varnothing | n_5 : n_6 : \varnothing)$ spiral saddle and $([n_1^{\rm m}, n_1^{\rm o}] : [n_2^{\rm m}, n_2^{\rm o}+1] : \varnothing : \varnothing | n_5 : n_6 : \varnothing)$ spiral saddle for the fixed point $(\mathbf{x}_k^*, \mathbf{p})$.

(iii) An $([n_1^{\rm m}, n_1^{\rm o}] : [\varnothing, \varnothing] : 1 : \varnothing | n_5 : \varnothing : \varnothing)$ state of the fixed point $(\mathbf{x}_{k0}^*, \mathbf{p}_0)$ is a stable saddle-node bifurcation of the $([n_1^{\rm m}+1, n_1^{\rm o}] : [\varnothing, \varnothing] : \varnothing : \varnothing | n_5 : \varnothing : \varnothing)$ spiral sink and $([n_1^{\rm m}, n_1^{\rm o}] : [1, \varnothing] : \varnothing : \varnothing | n_5 : \varnothing : \varnothing)$ spiral saddle for the fixed point $(\mathbf{x}_k^*, \mathbf{p})$.

(iv) An $([n_1^{\rm m}, n_1^{\rm o}] : [\varnothing, \varnothing] : \varnothing : 1 | n_5 : \varnothing : \varnothing)$ state of the fixed point $(\mathbf{x}_{k0}^*, \mathbf{p}_0)$ is a stable period-doubling bifurcation of the $([n_1^{\rm m}, n_1^{\rm o}+1] : [\varnothing, \varnothing] : \varnothing : \varnothing | n_5 : \varnothing : \varnothing)$ sink and $([n_1^{\rm m}, n_1^{\rm o}] : [\varnothing, 1] : \varnothing : \varnothing | n_5 : \varnothing : \varnothing)$ spiral saddle for the fixed point $(\mathbf{x}_k^*, \mathbf{p})$.

(v) An $([\varnothing, \varnothing] : [n_2^{\rm m}, n_2^{\rm o}] : 1 : \varnothing | \varnothing : n_6 : \varnothing)$ state of the fixed point $(\mathbf{x}_{k0}^*, \mathbf{p}_0)$ is an unstable saddle-node bifurcation of the $([\varnothing, \varnothing] : [n_2^{\rm m}+1, n_2^{\rm o}] : \varnothing : \varnothing | \varnothing : n_6 : \varnothing)$ spiral source and $([1, \varnothing] : [n_2^{\rm m}, n_2^{\rm o}] : \varnothing : \varnothing | \varnothing : n_6 : \varnothing)$ spiral saddle for the fixed point $(\mathbf{x}_k^*, \mathbf{p})$.

(vi) An $([\varnothing, \varnothing] : [n_2^{\rm m}, n_2^{\rm o}] : \varnothing : 1 | \varnothing : n_6 : \varnothing)$ state of the fixed point $(\mathbf{x}_{k0}^*, \mathbf{p}_0)$ is an unstable period-doubling bifurcation of the $([\varnothing, \varnothing] : [n_2^{\rm m}, n_2^{\rm o}+1] : \varnothing : \varnothing | \varnothing : n_6 : \varnothing)$ spiral source and $([\varnothing, 1] : [n_2^{\rm m}, n_2^{\rm o}] : \varnothing : \varnothing | \varnothing : n_6 : \varnothing)$ spiral saddle for the fixed point $(\mathbf{x}_k^*, \mathbf{p})$.

(vii) An $([n_1^{\rm m}, n_1^{\rm o}] : [n_2^{\rm m}, n_2^{\rm o}] : \varnothing : \varnothing | n_5 : n_6 : 1)$ state of the fixed point $(\mathbf{x}_{k0}^*, \mathbf{p}_0)$ is a switching of the $([n_1^{\rm m}, n_1^{\rm o}] : [n_2^{\rm m}, n_2^{\rm o}] : \varnothing : \varnothing | n_5+1 : n_6 : \varnothing)$ spiral saddle and $([n_1^{\rm m}, n_1^{\rm o}] : [n_2^{\rm m}, n_2^{\rm o}] : \varnothing : \varnothing | n_5 : n_6+1 : \varnothing)$ saddle for the fixed point $(\mathbf{x}_k^*, \mathbf{p})$.

(viii) An $([n_1^{\rm m}, n_1^{\rm o}] : [\varnothing, \varnothing] : \varnothing : \varnothing | n_5 : \varnothing : 1)$ state of the fixed point $(\mathbf{x}_{k0}^*, \mathbf{p}_0)$ is a Neimark bifurcation of the $([n_1^{\rm m}, n_1^{\rm o}] : [\varnothing, \varnothing] : \varnothing : \varnothing | n_5+1 : \varnothing : \varnothing)$ spiral sink and $([n_1^{\rm m}, n_1^{\rm o}] : [\varnothing, \varnothing] : \varnothing : \varnothing | n_5 : 1 : \varnothing)$ spiral saddle for the fixed point $(\mathbf{x}_k^*, \mathbf{p})$.

(ix) An $([\varnothing, \varnothing] : [n_2^{\rm m}, n_2^{\rm o}] : \varnothing : \varnothing | \varnothing : n_6 : 1)$ state of the fixed point $(\mathbf{x}_{k0}^*, \mathbf{p}_0)$ is an unstable Neimark bifurcation of the $([\varnothing, \varnothing] : [n_2^{\rm m}, n_2^{\rm o}] : \varnothing : \varnothing | \varnothing : n_6+1 : \varnothing)$ spiral source and $([\varnothing, \varnothing] : [n_2^{\rm m}, n_2^{\rm o}] : \varnothing : \varnothing | 1 : n_6 : \varnothing)$ spiral saddle for the fixed point $(\mathbf{x}_k^*, \mathbf{p})$.

(x) An $([n_1^{\rm m}, n_1^{\rm o}] : [n_2^{\rm m}, n_2^{\rm o}] : 1 : \varnothing | n_5 : n_6 : n_7)$ state of the fixed point $(\mathbf{x}_{k0}^*, \mathbf{p}_0)$ is a switching of the $([n_1^{\rm m}+1, n_1^{\rm o}] : [n_2^{\rm m}, n_2^{\rm o}] : \varnothing : \varnothing | n_5 : n_6 : n_7)$ state and $([n_1^{\rm m}, n_1^{\rm o}] : [n_2^{\rm m}+1, n_2^{\rm o}] : \varnothing : \varnothing | n_5 : n_6 : n_7)$ state for the fixed point $(\mathbf{x}_k^*, \mathbf{p})$.

(xi) An $([n_1^{\rm m}, n_1^{\rm o}] : [n_2^{\rm m}, n_2^{\rm o}] : \varnothing : 1 | n_5 : n_6 : n_7)$ state of the fixed point $(\mathbf{x}_{k0}^*, \mathbf{p}_0)$ is a switching of the $([n_1^{\rm m}, n_1^{\rm o}+1] : [n_2^{\rm m}, n_2^{\rm o}] : \varnothing : \varnothing | n_5 : n_6 : n_7)$ state and $([n_1^{\rm m}, n_1^{\rm o}] : [n_2^{\rm m}, n_2^{\rm o}+1] : \varnothing : \varnothing | n_5 : n_6 : n_7)$ state for the fixed point $(\mathbf{x}_k^*, \mathbf{p})$.

(xii) An $([n_1^{\rm m}, n_1^{\rm o}] : [n_2^{\rm m}, n_2^{\rm o}] : 1 : \varnothing | n_5 : n_6 : [n_7, l; \varnothing])$ state of the fixed point $(\mathbf{x}_{k0}^*, \mathbf{p}_0)$ is a switching of the $([n_1^{\rm m}+1, n_1^{\rm o}] : [n_2^{\rm m}, n_2^{\rm o}] : \varnothing : \varnothing | n_5 : n_6 : [n_7, l; \kappa_7])$ state and $([n_1^{\rm m}, n_1^{\rm o}] : [n_2^{\rm m}+1, n_2^{\rm o}] : \varnothing : \varnothing | n_5 : n_6 : [n_7, l; \kappa_7])$ state for the fixed point $(\mathbf{x}_k^*, \mathbf{p})$.

(xiii) An $([n_1^{\rm m}, n_1^{\rm o}] : [n_2^{\rm m}, n_2^{\rm o}] : \varnothing : 1|n_5 : n_6 : [n_7, l; \boldsymbol{\kappa}_7])$ state of the fixed point $(\mathbf{x}_{k0}^*, \mathbf{p}_0)$ is a switching of the $([n_1^{\rm m}, n_1^{\rm o}+1] : [n_2^{\rm m}, n_2^{\rm o}] : \varnothing : \varnothing|n_5 : n_6 : [n_7, l; \boldsymbol{\kappa}_7])$ state and $([n_1^{\rm m}, n_1^{\rm o}] : [n_2^{\rm m}, n_2^{\rm o}+1] : \varnothing : \varnothing|n_5 : n_6 : [n_7, l; \boldsymbol{\kappa}_7])$ state for the fixed point $(\mathbf{x}_k^*, \mathbf{p})$.

(xiv) An $([n_1^{\rm m}, n_1^{\rm o}] : [n_2^{\rm m}, n_2^{\rm o}] : \varnothing : \varnothing|n_5 : n_6 : n_7+1)$ state of the fixed point $(\mathbf{x}_{k0}^*, \mathbf{p}_0)$ is a switching of its $([n_1^{\rm m}, n_1^{\rm o}] : [n_2^{\rm m}, n_2^{\rm o}] : \varnothing : \varnothing|n_5+1 : n_6 : n_7)$ state and $([n_1^{\rm m}, n_1^{\rm o}] : [n_2^{\rm m}, n_2^{\rm o}] : \varnothing : \varnothing|n_5 : n_6+1 : n_7)$ state for the fixed point $(\mathbf{x}_k^*, \mathbf{p})$.

(xv) An $([n_1^{\rm m}, n_1^{\rm o}] : [n_2^{\rm m}, n_2^{\rm o}] : [n_3; \kappa_3] : [n_4; \kappa_4]|n_5 : n_6 : n_7+1)$ state of the fixed point $(\mathbf{x}_{k0}^*, \mathbf{p}_0)$ is a switching of the $([n_1^{\rm m}, n_1^{\rm o}] : [n_2^{\rm m}, n_2^{\rm o}] : [n_3; \kappa_3] : [n_4; \kappa_4]|\ n_5+1 : n_6 : n_7)$ state and $([n_1^{\rm m}, n_1^{\rm o}] : [n_2^{\rm m}, n_2^{\rm o}] : [n_3; \kappa_3] : [n_4; \kappa_4]|n_5 : n_6+1 : n_7)$ state for the fixed point $(\mathbf{x}_k^*, \mathbf{p})$.

II. *Complex switching*

(i) An $([n_1^{\rm m}, n_1^{\rm o}] : [n_2^{\rm m}, n_2^{\rm o}] : [n_3; \kappa_3] : \varnothing|n_5 : n_6 : n_7)$ state of the fixed point $(\mathbf{x}_{k0}^*, \mathbf{p}_0)$ is a switching of the $([n_1^{\rm m}+n_3, n_1^{\rm o}] : [n_2^{\rm m}, n_2^{\rm o}] : \varnothing : \varnothing|n_5 : n_6 : n_7)$ state and $([n_1^{\rm m}, n_1^{\rm o}] : [n_2^{\rm m}+n_3, n_2^{\rm o}] : \varnothing : \varnothing|n_5 : n_6 : n_7)$ state for the fixed point $(\mathbf{x}_k^*, \mathbf{p})$.

(ii) An $([n_1^{\rm m}, n_1^{\rm o}] : [n_2^{\rm m}, n_2^{\rm o}] : \varnothing : [n_4; \kappa_4]|n_5 : n_6 : n_7)$ state of the fixed point $(\mathbf{x}_{k0}^*, \mathbf{p}_0)$ is a switching of the $([n_1^{\rm m}, n_1^{\rm o}+n_4] : [n_2^{\rm m}, n_2^{\rm o}] : \varnothing : \varnothing|n_5 : n_6 : n_7)$ state and $([n_1^{\rm m}, n_1^{\rm o}] : [n_2^{\rm m}, n_2^{\rm o}+n_4] : \varnothing : \varnothing|n_5 : n_6 : n_7)$ state for the fixed point $(\mathbf{x}_k^*, \mathbf{p})$.

(iii) An $([n_1^{\rm m}, n_1^{\rm o}] : [n_2^{\rm m}, n_2^{\rm o}] : [n_3+k_3; \kappa_3] : \varnothing|n_5 : n_6 : n_7)$ state of the fixed point $(\mathbf{x}_{k0}^*, \mathbf{p}_0)$ is a switching of the $([n_1^{\rm m}+k_3, n_1^{\rm o}] : [n_2^{\rm m}, n_2^{\rm o}] : [n_3; \kappa_3'] : \varnothing|n_5 : n_6 : n_7)$ state and $([n_1^{\rm m}, n_1^{\rm o}] : [n_2^{\rm m}+k_3, n_2^{\rm o}] : [n_3; \kappa_3'] : \varnothing|n_5 : n_6 : n_7)$ state for the fixed point $(\mathbf{x}_k^*, \mathbf{p})$.

(iv) An $([n_1^{\rm m}, n_1^{\rm o}] : [n_2^{\rm m}, n_2^{\rm o}] : \varnothing : [n_4+k_4; \kappa_4]|n_5 : n_6 : n_7)$ state of the fixed point $(\mathbf{x}_{k0}^*, \mathbf{p}_0)$ is a switching of the $([n_1^{\rm m}, n_1^{\rm o}+k_4] : [n_2^{\rm m}, n_2^{\rm o}] : \varnothing : [n_4; \kappa_4']|n_5 : n_6 : n_7)$ state and $([n_1^{\rm m}, n_1^{\rm o}] : [n_2^{\rm m}, n_2^{\rm o}+k_4] : \varnothing : [n_4; \kappa_4']|n_5 : n_6 : n_7)$ state for the fixed point $(\mathbf{x}_k^*, \mathbf{p})$.

(v) An $([n_1^{\rm m}, n_1^{\rm o}] : [n_2^{\rm m}, n_2^{\rm o}] : [n_3+k_3; \kappa_3] : [n_4+k_4; \kappa_4]|n_5 : n_6 : n_7)$ state of the fixed point $(\mathbf{x}_{k0}^*, \mathbf{p}_0)$ is a switching of the $([n_1^{\rm m}+k_3, n_1^{\rm o}+k_4] : [n_2^{\rm m}, n_2^{\rm o}] : [n_3; \kappa_3'] : [n_4; \kappa_4']|n_5 : n_6 : n_7)$ state and $([n_1^{\rm m}, n_1^{\rm o}] : [n_2^{\rm m}+k_3, n_2^{\rm o}+k_4] : [n_3; \kappa_3'] : [n_4; \kappa_4']|n_5 : n_6 : n_7)$ state for the fixed point $(\mathbf{x}_k^*, \mathbf{p})$.

(vi) An $([n_1^{\rm m}, n_1^{\rm o}] : [n_2^{\rm m}, n_2^{\rm o}] : [n_3+k_3; \kappa_3] : \varnothing|n_5 : n_6 : [n_7, l; \boldsymbol{\kappa}_7])$ state of the fixed point $(\mathbf{x}_{k0}^*, \mathbf{p}_0)$ is a switching of the $([n_1^{\rm m}+k_3, n_1^{\rm o}] : [n_2^{\rm m}, n_2^{\rm o}] : [n_3; \kappa_3'] : \varnothing|n_5 : n_6 : [n_7, l; \boldsymbol{\kappa}_7])$ state and $([n_1^{\rm m}, n_1^{\rm o}] : [n_2^{\rm m}+k_3, n_2^{\rm o}] : [n_3, \kappa_3'] : \varnothing|n_5 : n_6 : [n_7, l; \boldsymbol{\kappa}_7])$ state for the fixed point $(\mathbf{x}_k^*, \mathbf{p})$.

(vii) An $([n_1^{\rm m}, n_1^{\rm o}] : [n_2^{\rm m}, n_2^{\rm o}] : \varnothing : [n_4+k_4; \kappa_4]|n_5 : n_6 : [n_7, l; \boldsymbol{\kappa}_7])$ state of the fixed point $(\mathbf{x}_{k0}^*, \mathbf{p}_0)$ is a switching of the $([n_1^{\rm m}, n_1^{\rm o}+k_4] : [n_2^{\rm m}, n_2^{\rm o}] : \varnothing : [n_4; \kappa_4']|n_5 : n_6 : [n_7, l; \boldsymbol{\kappa}_7])$ state and $([n_1^{\rm m}, n_1^{\rm o}] : [n_2^{\rm m}, n_2^{\rm o}+k_4] : \varnothing : [n_4; \kappa_4']|n_5 : n_6 : [n_7, l; \boldsymbol{\kappa}_7])$ state for the fixed point $(\mathbf{x}_k^*, \mathbf{p})$.

(viii) An $([n_1^{\rm m}, n_1^{\rm o}] : [n_2^{\rm m}, n_2^{\rm o}] : [n_3+k_3; \kappa_3] : [n_4+k_4; \kappa_4]|n_5 : n_6 : [n_7, l; \boldsymbol{\kappa}_7])$ state of the fixed point $(\mathbf{x}_{k0}^*, \mathbf{p}_0)$ is a switching of the $([n_1^{\rm m}+k_3, n_1^{\rm o}+k_4] : [n_2^{\rm m}, n_2^{\rm o}] :$

$[n_3;\kappa_3'] : [n_4;\kappa_4'] \| n_5 : n_6 : [n_7, l;\kappa_7])$ state and $([n_1^{\mathrm{m}}, n_1^{\mathrm{o}}] : [n_2^{\mathrm{m}}+k_3, n_2^{\mathrm{o}}+k_4] : [n_3;\kappa_3'] : [n_4;\kappa_4'] \| n_5 : n_6 : [n_7, l;\kappa_7])$ state for the fixed point $(\mathbf{x}_k^*, \mathbf{p})$.

(ix) An $([n_1^{\mathrm{m}}, n_1^{\mathrm{o}}] : [n_2^{\mathrm{m}}, n_2^{\mathrm{o}}] : [n_3;\kappa_3] : [n_4;\kappa_4] \| n_5 : n_6 : [n_7+k_7, l_1;\kappa_7])$ state of the fixed point $(\mathbf{x}_{k0}^*, \mathbf{p}_0)$ is a switching of the $([n_1^{\mathrm{m}}, n_1^{\mathrm{o}}] : [n_2^{\mathrm{m}}, n_2^{\mathrm{o}}] : [n_3;\kappa_3] : [n_4;\kappa_4] \| n_5+k_7 : n_6 : [n_7, l_2;\kappa_7'])$ state and $([n_1^{\mathrm{m}}, n_1^{\mathrm{o}}] : n_2^{\mathrm{o}}+k_4] : [n_3;\kappa_3] : [n_4;\kappa_4] \| n_5 : n_6+k_7 : [n_7, l_3;\kappa_7'])$ state for the fixed point $(\mathbf{x}_k^*, \mathbf{p})$.

Definition 2.22 Consider a discrete, nonlinear dynamical system $\mathbf{x}_{k+1} = \mathbf{f}(\mathbf{x}_k, \mathbf{p})$ in Eq. (2.4) with a fixed point $\mathbf{x}_k^*$. The corresponding solution is given by $\mathbf{x}_{k+j} = \mathbf{f}(\mathbf{x}_{k+j-1}, \mathbf{p})$ with $j \in \mathbb{Z}$. Suppose there is a neighborhood of the fixed point $\mathbf{x}_k^*$ (i.e., $U_k(\mathbf{x}_k^*) \subset \Omega$), and $\mathbf{f}(\mathbf{x}_k, \mathbf{p})$ is C^r $(r \geq 1)$-continuous in $U_k(\mathbf{x}_k^*)$ with Eq. (2.28). The linearized system is $\mathbf{y}_{k+j+1} = D\mathbf{f}(\mathbf{x}_k^*, \mathbf{p})\mathbf{y}_{k+j}$ $(\mathbf{y}_{k+j} = \mathbf{x}_{k+j} - \mathbf{x}_k^*)$ in $U_k(\mathbf{x}_k^*)$. The matrix $D\mathbf{f}(\mathbf{x}_k^*, \mathbf{p})$ possesses n eigenvalues λ_i $(i = 1, 2, \cdots, n)$. Set $N = \{1, 2, \cdots, n\}$, $N_p = \{p_1, p_2, \cdots, p_{n_p}\} \cup \varnothing$ with $p_q \in N$ $(q = 1, 2, \cdots, n_p, p = 1, 2, 3, 4)$ and $\Sigma_{p=1}^4 n_p = n$. $\cup_{p=1}^4 N_p = N$ and $N_p \cap N_l = \varnothing (l \neq p)$. $N_p = \varnothing$ if $n_p = 0$. $N_\alpha = N_\alpha^{\mathrm{m}} \cup N_\alpha^{\mathrm{o}}$ $(\alpha = 1, 2)$ and $N_\alpha^{\mathrm{m}} \cap N_\alpha^{\mathrm{o}} = \varnothing$ with $n_\alpha^{\mathrm{m}} + n_\alpha^{\mathrm{o}} = n_\alpha$ where superscripts "m" and "o" represent monotonic and oscillatory evolutions. The matrix $D\mathbf{f}(\mathbf{x}_k^*, \mathbf{p})$ possesses n_1-stable, n_2-unstable, n_3-invariant and n_4-flip real eigenvectors. An iterative response of $\mathbf{x}_{k+1} = \mathbf{f}(\mathbf{x}_k, \mathbf{p})$ is an $([n_1^{\mathrm{m}}, n_1^{\mathrm{o}}] : [n_2^{\mathrm{m}}, n_2^{\mathrm{o}}] : [n_3;\kappa_3] : [n_4;\kappa_4] \|$ flow in the neighborhood of the fixed point $\mathbf{x}_k^*$. $\kappa_p \in \{\varnothing, m_p\}$ $(p = 3, 4)$.

I. *Simple critical cases*

(i) An $([n_1^{\mathrm{m}}, n_1^{\mathrm{o}}] : [n_2^{\mathrm{m}}, n_2^{\mathrm{o}}] : 1 : \varnothing|$ state of the fixed point $(\mathbf{x}_{k0}^*, \mathbf{p}_0)$ is a switching of the $([n_1^{\mathrm{m}}+1, n_1^{\mathrm{o}}] : [n_2^{\mathrm{m}}, n_2^{\mathrm{o}}] : \varnothing : \varnothing|$ saddle and $([n_1^{\mathrm{m}}, n_1^{\mathrm{o}}] : [n_2^{\mathrm{m}}+1, n_2^{\mathrm{o}}] : \varnothing : \varnothing|$ saddle for the fixed point $(\mathbf{x}_k^*, \mathbf{p})$.

(ii) An $([n_1^{\mathrm{m}}, n_1^{\mathrm{o}}] : [n_2^{\mathrm{m}}, n_2^{\mathrm{o}}] : \varnothing : 1|$ state of the fixed point $(\mathbf{x}_{k0}^*, \mathbf{p}_0)$ is a switching of the $([n_1^{\mathrm{m}}, n_1^{\mathrm{o}}+1] : [n_2^{\mathrm{m}}, n_2^{\mathrm{o}}] : \varnothing : \varnothing|$ saddle and $([n_1^{\mathrm{m}}, n_1^{\mathrm{o}}] : [n_2^{\mathrm{m}}, n_2^{\mathrm{o}}+1] : \varnothing : \varnothing|$ saddle for the fixed point $(\mathbf{x}_k^*, \mathbf{p})$.

(iii) An $([n_1^{\mathrm{m}}, n_1^{\mathrm{o}}] : [\varnothing, \varnothing] : 1 : \varnothing|$ state of the fixed point $(\mathbf{x}_{k0}^*, \mathbf{p}_0)$ is a stable saddle-node bifurcation of the $([n_1^{\mathrm{m}}+1, n_1^{\mathrm{o}}] : [\varnothing, \varnothing] : \varnothing : \varnothing|$ sink and $([n_1^{\mathrm{m}}, n_1^{\mathrm{o}}] : [1, \varnothing] : \varnothing : \varnothing|$ saddle for the fixed point $(\mathbf{x}_k^*, \mathbf{p})$.

(iv) An $([n_1^{\mathrm{m}}, n_1^{\mathrm{o}}] : [\varnothing, \varnothing] : \varnothing : 1|$ state of the fixed point $(\mathbf{x}_{k0}^*, \mathbf{p}_0)$ is a stable period-doubling bifurcation of the $([n_1^{\mathrm{m}}, n_1^{\mathrm{o}}+1] : [\varnothing, \varnothing] : \varnothing : \varnothing|$ sink and $([n_1^{\mathrm{m}}, n_1^{\mathrm{o}}] : [\varnothing, 1] : \varnothing : \varnothing|$ saddle for the fixed point $(\mathbf{x}_k^*, \mathbf{p})$.

(v) An $([\varnothing, \varnothing] : [n_2^{\mathrm{m}}, n_2^{\mathrm{o}}] : 1 : \varnothing|$ state of the fixed point $(\mathbf{x}_{k0}^*, \mathbf{p}_0)$ is an unstable saddle-node bifurcation of the $([\varnothing, \varnothing] : [n_2^{\mathrm{m}}+1, n_2^{\mathrm{o}}] : \varnothing : \varnothing|$ source and $([1, \varnothing] : [n_2^{\mathrm{m}}, n_2^{\mathrm{o}}] : \varnothing : \varnothing|$ saddle for the fixed point $(\mathbf{x}_k^*, \mathbf{p})$.

(vi) An $([\varnothing, \varnothing] : [n_2^{\mathrm{m}}, n_2^{\mathrm{o}}] : \varnothing : 1|$ state of the fixed point $(\mathbf{x}_{k0}^*, \mathbf{p}_0)$ is an unstable period-doubling bifurcation of the $([\varnothing, \varnothing] : [n_2^{\mathrm{m}}, n_2^{\mathrm{o}}+1] : \varnothing : \varnothing|$ source and $([\varnothing, 1] : [n_2^{\mathrm{m}}, n_2^{\mathrm{o}}] : \varnothing : \varnothing|$ saddle for the fixed point $(\mathbf{x}_k^*, \mathbf{p})$.

(vii) An $([n_1^{\mathrm{m}}, n_1^{\mathrm{o}}] : [n_2^{\mathrm{m}}, n_2^{\mathrm{o}}] : 1 : \varnothing|$ state of the fixed point $(\mathbf{x}_{k0}^*, \mathbf{p}_0)$ is a switching of the $([n_1^{\mathrm{m}}+1, n_1^{\mathrm{o}}] : [n_2^{\mathrm{m}}, n_2^{\mathrm{o}}] : \varnothing : \varnothing|$ saddle and $([n_1^{\mathrm{m}}, n_1^{\mathrm{o}}] : [n_2^{\mathrm{m}}+1, n_2^{\mathrm{o}}] : \varnothing : \varnothing|$ saddle for the fixed point $(\mathbf{x}_k^*, \mathbf{p})$.

(viii) An $([n_1^{\mathrm{m}}, n_1^{\mathrm{o}}] : [n_2^{\mathrm{m}}, n_2^{\mathrm{o}}] : \varnothing : 1|$ state of the fixed point $(\mathbf{x}_{k0}^*, \mathbf{p}_0)$ is a switching of the $([n_1^{\mathrm{m}}, n_1^{\mathrm{o}}+1] : [n_2^{\mathrm{m}}, n_2^{\mathrm{o}}] : \varnothing : \varnothing|$ saddle and $([n_1^{\mathrm{m}}, n_1^{\mathrm{o}}] : [n_2^{\mathrm{m}}, n_2^{\mathrm{o}}+1] : \varnothing : \varnothing|$ saddle for the fixed point $(\mathbf{x}_k^*, \mathbf{p})$.

II. *Complex switching*

(i) An $([n_1^{\mathrm{m}}, n_1^{\mathrm{o}}] : [n_2^{\mathrm{m}}, n_2^{\mathrm{o}}] : [n_3; \kappa_3] : \varnothing|$ state of the fixed point $(\mathbf{x}_{k0}^*, \mathbf{p}_0)$ is a switching of the $([n_1^{\mathrm{m}}+n_3, n_1^{\mathrm{o}}] : [n_2^{\mathrm{m}}, n_2^{\mathrm{o}}] : \varnothing : \varnothing|$ saddle and $([n_1^{\mathrm{m}}, n_1^{\mathrm{o}}] : [n_2^{\mathrm{m}}+n_3, n_2^{\mathrm{o}}] : \varnothing : \varnothing|$ saddle for the fixed point $(\mathbf{x}_k^*, \mathbf{p})$.
(ii) An $([n_1^{\mathrm{m}}, n_1^{\mathrm{o}}] : [n_2^{\mathrm{m}}, n_2^{\mathrm{o}}] : \varnothing : [n_4; \kappa_4]|$ state of the fixed point $(\mathbf{x}_{k0}^*, \mathbf{p}_0)$ is a switching of the $([n_1^{\mathrm{m}}, n_1^{\mathrm{o}}+n_4] : [n_2^{\mathrm{m}}, n_2^{\mathrm{o}}] : \varnothing : \varnothing|$ saddle and $([n_1^{\mathrm{m}}, n_1^{\mathrm{o}}] : [n_2^{\mathrm{m}}, \ n_2^{\mathrm{o}}+n_4] : \varnothing : \varnothing|$ saddle for the fixed point $(\mathbf{x}_k^*, \mathbf{p})$.
(iii) An $([n_1^{\mathrm{m}}, n_1^{\mathrm{o}}] : [n_2^{\mathrm{m}}, n_2^{\mathrm{o}}] : [n_3+k_3; \kappa_3] : \varnothing|$ state of the fixed point $(\mathbf{x}_{k0}^*, \mathbf{p}_0)$ is a switching of the $([n_1^{\mathrm{m}}+k_3, n_1^{\mathrm{o}}] : [n_2^{\mathrm{m}}, n_2^{\mathrm{o}}] : [n_3; \kappa_3'] : \varnothing|$ state and $([n_1^{\mathrm{m}}, n_1^{\mathrm{o}}] : [n_2^{\mathrm{m}}+k_3, n_2^{\mathrm{o}}] : [n_3; \kappa_3'] : \varnothing|$ state for the fixed point $(\mathbf{x}_k^*, \mathbf{p})$.
(iv) An $([n_1^{\mathrm{m}}, n_1^{\mathrm{o}}] : [n_2^{\mathrm{m}}, n_2^{\mathrm{o}}] : \varnothing : [n_4+k_4; \kappa_4]|$ state of the fixed point $(\mathbf{x}_{k0}^*, \mathbf{p}_0)$ is a switching of the $([n_1^{\mathrm{m}}, n_1^{\mathrm{o}}+k_4] : [n_2^{\mathrm{m}}, n_2^{\mathrm{o}}] : \varnothing : [n_4; \kappa_4']|$ state and $([n_1^{\mathrm{m}}, n_1^{\mathrm{o}}] : [n_2^{\mathrm{m}}, n_2^{\mathrm{o}}+k_4] : \varnothing : [n_4; \kappa_4']|$ state for the fixed point $(\mathbf{x}_k^*, \mathbf{p})$.
(v) An $([n_1^{\mathrm{m}}, n_1^{\mathrm{o}}] : [n_2^{\mathrm{m}}, n_2^{\mathrm{o}}] : [n_3+k_3; \kappa_3] : [n_4+k_4; \kappa_4]|$ state of the fixed point $(\mathbf{x}_{k0}^*, \mathbf{p}_0)$ is a switching of the $([n_1^{\mathrm{m}}+k_3, n_1^{\mathrm{o}}+k_4] : [n_2^{\mathrm{m}}, n_2^{\mathrm{o}}] : [n_3; \kappa_3'] : [n_4; \kappa_4']|$ state and $([n_1^{\mathrm{m}}, n_1^{\mathrm{o}}] : [n_2^{\mathrm{m}}+k_3, n_2^{\mathrm{o}}+k_4] : [n_3; \kappa_3'] : [n_4; \kappa_4']|$ state for the fixed point $(\mathbf{x}_k^*, \mathbf{p})$.

Definition 2.23 Consider a discrete, nonlinear dynamical system $\mathbf{x}_{k+1} = \mathbf{f}(\mathbf{x}_k, \mathbf{p}) \in \mathscr{R}^{2n}$ in Eq. (2.4) with a fixed point $\mathbf{x}_k^*$. The corresponding solution is given by $\mathbf{x}_{k+j} = \mathbf{f}(\mathbf{x}_{k+j-1}, \mathbf{p})$ with $j \in \mathbb{Z}$. Suppose there is a neighborhood of the fixed point $\mathbf{x}_k^*$ (i.e., $U_k(\mathbf{x}_k^*) \subset \Omega$), and $\mathbf{f}(\mathbf{x}_k, \mathbf{p})$ is C^r $(r \geq 1)$-continuous in $U_k(\mathbf{x}_k^*)$ with Eq. (2.28). The linearized system is $\mathbf{y}_{k+j+1} = D\mathbf{f}(\mathbf{x}_k^*, \mathbf{p})\mathbf{y}_{k+j}$ $(\mathbf{y}_{k+j} = \mathbf{x}_{k+j} - \mathbf{x}_k^*)$ in $U_k(\mathbf{x}_k^*)$. The matrix $D\mathbf{f}(\mathbf{x}_k^*, \mathbf{p})$ possesses n-pairs of eigenvalues λ_i $(i = 1, 2, \cdots, n)$. Set $N = \{1, 2, \cdots, n\}$, $N_p = \{p_1, p_2, \cdots, p_{n_p}\} \cup \varnothing$ with $p_q \in N$ $(q = 1, 2, \cdots, n_p, p = 5, 6, 7)$ and $\Sigma_{p=5}^{7} n_p = n$. $\cup_{p=5}^{7} N_p = N$ and $N_p \cap N_l = \varnothing (l \neq p)$. $N_p = \varnothing$ if $n_p = 0$. The matrix $D\mathbf{f}(\mathbf{x}_k^*, \mathbf{p})$ possesses n_5-stable, n_6-unstable and n_7-center pairs of complex eigenvectors. Without repeated complex eigenvalues of $|\lambda_i| = 1 (i \in N_7)$, an iterative response of $\mathbf{x}_{k+1} = \mathbf{f}(\mathbf{x}_k, \mathbf{p})$ is an $|n_5 : n_6 : n_7)$ flow in the neighborhood of the fixed point $\mathbf{x}_k^*$. With repeated complex eigenvalues of $|\lambda_i| = 1$ $(i \in N_7)$, an iterative response of $\mathbf{x}_{k+1} = \mathbf{f}(\mathbf{x}_k, \mathbf{p})$ is an $|n_5 : n_6 : [n_7, l; \boldsymbol{\kappa}_7])$ flow in the neighborhood of the fixed point $\mathbf{x}_k^*$, $\boldsymbol{\kappa}_7 = (\kappa_{71}, \kappa_{72}, \cdots, \kappa_{7l})^{\mathrm{T}}$ with $\kappa_{7s} \in \{\varnothing, m_{7s}\} (s = 1, 2, \cdots, l)$.

I. *Simple switching and bifurcation*

(i) An $|n_5 : n_6 : 1)$ state of the fixed point $(\mathbf{x}_{k0}^*, \mathbf{p}_0)$ is a switching of the $|n_5+1 : n_6 : \varnothing)$ spiral saddle and $|n_5 : n_6+1 : \varnothing)$ saddle for the fixed point $(\mathbf{x}_k^*, \mathbf{p})$.
(ii) An $|n_5 : \varnothing : 1)$ state of the fixed point $(\mathbf{x}_{k0}^*, \mathbf{p}_0)$ is a stable Neimark bifurcation of the $|n_5+1 : \varnothing : \varnothing)$ spiral sink and $|n_5 : 1 : \varnothing)$ spiral saddle for the fixed point $(\mathbf{x}_k^*, \mathbf{p})$.

(iii) An $|\varnothing : n_6 : 1)$ state of the fixed point $(\mathbf{x}_{k0}^*, \mathbf{p}_0)$ is an unstable Neimark bifurcation of the $|\varnothing : n_6 + 1 : \varnothing)$ spiral source and $|1 : n_6 : \varnothing)$ spiral saddle for the fixed point $(\mathbf{x}_k^*, \mathbf{p})$.
(iv) An $|n_5 : n_6 : n_7 + 1)$ state of the fixed point $(\mathbf{x}_{k0}^*, \mathbf{p}_0)$ is a switching of the $|n_5 + 1 : n_6 : n_7)$ state and $|n_5 : n_6 + 1 : n_7)$ state for the fixed point $(\mathbf{x}_k^*, \mathbf{p})$.
(v) An $|\varnothing : n_6 : n_7 + 1)$ state of the fixed point $(\mathbf{x}_{k0}^*, \mathbf{p}_0)$ is a switching of the $|1 : n_6 : n_7)$ state and $|n_5 : n_6 + 1 : n_7)$ state for the fixed point $(\mathbf{x}_k^*, \mathbf{p})$.
(vi) An $|n_5 : \varnothing : n_7 + 1)$ state of the fixed point $(\mathbf{x}_{k0}^*, \mathbf{p}_0)$ is a switching of the $|n_5 + 1 : \varnothing : n_7)$ state and $|n_5 : 1 : n_7)$ state for the fixed point $(\mathbf{x}_k^*, \mathbf{p})$.

II. *Complex switching*

(i) An $|n_5 : n_6 : [n_7, l; \kappa_7])$ state of the fixed point $(\mathbf{x}_{k0}^*, \mathbf{p}_0)$ is a switching of the $|n_5 + n_7 : n_6 : \varnothing)$ spiral saddle and $|n_5 : n_6 + n_7 : \varnothing)$ spiral saddle for the fixed point $(\mathbf{x}_k^*, \mathbf{p})$.
(ii) An $|n_5 : n_6 : [n_7 + k_7, l_1; \kappa_7])$ state of the fixed point $(\mathbf{x}_{k0}^*, \mathbf{p}_0)$ is a switching of the $|n_5 + k_7 : n_6 : [n_7, l_1; \kappa_7'])$ state and $|n_5 : n_6 + k_7 : [n_7, l; \kappa_7'])$ state for the fixed point $(\mathbf{x}_k^*, \mathbf{p})$.
(iii) An $|n_5 : n_6 : [n_7 + k_5 - k_6, l_2; \kappa_7])$ state of the fixed point $(\mathbf{x}_{k0}^*, \mathbf{p}_0)$ is a switching of the $|n_5+k_5 : n_6 : [n_7, l_1; \kappa_7'])$ state and $|n_5 : n_6+k_6 : [n_7, l_3; \kappa_7''])$ state for the fixed point $(\mathbf{x}_k^*, \mathbf{p})$.

2.3.1 Stability and Switching

To extend the idea of Definitions 2.11 and 2.12, a new function will be defined to determine the stability and the stability state switching.

Definition 2.24 Consider a discrete, nonlinear dynamical system $\mathbf{x}_{k+1} = \mathbf{f}(\mathbf{x}_k, \mathbf{p}) \in \mathscr{R}^n$ in Eq. (2.4) with a fixed point $\mathbf{x}_k^*$. The corresponding solution is given by $\mathbf{x}_{k+j} = \mathbf{f}(\mathbf{x}_{k+j-1}, \mathbf{p})$ with $j \in \mathbb{Z}$. Suppose there is a neighborhood of the fixed point $\mathbf{x}_k^*$ (i.e., $U_k(\mathbf{x}_k^*) \subset \Omega$), and $\mathbf{f}(\mathbf{x}_k, \mathbf{p})$ is C^r ($r \geq 1$)-continuous in $U_k(\mathbf{x}_k^*)$ with Eq. (2.28). The linearized system is $\mathbf{y}_{k+j+1} = D\mathbf{f}(\mathbf{x}_k^*, \mathbf{p})\mathbf{y}_{k+j}$ ($\mathbf{y}_{k+j} = \mathbf{x}_{k+j} - \mathbf{x}_k^*$) in $U_k(\mathbf{x}_k^*)$ and there are n linearly independent vectors $\mathbf{v}_i$ ($i = 1, 2, \cdots, n$). For a perturbation of fixed point $\mathbf{y}_k = \mathbf{x}_k - \mathbf{x}_k^*$, let $\mathbf{y}_k^{(i)} = c_k^{(i)}\mathbf{v}_i$ and $\mathbf{y}_{k+1}^{(i)} = c_{k+1}^{(i)}\mathbf{v}_i$,

$$s_k^{(i)} = \mathbf{v}_i^{\mathrm{T}} \cdot \mathbf{y}_k = \mathbf{v}_i^{\mathrm{T}} \cdot (\mathbf{x}_k - \mathbf{x}_k^*) \tag{2.49}$$

where $s_k^{(i)} = c_k^{(i)}||\mathbf{v}_i||^2$. Define the following functions

$$G_i(\mathbf{x}_k, \mathbf{p}) = \mathbf{v}_i^{\mathrm{T}} \cdot [\mathbf{f}(\mathbf{x}_k, \mathbf{p}) - \mathbf{x}_k^*] \tag{2.50}$$

and

$$\begin{aligned} G^{(1)}_{s_k^{(i)}}(\mathbf{x},\mathbf{p}) &= \mathbf{v}_i^{\mathrm{T}} \cdot D_{s_k^{(i)}} \mathbf{f}(\mathbf{x}_k(s_k^{(i)}),\mathbf{p}) \\ &= \mathbf{v}_i^{\mathrm{T}} \cdot D_{\mathbf{x}_k} \mathbf{f}(\mathbf{x}_k(s_k^{(i)}),\mathbf{p}) \partial_{c_k^{(i)}} \mathbf{x}_k \partial_{s_k^{(i)}} c_k^{(i)} \\ &= \mathbf{v}_i^{\mathrm{T}} \cdot D_{\mathbf{x}} \mathbf{f}(\mathbf{x}_k(s_k^{(i)}),\mathbf{p}) \mathbf{v}_i ||\mathbf{v}_i||^{-2} \end{aligned} \tag{2.51}$$

$$\begin{aligned} G^{(m)}_{s_k^{(i)}}(\mathbf{x},\mathbf{p}) &= \mathbf{v}_i^{\mathrm{T}} \cdot D^{(m)}_{s_k^{(i)}} \mathbf{f}(\mathbf{x}_k(s_k^{(i)}),\mathbf{p}) \\ &= \mathbf{v}_i^{\mathrm{T}} \cdot D_{s_k^{(i)}} (D^{(m-1)}_{s_k^{(i)}} \mathbf{f}(\mathbf{x}_k(s_k^{(i)}),\mathbf{p})) \end{aligned} \tag{2.52}$$

where $D_{s_k^{(i)}}(\cdot) = \partial(\cdot)/\partial s_k^{(i)}$ and $D^{(m)}_{s_k^{(i)}}(\cdot) = D_{s_k^{(i)}}(D^{(m-1)}_{s_k^{(i)}}(\cdot))$.

Definition 2.25 Consider a discrete, nonlinear dynamical system $\mathbf{x}_{k+1} = \mathbf{f}(\mathbf{x}_k,\mathbf{p}) \in \mathscr{R}^n$ in Eq. (2.4) with a fixed point $\mathbf{x}_k^*$. The corresponding solution is given by $\mathbf{x}_{k+j} = \mathbf{f}(\mathbf{x}_{k+j-1},\mathbf{p})$ with $j \in \mathbb{Z}$. Suppose there is a neighborhood of the fixed point $\mathbf{x}_k^*$ (i.e., $U_k(\mathbf{x}_k^*) \subset \Omega$), and $\mathbf{f}(\mathbf{x}_k,\mathbf{p})$ is C^r $(r \geq 1)$-continuous in $U_k(\mathbf{x}_k^*)$ with Eq. (2.28). The linearized system is $\mathbf{y}_{k+j+1} = D\mathbf{f}(\mathbf{x}_k^*,\mathbf{p})\mathbf{y}_{k+j}$ $(\mathbf{y}_{k+j} = \mathbf{x}_{k+j} - \mathbf{x}_k^*)$ in $U_k(\mathbf{x}_k^*)$ and there are n linearly independent vectors $\mathbf{v}_i$ $(i = 1, 2, \cdots, n)$. For a perturbation of fixed point $\mathbf{y}_k = \mathbf{x}_k - \mathbf{x}_k^*$, let $\mathbf{y}_k^{(i)} = c_k^{(i)}\mathbf{v}_i$ and $\mathbf{y}_{k+1}^{(i)} = c_{k+1}^{(i)}\mathbf{v}_i$.

(i) $\mathbf{x}_{k+j} (j \in \mathbb{Z})$ at fixed point $\mathbf{x}_k^*$ on the direction $\mathbf{v}_i$ is stable if

$$|\mathbf{v}_i^{\mathrm{T}} \cdot (\mathbf{x}_{k+1} - \mathbf{x}_k^*)| < |\mathbf{v}_i^{\mathrm{T}} \cdot (\mathbf{x}_k - \mathbf{x}_k^*)| \tag{2.53}$$

for $\mathbf{x}_k \in U(\mathbf{x}_k^*) \subset \Omega_\alpha$. The fixed point $\mathbf{x}_k^*$ is called the sink (or stable node) on the direction $\mathbf{v}_i$.

(ii) $\mathbf{x}_{k+j} (j \in \mathbb{Z})$ at fixed point $\mathbf{x}_k^*$ on the direction $\mathbf{v}_i$ is unstable if

$$|\mathbf{v}_i^{\mathrm{T}} \cdot (\mathbf{x}_{k+1} - \mathbf{x}_k^*)| > |\mathbf{v}_i^{\mathrm{T}} \cdot (\mathbf{x}_k - \mathbf{x}_k^*)| \tag{2.54}$$

for $\mathbf{x}_k \in U(\mathbf{x}_k^*) \subset \Omega_\alpha$. The fixed point $\mathbf{x}_k^*$ is called the source (or unstable node) on the direction $\mathbf{v}_i$.

(iii) $\mathbf{x}_{k+j} (j \in \mathbb{Z})$ at fixed point $\mathbf{x}_k^*$ on the direction $\mathbf{v}_i$ is invariant if

$$\mathbf{v}_i^{\mathrm{T}} \cdot (\mathbf{x}_{k+1} - \mathbf{x}_k^*) = \mathbf{v}_i^{\mathrm{T}} \cdot (\mathbf{x}_k - \mathbf{x}_k^*) \tag{2.55}$$

for $\mathbf{x}_k \in U(\mathbf{x}_k^*) \subset \Omega_\alpha$. The fixed point $\mathbf{x}_k^*$ is called to be degenerate on the direction $\mathbf{v}_i$.

(iv) $\mathbf{x}_{k+j}^{(i)}(j \in \mathbb{Z})$ at fixed point $\mathbf{x}_k^*$ on the direction $\mathbf{v}_i$ is symmetrically flipped if

$$\mathbf{v}_i^{\mathrm{T}} \cdot (\mathbf{x}_{k+1} - \mathbf{x}_k^*) = -\mathbf{v}_i^{\mathrm{T}} \cdot (\mathbf{x}_k - \mathbf{x}_k^*) \tag{2.56}$$

for $\mathbf{x}_k \in U(\mathbf{x}_k^*) \subset \Omega_\alpha$. The fixed point $\mathbf{x}_k^*$ is called to be degenerate on the direction $\mathbf{v}_i$.

The stability of fixed points for a specific eigenvector is presented in Fig. 2.4. The solid curve is $\mathbf{v}_i^{\mathrm{T}} \cdot \mathbf{x}_{k+1} = \mathbf{v}_i^{\mathrm{T}} \cdot \mathbf{f}(\mathbf{x}_k, \mathbf{p})$. The circular symbol is the fixed point. The shaded regions are stable. The horizontal solid line is for a degenerate case. The vertical solid line is for a line with infinite slope. The monotonically stable node (sink) is presented in Fig. 2.4a. The dashed and dotted lines are for $\mathbf{v}_i^{\mathrm{T}} \cdot \mathbf{x}_k = \mathbf{v}_i^{\mathrm{T}} \cdot \mathbf{x}_{k+1}$ and $\mathbf{v}_i^{\mathrm{T}} \cdot \mathbf{y}_{k+1} = -\mathbf{v}_i^{\mathrm{T}} \cdot \mathbf{y}_k$, respectively. From the fixed point $\mathbf{x}_k^*$, $\mathbf{y}_k = \mathbf{x}_k - \mathbf{x}_k^*$ and $\mathbf{y}_{k+1} = \mathbf{x}_{k+1} - \mathbf{x}_k^*$. The iterative responses approach the fixed point. However, the monotonically unstable (source) is presented in Fig. 2.4b. The iterative responses go away from the fixed point. Similarly, the oscillatory stable node (sink) is presented in Fig. 2.4c. The dashed and dotted lines are for $\mathbf{v}_i^{\mathrm{T}} \cdot \mathbf{y}_{k+1} = -\mathbf{v}_i^{\mathrm{T}} \cdot \mathbf{y}_k$ and $\mathbf{v}_i^{\mathrm{T}} \cdot \mathbf{x}_k = \mathbf{v}_i^{\mathrm{T}} \cdot \mathbf{x}_{k+1}$, respectively. The oscillatory unstable node (source) is presented in Fig. 2.4d.

Theorem 2.5 *Consider a discrete, nonlinear dynamical system* $\mathbf{x}_{k+1} = \mathbf{f}(\mathbf{x}_k, \mathbf{p}) \in \mathscr{R}^n$ *in Eq.* (2.4) with a fixed point $\mathbf{x}_k^*$. *The corresponding solution is given by* $\mathbf{x}_{k+j} = \mathbf{f}(\mathbf{x}_{k+j-1}, \mathbf{p})$ *with* $j \in \mathbb{Z}$. *Suppose there is a neighborhood of the fixed point* $\mathbf{x}_k^*$ *(i.e.,* $U_k(\mathbf{x}_k^*) \subset \Omega$*), and* $\mathbf{f}(\mathbf{x}_k, \mathbf{p})$ *is* C^r $(r \geq 1)$*-continuous in* $U_k(\mathbf{x}_k^*)$ *with Eq.* (2.28). *The linearized system is* $\mathbf{y}_{k+j+1} = D\mathbf{f}(\mathbf{x}_k^*, \mathbf{p})\mathbf{y}_{k+j}$ $(\mathbf{y}_{k+j} = \mathbf{x}_{k+j} - \mathbf{x}_k^*)$ *in* $U_k(\mathbf{x}_k^*)$ *and there are n linearly independent vectors* $\mathbf{v}_i$ $(i = 1, 2, \cdots, n)$. *For a perturbation of fixed point* $\mathbf{y}_k = \mathbf{x}_k - \mathbf{x}_k^*$, *let* $\mathbf{y}_k^{(i)} = c_k^{(i)}\mathbf{v}_i$ *and* $\mathbf{y}_{k+1}^{(i)} = c_{k+1}^{(i)}\mathbf{v}_i$.

(i) $\mathbf{x}_{k+j}(j \in \mathbb{Z})$ *at fixed point* $\mathbf{x}_k^*$ *on the direction* $\mathbf{v}_i$ *is stable if and only if*

$$G_{s_k^{(i)}}^{(1)}(\mathbf{x}_k^*, \mathbf{p}) = \lambda_i \in (-1, 1) \tag{2.57}$$

for $\mathbf{x}_k \in U(\mathbf{x}_k^*) \subset \Omega_\alpha$.

(ii) $\mathbf{x}_{k+j}(j \in \mathbb{Z})$ *at fixed point* $\mathbf{x}_k^*$ *on the direction* $\mathbf{v}_i$ *is unstable if and only if*

$$G_{s_k^{(i)}}^{(1)}(\mathbf{x}_k^*, \mathbf{p}) = \lambda_i \in (1, \infty) \text{ and } (-\infty, -1) \tag{2.58}$$

for $\mathbf{x}_k \in U(\mathbf{x}_k^*) \subset \Omega_\alpha$.

(iii) $\mathbf{x}_{k+j}(j \in \mathbb{Z})$ *at fixed point* $\mathbf{x}_k^*$ *on the direction* $\mathbf{v}_i$ *is invariant if and only if*

$$G_{s_k^{(i)}}^{(1)}(\mathbf{x}_k^*, \mathbf{p}) = \lambda_i = 1 \text{ and } G_{s_k^{(i)}}^{(m_i)}(\mathbf{x}_k^*, \mathbf{p}) = 0 \quad \text{for } m_i = 2, 3, \cdots \tag{2.59}$$

for $\mathbf{x}_k \in U(\mathbf{x}_k^*) \subset \Omega_\alpha$.

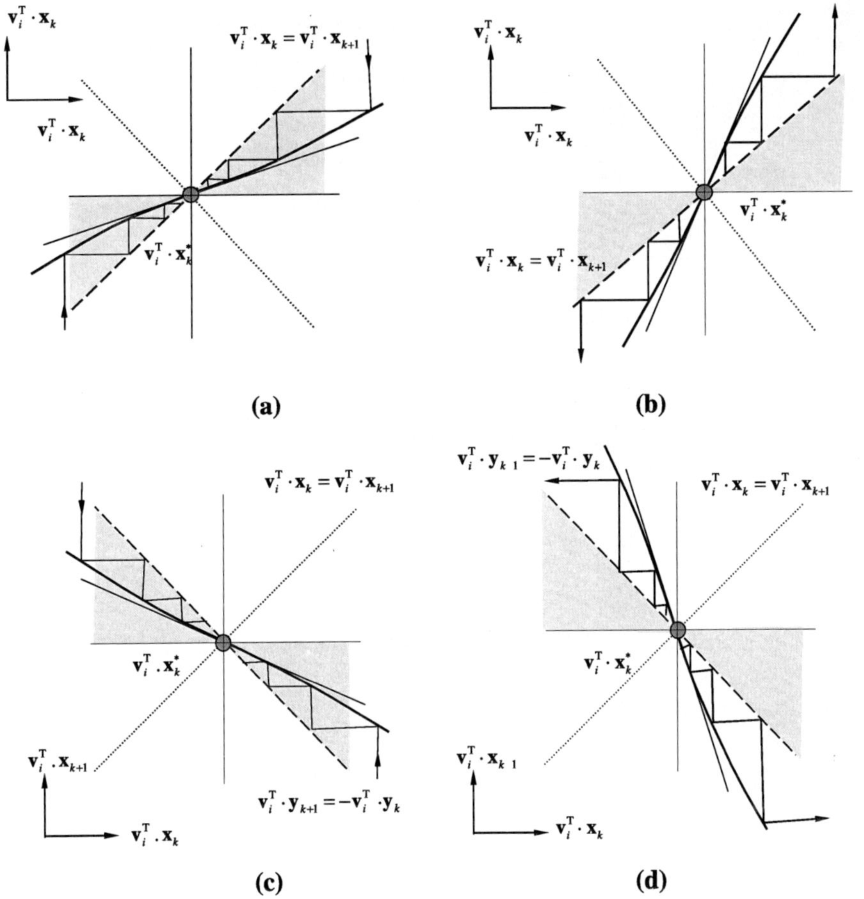

Fig. 2.4 Stability of fixed points: **a** monotonically stable node (sink), **b** monotonically unstable node (source), **c** oscillatory stable node (sink) and **d** oscillatory unstable node (sink). *Shaded areas* are stable zones. ($\mathbf{y}_k = \mathbf{x}_k - \mathbf{x}_k^*$ and $\mathbf{y}_{k+1} = \mathbf{x}_{k+1} - \mathbf{x}_k^*$)

(iv) $\mathbf{x}_{k+j}^{(i)} (j \in \mathbb{Z})$ *at fixed point* $\mathbf{x}_k^*$ *on the direction* $\mathbf{v}_i$ *is symmetrically flipped if and only if*

$$G_{s_k^{(i)}}^{(1)}(\mathbf{x}_k^*, \mathbf{p}) = \lambda_i = -1 \text{ and } G_{s_k^{(i)}}^{(m_i)}(\mathbf{x}_k^*, \mathbf{p}) = 0 \quad \text{for } m_i = 2, 3, \cdots \tag{2.60}$$

for $\mathbf{x}_k \in U(\mathbf{x}_k^*) \subset \Omega_\alpha$.

Proof Because

$$\begin{aligned} s_{k+1}^{(i)} &= \mathbf{v}_i^{\mathrm{T}} \cdot (\mathbf{x}_{k+1} - \mathbf{x}_k^*) = \mathbf{v}_i^{\mathrm{T}} \cdot \mathbf{x}_k^* + G_{s_k^{(i)}}^{(1)}(\mathbf{x}_k^*, \mathbf{p}) s_k^{(i)} + o(s_k^{(i)}) - \mathbf{v}_i^{\mathrm{T}} \cdot \mathbf{x}_k^* \\ &= G_{s_k^{(i)}}^{(1)}(\mathbf{x}_k^*, \mathbf{p}) s_k^{(i)} + o(s_k^{(i)}) \end{aligned}$$

due to any selection of $s_k^{(i)}$ and $s_{k+1}^{(i)}$ as an infinitesimal, we have

$$\begin{aligned} s_{k+1}^{(i)} &= G_{s_k^{(i)}}^{(1)}(\mathbf{x}_k^*, \mathbf{p}) s_k^{(i)}, \\ G_{s_k^{(i)}}^{(1)}(\mathbf{x}_k^*, \mathbf{p}) &= \mathbf{v}_i^{\mathrm{T}} \cdot D_{\mathbf{x}} \mathbf{f}(\mathbf{x}_k^*, \mathbf{p}) \mathbf{v}_i ||\mathbf{v}_i||^{-2} \\ &= \mathbf{v}_i^{\mathrm{T}} \cdot \lambda_i \mathbf{v}_i ||\mathbf{v}_i||^{-2} = \lambda_i. \end{aligned}$$

(i) From the definition in Eq. (2.53), we have

$$|\mathbf{v}_i^{\mathrm{T}} \cdot (\mathbf{x}_{k+1} - \mathbf{x}_k^*)| < |\mathbf{v}_i^{\mathrm{T}} \cdot (\mathbf{x}_k - \mathbf{x}_k^*)| \Rightarrow |s_{k+1}^{(i)}| < |s_k^{(i)}|$$

which gives

$$|G_{s_k^{(i)}}^{(1)}(\mathbf{x}_k^*, \mathbf{p}) s_k^{(i)}| < |s_k^{(i)}|.$$

Thus,

$$|G_{s_k^{(i)}}^{(1)}(\mathbf{x}_k^*, \mathbf{p})| < 1 \Rightarrow G_{s_k^{(i)}}^{(1)}(\mathbf{x}_k^*, \mathbf{p}) = \lambda_i \in (-1, 1).$$

Therefore, $\mathbf{x}_{k+j} (j \in \mathbb{Z})$ at fixed point $\mathbf{x}_k^*$ on the direction $\mathbf{v}_i$ is stable and vice versa.

(ii) From the definition in Eq. (2.54), we have

$$|\mathbf{v}_i^{\mathrm{T}} \cdot (\mathbf{x}_{k+1} - \mathbf{x}_k^*)| > |\mathbf{v}_i^{\mathrm{T}} \cdot (\mathbf{x}_k - \mathbf{x}_k^*)| \Rightarrow |s_{k+1}^{(i)}| > |s_k^{(i)}|$$

which gives

$$|G_{s_k^{(i)}}^{(1)}(\mathbf{x}_k^*, \mathbf{p}) s_k^{(i)}| > |s_k^{(i)}|.$$

Thus,

$$|G_{s_k^{(i)}}^{(1)}(\mathbf{x}_k^*, \mathbf{p})| > 1 \Rightarrow G_{s_k^{(i)}}^{(1)}(\mathbf{x}_k^*, \mathbf{p}) = \lambda_i \in (-\infty, -1) \text{ and } (1, \infty).$$

Therefore, $\mathbf{x}_{k+j} (j \in \mathbb{Z})$ at fixed point $\mathbf{x}_k^*$ on the direction $\mathbf{v}_i$ is unstable and vice versa.

(iii) Because

$$\begin{aligned} s_{k+1}^{(i)} &= \mathbf{v}_i^{\mathrm{T}} \cdot (\mathbf{x}_{k+1} - \mathbf{x}_k^*) \\ &= \mathbf{v}_i^{\mathrm{T}} \cdot \mathbf{x}_k^* + G_{s_k^{(i)}}^{(1)}(\mathbf{x}_k^*, \mathbf{p}) s_k^{(i)} + \sum_{m_i=2}^{\infty} \frac{1}{m_i!} G_{s_k^{(i)}}^{(m_i)}(\mathbf{x}_k^*, \mathbf{p}) (s_k^{(i)})^{m_i} - \mathbf{v}_i^{\mathrm{T}} \cdot \mathbf{x}_k^* \\ &= G_{s_k^{(i)}}^{(1)}(\mathbf{x}_k^*, \mathbf{p}) s_k^{(i)} + \sum_{m_i=2}^{\infty} \frac{1}{m_i!} G_{s_k^{(i)}}^{(m_i)}(\mathbf{x}_k^*, \mathbf{p}) (s_k^{(i)})^{m_i}, \end{aligned}$$

from the definition in Eq. (2.55)

$$\mathbf{v}_i^{\mathrm{T}} \cdot (\mathbf{x}_{k+1} - \mathbf{x}_k^*) = \mathbf{v}_i^{\mathrm{T}} \cdot (\mathbf{x}_k - \mathbf{x}_k^*) \Rightarrow s_{k+1}^{(i)} = s_k^{(i)}.$$

Due to any selection of $s_k^{(i)}$ and $s_{k+1}^{(i)}$ as an infinitesimal, we have

$$G_{s_k^{(i)}}^{(1)}(\mathbf{x}_k^*, \mathbf{p}) = \lambda_i = 1 \text{ and } G_{s_k^{(i)}}^{(m_i)}(\mathbf{x}_k^*, \mathbf{p}) = 0 \quad \text{for } m_i = 2, 3, \cdots.$$

Therefore, $\mathbf{x}_{k+j} (j \in \mathbb{Z})$ at fixed point $\mathbf{x}_k^*$ on the direction $\mathbf{v}_i$ is invariant and vice versa.

(iv) From the definition in Eq. (2.56)

$$\mathbf{v}_i^{\mathrm{T}} \cdot (\mathbf{x}_{k+1} - \mathbf{x}_k^*) = -\mathbf{v}_i^{\mathrm{T}} \cdot (\mathbf{x}_k - \mathbf{x}_k^*) \Rightarrow s_{k+1}^{(i)} = -s_k^{(i)}.$$

Due to any selection of $s_k^{(i)}$ and $s_{k+1}^{(i)}$ as an infinitesimal, we have

$$G_{s_k^{(i)}}^{(1)}(\mathbf{x}_k^*, \mathbf{p}) = \lambda_i = -1 \text{ and } G_{s_k^{(i)}}^{(m_i)}(\mathbf{x}_k^*, \mathbf{p}) = 0 \quad \text{for } m_i = 2, 3, \cdots.$$

Therefore, $\mathbf{x}_{k+j} (j \in \mathbb{Z})$ at fixed point $\mathbf{x}_k^*$ on the direction $\mathbf{v}_i$ is flipped and vice versa. The theorem is proved. ■

Definition 2.26 Consider a discrete, nonlinear dynamical system $\mathbf{x}_{k+1} = \mathbf{f}(\mathbf{x}_k, \mathbf{p}) \in \mathscr{R}^n$ in Eq. (2.4) with a fixed point $\mathbf{x}_k^*$. The corresponding solution is given by $\mathbf{x}_{k+j} = \mathbf{f}(\mathbf{x}_{k+j-1}, \mathbf{p})$ with $j \in \mathbb{Z}$. Suppose there is a neighborhood of the fixed point $\mathbf{x}_k^*$ (i.e., $U_k(\mathbf{x}_k^*) \subset \Omega$), and $\mathbf{f}(\mathbf{x}_k, \mathbf{p})$ is C^r $(r \geq 1)$-continuous in $U_k(\mathbf{x}_k^*)$ with Eq. (2.28). The linearized system is $\mathbf{y}_{k+j+1} = D\mathbf{f}(\mathbf{x}_k^*, \mathbf{p})\mathbf{y}_{k+j}$ $(\mathbf{y}_{k+j} = \mathbf{x}_{k+j} - \mathbf{x}_k^*)$ in $U_k(\mathbf{x}_k^*)$ and there are n linearly independent vectors $\mathbf{v}_i$ $(i = 1, 2, \cdots, n)$. For a perturbation of the fixed point $\mathbf{y}_k = \mathbf{x}_k - \mathbf{x}_k^*$, let $\mathbf{y}_k^{(i)} = c_k^{(i)}\mathbf{v}_i$ and $\mathbf{y}_{k+1}^{(i)} = c_{k+1}^{(i)}\mathbf{v}_i$.

(i) $\mathbf{x}_{k+j} (j \in \mathbb{Z})$ at fixed point $\mathbf{x}_k^*$ on the direction $\mathbf{v}_i$ is monotonically stable of the $(2m_i + 1)$th-order if

$$\begin{aligned} &G_{s_k^{(i)}}^{(1)}(\mathbf{x}_k^*, \mathbf{p}) = \lambda_i = 1, \\ &G_{s_k^{(i)}}^{(r_i)}(\mathbf{x}_k^*, \mathbf{p}) = 0 \text{ for } r_i = 2, 3, \cdots, 2m_i, \\ &G_{s_k^{(i)}}^{(2m_i+1)}(\mathbf{x}_k^*, \mathbf{p}) \neq 0, \\ &|\mathbf{v}_i^{\mathrm{T}} \cdot (\mathbf{x}_{k+1} - \mathbf{x}_k^*)| < |\mathbf{v}_i^{\mathrm{T}} \cdot (\mathbf{x}_k - \mathbf{x}_k^*)| \end{aligned} \quad (2.61)$$

for $\mathbf{x}_k \in U(\mathbf{x}_k^*) \subset \Omega_\alpha$. The fixed point $\mathbf{x}_k^*$ is called the monotonic sink (or stable node) of the $(2m_i + 1)$th-order on the direction $\mathbf{v}_i$.

(ii) $\mathbf{x}_{k+j} (j \in \mathbb{Z})$ at fixed point $\mathbf{x}_k^*$ on the direction $\mathbf{v}_i$ is monotonically unstable of the $(2m_i + 1)$th-order if

$$
\begin{aligned}
&G^{(1)}_{s_k^{(i)}}(\mathbf{x}_k^*, \mathbf{p}) = \lambda_i = 1,\\
&G^{(r_i)}_{s_k^{(i)}}(\mathbf{x}_k^*, \mathbf{p}) = 0 \text{ for } r_i = 2, 3, \cdots, 2m_i;\\
&G^{(2m_i+1)}_{s_k^{(i)}}(\mathbf{x}_k^*, \mathbf{p}) \neq 0;\\
&|\mathbf{v}_i^{\mathrm{T}} \cdot (\mathbf{x}_{k+1} - \mathbf{x}_k^*)| > |\mathbf{v}_i^{\mathrm{T}} \cdot (\mathbf{x}_k - \mathbf{x}_k^*)|
\end{aligned} \tag{2.62}
$$

for $\mathbf{x}_k \in U(\mathbf{x}_k^*) \subset \Omega_\alpha$. The fixed point $\mathbf{x}_k^*$ is called the monotonic source (or unstable node) of the $(2m_i + 1)$th-order on the direction $\mathbf{v}_i$.

(iii) $\mathbf{x}_{k+j} (j \in \mathbb{Z})$ at fixed point $\mathbf{x}_k^*$ on the direction $\mathbf{v}_i$ is monotonically unstable of the $(2m_i)$th-order, lower saddle if

$$
\begin{aligned}
&G^{(1)}_{s_k^{(i)}}(\mathbf{x}_k^*, \mathbf{p}) = \lambda_i = 1,\\
&G^{(r_i)}_{s_k^{(i)}}(\mathbf{x}_k^*, \mathbf{p}) = 0 \text{ for } r_i = 2, 3, \cdots, 2m_i - 1;\\
&G^{(2m_i)}_{s_k^{(i)}}(\mathbf{x}_k^*, \mathbf{p}) \neq 0,\\
&|\mathbf{v}_i^{\mathrm{T}} \cdot (\mathbf{x}_{k+1} - \mathbf{x}_k^*)| < |\mathbf{v}_i^{\mathrm{T}} \cdot (\mathbf{x}_k - \mathbf{x}_k^*)| \text{ for } s_k^{(i)} > 0\\
&|\mathbf{v}_i^{\mathrm{T}} \cdot (\mathbf{x}_{k+1} - \mathbf{x}_k^*)| > |\mathbf{v}_i^{\mathrm{T}} \cdot (\mathbf{x}_k - \mathbf{x}_k^*)| \text{ for } s_k^{(i)} < 0
\end{aligned} \tag{2.63}
$$

for $\mathbf{x}_k \in U(\mathbf{x}_k^*) \subset \Omega_\alpha$. The fixed point $\mathbf{x}_k^*$ is called the monotonic, lower saddle of the $(2m_i)$th-order on the direction $\mathbf{v}_i$.

(iv) $\mathbf{x}_{k+j} (j \in \mathbb{Z})$ at fixed point $\mathbf{x}_k^*$ on the direction $\mathbf{v}_i$ is monotonically unstable of the $(2m_i)$th-order, upper saddle if

$$
\begin{aligned}
&G^{(1)}_{s_k^{(i)}}(\mathbf{x}_k^*, \mathbf{p}) = \lambda_i = 1,\\
&G^{(r_i)}_{s_k^{(i)}}(\mathbf{x}_k^*, \mathbf{p}) = 0 \text{ for } r_i = 2, 3, \cdots, 2m_i - 1;\\
&G^{(2m_i)}_{s_k^{(i)}}(\mathbf{x}_k^*, \mathbf{p}) \neq 0,\\
&|\mathbf{v}_i^{\mathrm{T}} \cdot (\mathbf{x}_{k+1} - \mathbf{x}_k^*)| > |\mathbf{v}_i^{\mathrm{T}} \cdot (\mathbf{x}_k - \mathbf{x}_k^*)| \text{ for } s_k^{(i)} > 0\\
&|\mathbf{v}_i^{\mathrm{T}} \cdot (\mathbf{x}_{k+1} - \mathbf{x}_k^*)| < |\mathbf{v}_i^{\mathrm{T}} \cdot (\mathbf{x}_k - \mathbf{x}_k^*)| \text{ for } s_k^{(i)} < 0
\end{aligned} \tag{2.64}
$$

for $\mathbf{x}_k \in U(\mathbf{x}_k^*) \subset \Omega_\alpha$. The fixed point $\mathbf{x}_k^*$ is called the monotonic, upper saddle of the $(2m_i)$th-order on the direction $\mathbf{v}_i$.

(v) $\mathbf{x}_{k+j} (j \in \mathbb{Z})$ at fixed point $\mathbf{x}_k^*$ on the direction $\mathbf{v}_i$ is oscillatory stable of the $(2m_i + 1)$th-order if

$$
\begin{aligned}
&G^{(1)}_{s_k^{(i)}}(\mathbf{x}_k^*, \mathbf{p}) = \lambda_i = -1,\\
&G^{(r_i)}_{s_k^{(i)}}(\mathbf{x}_k^*, \mathbf{p}) = 0 \text{ for } r_i = 2, 3, \cdots, 2m_i;\\
&G^{(2m_i+1)}_{s_k^{(i)}}(\mathbf{x}_k^*, \mathbf{p}) \neq 0;\\
&|\mathbf{v}_i^{\mathrm{T}} \cdot (\mathbf{x}_{k+1} - \mathbf{x}_k^*)| < |\mathbf{v}_i^{\mathrm{T}} \cdot (\mathbf{x}_k - \mathbf{x}_k^*)|
\end{aligned} \tag{2.65}
$$

for $\mathbf{x}_k \in U(\mathbf{x}_k^*) \subset \Omega_\alpha$. The fixed point $\mathbf{x}_k^*$ is called the oscillatory sink (or stable node) of the $(2m_i + 1)$th-order in the direction $\mathbf{v}_i$.

(vi) $\mathbf{x}_{k+j} (j \in \mathbb{Z})$ at fixed point $\mathbf{x}_k^*$ on the direction $\mathbf{v}_i$ is oscillatory unstable of the $(2m_i + 1)$th-order if

$$\begin{aligned}
&G^{(1)}_{s_k^{(i)}}(\mathbf{x}_k^*, \mathbf{p}) = \lambda_i = -1;\\
&G^{(r_i)}_{s_k^{(i)}}(\mathbf{x}_k^*, \mathbf{p}) = 0 \text{ for } r_i = 2, 3, \cdots, 2m_i;\\
&G^{(2m_i+1)}_{s_k^{(i)}}(\mathbf{x}_k^*, \mathbf{p}) \neq 0;\\
&|\mathbf{v}_i^{\mathrm{T}} \cdot (\mathbf{x}_{k+1} - \mathbf{x}_k^*)| > |\mathbf{v}_i^{\mathrm{T}} \cdot (\mathbf{x}_k - \mathbf{x}_k^*)|
\end{aligned} \tag{2.66}$$

for $\mathbf{x}_k \in U(\mathbf{x}_k^*) \subset \Omega_\alpha$. The fixed point $\mathbf{x}_k^*$ is called the oscillatory source (or unstable node) of the $(2m_i + 1)$th-order in the direction $\mathbf{v}_i$.

(vii) $\mathbf{x}_{k+j} (j \in \mathbb{Z})$ at fixed point $\mathbf{x}_k^*$ on the direction $\mathbf{v}_i$ is oscillatory unstable of the $(2m_i)$th-order, lower saddle-node if

$$\begin{aligned}
&G^{(1)}_{s_k^{(i)}}(\mathbf{x}_k^*, \mathbf{p}) = \lambda_i = -1,\\
&G^{(r_i)}_{s_k^{(i)}}(\mathbf{x}_k^*, \mathbf{p}) = 0 \text{ for } r_i = 2, 3, \cdots, 2m_i - 1;\\
&G^{(2m_i)}_{s_k^{(i)}}(\mathbf{x}_k^*, \mathbf{p}) \neq 0,\\
&|\mathbf{v}_i^{\mathrm{T}} \cdot (\mathbf{x}_{k+1} - \mathbf{x}_k^*)| > |\mathbf{v}_i^{\mathrm{T}} \cdot (\mathbf{x}_k - \mathbf{x}_k^*)| \text{ for } s_k^{(i)} > 0,\\
&|\mathbf{v}_i^{\mathrm{T}} \cdot (\mathbf{x}_{k+1} - \mathbf{x}_k^*)| < |\mathbf{v}_i^{\mathrm{T}} \cdot (\mathbf{x}_k - \mathbf{x}_k^*)| \text{ for } s_k^{(i)} < 0,
\end{aligned} \tag{2.67}$$

for $\mathbf{x}_k \in U(\mathbf{x}_k^*) \subset \Omega_\alpha$. The fixed point $\mathbf{x}_k^*$ is called the oscillatory lower saddle of the $(2m_i)$th-order on the direction $\mathbf{v}_i$.

(viii) $\mathbf{x}_{k+j} (j \in \mathbb{Z})$ at fixed point $\mathbf{x}_k^*$ on the direction $\mathbf{v}_i$ is oscillatory unstable of the $(2m_i)$th-order, upper saddle-node if

$$\begin{aligned}
&G^{(1)}_{s_k^{(i)}}(\mathbf{x}_k^*, \mathbf{p}) = \lambda_i = -1,\\
&G^{(r_i)}_{s_k^{(i)}}(\mathbf{x}_k^*, \mathbf{p}) = 0 \text{ for } r_i = 2, 3, \cdots, 2m_i - 1;\\
&G^{(2m_i)}_{s_k^{(i)}}(\mathbf{x}_k^*, \mathbf{p}) \neq 0,\\
&|\mathbf{v}_i^{\mathrm{T}} \cdot (\mathbf{x}_{k+1} - \mathbf{x}_k^*)| < |\mathbf{v}_i^{\mathrm{T}} \cdot (\mathbf{x}_k - \mathbf{x}_k^*)| \text{ for } s_k^{(i)} > 0,\\
&|\mathbf{v}_i^{\mathrm{T}} \cdot (\mathbf{x}_{k+1} - \mathbf{x}_k^*)| > |\mathbf{v}_i^{\mathrm{T}} \cdot (\mathbf{x}_k - \mathbf{x}_k^*)| \text{ for } s_k^{(i)} < 0
\end{aligned} \tag{2.68}$$

for $\mathbf{x}_k \in U(\mathbf{x}_k^*) \subset \Omega_\alpha$. The fixed point $\mathbf{x}_k^*$ is called the oscillatory, upper saddle of the $(2m_i)$th-order on the direction $\mathbf{v}_i$.

The monotonic stability of fixed points with higher order singularity for a specific eigenvector is presented in Fig. 2.5. The solid curve is $\mathbf{v}_i^{\mathrm{T}} \cdot \mathbf{x}_{k+1} = \mathbf{v}_i^{\mathrm{T}} \cdot \mathbf{f}(\mathbf{x}_k, \mathbf{p})$.

The circular symbol is fixed pointed. The shaded regions are stable. The horizontal solid line is also for the degenerate case. The vertical solid line is for a line with infinite slope. The monotonically stable node (sink) of the $(2m_i + 1)$th order is sketched in Fig. 2.5a. The dashed and dotted lines are for $\mathbf{v}_i^{\mathrm{T}} \cdot \mathbf{x}_k = \mathbf{v}_i^{\mathrm{T}} \cdot \mathbf{x}_{k+1}$ and $\mathbf{v}_i^{\mathrm{T}} \cdot \mathbf{y}_{k+1} = -\mathbf{v}_i^{\mathrm{T}} \cdot \mathbf{y}_k$, respectively. The nonlinear curve lies in the unstable zone, and the iterative responses approach the fixed point. However, the monotonically unstable (source) of the $(2m_i + 1)$th order is presented in Fig. 2.5b. The nonlinear curve lies in the unstable zone, and the iterative responses go away from the fixed point. The monotonically lower saddle of the $(2m_i)$th order is presented in Fig. 2.5c. The nonlinear curve is tangential to the line of $\mathbf{v}_i^{\mathrm{T}} \cdot \mathbf{x}_k = \mathbf{v}_i^{\mathrm{T}} \cdot \mathbf{x}_{k+1}$ with the $(2m_i)$th order, and one branch is in the stable zone while the other branch is in the unstable zone. Similarly, the monotonically upper saddle of the $(2m_i)$th order is presented in Fig. 2.5d.

Similar to Fig. 2.5, the oscillatory stability of fixed points with higher order singularity for a specific eigenvector is presented in Fig. 2.6. The oscillatory stable node (sink) of the $(2m_i + 1)$th order is sketched in Fig. 2.6a. The dashed and dotted lines are for $\mathbf{v}_i^{\mathrm{T}} \cdot \mathbf{y}_{k+1} = -\mathbf{v}_i^{\mathrm{T}} \cdot \mathbf{y}_k$ and $\mathbf{v}_i^{\mathrm{T}} \cdot \mathbf{x}_k = \mathbf{v}_i^{\mathrm{T}} \cdot \mathbf{x}_{k+1}$, respectively. The nonlinear curve lies in the unstable zone, and the iterative responses approach the fixed point. However, the oscillatory unstable (source) of the $(2m_i + 1)$th order is presented in Fig. 2.6b. The nonlinear curve lies in the unstable zone, and the iterative responses go away from the fixed point. The oscillatory lower saddle of the $(2m_i)$th order is presented in Fig. 2.6c. The nonlinear curve is tangential to and below the line of $\mathbf{v}_i^{\mathrm{T}} \cdot \mathbf{y}_{k+1} = -\mathbf{v}_i^{\mathrm{T}} \cdot \mathbf{y}_k$ with the $(2m_i)$th order, and one branch is in the stable zone while the branch is in the unstable zone. Finally, the oscillatory upper saddle of the $(2m_i)$th order is presented in Fig. 2.6d.

Theorem 2.6 *Consider a discrete, nonlinear dynamical system* $\mathbf{x}_{k+1} = \mathbf{f}(\mathbf{x}_k, \mathbf{p}) \in \mathscr{R}^n$ *in Eq.* (2.4) with a fixed point $\mathbf{x}_k^*$. *The corresponding solution is given by* $\mathbf{x}_{k+j} = \mathbf{f}(\mathbf{x}_{k+j-1}, \mathbf{p})$ *with* $j \in \mathbb{Z}$. *Suppose there is a neighborhood of the fixed point* $\mathbf{x}_k^*$ *(i.e.,* $U_k(\mathbf{x}_k^*) \subset \Omega$*), and* $\mathbf{f}(\mathbf{x}_k, \mathbf{p})$ *is* C^r $(r \geq 1)$ *-continuous in* $U_k(\mathbf{x}_k^*)$ *with Eq.* (2.28). *The linearized system is* $\mathbf{y}_{k+j+1} = D\mathbf{f}(\mathbf{x}_k^*, \mathbf{p})\mathbf{y}_{k+j}$ $(\mathbf{y}_{k+j} = \mathbf{x}_{k+j} - \mathbf{x}_k^*)$ *in* $U_k(\mathbf{x}_k^*)$ *and there are n linearly independent vectors* $\mathbf{v}_i$ $(i = 1, 2, \cdots, n)$. *For a perturbation of fixed point* $\mathbf{y}_k = \mathbf{x}_k - \mathbf{x}_k^*$, *let* $\mathbf{y}_k^{(i)} = c_k^{(i)}\mathbf{v}_i$ *and* $\mathbf{y}_{k+1}^{(i)} = c_{k+1}^{(i)}\mathbf{v}_i$.

(i) $\mathbf{x}_{k+j} (j \in \mathbb{Z})$ *at fixed point* $\mathbf{x}_k^*$ *on the direction* $\mathbf{v}_i$ *is monotonically stable of the* $(2m_i + 1)$th *-order if and only if*

$$
\begin{aligned}
&G_{s_k^{(i)}}^{(1)}(\mathbf{x}_k^*, \mathbf{p}) = \lambda_i = 1,\\
&G_{s_k^{(i)}}^{(r_i)}(\mathbf{x}_k^*, \mathbf{p}) = 0 \text{ for } r_i = 2, 3, \cdots, 2m_i,\\
&G_{s_k^{(i)}}^{(2m_i+1)}(\mathbf{x}_k^*, \mathbf{p}) < 0
\end{aligned}
\tag{2.69}
$$

for $\mathbf{x}_k \in U(\mathbf{x}_k^*) \subset \Omega_\alpha$.

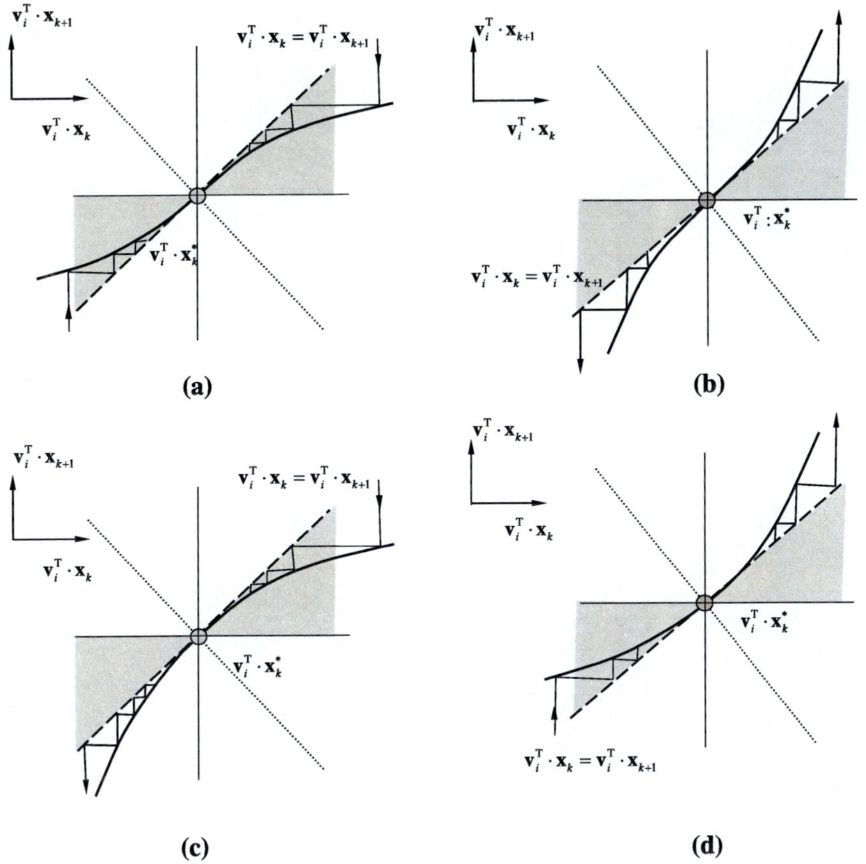

Fig. 2.5 Monotonic stability of fixed points with higher order singularity: **a** monotonically stable node (sink) of $(2m_i + 1)$th-order, **b** monotonically unstable node (source) of $(2m_i + 1)$th-order, **c** monotonically *lower* saddle of $(2m_i)$th-order and **d** monotonically upper saddle of $(2m_i)$th-order. *Shaded areas* are stable zones

(ii) $\mathbf{x}_{k+j} (j \in \mathbb{Z})$ *at fixed point* $\mathbf{x}_k^*$ *on the direction* $\mathbf{v}_i$ *is monotonically unstable of the* $(2m_i + 1)$th *-order if and only if*

$$
\begin{aligned}
&G^{(1)}_{s_k^{(i)}}(\mathbf{x}_k^*, \mathbf{p}) = \lambda_i = 1,\\
&G^{(r_i)}_{s_k^{(i)}}(\mathbf{x}_k^*, \mathbf{p}) = 0 \text{ for } r_i = 2, 3, \cdots, 2m_i,\\
&G^{(2m_i+1)}_{s_k^{(i)}}(\mathbf{x}_k^*, \mathbf{p}) > 0
\end{aligned}
\tag{2.70}
$$

for $\mathbf{x}_k \in U(\mathbf{x}_k^*) \subset \Omega_\alpha$.

(iii) $\mathbf{x}_{k+j} (j \in \mathbb{Z})$ *at fixed point* $\mathbf{x}_k^*$ *on the direction* $\mathbf{v}_i$ *is monotonically stable of the* $(2m_i)$th *-order, lower saddle if and only if*

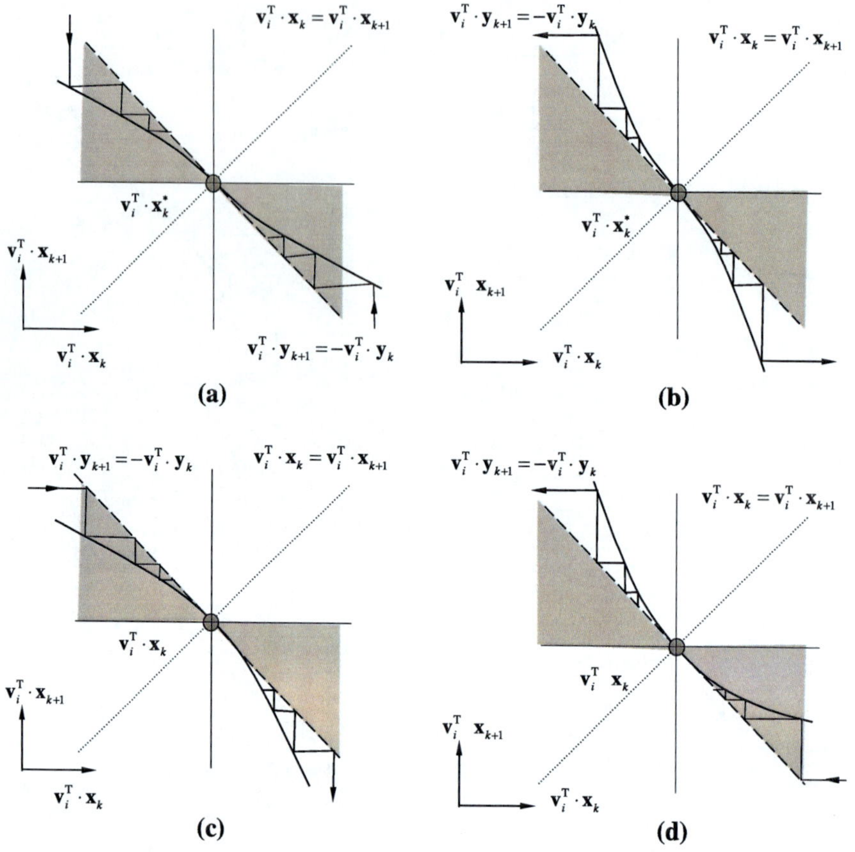

Fig. 2.6 Oscillatory stability of fixed points with higher order singularity: **a** oscillatory stable node (sink) of $(2m_i+1)$th-order, **b** oscillatory unstable node (source) of $(2m_i+1)$th-order, **c** oscillatory *lower* saddle of $(2m_i)$th-order and **d** oscillatory *upper* saddle of $(2m_i)$th-order. *Shaded areas* are stable zones ($\mathbf{y}_k = \mathbf{x}_k - x_k^*$ and $\mathbf{y}_{k+1} = \mathbf{x}_{k+1} - \mathbf{x}_k^*$)

$$
\begin{aligned}
&G^{(1)}_{s_k^{(i)}}(\mathbf{x}_k^*, \mathbf{p}) = \lambda_i = 1,\\
&G^{(r_i)}_{s_k^{(i)}}(\mathbf{x}_k^*, \mathbf{p}) = 0 \text{ for } r_i = 2, 3, \cdots, 2m_i - 1,\\
&G^{(2m_i)}_{s_k^{(i)}}(\mathbf{x}_k^*, \mathbf{p}) < 0 \text{ stable for } s_k^{(i)} > 0;\\
&G^{(2m_i)}_{s_k^{(i)}}(\mathbf{x}_k^*, \mathbf{p}) < 0 \text{ unstable for } s_k^{(i)} < 0
\end{aligned}
\tag{2.71}
$$

for $\mathbf{x}_k \in U(\mathbf{x}_k^*) \subset \Omega_\alpha$.

(iv) $\mathbf{x}_{k+j}(j \in \mathbb{Z})$ *at fixed point* $\mathbf{x}_k^*$ *on the direction* $\mathbf{v}_i$ *is monotonically unstable of the* $(2m_i)$th *order, upper saddle if and only if*

$$\begin{aligned} &G_{s_k^{(i)}}^{(1)}(\mathbf{x}_k^*, \mathbf{p}) = \lambda_i = 1, \\ &G_{s_k^{(i)}}^{(r_i)}(\mathbf{x}_k^*, \mathbf{p}) = 0 \text{ for } r_i = 2, 3, \cdots, 2m_i - 1, \\ &G_{s_k^{(i)}}^{(2m_i)}(\mathbf{x}_k^*, \mathbf{p}) > 0 \text{ unstable for } s_k^{(i)} > 0; \\ &G_{s_k^{(i)}}^{(2m_i)}(\mathbf{x}_k^*, \mathbf{p}) > 0 \text{ stable for } s_k^{(i)} < 0 \end{aligned} \tag{2.72}$$

for $\mathbf{x}_k \in U(\mathbf{x}_k^*) \subset \Omega_\alpha$.

(v) $\mathbf{x}_{k+j}(j \in \mathbb{Z})$ *at fixed point* $\mathbf{x}_k^*$ *on the direction* $\mathbf{v}_i$ *is oscillatory stable of the* $(2m_i + 1)$th *-order if and only if*

$$\begin{aligned} &G_{s_k^{(i)}}^{(1)}(\mathbf{x}_k^*, \mathbf{p}) = \lambda_i = -1, \\ &G_{s_k^{(i)}}^{(r_i)}(\mathbf{x}_k^*, \mathbf{p}) = 0 \text{ for } r_i = 2, 3, \cdots, 2m_i, \\ &G_{s_k^{(i)}}^{(2m_i+1)}(\mathbf{x}_k^*, \mathbf{p}) > 0 \end{aligned} \tag{2.73}$$

for $\mathbf{x}_k \in U(\mathbf{x}_k^*) \subset \Omega_\alpha$.

(vi) $\mathbf{x}_{k+j}(j \in \mathbb{Z})$ *at fixed point* $\mathbf{x}_k^*$ *on the direction* $\mathbf{v}_i$ *is oscillatory unstable of the* $(2m_i + 1)$th *-order if and only if*

$$\begin{aligned} &G_{s_k^{(i)}}^{(1)}(\mathbf{x}_k^*, \mathbf{p}) = \lambda_i = -1, \\ &G_{s_k^{(i)}}^{(r_i)}(\mathbf{x}_k^*, \mathbf{p}) = 0 \text{ for } r_i = 2, 3, \cdots, 2m_i, \\ &G_{s_k^{(i)}}^{(2m_i+1)}(\mathbf{x}_k^*, \mathbf{p}) < 0 \end{aligned} \tag{2.74}$$

for $\mathbf{x}_k \in U(\mathbf{x}_k^*) \subset \Omega_\alpha$.

(vii) $\mathbf{x}_{k+j}(j \in \mathbb{Z})$ *at fixed point* $\mathbf{x}_k^*$ *on the direction* $\mathbf{v}_i$ *is oscillatory unstable of the* $(2m_i)$th *-order, upper saddle if and only if*

$$\begin{aligned} &G_{s_k^{(i)}}^{(1)}(\mathbf{x}_k^*, \mathbf{p}) = \lambda_i = -1, \\ &G_{s_k^{(i)}}^{(r_i)}(\mathbf{x}_k^*, \mathbf{p}) = 0 \text{ for } r_i = 2, 3, \cdots, 2m_i - 1, \\ &G_{s_k^{(i)}}^{(2m_i)}(\mathbf{x}_k^*, \mathbf{p}) > 0 \text{ stable for } s_k^{(i)} > 0; \\ &G_{s_k^{(i)}}^{(2m_i)}(\mathbf{x}_k^*, \mathbf{p}) > 0 \text{ unstable for } s_k^{(i)} < 0 \end{aligned} \tag{2.75}$$

for $\mathbf{x}_k \in U(\mathbf{x}_k^*) \subset \Omega_\alpha$.

(viii) $\mathbf{x}_{k+j} (j \in \mathbb{Z})$ *at fixed point* $\mathbf{x}_k^*$ *in the direction* $\mathbf{v}_i$ *is oscillatory unstable of the* $(2m_i)$th *-order lower saddle if and only if*

$$\begin{aligned} &G^{(1)}_{s_k^{(i)}}(\mathbf{x}_k^*, \mathbf{p}) = \lambda_i = -1, \\ &G^{(r_i)}_{s_k^{(i)}}(\mathbf{x}_k^*, \mathbf{p}) = 0 \text{ for } r_i = 2, 3, \cdots, 2m_i - 1, \\ &G^{(2m_i)}_{s_k^{(i)}}(\mathbf{x}_k^*, \mathbf{p}) < 0 \text{ stable for } s_k^{(i)} < 0; \\ &G^{(2m_i)}_{s_k^{(i)}}(\mathbf{x}_k^*, \mathbf{p}) < 0 \text{ unstable for } s_k^{(i)} > 0 \end{aligned} \tag{2.76}$$

for $\mathbf{x}_k \in U(\mathbf{x}_k^*) \subset \Omega_\alpha$.

Proof Because

$$\begin{aligned} s_{k+1}^{(i)} =& \mathbf{v}_i^{\mathrm{T}} \cdot (\mathbf{x}_{k+1} - \mathbf{x}_k^*) \\ =& \mathbf{v}_i^{\mathrm{T}} \cdot \mathbf{x}_k^* + G^{(1)}_{s_k^{(i)}}(\mathbf{x}_k^*, \mathbf{p}) s_k^{(i)} + \sum_{r_i=2}^{2m_i+1} \frac{1}{r_i!} G^{(r_i)}_{s_k^{(i)}}(\mathbf{x}_k^*, \mathbf{p}) (s_k^{(i)})^{r_i} \\ &- \mathbf{v}_i^{\mathrm{T}} \cdot \mathbf{x}_k^* + o((s_k^{(i)})^{2m_i+1}) \\ =& G^{(1)}_{s_k^{(i)}}(\mathbf{x}_k^*, \mathbf{p}) s_k^{(i)} + \sum_{r_i=2}^{2m_i} \frac{1}{r_i!} G^{(r_i)}_{s_k^{(i)}}(\mathbf{x}_k^*, \mathbf{p}) (s_k^{(i)})^{r_i} \\ &+ \frac{1}{(2m_i+1)!} G^{(2m_i+1)}_{s_k^{(i)}}(\mathbf{x}_k^*, \mathbf{p}) (s_k^{(i)})^{2m_i+1} + o((s_k^{(i)})^{2m_i+1}) \end{aligned}$$

and

$$s_k^{(i)} = \mathbf{v}_i^{\mathrm{T}} \cdot (\mathbf{x}_k - \mathbf{x}_k^*).$$

(i) From the first two equations of Eq. (2.69), for the infinitesimal $s_k^{(i)}$, one obtains

$$s_{k+1}^{(i)} = [G^{(1)}_{s_k^{(i)}}(\mathbf{x}_k^*, \mathbf{p}) + \frac{1}{(2m_i+1)!} G^{(2m_i+1)}_{s_k^{(i)}}(\mathbf{x}_k^*, \mathbf{p}) (s_k^{(i)})^{2m_i}] s_k^{(i)}.$$

Since

$$|\mathbf{v}_i^{\mathrm{T}} \cdot (\mathbf{x}_{k+1} - \mathbf{x}_k^*)| < |\mathbf{v}_i^{\mathrm{T}} \cdot (\mathbf{x}_k - \mathbf{x}_k^*)|,$$

we have

$$
\begin{aligned}
|s_{k+1}^{(i)}| &= \left|[G_{s_k^{(i)}}^{(1)}(\mathbf{x}_k^*, \mathbf{p}) + \frac{1}{(2m_i+1)!} G_{s_k^{(i)}}^{(2m_i+1)}(\mathbf{x}_k^*, \mathbf{p})(s_k^{(i)})^{2m_i}] s_k^{(i)}\right| \\
&= \left| G_{s_k^{(i)}}^{(1)}(\mathbf{x}_k^*, \mathbf{p}) \frac{1}{(2m_i+1)!} + G_{s_k^{(i)}}^{(2m_i+1)}(\mathbf{x}_k^*, \mathbf{p})(s_k^{(i)})^{2m_i} \right| |s_k^{(i)}| \\
&< |s_k^{(i)}|.
\end{aligned}
$$

For $G_{s_k^{(i)}}^{(1)}(\mathbf{x}_k^*, \mathbf{p}) = 1$, we have

$$
|1 + \frac{1}{(2m_i+1)!} G_{s_k^{(i)}}^{(2m_i+1)}(\mathbf{x}_k^*, \mathbf{p})(s_k^{(i)})^{2m_i}| < 1.
$$

Since the infinitesimal $s_k^{(i)}$ is arbitrarily selected, the foregoing equation gives

$$
G_{s_k^{(i)}}^{(2m_i+1)}(\mathbf{x}_k^*, \mathbf{p}) < 0.
$$

Therefore, $\mathbf{x}_{k+j}$ ($j \in \mathbb{Z}$) at fixed point $\mathbf{x}_k^*$ on the direction $\mathbf{v}_i$ is monotonically stable of the $(2m_i + 1)$th-order, vice versa.

(ii) Similarly, since

$$
|\mathbf{v}_i^{\mathrm{T}} \cdot (\mathbf{x}_{k+1} - \mathbf{x}_k^*)| < |\mathbf{v}_i^{\mathrm{T}} \cdot (\mathbf{x}_k - \mathbf{x}_k^*)|,
$$

we have

$$
\left|1 + \frac{1}{(2m_i+1)!} G_{s_k^{(i)}}^{(2m_i+1)}(\mathbf{x}_k^*, \mathbf{p})(s_k^{(i)})^{2m_i}\right| > 1.
$$

For the arbitrarily infinitesimal $s_k^{(i)}$, the foregoing equation requires

$$
G_{s_k^{(i)}}^{(2m_i+1)}(\mathbf{x}_k^*, \mathbf{p}) > 0.
$$

Therefore, $\mathbf{x}_{k+j}(j \in \mathbb{Z})$ at fixed point $\mathbf{x}_k^*$ on the direction $\mathbf{v}_i$ is monotonically unstable of the $(2m_i + 1)$th-order and vice versa.

(iii) The Taylor expansion of $s_{k+1}^{(i)}$ keeps up to the $(2m_i)$th term of $s_k^{(i)}$

$$
\begin{aligned}
s_{k+1}^{(i)} =& \mathbf{v}_i^{\mathrm{T}} \cdot (\mathbf{x}_{k+1} - \mathbf{x}_k^*) \\
=& \mathbf{v}_i^{\mathrm{T}} \cdot \mathbf{x}_k^* + G_{s_k^{(i)}}^{(1)}(\mathbf{x}_k^*, \mathbf{p}) s_k^{(i)} + \sum_{r_i=2}^{2m_i} \frac{1}{r_i!} G_{s_k^{(i)}}^{(r_i)}(\mathbf{x}_k^*, \mathbf{p})(s_k^{(i)})^{r_i} \\
& - \mathbf{v}_i^{\mathrm{T}} \cdot \mathbf{x}_k^* + o((s_k^{(i)})^{2m_i}) \\
=& G_{s_k^{(i)}}^{(1)}(\mathbf{x}_k^*, \mathbf{p}) s_k^{(i)} + \sum_{r_i=2}^{2m_i-1} \frac{1}{r_i!} G_{s_k^{(i)}}^{(r_i)}(\mathbf{x}_k^*, \mathbf{p})(s_k^{(i)})^{r_i} \\
& + \frac{1}{(2m_i)!} G_{s_k^{(i)}}^{(2m_i)}(\mathbf{x}_k^*, \mathbf{p})(s_k^{(i)})^{2m_i} + o((s_k^{(i)})^{2m_i}).
\end{aligned}
$$

From the first two equations of Eq. (2.71), for the infinitesimal $s_k^{(i)}$, one obtains

$$s_{k+1}^{(i)} = s_k^{(i)} + \frac{1}{(2m_i)!} G_{s_k^{(i)}}^{(2m_i)}(\mathbf{x}_k^*, \mathbf{p})(s_k^{(i)})^{2m_i}.$$

Thus

$$\begin{aligned} |s_{k+1}^{(i)}| &= \left| [1 + \frac{1}{(2m_i)!} G_{s_k^{(i)}}^{(2m_i)}(\mathbf{x}_k^*, \mathbf{p})(s_k^{(i)})^{2m_i-1}] s_k^{(i)} \right| \\ &= \left| 1 + \frac{1}{(2m_i)!} G_{s_k^{(i)}}^{(2m_i)}(\mathbf{x}_k^*, \mathbf{p})(s_k^{(i)})^{2m_i-1} \right| |s_k^{(i)}|. \end{aligned}$$

For $G_{s_k^{(i)}}^{(2m_i)}(\mathbf{x}_k^*, \mathbf{p}) < 0$, if $s_k^{(i)} > 0$, we have

$$|s_{k+1}^{(i)}| < |s_k^{(i)}| \Rightarrow |\mathbf{v}_i^{\mathrm{T}} \cdot (\mathbf{x}_{k+1} - \mathbf{x}_k^*)| < |\mathbf{v}_i^{\mathrm{T}} \cdot (\mathbf{x}_k - \mathbf{x}_k^*)|,$$

if $s_k^{(i)} < 0$, we have

$$|s_{k+1}^{(i)}| > |s_k^{(i)}| \Rightarrow |\mathbf{v}_i^{\mathrm{T}} \cdot (\mathbf{x}_{k+1} - \mathbf{x}_k^*)| > |\mathbf{v}_i^{\mathrm{T}} \cdot (\mathbf{x}_k - \mathbf{x}_k^*)|,$$

Thus, $\mathbf{x}_{k+j} (j \in \mathbb{Z})$ at fixed point $\mathbf{x}_k^*$ on the direction $\mathbf{v}_i$ is monotonically unstable of the $(2m_i)$th-order, lower saddle and vice versa.

(iv) Similar to (iii), for $G_{s_k^{(i)}}^{(2m_i)}(\mathbf{x}_k^*, \mathbf{p}) > 0$, if $s_k^{(i)} > 0$, we have

$$|s_{k+1}^{(i)}| > |s_k^{(i)}| \Rightarrow |\mathbf{v}_i^{\mathrm{T}} \cdot (\mathbf{x}_{k+1} - \mathbf{x}_k^*)| > |\mathbf{v}_i^{\mathrm{T}} \cdot (\mathbf{x}_k - \mathbf{x}_k^*)|,$$

if $s_k^{(i)} < 0$, we have

$$|s_{k+1}^{(i)}| < |s_k^{(i)}| \Rightarrow |\mathbf{v}_i^{\mathrm{T}} \cdot (\mathbf{x}_{k+1} - \mathbf{x}_k^*)| < |\mathbf{v}_i^{\mathrm{T}} \cdot (\mathbf{x}_k - \mathbf{x}_k^*)|.$$

Thus, $\mathbf{x}_{k+j} (j \in \mathbb{Z})$ at fixed point $\mathbf{x}_k^*$ on the direction $\mathbf{v}_i$ is monotonically unstable of the $(2m_i)$th-order, upper saddle and vice versa.

(v) Similar to case (i), consider

$$|\mathbf{v}_i^{\mathrm{T}} \cdot (\mathbf{x}_{k+1} - \mathbf{x}_k^*)| < |\mathbf{v}_i^{\mathrm{T}} \cdot (\mathbf{x}_k - \mathbf{x}_k^*)|,$$

For $G_{s_k^{(i)}}^{(1)}(\mathbf{x}_k^*, \mathbf{p}) = -1$, we have

$$|-1 + \frac{1}{(2m_i+1)!} G_{s_k^{(i)}}^{(2m_i+1)}(\mathbf{x}_k^*, \mathbf{p})(s_k^{(i)})^{2m_i}| < 1.$$

Since the infinitesimal $s_k^{(i)}$ is arbitrarily selected, the foregoing equation gives

$$G^{(2m_i+1)}_{s_k^{(i)}}(\mathbf{x}_k^*, \mathbf{p}) > 0.$$

Therefore, $\mathbf{x}_{k+j} (j \in \mathbb{Z})$ at fixed point $\mathbf{x}_k^*$ on the direction $\mathbf{v}_i$ is oscillatory stable of the $(2m_i + 1)$th-order, vice versa.

(vi) Similar to case (ii), consider

$$|\mathbf{v}_i^{\mathrm{T}} \cdot (\mathbf{x}_{k+1} - \mathbf{x}_k^*)| > |\mathbf{v}_i^{\mathrm{T}} \cdot (\mathbf{x}_k - \mathbf{x}_k^*)|,$$

For $G^{(1)}_{s_k^{(i)}}(\mathbf{x}_k^*, \mathbf{p}) = -1$, we have

$$\left|-1 + \frac{1}{(2m_i + 1)!} G^{(2m_i+1)}_{s_k^{(i)}}(\mathbf{x}_k^*, \mathbf{p})(s_k^{(i)})^{2m_i}\right| > 1.$$

Since the infinitesimal $s_k^{(i)}$ is arbitrarily selected, the foregoing equation gives

$$G^{(2m_i+1)}_{s_k^{(i)}}(\mathbf{x}_k^*, \mathbf{p}) < 0.$$

Therefore, $\mathbf{x}_{k+j} (j \in \mathbb{Z})$ at fixed point $\mathbf{x}_k^*$ on the direction $\mathbf{v}_i$ is oscillatory unstable of the $(2m_i + 1)$th-order and vice versa.

(vii) Similar to (iii), from the first two equations of Eq. (2.75), for the infinitesimal $s_k^{(i)}$, one obtains

$$s_{k+1}^{(i)} = -s_k^{(i)} + \frac{1}{(2m_i)!} G^{(2m_i)}_{s_k^{(i)}}(\mathbf{x}_k^*, \mathbf{p})(s_k^{(i)})^{2m_i}$$

Thus

$$\begin{aligned} |s_{k+1}^{(i)}| &= \left|[-1 + \frac{1}{(2m_i)!} G^{(2m_i)}_{s_k^{(i)}}(\mathbf{x}_k^*, \mathbf{p})(s_k^{(i)})^{2m_i-1}] s_k^{(i)}\right| \\ &= \left|-1 + \frac{1}{(2m_i)!} G^{(2m_i)}_{s_k^{(i)}}(\mathbf{x}_k^*, \mathbf{p})(s_k^{(i)})^{2m_i-1}\right| |s_k^{(i)}|. \end{aligned}$$

For $G^{(2m_i)}_{s_k^{(i)}}(\mathbf{x}_k^*, \mathbf{p}) > 0$, if $s_k^{(i)} < 0$, we have

$$|s_{k+1}^{(i)}| < |s_k^{(i)}| \Rightarrow |\mathbf{v}_i^{\mathrm{T}} \cdot (\mathbf{x}_{k+1} - \mathbf{x}_k^*)| < |\mathbf{v}_i^{\mathrm{T}} \cdot (\mathbf{x}_k - \mathbf{x}_k^*)|,$$

if $s_k^{(i)} < 0$, we have

$$|s_{k+1}^{(i)}| > |s_k^{(i)}| \Rightarrow |\mathbf{v}_i^{\mathrm{T}} \cdot (\mathbf{x}_{k+1} - \mathbf{x}_k^*)| > |\mathbf{v}_i^{\mathrm{T}} \cdot (\mathbf{x}_k - \mathbf{x}_k^*)|.$$

Thus, $\mathbf{x}_{k+j} (j \in \mathbb{Z})$ at fixed point $\mathbf{x}_k^*$ on the direction $\mathbf{v}_i$ is oscillatory unstable of the $(2m_i)$th-order, upper saddle and vice versa.

(viii) Similar to (vii), for $G^{(2m_i)}_{s_k^{(i)}}(\mathbf{x}_k^*, \mathbf{p}) < 0$, if $s_k^{(i)} > 0$, we have

$$|s_{k+1}^{(i)}| > |s_k^{(i)}| \Rightarrow |\mathbf{v}_i^{\mathrm{T}} \cdot (\mathbf{x}_{k+1} - \mathbf{x}_k^*)| > |\mathbf{v}_i^{\mathrm{T}} \cdot (\mathbf{x}_k - \mathbf{x}_k^*)|,$$

if $s_k^{(i)} > 0$, we have

$$|s_{k+1}^{(i)}| < |s_k^{(i)}| \Rightarrow |\mathbf{v}_i^{\mathrm{T}} \cdot (\mathbf{x}_{k+1} - \mathbf{x}_k^*)| < |\mathbf{v}_i^{\mathrm{T}} \cdot (\mathbf{x}_k - \mathbf{x}_k^*)|.$$

Thus, $\mathbf{x}_{k+j}(j \in \mathbb{Z})$ at fixed point $\mathbf{x}_k^*$ in the direction $\mathbf{v}_i$ is monotonically unstable of the $(2m_i)$th-order, lower saddle and vice versa. This theorem is proved. ■

Definition 2.27 Consider a discrete, nonlinear dynamical system $\mathbf{x}_{k+1} = \mathbf{f}(\mathbf{x}_k, \mathbf{p}) \in \mathscr{R}^n$ in Eq. (2.4) with a fixed point $\mathbf{x}_k^*$. The corresponding solution is given by $\mathbf{x}_{k+j} = \mathbf{f}(\mathbf{x}_{k+j-1}, \mathbf{p})$ with $j \in \mathbb{Z}$. Suppose there is a neighborhood of the fixed point $\mathbf{x}_k^*$ (i.e., $U_k(\mathbf{x}_k^*) \subset \Omega$), and $\mathbf{f}(\mathbf{x}_k, \mathbf{p})$ is C^r $(r \geq 1)$-continuous in $U_k(\mathbf{x}_k^*)$ with Eq. (2.28). The linearized system is $\mathbf{y}_{k+j+1} = D\mathbf{f}(\mathbf{x}_k^*, \mathbf{p})\mathbf{y}_{k+j}$ $(\mathbf{y}_{k+j} = \mathbf{x}_{k+j} - \mathbf{x}_k^*)$ in $U_k(\mathbf{x}_k^*)$. Consider a pair of complex eigenvalue $\alpha_i \pm \mathbf{i}\beta_i$ $(i \in N = \{1, 2, \cdots, n\},\ \mathbf{i} = \sqrt{-1})$ of matrix $D\mathbf{f}(\mathbf{x}^*, \mathbf{p})$ with a pair of eigenvectors $\mathbf{u}_i \pm \mathbf{i}\mathbf{v}_i$. On the invariant plane of $(\mathbf{u}_i, \mathbf{v}_i)$, consider $\mathbf{r}_k^{(i)} = \mathbf{y}_k = \mathbf{y}_{k+}^{(i)} + \mathbf{y}_{k-}^{(i)}$ with

$$\begin{aligned} \mathbf{r}_k^{(i)} &= c_k^{(i)}\mathbf{u}_i + d_k^{(i)}\mathbf{v}_i, \\ \mathbf{r}_{k+1}^{(i)} &= c_{k+1}^{(i)}\mathbf{u}_i + d_{k+1}^{(i)}\mathbf{v}_i; \end{aligned} \tag{2.77}$$

and

$$\begin{aligned} c_k^{(i)} &= \frac{1}{\Delta}[\Delta_2(\mathbf{u}_i^{\mathrm{T}} \cdot \mathbf{y}_k) - \Delta_{12}(\mathbf{v}_i^{\mathrm{T}} \cdot \mathbf{y}_k)], \\ d_k^{(i)} &= \frac{1}{\Delta}[\Delta_1(\mathbf{v}_i^{\mathrm{T}} \cdot \mathbf{y}_k) - \Delta_{12}(\mathbf{u}_i^{\mathrm{T}} \cdot \mathbf{y}_k)]; \\ \Delta_1 &= ||\mathbf{u}_i||^2, \Delta_2 = ||\mathbf{v}_i||^2, \Delta_{12} = \mathbf{u}_i^{\mathrm{T}} \cdot \mathbf{v}_i; \\ \Delta &= \Delta_1\Delta_2 - \Delta_{12}^2. \end{aligned} \tag{2.78}$$

Consider a polar coordinate of (r_k, θ_k) defined by

$$\begin{aligned} & c_k^{(i)} = r_k^{(i)} \cos\theta_k^{(i)}, \text{ and } d_k^{(i)} = r_k^{(i)} \sin\theta_k^{(i)}; \\ & r_k^{(i)} = \sqrt{(c_k^{(i)})^2 + (d_k^{(i)})^2}, \text{ and } \theta_k^{(i)} = \arctan d_k^{(i)}/c_k^{(i)}. \end{aligned} \tag{2.79}$$

Thus

$$\begin{aligned} c_{k+1}^{(i)} &= \frac{1}{\Delta}[\Delta_2 G_{c_k^{(i)}}(\mathbf{x}_k, \mathbf{p}) - \Delta_{12} G_{d_k^{(i)}}(\mathbf{x}_k, \mathbf{p})] \\ d_{k+1}^{(i)} &= \frac{1}{\Delta}[\Delta_1 G_{d_k^{(i)}}(\mathbf{x}_k, \mathbf{p}) - \Delta_{12} G_{c_k^{(i)}}(\mathbf{x}_k, \mathbf{p})] \end{aligned} \tag{2.80}$$

where

$$G_{c_k^{(i)}}(\mathbf{x}_k,\mathbf{p}) = \mathbf{u}_i^{\mathrm{T}} \cdot [\mathbf{f}(\mathbf{x}_k,\mathbf{p}) - \mathbf{x}_k^*] = \sum\nolimits_{m_i=1}^{\infty} \frac{1}{m_i!} G_{c_k^{(i)}}^{(m_i)}(\theta_k^{(i)})(r_k^{(i)})^{m_i},$$
$$G_{d_k^{(i)}}(\mathbf{x}_k,\mathbf{p}) = \mathbf{v}_i^{\mathrm{T}} \cdot [\mathbf{f}(\mathbf{x}_k,\mathbf{p}) - \mathbf{x}_k^*] = \sum\nolimits_{m_i=1}^{\infty} \frac{1}{m_i!} G_{d_k^{(i)}}^{(m_i)}(\theta_k^{(i)})(r_k^{(i)})^{m_i}; \tag{2.81}$$

$$G_{c_k^{(i)}}^{(m_i)}(\theta_k^{(i)}) = \mathbf{u}_i^{\mathrm{T}} \cdot \partial_{\mathbf{x}_k}^{(m_i)} \mathbf{f}(\mathbf{x}_k,\mathbf{p})[\mathbf{u}_i \cos\theta_k^{(i)} + \mathbf{v}_i \sin\theta_k^{(i)}]^{m_i}\Big|_{(\mathbf{x}_k^*,\mathbf{p})},$$
$$G_{d_k^{(i)}}^{(m_i)}(\theta_k^{(i)}) = \mathbf{v}_i^{\mathrm{T}} \cdot \partial_{\mathbf{x}_k}^{(m)} \mathbf{f}(\mathbf{x}_k,\mathbf{p})[\mathbf{u}_i \cos\theta_k^{(i)} + \mathbf{v}_i \sin\theta_k^{(i)}]^{m_i}\Big|_{(\mathbf{x}_k^*,\mathbf{p})}. \tag{2.82}$$

Thus

$$\begin{aligned} r_{k+1}^{(i)} &= \sqrt{(c_{k+1}^{(i)})^2 + (d_{k+1}^{(i)})^2} = \sqrt{\sum\nolimits_{m_i=2}^{\infty} (r_k^{(i)})^{m_i} G_{r_{k+1}^{(i)}}^{(m_i)}(\theta_k^{(i)})} \\ &= \sqrt{G_{r_{k+1}^{(i)}}^{(2)} r_k^{(i)}} \sqrt{1 + (G_{r_{k+1}^{(i)}}^{(2)})^{-1} \sum\nolimits_{m_i=3}^{\infty} (r_k^{(i)})^{m_i-2} G_{r_{k+1}^{(i)}}^{(m_i)}(\theta_k^{(i)})} \\ \theta_{k+1}^{(i)} &= \arctan(d_{k+1}^{(i)}/c_{k+1}^{(i)}) \end{aligned} \tag{2.83}$$

where

$$\begin{aligned} G_{r_{k+1}^{(i)}}^{(m_i)}(\theta_k^{(i)}) &= \sum_{r_i=1}^{\infty}\sum_{s_i=1}^{\infty} \frac{1}{r_i!}\frac{1}{s_i!} G_{c_{k+1}^{(i)}}^{(r_i)}(\theta_k^{(i)}) G_{c_{k+1}^{(i)}}^{(s_i)}(\theta_k^{(i)}) + G_{d_{k+1}^{(i)}}^{(r_i)}(\theta_k^{(i)}) G_{d_{k+1}^{(i)}}^{(s_i)}(\theta_k^{(i)}) \delta_{m_i}^{r_i+s_i} \\ &= \frac{1}{m_i!} \sum_{q=1}^{m_i-1} C_{m_i}^q [G_{c_{k+1}^{(i)}}^{(q)}(\theta_k^{(i)}) G_{c_{k+1}^{(i)}}^{(m_i-q)}(\theta_k^{(i)}) + G_{d_{k+1}^{(i)}}^{(q)}(\theta_k^{(i)}) G_{d_{k+1}^{(i)}}^{(m_i-q)}(\theta_k^{(i)})], \end{aligned} \tag{2.84}$$

and

$$\begin{aligned} G_{c_{k+1}^{(i)}}^{(m_i)}(\theta_k^{(i)}) &= \frac{1}{\Delta}[\Delta_2 G_{c_k^{(i)}}^{(m_i)}(\theta_k^{(i)}) - \Delta_{12} G_{d_k^{(i)}}^{(m_i)}(\theta_k^{(i)})], \\ G_{d_{k+1}^{(i)}}^{(m_i)}(\theta_k) &= \frac{1}{\Delta}[\Delta_1 G_{d_k^{(i)}}^{(m_i)}(\theta_k^{(i)}) - \Delta_{12} G_{c_k^{(i)}}^{(m_i)}(\theta_k^{(i)})], \\ C_{m_i}^q &= \frac{m_i!}{q!(m_i-q)!}. \end{aligned} \tag{2.85}$$

From the foregoing definition, consider the first order terms of G-function

$$\begin{aligned} G_{c_k^{(i)}}^{(1)}(\mathbf{x}_k,\mathbf{p}) &= G_{c_k^{(i)}1}^{(1)}(\mathbf{x}_k,\mathbf{p}) + G_{c_k^{(i)}2}^{(1)}(\mathbf{x}_k,\mathbf{p}), \\ G_{d_k^{(i)}}^{(1)}(\mathbf{x}_k,\mathbf{p}) &= G_{d_k^{(i)}1}^{(1)}(\mathbf{x}_k,\mathbf{p}) + G_{d_k^{(i)}2}^{(1)}(\mathbf{x}_k,\mathbf{p}), \end{aligned} \tag{2.86}$$

where

$$\begin{aligned} G^{(1)}_{c_k^{(i)}1}(\mathbf{x}_k,\mathbf{p}) &= \mathbf{u}_i^{\mathrm{T}}\cdot D_{\mathbf{x}_k}\mathbf{f}(\mathbf{x}_k,\mathbf{p})\partial_{c_k^{(i)}}\mathbf{x}_k = \mathbf{u}_i^{\mathrm{T}}\cdot D_{\mathbf{x}_k}\mathbf{f}(\mathbf{x}_k,\mathbf{p})\mathbf{u}_i \\ &= \mathbf{u}_i^{\mathrm{T}}\cdot(-\beta_i\mathbf{v}_i+\alpha_i\mathbf{u}_i) = \alpha_i\Delta_1-\beta_i\Delta_{12}, \\ G^{(1)}_{c_k^{(i)}2}(\mathbf{x}_k,\mathbf{p}) &= \mathbf{u}_i^{\mathrm{T}}\cdot D_{\mathbf{x}_k}\mathbf{f}(\mathbf{x}_k,\mathbf{p})\partial_{d_k^{(i)}}\mathbf{x}_k = \mathbf{u}_i^{\mathrm{T}}\cdot D_{\mathbf{x}_k}\mathbf{f}(\mathbf{x}_k,\mathbf{p})\mathbf{v}_i \\ &= \mathbf{u}_i^{\mathrm{T}}\cdot(\beta_i\mathbf{u}_i+\alpha_i\mathbf{v}_i) = \alpha_i\Delta_{12}+\beta_i\Delta_1; \end{aligned} \tag{2.87}$$

and

$$\begin{aligned} G^{(1)}_{d_k^{(i)}1}(\mathbf{x}_k,\mathbf{p}) &= \mathbf{v}_i^{\mathrm{T}}\cdot D_{\mathbf{x}_k}\mathbf{f}(\mathbf{x}_k,\mathbf{p})\partial_{c_k^{(i)}}\mathbf{x}_k = \mathbf{v}_i^{\mathrm{T}}\cdot D_{\mathbf{x}_k}\mathbf{f}(\mathbf{x}_k,\mathbf{p})\mathbf{u}_i \\ &= \mathbf{v}_i^{\mathrm{T}}\cdot(-\beta_i\mathbf{v}_i+\alpha_i\mathbf{u}_i) = -\beta_i\Delta_2+\alpha_i\Delta_{12}, \\ G^{(1)}_{d_k^{(i)}2}(\mathbf{x},\mathbf{p}) &= \mathbf{v}_i^{\mathrm{T}}\cdot D_{\mathbf{x}_k}\mathbf{f}(\mathbf{x}_k,\mathbf{p})\partial_{d_k^{(i)}}\mathbf{x}_k = \mathbf{v}_i^{\mathrm{T}}\cdot D_{\mathbf{x}_k}\mathbf{f}(\mathbf{x}_k,\mathbf{p})\mathbf{v}_i \\ &= \mathbf{v}_i^{\mathrm{T}}\cdot(\beta_i\mathbf{u}_i+\alpha_i\mathbf{v}_i) = \alpha_i\Delta_2+\beta_i\Delta_{12}. \end{aligned} \tag{2.88}$$

Substitution of Eqs. (2.86)–(2.88) into Eq. (2.82) gives

$$\begin{aligned} G^{(1)}_{c_k^{(i)}}(\theta_k^{(i)}) &= G^{(1)}_{c_k^{(i)}1}(\mathbf{x}_k,\mathbf{p})\cos\theta_k^{(i)} + G^{(1)}_{c_k^{(i)}2}(\mathbf{x}_k,\mathbf{p})\sin\theta_k^{(i)} \\ &= (\alpha_i\Delta_1-\beta_i\Delta_{12})\cos\theta_k^{(i)} + (\alpha_i\Delta_{12}+\beta_i\Delta_1)\sin\theta_k^{(i)}, \\ G^{(1)}_{d_k^{(i)}}(\theta_k^{(i)}) &= G^{(1)}_{d_k^{(i)}1}(\mathbf{x}_k,\mathbf{p})\cos\theta_k^{(i)} + G^{(1)}_{d_k^{(i)}2}(\mathbf{x}_k,\mathbf{p})\sin\theta_k^{(i)} \\ &= (-\beta_i\Delta_2+\alpha_i\Delta_{12})\cos\theta_k^{(i)} + (\alpha_i\Delta_2+\beta_i\Delta_{12})\sin\theta_k^{(i)}. \end{aligned} \tag{2.89}$$

From Eq. (2.85), we have

$$\begin{aligned} G^{(1)}_{c_{k+1}^{(i)}}(\theta_k^{(i)}) &= \frac{1}{\Delta}[\Delta_2 G^{(1)}_{c_k^{(i)}}(\theta_k^{(i)}) - \Delta_{12}G^{(1)}_{d_k^{(i)}}(\theta_k^{(i)})] \\ &= \alpha_i\cos\theta_k^{(i)} + \beta_i\sin\theta_k^{(i)} \\ G^{(1)}_{d_{k+1}^{(i)}}(\theta_k^{(i)}) &= \frac{1}{\Delta}[\Delta_1 G^{(1)}_{d_k^{(i)}}(\theta_k^{(i)}) - \Delta_{12}G^{(1)}_{c_k^{(i)}}(\theta_k^{(i)})] \\ &= \alpha_i\sin\theta_k^{(i)} - \beta_i\cos\theta_k^{(i)}. \end{aligned} \tag{2.90}$$

Thus

$$\begin{aligned} G^{(2)}_{r_{k+1}^{(i)}}(\theta_k^{(i)}) &= [G^{(1)}_{c_{k+1}^{(i)}}(\theta_k^{(i)})G^{(1)}_{c_{k+1}^{(i)}}(\theta_k^{(i)}) + G^{(1)}_{d_{k+1}^{(i)}}(\theta_k^{(i)})G^{(1)}_{d_{k+1}^{(i)}}(\theta_k^{(i)})] \\ &= \alpha_i^2+\beta_i^2. \end{aligned} \tag{2.91}$$

Furthermore, Eq. (2.83) gives

$$r_{k+1}^{(i)} = \rho_i r_k^{(i)} + o(r_k^{(i)}) \text{ and } \theta_{k+1}^{(i)} = \theta_k^{(i)} - \vartheta_i + o(r_k^{(i)}), \tag{2.92}$$

where

$$\vartheta_i = \arctan(\beta_i/\alpha_i) \text{ and } \rho_i = \sqrt{\alpha_i^2 + \beta_i^2}. \tag{2.93}$$

As $r_k^{(i)} \ll 1$ and $r_k^{(i)} \to 0$, we have

$$r_{k+1}^{(i)} = \rho_i r_k^{(i)} \text{ and } \theta_{k+1}^{(i)} = \vartheta_i - \theta_k^{(i)}. \tag{2.94}$$

With an initial condition of $r_k^{(i)} = r_k^0$ and $\theta_k^{(i)} = \theta_k^{(i)}$, the corresponding solution of Eq. (2.94) is

$$r_{k+j}^{(i)} = (\rho_i)^j r_k^0 \text{ and } \theta_{k+j}^{(i)} = j\vartheta_i - \theta_k^{(i)}. \tag{2.95}$$

From Eqs. (2.80), (2.81) and (2.90), we have

$$\begin{aligned} c_{k+1}^{(i)} &= \alpha_i r_k^{(i)} \cos\theta_k^{(i)} + \beta_i r_k^{(i)} \sin\theta_k^{(i)} = \alpha_i c_k^{(i)} + \beta_i d_k^{(i)}, \\ d_{k+1}^{(i)} &= \alpha_i r_k^{(i)} \sin\theta_k^{(i)} - \beta_i r_k^{(i)} \cos\theta_k^{(i)} = -\beta_i c_k^{(i)} + \alpha_i d_k^{(i)}. \end{aligned} \tag{2.96}$$

That is,

$$\begin{Bmatrix} c_{k+1}^{(i)} \\ d_{k+1}^{(i)} \end{Bmatrix} = \begin{bmatrix} \alpha_i & \beta_i \\ -\beta_i & \alpha_i \end{bmatrix} \begin{Bmatrix} c_k^{(i)} \\ d_k^{(i)} \end{Bmatrix} = \rho_i \begin{bmatrix} \cos\vartheta_i & \sin\vartheta_i \\ -\sin\vartheta_i & \cos\vartheta_i \end{bmatrix} \begin{Bmatrix} c_k^{(i)} \\ d_k^{(i)} \end{Bmatrix}. \tag{2.97}$$

From the foregoing equation, we have

$$\begin{Bmatrix} c_{k+j}^{(i)} \\ d_{k+j}^{(i)} \end{Bmatrix} = \begin{bmatrix} \alpha_i & \beta_i \\ -\beta_i & \alpha_i \end{bmatrix}^j \begin{Bmatrix} c_k^{(i)} \\ d_k^{(i)} \end{Bmatrix} = (\rho_i)^j \begin{bmatrix} \cos j\vartheta_i & \sin j\vartheta_i \\ -\sin j\vartheta_i & \cos j\vartheta_i \end{bmatrix} \begin{Bmatrix} c_k^{(i)} \\ d_k^{(i)} \end{Bmatrix}. \tag{2.98}$$

Definition 2.28 Consider a discrete, nonlinear dynamical system $\mathbf{x}_{k+1} = \mathbf{f}(\mathbf{x}_k, \mathbf{p}) \in \mathscr{R}^n$ in Eq. (2.4) with a fixed point $\mathbf{x}_k^*$. The corresponding solution is given by $\mathbf{x}_{k+j} = \mathbf{f}(\mathbf{x}_{k+j-1}, \mathbf{p})$ with $j \in \mathbb{Z}$. Suppose there is a neighborhood of the fixed point $\mathbf{x}_k^*$ (i.e., $U_k(\mathbf{x}_k^*) \subset \Omega$), and $\mathbf{f}(\mathbf{x}_k, \mathbf{p})$ is C^r $(r \geq 1)$-continuous in $U_k(\mathbf{x}_k^*)$ with Eq. (2.28). The linearized system is $\mathbf{y}_{k+j+1} = D\mathbf{f}(\mathbf{x}_k^*, \mathbf{p})\mathbf{y}_{k+j}$ $(\mathbf{y}_{k+j} = \mathbf{x}_{k+j} - \mathbf{x}_k^*)$ in $U_k(\mathbf{x}_k^*)$. Consider a pair of complex eigenvalues $\alpha_i \pm \mathbf{i}\beta_i$ $(i \in N = \{1, 2, \cdots, n\},\ \mathbf{i} = \sqrt{-1})$ of matrix $D\mathbf{f}(\mathbf{x}^*, \mathbf{p})$ with a pair of eigenvectors $\mathbf{u}_i \pm \mathbf{i}\mathbf{v}_i$. On the invariant plane of $(\mathbf{u}_i, \mathbf{v}_i)$, consider $\mathbf{r}_k^{(i)} = \mathbf{y}_k = \mathbf{y}_{k+}^{(i)} + \mathbf{y}_{k-}^{(i)}$ with Eqs. (2.77) and (2.79). For any arbitrarily small $\varepsilon > 0$, the stability of the fixed point $\mathbf{x}_k^*$ on the invariant plane of $(\mathbf{u}_i, \mathbf{v}_i)$ can be determined.

(i) $\mathbf{x}^{(k)}$ at the fixed point $\mathbf{x}_k^*$ on the plane of $(\mathbf{u}_i, \mathbf{v}_i)$ is spirally stable if

$$r_{k+1}^{(i)} - r_k^{(i)} < 0. \tag{2.99}$$

(ii) $\mathbf{x}^{(k)}$ at the fixed point $\mathbf{x}_k^*$ on the plane of $(\mathbf{u}_i, \mathbf{v}_i)$ is spirally unstable if

$$r_{k+1}^{(i)} - r_k^{(i)} > 0. \tag{2.100}$$

(iii) $\mathbf{x}^{(k)}$ at the fixed point $\mathbf{x}_k^*$ on the plane of $(\mathbf{u}_i, \mathbf{v}_i)$ is spirally stable with the m_ith-order singularity if for $\theta_k^{(i)} \in [0, 2\pi]$

$$\begin{aligned} &\rho_i = \sqrt{\alpha_i^2 + \beta_i^2} = 1, \\ &G_{r_{k+1}^{(i)}}^{(s_k^{(i)})}(\theta_k) = 0 \quad \text{for } s_k^{(i)} = 1, 2, \cdots, m_i - 1; \\ &r_{k+1}^{(i)} - r_k^{(i)} < 0. \end{aligned} \tag{2.101}$$

(iv) $\mathbf{x}^{(k)}$ at the fixed point $\mathbf{x}_k^*$ on the plane of $(\mathbf{u}_i, \mathbf{v}_i)$ is spirally unstable with the m_ith-order singularity if for $\theta_k^{(i)} \in [0, 2\pi]$

$$\begin{aligned} &\rho_i = \sqrt{\alpha_i^2 + \beta_i^2} = 1, \\ &G_{r_{k+1}^{(i)}}^{(s_k^{(i)})}(\theta_k) = 0 \quad \text{for } s_k^{(i)} = 0, 1, 2, \cdots, m_i - 1; \\ &r_{k+1}^{(i)} - r_k^{(i)} > 0. \end{aligned} \tag{2.102}$$

(v) $\mathbf{x}^{(k)}$ at the fixed point $\mathbf{x}_k^*$ on the plane of $(\mathbf{u}_i, \mathbf{v}_i)$ is circular if for $\theta_k^{(i)} \in [0, 2\pi]$

$$r_{k+1}^{(i)} - r_k^{(i)} = 0. \tag{2.103}$$

(vi) $\mathbf{x}^{(k)}$ at the fixed point $\mathbf{x}^*$ on the plane of $(\mathbf{u}_i, \mathbf{v}_i)$ is degenerate in the direction of $\mathbf{u}_k$ if

$$\beta_i = 0 \text{ and } \theta_{k+1}^{(i)} - \theta_k^{(i)} = 0. \tag{2.104}$$

Theorem 2.7 *Consider a discrete, nonlinear dynamical system $\mathbf{x}_{k+1} = \mathbf{f}(\mathbf{x}_k, \mathbf{p}) \in \mathscr{R}^n$ in Eq. (2.4) with a fixed point $\mathbf{x}_k^*$. The corresponding solution is given by $\mathbf{x}_{k+j} = \mathbf{f}(\mathbf{x}_{k+j-1}, \mathbf{p})$ with $j \in \mathbb{Z}$. Suppose there is a neighborhood of the fixed point $\mathbf{x}_k^*$ (i.e., $U_k(\mathbf{x}_k^*) \subset \Omega$), and $\mathbf{f}(\mathbf{x}_k, \mathbf{p})$ is C^r $(r \geq 1)$-continuous in $U_k(\mathbf{x}_k^*)$ with Eq. (2.28). The linearized system is $\mathbf{y}_{k+j+1} = D\mathbf{f}(\mathbf{x}_k^*, \mathbf{p})\mathbf{y}_{k+j}$ $(\mathbf{y}_{k+j} = \mathbf{x}_{k+j} - \mathbf{x}_k^*)$ in $U_k(\mathbf{x}_k^*)$. Consider a pair of complex eigenvalues $\alpha_i \pm \mathbf{i}\beta_i$ $(i \in N = \{1, 2, \cdots, n\},\ \mathbf{i} = \sqrt{-1})$ of matrix $D\mathbf{f}(\mathbf{x}^*, \mathbf{p})$ with a pair of eigenvectors $\mathbf{u}_i \pm \mathbf{i}\mathbf{v}_i$. On the invariant plane of $(\mathbf{u}_i, \mathbf{v}_i)$, consider $\mathbf{r}_k^{(i)} = \mathbf{y}_k = \mathbf{y}_{k+}^{(i)} + \mathbf{y}_{k-}^{(i)}$ with Eqs. (2.77) and (2.79). For any arbitrarily small $\varepsilon > 0$, the stability of the fixed point $\mathbf{x}_k^*$ on the invariant plane of $(\mathbf{u}_i, \mathbf{v}_i)$ can be determined.*

(i) $\mathbf{x}^{(k)}$ *at the fixed point* $\mathbf{x}_k^*$ *on the plane of* $(\mathbf{u}_i, \mathbf{v}_i)$ *is spirally stable if and only if*

$$\rho_i < 1. \tag{2.105}$$

(ii) $\mathbf{x}^{(k)}$ *at the fixed point* $\mathbf{x}_k^*$ *on the plane of* $(\mathbf{u}_i, \mathbf{v}_i)$ *is spirally unstable if*

$$\rho_i > 1. \tag{2.106}$$

(iii) $\mathbf{x}^{(k)}$ *at the fixed point* $\mathbf{x}_k^*$ *on the plane of* $\mathbf{u}_i, \mathbf{v}_i$ *is spirally stable with the* (m_k)th-*order singularity if and only if for* $\theta_k^{(i)} \in [0, 2\pi]$

$$\begin{aligned} &\rho_i = \sqrt{\alpha_i^2 + \beta_i^2} = 1, \\ &G_{r_{k+1}^{(i)}}^{(s_k^{(i)})}(\theta_k^{(i)}) = 0 \quad \text{for } s_k^{(i)} = 1, 2, \cdots, m_i - 1; \\ &G_{r_{k+1}^{(i)}}^{(m_i)}(\theta_k^{(i)}) < 0. \end{aligned} \tag{2.107}$$

(iv) $\mathbf{x}^{(k)}$ *at the fixed point* $\mathbf{x}_k^*$ *on the plane of* $(\mathbf{u}_i, \mathbf{v}_i)$ *is spirally unstable with the* (m_i)th-*order singularity if and only if for* $\theta_k^{(i)} \in [0, 2\pi]$

$$\begin{aligned} &\rho_i = \sqrt{\alpha_i^2 + \beta_i^2} = 1, \\ &G_{r_{k+1}^{(i)}}^{(s_k^{(i)})}(\theta_k^{(i)}) = 0 \quad \text{for } s_k^{(i)} = 0, 1, 2, \cdots, m_i - 1; \\ &G_{r_{k+1}^{(i)}}^{(m_i)}(\theta_k^{(i)}) > 0. \end{aligned} \tag{2.108}$$

(v) $\mathbf{x}^{(k)}$ *at the fixed point* $\mathbf{x}_k^*$ *on the plane of* $(\mathbf{u}_i, \mathbf{v}_i)$ *is circular if and only if for* $\theta_k^{(i)} \in [0, 2\pi]$

$$\begin{aligned} &\rho_i = \sqrt{\alpha_i^2 + \beta_i^2} = 1, \\ &G_{r_{k+1}^{(i)}}^{(s_k^{(i)})}(\theta_k^{(i)}) = 0 \quad \text{for } s_k^{(i)} = 0, 1, 2, \cdots. \end{aligned} \tag{2.109}$$

Proof Since

$$\begin{aligned} c_{k+1}^{(i)} &= \frac{1}{\Delta}[\Delta_2 G_{c_k^{(i)}}(\mathbf{x}_k, \mathbf{p}) - \Delta_{12} G_{d_k^{(i)}}(\mathbf{x}_k, \mathbf{p})] \\ d_{k+1}^{(i)} &= \frac{1}{\Delta}[\Delta_1 G_{d_k^{(i)}}(\mathbf{x}_k, \mathbf{p}) - \Delta_{12} G_{c_k^{(i)}}(\mathbf{x}_k, \mathbf{p})] \end{aligned}$$

For $\mathbf{x}_{k+1} = \mathbf{x}_k = \mathbf{x}_k^*$, $r_k = 0$. The first order approximation of $c_{k+1}^{(i)}$ and $d_{k+1}^{(i)}$ in the Taylor series expansion gives

$$c_{k+1}^{(i)} = G_{c_{k+1}^{(i)}}^{(1)}(\theta_k^{(i)})r_k^{(i)} + o(r_k^{(i)}),$$
$$d_{k+1}^{(i)} = G_{d_{k+1}^{(i)}}^{(1)}(\theta_k^{(i)})r_k^{(i)} + o(r_k^{(i)})$$

where $r_k^{(i)} = \sqrt{(c_k^{(i)})^2 + (d_k^{(i)})^2}$ and $\theta_k^{(i)} = \arctan(d_k^{(i)}/c_k^{(i)})$

$$G_{c_{k+1}^{(i)}}^{(1)}(\theta_k^{(i)}) = \frac{1}{\Delta}[\Delta_2 G_{c_k^{(i)}}^{(1)}(\theta_k^{(i)}) - \Delta_{12} G_{d_k^{(i)}}^{(1)}(\theta_k^{(i)})]$$
$$G_{d_{k+1}^{(i)}}^{(1)}(\theta_k^{(i)}) = \frac{1}{\Delta}[\Delta_1 G_{d_k^{(i)}}^{(1)}(\theta_k^{(i)}) - \Delta_{12} G_{c_k^{(i)}}^{(1)}(\theta_k^{(i)})]$$

and

$$G_{c_k^{(i)}}^{(1)}(\theta_k^{(i)}) = (\alpha_i\Delta_1 - \beta_i\Delta_{12})\cos\theta_k^{(i)} + (\alpha_i\Delta_{12} + \beta_i\Delta_1)\sin\theta_k^{(i)},$$
$$G_{d_k^{(i)}}^{(1)}(\theta_k^{(i)}) = (-\beta_i\Delta_2 + \alpha_i\Delta_{12})\cos\theta_k^{(i)} + (\alpha_i\Delta_2 + \beta_i\Delta_{12})\sin\theta_k^{(i)}.$$

Therefore,

$$G_{c_{k+1}^{(i)}}^{(1)}(\theta_k^{(i)}) = \alpha_i\cos\theta_k^{(i)} + \beta_i\sin\theta_k^{(i)},$$
$$G_{d_{k+1}^{(i)}}^{(1)}(\theta_k^{(i)}) = \alpha_i\sin\theta_k^{(i)} - \beta_i\cos\theta_k^{(i)}.$$

Further

$$c_{k+1}^{(i)} = \alpha_i r_k^{(i)}\cos\theta_k^{(i)} + \beta_i r_k^{(i)}\sin\theta_k^{(i)} = \alpha_i c_k^{(i)} + \beta_i d_k^{(i)},$$
$$d_{k+1}^{(i)} = \alpha_i r_k^{(i)}\sin\theta_k^{(i)} - \beta_i r_k^{(i)}\cos\theta_k^{(i)} = -\beta_i c_k^{(i)} + \alpha_i d_k^{(i)}.$$

That is,

$$\begin{Bmatrix} c_{k+1}^{(i)} \\ d_{k+1}^{(i)} \end{Bmatrix} = \begin{bmatrix} \alpha_i & \beta_i \\ -\beta_i & \alpha_i \end{bmatrix}\begin{Bmatrix} c_k^{(i)} \\ d_k^{(i)} \end{Bmatrix} = \rho_i\begin{bmatrix} \cos\vartheta_i & \sin\vartheta_i \\ -\sin\vartheta_i & \cos\vartheta_i \end{bmatrix}\begin{Bmatrix} c_k^{(i)} \\ d_k^{(i)} \end{Bmatrix}.$$

From the foregoing equation, we have

$$r_{k+1}^{(i)} = \rho_i r_k^{(i)} + o(r_k^{(i)}) \text{ and } \theta_{k+1}^{(i)} = \theta_k^{(i)} - \vartheta_i + o(r_k^{(i)}).$$

where

$$\vartheta_i = \arctan(\beta_i/\alpha_i) \text{ and } \rho_i = \sqrt{\alpha_i^2 + \beta_i^2}.$$

As $r_k^{(i)} \ll 1$ and $r_k^{(i)} \to 0$, we have

$$r_{k+1}^{(i)} = \rho_i r_k^{(i)} \text{ and } \theta_{k+1}^{(i)} = \vartheta_i - \theta_k^{(i)}.$$

(i) For fixed point stability, if $\rho_i < 1$, then

$$r_{k+1}^{(i)} < r_k^{(i)}$$

which implies that $\mathbf{x}_k^{(i)}$ at the fixed point $\mathbf{x}_k^*$ on the plane of $(\mathbf{u}_i, \mathbf{v}_i)$ is spirally stable and vice versa.

(ii) If $\rho_i > 1$, then

$$r_{k+1}^{(i)} > r_k^{(i)}$$

which implies that $\mathbf{x}_k^{(i)}$ at the fixed point $\mathbf{x}_k^*$ on the plane of $(\mathbf{u}_i, \mathbf{v}_i)$ is spirally stable and vice versa.

(iii) If for $\theta_k^{(i)} \in [0, 2\pi]$ the following conditions exist

$$\rho_i = \sqrt{G_{r_{k+1}^{(i)}}^{(2)}} = \sqrt{\alpha_i^2 + \beta_i^2} = 1,$$

$$G_{r_{k+1}^{(i)}}^{(s_k^{(i)})}(\theta_k^{(i)}) = 0 \quad \text{for } s_k^{(i)} = 1, 2, \cdots, m_i - 1;$$

$$G_{r_{k+1}^{(i)}}^{(m_i)}(\theta_k^{(i)}) \neq 0, \quad \text{and } |G_{r_{k+1}^{(i)}}^{(s_k^{(i)})}(\theta_k^{(i)})| < \infty \text{ for } s_k^{(i)} = m_i + 1, m + 2 \cdots .$$

then the higher terms can be ignored, i.e.,

$$r_{k+1}^{(i)} = r_k^{(i)} \sqrt{1 + \sum\nolimits_{m_i=3}^{\infty} (r_k^{(i)})^{m_i - 2} G_{r_{k+1}^{(i)}}^{(m_i)}(\theta_k^{(i)})}.$$

If $G_{r_{k+1}^{(i)}}^{(m_i)}(\theta_k^{(i)})$ is independent of θ_k (i.e., $G_{r_{k+1}^{(i)}}^{(m_i)}(\theta_k^{(i)}) = \text{const}$), it can be used to determine the stability of the fixed point. If $G_{r_{k+1}^{(i)}}^{(m_i)}(\theta_k^{(i)}) < 0$, then

$$r_{k+1}^{(i)} < r_k^{(i)}.$$

In other words this implies $\mathbf{x}_k^{(i)}$ at the fixed point $\mathbf{x}_k^*$ on the plane of $(\mathbf{u}_i, \mathbf{v}_i)$ is spirally stable with m_kth order singularity, and vice versa.

(iv) If $G_{r_{k+1}^{(i)}}^{(m_i)}(\theta_k^{(i)}) > 0$, then

$$r_{k+1}^{(i)} > r_k^{(i)}.$$

That is, $\mathbf{x}^{(k)}$ at the fixed point $\mathbf{x}_k^*$ on the plane of $(\mathbf{u}_i, \mathbf{v}_i)$ is spirally unstable with the m_kth-order singularity, and vice versa.

(v) If for $\theta_k^{(i)} \in [0, 2\pi]$ the following conditions exist

$$G_{r_{k+1}^{(i)}}^{(s_k^{(i)})}(\theta_k^{(i)}) = 0 \text{ for } s_k^{(i)} = 1, 2, \cdots,$$

then

$$r_{k+1}^{(i)} = r_k^{(i)}$$

and vice versa. Therefore $\mathbf{x}^{(k)}$ at the fixed point $\mathbf{x}_k^*$ on the plane of $(\mathbf{u}_i, \mathbf{v}_i)$ is circular. This theorem is proved. ■

2.3.2 Bifurcations

Definition 2.29 Consider a discrete, nonlinear dynamical system $\mathbf{x}_{k+1} = \mathbf{f}(\mathbf{x}_k, \mathbf{p}) \in \mathscr{R}^n$ in Eq. (2.4) with a fixed point $\mathbf{x}_k^*$. The corresponding solution is given by $\mathbf{x}_{k+j} = \mathbf{f}(\mathbf{x}_{k+j-1}, \mathbf{p})$ with $j \in \mathbb{Z}$. Suppose there is a neighborhood of the fixed point $\mathbf{x}_k^*$ (i.e., $U_k(\mathbf{x}_k^*) \subset \Omega$), and $\mathbf{f}(\mathbf{x}_k, \mathbf{p})$ is C^r $(r \geq 1)$-continuous in $U_k(\mathbf{x}_k^*)$ with Eq. (2.28). The linearized system is $\mathbf{y}_{k+j+1} = D\mathbf{f}(\mathbf{x}_k^*, \mathbf{p})\mathbf{y}_{k+j}$ $(\mathbf{y}_{k+j} = \mathbf{x}_{k+j} - \mathbf{x}_k^*)$ in $U_k(\mathbf{x}_k^*)$ and there are n linearly independent vectors $\mathbf{v}_i$ $(i = 1, 2, \cdots, n)$. For a perturbation of the fixed point $\mathbf{y}_k = \mathbf{x}_k - \mathbf{x}_k^*$, let $\mathbf{y}_k^{(i)} = c_k^{(i)}\mathbf{v}_i$ and $\mathbf{y}_{k+1}^{(i)} = c_{k+1}^{(i)}\mathbf{v}_i$.

$$s_i^{(i)} = \mathbf{v}_i^{\mathrm{T}} \cdot \mathbf{y}_k = \mathbf{v}_i^{\mathrm{T}} \cdot (\mathbf{x}_k - \mathbf{x}_k^*) \tag{2.110}$$

where $s_k^{(i)} = c_k^{(i)}||\mathbf{v}_i||^2$.

$$s_{k+1}^{(i)} = \mathbf{v}_i^{\mathrm{T}} \cdot \mathbf{y}_{k+1} = \mathbf{v}_i^{\mathrm{T}} \cdot [\mathbf{f}(\mathbf{x}_k, \mathbf{p}) - \mathbf{x}_k^*]. \tag{2.111}$$

In the vicinity of point $(\mathbf{x}_{k(0)}^*, \mathbf{p}_0)$, $\mathbf{v}_i^{\mathrm{T}} \cdot \mathbf{f}(\mathbf{x}_k, \mathbf{p})$ can be expended for $(0 < \theta < 1)$ as

$$\begin{aligned}
\mathbf{v}_i^{\mathrm{T}} \cdot [\mathbf{f}(\mathbf{x}_k, \mathbf{p}) - \mathbf{x}_{k(0)}^*] &= a_i(s_k^{(i)} - s_{k(0)}^{(i)*}) + \mathbf{b}_i^{\mathrm{T}} \cdot (\mathbf{p} - \mathbf{p}_0) \\
&\quad + \sum_{q=2}^{m} \sum_{r=0}^{q} \frac{1}{q!} C_q^r \mathbf{a}_i^{(q-r,r)} (s_k^{(i)} - s_{k(0)}^{(i)*})^{q-r} (\mathbf{p} - \mathbf{p}_0)^r \\
&\quad + \frac{1}{(m+1)!} [(s_k^{(i)} - s_{k(0)}^{(i)*})\partial_{s_k^{(i)}} + (\mathbf{p} - \mathbf{p}_0)\partial_{\mathbf{p}}]^{m+1} \\
&\quad \times (\mathbf{v}_i^{\mathrm{T}} \cdot \mathbf{f}(\mathbf{x}_{k(0)}^* + \theta\Delta\mathbf{x}_k, \mathbf{p}_0 + \theta\Delta\mathbf{p}))
\end{aligned} \tag{2.112}$$

where

$$
\begin{aligned}
a_i &= \mathbf{v}_i^{\mathrm{T}} \cdot \partial_{s_k^{(i)}} \mathbf{f}(\mathbf{x}_k, \mathbf{p})\Big|_{(\mathbf{x}_{k(0)}^*, \mathbf{p}_0)}, \\
\mathbf{b}_i^{\mathrm{T}} &= \mathbf{v}_i^{\mathrm{T}} \cdot \partial_{\mathbf{p}} \mathbf{f}(\mathbf{x}_k, \mathbf{p})\Big|_{(\mathbf{x}_{k(0)}^*, \mathbf{p}_0)}, \\
\mathbf{a}_i^{(r,s)} &= \mathbf{v}_i^{\mathrm{T}} \cdot \partial_{s_k^{(i)}}^{(r)} \partial_{\mathbf{p}}^{(s)} \mathbf{f}(\mathbf{x}_k, \mathbf{p})\Big|_{(\mathbf{x}_{k(0)}^*, \mathbf{p}_0)}.
\end{aligned} \tag{2.113}
$$

If $a_i = 1$ and $\mathbf{p} = \mathbf{p}_0$, the stability of fixed point $\mathbf{x}_k^*$ on an eigenvector $\mathbf{v}_i$ changes from stable to unstable state (or from unstable to stable state). The bifurcation manifold in the direction of $\mathbf{v}_i$ is determined by

$$
\mathbf{b}_i^{\mathrm{T}} \cdot (\mathbf{p} - \mathbf{p}_0) + \sum_{q=2}^{m} \sum_{r=0}^{q} \frac{1}{q!} C_q^r \mathbf{a}_i^{(r,q,-r)} (\mathbf{p} - \mathbf{p}_0)^{q-r} (s_k^{(i)} - s_{k(0)}^{(i)*})^r = 0. \tag{2.114}
$$

In the neighborhood of $(\mathbf{x}_{k(0)}^*, \mathbf{p}_0)$, when other components of fixed point $\mathbf{x}_k^*$ on the eigenvector of $\mathbf{v}_j$ for all $j \neq i$, $(i, j \in N)$ do not change their stability states, Eq. (2.114) possesses l-branch solutions of equilibrium $s_k^{(i)*}$ $(0 < l \leq m)$ with l_1-stable and l_2-unstable solutions $(l_1, l_2 \in \{0, 1, 2, \cdots, l\})$. Such l-branch solutions are called the bifurcation solutions of fixed point $\mathbf{x}_k^*$ on the eigenvector of $\mathbf{v}_i$ in the neighborhood of $(\mathbf{x}_{k(0)}^*, \mathbf{p}_0)$. Such a bifurcation at point $(\mathbf{x}_{k(0)}^*, \mathbf{p}_0)$ is called the hyperbolic bifurcation of mth-order on the eigenvector of $\mathbf{v}_i$. Consider two special cases herein.

(i) If

$$
\mathbf{a}_i^{(1,1)} = \mathbf{0} \text{ and } \mathbf{b}_i^{\mathrm{T}} \cdot (\mathbf{p} - \mathbf{p}_0) + \frac{1}{2!} a_i^{(2,0)} (s_k^{(i)*} - s_{k0}^{(i)*})^2 = 0 \tag{2.115}
$$

where

$$
\begin{aligned}
a_i^{(2,0)} &= \mathbf{v}_i^{\mathrm{T}} \cdot \partial_{s_k^{(i)}}^{(2)} \partial_{\mathbf{p}}^{(0)} \mathbf{f}(\mathbf{x}_k, \mathbf{p})\Big|_{(\mathbf{x}_{k(0)}^*, \mathbf{p}_0)} = \mathbf{v}_i^{\mathrm{T}} \cdot \partial_{s_k^{(i)}}^{(2)} \mathbf{f}(\mathbf{x}_k, \mathbf{p})\Big|_{(\mathbf{x}_{k(0)}^*, \mathbf{p}_0)} \\
&= \mathbf{v}_i^{\mathrm{T}} \cdot \partial_{\mathbf{x}}^{(2)} \mathbf{f}(\mathbf{x}_k, \mathbf{p})(\mathbf{v}_k \mathbf{v}_k)\Big|_{(\mathbf{x}_{k(0)}^*, \mathbf{p}_0)} = G_{s_k^{(i)}}^{(2)}(\mathbf{x}_{k(0)}^*, \mathbf{p}_0) \neq 0,
\end{aligned} \tag{2.116}
$$

$$
\mathbf{b}_i^{\mathrm{T}} = \mathbf{v}_i^{\mathrm{T}} \cdot \partial_{\mathbf{p}} \mathbf{f}(\mathbf{x}_k, \mathbf{p})\Big|_{(\mathbf{x}_{k(0)}^*, \mathbf{p}_0)} \neq \mathbf{0},
$$

$$
a_i^{(2,0)} \times [\mathbf{b}_i^{\mathrm{T}} \cdot (\mathbf{p} - \mathbf{p}_0)] < 0, \tag{2.117}
$$

such a bifurcation at point $(\mathbf{x}_0^*, \mathbf{p}_0)$ is called the *saddle-node* bifurcation on the eigenvector of $\mathbf{v}_i$.

(ii) If

$$\mathbf{b}_i^{\mathrm{T}} \cdot (\mathbf{p} - \mathbf{p}_0) = 0 \text{ and}$$

$$\mathbf{a}_i^{(1,1)} \cdot (\mathbf{p} - \mathbf{p}_0)(s_k^{(i)*} - s_{k(0)}^{(i)*}) + \frac{1}{2!} a_i^{(2,0)} (s_k^{(i)*} - s_{k(0)}^{(i)*})^2 = 0 \tag{2.118}$$

where

$$\begin{aligned} a_i^{(2,0)} &= \mathbf{v}_i^{\mathrm{T}} \cdot \partial_{s_k^{(i)}}^{(2)} \partial_{\mathbf{p}}^{(0)} \mathbf{f}(\mathbf{x}_k, \mathbf{p}) \Big|_{(\mathbf{x}_{k(0)}^*, \mathbf{p}_0)} = \mathbf{v}_i^{\mathrm{T}} \cdot \partial_{s_k^{(i)}}^{(2)} \mathbf{f}(\mathbf{x}_k, \mathbf{p}) \Big|_{(\mathbf{x}_0^*, \mathbf{p}_0)} \\ &= \mathbf{v}_i^{\mathrm{T}} \cdot \partial_{\mathbf{x}_k}^{(2)} \mathbf{f}(\mathbf{x}_k, \mathbf{p})(\mathbf{v}_i \mathbf{v}_i) \Big|_{(\mathbf{x}_{k(0)}^*, \mathbf{p}_0)} = G_{s_k^{(i)}}^{(2)}(\mathbf{x}_{k(0)}^*, \mathbf{p}_0) \neq 0, \\ \mathbf{a}_i^{(1,1)} &= \mathbf{v}_i^{\mathrm{T}} \cdot \partial_{s_k^{(i)}}^{(1)} \partial_{\mathbf{p}}^{(1)} \mathbf{f}(\mathbf{x}_k, \mathbf{p}) \Big|_{(\mathbf{x}_{k(0)}^*, \mathbf{p}_0)} = \mathbf{v}_i^{\mathrm{T}} \cdot \partial_{s_k^{(i)}} \partial_{\mathbf{p}} \mathbf{f}(\mathbf{x}_k, \mathbf{p}) \Big|_{(\mathbf{x}_{k(0)}^*, \mathbf{p}_0)} \\ &= \mathbf{v}_i^{\mathrm{T}} \cdot \partial_{\mathbf{x}_k} \partial_{\mathbf{p}} \mathbf{f}(\mathbf{x}_k, \mathbf{p}) \mathbf{v}_i \Big|_{(\mathbf{x}_{k(0)}^*, \mathbf{p}_0)} \neq \mathbf{0}, \end{aligned} \tag{2.119}$$

$$a_i^{(2,0)} \times [\mathbf{a}_i^{(1,1)} \cdot (\mathbf{p} - \mathbf{p}_0)] < 0, \tag{2.120}$$

such a bifurcation at point $(\mathbf{x}_{k(0)}^*, \mathbf{p}_0)$ is called the *transcritical* bifurcation on the eigenvector of $\mathbf{v}_i$.

Definition 2.30 Consider a discrete, nonlinear dynamical system $\mathbf{x}_{k+1} = \mathbf{f}(\mathbf{x}_k, \mathbf{p}) \in \mathscr{R}^n$ in Eq. (2.4) with a fixed point $\mathbf{x}_k^*$. The corresponding solution is given by $\mathbf{x}_{k+j} = \mathbf{f}(\mathbf{x}_{k+j-1}, \mathbf{p})$ with $j \in \mathbb{Z}$. Suppose there is a neighborhood of the fixed point $\mathbf{x}_k^*$ (i.e., $U_k(\mathbf{x}_k^*) \subset \Omega$), and $\mathbf{f}(\mathbf{x}_k, \mathbf{p})$ is C^r $(r \geq 1)$-continuous in $U_k(\mathbf{x}_k^*)$ with Eq. (2.28). The linearized system is $\mathbf{y}_{k+j+1} = D\mathbf{f}(\mathbf{x}_k^*, \mathbf{p})\mathbf{y}_{k+j}$ $(\mathbf{y}_{k+j} = \mathbf{x}_{k+j} - \mathbf{x}_k^*)$ in $U_k(\mathbf{x}_k^*)$ and there are n linearly independent vectors $\mathbf{v}_i$ $(i = 1, 2, \cdots, n)$. For a perturbation of fixed point $\mathbf{y}_k = \mathbf{x}_k - \mathbf{x}_k^*$, let $\mathbf{y}_k^{(i)} = c_k^{(i)} \mathbf{v}_i$ and $\mathbf{y}_{k+1}^{(i)} = c_{k+1}^{(i)} \mathbf{v}_i$. Equations (2.110), (2.111) and (2.113) hold. In the vicinity of point $(\mathbf{x}_{k0}^*, \mathbf{p}_0)$, $\mathbf{v}_i^{\mathrm{T}} \cdot \mathbf{f}(\mathbf{x}_k, \mathbf{p})$ can be expended for $(0 < \theta < 1)$ as

$$\begin{aligned} \mathbf{v}_i^{\mathrm{T}} \cdot [\mathbf{f}(\mathbf{x}_k, \mathbf{p}) - \mathbf{x}_{k+1(0)}^*] &= a_i (s_k^{(i)} - s_{k(0)}^{(i)*}) + \mathbf{b}_i^{\mathrm{T}} \cdot (\mathbf{p} - \mathbf{p}_0) \\ &\quad + \sum_{q=2}^{m} \sum_{r=0}^{q} \frac{1}{q!} C_q^r \mathbf{a}_i^{(q-r,r)} (s_k^{(i)} - s_{k(0)}^{(i)*})^{q-r} (\mathbf{p} - \mathbf{p}_0)^r \\ &\quad + \frac{1}{(m+1)!} [(s_k^{(i)} - s_{k(0)}^{(i)*}) \partial_{s_k^{(i)}} + (\mathbf{p} - \mathbf{p}_0) \partial_{\mathbf{p}}]^{m+1} \\ &\quad \times (\mathbf{v}_k^{\mathrm{T}} \cdot \mathbf{f}(\mathbf{x}_{k(0)}^* + \theta \Delta \mathbf{x}_k, \mathbf{p}_0 + \theta \Delta \mathbf{p})) \end{aligned} \tag{2.121}$$

and

$$
\begin{aligned}
\mathbf{v}_i^{\mathrm{T}} \cdot [\mathbf{f}(\mathbf{x}_{k+1}, \mathbf{p}) - \mathbf{x}_{k(0)}^*] &= a_i (s_{k+1}^{(i)} - s_{k+1(0)}^{(i)*}) + \mathbf{b}_i^{\mathrm{T}} \cdot (\mathbf{p} - \mathbf{p}_0) \\
&\quad + \sum_{q=2}^{m} \sum_{r=0}^{q} \frac{1}{q!} C_q^r \mathbf{a}_i^{(q-r,r)} (s_{k+1}^{(i)} - s_{k+1(0)}^{(i)*})^{q-r} (\mathbf{p} - \mathbf{p}_0)^r \\
&\quad + \frac{1}{(m+1)!} [(s_{k+1}^{(i)} - s_{k+1(0)}^{(i)*}) \partial_{s_{k+1}^{(i)}} + (\mathbf{p} - \mathbf{p}_0) \partial_{\mathbf{p}}]^{m+1} \\
&\quad \times (\mathbf{v}_i^{\mathrm{T}} \cdot \mathbf{f}(\mathbf{x}_{k+1(0)}^* + \theta \Delta \mathbf{x}_{k+1}, \mathbf{p}_0 + \theta \Delta \mathbf{p})).
\end{aligned}
\tag{2.122}
$$

If $a_i = -1$ and $\mathbf{p} = \mathbf{p}_0$, the stability of current equilibrium $\mathbf{x}_k^*$ on an eigenvector $\mathbf{v}_i$ changes from stable to unstable state (or from unstable to stable state). The bifurcation manifold in the direction of $\mathbf{v}_i$ is determined by

$$
\left.
\begin{aligned}
&\mathbf{b}_i^{\mathrm{T}} \cdot (\mathbf{p} - \mathbf{p}_0) + a_i (s_k^{(i)*} - s_{k(0)}^{(i)*}) \\
&\quad + \sum_{q=2}^{m} \sum_{r=0}^{q} \frac{1}{q!} C_q^r \mathbf{a}_i^{(q-r,r)} (s_k^{(i)} - s_{k(0)}^{(i)*})^{q-r} (\mathbf{p} - \mathbf{p}_0)^r = (s_{k+1}^{(i)*} - s_{k+1(0)}^{(i)*}); \\
&\mathbf{b}_i^{\mathrm{T}} \cdot (\mathbf{p} - \mathbf{p}_0) + a_i (s_{k+1}^{(i)*} - s_{k+1(0)}^{(i)*}) \\
&\quad + \sum_{q=2}^{m} \sum_{r=0}^{q} \frac{1}{q!} C_q^r \mathbf{a}_i^{(q-r,r)} (s_{k+1}^{(i)} - s_{k+1(0)}^{(i)*})^{q-r} (\mathbf{p} - \mathbf{p}_0)^r = (s_k^{(i)*} - s_{k(0)}^{(i)*}).
\end{aligned}
\right\}
\tag{2.123}
$$

In the neighborhood of $(\mathbf{x}_{k(0)}^*, \mathbf{p}_0)$, when other components of fixed point $\mathbf{x}_{k(0)}^*$ on the eigenvector of $\mathbf{v}_j$ for all $j \neq i,\ (j, i \in N)$ do not change their stability states, Eq. (2.123) possesses l-branch solutions of equilibrium $s_k^{(i)*}$ $(0 < l \leq m)$ with l_1-stable and l_2-unstable solutions $(l_1, l_2 \in \{0, 1, 2, \cdots, l\})$. Such l-branch solutions are called the bifurcation solutions of fixed point $\mathbf{x}_k^*$ on the eigenvector of $\mathbf{v}_i$ in the neighborhood of $(\mathbf{x}_{k(0)}^*, \mathbf{p}_0)$. Such a bifurcation at point $(\mathbf{x}_{k(0)}^*, \mathbf{p}_0)$ is called the *hyperbolic bifurcation* of mth-order with doubling iterations on the eigenvector of $\mathbf{v}_i$. Consider a special case. If

$$
\left.
\begin{aligned}
&\mathbf{b}_i^{\mathrm{T}} \cdot (\mathbf{p} - \mathbf{p}_0) = 0,\ a_i = -1,\ a_i^{(2,0)} = 0,\ \mathbf{a}_i^{(2,1)} = 0,\ \mathbf{a}_i^{(1,2)} = 0, \\
&[\mathbf{a}^{(1,1)} \cdot (\mathbf{p} - \mathbf{p}_0) + a_i](s_k^{(i)*} - s_{k(0)}^{(i)*}) + \frac{1}{3!} a_i^{(3,0)} (s_k^* - s_{k(0)}^*)^3 \\
&= (s_{k+1}^{(i)*} - s_{k+1(0)}^{(i)*}), \\
&[\mathbf{a}^{(1,1)} \cdot (\mathbf{p} - \mathbf{p}_0) + a_i](s_{k+1}^{(i)*} - s_{k+1(0)}^{(i)*})] + \frac{1}{3!} a_i^{(3,0)} (s_{k+1}^* - s_{k+1(0)}^*)^3 \\
&= (s_k^{(i)*} - s_{k(0)}^{(i)*})
\end{aligned}
\right\}
\tag{2.124}
$$

where

$$
\begin{aligned}
a_i^{(3,0)} &= \mathbf{v}_i^{\mathrm{T}} \cdot \partial_{s_k^{(i)}}^{(3)} \partial_{\mathbf{p}}^{(0)} \mathbf{f}(\mathbf{x}_k, \mathbf{p})\Big|_{(\mathbf{x}_{k(0)}^*, \mathbf{p}_0)} = \mathbf{v}_i^{\mathrm{T}} \cdot \partial_{s_k^{(i)}}^{(3)} \mathbf{f}(\mathbf{x}_k, \mathbf{p})\Big|_{(\mathbf{x}_{k(0)}^*, \mathbf{p}_0)} \\
&= \mathbf{v}_i^{\mathrm{T}} \cdot \partial_{\mathbf{x}_k}^{(3)} \mathbf{f}(\mathbf{x}_k, \mathbf{p})(\mathbf{v}_i \mathbf{v}_i \mathbf{v}_i)\Big|_{(\mathbf{x}_{k(0)}^*, \mathbf{p}_0)} = G_{s_k^{(i)}}^{(3)}(\mathbf{x}_{k(0)}^*, \mathbf{p}_0) \neq 0, \\
\mathbf{a}_i^{(1,1)} &= \mathbf{v}_i^{\mathrm{T}} \cdot \partial_{s_k^{(i)}}^{(1)} \partial_{\mathbf{p}}^{(1)} \mathbf{f}(\mathbf{x}_k, \mathbf{p})\Big|_{(\mathbf{x}_{k(0)}^*, \mathbf{p}_0)} = \mathbf{v}_i^{\mathrm{T}} \cdot \partial_{s_k^{(i)}} \partial_{\mathbf{p}} \mathbf{f}(\mathbf{x}_k, \mathbf{p})\Big|_{(\mathbf{x}_{k(0)}^*, \mathbf{p}_0)} \\
&= \mathbf{v}_i^{\mathrm{T}} \cdot \partial_{\mathbf{x}_k} \partial_{\mathbf{p}} \mathbf{f}(\mathbf{x}_k, \mathbf{p}) \mathbf{v}_i\Big|_{(\mathbf{x}_{k(0)}^*, \mathbf{p}_0)} \neq \mathbf{0},
\end{aligned}
\tag{2.125}
$$

$$
a_i^{(3,0)} \times [\mathbf{a}_i^{(1,1)} \cdot (\mathbf{p} - \mathbf{p}_0)] < 0, \tag{2.126}
$$

such a bifurcation at point $(\mathbf{x}_{k(0)}^*, \mathbf{p}_0)$ is called the *pitchfork* bifurcation (or period-doubling bifurcation) on the eigenvector of $\mathbf{v}_i$.

The three types of special cases can be discussed through 1-D systems and intuitive illustrations are presented in Fig. 2.7 for a better understanding of bifurcation for nonlinear discrete maps. Similarly, other cases on the eigenvector of $\mathbf{v}_i$ can be discussed from Eqs. (2.114) and (2.123). In Fig. 2.7, the bifurcation point is also represented by a solid circular symbol. The stable and unstable fixed point branches are given by solid and dashed curves, respectively. The vector fields are represented by lines with arrows. If no fixed points exist, such a region is shaded.

Consider a saddle-node bifurcation in 1-D system

$$
x_{k+1} = f(x_k, p) \equiv x_k + p - x_k^2. \tag{2.127}
$$

For $x_{k+1} = x_k$, the fixed points of the foregoing equation are $x_k^* = \pm\sqrt{p}$ $(p > 0)$ and no fixed points exist for $p < 0$. From Eq. (2.127), the linearized equation in the vicinity of the fixed points with $y_k = x_k - x_k^*$ is

$$
y_{k+1} = Df(x_k^*, p) y_k = (1 - 2x_k^*) y_k. \tag{2.128}
$$

For the branch of $x_k^* = +\sqrt{p}$ $(p > 0)$, the fixed point is stable due to $|y_{k+1}| < |y_k|$. However, for the branch of $x_k^* = -\sqrt{p}$ $(p > 0)$, such a fixed point is unstable due to $|y_{k+1}| > |y_k|$, For $p = p_0 = 0$, we have $x_k^* = x_{k(0)}^* = 0$ and $Df(x_{k(0)}^*, p_0) = 1$. Since $D^2 f(x_{k(0)}^*, p_0) = -2 < 0$, we have

$$
y_{k+1} = y_k + D^2 f(x_{k(0)}^*, p_0) y_k^2 = (1 - 2y_k) y_k. \tag{2.129}
$$

At $(x_{k(0)}^*, p_0) = (0, 0)$, $|y_{k+1}| < |y_k|$ for $y_k > 0$ and $|y_{k+1}| > |y_k|$ for $y_k < 0$. The fixed point $(x_{k(0)}^*, p_0) = (0, 0)$ is bifurcation point, which is a decreasing saddle of the second order. For $p < 0$, from Eq. (2.127), $p - (x_k^*)^2 < 0$. Thus, no fixed point exists. The fixed point x_k^* varying with parameter p is sketched in Fig. 2.7a. On the left side of x_k-axes, no fixed point exists. So only the vector field of the map in Eq. (2.127) is presented.

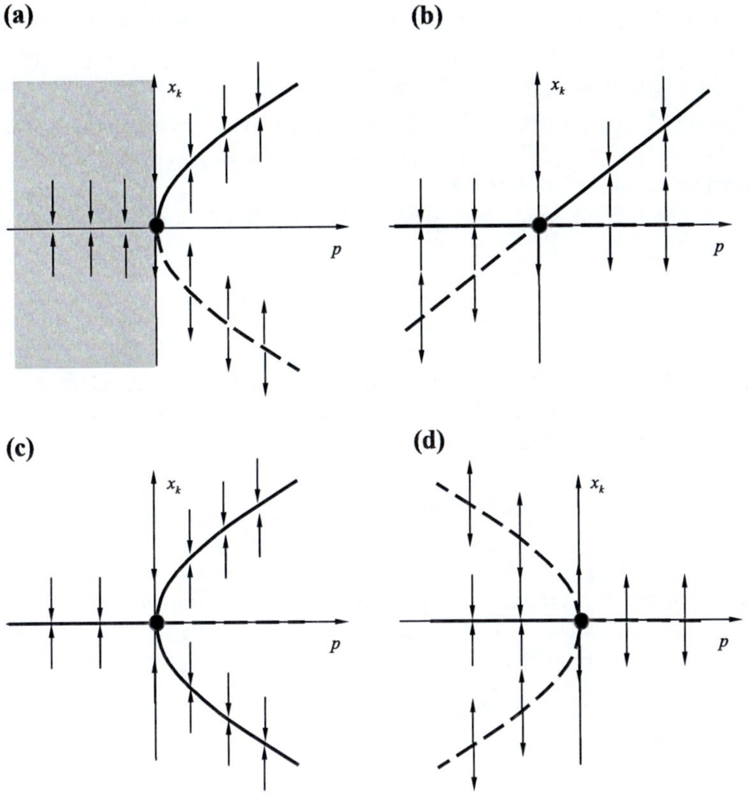

Fig. 2.7 Bifurcation diagrams: **a** saddle-node bifurcation of the first kind, **b** transcritical bifurcation, **c** pitchfork bifurcation for stable-symmetry (or saddle-node bifurcation of the second kind) and **d** pitchfork bifurcation for unstable-symmetry (or unstable saddle-node bifurcation of the second kind)

Consider a transcritical bifurcation through a 1-D discrete system as

$$x_{k+1} = f(x_k, p) \equiv x_k + px_k - x_k^2. \tag{2.130}$$

The fixed points of the map in the foregoing equation are $x_k^* = 0, p$. From Eq. (2.130), the linearized equation in the vicinity of the fixed points with $y_k = x_k - x_k^*$ is

$$y_{k+1} = Df(x_k^*, p)y_k = (1 + p - 2x_k^*)y_k. \tag{2.131}$$

For the branch of $x_k^* = 0$ $(p > 0)$, the fixed point is unstable due to $|y_{k+1}| < |y_k|$. For the branch of $x_k^* = p$ $(p > 0)$, such a fixed point is stable because of $|y_{k+1}| < |y_k|$. However, for the branch of $x_k^* = 0$ $(p < 0)$, the fixed point is stable due to $|y_{k+1}| < |y_k|$. For the branch of $x_k^* = p$ $(p < 0)$, such a fixed point is unstable owing to $|y_{k+1}| > |y_k|$. For $p = p_0 = 0$, $x_k^* = x_{k(0)}^* = 0$ and $Df(x_{k(0)}^*, p_0) = 1$

are obtained. $D^2f(x^*_{k(0)}, p_0) = -2$ is needed. Thus the variational equation at the fixed point is given by $y_{k+1} = y_k - 2y_k^2 = (1 - 2y_k)y_k$. From this equation, at $(x^*_{k(0)}, p_0) = (0, 0)$, $|y_{k+1}| < |y_k|$ for $y_k > 0$ and $|y_{k+1}| > |y_k|$ for $y_k < 0$. The fixed point $(x^*_{k(0)}, p_0) = (0, 0)$ is a bifurcation point, which is a decreasing saddle of the second order. The fixed point varying with parameter p is sketched in Fig. 2.7b.

Consider the pitchfork bifurcation with stable-symmetry (or saddle-node bifurcation of the second kind, or period-doubling bifurcation) with a 1-D system as

$$x_{k+1} = (-1 - p)x_k + x_k^3. \tag{2.132}$$

For $-x_{k+1} = x_k = x_k^*$, the corresponding fixed point are $x_k^* = 0, \pm\sqrt{p}$ $(p > 0)$ and $x_k^* = 0$ $(p \leq 0)$. From Eq. (2.132), the linearized equation in the vicinity of the fixed point with $y_k = x_k - x_k^*$ is

$$y_{k+1} = Df(x_k^*, p)y_k = [-1 - p + 3(x_k^*)^2]y_k. \tag{2.133}$$

For the branch of $x_k^* = 0$ $(p > 0)$, the fixed point is unstable due to $|y_{k+1}| > |y_k|$. For the branches of $x_k^* = \pm\sqrt{p}$ $(p > 0)$, such two fixed points are stable because of $|y_{k+1}| < |y_k|$. However, for the branch of $x_k^* = 0$ $(p < 0)$, the fixed point is stable due to $|y_{k+1}| < |y_k|$. For $p = p_0 = 0$, $x_k^* = x^*_{k(0)} = 0$ and $Df(x^*_{k(0)}, p_0) = -1$ are obtained. However, $D^2f(x^*_{k(0)}, p_0) = 6x^*_{k(0)} = 0$ is also obtained. Further, $D^3f(x^*_{k(0)}, p_0) = 6 > 0$ is computed. Thus the variational equation at the fixed point is

$$y_{k+1} = -y_k + D^3f(x^*_{k(0)}, p_0)y_k^3 = (-1 + 6y_k^2)y_k. \tag{2.134}$$

At $(x^*_{k(0)}, p_0) = (0, 0)$, $|y_{k+1}| < |y_k|$ exists always. The fixed point $(x^*_{k0}, p_0) = (0, 0)$ is a bifurcation point, which is an oscillatory sink of the third order due to $D^3f > 0$. The fixed point varying with parameter p is sketched in Fig. 2.7c.

Consider the pitchfork bifurcation for unstable-symmetry (or unstable saddle-node bifurcation of the second kind, or unstable period-doubling bifurcation) with a 1-D system as

$$x_{k+1} = (-1 - p)x_k - x_k^3. \tag{2.135}$$

For $-x_{k+1} = x_k = x_k^*$, the fixed points are $x_k^* = 0, \pm\sqrt{-p}$ $(p < 0)$ and $x_k^* = 0$ $(p \geq 0)$. From Eq. (2.135), the linearized equation in vicinity of fixed points with $y_k = x_k - x_k^*$ is

$$y_{k+1} = Df(x_k^*, p)y_k = [-1 - p - 3(x_k^*)^2]y_k. \tag{2.136}$$

For the branch of $x_k^* = 0$ $(p < 0)$, the fixed point is stable due to $|y_{k+1}| < |y_k|$. For the branches of $x_k^* = \pm\sqrt{-p}$ $(p < 0)$, such two fixed points are unstable due to $|y_{k+1}| >$

$|y_k|$. However, for the branch of $x_k^* = 0$ $(p > 0)$, the fixed point is unstable due to $|y_{k+1}| > |y_k|$. For $p = p_0 = 0$, $x_k^* = x_{k(0)}^* = 0$ and $Df(x_{k(0)}^*, p_0) = -1$ are obtained. $D^2 f(x_{k(0)}^*, p_0) = -6x_{k(0)}^* = 0$ are obtained. Furthermore, $D^3 f(x_{k(0)}^*, p_0) = -6 < 0$. Thus the variational equation at the fixed point is

$$y_{k+1} = -y_k + D^3 f(x_{k(0)}^*, p_0) y_k^3 = (-1 - 6y_k^2) y_k. \tag{2.137}$$

At $(x_{k(0)}^*, p_0) = (0, 0)$, $|y_{k+1}| > |y_k|$ exists always. The fixed point $(x_0^*, p_0) = (0, 0)$ is a bifurcation point, which is an oscillatory source of the third order. The fixed point varying with parameter p is sketched in Fig. 2.7d.

From the proceeding analysis, the bifurcation points possess the higher-order singularity of the flow in discrete dynamical system. For the saddle-node bifurcation of the first kind, the $(2m)$th order singularity of the flow at the bifurcation point exists as a saddle of the $(2m)$th order. For the transcritical bifurcation, the $(2m)$th order singularity of the flow at the bifurcation point exists as a saddle of of the $(2m)$th order. However, for the stable pitchfork bifurcation (or saddle-node bifurcation of the second kind, or period-doubling bifurcation), the $(2m+1)$th order singularity of the flow at the bifurcation point exists as an oscillatory sink of the $(2m+1)$th order. For the unstable pitchfork bifurcation (or the unstable saddle- node bifurcation of the second kind, or unstable period-doubling bifurcation), the $(2m+1)$th order singularity of the flow at the bifurcation point exists as an oscillatory source of the $(2m+1)$th order.

Definition 2.31 Consider a discrete, nonlinear dynamical system $\mathbf{x}_{k+1} = \mathbf{f}(\mathbf{x}_k, \mathbf{p}) \in R^{2n}$ in Eq. (2.4) with a fixed point $\mathbf{x}_k^*$. The corresponding solution is given by $\mathbf{x}_{k+j} = \mathbf{f}(\mathbf{x}_{k+j-1}, \mathbf{p})$ with $j \in \mathbb{Z}$. Suppose there is a neighborhood of the fixed point $\mathbf{x}_k^*$ (i.e., $U_k(\mathbf{x}_k^*) \subset \Omega$), and $\mathbf{f}(\mathbf{x}_k, \mathbf{p})$ is C^r $(r \geq 1)$-continuous in $U_k(\mathbf{x}_k^*)$ with Eq. (2.28). The linearized system is $\mathbf{y}_{k+j+1} = D\mathbf{f}(\mathbf{x}_k^*, \mathbf{p})\mathbf{y}_{k+j}$ $(\mathbf{y}_{k+j} = \mathbf{x}_{k+j} - \mathbf{x}_k^*)$ in $U_k(\mathbf{x}_k^*)$. Consider a pair of complex eigenvalues $\alpha_i \pm \mathbf{i}\beta_i$ $(i \in N = \{1, 2, \cdots, n\}$, $\mathbf{i} = \sqrt{-1})$ of matrix $D\mathbf{f}(\mathbf{x}^*, \mathbf{p})$ with a pair of eigenvectors $\mathbf{u}_i \pm \mathbf{i}\mathbf{v}_i$. On the invariant plane of $(\mathbf{u}_i, \mathbf{v}_i)$, consider $\mathbf{r}_k^{(i)} = \mathbf{y}_k^{(i)} = \mathbf{y}_{k+}^{(i)} + \mathbf{y}_{k-}^{(i)}$ with

$$\begin{aligned} \mathbf{r}_k^{(i)} &= c_k^{(i)} \mathbf{u}_i + d_k^{(i)} \mathbf{v}_i, \\ \mathbf{r}_{k+1}^{(i)} &= c_{k+1}^{(i)} \mathbf{u}_i + d_{k+1}^{(i)} \mathbf{v}_i. \end{aligned} \tag{2.138}$$

and

$$\begin{aligned} c_k^{(i)} &= \frac{1}{\Delta}[\Delta_2(\mathbf{u}_i^{\mathrm{T}} \cdot \mathbf{y}_k) - \Delta_{12}(\mathbf{v}_i^{\mathrm{T}} \cdot \mathbf{y}_k)], \\ d_k^{(i)} &= \frac{1}{\Delta}[\Delta_1(\mathbf{v}_i^{\mathrm{T}} \cdot \mathbf{y}_k) - \Delta_{12}(\mathbf{u}_i^{\mathrm{T}} \cdot \mathbf{y}_k)]; \\ \Delta_1 &= ||\mathbf{u}_i||^2, \Delta_2 = ||\mathbf{v}_i||^2, \Delta_{12} = \mathbf{u}_i^{\mathrm{T}} \cdot \mathbf{v}_i; \\ \Delta &= \Delta_1 \Delta_2 - \Delta_{12}^2. \end{aligned} \tag{2.139}$$

Consider a polar coordinate of (r_k, θ_k) defined by

$$\begin{aligned} c_k^{(i)} &= r_k^{(i)} \cos\theta_k^{(i)}, \text{ and } d_k^{(i)} = r_k^{(i)} \sin\theta_k^{(i)}; \\ r_k^{(i)} &= \sqrt{(c_k^{(i)})^2 + (d_k^{(i)})^2}, \text{ and } \theta_k^{(i)} = \arctan d_k^{(i)}/c_k^{(i)}. \end{aligned} \tag{2.140}$$

Thus

$$\begin{aligned} c_{k+1}^{(i)} &= \frac{1}{\Delta}[\Delta_2 G_{c_k^{(i)}}(\mathbf{x}_k, \mathbf{p}) - \Delta_{12} G_{d_k^{(i)}}(\mathbf{x}_k, \mathbf{p})] \\ d_{k+1}^{(i)} &= \frac{1}{\Delta}[\Delta_1 G_{d_k^{(i)}}(\mathbf{x}_k, \mathbf{p}) - \Delta_{12} G_{c_k^{(i)}}(\mathbf{x}_k, \mathbf{p})] \end{aligned} \tag{2.141}$$

where

$$\begin{aligned} G_{c_k^{(i)}}(\mathbf{x}_k, \mathbf{p}) &= \mathbf{u}_i^{\mathrm{T}} \cdot [\mathbf{f}(\mathbf{x}_k, \mathbf{p}) - \mathbf{x}_{k(0)}^*] \\ &= \mathbf{a}_i^{\mathrm{T}} \cdot (\mathbf{p} - \mathbf{p}_0) + a_{i11}(c_k^{(i)} - c_{k(0)}^{(i)*}) + a_{i12}(d_k^{(i)} - d_{k(0)}^{(i)*}) \\ &\quad + \sum_{q=2}^{m_i} \sum_{r_1=0}^{q} \frac{1}{q!} C_q^{r_i} \mathbf{G}_{c_k^{(i)}}^{(q-r_i, r_i)}(\mathbf{x}_k^*, \mathbf{p}_0)(\mathbf{p} - \mathbf{p}_0)^{r_i} (r_k^{(i)})^{q-r_i} \\ &\quad + \frac{1}{(m_i+1)!}[(c_k^{(i)} - c_{k(0)}^{(i)*})\partial_{c_k^{(i)}} + (d_k^{(i)} - d_{k(0)}^{(i)*})\partial_{d_k^{(i)}} + (\mathbf{p} - \mathbf{p}_0)\partial_{\mathbf{p}}]^{m_i+1} \\ &\quad \times (\mathbf{u}_i^{\mathrm{T}} \cdot \mathbf{f}(\mathbf{x}_{k0}^* + \theta \Delta \mathbf{x}_k, \mathbf{p}_0 + \theta \Delta \mathbf{p})), \\ G_{d_k^{(i)}}(\mathbf{x}_k, \mathbf{p}) &= \mathbf{v}_i^{\mathrm{T}} \cdot [\mathbf{f}(\mathbf{x}_k, \mathbf{p}) - \mathbf{x}_{k(0)}^*] \\ &= \mathbf{b}_i^{\mathrm{T}} \cdot (\mathbf{p} - \mathbf{p}_0) + a_{i21}(c_k^{(i)} - c_{k(0)}^{(i)*}) + a_{i22}(d_k^{(i)} - d_{k(0)}^{(i)*}) \\ &\quad + \sum_{q=2}^{m_i} \sum_{r_i=0}^{q} \frac{1}{q!} C_q^{r_i} \mathbf{G}_{d_k^{(i)}}^{(q-r_i, r_i)}(\mathbf{x}_k^*, \mathbf{p}_0)(\mathbf{p} - \mathbf{p}_0)^{r_i} (r_k^{(i)})^{q-r_i} \\ &\quad + \frac{1}{(m_i+1)!}[(c_k^{(i)} - c_{k(0)}^{(i)*})\partial_{c_k^{(i)}} + (d_k^{(i)} - d_{k(0)}^{(i)*})\partial_{d_k^{(i)}} + (\mathbf{p} - \mathbf{p}_0)\partial_{\mathbf{p}}]^{m+1} \\ &\quad \times (\mathbf{v}_i^{\mathrm{T}} \cdot \mathbf{f}(\mathbf{x}_{k(0)}^* + \theta \Delta \mathbf{x}, \mathbf{p}_0 + \theta \Delta \mathbf{p})); \end{aligned} \tag{2.142}$$

and

$$\begin{aligned} &\mathbf{G}_{c_k^{(i)}}^{(s,r)}(\mathbf{x}_{k(0)}^*, \mathbf{p}_0) \\ &= \mathbf{u}_i^{\mathrm{T}} \cdot [\partial_{\mathbf{x}_k}()\mathbf{u}_i \cos\theta_k^{(i)} + \partial_{\mathbf{x}_k}()\mathbf{v}_i \sin\theta_k^{(i)}]^s \partial_{\mathbf{p}}^{(r)} \mathbf{f}(\mathbf{x}_k, \mathbf{p})\Big|_{(\mathbf{x}_{k(0)}^*, \mathbf{p}_0)}, \\ &\mathbf{G}_{d_k^{(i)}}^{(s,r)}(\mathbf{x}_{k(0)}^*, \mathbf{p}_0) \\ &= \mathbf{v}_i^{\mathrm{T}} \cdot [\partial_{\mathbf{x}_k}()\mathbf{u}_i \cos\theta_k^{(i)} + \partial_{\mathbf{x}_k}()\mathbf{v}_i \sin\theta_k^{(i)}]^s \partial_{\mathbf{p}}^{(r)} \mathbf{f}(\mathbf{x}_k, \mathbf{p})\Big|_{(\mathbf{x}_{k(0)}^*, \mathbf{p}_0)}; \end{aligned} \tag{2.143}$$

$$\begin{aligned} &\mathbf{a}_i^{\mathrm{T}} = \mathbf{u}_i^{\mathrm{T}} \cdot \partial_{\mathbf{p}} \mathbf{f}(\mathbf{x}_k, \mathbf{p}), \mathbf{b}_i^{\mathrm{T}} = \mathbf{v}_i^{\mathrm{T}} \cdot \partial_{\mathbf{p}} \mathbf{f}(\mathbf{x}_k, \mathbf{p}); \\ &a_{i11} = \mathbf{u}_i^{\mathrm{T}} \cdot \partial_{\mathbf{x}_k} \mathbf{f}(\mathbf{x}_k, \mathbf{p})\mathbf{u}_i, a_{i12} = \mathbf{u}_i^{\mathrm{T}} \cdot \partial_{\mathbf{x}_k} \mathbf{f}(\mathbf{x}_k, \mathbf{p})\mathbf{u}_i; \\ &a_{i21} = \mathbf{v}_i^{\mathrm{T}} \cdot \partial_{\mathbf{x}_k} \mathbf{f}(\mathbf{x}_k, \mathbf{p})\mathbf{u}_i, a_{i22} = \mathbf{v}_i^{\mathrm{T}} \cdot \partial_{\mathbf{x}_k} \mathbf{f}(\mathbf{x}_k, \mathbf{p})\mathbf{v}_i. \end{aligned} \tag{2.144}$$

Suppose

$$\mathbf{a}_i = \mathbf{0} \text{ and } \mathbf{b}_i = \mathbf{0} \tag{2.145}$$

then

$$\begin{aligned} r_{k+1}^{(i)} &= \sqrt{(c_{k+1}^{(i)})^2 + (d_{k+1}^{(i)})^2} = \sqrt{\sum\nolimits_{m=2}^{\infty} (r_k^{(i)})^m G_{r_{k+1}^{(i)}}^{(m)}} \\ &= \sqrt{G_{r_{k+1}^{(i)}}^{(2,0)}} r_k^{(i)} \sqrt{1 + \lambda^{(i)} + \sum\nolimits_{m=3}^{\infty} \lambda_m^{(i)} (r_k^{(i)})^{m-2}} \\ \theta_{k+1}^{(i)} &= \arctan(d_{k+1}^{(i)}/c_{k+1}^{(i)}) \end{aligned} \tag{2.146}$$

where

$$\begin{aligned} G_{r_{k+1}^{(i)}}^{(2)} &= G_{r_{k+1}^{(i)}}^{(2,0)} + G_{r_{k+1}^{(i)}}^{(1,1)} \text{ and } \lambda^{(i)} = G_{r_{k+1}^{(i)}}^{(1,1)} / G_{r_{k+1}^{(i)}}^{(2,0)} \text{ with} \\ G_{r_{k+1}^{(i)}}^{(2,0)} &= [G_{c_{k+1}^{(i)}}^{(1,0)}(\theta_k^{(i)}, \mathbf{p}_0)]^2 + [G_{d_{k+1}^{(i)}}^{(1,0)}(\theta_k^{(i)}, \mathbf{p}_0)]^2, \\ G_{r_{k+1}^{(i)}}^{(1,1)} &= \sum\nolimits_{m_i=2}^{M} \sum\nolimits_{m_j=2}^{M} \frac{1}{(m_i + m_j - 2)!} C_{m_i+m_j-2}^{m_i-1} \\ &\quad ([G_{c_{k+1}^{(i)}}^{(1,m_i-1)}(\theta_k^{(i)}, \mathbf{p}_0) \cdot (\mathbf{p} - \mathbf{p}_0)^{m_i-1}][G_{c_{k+1}^{(i)}}^{(1,m_j-1)}(\theta_k^{(i)}, \mathbf{p}_0) \cdot (\mathbf{p} - \mathbf{p}_0)^{m_j-1}] \\ &\quad + [G_{d_{k+1}^{(i)}}^{(1,m_i-1)}(\theta_k^{(i)}, \mathbf{p}_0) \cdot (\mathbf{p} - \mathbf{p}_0)^{m_i-1}][G_{d_{k+1}^{(i)}}^{(1,m_j-1)}(\theta_k^{(i)}, \mathbf{p}_0) \cdot (\mathbf{p} - \mathbf{p}_0)^{m_j-1}]) \end{aligned} \tag{2.147}$$

and

$$\begin{aligned} \lambda_m^{(i)} &= G_{r_{k+1}^{(i)}}^{(m)} / G_{r_{k+1}^{(i)}}^{(2,0)} \text{ with} \\ G_{r_{k+1}^{(i)}}^{(m)} &= \sum\nolimits_{m_i=0}^{\infty} \sum\nolimits_{m_j=0}^{\infty} \frac{1}{m_i!} \frac{1}{m_j!} [\mathbf{G}_{c_{k+1}^{(i)}}^{(m_i-r_i,r_i)}(\theta_k^{(i)}, \mathbf{p}_0) \cdot (\mathbf{p} - \mathbf{p}_0)^{m_i-r_i}] \\ &\quad \times \mathbf{G}_{c_{k+1}^{(i)}}^{(m_j-s_j,s_j)}(\theta_k^{(i)}, \mathbf{p}_0) \cdot (\mathbf{p} - \mathbf{p}_0)^{m_j-s_j} \\ &\quad + \mathbf{G}_{d_{k+1}^{(i)}}^{(m_i-r_i,r_i)}(\theta_k^{(i)}, \mathbf{p}_0) \cdot (\mathbf{p} - \mathbf{p}_0)^{m_i-r_i}] \\ &\quad \times \mathbf{G}_{d_{k+1}^{(i)}}^{(m_j-s_j,s_j)}(\theta_k^{(i)}, \mathbf{p}_0) \cdot (\mathbf{p} - \mathbf{p}_0)^{m_j-s_j}] \delta_m^{(r_i+s_j)} \\ &= \sum_{s=1}^{m-1} \sum_{m_i=1}^{M} \sum_{m_j=1}^{M} \frac{1}{m!} \frac{1}{(m_i + m_j - m)!} C_m^s C_{m_i+m_j-m}^{m_i-s} \\ &\quad [\mathbf{G}_{c_{k+1}^{(i)}}^{(s,m_i-s)}(\theta_k^{(i)}, \mathbf{p}_0) \cdot (\mathbf{p} - \mathbf{p}_0)^{m_i-s} \mathbf{G}_{c_{k+1}^{(i)}}^{(m-s,m_j-m+s)}(\theta_k^i, \mathbf{p}_0) \cdot (\mathbf{p} - \mathbf{p}_0)^{m_j-m+s} \\ &\quad + \mathbf{G}_{d_{k+1}^{(i)}}^{(s,m_i-s)}(\theta_k^{(i)}, \mathbf{p}_0) \cdot (\mathbf{p} - \mathbf{p}_0)^{m_i-s} \mathbf{G}_{d_{k+1}^{(i)}}^{(m-s,m_j-m+s)}(\theta_k^i, \mathbf{p}_0) \cdot (\mathbf{p} - \mathbf{p}_0)^{m_j-m+s}], \end{aligned} \tag{2.148}$$

$$\begin{aligned}
\mathbf{G}^{(m-r,r)}_{c^{(i)}_{k+1}}(\theta_k, \mathbf{p}_0) &= \frac{1}{\Delta}[\Delta_2 \mathbf{G}^{(m-r,r)}_{c^{(i)}_k}(\mathbf{x}^*_{k(0)}, \mathbf{p}_0) - \Delta_{12}\mathbf{G}^{(m-r,r)}_{d^{(i)}_k}(\mathbf{x}^*_{k(0)}, \mathbf{p}_0)], \\
\mathbf{G}^{(m-r,r)}_{d^{(i)}_{k+1}}(\theta_k, \mathbf{p}_0) &= \frac{1}{\Delta}[\Delta_1 \mathbf{G}^{(m-r,r)}_{d^{(i)}_k}(\mathbf{x}^*_{k(0)}, \mathbf{p}_0) - \Delta_{12}\mathbf{G}^{(m-r,r)}_{c^{(i)}_k}(\mathbf{x}^*_{k(0)}, \mathbf{p}_0)].
\end{aligned} \tag{2.149}$$

If $G^{(2,0)}_{r^{(i)}_{k+1}} = 1$ and $\mathbf{p} = \mathbf{p}_0$, the stability of current fixed point $\mathbf{x}^*_k$ on an eigenvector plane of $(\mathbf{u}_i, \mathbf{v}_i)$ changes from stable to unstable state (or from unstable to stable state). The bifurcation manifold in the direction of $\mathbf{v}_k$ is determined by

$$\lambda^{(i)} + \sum_{m=3}^{\infty} \lambda^{(i)}_m (r^{(i)}_k)^{m-2} = 0. \tag{2.150}$$

Such a bifurcation at the fixed point $(\mathbf{x}^*_{k(0)}, \mathbf{p}_0)$ is called the generalized Neimark bifurcation on the eigenvector plane of $(\mathbf{u}_i, \mathbf{v}_i)$.

For a special case, if

$$\lambda^{(i)} + \lambda^{(i)}_4 (r^{(i)}_k)^2 = 0, \text{ for } \lambda^{(i)} \times \lambda^{(i)}_4 < 0 \text{ and } \lambda^{(i)}_3 = 0 \tag{2.151}$$

such a bifurcation at the fixed point $(\mathbf{x}^*_{k(0)}, \mathbf{p}_0)$ is called the Neimark bifurcation on the eigenvector plane of $(\mathbf{u}_i, \mathbf{v}_i)$.

For the repeating eigenvalues of $DP(\mathbf{x}^*_k, \mathbf{p})$, the bifurcation of fixed point $\mathbf{x}^*_k$ can be similarly discussed in the foregoing two Theorems 2.5 and 2.6. Herein, such a procedure will not be repeated.

Consider a dynamical system

$$\begin{aligned}
x_{k+1} &= \alpha[1 + \lambda + a(x_k^2 + y_k^2)]x_k + \beta[1 + \lambda + a(x_k^2 + y_k^2)]y_k, \\
y_{k+1} &= -\beta[1 + \lambda + a(x_k^2 + y_k^2)]x_k + \alpha[1 + \lambda + a(x_k^2 + y_k^2)]y_k.
\end{aligned} \tag{2.152}$$

Setting

$$r_k^2 = x_k^2 + y_k^2 \text{ with } x_k = r_k \cos\theta_k \text{ and } y_k = r_k \sin\theta_k, \tag{2.153}$$

we have

$$\begin{aligned}
r_{k+1} &= \sqrt{x_{k+1}^2 + y_{k+1}^2} = \rho r_k (1 + \lambda + a r_k^2), \\
\theta_{k+1} &= \arctan\frac{-\beta\cos\theta_k + \alpha\sin\theta_k}{\beta\sin\theta_k + \alpha\cos\theta_k} = \theta_k - \vartheta, \\
\rho &= \sqrt{\alpha^2 + \beta^2} \text{ and } \vartheta = \arctan\frac{\beta}{\alpha}.
\end{aligned} \tag{2.154}$$

If $\rho = 1$, the fixed point is

$$\begin{aligned}
r^*_{k(1)} &= 0 \text{ for } \lambda \in (-\infty, +\infty), \\
r^*_{k(2)} &= (-\lambda/a)^{1/2} \text{ for } \lambda \times a < 0.
\end{aligned} \tag{2.155}$$

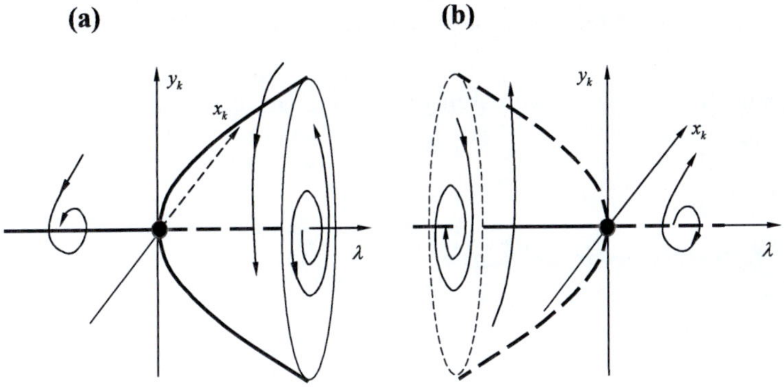

Fig. 2.8 Neimark bifurcations: **a** supercritical ($a < 0$) and **b** subcritical ($a > 0$)

If $\lambda \neq 0$, we have $Df_r(r_k^*, \lambda) = 1 + \lambda + 3a(r_k^*)^2$, the variational equation is

$$s_{k+1} = Df_r(r_k^*, \lambda)s_k = [1 + \lambda + 3a(r_k^*)^2]s_k \text{ with } s_k = r_k - r_k^*. \tag{2.156}$$

For $r_{k(1)}^* = 0,\ Df_r = 1 + \lambda$. This fixed point is stable for $\lambda < 0$ owing to $s_{k+1} < s_k$ or unstable for $\lambda > 0$ owing to $s_{k+1} > s_k$. The fixed point is a critical point for $\lambda = 0$. The fixed point of $r_{k(2)}^* = (-\lambda/a)^{1/2}$ requires $a\lambda < 0$. For $a > 0$, such a fixed point exists for $\lambda < 0$. For $a < 0$, the fixed point existence condition is $\lambda > 0$. From $Df_r = 1 - 2\lambda$, the fixed point is stable for $\lambda > 0$ owing to $s_{k+1} < s_k$ and unstable for $\lambda < 0$ owing to $s_{k+1} > s_k$. For $\lambda = 0$, we have $r_{k(0)}^* = 0$ and

$$Df_r(r_k^*, \lambda) = 1 \text{ and } D_\lambda Df_r(r^*, \alpha) = 1 \neq 0. \tag{2.157}$$

For $r_{k(0)}^* = 0$ and $\lambda = 0,\ Df_r(r_k^*, \lambda) = 1$ and $D^2 f_r(r_k^*, \lambda) = 6ar_k^* = 0$ exists. So we have $D^3 f_r(r_k^*, \lambda) = 6a$. The variational equation is given by $s_{k+1} = (1 + 6as_k^2)s_k$. For $a < 0,\ s_{k+1} < s_k$, the fixed point $(r_{k(0)}^*, \lambda) = (0, 0)$ is sprially stable of the third order. The bifurcation of the fixed point $(r_{k(0)}^*, \lambda) = (0, 0)$ is the Neimark bifurcation. The Neimark bifurcation with stable focus ($a < 0$) is called a supercritical case. For $a > 0,\ s_{k+1} > s_k$, the fixed point $(r_{k(0)}^*, \lambda) = (0, 0)$ is spirally unstable of the third order. The bifurcation of the fixed point $(r_{k(0)}^*, \lambda) = (0, 0)$ is the Neimark bifurcation. The Neimark bifurcation with unstable focus ($a > 0$) is called a subcritical case. The supercritical and subcritical Neimark bifurcation is shown in Fig. 2.8a and b. The solid lines and curves represent stable fixed point. The dashed lines and curves represent unstable fixed point. The phase shift is determined by $\theta_{k+1} = \theta_k - \vartheta$ and $r_{k(2)}^* \neq 0$, one get a unstable or unstable periodic solution on the circle.

From the foregoing analysis of the Neimark bifurcation, the Neimark bifurcation points possess the higher-order singularity of the flow in discrete dynamical system in the radial direction. For the stable Neimark bifurcation, the mth order singularity of the flow at the bifurcation point exists as a sink of the mth order in the radial direction. For the unstable Neimark bifurcation, the mth order singularity of the flow at the bifurcation point exists as a source of the mth order in the radial direction.

2.4 Lower Dimensional Discrete Systems

For a better understanding, the stability and bifurcation of 1-D and 2-D maps will be discussed.

2.4.1 *One-Dimensional Maps*

Consider a 1-D map,

$$P : x_k \rightarrow x_{k+1} \text{ with } x_{k+1} = f(x_k, \mathbf{p}) \tag{2.158}$$

where $\mathbf{p}$ is a parameter vector. To determine the period-1 solution (fixed point) of Eq. (2.158), substitution of $x_{k+1} = x_k$ into Eq. (2.158) yields the periodic solution $x_k = x_k^*$. The stability and bifurcation of the period-1solution is presented.

(i) Pitchfork bifurcation (period-doubling bifurcation)

$$\frac{dx_{k+1}}{dx_k} = \left.\frac{df(x_k, \mathbf{p})}{dx_k}\right|_{x_k = x_k^*} = -1. \tag{2.159}$$

(ii) Tangent (saddle-node) bifurcation

$$\frac{dx_{k+1}}{dx_k} = \left.\frac{df(x_k, \mathbf{p})}{dx_k}\right|_{x_k = x_k^*} = 1. \tag{2.160}$$

With two such conditions and fixed points $x_k = x_k^*$, the critical parameter vector $\mathbf{p}_0$ on the corresponding parameter manifolds can be determined. The two kinds of bifurcations for 1-D iterative maps are depicted in Fig. 2.9. Note that the most common pitchfork bifurcation involves an infinite cascade of period-doubling bifurcations with universal scalings. An exact renormalization theory for period-doubling bifurcation was developed in terms of a functional equation by Feigenbaum (1978), and Collet and Eckmann (1980). Helleman (1980a, b) employed an *algebraic* renormalization procedure to determine the rescaling constants. It is assumed that $f(x_k, \mathbf{p})$ has a quadratic maximum at $x_k = x_k^0$. If chaotic solution ensues at $\mathbf{p}_\infty$ via the period-doubling bifurcation, the function $x_{k+1} = f(x_k, \mathbf{p}_\infty)$ is rescaled by a scale factor α and a self-similar structure exists near $x_k = x_k^0$. Under the transition to chaos, the period doubling bifurcation will be discussed where two renormalization procedures will be presented in next section, namely, the renormalization group approach via the functional equation method as outlined by Feigenbaum (1978) (see also, Schuster 1988; Lichtenberg and Lieberman 1992), and the algebraic renormalization technique as described by Helleman (1980a, b). In next section, the quasiperiodicity route to chaos and the intermittency route to chaos will be discussed.

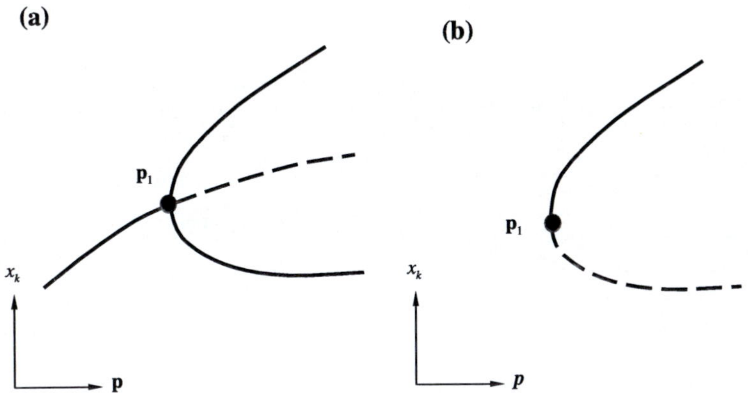

Fig. 2.9 Bifurcation types: **a** period-doubling and **b** saddle-node

2.4.2 Two-Dimensional Maps

Consider a 2-D map

$$P : \mathbf{x}_k \to \mathbf{x}_{k+1} \text{ with } \mathbf{x}_{k+1} = \mathbf{f}(\mathbf{x}_k, \mathbf{p}), \tag{2.161}$$

where $\mathbf{x}_k = (x_k, y_k)^{\mathrm{T}}$ and $\mathbf{f} = (f_1, f_2)^{\mathrm{T}}$ with a parameter vector $\mathbf{p}$. The period-n fixed point for Eq. (2.161) is $(\mathbf{x}_k^*, \mathbf{p})$, i.e., $P^{(n)}\mathbf{x}_k^* = \mathbf{x}_{k+n}^*$, where $P^{(n)} = P \circ P^{(n-1)}$ and $P^{(0)} = \mathbf{I}$, and its stability and bifurcation conditions are given as follows:

(i) period-doubling (flip or pitchfork) bifurcation

$$\mathrm{tr}(DP^{(n)}) + \det(DP^{(n)}) + 1 = 0; \tag{2.162}$$

(ii) saddle-node bifurcation

$$\det(DP^{(n)}) + 1 = \mathrm{tr}(DP^{(n)}); \tag{2.163}$$

(iii) Neimark bifurcation

$$\det(DP^{(n)}) = 1, \tag{2.164}$$

where

$$DP^{(n)}(\mathbf{x}_k^*) = \prod\nolimits_{j=n-1}^{0} DP(\mathbf{x}_{k+j}^*) = \underbrace{\left[\frac{\partial \mathbf{x}_{k+n}}{\partial \mathbf{x}_{k+n-1}}\right]_{\mathbf{x}_{k+n-1}^*} \cdot \cdots \cdot \left[\frac{\partial \mathbf{x}_{k+1}}{\partial \mathbf{x}_k}\right]_{\mathbf{x}_k^*}}_{n}. \tag{2.165}$$

For $n = 1$, we have

$$DP(\mathbf{x}_k^*) = \left[\frac{\partial \mathbf{x}_{k+1}}{\partial \mathbf{x}_k}\right]_{\mathbf{x}_k^*} = \begin{bmatrix} \partial_{x_k} f_1 & \partial_{y_k} f_1 \\ \partial_{x_k} f_2 & \partial_{y_k} f_2 \end{bmatrix}_{\mathbf{x}_k^*} \tag{2.166}$$

$$\begin{aligned} \operatorname{tr}(DP) &= \partial_{x_k} f_1 + \partial_{y_k} f_2 \\ \det(DP) &= \partial_{x_k} f_1 \cdot \partial_{y_k} f_2 - \partial_{y_k} f_1 \cdot \partial_{x_k} f_2 \end{aligned} \tag{2.167}$$

and

$$\begin{aligned} \partial_{x_k} f_1 &= \partial f_1(\mathbf{x}_k, p)/\partial x_k|_{\mathbf{x}_k=\mathbf{x}_k^*}, \\ \partial_{y_k} f_1 &= \partial f_1(\mathbf{x}_k, p)/\partial y_k|_{\mathbf{x}_k=\mathbf{x}_k^*}, \\ \partial_{x_k} f_2 &= \partial f_2(\mathbf{x}_k, p)/\partial x_k|_{\mathbf{x}_k=\mathbf{x}_k^*}, \\ \partial_{y_k} f_2 &= \partial f_2(\mathbf{x}_k, p)/\partial y_k|_{\mathbf{x}_k=\mathbf{x}_k^*}. \end{aligned} \tag{2.168}$$

The bifurcation and stability conditions for the solution of period-n for Eq. (2.161) are summarized in Fig. 2.10 with $\det(DP^{(n)}) = \det(DP^{(n)}(\mathbf{x}_{k(0)}^*, \mathbf{p}_0))$ and $\operatorname{tr}(DP^{(n)}) = \operatorname{tr}(DP^{(n)}(\mathbf{x}_{k(0)}^*, \mathbf{p}_0))$. The thick dashed lines are bifurcation lines. The stability of fixed point is given by the eigenvalues in complex plane. The stability of fixed point for higher dimensional systems can be identified by using a naming of stability for linear dynamical systems in Appendix B. The saddle-node bifurcation possesses stable saddle-node bifurcation (critical) and unstable saddle-node bifurcation (degenerate).

2.4.3 Finite-Dimensional Maps

Consider an m-D map

$$P : \mathbf{x}_k \to \mathbf{x}_{k+1} \text{ with } \mathbf{x}_{k+1} = \mathbf{f}(\mathbf{x}_k, \mathbf{p}), \tag{2.169}$$

where $\mathbf{x}_k = (x_{1k}, x_{2k}, \cdots x_{mk})^{\mathrm{T}}$ and $\mathbf{f} = (f_1, f_2, \cdots, f_m)^{\mathrm{T}}$ with a parameter $\mathbf{p}$. The period-n fixed point for Eq. (2.169) is $(\mathbf{x}_k^*, \mathbf{p})$, and its stability and bifurcation conditions are given as follows. Similarly, $P^{(n)}\mathbf{x}_k^* = \mathbf{x}_{k+n}^*$, where $P^{(n)} = P \circ P^{(n-1)}$ and $P^{(0)} = \mathbf{I}$.

(i) period-doubling (flip or pitchfork) bifurcation

$$|DP^{(n)} + \mathbf{I}_{m\times m}| = 0, \tag{2.170}$$

(ii) saddle-node bifurcation

$$|DP^{(n)} - \mathbf{I}_{m\times m}| = 0, \tag{2.171}$$

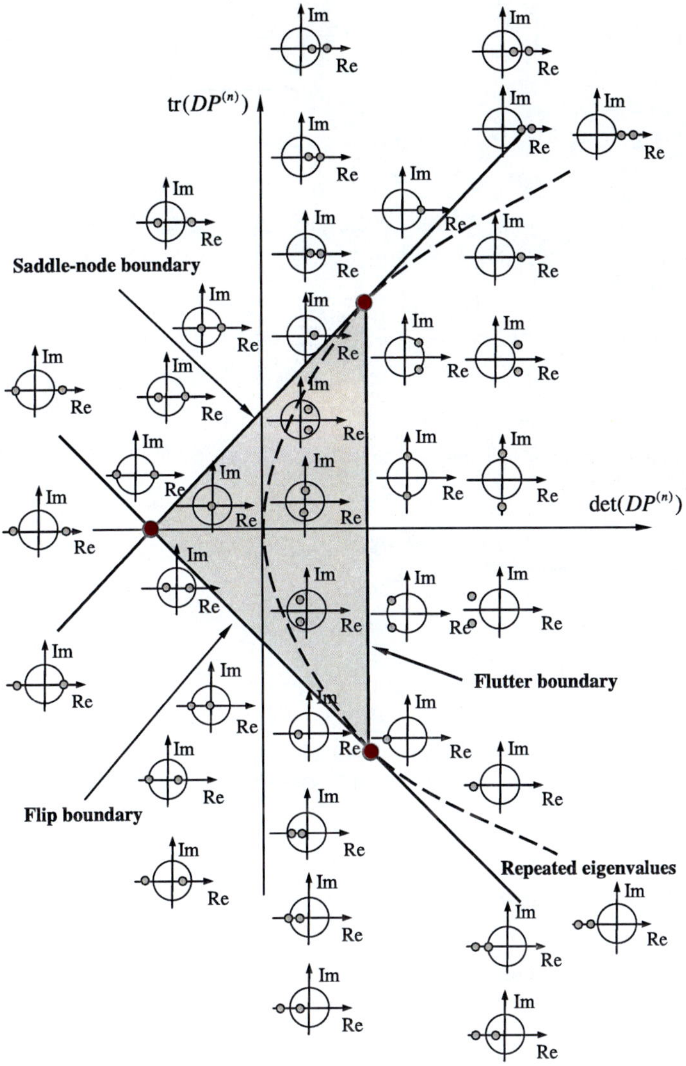

Fig. 2.10 Stability and bifurcation diagrams through the complex plane of eigenvalues for 2D-discrete dynamical systems

(iii) Neimark bifurcation

$$\begin{vmatrix} DP^{(n)} - \alpha \mathbf{I}_{m\times m} & -\beta \mathbf{I}_{m\times m} \\ \beta \mathbf{I}_{m\times m} & DP^{(n)} - \alpha \mathbf{I}_{m\times m} \end{vmatrix} = 0 \text{ with } \alpha^2 + \beta^2 = 1 \tag{2.172}$$

where

$$DP^{(n)}(\mathbf{x}_k^*) = \prod_{j=n-1}^{0} DP(\mathbf{x}_{k+j}^*) = \underbrace{\left[\frac{\partial \mathbf{x}_{k+n}}{\partial \mathbf{x}_{k+n-1}}\right]_{\mathbf{x}_{k+n-1}^*} \cdot \cdots \cdot \left[\frac{\partial \mathbf{x}_{k+1}}{\partial \mathbf{x}_k}\right]_{\mathbf{x}_k^*}}_{n} \tag{2.173}$$

with

$$DP(\mathbf{x}_{k+j}^*) = \left[\frac{\partial \mathbf{x}_{k+j+1}}{\partial \mathbf{x}_{k+j}}\right]_{\mathbf{x}_{k+j}^*} = \begin{bmatrix} \partial_{x_{1(k+j)}} f_1 & \partial_{x_{2(k+j)}} f_1 & \cdots & \partial_{x_{m(k+j)}} f_1 \\ \partial_{x_{1(k+j)}} f_2 & \partial_{x_{2(k+j)}} f_2 & \cdots & \partial_{x_{m(k+j)}} f_2 \\ \vdots & \vdots & \vdots & \vdots \\ \partial_{x_{1(k+j)}} f_m & \partial_{x_{2(k+j)}} f_m & \cdots & \partial_{x_{m(k+j)}} f_m \end{bmatrix}_{\mathbf{x}_{k+j}^*} \tag{2.174}$$

for $j = 0, 1, \cdots, n-1$ and

$$\partial_{x_{\alpha(k+j)}} f_\beta = \frac{\partial f_\beta(\mathbf{x}_{k+j}, \mathbf{p})}{\partial x_{\alpha(k+j)}} \text{ for } \alpha, \beta = 1, 2, \cdots, m. \tag{2.175}$$

2.5 Routes to Chaos

The routes to chaos will be discussed. The 1-D discrete system will be discussed first, then we will discuss the 2-D discrete systems.

2.5.1 *One-Dimensional Maps*

(A) Period doubling route to chaos

Herein, two renormalization group methods will be discussed as follows.

(i) *Functional renormalization theory*. Consider a universal function as

$$g^*(x) = \lim_{n\to\infty} \alpha^n f^{(2^n)}(x/\alpha^n, \mathbf{p}_\infty) \tag{2.176}$$

where g^* must satisfy the rescaling equation of the geometry, that is,

$$g^* = \alpha g^*(g^*(x/\alpha)) = Tg^* \tag{2.177}$$

in which T is a period-doubling operator. From Eq. (2.177), the universality of the scale factor α is obtained. The linearization of $f(x, \mathbf{p}_n)$ at $\mathbf{p}_n = \mathbf{p}_\infty$ yields the universal constant δ.

$$f(x,\mathbf{p}_n) = f(x,\mathbf{p}_\infty) + \left.\frac{\partial f(x,\mathbf{p}_n)}{\partial \mathbf{p}_n}\right|_{\mathbf{p}_n=\mathbf{p}_\infty} (\mathbf{p}_n - \mathbf{p}_\infty) + o(\| \mathbf{p}_n - \mathbf{p}_\infty \|). \tag{2.178}$$

Applying the period-doubling operator n times to Eq. (2.178) yields,

$$\lim_{n\to\infty} T^n f(x,\mathbf{p}_n) = g^*(x) + L^n_{g^*}\left.\left(\frac{\partial f(x,\mathbf{p}_n)}{\partial \mathbf{p}_n}\right)\right|_{\mathbf{p}_n=\mathbf{p}_\infty} (\mathbf{p}_n - \mathbf{p}_\infty). \tag{2.179}$$

Substitution of the unstable eigenvalue of L_{g^*} into Eq. (2.179) gives

$$\lim_{n\to\infty} T^n f(x,\mathbf{p}_n) = g^*(x) + \delta^n\left.\left(\frac{\partial f(x,\mathbf{p}_n)}{\partial \mathbf{p}_n}\right)\right|_{\mathbf{p}_n=\mathbf{p}_\infty} (\mathbf{p}_n - \mathbf{p}_\infty). \tag{2.180}$$

Transformation of the point of origin to $x = x_0$ and normalization of Eq. (2.158) by setting $g^*(0) = 1$, the condition is

$$f^{(2^n)}(0,\mathbf{p}_n) = 0. \tag{2.181}$$

From Eqs. (2.180) and (2.181), the universal constant is proportional to

$$||\mathbf{p}_n - \mathbf{p}_\infty|| \sim \delta^{-n}. \tag{2.182}$$

(ii) *Algebraic renormalization theory.* Taking into account the period-2 solutions of Eq. (2.158), we can solve for $x_{1\pm}, x_{2\pm}$ at $x_k = x_{k+2}$:

$$f(x_k,\mathbf{p}) = x_{k+1} \text{ and } f(x_{k+1},\mathbf{p}) = x_{k+2}. \tag{2.183}$$

Using a Taylor series expansion, we can apply a perturbation to Eq. (2.183) at $x_k = x_{k(2)\pm} + \Delta x_k$, $x_{k+1} = x_{k(1)\pm} + \Delta x_{k+1}$ and $x_{k+2} = x_{k(2)\pm} + \Delta x_{k+2}$, that is,

$$\Delta x_{k+1} = f_1(\Delta x_k, \mathbf{p}), \tag{2.184}$$

$$\Delta x_{k+2} = f_2(\Delta x_{k+1}, \mathbf{p}). \tag{2.185}$$

Substitution of Eq. (2.184) into Eq. (2.185) yields

$$\Delta x_{k+2} = f_2(f_1(\Delta x_k, \mathbf{p}), \mathbf{p}) = f(\Delta x_k, \bar{\mathbf{p}}). \tag{2.186}$$

Rescaling Eq. (2.186) with

$$x'_k = \alpha \Delta x_k \tag{2.187}$$

gives the corresponding renormalized equation, i.e.,

$$x'_{k+2} = f(x'_k, \bar{\mathbf{p}}_{2^{k+1}}), \tag{2.188}$$

where

$$\bar{\mathbf{p}}_{2^{k+1}} = \mathbf{g}(\bar{\mathbf{p}}_{2^k}). \tag{2.189}$$

Equation (2.189) presents a relationship of the bifurcation values between two period-doubling bifurcations. The rescaling factor α is determined by comparing Eq (2.161) with Eq. (2.158). If chaos appears via the period-doubling cascade, i. e., $\bar{\mathbf{p}}_{2^{k+1}} = \bar{\mathbf{p}}_{2^k} = \mathbf{p}_\infty$, the universal parameter manifolds are determined.

(B) Quasiperiodicity route to chaos
Consider a mapping defined on the unit interval $0 \leq x \leq 1$, that is,

$$x_{k+1} = x_k + \Omega + f(x_k, \mathbf{p}) = F(x_k, \Omega, \mathbf{p}), \tag{2.190}$$

where $f(x_k, \mathbf{p})$ is a periodic modulo, i. e., $f(x_k + 1, \mathbf{p}) = f(x_k, \mathbf{p})$; and Ω is a prescribed parameter defined in the interval $0 \leq \Omega \leq 1$. In Eq. (2.190), parameters $(\Omega, \mathbf{p})$ can be adjusted to generate a transition from quasiperiodicity to chaos. We can increase the parameter vector amplitude $||\mathbf{p}||$ first under a rational winding number $w = p/q$ fixed to a selected value, and we will have to increase Ω as well. The winding number w is an important quantity for describing the dynamics, which is defined by

$$w(\Omega, \mathbf{p}) = \lim_{k\to\infty} \frac{x_k - x_0}{k}. \tag{2.191}$$

Define a quantity $\Omega_{p,q}(\mathbf{p})$ which belongs to a q-cycle of the map $f(x_k, \mathbf{p})$ and shifted by p. This quantity generates a rational winding number $w = p/q$ and for a fixed value of $\mathbf{p}$, it can be determined from

$$F^{(q)}(0, \Omega_{p,q}, \mathbf{p}) = p, \tag{2.192}$$

where $F^{(q)} = F(F^{(q-1)})$. Choosing the winding number equal to the golden mean $w^* = (\sqrt{5} - 1)/2$, the universal constants for chaos can be computed.

(C) Intermittency route to chaos
There are three types of intermittencies, Types I, II and III. In this section, we will present only Type I and III intermittencies. The Type II intermittency will be discussed in a later section under 2-D maps.

(i) *Type I intermittency*. Consider an iterative map with a small perturbation defined by

$$x_{k+1} = f(x_k, \varepsilon) = \varepsilon + x_k + \eta x_k^2, \tag{2.193}$$

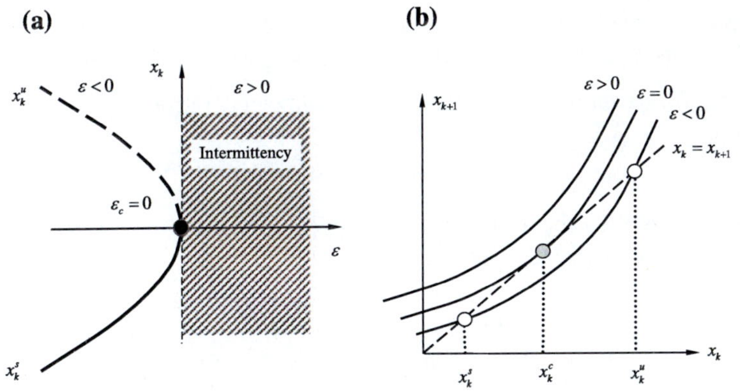

Fig. 2.11 **a** bifurcation and **b** iterative map for Eq. (2.193)

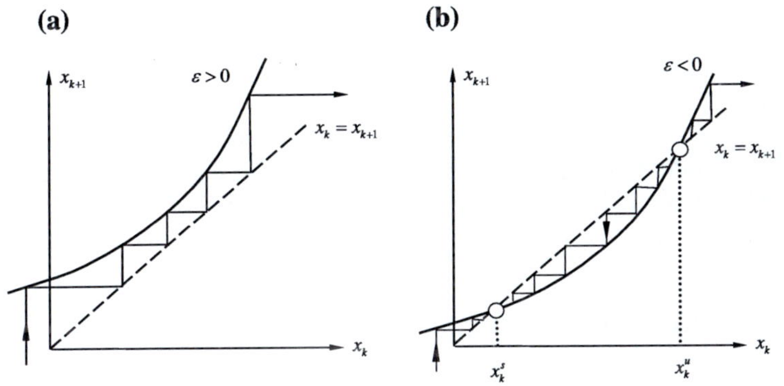

Fig. 2.12 **a** Intermittency and **b** stable and unstable fixed points for Eq. (2.193)

where ε is a control parameter and η is a prescribed parameter. This mapping results in the Type I intermittency caused by the tangent bifurcation which occurs when a real eigenvalue of Eq. (2.193) crosses the unit circle at +1. In other words,

$$x_k^* = \pm(-\varepsilon/\eta)^{1/2} \text{ and } Df(x_k^*) = df/dx_k|_{x_k=x_k^*} = 1 + 2\eta x_k^*. \qquad (2.194)$$

For $\eta > 0$, if $\varepsilon > 0$, no fixed point exists. If $\varepsilon = 0$, $x_k^* = x_k^c = 0$ with $Df(x_k^*) = 1$. Since $D^2 f(x_k^*) = 2\eta \neq 0$, the saddle-node bifurcation occurs. For $\eta > 0$, if $\varepsilon > 0$, $x_k^* = (-\varepsilon/\eta)^{1/2} \equiv x_k^u$ with $Df(x_k^*) > 1$ and $x_k^* = -(-\varepsilon/\eta)^{1/2} \equiv x_k^s$ with $Df(x_k^*) < 1$. The tangent bifurcation and iterative map for the Type I intermittency is shown in Fig. 2.11. The intermittency and the stable and unstable fixed points are presented in Fig. 2.12. This case includes the Poincare map for the Lorenz model and the iterative map for the window of period-3 solution in the chaotic band. The renormalization procedure of Eq. (2.193) has been presented in Hu and Rudnick (1982). Also, interested readers can refer to Guckenheimer and Holmes (1990), and Schuster (1988) for details.

(ii) *Type III intermittency*. Consider the following an iterative map

$$x_{k+1} = f(x_k, \varepsilon) = -(1+\varepsilon)x_k - \eta x_k^3, \tag{2.195}$$

which produces the Type III intermittency caused by the inverse pitchfork bifurcation. The fixed points are

$$\begin{aligned} &x_k^* = 0 \text{ and } x_k^* = \pm(-\varepsilon/\eta)^{1/2} \text{ with} \\ &Df(x_k^*) = df/dx_k|_{x_k = x_k^*} = -(1+\varepsilon) - 3\eta x_k^2. \end{aligned} \tag{2.196}$$

For $\eta > 0$, if $\varepsilon > 0$, only one unstable fixed point exists. $x_k^* = x_k^u = 0$ because of $Df(x_k^*) < -1$. If $\varepsilon < 0$, there are three fixed points. $x_k^* = x_k^s = 0$ and $Df(x_k^*) > -1$; $x_k^* \equiv x_k^s = \pm(-\varepsilon/\eta)^{1/2}$ are stable with $Df(x_k^*) < -1$. If $\varepsilon = 0$, $x_k^* = x_k^c = 0$ with $Df(x_k^*) = -1$. $D^2 f(x_k^*) = -6\eta x_k^* = 0$, however, $D^3 f(x_k^*) = -6\eta < 0$. This is an inverse pitchfork bifurcation (unstable period-doubling bifurcation). The bifurcation diagram and the iterative map for the Type III intermittency is presented in Fig. 2.13. In addition, the intermittency and the stable and unstable fixed points are presented in Fig. 2.14.

2.5.2 *Two-Dimensional Systems*

For 2-D invertible maps, the transition from regular motion to chaos takes place via a series of cascades of period-doubling bifurcations. The renormalization procedure of the period-doubling route to chaos for a 2-D map appears in next section through an example. The quasi-periodic transition to chaos and the intermittence to chaos are briefly presented through an example. The quasiperiodic transition to chaos and the intermittence to chaos are presented briefly.

(A) Quasiperiodic transition to chaos

This route to chaos is presented via the standard map as

$$x_{k+1} = x_k + K \sin\theta_k \text{ and } \theta_{k+1} = \theta_k + x_{k+1}. \tag{2.197}$$

The critical condition of Eq. (2.197) for transition from local to global stochasticity is $K_{\text{cr}} \approx 0.9716\cdots$. For a dissipative standard map, consider

$$x_{k+1} = (1-\delta)x_k + K \sin\theta_k \text{ and } \theta_{k+1} = \theta_k + x_{k+1}, \tag{2.198}$$

where δ is the dissipative coefficient. Some results are given in Lichtenberg and Lieberman (1992).

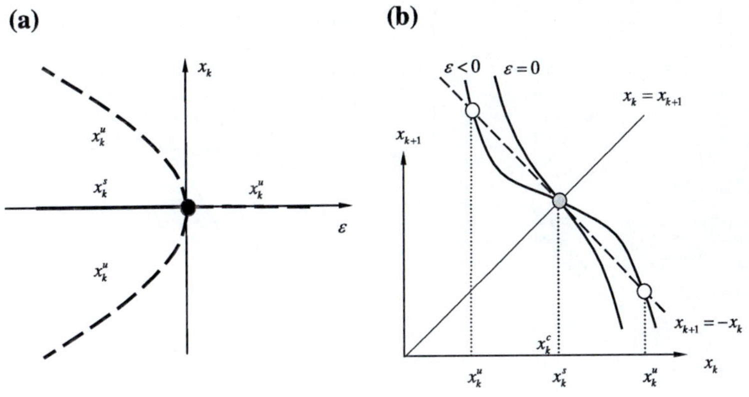

Fig. 2.13 **a** Inverse pitchfork bifurcation and **b** iterative map for Eq. (2.195)

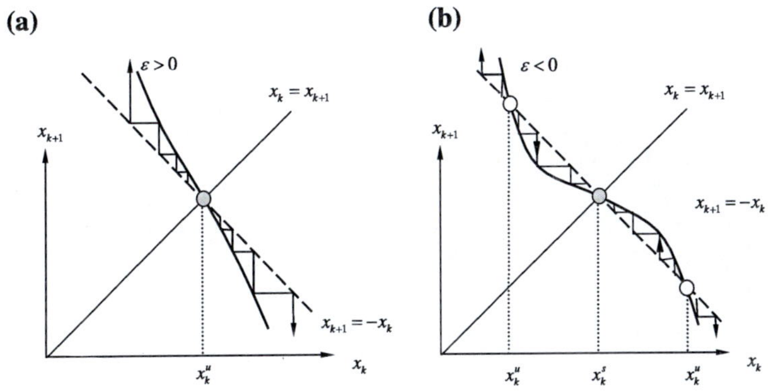

Fig. 2.14 **a** Intermittency and **b** stable and unstable fixed points for Eq. (2.195)

(B) Type II intermittency to chaos
Consider the following mapping which represents Type II intermittency to chaos,

$$r_{k+1} = (1+\varepsilon)r_k + \eta r_k^3 \text{ and } \theta_{k+1} = \theta_k + \Omega, \tag{2.199}$$

$$x_k = r_k \cos\theta_k \text{ and } y_k = r_k \sin\theta_k. \tag{2.200}$$

When a pair of complex eigenvalues of Eq. (2.199) passes over the unit circle, the subcritical Neimark bifurcation occurs. Hence, Type II intermittency results from the subcritical Neimark bifurcation as shown in Fig. 2.15, and the corresponding intermittency and stable and unstable fixed points are presented in Fig. 2.16.

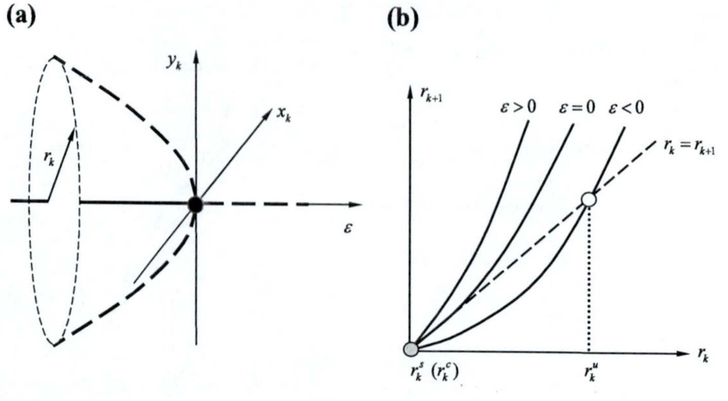

Fig. 2.15 **a** Subcritical Neimark bifurcation and **b** iterative map for Eq. (2.199)

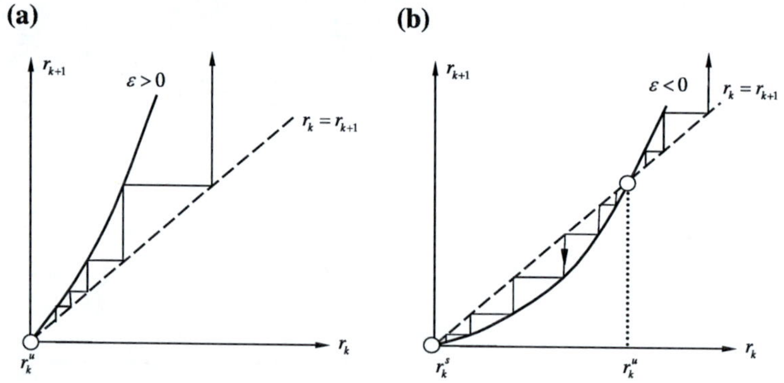

Fig. 2.16 **a** Intermittency and **b** stable and unstable fixed points for Eq. (2.199)

2.6 Universality for Discrete Duffing Systems

Consider a Duffing oscillator

$$\ddot{x} + \delta\dot{x} + \alpha_1 x + \alpha_2 x^3 = Q_0 \cos \Omega t, \tag{2.201}$$

where system parameters are $\delta, \alpha_1, \alpha_2, Q_0$ and Ω. Discretizing it with respect to time yields a discrete map to investigate qualitatively its universal behavior. Here, by means of the Naive discretization of the time derivative, Eq. (2.201) is discretized at $x_k = x(t_k)$ and $t_k = 2k\pi/\Omega$ using time difference $\Delta t = 2\pi/\Omega$ for external excitation. Therefore

$$x_{k+1} - 2x_k + x_{k-1} + (1-b)(x_k - x_{k-1}) + cx_k + dx_k^3 = (1-b)\varpi, \tag{2.202}$$

where all parameters are defined as

$$b = 1 - \frac{2\pi}{\Omega}\delta,\ c = (\frac{2\pi}{\Omega})^2\alpha_1,\ d = (\frac{2\pi}{\Omega})^2\alpha_2,\ \varpi = \frac{Q_0}{\alpha_1}\frac{2\pi}{\Omega}. \tag{2.203}$$

From Eq. (2.202), a Duffing map is constructed as

$$P: x_{k+1} = x_k + \varpi + y_{k+1} \text{ and } y_{k+1} = by_k - cx_k - dx_k^3. \tag{2.204}$$

To qualitatively investigate the Feigenbaum cascade of Eq. (2.201), the renormalization of Eq. (2.204) will be presented via period-doubling bifurcation cascade. Consider a transformation as

$$x_k = X_k + \bar{\delta} \text{ and } y_k = Y_k - \varpi. \tag{2.205}$$

Substitution of the foregoing equation into Eq. (2.204) yields

$$X_{k+1} = X_k + Y_{k+1} \text{ and } Y_{k+1} = BY_k + CX_k + DX_k^2 - EX_k^3, \tag{2.206}$$

where

$$-d\bar{\delta}^3 + c\bar{\delta} + (-1 + b)\varpi = 0, \tag{2.207}$$

$$B = b,\ C = -(c + 3d\bar{\delta}^2),\ D = -3d\bar{\delta},\ \text{and } E = d. \tag{2.208}$$

From Eq. (2.207), the parameter $\bar{\delta}$ is determined. Using Eq. (2.206), its period-1 solution is determined by

$$\begin{aligned} &Y_{k(1)}^* = 0 \text{ and } X_{k(1)}^* = 0; \\ &Y_{k(2,3)}^* = 0 \text{ and } X_{k(2,3)}^* = \frac{D \pm \sqrt{D^2 + 4CE}}{2E}. \end{aligned} \tag{2.209}$$

Deformation of Eq. (2.206) is

$$X_{k+1} + BX_{k-1} = (1 + B + C)X_k + DX_k^2 - EX_k^3. \tag{2.210}$$

The second iteration of Eq. (2.206) gives

$$X_{k+2} + BX_k = (1 + B + C)X_{k+1} + DX_{k+1}^2 - EX_{k+1}^3. \tag{2.211}$$

The period-2 of Eq. (2.206) requires $X_{k+2} = X_k$ and $X_{k+1} = X_{k-1}$. Thus simplification of Eqs. (2.210) and (2.211) gives

$$a_6X_k^6 + a_5X_k^5 + a_4X_k^4 + a_3X_k^3 + a_2X_k^2 + a_1X_k + a_0 = 0, \tag{2.212}$$

where

$$\begin{aligned} &a_0 = (1 + B)^2[C + 2(1 + B)],\ a_1 = D(1 + B)[C + 2(1 + B)], \\ &a_2 = D^2(1 + B) + EC^2 - 3(1 + B)E[C + 2(1 + B)], \\ &a_3 = 2DE[C - (1 + B)],\ a_4 = E^2[3(1 + B) + 2C] - D^2E,\ a_6 = -E^3. \end{aligned} \tag{2.213}$$

For all the given parameters, solving Eq. (2.212) numerically obtains $X_k = X_k^*$, meanwhile, using one of Eqs. (2.210) or (2.211) and $X_{k+2} = X_k$ and $X_{k+1} = X_{k-1}$, the second solution $X_{k+1} = X_{k+1}^*$ can be determined. With the periodicity of period-2, solving of Eqs.(2.210) and (2.211) directly gives the period-2 solutions via Newton-Raphson method. In the neighborhoods of solutions X_{k+j}^*, consider a perturbation as

$$X_{k+j} = X_{k+j}^* + \Delta X_{k+j} \text{ for } j = -1, 0, 1, 2. \tag{2.214}$$

Substitution of them into Eq. (2.210) yields a group of iterative equations as:

$$\begin{aligned} \Delta X_k + B\Delta X_{k-2} &= e_{11}\Delta X_{k-1} + e_{12}\Delta X_{k-1}^2 - E\Delta X_{k-1}^3, \\ \Delta X_{k+1} + B\Delta X_{k-1} &= e_{21}\Delta X_k + e_{22}\Delta X_k^2 - E\Delta X_k^3, \\ \Delta X_{k+2} + B\Delta X_k &= e_{11}\Delta X_{k+1} + e_{12}\Delta X_{k+1}^2 - E\Delta X_{k+1}^3 \end{aligned} \tag{2.215}$$

where e_{11}, e_{12} and e_{21}, e_{22} are

$$\begin{aligned} e_{11} &= 1 + B + C + 2DX_n^* - 3E(X_k^*)^2, e_{12} = D - 3EX_k^* \\ e_{21} &= 1 + B + C + 2DX_{n+1}^* - 3E(X_{k+1}^*)^2, e_{22} = D - 3EX_{k+1}^*. \end{aligned} \tag{2.216}$$

Multiplication of the first equation by B and *the second equation* by e_{11} of Eq. (2.215), and adding both equations into the third equation yields

$$\begin{aligned} \Delta X_{k+2} + B^2\Delta X_{k-2} =&(e_{11}e_{21} - 2B)\Delta X_k + e_{11}e_{22}\Delta X_k^2 \\ &-e_{11}\Delta X_k^3 + e_{12}(\Delta X_{k+1}^2 + B\Delta X_{k-1}^2) \\ &+e_{13}(\Delta X_{k+1}^3 + B\Delta X_{k-1}^3). \end{aligned} \tag{2.217}$$

For a small vicinity of the bifurcation of period-2, ΔX_{k+1} and ΔX_{k-1} are quite close. Therefore a similar linear scale ratio is introduced as

$$r = \Delta X_{k+1}/\Delta X_{k-1}. \tag{2.218}$$

The nonlinear terms can be ignored because ΔX_{k+j} ($j = -1, 0, 1, 2$) is the infinitesimal quantity. The second equation of Eq. (2.215) gives an approximate relationship as

$$\Delta X_{k-1} \approx \frac{e_{21}}{r + B}\Delta X_k. \tag{2.219}$$

From Eqs. (2.212) and (2.219), equation (2.217) becomes

$$\begin{aligned} \Delta X_{k+2} + B^2\Delta X_{k-2} = &(e_{11}e_{21} - 2B)\Delta X_k \\ &+ [e_{11}e_{22} + e_{12}e_{21}^2(r^2 + B^2)(r + B)^{-2}]\Delta X_k^2 \\ &- [e_{11}E - e_{13}e_{21}^3(r^3 + B^3)(r + B)^{-3}]\Delta X_k^3. \end{aligned} \tag{2.220}$$

From the renormalization theory, the rescaling length of the variables should be adopted.

$$X'_k = \varepsilon \Delta X_k \text{ and } X'_{k\pm1} = \varepsilon \Delta X_{k\pm2}, \tag{2.221}$$

where ε is a scaling constant. The foregoing equation makes Eq. (2.220) have an algebraically similar structure to Eq. (2.210), i.e.,

$$X'_{k+1} + B'X'_{k-1} = C'_1 X'_k + D'(X'_k)^2 - E'(X'_k)^3, \tag{2.222}$$

where

$$\begin{aligned} B' &= B^2,\ C'_1 = e_{11}e_{21} - 2B,\ C_1 = 1 + B + C, \\ D' &= \varepsilon[e_{11}e_{22} + e_{12}e_{21}^2(r^2 + B^2)(r + B)^{-2}], \\ E' &= \varepsilon^2[e_{11}E - e_{13}e_{21}^3(r^3 + B^3)(r + B)^{-3}]. \end{aligned} \tag{2.223}$$

If Eq. (2.222) has a self-similarity with Eq. (2.210), the similar scaling ratio r in Eq. (2.218) should be one, i.e., $r = 1$ in Eq. (2.218). The similar parameters have the same property as the scaling of variable Δx_k, and this property indicates that the cascade of bifurcations will be accumulated. Therefore the chaos is generated via period-doubling bifurcation at $B = B' = B_\infty,\ C_1 = C'_1 = C_{1\infty},\ D = D' = D_\infty$ and $E = E' = E_\infty$. Thus Eq. (2.223) becomes

$$\begin{aligned} B &= 0 \text{ or } 1, \\ C_1 + 2B &= [C_1 + 2DX_k^* - 3E(X_k^*)^2][C_1 + 2DX_{k+1}^* - 3E(X_{k+1}^*)^2], \\ D &= \varepsilon[e_{11}e_{22} + e_{12}e_{21}^2(r^2 + B^2)(r + B)^{-2}], \\ E &= \varepsilon^2[e_{11}E - e_{13}e_{21}^3(r^3 + B^3)(r + B)^{-3}]. \end{aligned} \tag{2.224}$$

where $C_1 = 1 + B + C$. To solve five parameters from four equations, one parameter should be given. However, from Eqs. (2.207) and (2.207), the parameter D is determined if parameters B, C and E have already been computed. Employing Eqs. (2.212), (2.213), (2.223) and (2.224), the universal parameter values for Eq. (2.206) are determined. To verify these values, the corresponding universal values are determined numerically via iteration of Eq. (2.206). Taking the parameters $D = 1.0$ and $E = 1.0$ into account, the universalized parameter C_1, computed via *R*enormalization *G*roup (*RG*) and *N*umerical *S*imulation (*NS*), versus the damping parameter B are shown in Fig. 2.17a with solid and circular-symbol curves, respectively. *RG* values are close to the *NS* values. For the parameters $D = -1.0$ and $E = 1.0$, the universalized parameter C_1 values *RG*(+), *NS*(+), *RG*(−) and *NS*(−) are the *RG*(−), *NS*(−), *RG*(+) and *NS*(+) of the situation of $D = 1.0$ and $E = 1.0$, respectively. As for the situation of $D = -1.0$ and $E = -1.0$, the universalized parameter C_1 versus the damping parameter B is also plotted in Fig. 2.17b. This renormalization group can provide a good prediction for $D = 1.0$ and $E = 1.0$ as the parameter B corresponding to the damping is in the range of $B = (-0.5, 0.9)$. However, a good prediction for

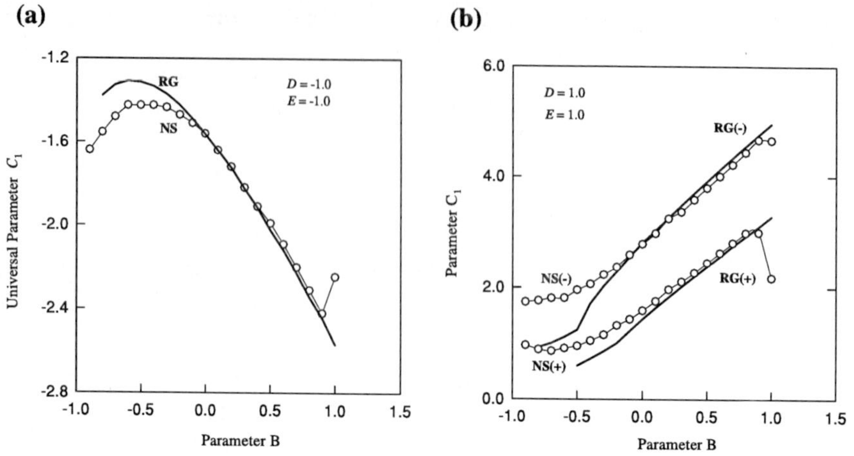

Fig. 2.17 Universal parameters via renormalization: **a** Duffing map with a soft spring and **b** Duffing map with a double-potential well

$D = -1.0$ and $E = -1.0$ is given for $B = (-0.8, 0.9)$. $B = 1$ implies the conservative system, and $B > 1$ implies the negative damping system. From such universal parameters for chaos generated by the period-doubling, the system parameters are $\delta, \alpha_1, \alpha_2, Q_0$ and Ω can be determined.

References

Collet, P. and Eckmann, J. P., 1980, *Iterated Maps on the Interval as Dynamical Systems*, Progress on Physics, **I**, Birkhauser, Boston.

Feigenbaum M. J., 1978, Quantitative universality for a class of nonlinear transformations, *Journal of Statistical Physics*, **19**, 25–52.

Guckenhiemer J. and Holmes P., 1990, *Nonlinear Oscillations, Dynamical Systems, and Bifurcations of Vector Fields*, Springer, New York.

Helleman R. H. G., 1980a, Self-generated chaotic behavior in nonlinear mechanics, *Fundamental Problems in Statistical Mechanics*, E. G. D. Cohen (ed.), **5**, 165–233 North-Holland: Amsterdam.

Helleman, R. H. G.(ed.), 1980b, *Nonlinear Dynamics*, (Annals of the New York Academy of Science) **357**, New York Academy of Sciences: New York.

Hu, B. and Rudnick, J.,1982, Exact solutions of the Feigenbaum renormalization group equations for intermittence, *Physical Review Letters*, **48**, 1645-1648.

Nitecki Z., 1971, *Differentiable Dynamics: An Introduction to the Orbit Structures of Diffeomorphisms*, MIT Press: Cambridge. MA.

Lichtenberg A. J. and Lieberman M. A., 1992, *Regular and Chaotic Dynamics*, 2nd ed., Springer-Verlag: New York.

Schuster H. G., 1988, *Deterministic Chaos, An Introduction*, 2nd ed., VCH Verlagsgesellschaft MBH: Weinheim, Germany.

Chapter 3
Chaos and Multifractality

In this Chapter, basic concepts of fractal in nonlinear dynamical systems will be presented as an introduction. The fractal generation rules will be presented for nonrandom and random fractals. The multifractals based on the single- and joint-multifractal measures will be presented. Multifractality of chaos generated by period-doubling bifurcation will be presented via a geometrical approach and self-similarity. Fractality of hyperbolic chaos will be discussed.

3.1 Introduction to Fractals

Chaos possesses self-similar structures which imply the presence of fractals. By self-similarity, we mean that no matter how much the view is zoomed, the same basic shape is retained. Therefore, whether an object is viewed globally or locally, the same basic structure is observed. It is also possible to use fractal dimension measurement to describe chaotic or strange attractors in a dissipative dynamical system. Most fractals in chaotic dynamics have multiscales and multimeasures, and thus, they are nonuniform fractals or multifractals. Unlike fractals which are geometrically self-similar, multifractals are statistically self-similar. Some basic concepts of fractals necessary to study their characteristics in chaos will be presented next.

3.1.1 Basic Concepts

What are fractals? Simply speaking, fractals are geometric objects that possess non-integer dimension and self-similarity. They do not necessarily have characteristic sizes, namely, we cannot measure dimensional quantities such as length, area and volume. The geometry can only be realized using a recursion of the iterative map. Before definition of fractals is given, the following two examples known as the Sierpinski gasket fractals are presented. In Fig. 3.1, the similar structures are generated

A. C. J. Luo, *Regularity and Complexity in Dynamical Systems*,
DOI: 10.1007/978-1-4614-1524-4_3, © Springer Science+Business Media, LLC 2012

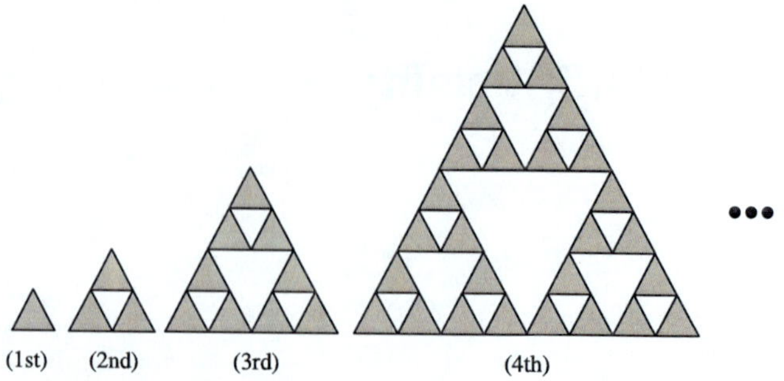

Fig. 3.1 The first four generations of the Sierpinski gasket fractals based on an expansion rule

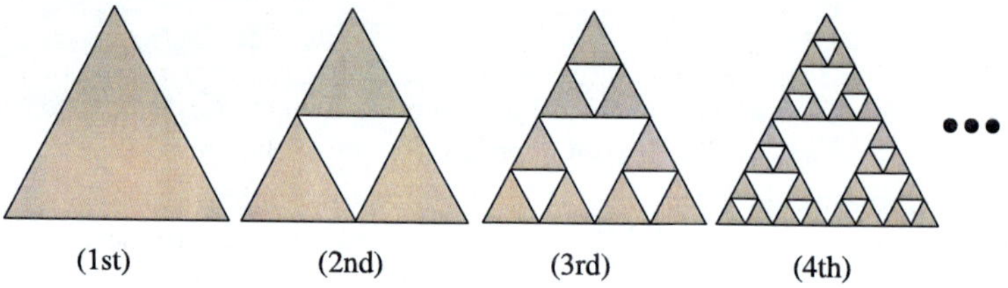

Fig. 3.2 The first four generations of the Sierpinski gasket fractals based on a reduction rule

by an expansion rule, whereas in Fig. 3.2, the similar structures are produced by a contraction rule. Both of the similar structure possess the characteristics of self-similarity, and have a non-integer dimension of 1.58. Thus, the Sierpinski gasket is a fractal. From Figs. 3.1 and 3.2, the scaling size L and number of self-similar structures $N(L)$ can be summarized in Tables 3.1 and 3.2. Note that r_0 is the original size.

The self-similar law in Fig. 3.1, from Table 3.1 and the expansion rule, is:

$$N(L) = (L/r_0)^D . \tag{3.1}$$

whereas, the self-similar law in Fig. 3.2, from Table 3.2 and the reduction rule, is:

$$N(L) = (r_0/L)^D . \tag{3.2}$$

From Eq. (3.1), the scaling size L grows rapidly as expected under the aggregation rule. On the other hand, from Eq. (3.2), the scaling size L shrinks rapidly as expected under the reduction rule. To characterize a fractal, the Hausdorff dimension based on a uniform formula is defined as follows.

Definition 3.1 For any object with non-empty N parts which are scaled by a ratio r in the m-D Euclidean space, there is a self-similarity satisfying

$$N(L)r^D = 1, \tag{3.3}$$

Table 3.1 Fractal distribution of the Sierpinski gasket with an expansion rule

Generation no.	Scaling size L	No. of self-similar structure $N(L)$
1st	$2^0 r_0$	3^0
2nd	$2^1 r_0$	3^1
3rd	$2^2 r_0$	3^2
4th	$2^3 r_0$	3^3
⋮	⋮	⋮
nth	$2^n r_0$	3^n

Table 3.2 Fractal distribution of the Sierpinski gasket with a reduction rule

Generation no.	Scaling size L	No. of self-similar structure $N(L)$
1st	$(1/2)^0 r_0$	3^0
2nd	$(1/2)^1 r_0$	3^1
3rd	$(1/2)^2 r_0$	3^2
4th	$(1/2)^3 r_0$	3^3
⋮	⋮	⋮
nth	$(1/2)^n r_0$	3^n

the Hausdorff dimension D in Eq. (3.3) is defined as

$$D = -\frac{\log N}{\log r}. \tag{3.4}$$

From the measure theory, mathematically, the Hausdorff dimension can be defined (e.g., Falconer 1990).

Definition 3.2 Consider a map $S : E \to E$ in $\mathscr{R}^n$, where $E \in \mathscr{R}^n$ is a closed set. If there is a number r with $0 < r < 1$ such that $|S(\mathbf{x}) - S(\mathbf{y})| = r||\mathbf{x} - \mathbf{y}||$ for all $\mathbf{x}, \mathbf{y} \in E$, then the mapping S is termed a similarity. Suppose a self-similar set $F \subseteq E$ is invariant under the mapping S. The Hausdorff dimension measure $H^D(F)$ is defined for any $\delta > 0$:

$$\begin{aligned} \mathrm{H}^D(F) &= \lim_{\delta\to 0} \mathrm{H}^D_\delta(F) \\ &= \lim_{\delta\to 0} \inf \left\{ \sum_{j=1}^{\infty} |U_j|^D \,\middle|\, F \subset \cup_{j=1}^{\infty} U_j \text{ and } 0 < |U_j| \le \delta \text{ for all } j \right\} \end{aligned} \tag{3.5}$$

where U_j is any non-empty δ-cover of F in $\mathscr{R}^n$.

Theorem 3.1 *If* $F \subset \mathscr{R}^n$ *and* $r > 0$ *then*

$$\mathrm{H}^D(rF) = r^D \mathrm{H}^D(F) \tag{3.6}$$

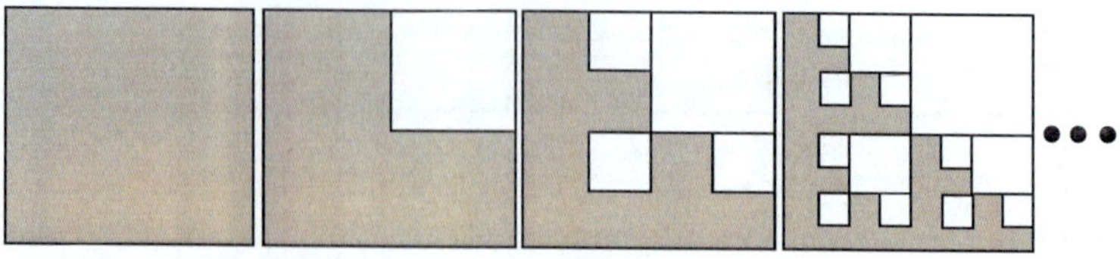

Fig. 3.3 Fractals generated by an iterative map in 2-D space

where $rF = \{rx | x \in F\}$ *means that the set F is scaled by a similar factor r.*

Proof Consider $\{U_j\}$ to be a non-empty δ-cover of F in $\mathscr{R}^n$ and $\{rU_j\}$ to be a non-empty δ-cover of rF in $\mathscr{R}^n$. Thus

$$\mathrm{H}^D_{r\delta}(rF) \le \sum_j |rU_j|^D = r^D \sum_j |U_j|^D \le r^D \mathrm{H}^D_\delta(F).$$

This is because U_j is any δ-cover of F in $\mathscr{R}^n$. If $\delta \to 0$, $\mathrm{H}^D(rF) \le r^D \mathrm{H}^D(F)$. Similarly, if the scaling factor r is replaced by $1/r$ and F is replaced by rF, then $r^D \mathrm{H}^D(F) \le \mathrm{H}^D(rF)$. Therefore, $\mathrm{H}^D(rF) = r^D \mathrm{H}^D(F)$. ∎

Application of $\mathrm{H}^D(F)$ in Eq. (3.5) to the self-similar set F gives

$$\begin{aligned} \mathrm{H}^D(F) &= \sum_{i=1}^{N(E)} \mathrm{H}^D(S(F_i)) = N(E) r^D \mathrm{H}^D(F) \\ \text{with } F &= \cup_{i=1}^{N(E)} F_i \text{ and } \mathrm{H}^D(rF) = r^D \mathrm{H}^D(F). \end{aligned} \tag{3.7}$$

The scaling ratios are $r = r_0/(2^n r_0) = 2^{-n}$ in Fig. 3.1 and $r = (2^{-n} r_0)/r_0 = 2^{-n}$ in Fig. 3.2. Thus, the Hausdorff dimension for the Sierpinski gasket fractals is

$$D = \frac{\log(3^n)}{\log(2^n)} = \frac{\log(3)}{\log(2)} = 1.58\cdots. \tag{3.8}$$

Another interesting point about fractals is that different fractals can have the same fractal dimension as for example, the Sierpinski gasket fractals. Consider a fractal shown in Fig. 3.3. Using Eq. (3.4), the Hausdorff dimension is $D = 1.58\cdots$.

Fractals can be classified as *nonrandom* and *random*. In addition, from scaling and measures, fractal can also be uniform or nonuniform. Uniform fractals are called simply as 'fractals' whereas nonuniform fractals are known as 'multifractals'. As expected, a nonrandom fractal is generated by a deterministic rule, such as a given iterative map. However, a random fractal is generated by a stochastic rule. Random fractals, whether uniform or otherwise, are always statistically self-similar. In other words, such random fractals cannot be geometrically self-similar. However, random fractals can represent natural phenomena such as coastlines, land surfaces, roughness, cloud boundaries and so on.

3.1.2 Fractal Generation Rules

The generation of fractals based on multiple generators will be briefly presented in this section, and the detailed discussion can be referred to Luo (1991) (also see, Leung and Luo 1992). The generation of nonrandom fractals will be discussed first by use of a single-scale single generator. Next, the generation of random fractals will be discussed through a single-scale multigenerator.

3.1.2.1 Nonrandom Fractals

If there are K generators in a fractal structure, and the ith generator has N_i non-empty sets with a linear scaling ratio r_i in $\mathscr{R}^n$, then for all K generators, the total number of equivalent nonempty sets N and linear scaling ratio r are determined by

$$N = \prod_{i=1}^{K} N_i \quad \text{and} \quad r = \prod_{i=1}^{K} r_i. \tag{3.9}$$

From Eq. (3.3), the Hausdorff dimension of this fractal is

$$D = -\frac{\sum_{i=1}^{K} \log N_i}{\sum_{i=1}^{K} \log r_i}. \tag{3.10}$$

To extend the above idea, consider the ith generator has m_i-time action on the fractal. The corresponding Hausdorff dimension is computed by

$$D = -\frac{\sum_{i=1}^{K} m_i \log N_i}{\sum_{i=1}^{K} m_i \log r_i} \tag{3.11}$$

However, from Eq. (3.9), the Hausdorff dimension is independent of the action order of the generators, which implies that the different fractals can have the same Hausdorff dimension.

3.1.2.2 Random Fractals

Basically, there are three possible methods of generation: random action, random generators and combined random action and generators.

(A) *Random action*: From Eq. (3.11), setting $M = \sum_{i=1}^{K} m_i$ and manipulation gives

$$D = -\frac{\sum_{i=1}^{K} \frac{m_i}{M} \log N_i}{\sum_{i=1}^{K} \frac{m_i}{M} \log r_i} = -\frac{\sum_{i=1}^{K} p_i \log N_i}{\sum_{i=1}^{K} p_i \log r_i} \tag{3.12}$$

where p_i is the action probability of the ith generator with the properties of $\sum_{i=1}^{K} p_i = 1$ and $p_i = m_i/M$.

(B) *Random generator*: If the ith generator has L_i subgenerators with action probability of p_{ij} ($\sum_{j=1}^{L_i} p_{ij} = 1$), linear scaling ratio r_{ij} and, N_{ij} non-empty subsets ($j = 1, 2, \ldots, L_i$), then the corresponding mean values are

$$\langle N_i \rangle = \sum_{j=1}^{L_i} p_{ij} N_{ij} \quad \text{and} \quad \langle r_i \rangle = \sum_{j=1}^{L_i} p_{ij} r_{ij}, \tag{3.13}$$

and the corresponding Hausdorff dimension is given by

$$D = -\frac{\sum_{i=1}^{K} m_i \log \langle N_i \rangle}{\sum_{i=1}^{K} m_i \log \langle r_i \rangle} = -\frac{\sum_{i=1}^{K} m_i \log(\sum_{j=1}^{L_i} p_{ij} N_{ij})}{\sum_{i=1}^{K} m_i \log(\sum_{j=1}^{L_i} p_{ij} N_{ij})}. \tag{3.14}$$

(C) *Random action and random generator*: The Hausdorff dimension for a fractal generated by the combination of random action and random generators is given by

$$D = -\frac{\sum_{i=1}^{K} p_i \log \langle N_i \rangle}{\sum_{i=1}^{K} p_i \log \langle r_i \rangle} = -\frac{\sum_{i=1}^{K} p_i \log(\sum_{j=1}^{L_i} p_{ij} N_{ij})}{\sum_{i=1}^{K} p_i \log(\sum_{j=1}^{L_i} p_{ij} N_{ij})}. \tag{3.15}$$

Readers are interested in the other discussion of random fractals which can be referred to Falconer (1990).

3.1.3 *Multifractals*

In the foregoing section, the fractal generators are based on the uniformal scaling. In this section, the multifractals with nonuniform scales and measures will be discussed. First, we introduce the concept of fractal measures and then fractal scales.

3.1.3.1 Single Multifractal Measure

(A) *One scaling multifractal*: Multifractal distributions can be described using the scaling properties of the coarse-grained measures. Consider $p_r(x_i)$ to be a probability measure in a box of size l_i centered at point x_i, and this box has a scaling ratio $r_i = l_i/L$, where L denotes its largest scale. The scaling index α can be defined as a local singularity strength at position x_i (e.g., Halsey et al. 1986),

$$p_{r_i}(x_i) \sim r_i^{\alpha}. \tag{3.16}$$

Consider the scaling of the qth order moment of $p_{r_i}(x_i)$ with box size l_i. A new auxiliary parameter $\tau(q)$ is introduced by

$$\sum_i [p_{r_i}(x_i)]^q \sim r_i^{\tau(q)}. \tag{3.17}$$

From Eq. (3.17), the auxiliary parameter $\tau(q)$ is defined as

$$\tau(q) = \lim_{r_i \to \infty} \frac{\log \sum_i [p_{r_i}(x_i)]^q}{\log(r_i)}. \tag{3.18}$$

The generalized dimension is introduced by

$$D_q = \frac{\tau(q)}{q-1}. \tag{3.19}$$

Consider $N_{r_i}(\alpha)$ to be a number of a box of size r_i with a value of α in the band $d\alpha$. The fractal spectrum $f(\alpha)$ is defined though

$$N_{r_i}(x_i) = \rho(\alpha) r_i^{-f(\alpha)} d\alpha \tag{3.20}$$

where $\rho(\alpha)$ is a nonsingular weighting function. From Eqs. (3.16), (3.17) and (3.20), for all α, one achieves

$$\sum_i [p_{r_i}(x_i)]^q = \int_0^\alpha \rho(\xi) r_i^{q\xi - f(\xi)} d\xi \sim r_i^{\tau(q)} \tag{3.21}$$

For $r = \max_i(r_i) \to 0$, the foregoing equation gives

$$\tau(q) = q\alpha - f(\alpha). \tag{3.22}$$

Differentiation of Eq. (3.22) with respect to q yields

$$\alpha = \frac{d\tau(q)}{dq} \quad \text{and} \quad q = \frac{df(\alpha)}{d\alpha}. \tag{3.23}$$

(B) *Multiscaling fractals*: Before discussion of the multiscaling fractals, the Hausdorff dimension of fractals with multiscales will be introduced as in Sect. 3.1.1.

Definition 3.3 Consider a group of mappings $S_i : E \to E$ $(i = 1, 2, \ldots)$ in $\mathscr{R}^n$ which $E \in \mathscr{R}^n$ is a closed set. For a number r_i with $0 < r_i < 1$. If $|S_i(\mathbf{x}) - S_i(\mathbf{y})| = r_i |\mathbf{x} - \mathbf{y}|$ for all $\mathbf{x}, \mathbf{y} \in E$, the mapping S_i is called the similarity. Suppose a self-similar set $F \subseteq E$ is invariant under the mapping S_i with$F = \cup_{i=1}^{N(E)} S_i(F)$. After mapping S_i has acted on F, it produced $N(E)$ similar sets. The Hausdorff dimension $H^D(F)$ for the self-similar set F is defined as for any $\delta > 0$,

$$\begin{aligned} \mathrm{H}^D(F) &= \sum_{i=1}^{N(E)} \lim_{\delta_i \to 0} \mathrm{H}^D_{\delta_i}(F) \\ &= \sum_{i=1}^{N(E)} \lim_{\delta_i \to 0} \inf \left\{ \sum_{j=1}^{\infty} |U_j^{(i)}|^D \,\middle|\, \begin{array}{l} F \subset \cup_i^{N(E)} \cup_{j=1}^{\infty} U_j^{(i)} \text{ and} \\ 0 < |U_j^{(i)}| \le \delta \text{ for all } j \end{array} \right\}. \end{aligned} \tag{3.24}$$

Theorem 3.2 *If $F_i \subset \mathscr{R}^n$ and $r_i > 0$ then*

$$\mathrm{H}^D(r_i F) = r_i^D \mathrm{H}^D(F) \tag{3.25}$$

where $r_i F = \{r_i x | x \in F\}$ means that the set F is scaled by a similar factor r_i.

Proof The proof is the same as the proof of Theorem 3.1. ■

Application of $H^D(F)$ in Eq. (3.24) to the self-similar set F gives

$$\mathrm{H}^D(F) = \sum_{i=1}^{N(E)} \mathrm{H}^D(S_i(F)) = \sum_{i=1}^{N(E)} r_i^D \mathrm{H}^D(F). \tag{3.26}$$

Simplification of the foregoing equation leads to

$$\sum_{i=1}^{N(E)} r_i^D = 1 \tag{3.27}$$

for the Hausdorff dimension of multifractals.

For the measure of a multifractal, consider a multiscaling box as a measure of fractals. As in Grassberger (1983a, b, c), and Halsey et al. (1986), a general spectrum of fractal dimensions is introduced. If the scaling ratio $r_i = l_i/L$ of every box is variable, a partition sum can be similarly defined as

$$\Gamma(q, \tau, r_i) = \sum_i \frac{p_i^q}{r_i^{\tau(q)}} \tag{3.28}$$

where the auxiliary parameter τ is

$$\tau(q) = (q-1)D_q. \tag{3.29}$$

For a chosen value of q, for $r = \max_i\{r_i\} \to 0$, the partition sum goes from zero to infinity, i.e.,

$$\Gamma(q, \tau(q), r_i) = \begin{cases} 0 & \tau < \tau(q) \\ \infty & \tau > \tau(q) \\ \text{constant} & \tau = \tau(q) \end{cases} \tag{3.30}$$

and Eqs. (3.22) and (3.23) can be used for the scaling index and fractal spectrum.

3.1.3.2 Joint Multifractal Measure

As in Luo (1995), consider $p_{ij}(x_i)$ denotes the jth measure in the total m-probability measures for a box of size l_i centered at the point x_i for $j = 1, 2, \ldots, m$. The box has a scaling ratio $r_i = l_i/L$, where L is the largest scale. The scaling index

$\alpha_j(q_1, q_2 \cdots q_m)$ is defined as a local singularity strength at position x_i for probability $p_{ij}(x_i)$, i.e.,

$$p_{ij}(x_i) \sim r_i^{\alpha_j}. \tag{3.31}$$

Similarly, a partition sum is defined as

$$\Gamma(\{q_1, q_2, \ldots, q_m\}, \tau(q_1, q_2, \ldots, q_m), \{r_1, r_2, \ldots, r_m\}) = \sum_i \frac{\prod_{j=1}^m p_{ij}^{q_j}}{r_i^{\tau(q_1, q_2, \ldots, q_m)}} \tag{3.32}$$

where the auxiliary parameter $\tau(q_1, q_2, \ldots, q_m)$ is now given by

$$\tau(0, \ldots, q_j, \ldots, 0) = (q_j - 1)D_{q_j}. \tag{3.33}$$

For all q_i with $r = \max_i\{r_i\} \to 0$, the partition sum goes from zero to infinity.

$$\Gamma(\{q_1, q_2, \ldots, q_m\}, \tau(q_1, q_2, \ldots, q_m), \{r_1, r_2, \ldots, r_m\}) = \begin{cases} 0 & \tau < \tau(q_1, q_2, \ldots, q_m), \\ \infty & \tau > \tau(q_1, q_2, \ldots, q_m), \\ \text{constant} & \tau = \tau(q_1, q_2, \ldots, q_m). \end{cases} \tag{3.34}$$

Consider $N_{r_i}(\alpha_1, \alpha_2, \ldots, \alpha_m)$ boxes of size r_i, with values of $(\alpha_1, \alpha_2, \ldots, \alpha_m)$ in a volume $\prod_{i=1}^m d\alpha_j$, the fractal spectrum $f(\alpha_1, \alpha_2, \ldots, \alpha_m)$ is defined through

$$N_{r_i}(\alpha_1, \alpha_2, \ldots, \alpha_m) = \rho(\alpha_1, \alpha_2, \ldots, \alpha_m) r_i^{-f(\alpha_1, \alpha_2, \ldots, \alpha_m)} \prod_{i=1}^m d\alpha_j. \tag{3.35}$$

For all $(\alpha_1, \alpha_2, \ldots, \alpha_m)$, with Eqs. (3.31) and (3.35), equation (3.32) becomes

$$\sum_i \frac{\prod_{j=1}^m p_{ij}^{q_j}}{r_i^{\tau(q_1, q_2, \ldots q_m)}} = \int \rho(\alpha_1, \alpha_2, \ldots, \alpha_m) r_i^{-\tau(q_1, q_2, \ldots, q_m) + \sum_{j=1}^m q_j \alpha_j - f(\alpha_1, \alpha_2, \ldots, \alpha_m)} \prod_{i=1}^m d\alpha_j. \tag{3.36}$$

For $r = \max_i (r_i) \to 0$,

$$\tau(q_1, q_2, \ldots, q_m) = \sum_{j=1}^m q_j \alpha_j - f(\alpha_1, \alpha_2, \ldots, \alpha_m). \tag{3.37}$$

Differentiation of Eq. (3.37) with respect to q_i and α_i yields

$$\alpha_j(q_1, q_2, \ldots, q_m) = \frac{\partial \tau(q_1, q_2, \ldots, q_m)}{\partial q_j}, \quad q_j = \frac{\partial f(\alpha_1, \alpha_2, \ldots, \alpha_m)}{\partial \alpha_j}. \tag{3.38}$$

3.2 Multifractals in 1D Iterative Maps

Consider the 1D iterative map

$$x_{k+1} = f(x_k, \mu) \text{ for } k \in \mathbb{N}, \tag{3.39}$$

where $\mathbb{N}$ is the natural number set. The kth iteration of $f(x_k)$ is given by

$$f^{(k)}(x, \mu) = f(f^{(k-1)}(x, \mu)), \;\; f^{(0)}(x, \mu) = x. \tag{3.40}$$

For a 1D discrete process, the simplest nonlinear difference equation has rich dynamical behaviors. Such a mathematical model has been studied extensively. For instance, May (1976) gave an interesting discrete model for dynamical processes in biological, economic and social sciences. The metric universality for such a discrete model is a very important characteristic (e.g., Derrida et al. (1979)). Feigenbaum (1978, 1980) studied the universal behavior of 1D systems and quantitatively determined the universal numbers. Such numbers give the threshold values from period doubling bifurcation to chaos. In 1981, Nauenberg and Rudnick (1981) discussed the universality and the power spectrum at the onset of chaos for 1D iterative maps. Collet et al. (1981) generalized the period doubling theory to higher dimensions. Zisook (1981) studied the universal effects of dissipation in the 2D mapping. The computation of the universal rescaling factors for both 1-D and 2-D maps has been carried to a very high precision by Hu and Mao (1985). Halsey et al. (1986) provided fractal measures and their singularities, and applied them to characterize strange sets. They studied the fractal of the 2^{∞}-cycle of period doubling by choosing fractal scales $l_1 = 1/\alpha_{\mathrm{PD}}, l_2 = 1/\alpha_{\mathrm{PD}}^2$, where $\alpha_{\mathrm{PD}} = 2.502\,907\,875$ is the factor in the period doubling for the iterate map $x_{k+1} = \mu x_k(1 - x_k)$. They obtained the following dimensions:

$$\begin{aligned} D_0 &= 0.537\ldots, \\ D_{-\infty} &= \frac{\log 2}{\log \alpha_{\mathrm{PD}}} = 0.755\,51\ldots, \\ D_{+\infty} &= \frac{\log 2}{2\log \alpha_{\mathrm{PD}}} = 0.377\,75\ldots, \end{aligned} \tag{3.41}$$

where D_0 is the Hausdorff dimension and $D_{-\infty}$, and $D_{+\infty}$ are the limit dimensions. A more accurate Hausdorff dimension was given by Rasband (1989):

$$D_0 = -\frac{\log 2}{\log\left[(\alpha_{\mathrm{PD}}^{-1} + (\alpha_{\mathrm{PD}}^{-2})\right] - \log 2} = 0.543\,87\cdots. \tag{3.42}$$

In this section, a method to compute the period doubling solutions of a general 1D iterative map is presented. An example is presented for demonstration and showing procedure.

3.2.1 Similar Structures in Period Doubling

Consider a 1D map of one parameter as

$$x_{k+1} = f(x_k, \mu) \tag{3.43}$$

where is the parameter. The fixed point x_k^* is determined from Eq. (3.43) by setting $x_{k+1} = x_k$. If

$$\frac{dx_{k+1}}{dx_k} = \left.\frac{df(x_k, \mu)}{dx_k}\right|_{x_k = x_k^*} = -1 \tag{3.44}$$

then this fixed point is the critical point of bifurcation. Suppose the solution of Eq. (3.43) exist for $x_{k+1} = x_k = x_k^*$. The minimum value μ_0^* can be determined for the onset of the fixed point x_k^*, and the maximum value of μ just before the first bifurcation is μ_1^*. Thus, the stable solution of Eq. (3.43) for the period-1 solution can be determined for $\mu \in (\mu_0^*, \mu_1^*)$. For $\mu = \mu_1^*$, the period-doubling bifurcation exists. For $\mu > \mu_1^*$, the period-2 solution of Eq. (3.43) is determined by

$$x_{k+2} = f^{(2)}(x_k, \mu) \quad \text{and} \quad x_{k+2} = x_k = x_k^*. \tag{3.45}$$

If there is a critical parameter of μ_2^* for $x_{k+2} = x_k = x_k^*$, and the following equation holds

$$\frac{dx_{k+2}}{dx_k} = \left.\frac{df^{(2)}(x_k, \mu)}{dx_k}\right|_{x_k = x_k^*} = \left.\frac{dx_{k+2}}{dx_{k+1}}\frac{dx_{k+1}}{dx_k}\right|_{(x_k^*, x_{k+1}^*)} = -1, \tag{3.46}$$

then the critical parameter $\mu = \mu_2^*$ is for the period-doubling bifurcation of the period-2 solution. The stable period-2 solution of Eq. (3.43) exists at $\mu \in (\mu_1^*, \mu_2^*)$. For $\mu > \mu_2^*$, the period-4 solution of Eq. (3.43) is determined by

$$x_{k+4} = f^{(4)}(x_k, \mu) \quad \text{and} \quad x_{k+4} = x_k = x_k^*. \tag{3.47}$$

In general, the period-2^m solution of Eq. (3.43) is determined by

$$x_{k+2^m} = f^{(2^m)}(x_k, \mu) \quad \text{and} \quad x_{k+2^m} = x_k = x_k^*. \tag{3.48}$$

To illustrate such iteration process, consider solutions caused by the period-doubling bifurcation as in Fig. 3.4(a)–(d). The similar structures of solution existence intervals are very clear with different scalings. From the horizontal direction, the similar structures are extracted as summarized in Fig. 3.5. From the similar structure construction in Fig. 3.5, Eq. (3.48) can be renormalized by rescaling its map. That is, the origin is moved to the fixed point in Eq. (3.43), by letting $\bar{z} = x - x^*$, and $z = a\bar{z}$, where a is the scaling factor of renormalization. Equation (3.45) then becomes

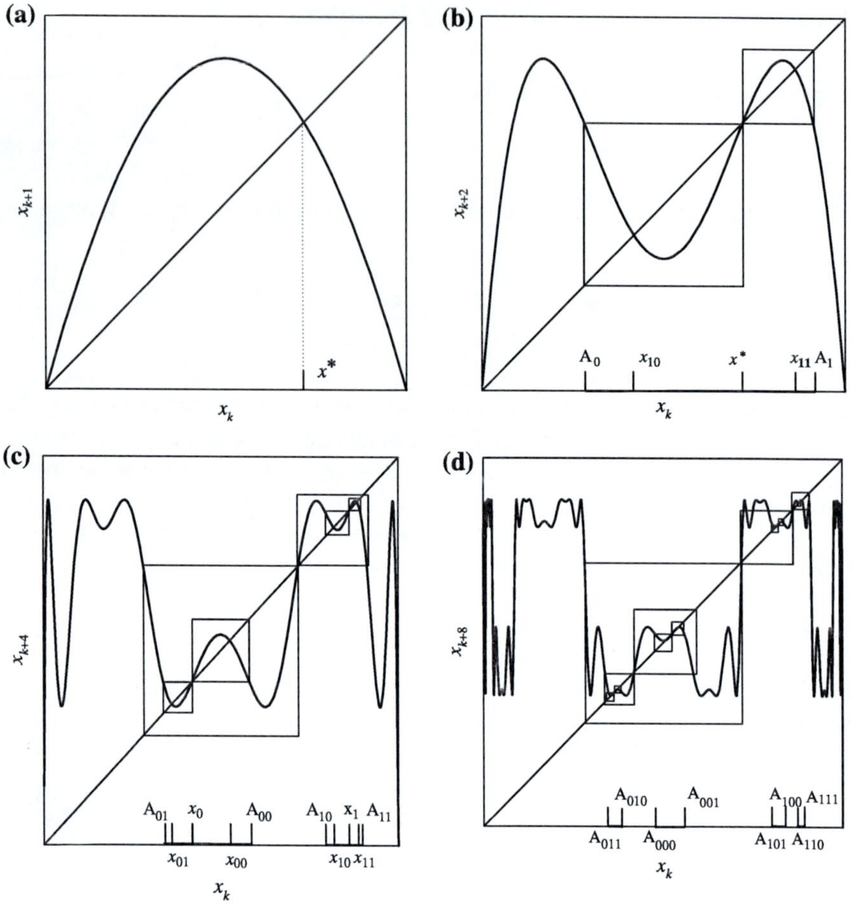

Fig. 3.4 The iteration of iterative map ($x_{k+1} = f(x_k, \mu) = \mu x_k(1 - x_k)$) : **a** the first iteration, **b** the second iteration, **c** the third iteration and **d** the fourth iteration

$$z_{k+1} = f(z_k, \mu_1) \tag{3.49}$$

where the new parameter μ_1 is given by the function

$$\mu_1 = g(\mu). \tag{3.50}$$

Equation (3.49) is similar to Eq. (3.43). If the period doubling bifurcation occurs again, we obtain

$$\mu_2 = g(\mu_1) = g(g(\mu)). \tag{3.51}$$

After m-cycle period-doubling bifurcations, we have

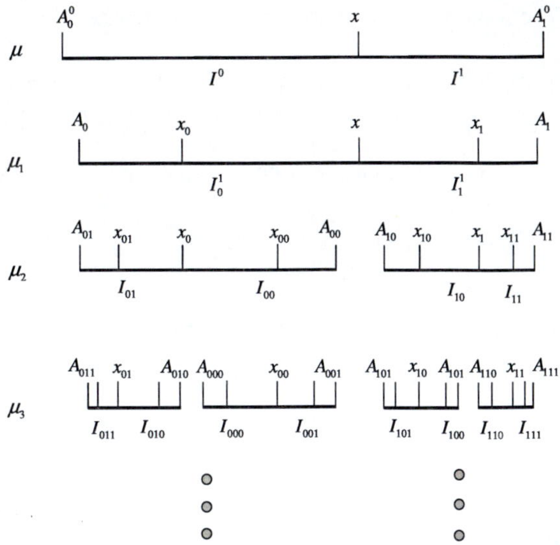

Fig. 3.5 Period-doubling construction of the 1-D iterative map $x_{k+1} = f(x_k, \mu)$

$$\mu_m = g(\mu_{m-1}) = g^{(m)}(\mu). \tag{3.52}$$

If $\mu_m = \mu_{n-1} = \mu_\infty$, the period-doubling process approaches chaos. For $\mu_m < \mu_\infty$, the iterative map, $x_{k+1} = f(x_k, \mu)$, will have m-cycles period-doubling bifurcations. The period-doubling length scaling factors are defined as

$$I_{s_i}^i = |z_{i-1}^* - z_{A_{s_0 s_1 \cdots s_i}}|, \quad I^i = |z_i^* - z_{A_{s_0 s_1 \cdots s_i}}|. \tag{3.53}$$

in which the index $i \in \{1, 2, \ldots, m\}$ refers to the ith bifurcation of the iterative map and $s_i \in \{0, 1\}$. The terms $z_{i-1}^*, z_{A_{s_0 s_1 \cdots s_i}}$ are computed, respectively, from:

$$z_{i-1}^* = f(z_{i-1}^*, \mu_{i-1}), \quad z_{i-1}^* = f^{(2)}(z_{i-1}, \mu_{i-1}). \tag{3.54}$$

In determining $z_{A_{s_0 s_1 \cdots s_i}}$, only two of its three nonzero z_{i-1} are selected which results in minimum $|z_{i-1}^* - z_{i-1}|$. In particular, the length scaling factors of the *first* period-doubling bifurcation are given by

$$I_0^1 = x^* - x_{A_0}, \quad I_1^1 = x_{A_1} - x^*, \quad I^0 = x^* - x_{A_0^0} \tag{3.55}$$

where as shown in Fig. 3.5, x_{A_0} and x_{A_1} are determined from

$$x^* = f^{(2)}(x, \mu). \tag{3.56}$$

In general, for the mth-cycle period-doubling bifurcations, the associated length scaling factor of the similar structure is defined as

$$I_{s_1 s_2 \cdots s_m} = I_{s_1 s_2 \cdots s_{m-1}} I_{s_m}^m = \prod_{i=1}^{m} I_{s_i}^i. \tag{3.57}$$

The solution of the period doubling for the iterative map in Eq. (3.43) is

$$x_{s_1 s_2 \cdots s_m} = x_{s_1 s_2 \cdots s_{m-1}} + (-1)^{(m - \sum_{i=1}^{m} s_i)} I_{s_1 s_2 \cdots s_m} I^m, \tag{3.58}$$

where $x_{s_0} = x^*$ is its fixed point. Equation (3.58) becomes

$$x_{s_1 s_2 \cdots s_m} = x^* + \sum_{i=1}^{m} (-1)^{(i - \sum_{j=1}^{i} s_j)} I_{s_1 s_2 \cdots s_i} I^i \tag{3.59}$$

or

$$x_{s_1 s_2 \cdots s_m} = x^* + \sum_{i=1}^{m} (-1)^{(i - \sum_{j=1}^{i} s_j)} \prod_{i=1}^{m} I_{s_i}^i I^i. \tag{3.60}$$

Since Eq. (3.60) gives all the solutions of the mth-cycle period-doubling bifurcation of the iterative map, the mth-solutions $x_{s_1 s_2 \cdots s_m}$ are stable just before mth cycle period-doubling bifurcation. All other solutions up to the $(m-1)$-cycle period-doubling bifurcation $(x_{s_1}, x_{s_1 s_2}, \ldots, x_{s_1 s_2 \cdots s_{m-1}})$ are unstable. From the stable and unstable solutions, the chaotic solutions caused by the period-doubling bifurcation of the iterative map can be written as

$$x_{s_1 s_2 \cdots s_m} = x_{s_1 s_2 \cdots s_{m-1}} + (-1)^{(m-k)} (I_1^1)^k (I_0^1)^{(m-k)} I^0, \tag{3.61}$$

where k is the total number of $s_i = 1, i \in \{1, 2, 3, \ldots, m\}$ as $m \to \infty$. The length scaling factors $I_1^1 = I_1^i$, $I_0^1 = I_0^i$ and $I^i = I^0$ remain constant. The foregoing equation can be expressed by

$$x_{s_1 s_2 \cdots s_m} = x^* + \sum_{i=1}^{m} (-1)^{(i-k)} (I_1^1)^k (I_0^1)^{(i-k)} I^i, \tag{3.62}$$

where k is the number of $s_j = 1, j \in \{1, 2, 3, \ldots, i\}$ for every i, as $m \to \infty$. This similar structure analysis can be done by the symbolic dynamics approach.

3.2.2 Fractality of Chaos Via PD Bifurcation

For the period-doubling bifurcation of 1D iterative map leading to chaos, the fractal is a multifractal as shown in Fig. 3.5. From Eq. (3.52), μ_m is constant at chaos, i.e., $\mu_m = \mu_\infty$, and the similar structure of iterative map will become the self-similar structure. Thus

$$I_{s_i}^i = I_{s_1}^1 = I_{s_1}, s_i \in \{0, 1\} \quad \text{for } i \in \mathbb{N}. \tag{3.63}$$

The chaotic fractal scalings of period doubling are constant, i.e.,

$$l_1 = I_0, \quad l_2 = I_1. \tag{3.64}$$

From Halsey et al. (1986) the multifractal partition sum function is

$$\Gamma = \sum_{i=1}^{n} \frac{p_i^q}{l_i^\tau} = 1, \tag{3.65}$$

where for the two-scale fractal ($n = 2$), τ is a weight parameter and $p_i = 1/2$ is the action probability. For the same action in period-2 bifurcation similar structure, we have:

$$\frac{2^{-q}}{I_0^\tau} + \frac{2^{-q}}{I_1^\tau} = 1. \tag{3.66}$$

The weight parameter is

$$\tau(q) = \frac{\log[1 + (I_0/I_1)^\tau] - q \log 2}{\log I_0}. \tag{3.67}$$

Since $\tau(q) = (q-1)D_q$, the generalized fractal dimension D_q becomes

$$D_q = \frac{\log\left[1 + (I_0/I_1)^{(q-1)D_q}\right] - q \log 2}{(q-1)\log I_0}. \tag{3.68}$$

Several special cases of the generalized fractal dimensions are given as follows. The Hausdorff dimension is

$$D_0 = -\frac{\log\left[1 + (I_1/I_0)^{D_0}\right]}{\log I_0}. \tag{3.69}$$

The information dimension is

$$D_1 = -\frac{2 \log 2}{\log I_0 + \log I_1}. \tag{3.70}$$

The two limit dimensions are

$$D_{-\infty} = -\frac{\log 2}{\log I_0}, \quad D_{+\infty} = -\frac{\log 2}{\log I_1}. \tag{3.71}$$

The scaling index is

$$\alpha = \frac{d\tau(q)}{dq} = \frac{-\log 2\,[1 + (I_1/I_0)^\tau]}{(I_1/I_0)^\tau \log I_0 + \log I_1}. \tag{3.72}$$

The singular fractal spectrum function is

$$f(\alpha) = \alpha q - \tau(q). \tag{3.73}$$

For the correlation dimension D_2, we have

$$D_2 = 2\alpha(q) - f(\alpha(q))|_{q=2}\,. \tag{3.74}$$

The characteristic parameters of the multifractal can be determined using Eqs. (3.67)–(3.74). The relationships are different from those in Halsey et al. (1986) and Cosenza et al. (1989). Since the fractal is constructed from the similar structure of the period-doubling solutions of the iterative map, the scaling factors derived herein are based on a geometric approach.

3.2.3 An Example

To demonstrate the similar structure approach for the iterative map at periodic doubling, consider

$$x_{k+1} = \mu x_k(1 - x_k). \tag{3.75}$$

Renormalization of the ith-period-doubling bifurcation equation in Eq. (3.75) yields

$$x^i_{k+1} = \mu_i x^i_k(1 - x^i_k), \tag{3.76}$$

where the parameter relation is given by

$$\mu_i = \mu^2_{i-1} - 2\mu_{i-1} - 2. \tag{3.77}$$

Let $\mu = 3.5$ and from Eq. (3.77), the renormalized parameter $\mu_1 = 3.25$. Since $\mu_1 > 3$ which is the threshold value determined from Eq. (3.44), the new iterative map of Eq. (3.75) will exhibit period-doubling bifurcations. Invoking Eq. (3.77) once again for $\mu_1 = 3.25$ yields the renormalized parameter $\mu_2 = 2.06$, associated with the period-2 doubling bifurcation. However, since $\mu_2 < 3$, this map will not exhibit period-doubling bifurcations, and thus for this case, its solutions are stable. The first fixed point of the iterative map in Eq. (3.75) is $x^* = 1 - 1/\mu$ and its period-doubling factors are

Table 3.3 Solutions of $x_{k+1} = \mu x_k(1 - x_k)$ at $\mu = 3.5$

Sampling point	Doubling time	Stability status	Similar structure solution	Exact result	Relative error (%)
x^*	0	unstable	0.714 285 143	0.714 285 143	0.00
x_0	1	unstable	0.417 582 417	0.428 571 428	2.56
x_1	1	unstable	0.850 005 845	0.857 142 857	0.83
x_{01}	2	stable	0.373 027 890	–	–
x_{00}	2	stable	0.502 497 502	–	–
x_{10}	2	stable	0.811 163 383	–	–
x_{11}	2	stable	0.870 386 293	–	–

$$
\begin{aligned}
I^i &= z_i^* = 1 - \frac{1}{\mu_i}; \\
z^i_{A_{s_0 s_1 \ldots s_{i-1} 0}} &= \frac{1}{\mu_{i-1}}, \quad z^i_{A_{s_0 s_1 \ldots s_{i-1} 1}} = \frac{\mu_{i-1} + \sqrt{\mu_{i-1}^2 - 4}}{2\mu_{i-1}}, \\
I_0^i &= 1 - \frac{2}{\mu_{i-1}}, \quad I_0^i = \frac{2 - \mu_{i-1} + \sqrt{\mu_{i-1}^2 - 4}}{2\mu_{i-1}}.
\end{aligned}
\tag{3.78}
$$

For $\mu = 3.5$, the solutions of Eq. (3.75) are

$$
\begin{aligned}
& x^* = 1 - \frac{1}{\mu}; \\
& x_0 = x^* - I_0^1 I^1, x_1 = x^* + I_1^1 I^1; \\
& \left.\begin{aligned} x_{01} = x_0 - I_0^1 I_1^2 I^2, \; x_{01} = x_0 + I_0^1 I_0^2 I^2, \\ x_{10} = x_1 - I_1^1 I_0^2 I^2, \; x_{10} = x_1 - I_1^1 I_1^2 I^2. \end{aligned}\right\}
\end{aligned}
\tag{3.79}
$$

According to the above analysis, the solutions of iterative map x^*, x_0 and x_1 are unstable at $\mu = 3.5$ but the period-2 bifurcation solutions x_{01}, x_{00}, x_{10} and x_{11} are stable. These results are tabulated in Table 3.1. For comparison, the exact period-2 solutions of Eq. (3.75) are

$$
x_0 = \frac{1 + \mu - \sqrt{\mu^2 - 2\mu - 3}}{2\mu} \quad \text{and} \quad x_1 = \frac{1 + \mu + \sqrt{\mu^2 - 2\mu - 3}}{2\mu}. \tag{3.80}
$$

In Table 3.3, the similar structure technique for computing the period-doubling solutions for the 1D iterative map yields a good agreement with the exact solutions. If the period-doubling solutions are chaotic at $\mu = \mu_\infty$, this structure will be a similar structure, and its solutions can be determined from Eq. (3.62). Note that the scaling factors of period doubling for these solutions are constant. The period-doubling solutions of the iterative map in Eq. (3.75) at $\mu = \mu_\infty = 3.569\,945\,6\cdots$ in a binary tree format is presented in Fig. 3.6.

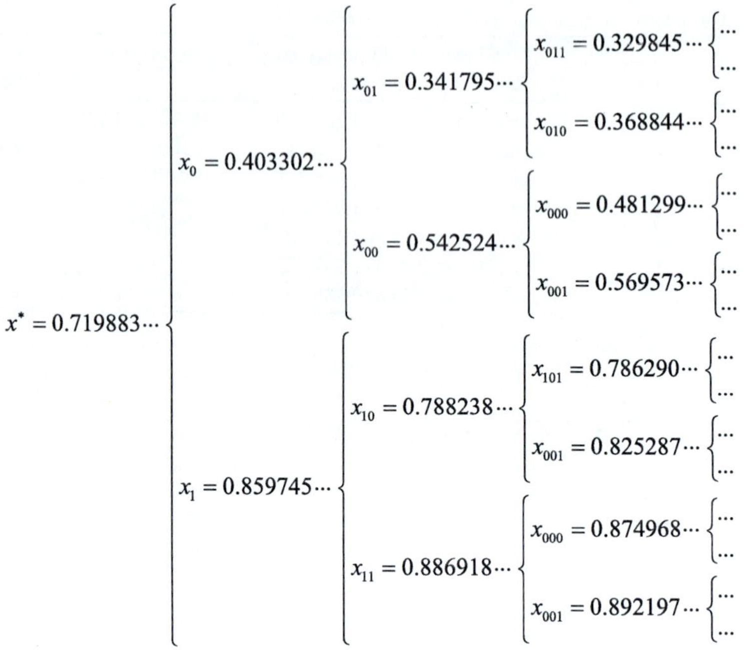

Fig. 3.6 Binary tree for the chaotic solution at $\mu = \mu_\infty = 3.569\,945\,6\cdots$

Taking $\mu_i = \mu_{i-1}$ in Eq. (3.77), the critical chaos parameter of the period-doubling solutions can be calculated to yield, $\mu = \mu_\infty = 3.561\,552\,8\cdots$ and the length scaling factors are

$$\begin{aligned} I_0 = I_0^1 &= 0.438\,447\,185\ \ldots, \\ I_1 = I_1^1 &= 0.194\,496\,855\ldots, \\ I_1/I_0 &= 0.433\,603\,840\ \ldots. \end{aligned} \tag{3.81}$$

Substitution of these length scaling factors into Eqs. (3.69)–(3.71), several of the generalized fractal dimensions can be computed and the results are listed in the Table 3.4. To assess the accuracy of these results obtained through renormalization, the length scaling factors with the critical parameter $\mu_\infty = 3.5699456\cdots$ for chaos are

$$\begin{aligned} I_0 = I_0^1 &= 0.439\,767\,373\ \ldots, \\ I_1 = I_1^1 &= 0.194\,283\,973\ldots, \\ I_1/I_0 &= 0.441\,788\,057\ \cdots \end{aligned} \tag{3.82}$$

Table 3.4 Comparison of the computed generalized fractal dimension D_q for a iterative mapping $x_{k+1} = \mu x_k(1 - x_k)$

D_q	Renormalization results	Exact solution	Halsey et al. (1986), Cosenza et al. (1989)	Rasband (1989)
D_0	0.585 286 432	0.586 670 729	0.537	0.543 87
D_1	0.563 109 625	0.563 547 168	–	
$D_{-\infty}$	0.840 671 676	0.843 748 337	0.755 51	–
$D_{+\infty}$	0.423 337 537	0.423 054 580	0.377 75	–

are exact, and when substitution of the accurate length scaling factors into Eqs. (3.69)–(3.71) yield the "exact" results of the generalized fractal dimensions. It is observed that the renormalization technique gives very good results compared to the exact resluts. For comparison, some available solutions are also tabulated in Table 3.4. As shown, the results of Rasband (1989) are not only slightly larger than those of Halsey et al. (1986) and Cosenza et al. (1989). However, the existing results does not match very well with exact solutions because the approximate models were adopted.

The fractality characteristics of chaos via period-doubling bifurcation are presented in Fig. 3.7. The weight parameter function $\tau(q)$ in Eq. (3.67) is computed using the two different sets of length scaling factors, as shown in Fig. 3.7(a). The generalized fractal dimension D_q is sketched in Fig. 3.7(b), the scaling index $\alpha(q)$ in Fig. 3.7(c), and the fractal spectrum function in Fig. 3.7(d). The "exact" and renormalization results are by solid and circular symbol curves, respectively. The results of Halsey et al. (1986) are denoted by dotted curves. The experimental results of Glazier et al. (1986) are denoted by solid circle symbols. The analytical solutions presented herein (also see, Luo and Han, 1992) agree well with the experimental results.

The complexity of chaos caused by tangential bifurcation is still unsolved. How to construct the fractal structure of chaos should be further investigated, and the fractality of chaos to measure the corresponding complexity should be completed.

3.3 Fractals in Hyperbolic Chaos

In this section, multifractals in chaotic dynamics of m-D horseshoe maps will be discussed. In chaotic dynamics, characterizing the complexity of chaos is very important. One adopted Poincare mapping sections, power spectrum analysis, Lyapunov exponent and generalized Hausdorff dimension, statistical thermodynamic approach, and ergodic theory.

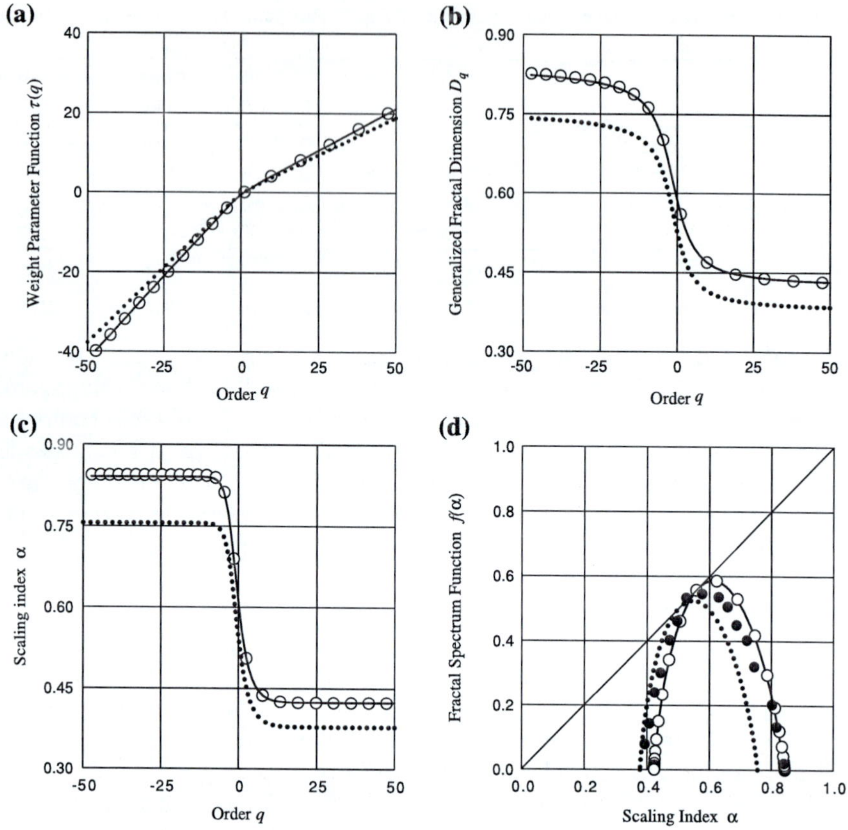

Fig. 3.7 Fractality characteristics for the iterative map $x_{k+1} = \mu x_k(1 - x_k)$: **a** weight parameter function τ_q, **b** generalized fractal dimensions D_q, **c** scaling index $\alpha(q)$ and **d** fractal spectrum function $f(\alpha)$. The solid, circular symbol and dotted curves are for exact and renormalization solutions, and Hasley et al., respectively. The filled circular symbols are experimental results

3.3.1 Fractal Theory for Hyperbolic Chaos

In this section, a new theory for describing multifractals of the hyperbolic invariant sets is established. Consider a 1D unit interval divided into M parts with a scaling ratio of $r = 1/M$. Repeating this process ad infinitum for N non-empty parts among M parts yields a fractal. As in Eq. (3.3), Mandelbrot (1977) presented the following definition of a 1D fractal

$$Nr^D = 1. \tag{3.83}$$

Fig. 3.8 2-D fractal object

The exponent D is called the Hausdorff dimension

$$D = \frac{\log N}{\log M} = -\frac{\log N}{\log r}. \tag{3.84}$$

Extending the 1D fractal concept to a 2D Euclidean fractal body, consider a non-fractal body of unit square divided into $M = M_h \times M_v$ parts in two directions with $r_h = 1/M_h$ and $r_v = 1/M_v$ where M_h and M_v pieces are in the *horizontal* and *vertical* directions, respectively. Repeating the process ad infinitum for non-empty parts $N = N_h \times N_v$ generate a 2D fractal object, as shown in Fig. 3.8. It is assumed that the fractals in each of the directions are generated independently. Thus,

$$N_h N_v r_h^{D_h} r_v^{D_v} = 1. \tag{3.85}$$

For each of the directions, $N_h r_h^{D_h} = 1$ and $N_v r_v^{D_v} = 1$ leads to

$$D_h = \frac{\log N_h}{\log M_h} = -\frac{\log N_h}{\log r_h} \quad \text{and} \quad D_v = \frac{\log N_v}{\log M_v} = -\frac{\log N_v}{\log r_v}. \tag{3.86}$$

Thus the fractal dimension of the 2D fractal object is

$$D = D_h + D_v. \tag{3.87}$$

Generalizing the concept to handle the computation of the m-D fractal dimension of an m-D fractal body where $m \le n$. Dividing an n-D unit nonfractal geometric object into M subobjects in the m-D Euclidean space leads to $M = \prod_{i=1}^{m} M_i$ with the scaling ratio $r_i = 1/M_i$. If there are N nonempty subobjects corresponding to the m-D Euclidean space, then $N = \prod_{i=1}^{m} N_i$. If the fractals in each of the directions are generated independently, then

$$N_i r_i^{D_i} = 1, \quad D_i = \frac{\log N_i}{\log M_i} = -\frac{\log N_i}{\log r_i}. \tag{3.88}$$

The fractal dimension of the n-D objects is

$$D = (n - m) + \sum_{i=1}^{m} D_i. \tag{3.89}$$

The foregoing idea will be extended to m-D *nonuniform* fractals (or simply, multifractals) in the n-D Euclidean space. Suppose there are N_i subobjects having n_i scales in the ith direction. The jth-scale has a measured length r_{ij}, its scale probability weight p_{ij} and its scale number n_{ij} $(j = 1, 2, \ldots, n_i)$. The multifractal partition function is given by,

$$\Gamma_i(\tau_i, q) = \sum_{j=1}^{n_i} n_{ij} \frac{p_{ij}^q}{l_{ij}^{\tau_i}} = 1, \tag{3.90}$$

where $\tau_i(q)$ is the weight parameter for multifractals in the ith direction, given by

$$\tau_i(q) = (q-1)D_i(q) \tag{3.91}$$

and the partition function is defined as

$$\Gamma_i(\tau_i, q) = \begin{cases} 0 & \text{at } \tau_{\text{i}} < \tau_i(q), \\ \infty & \text{at } \tau_{\text{i}} > \tau_i(q), \\ \text{constant} & \text{at } \tau_{\text{i}} = \tau_i(q). \end{cases} \tag{3.92}$$

Furthermore, from thermodynamics, the scaling index in the ith direction is

$$\alpha_i = \frac{d\tau_i}{dq} \tag{3.93}$$

and applying the Legendre transform yields,

$$\tau_i(q) = \alpha_i q - f_i(\alpha_i) \tag{3.94}$$

in which the $f_i(\alpha_i)$ is a fractal spectrum in the ith direction given by

$$\frac{df_i}{d\alpha_i} = q. \tag{3.95}$$

Summarizing the results for all the directions, the following equations for the m-D multifractal theory in n-D Euclidean space are achieved,

$$\tau(q) = (n-m)(q-1) + \sum_{i=1}^{m} \tau_i(q), \tag{3.96}$$

$$\alpha = \sum_{i=1}^{m} \alpha_i, \tag{3.97}$$

$$f(\alpha) = \sum_{i=1}^{m} f(\alpha_i). \tag{3.98}$$

3.3.2 A 1D Horseshoe Iterative Map

In this section, fractals generated by a 1D horseshoe iterative map in chaotic dynamics are studied.

(A) *A uniform 1D Cantor-horseshoe*: Consider a 1D iterative map that possesses a uniform horseshoe structure. In other words, a uniform cantor structure in phase space $x_{k+1} = f(x_k, \mu)$ $(k \in \mathbb{N})$ and μ is a control parameter. Note that $\mathbb{N}$ is a natural number set. Consider a tent map *f* in the unit interval $I = [0, 1]$ as an example,

$$f: x_{k+1} = \begin{cases} \mu x_k & \text{for } 0 \le x_k \le 1/2 \\ \mu(1 - x_k) & \text{for } 1/2 \le x_k \le 1 \end{cases} \tag{3.99}$$

where $\mu \ge 2$. The phase graph and fractal structure are procreated using Eq. (3.99) for a unit interval I as shown in Fig. 3.9. The two subintervals, I_0 and I_1 in Fig. 3.9(a), are obtained from the first iteration of Eq. (3.99) with $x_{k+1} \le 1$. That is, $I_0 = I_1 = 1/\mu$. Therefore, for the first iteration of Eq. (3.99), its invariant set is

$$f(I) = I_0 \cup I_1. \tag{3.100}$$

Similarly for the second iteration, we have $f^{(2)}(I) = \cup I_{\sigma_1\sigma_2}$, and $\sigma_i \in \{0, 1\}$ for $i \in \{1, 2\}$. Repeating such an iteration ad infinitum leads to its invariant set as

$$\Lambda = \cap_{k=0}^{\infty} f^{(k)}(I) \tag{3.101}$$

where $f^{(k)}(I) = \cup_k I_{\sigma_1\sigma_2\cdots\sigma_k}$, and $\sigma_i \in \{1, 2\}$ for $i \in \{1, 2, \ldots, k\}$.

For any value *k*, the scale ratio and the number of the non-empty interval are

$$r = \left| I_{\sigma_1\sigma_2\cdots\sigma_k} \right| = \frac{1}{\mu^k} \quad \text{and} \quad N = 2^k. \tag{3.102}$$

The Hausdorff dimension of the invariant set shown in Eq. (3.101) is

$$D_0 = \lim_{k\to\infty} \frac{\log N}{\log r} = \frac{\log 2}{\log \mu}. \tag{3.103}$$

(B) *A nonuniform 1D Cantor-horseshoe*: Consider a 1D iterative map experiencing the multiscale Cantor-horseshoe structure. For instance, an asymmetric tent map is

$$f: x_{k+1} = \begin{cases} \mu_1 x_k & \text{for } 0 \le x_k \le \frac{\mu_2}{\mu_1+\mu_2}, \\ \mu_2(1 - x_k) & \text{for } \frac{\mu_2}{\mu_1+\mu_2} \le x_k \le 1. \end{cases} \tag{3.104}$$

where the control parameters μ_1 and μ_2 satisfy $\mu_1\mu_2 \ge (\mu_1 + \mu_2)$. In Fig. 3.10, the phase graph and fractal structure are procreated using Eq. (3.104). Due to the nonuniform structure, we now have a two-scale multifractal. Thus, after the first iteration of Eq. (3.104) on the original interval $I = [0, 1]$, i.e., $f(I) = I_0 \cup I_1$; the lengths of two new subintervals are not identical, namely, $r_1 = |I_0| = 1/\mu_1$,

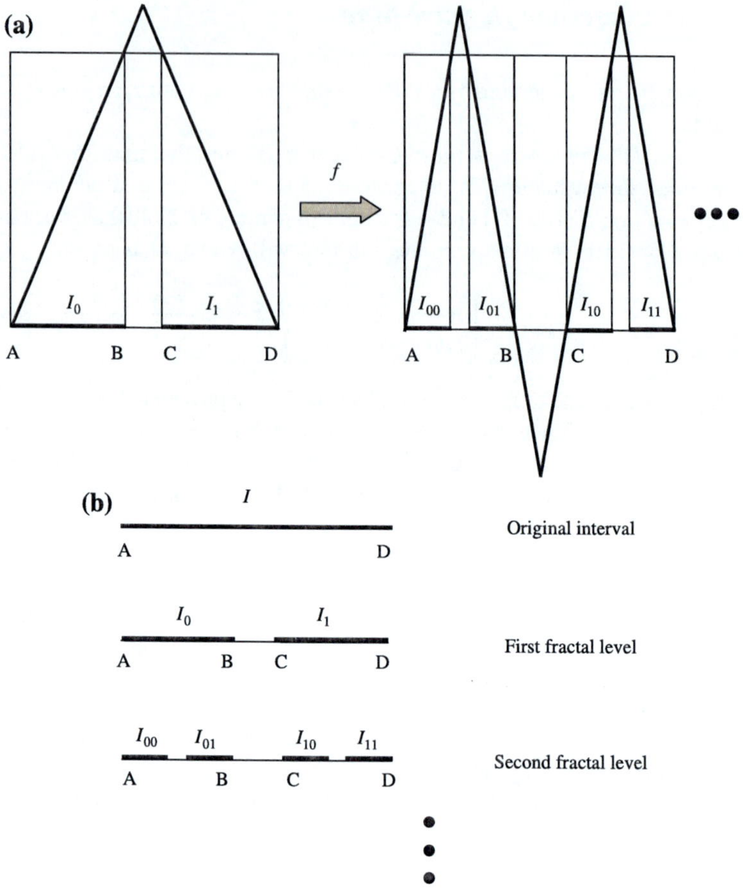

Fig. 3.9 Phase graphs and fractal structures generated via Eq. (3.99)

$r_2 = |I_1| = 1/\mu_2$. Repeating this iterative process ad infinitum results in an invariant set,

$$\Lambda = \bigcap_{k=0}^{\infty} f^{(k)}(I) \tag{3.105}$$

where $f^{(k)}(I) = \cup_k I_{\sigma_1\sigma_2\cdots\sigma_k}$, and $\sigma_i \in \{1, 2\}$ for $i \in \{1, 2, \ldots, k\}$. From the iteration process, the probability of appearance for the two scales is

$$p_0 = p_1 = \frac{1}{2}. \tag{3.106}$$

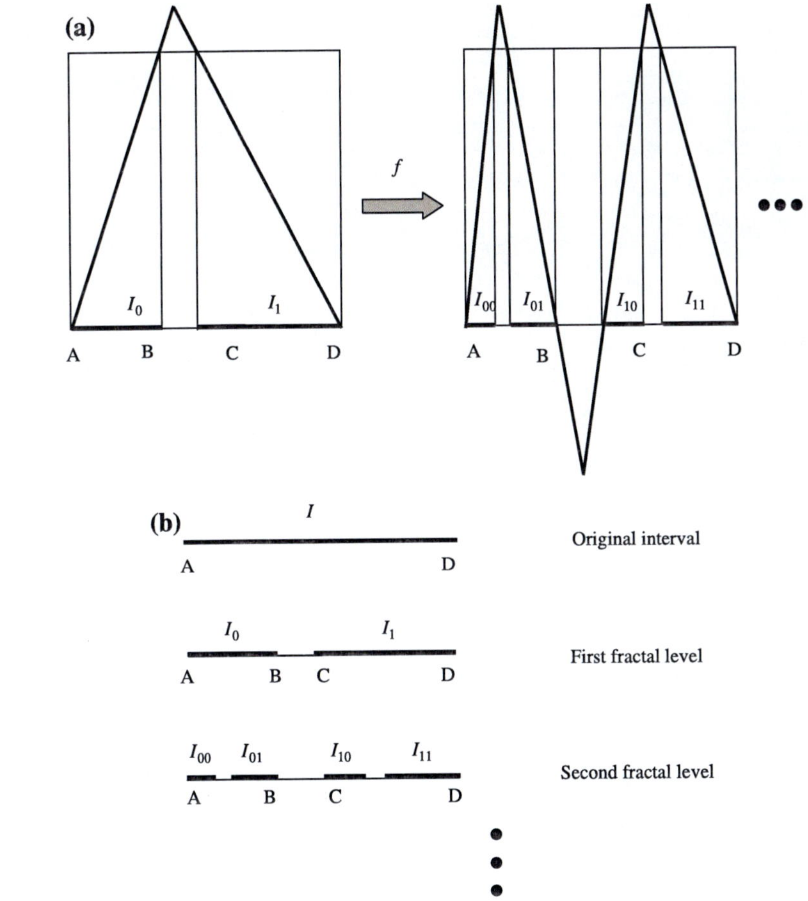

Fig. 3.10 Phase graphs and fractal structures generated via Eq. (3.104)

Applying Eq. (3.91), a partition function for the horseshoe invariant set of Eq. (3.105) is

$$\Gamma = (\frac{\mu_1^{\tau}}{2^q} + \frac{\mu_2^{\tau}}{2^q})^n = 1 \tag{3.107}$$

from which, we obtain

$$q = \frac{\log(\mu_1^{\tau} + \mu_2^{\tau})}{\log 2}. \tag{3.108}$$

From Eqs. (3.91)–(3.94) and (3.108), the fractal dimension, scaling index and fractal spectrum are given, respectively, by

$$D_q = \frac{\tau \log 2}{\log(\mu_1^{\tau} + \mu_2^{\tau}) - \log 2}, \tag{3.109}$$

$$\alpha = \frac{(\mu_1^{\tau} + \mu_2^{\tau}) \log 2}{\mu_1^{\tau} \log \mu_1 + \mu_2^{\tau} \log \mu_2}, \tag{3.110}$$

$$f(\alpha) = \alpha q - \tau(q). \tag{3.111}$$

Imposing $\mu_1 \leq \mu_2$ in Eqs. (3.109)–(3.111), the following specific fractal dimensions are obtained.

$$D_1 = \alpha, D_{-\infty} = \frac{\log 2}{\log \mu_1}, \quad D_{\infty} = \frac{\log 2}{\log \mu_2}; \tag{3.112}$$

and the Hausdorff dimension D_0 is determined using

$$\mu_1^{-D_0} + \mu_2^{-D_0} = 1. \tag{3.113}$$

3.3.3 Fractals of the 2D Horseshoe Chaos

In this section, for 2D fractals based on the Smale horseshoe map, consider fractals of the uniform horseshoe sets, and then fractals of the nonuniform horseshoe sets.

(A) *A uniform Smale horseshoe*: The Smale horseshoe arising from the transversely homoclinic orbits via the Poincare map is very important for describing chaotic dynamics in neighborhood of the saddle. To analyze the fractality of this 2D invariant set, consider the original 2D unit square $D = \{(x, y) \in \mathscr{R}^2 \,|0 \leq x \leq 1, 0 \leq y \leq 1\}$ and define a mapping $f : D \to \mathscr{R}^2$. Therefore,

$$f : \begin{cases} \begin{Bmatrix} x_{k+1} \\ y_{k+1} \end{Bmatrix} = \begin{bmatrix} \lambda & 0 \\ 0 & \mu \end{bmatrix} \begin{Bmatrix} x_k \\ y_k \end{Bmatrix} & \text{on } H_0, \\ \begin{Bmatrix} x_{k+1} \\ y_{k+1} \end{Bmatrix} = \begin{bmatrix} -\lambda & 0 \\ 0 & -\mu \end{bmatrix} \begin{Bmatrix} x_k \\ y_k \end{Bmatrix} + \begin{Bmatrix} 1 \\ \mu \end{Bmatrix} & \text{on } H_1. \end{cases} \tag{3.114}$$

where $0 < \lambda \leq 1/2, \mu \geq 2$. From Eq. (3.114), two rectangles in the horizontal direction are defined as

$$\left. \begin{array}{ll} H_0 = & \{(x, y) \in \mathscr{R}^2 \,|\, 0 \leq x \leq 1, 0 \leq y \leq 1/\mu\}, \\ H_1 = & \{(x, y) \in \mathscr{R}^2 \,|\, 0 \leq x \leq 1, 1 - 1/\mu \leq y \leq 1\}. \end{array} \right\} \tag{3.115}$$

Application of f to the two horizontal rectangles produces two vertical rectangles. That is,

$$\begin{aligned} f(H_0) \equiv V_0 = & \ \{(x, y) \in \mathscr{R}^2 \mid 0 \leq x \leq \lambda, 0 \leq y \leq 1\}, \\ f(H_1) \equiv V_1 = & \ \{(x, y) \in \mathscr{R}^2 \mid 1 - \lambda \leq x \leq 1, 0 \leq y \leq 1\}. \end{aligned} \tag{3.116}$$

To construct the Smale horseshoe that intersects between a vertical invariant set and a horizontal invariant set, application of f ad infinitum to the unit square S, namely,

$$\begin{aligned} \Lambda_V &= \bigcap_{k=1}^{\infty} f^{(k)}(D) = \bigcup_{\substack{s_{-i}\in S \\ i=1,2,\cdots}} \left(f(V_{s_{-2}\cdots s_{-k}\cdots}) \cap V_{s_{-1}}\right) = \bigcup_{\substack{s_{-i}\in S \\ i=1,2,\cdots}} V_{s_{-1}\cdots s_{-k}\cdots} \\ &= \left\{(x, y) \in D \,\middle|\, f^{(-i+1)}(p) \in V_{s_{-i}}, p = (x, y), s_{-i} \in S, i = 1, 2, \cdots\right\}, \end{aligned} \tag{3.117}$$

where $S = \{0, 1\}$. The results in the vertical invariant set are shown in Fig. 3.11. The fractal has a scaling ratio $r_x = \lambda$ and the corresponding Hausdorff dimension can be computed by

$$D_{0x} = -\frac{\log 2}{\log \lambda}. \tag{3.118}$$

Since the vertical invariant set does not have fractals in the y-direction, $D_{0y} = 1$. Therefore, the resultant Hausdorff dimension for the vertical invariant set is

$$D = D_{0x} + D_{0y} = 1 - \frac{\log 2}{\log \lambda}. \tag{3.119}$$

Note that Eq. (3.119) is identical to the expression in Guckenheimer and Holmes (1983). In a similar manner, the horizontal invariant set can be reprocreated via the inverse map $f^{(-1)}$ acting on the unit square S, that is,

$$\begin{aligned} \Lambda_H &= \bigcap_{k=1}^{\infty} f^{(-k)}(D) = \bigcup_{\substack{s_{-i}\in S \\ i=1,2,\cdots}} \left(f(H_{s_1\cdots s_k\cdots}) \cap H_{s_0}\right) = \bigcup_{\substack{s_{-i}\in S \\ i=1,2,\cdots}} H_{s_0\cdots s_k\cdots} \\ &= \left\{(x, y) \in D \,\middle|\, f^{(i)}(p) \in H_{s_i}, p = (x, y), s_i \in S, i = 0, 1, 2, \cdots\right\}. \end{aligned} \tag{3.120}$$

The result is sketched in Fig. 3.12 with its scaling ratio $r_y = 1/\mu$. The fractal of the horizontal invariant set in the y-direction has a Hausdorff dimension of

$$D_{0y} = \frac{\log 2}{\log \mu}. \tag{3.121}$$

The intersection of the vertical and the horizontal invariant sets yields the Smale horseshoe which is shown in Fig. 3.13. That is, we have:

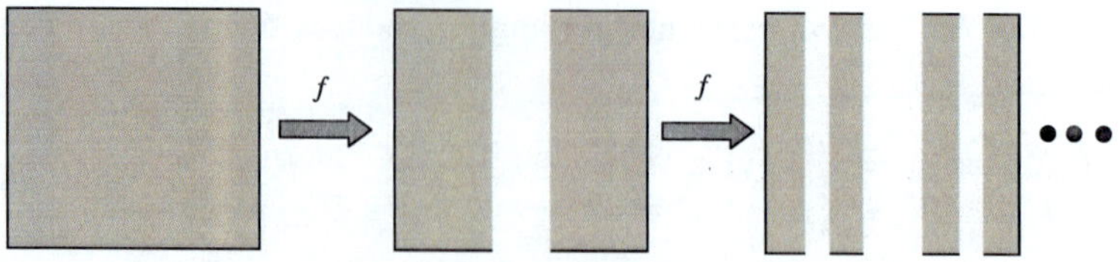

Fig. 3.11 Vertical invariant set procreated via Eq. (3.114)

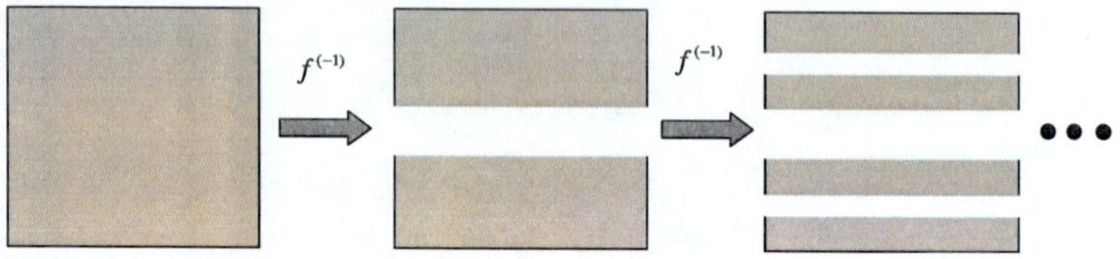

Fig. 3.12 Horizontal invariant set procreated via an inverse map of Eq. (3.114)

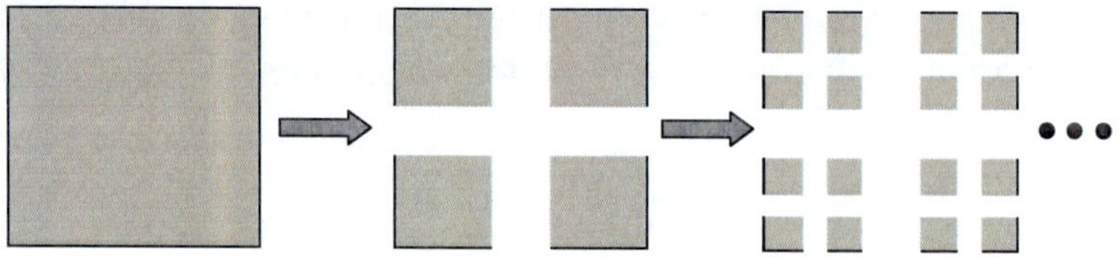

Fig. 3.13 Smale horseshoe generated by iteration of Eq. (3.114).

$$\Lambda = \Lambda_V \cap \Lambda_H = \bigcap_{k=-\infty}^{\infty} f^{(k)}(D). \tag{3.122}$$

The Hausdorff dimension of the Smale horseshoe procreated via ad infinitum n iterations of the 2D map f on the unit square S is

$$D = D_{0x} + D_{0y} = \log 2 \left[\frac{1}{\log \mu} - \frac{1}{\log \lambda} \right]. \tag{3.123}$$

(B) *A nonuniform Smale horseshoe*: The multifractality of the nonuniform Smale horseshoe will be discussed through a map of nonuniform Smale horseshoe given by

$$f : \begin{cases} \begin{Bmatrix} x_{k+1} \\ y_{k+1} \end{Bmatrix} = \begin{bmatrix} \lambda_1 & 0 \\ 0 & \mu_1 \end{bmatrix} \begin{Bmatrix} x_k \\ y_k \end{Bmatrix} & \text{on } H_0, \\ \begin{Bmatrix} x_{k+1} \\ y_{k+1} \end{Bmatrix} = \begin{bmatrix} -\lambda_2 & 0 \\ 0 & -\mu_2 \end{bmatrix} \begin{Bmatrix} x_k \\ y_k \end{Bmatrix} + \begin{Bmatrix} 1 \\ \mu_2 \end{Bmatrix} & \text{on } H_1. \end{cases} \tag{3.124}$$

where $0 < \lambda_i \leq 1/2$ and $\mu_i \geq 2$ and $i = \{1, 2\}$. From Eq. (3.124), two separate rectangles in the horizontal direction can be defined as

$$\left.\begin{aligned} H_0 &= \left\{(x, y) \in \mathscr{R}^2 \,|0 \leq x \leq 1, 0 \leq y \leq 1/\mu_1\right\}, \\ H_1 &= \left\{(x, y) \in \mathscr{R}^2 \,|0 \leq x \leq 1, 1 - 1/\mu_2 \leq y \leq 1\right\}. \end{aligned}\right\} \tag{3.125}$$

Application of f in Eq. (3.124) to the two horizontal rectangles in Eq. (3.125) produces two vertical rectangles, i.e.,

$$\left.\begin{aligned} f(H_0) \equiv V_0 &= \left\{(x, y) \in \mathscr{R}^2 \,|0 \leq x \leq \lambda_1, 0 \leq y \leq 1\right\}, \\ f(H_1) \equiv V_1 &= \left\{(x, y) \in \mathscr{R}^2 \,|1 - \lambda_2 \leq x \leq 1, 0 \leq y \leq 1\right\}. \end{aligned}\right\} \tag{3.126}$$

Application of maps f and $f^{(-1)}$ ad infinitum to the unit square S yields the Smale horseshoe, that is,

$$\Lambda = \Lambda_V \cap \Lambda_H = \bigcap_{k=-\infty}^{\infty} f^{(k)}(D) = \left[\bigcap_{k=0}^{\infty} f^{(-k)}(D)\right] \cap \left[\bigcap_{k=1}^{\infty} f^{(k)}(D)\right]. \tag{3.127}$$

The vertical invariant set of Eq. (3.124) has two scaling ratios $r_{1x} = \lambda_1$ and $r_{2x} = \lambda_2$ and probability weight $p_{1x} = p_{1x} = 1/2$. Therefore, from the 1D multifractal theory, its partition function in x-direction is

$$\Gamma_x = \left[\frac{\lambda_1^{-\tau_x}}{2^q} + \frac{\lambda_2^{-\tau_x}}{2^q}\right]^n = 1. \tag{3.128}$$

Re-expressing Eq. (3.128), we have

$$q = \frac{\log(\lambda_1^{-\tau_x} + \lambda_2^{-\tau_x})}{\log 2}. \tag{3.129}$$

The multifractal dimension, scaling index and fractal spectrum for the vertical invariant set in the x-direction are

$$D_{qx} = \frac{\tau_x \log 2}{\log(\lambda_1^{-\tau_x} + \lambda_2^{-\tau_x}) - \log 2}, \tag{3.130}$$

$$\alpha_x = \frac{(\lambda_1^{-\tau_x} + \lambda_1^{-\tau_x}) \log 2}{\lambda_1^{-\tau_x} \log \lambda_1 + \lambda_2^{-\tau_x} \log \lambda_2}, \tag{3.131}$$

$$f_x(\alpha_x) = \alpha_x q - \tau_x(q). \tag{3.132}$$

Next, consider the multifractality of the horizontal invariant set of Eq. (3.124) in the y-direction and the results are

$$q = \frac{\log(\mu_1^{\tau_y} + \mu_2^{\tau_y})}{\log 2}, \tag{3.133}$$

$$D_{qy} = \frac{\tau_y \log 2}{\log(\mu_1^{\tau_y} + \mu_2^{\tau_y}) - \log 2}, \tag{3.134}$$

$$\alpha_y = \frac{(\mu_1^{\tau_y} + \mu_2^{\tau_y}) \log 2}{\mu_1^{\tau_y} \log \mu_1 + \mu_2^{\tau_y} \log \mu_2}, \tag{3.135}$$

$$f(\alpha_y) = \alpha_y q - \tau_y(q). \tag{3.136}$$

Therefore, the multifractal characteristics for the vertical invariant set, the horizontal invariant set and the nonuniform Smale horseshoe are

$$\tau_q = q - 1 + \tau_{qx},\ D_q = 1 + D_{qx},\ \alpha = 1 + \alpha_x,\ f(\alpha) = 1 + f(\alpha_x). \tag{3.137}$$

for vertical invariant set,

$$\tau_q = q - 1 + \tau_{qy},\ D_q = 1 + D_{qy},\ \alpha = 1 + \alpha_y,\ f(\alpha) = 1 + f(\alpha_y). \tag{3.138}$$

for horizontal invariant set,

$$\tau = \tau_x + \tau_y,\ D_q = D_{qx} + D_{qy},\ \alpha = \alpha_x + \alpha_y,\ f(\alpha) = f(\alpha_x) + f(\alpha_y). \tag{3.139}$$

for the Smale horseshoe set.

References

Collet, P., Eckmann, J. P. and Koch, H., 1981, Period doubling bifurcations for families of maps on R^n, *Journal of Statistical Physics*, **25**, 1–14.

Cosenza, M. G., McCormick, W. D. and Swift, J. B., 1989, Finite-size effects on the fractal spectrum of the period-doubling attractor, *Physical Review*, **39A**, 2734–2737.

Derrida D., Gervois, A. and Pomeau, Y., 1979, Universal metric properties of bifurcations of endomorphisms, *Journal of Physics*, **12A**, 269–296.

Falconer, K., 1990, *Fractal Geometry: Mathematical Foundations and Applications*, John Wiley and Sons: Chichester.

Feigenbaum, M. J., 1978, Quantitative universality for a class of nonlinear transformations, *Journal of Statistical Physics*, **19**, 25–52.

Feigenbaum, M. J., 1980, Universal behavior in nonlinear systems, *Los Alamos Science*, **1**, 4–27.

Glazier, J. A., Jensen, M. H., Libchaber, A., and Stavens, J., 1986, Structure of Arnold tongues and $f(\alpha)$ spectrum for period doubling: experimental results, *Physical Review*, **34A**, 1621–1625.

Grassberger, P. and Procaccia, I., 1983a, On the characterization of strange attractors, *Physical Review Letter*, **50**, 346–349.

Grassberger, P. and Procaccia, I., 1983b, Estimation of the Kolmogorov entropy from a chaotic signal, *Phys. Rev.*, **29A**, 2591–2593.

Grassberger, P. and Procaccia, I., 1983c, Measuring the strangeness of strange attractors, *Physica*, **9D**, 189–208.

Guckenheimer, J. and Holmes, P., 1983, *Nonlinear Oscillations, Dynamical Systems, and Bifurcations of Vector Fields*, Springer-Verlag: New York.

Halsey, T.C., Jensen, M.H., Kadanoff, L.P., Procaccia, I. and Shraiman, B.I. 1986, Fractal Measures and their Singularities: the Characterization of Stranger Sets, *Physical Review* **33A**, 1141–1151.

Hu, B. and Mao, J., 1985, The eigenvalue matching renormalization group, *Physical Letter*, **108A**, 305–307.

Leung, A.Y.T. and Luo, A.C.J., 1992, Fractals by Multigenerator, *Microcomputer in Civil Engineering*, **7**, 257–264.

Luo, A.C.J., 1991, Generalized Fractal Theory, *Proceeding on Chinese Conference on Turbulence and Stability*, Tianjing, P. R. China (in Chinese).

Luo, A.C.J. and Han, R.P.S., 1992, Period doubling and multifractal in 1-D iterative maps, *Chaos, Solitons & Fractals*, **2**, 335–348.

Luo, A.C.J., 1995, *Analytical Modeling of Bifurcation, Chaos and Multifractals in Nonlinear Dynamics*, Ph.D. Dissertation, University of Manitoba: Winnipeg.

Mandelbrot, B.B., 1977, *Fractals: Form, Chance and Dimension*, W. H. Freeman and Company: San Francisco.

May, R.M., 1976, Simple mathematical models with very complicated dynamics, *Nature*, **261**, 459–467.

Nauenberg, M. and Rudnick, J., 1981, Universality and power spectrum at the onset of chaos, *Physical Review*, **24B**, 493–495.

Rasband, S. N., 1989, *Chaotic Dynamics of Nonlinear System*, John Wiley and Sons: New York.

Zisook, A.Z., 1981, Universal effects of dissipation in two-dimensional mappings, *Physical Review*, **24A**, 1640–1642.

Chapter 4
Complete Dynamics and Synchronization

This chapter will present a Ying–Yang theory for nonlinear discrete dynamical systems with consideration of positive and negative iterations of discrete iterative maps. In existing analysis, the solutions relative to "Yang" in nonlinear dynamical systems are extensively investigated. However, the solutions pertaining to "Ying" in nonlinear dynamical systems will be presented. A set of concepts on "Ying" and "Yang" in discrete dynamical systems will be introduced. Based on the Ying–Yang theory, the complete dynamics of discrete dynamical systems can be discussed. A discrete dynamical system with the Henon map will be presented as an example. The companion and synchronization of discrete dynamical systems will be introduced, and the corresponding conditions are developed. The synchronization dynamics of Duffing and Henon maps will be discussed.

4.1 Discrete Systems with a Single Nonlinear Map

Definition 4.1 Consider an implicit vector function $\mathbf{f} : D \to D$ on an open set $D \subset \mathscr{R}^n$ in an n-dimensional discrete dynamical system. For $\mathbf{x}_k, \mathbf{x}_{k+1} \in D$, there is a discrete relation as

$$\mathbf{f}(\mathbf{x}_k, \mathbf{x}_{k+1}, \mathbf{p}) = \mathbf{0} \tag{4.1}$$

where the vector function is $\mathbf{f} = (f_1, f_2, \cdots, f_n)^{\mathrm{T}} \in \mathscr{R}^n$ and discrete variable vector is $\mathbf{x}_k = (x_{k1}, x_{k2}, \cdots, x_{kn})^{\mathrm{T}} \in D$ with a parameter vector $\mathbf{p} = (p_1, p_2, \cdots, p_m)^{\mathrm{T}} \in \mathscr{R}^m$.

As in Luo (2010), to symbolically describe the discrete dynamical systems, introduce two discrete sets.

Definition 4.2 For a discrete dynamical system in Eq. (4.1), the positive and negative discrete sets are defined by

A. C. J. Luo, *Regularity and Complexity in Dynamical Systems*,
DOI: 10.1007/978-1-4614-1524-4_4,

$$\left.\begin{array}{l}\Sigma_{+} = \{\mathbf{x}_{k+i} | \mathbf{x}_{k+i} \in \mathscr{R}^n, i \in \mathbb{Z}_{+}\} \subset D \text{ and} \\ \Sigma_{-} = \{\mathbf{x}_{k-i} | \mathbf{x}_{k-i} \in \mathscr{R}^n, i \in \mathbb{Z}_{+}\} \subset D\end{array}\right\} \tag{4.2}$$

respectively. The discrete set is

$$\Sigma = \Sigma_{+} \cup \Sigma_{-}. \tag{4.3}$$

A positive mapping is defined as

$$P_{+} : \Sigma \to \Sigma_{+} \quad \Rightarrow \quad P_{+} : \mathbf{x}_k \to \mathbf{x}_{k+1} \tag{4.4}$$

and a negative mapping is defined by

$$P_{-} : \Sigma \to \Sigma_{-} \quad \Rightarrow \quad P_{-} : \mathbf{x}_k \to \mathbf{x}_{k-1}. \tag{4.5}$$

Definition 4.3 For a discrete dynamical system in Eq. (4.1), consider two points $\mathbf{x}_k \in D$ and $\mathbf{x}_{k+1} \in D$, and there is a specific, differentiable, vector function $\mathbf{g} \in \mathscr{R}^n$ to make $\mathbf{g}(\mathbf{x}_k, \mathbf{x}_{k+1}, \boldsymbol{\lambda}) = \mathbf{0}$.

(i) The stable solution based on $\mathbf{x}_{k+1} = P_{+}\mathbf{x}_k$ for the positive mapping P_{+} is called the "Yang" of the discrete dynamical system in Eq. (4.1) in sense of $\mathbf{g}(\mathbf{x}_k, \mathbf{x}_{k+1}, \boldsymbol{\lambda}) = \mathbf{0}$ if solutions $(\mathbf{x}_k^*, \mathbf{x}_{k+1}^*)$ of $\mathbf{f}(\mathbf{x}_k, \mathbf{x}_{k+1}, \mathbf{p}) = \mathbf{0}$ and $\mathbf{g}(\mathbf{x}_k, \mathbf{x}_{k+1}, \boldsymbol{\lambda}) = \mathbf{0}$ exist.
(ii) The stable solution based on $\mathbf{x}_k = P_{-}\mathbf{x}_{k+1}$ for the negative mapping P_{-} is called the "Ying" of the discrete dynamical system in Eq. (4.1) in sense of $\mathbf{g}(\mathbf{x}_k, \mathbf{x}_{k+1}, \boldsymbol{\lambda}) = \mathbf{0}$ if solutions $(\mathbf{x}_k^*, \mathbf{x}_{k+1}^*)$ of $\mathbf{f}(\mathbf{x}_k, \mathbf{x}_{k+1}, \mathbf{p}) = \mathbf{0}$ and $\mathbf{g}(\mathbf{x}_k, \mathbf{x}_{k+1}, \boldsymbol{\lambda}) = \mathbf{0}$ exist.
(iii) The solution based on $\mathbf{x}_{k+1} = P_{+}\mathbf{x}_k$ is called the "Ying–Yang" for the positive mapping P_{+} of the discrete dynamical system in Eq. (4.1) in sense of $\mathbf{g}(\mathbf{x}_k, \mathbf{x}_{k+1}, \boldsymbol{\lambda}) = \mathbf{0}$ if solutions $(\mathbf{x}_k^*, \mathbf{x}_{k+1}^*)$ of $\mathbf{f}(\mathbf{x}_k, \mathbf{x}_{k+1}, \mathbf{p}) = \mathbf{0}$ and $\mathbf{g}(\mathbf{x}_k, \mathbf{x}_{k+1}, \boldsymbol{\lambda}) = \mathbf{0}$ exist and the eigenvalues of $DP_{+}(\mathbf{x}_k^*)$ are distributed inside and outside the unit cycle.
(iv) The solution based on $\mathbf{x}_k = P_{-}\mathbf{x}_{k+1}$ is called the "Ying–Yang" for the negative mapping P_{-} of the discrete dynamical system in Eq. (4.1) in sense of $\mathbf{g}(\mathbf{x}_k, \mathbf{x}_{k+1}, \boldsymbol{\lambda}) = \mathbf{0}$ if solutions $(\mathbf{x}_k^*, \mathbf{x}_{k+1}^*)$ of $\mathbf{f}(\mathbf{x}_k, \mathbf{x}_{k+1}, \mathbf{p}) = \mathbf{0}$ and $\mathbf{g}(\mathbf{x}_k, \mathbf{x}_{k+1}, \boldsymbol{\lambda}) = \mathbf{0}$ exist and the eigenvalues of $DP_{-}(\mathbf{x}_{k+1}^*)$ are distributed inside and outside unit cycle.

Consider the positive and negative mappings are

$$\mathbf{x}_{k+1} = P_{+}\mathbf{x}_k \text{ and } \mathbf{x}_k = P_{-}\mathbf{x}_{k+1}. \tag{4.6}$$

For the simplest case, consider the constraint condition of $\mathbf{g}(\mathbf{x}_k, \mathbf{x}_{k+1}, \boldsymbol{\lambda}) = \mathbf{x}_{k+1} - \mathbf{x}_k = \mathbf{0}$. Thus, the positive and negative mappings have, respectively, the constraints

$$\mathbf{x}_{k+1} = \mathbf{x}_k \text{ and } \mathbf{x}_k = \mathbf{x}_{k+1}. \tag{4.7}$$

Both positive and negative mappings are governed by the discrete relation in Eq. (4.1). In other words, equation (4.6) gives

$$\mathbf{f}(\mathbf{x}_k, \mathbf{x}_{k+1}, \mathbf{p}) = \mathbf{0} \text{ and } \mathbf{f}(\mathbf{x}_k, \mathbf{x}_{k+1}, \mathbf{p}) = \mathbf{0}. \tag{4.8}$$

Setting the period-1 solution $\mathbf{x}_k^*$ and substitution of Eqs. (4.7) into (4.8) gives

$$\mathbf{f}(\mathbf{x}_k^*, \mathbf{x}_k^*, \mathbf{p}) = \mathbf{0} \text{ and } \mathbf{f}(\mathbf{x}_k^*, \mathbf{x}_k^*, \mathbf{p}) = \mathbf{0}. \tag{4.9}$$

It is observed that the period-1 solutions for the positive and negative mappings are identical. The two relations for positive and negative mappings are illustrated in Figs. 4.1a, b, respectively. To determine the period-1 solution, the fixed points of Eq. (4.7) exist under constraints in Eq. (4.8), which are also shown in Fig. 4.1. The two thick lines on the axis are two sets for the mappings from the starting to final states. The relation in Eq. (4.7) is presented by a solid curve. The intersection points of the curves and straight lines for relations in Eqs. (4.7) and (4.8) give the fixed points of Eq. (4.9), which are period-1 solutions, labeled by the circular symbols. However, their stability and bifurcation for the period-1 solutions are different. To determine the stability and bifurcation of the positive and negative mappings, the following theorem is stated.

Theorem 4.1 *For a discrete dynamical system in Eq.* (4.1), *there are two points* $\mathbf{x}_k \in D$ *and* $\mathbf{x}_{k+1} \in D$, *and two positive and negative mappings are*

$$\mathbf{x}_{k+1} = P_+\mathbf{x}_k \text{ and } \mathbf{x}_k = P_-\mathbf{x}_{k+1} \tag{4.10}$$

with

$$\mathbf{f}(\mathbf{x}_k, \mathbf{x}_{k+1}, \mathbf{p}) = \mathbf{0} \text{ and } \mathbf{f}(\mathbf{x}_k, \mathbf{x}_{k+1}, \mathbf{p}) = \mathbf{0}. \tag{4.11}$$

Suppose a specific, differentiable, vector function $\mathbf{g} \in \mathscr{R}^n$ *makes* $\mathbf{g}(\mathbf{x}_k, \mathbf{x}_{k+1}, \boldsymbol{\lambda}) = \mathbf{0}$ *hold. If the solutions* $(\mathbf{x}_k^*, \mathbf{x}_{k+1}^*)$ *of both* $\mathbf{f}(\mathbf{x}_k, \mathbf{x}_{k+1}, \mathbf{p}) = \mathbf{0}$ *and* $\mathbf{g}(\mathbf{x}_k, \mathbf{x}_{k+1}, \boldsymbol{\lambda}) = \mathbf{0}$ *exist, then the following conclusions in the sense of* $\mathbf{g}(\mathbf{x}_k, \mathbf{x}_{k+1}, \boldsymbol{\lambda}) = \mathbf{0}$ *hold.*

(i) *The stable* P_+*-1 solutions are the unstable* P_-*-1 solutions with all eigenvalues of* $DP_-(\mathbf{x}_k^*)$ *outside the unit cycle, vice versa.*
(ii) *The unstable* P_+*-1 solutions with all eigenvalues of* $DP_+(\mathbf{x}_k^*)$ *outside the unit cycle are the stable* P_-*-1 solutions, vice versa.*
(iii) *For the unstable* P_+*-1 solutions with eigenvalue distribution of* $DP_+(\mathbf{x}_k^*)$ *inside and outside the unit cycle, the corresponding* P_-*-1 solution is also unstable with switching the eigenvalue distribution of* $DP_-(\mathbf{x}_k^*)$ *inside and outside the unit cycle, vice versa.*
(iv) *All the bifurcations of the stable and unstable* P_+*-1 solutions are all the bifurcations of the unstable and stable* P_-*-1 solutions, respectively.*

Proof Consider the positive and negative mappings with relations in Eq. (4.9). The periodic solution in sense of $\mathbf{g}(\mathbf{x}_k, \mathbf{x}_{k+1}, \boldsymbol{\lambda}) = \mathbf{0}$ is given by

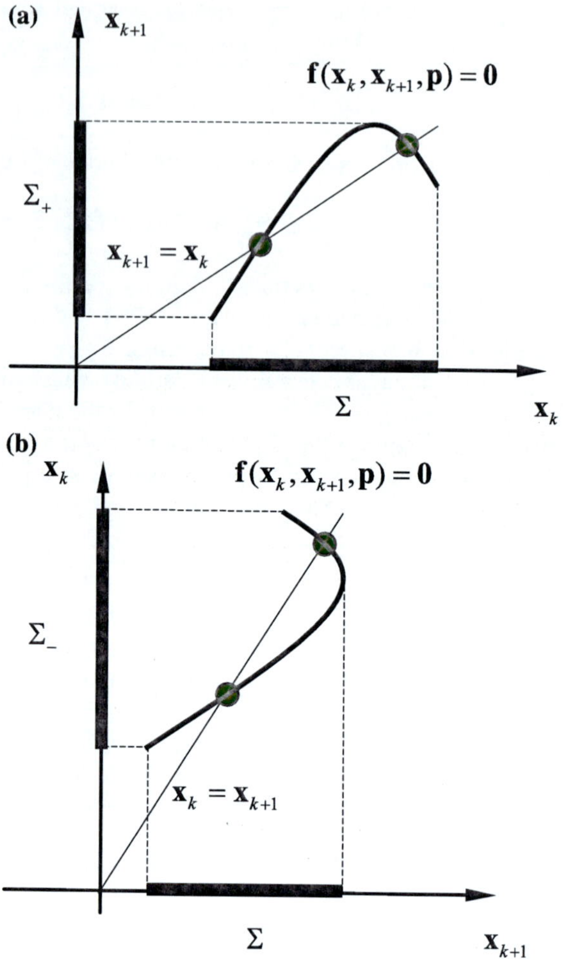

Fig. 4.1 Period-1 solution for **a** positive mapping and **b** negative mapping. The two *thick lines* on the axis are two sets for the mappings from the starting to final states. The mapping relation is presented by a *solid curve*. The circular symbols give period-1 solutions for the positive and negative mappings

$$\mathbf{f}(\mathbf{x}_k, \mathbf{x}_{k+1}, \mathbf{p}) = \mathbf{0} \text{ and } \mathbf{g}(\mathbf{x}_k, \mathbf{x}_{k+1}, \boldsymbol{\lambda}) = \mathbf{0}$$

from which the fixed points $(\mathbf{x}_k^*, \mathbf{x}_{k+1}^*)$ can be determined. Consider a small perturbation

$$\mathbf{x}_{k+1} = \mathbf{x}_{k+1}^* + \delta\mathbf{x}_{k+1} \text{ and } \mathbf{x}_k = \mathbf{x}_k^* + \delta\mathbf{x}_k.$$

The linearization of mappings in Eq. (4.6) gives

$$\delta\mathbf{x}_{k+1} = DP_+(\mathbf{x}_k^*)\delta\mathbf{x}_k \text{ and } \delta\mathbf{x}_k = DP_-(\mathbf{x}_{k+1}^*)\delta\mathbf{x}_{k+1}$$

where

$$DP_{+}(\mathbf{x}_k^*) = [\frac{\partial \mathbf{x}_{k+1}}{\partial \mathbf{x}_k}]_{\mathbf{x}_k^*} \text{ and } DP_{-}(\mathbf{x}_{k+1}^*) = [\frac{\partial \mathbf{x}_k}{\partial \mathbf{x}_{k+1}}]_{\mathbf{x}_{k+1}^*}.$$

From Eq. (4.1), one obtains

$$\begin{aligned}
&([\frac{\partial \mathbf{f}}{\partial \mathbf{x}_k}] + [\frac{\partial \mathbf{f}}{\partial \mathbf{x}_{k+1}}][\frac{\partial \mathbf{x}_{k+1}}{\partial \mathbf{x}_k}])_{(\mathbf{x}_k^*, \mathbf{x}_{k+1}^*)} = \mathbf{0}, \\
&\mathbf{g}(\mathbf{x}_k, \mathbf{x}_{k+1}, \boldsymbol{\lambda}) = \mathbf{g}(\mathbf{x}_k^*, \mathbf{x}_{k+1}^*, \boldsymbol{\lambda}) = \mathbf{0}; \\
&([\frac{\partial \mathbf{f}}{\partial \mathbf{x}_{k+1}}] + [\frac{\partial \mathbf{f}}{\partial \mathbf{x}_k}][\frac{\partial \mathbf{x}_k}{\partial \mathbf{x}_{k+1}}])_{(\mathbf{x}_k^*, \mathbf{x}_{k+1}^*)} = \mathbf{0}, \\
&\mathbf{g}(\mathbf{x}_k^*, \mathbf{x}_{k+1}^*, \boldsymbol{\lambda}) = \mathbf{g}(\mathbf{x}_k, \mathbf{x}_{k+1}, \boldsymbol{\lambda}) = \mathbf{0}.
\end{aligned}$$

That is,

$$\begin{aligned}
DP_{+}(\mathbf{x}_k^*) &= [\frac{\partial \mathbf{x}_{k+1}}{\partial \mathbf{x}_k}]|_{\mathbf{x}_k^*} = -([\frac{\partial \mathbf{f}}{\partial \mathbf{x}_{k+1}}]^{-1}[\frac{\partial \mathbf{f}}{\partial \mathbf{x}_k}])_{(\mathbf{x}_k^*, \mathbf{x}_{k+1}^*)}, \\
DP_{-}(\mathbf{x}_{k+1}^*) &= [\frac{\partial \mathbf{x}_k}{\partial \mathbf{x}_{k+1}}]_{\mathbf{x}_{k+1}^*} = -([\frac{\partial \mathbf{f}}{\partial \mathbf{x}_k}]^{-1}[\frac{\partial \mathbf{f}}{\partial \mathbf{x}_{k+1}}])_{(\mathbf{x}_k^*, \mathbf{x}_{k+1}^*)}.
\end{aligned}$$

Taking the inverse of the second equation in the foregoing equation gives

$$DP_{-}^{-1}(\mathbf{x}_{k+1}^*) = [\frac{\partial \mathbf{x}_k}{\partial \mathbf{x}_{k+1}}]^{-1}|_{\mathbf{x}_{k+1}^*} = -[\frac{\partial \mathbf{f}}{\partial \mathbf{x}_{k+1}}]^{-1}[\frac{\partial \mathbf{f}}{\partial \mathbf{x}_k}]_{(\mathbf{x}_{k+1}^*, \mathbf{x}_k^*)}$$

which is identical to $DP_{+}(\mathbf{x}_k^*)$. Therefore, one obtains

$$DP_{-}^{-1}(\mathbf{x}_{k+1}^*) = DP_{+}(\mathbf{x}_k^*).$$

In other words, $DP_{+}(\mathbf{x}_k^*)$ is the inverse of $DP_{-}(\mathbf{x}_{k+1}^*)$.

Consider the eigenvalues λ_{-} and λ_{+} of $DP_{-}(\mathbf{x}_{k+1}^*)$ and $DP_{+}(\mathbf{x}_k^*)$, accordingly. The following relations hold

$$\begin{aligned}
(DP_{-}(\mathbf{x}_{k+1}^*) - \lambda_{-}\mathbf{I})\delta\mathbf{x}_{k+1} &= \mathbf{0}, \\
(DP_{+}(\mathbf{x}_k^*) - \lambda_{+}\mathbf{I})\delta\mathbf{x}_k &= \mathbf{0}.
\end{aligned}$$

Left multiplication of $DP_{+}(\mathbf{x}_k^*)$ in the first equation of the foregoing equation and $\delta\mathbf{x}_{k+1} = \lambda_{-}^{-1} d\mathbf{x}_k$ gives

$$[DP_{+}(\mathbf{x}_k^*) - \lambda_{-}^{-1}\mathbf{I}]\delta\mathbf{x}_k = \mathbf{0}.$$

Thus, one can obtain

$$\lambda_{+} = \lambda_{-}^{-1}.$$

From the stability and bifurcation theory for P_{+}-1 and P_{-}-1 solutions for discrete dynamical system in Eq. (4.1), the following conclusions can be given as follows:

(i) The stable P_+-1 solutions are the unstable P_--1 solutions with all eigenvalues of $DP_-(\mathbf{x}_k^*)$ outside the unit cycle, vice versa.
(ii) The unstable P_+-1 solutions with all eigenvalues of $DP_+(\mathbf{x}_k^*)$ outside the unit cycle are the stable P_--1 solutions, vice versa.
(iii) For the unstable P_+-1 solutions with eigenvalue distribution of $DP_+(\mathbf{x}_k^*)$ inside and outside the unit cycle, the corresponding P_--1 solution is also unstable with switching the eigenvalue distribution of $DP_-(\mathbf{x}_k^*)$ inside and outside the unit cycle, vice versa.
(iv) All the bifurcations of the *stable* and *unstable* P_+-1 solutions are all the bifurcations of the *unstable* and *stable* P_--1 solutions, respectively.

This theorem is proved. ■

From the foregoing theorem, the *Ying*, *Yang* and *Ying-Yang* states in discrete dynamical systems exist. To generate the above ideas to $P_+^{(N)}$-1 and $P_-^{(N)}$-1 solutions in discrete dynamical systems in sense of $\mathbf{g}(\mathbf{x}_k, \mathbf{x}_{k+N}, \boldsymbol{\lambda}) = \mathbf{0}$, the mapping structure consisting of *N*- positive or negative mappings is considered.

Definition 4.4 For a discrete dynamical system in Eq. (4.1), the mapping structures of *N*-mappings for the positive and negative mappings are defined as

$$\mathbf{x}_{k+N} = \underbrace{P_+ \circ P_+ \circ \cdots \circ P_+}_{N} \mathbf{x}_k = P_+^{(N)} \mathbf{x}_k, \tag{4.12}$$

$$\mathbf{x}_k = \underbrace{P_- \circ P_- \circ \cdots \circ P_-}_{N} \mathbf{x}_{k+N} = P_-^{(N)} \mathbf{x}_{k+N}. \tag{4.13}$$

with

$$\mathbf{f}(\mathbf{x}_{k+i-1}, \mathbf{x}_{k+i}, \mathbf{p}) = \mathbf{0} \text{ for } i = 1, 2, \ldots, N \tag{4.14}$$

where $P_+^{(0)} = 1$ and $P_-^{(0)} = 1$ for $N = 0$.

Definition 4.5 For a discrete dynamical system in Eq. (4.1), consider two points $\mathbf{x}_{k+i-1} \in D$ $(i = 1, 2, \ldots, N)$ and $\mathbf{x}_{k+N} \in D$, and there is a specific, differentiable, vector function $\mathbf{g} \in \mathscr{R}^n$ to make $\mathbf{g}(\mathbf{x}_k, \mathbf{x}_{k+N}, \boldsymbol{\lambda}) = \mathbf{0}$.

(i) The stable solution based on $\mathbf{x}_{k+N} = P_+^{(N)} \mathbf{x}_k$ for the positive mapping P_+ is called the "Yang" of the discrete dynamical system in Eq. (4.1) in sense of $\mathbf{g}(\mathbf{x}_k, \mathbf{x}_{k+N}, \boldsymbol{\lambda}) = \mathbf{0}$ if the solutions $(\mathbf{x}_k^*, \mathbf{x}_{k+1}^*, \ldots, \mathbf{x}_{k+N}^*)$ of Eq. (4.14) with $\mathbf{g}(\mathbf{x}_k, \mathbf{x}_{k+N}, \boldsymbol{\lambda}) = \mathbf{0}$ exist.
(ii) The stable solution based on $\mathbf{x}_k = P_-^{(N)} \mathbf{x}_{k+N}$ for the negative mapping P_- is called the "Ying" of the discrete dynamical system in Eq. (4.1) in sense of $\mathbf{g}(\mathbf{x}_k, \mathbf{x}_{k+N}, \boldsymbol{\lambda}) = \mathbf{0}$ if the solutions $(\mathbf{x}_k^*, \mathbf{x}_{k+1}^*, \cdots, \mathbf{x}_{k+N}^*)$ of Eq. (4.14) with $\mathbf{g}(\mathbf{x}_k, \mathbf{x}_{k+N}, \boldsymbol{\lambda}) = \mathbf{0}$ exist.
(iii) The solution based on $\mathbf{x}_{k+N} = P_+^{(N)} \mathbf{x}_k$ is called the "Ying–Yang" for the positive mapping P_+ of the discrete dynamical system in Eq. (4.1) in sense of

$\mathbf{g}(\mathbf{x}_k, \mathbf{x}_{k+N}, \boldsymbol{\lambda}) = \mathbf{0}$ if the solutions $(\mathbf{x}_k^*, \mathbf{x}_{k+1}^*, \ldots, \mathbf{x}_{k+N}^*)$ of Eq. (4.14) with $\mathbf{g}(\mathbf{x}_k, \mathbf{x}_{k+N}, \boldsymbol{\lambda}) = \mathbf{0}$ exist and the eigenvalues of $DP_+^{(N)}(\mathbf{x}_k^*)$ are distributed inside and outside the unit cycle.

(iv) The solution based on $\mathbf{x}_k = P_-^{(N)}\mathbf{x}_{k+N}$ is called the "Ying–Yang" for the negative mapping P_- of the discrete dynamical system in Eq. (4.1) in sense of $\mathbf{g}(\mathbf{x}_k, \mathbf{x}_{k+N}, \boldsymbol{\lambda}) = \mathbf{0}$ if the solutions $(\mathbf{x}_k^*, \mathbf{x}_{k+1}^*, \ldots, \mathbf{x}_{k+N}^*)$ of Eq. (4.14) with $\mathbf{g}(\mathbf{x}_k, \mathbf{x}_{k+N}, \boldsymbol{\lambda}) = \mathbf{0}$ exist and the eigenvalues of $DP_-^{(N)}(\mathbf{x}_{k+N}^*)$ are distributed inside and outside unit cycle.

To determine the Ying–Yang properties of $P_+^{(N)}$-1 and $P_-^{(N)}$-1 in the discrete mapping system in Eq. (4.1), the corresponding theorem is presented as follows.

Theorem 4.2 *For a discrete dynamical system in Eq.* (4.1), *there are two points* $\mathbf{x}_k \in D$ *and* $\mathbf{x}_{k+N} \in D$, *and two positive and negative mappings are*

$$\mathbf{x}_{k+N} = P_+^{(N)}\mathbf{x}_k \text{ and } \mathbf{x}_k = P_-^{(N)}\mathbf{x}_{k+N}. \tag{4.15}$$

and $\mathbf{x}_{k+i} = P_+\mathbf{x}_{k+i-1}$ *and* $\mathbf{x}_{k+i-1} = P_-\mathbf{x}_{k+i}$ *can be governed by*

$$\mathbf{f}(\mathbf{x}_{k+i-1}, \mathbf{x}_{k+i}, \mathbf{p}) = \mathbf{0} \text{ for } i = 1, 2, \ldots, N. \tag{4.16}$$

Suppose a specific, differentiable, vector function of $\mathbf{g} \in \mathscr{R}^n$ *makes* $\mathbf{g}(\mathbf{x}_k, \mathbf{x}_{k+N}, \boldsymbol{\lambda}) = \mathbf{0}$ *hold. If the solutions* $(\mathbf{x}_k^*, \ldots, \mathbf{x}_{k+i}^*)$ *of Eq.* (4.16) *with* $\mathbf{g}(\mathbf{x}_k, \mathbf{x}_{k+N}, \boldsymbol{\lambda}) \equiv \mathbf{0}$ *exist, then the following conclusions in the sense of* $\mathbf{g}(\mathbf{x}_k, \mathbf{x}_{k+N}, \boldsymbol{\lambda}) = \mathbf{0}$ *hold.*

(i) *The stable* $P_+^{(N)}$*-1 solution is the unstable* $P_-^{(N)}$*-1 solution with all eigenvalues of* $DP_-^{(N)}(\mathbf{x}_{k+N}^*)$ *outside the unit cycle, vice versa.*

(ii) *The unstable* $P_+^{(N)}$*-1 solution with all eigenvalues of* $DP_+^{(N)}(\mathbf{x}_k^*)$ *outside the unit cycle is the stable* $P_-^{(N)}$*-1 solution, vice versa.*

(iii) *For the unstable* $P_+^{(N)}$*-1 solution with eigenvalue distribution of* $DP_+^{(N)}(\mathbf{x}_k^*)$ *inside and outside the unit cycle, the corresponding* $P_-^{(N)}$*-1 solution is also unstable with switching eigenvalue distribution of* $DP_-^{(N)}(\mathbf{x}_{k+N}^*)$ *inside and outside the unit cycle, vice versa.*

(iv) *All the bifurcations of the stable and unstable* $P_+^{(N)}$*-1 solution are all the bifurcations of the unstable and stable* $P_-^{(N)}$*-1 solution, respectively.*

Proof Consider positive and negative mappings with relations in Eq. (4.16), i.e.,

$$\mathbf{f}(\mathbf{x}_{k+i-1}, \mathbf{x}_{k+i}, \mathbf{p}) = \mathbf{0} \text{ for } i = 1, 2, \ldots, N$$

from which $\mathbf{x}_{k+i}$ is a function of $\mathbf{x}_{k+i-1}$ in the positive mapping iteration and $\mathbf{x}_{k+i-1}$ is a function of $\mathbf{x}_{k+i}$ in the negative mapping iteration. The periodic solution in sense of $\mathbf{g}(\mathbf{x}_k, \mathbf{x}_{k+N}, \boldsymbol{\lambda}) = \mathbf{0}$ is given by

$$\begin{aligned}&\mathbf{f}(\mathbf{x}_{k+i-1}, \mathbf{x}_{k+i}, \mathbf{p}) = \mathbf{0} \text{ for } i = 1, 2, \ldots, N\\&\mathbf{g}(\mathbf{x}_k, \mathbf{x}_{k+N}, \boldsymbol{\lambda}) = \mathbf{0}.\end{aligned}$$

Setting the period-1 solution be $\mathbf{x}^*_{k+i-1}$ or $\mathbf{x}^*_{k+i}$ $(i = 1, 2, \ldots, N)$ and the foregoing equation gives

$$\mathbf{f}(\mathbf{x}^*_{k+i-1}, \mathbf{x}^*_{k+i}, \mathbf{p}) = \mathbf{0} \text{ for } i = 0, 1, \ldots, N;$$
$$\mathbf{g}(\mathbf{x}^*_k, \mathbf{x}^*_{k+N}, \boldsymbol{\lambda}) = \mathbf{0}$$

for both the positive and negative mapping iterations. The existence condition of the foregoing equation requires

$$\det[(D_{ij})_{N\times N}] \neq 0$$

where

$$\begin{aligned}
D_{N1} &= -\left([\frac{\partial \mathbf{f}(\mathbf{x}_{k+N-1}, \mathbf{x}_{k+N}, \mathbf{p})}{\partial \mathbf{x}_{k+N}}]_{n\times n}[\frac{\partial \mathbf{x}_k}{\partial \mathbf{x}_{k+N}}]^{-1}_{n\times n}\right)_{(\mathbf{x}^*_{k+N-1}, \mathbf{x}^*_k)}\\
&= -\left([\frac{\partial \mathbf{f}(\mathbf{x}_{k+N-1}, \mathbf{x}_{k+N}, \mathbf{p})}{\partial \mathbf{x}_{k+N}}]_{n\times n}[\frac{\partial \mathbf{g}(\mathbf{x}_{k+N}, \mathbf{x}_k, \mathbf{p})}{\partial \mathbf{x}_{k+N}}]^{-1}_{n\times n}\right.\\
&\quad \left.\times [\frac{\partial \mathbf{g}(\mathbf{x}_{k+N}, \mathbf{x}_k, \mathbf{p})}{\partial \mathbf{x}_k}]_{n\times n}\right)\Big|_{(\mathbf{x}^*_{k+N-1}, \mathbf{x}^*_k)},\\
D_{NN} &= [\frac{\partial \mathbf{f}(\mathbf{x}_{k+N-1}, \mathbf{x}_k, \mathbf{p})}{\partial \mathbf{x}_{k+N-1}}]_{n\times n}|_{(\mathbf{x}^*_{k+N-1}, \mathbf{x}^*_k)},\\
D_{Nj} &= [\mathbf{0}]_{n\times n} \text{ for } j = 2, 3, \ldots, N-1;
\end{aligned}$$

$$\begin{aligned}
D_{ii} &= [\frac{\partial \mathbf{f}(\mathbf{x}_{k+i-1}, \mathbf{x}_{k+i}, \mathbf{p})}{\partial \mathbf{x}_{k+i-1}}]_{n\times n}|_{(\mathbf{x}^*_{k+i-1}, \mathbf{x}^*_{k+i})},\\
D_{i(i+1)} &= [\frac{\partial \mathbf{f}(\mathbf{x}_{k+i-1}, \mathbf{x}_{k+i}, \mathbf{p})}{\partial \mathbf{x}_{k+i}}]_{n\times n}|_{(\mathbf{x}^*_{k+i-1}, \mathbf{x}^*_{k+i})}\\
D_{ij} &= [\mathbf{0}]_{n\times n} \text{ for } i = 1, 2, \ldots, N-1;\\
j &= 1, 2, \ldots, i-1; i+2, i+3, \ldots, N.
\end{aligned}$$

Once $\mathbf{x}^*_{k+i-1}$ or $\mathbf{x}^*_{k+i}$ $(i = 1, 2, \ldots, N)$ is obtained in sense of $\mathbf{g}(\mathbf{x}_k, \mathbf{x}_{k+N}, \boldsymbol{\lambda}) \equiv \mathbf{0}$, the corresponding stability and bifurcation of the periodic solutions can be determined. However, the stability and bifurcation of the $P^{(N)}_+$-1 and $P^{(N)}_-$-1 solutions will be different. Herein, consider a small perturbation from the periodic solution

$$\left.\begin{aligned}
\mathbf{x}_{k+i} &= \mathbf{x}^*_{k+i} + \delta\mathbf{x}_{k+i}\\
\mathbf{x}_{k+i+1} &= \mathbf{x}^*_{k+i+1} + \delta\mathbf{x}_{k+i+1}
\end{aligned}\right\} \text{ for } i = 0, 1, \ldots, N.$$

With the foregoing equation, linearization of Eq. (4.15) gives

$$\begin{aligned}
\delta\mathbf{x}_{k+N} &= \underbrace{DP_+ \cdot DP_+ \cdot \ldots \cdot DP_+}_{N}|_{\mathbf{x}^*_k}\delta\mathbf{x}_k\\
&= DP^{(N)}_+(\mathbf{x}^*_k)\delta\mathbf{x}_k,
\end{aligned}$$

$$\delta \mathbf{x}_k = \underbrace{DP_- \cdot DP_- \cdot \ldots \cdot DP_-}_{N} |_{\mathbf{x}^*_{k+N}} \delta \mathbf{x}_{k+N}$$
$$= DP_-^{(N)}(\mathbf{x}^*_{k+N})\delta \mathbf{x}_{k+N}.$$

On the other hand, for each single positive and negative mappings gives

$$\delta \mathbf{x}_{k+i} = DP_+(\mathbf{x}^*_{k+i-1})\delta \mathbf{x}_{k+i-1} \text{ for } i = 1, 2, \ldots, N$$
$$\delta \mathbf{x}_{k+i-1} = DP_-(\mathbf{x}^*_{k+i})\delta \mathbf{x}_{k+i} \text{ for } i = 1, 2, \ldots, N$$

where

$$DP_+(\mathbf{x}^*_{k+i-1}) = [\frac{\partial \mathbf{x}_{k+i}}{\partial \mathbf{x}_{k+i-1}}]_{\mathbf{x}^*_{k+i-1}} \quad \text{for } i = 1, 2, \ldots, N$$
$$DP_-(\mathbf{x}^*_{k+i}) = [\frac{\partial \mathbf{x}_{k+i-1}}{\partial \mathbf{x}_{k+i}}]_{\mathbf{x}^*_{k+i}} \quad \text{for } i = 1, 2, \ldots, N$$

and for $i = 1, 2, \ldots N$, linearization of Eq. (4.15) gives

$$DP_+(\mathbf{x}^*_{k+i-1}) = [\frac{\partial \mathbf{x}_{k+i}}{\partial \mathbf{x}_{k+i-1}}]_{\mathbf{x}^*_{k+i-1}} = -([\frac{\partial \mathbf{f}}{\partial \mathbf{x}_{k+i}}]^{-1}[\frac{\partial \mathbf{f}}{\partial \mathbf{x}_{k+i-1}}])_{(\mathbf{x}^*_{k+i},\mathbf{x}^*_{k+i-1})},$$
$$DP_-(\mathbf{x}^*_{k+i}) = [\frac{\partial \mathbf{x}_{k+i-1}}{\partial \mathbf{x}_{k+i}}]_{\mathbf{x}^*_{k+i}} = -([\frac{\partial \mathbf{f}}{\partial \mathbf{x}_{k+i-1}}]^{-1}[\frac{\partial \mathbf{f}}{\partial \mathbf{x}_{k+i}}])_{(\mathbf{x}^*_{k+i},\mathbf{x}^*_{k+i-1})}.$$

Therefore, the resultant Jacobian matrices for $P_+^{(N)}$-1 and $P_-^{(N)}$-1 are

$$\begin{aligned} DP_+^{(N)}(\mathbf{x}^*_k) &= DP_+(\mathbf{x}^*_{k+N-1}) \cdot DP_+(\mathbf{x}^*_{k+N-2}) \cdot \ldots \cdot DP_+(\mathbf{x}^*_{k+1}) \cdot DP_+(\mathbf{x}^*_k) \\ &= [\frac{\partial \mathbf{x}_{k+N}}{\partial \mathbf{x}_{k+N-1}}]_{\mathbf{x}^*_{k+N-1}} \cdot [\frac{\partial \mathbf{x}_{k+N}}{\partial \mathbf{x}_{k+N-1}}]_{\mathbf{x}^*_{k+N-2}} \cdot \ldots \cdot [\frac{\partial \mathbf{x}_{k+2}}{\partial \mathbf{x}_{k+1}}]_{\mathbf{x}^*_{k+1}} \cdot [\frac{\partial \mathbf{x}_{k+1}}{\partial \mathbf{x}_k}]_{\mathbf{x}^*_k} \\ &= (-1)^N([\frac{\partial \mathbf{f}}{\partial \mathbf{x}_{k+N}}]^{-1}[\frac{\partial \mathbf{f}}{\partial \mathbf{x}_{k+N-1}}])_{(\mathbf{x}^*_{k+N},\mathbf{x}^*_{k+N-1})} \cdot \ldots \cdot \\ &\quad ([\frac{\partial \mathbf{f}}{\partial \mathbf{x}_{k+1}}]^{-1}[\frac{\partial \mathbf{f}}{\partial \mathbf{x}_k}])_{(\mathbf{x}^*_{k+1},\mathbf{x}^*_k)}, \end{aligned}$$

$$\begin{aligned} &DP_-^{(N)}(\mathbf{x}^*_{k+N}) \\ &= DP_-(\mathbf{x}^*_{k+1}) \cdot DP_-(\mathbf{x}^*_{k+2}) \cdot \ldots \cdot DP_-(\mathbf{x}^*_{k+N-1}) \cdot DP_-(\mathbf{x}^*_{k+N}) \\ &= [\frac{\partial \mathbf{x}_k}{\partial \mathbf{x}_{k+1}}]_{\mathbf{x}^*_{k+1}} \cdot [\frac{\partial \mathbf{x}_{k+1}}{\partial \mathbf{x}_{k+2}}]_{\mathbf{x}^*_{k+2}} \cdot \ldots \cdot [\frac{\partial \mathbf{x}_{k+N-2}}{\partial \mathbf{x}_{k+N-1}}]_{\mathbf{x}^*_{k+N-1}} \cdot [\frac{\partial \mathbf{x}_{k+N-1}}{\partial \mathbf{x}_{k+N}}]_{\mathbf{x}^*_{k+N}} \\ &= (-1)^N([\frac{\partial \mathbf{f}}{\partial \mathbf{x}_k}]^{-1}[\frac{\partial \mathbf{f}}{\partial \mathbf{x}_{k+1}}])_{(\mathbf{x}^*_{k+1},\mathbf{x}^*_k)} \cdot \ldots \cdot \\ &([\frac{\partial \mathbf{f}}{\partial \mathbf{x}_{k+N-1}}]^{-1}[\frac{\partial \mathbf{f}}{\partial \mathbf{x}_{k+N}}])_{(\mathbf{x}^*_{k+N},\mathbf{x}^*_{k+N-1})}. \end{aligned}$$

From the two equations, it is very easily proved that the two resultant Jacobian matrices are inverse each other, i.e.,

$$DP_{+}^{(N)}(\mathbf{x}_k^*)\cdot DP_{-}^{(N)}(\mathbf{x}_{k+N}^*)=\mathbf{I}_{n\times n}.$$

Similarly, consider eigenvalues λ_- and λ_+ of $DP_{-}^{(N)}(\mathbf{x}_{k+N}^*)$ and $DP_{+}^{(N)}(\mathbf{x}_k^*)$, accordingly. The following relations hold

$$(DP_{-}^{(N)}(\mathbf{x}_{k+N}^*)-\lambda_-\mathbf{I})\delta\mathbf{x}_{k+N}=\mathbf{0},$$
$$(DP_{+}^{(N)}(\mathbf{x}_k^*)-\lambda_+\mathbf{I})\delta\mathbf{x}_k=\mathbf{0}.$$

Left multiplication of $DP_{+}^{(N)}(\mathbf{x}_k^*)$ in the first equation of the foregoing equation and $\delta\mathbf{x}_{k+N}=\lambda_-^{-1}d\mathbf{x}_k$ gives

$$(DP_{+}^{(N)}(\mathbf{x}_k^*)-\lambda_-^{-1}\mathbf{I})\delta\mathbf{x}_k=\mathbf{0}.$$

Compared to $(DP_{+}^{(N)}(\mathbf{x}_k^*)-\lambda_+\mathbf{I})\delta\mathbf{x}_k=\mathbf{0}$, one obtains

$$\lambda_+=\lambda_-^{-1}$$

in the sense of $\mathbf{g}(\mathbf{x}_k,\mathbf{x}_{k+N},\boldsymbol{\lambda})=\mathbf{0}$ hold. From the stability and bifurcation theory for discrete dynamical systems, the following conclusions can be summarized as

(i) The stable $P_{+}^{(N)}$-1 solution is the unstable $P_{-}^{(N)}$-1 solution with all eigenvalues of $DP_{-}^{(N)}(\mathbf{x}_{k+N}^*)$ outside the unit cycle, vice versa.
(ii) The unstable $P_{+}^{(N)}$-1 solution with all eigenvalues of $DP_{+}^{(N)}(\mathbf{x}_k^*)$ outside the unit cycle is the stable $P_{-}^{(N)}$-1 solution, vice versa.
(iii) For the unstable $P_{+}^{(N)}$-1 solution with eigenvalue distribution of $DP_{+}^{(N)}(\mathbf{x}_k^*)$ inside and outside the unit cycle, the corresponding $P_{-}^{(N)}$-1 solution is also unstable with switching eigenvalue distribution of $DP_{-}^{(N)}(\mathbf{x}_{k+N}^*)$ *inside* and *outside* the unit cycle, vice versa.
(iv) All the bifurcations of the *stable* and *unstable* $P_{+}^{(N)}$-1 solution are all the bifurcations of the *unstable* and *stable* $P_{-}^{(N)}$-1 solution, respectively.

This theorem is proved. ■

Notice that the number N for the $P_{+}^{(N)}$-1 and $P_{-}^{(N)}$-1 solutions in the discrete dynamical system can be any integer if such a solution exists in sense of $\mathbf{g}(\mathbf{x}_k,\mathbf{x}_{k+N},\boldsymbol{\lambda})=\mathbf{0}$.

Theorem 4.3 *For a discrete dynamical system in Eq.* (4.1), *there are two points* $\mathbf{x}_k\in D$ *and* $\mathbf{x}_{k+N}\in D$. *If the period-doubling cascade of the* $P_{+}^{(N)}$-1 *and* $P_{-}^{(N)}$-1 *solution occurs, the corresponding mapping structures are given by*

$$
\begin{aligned}
&\mathbf{x}_{k+2N} = P_{+}^{(N)} \circ P_{+}^{(N)}\mathbf{x}_k = P_{+}^{(2N)}\mathbf{x}_k \text{ and } \mathbf{g}(\mathbf{x}_k, \mathbf{x}_{k+2N}, \boldsymbol{\lambda}) = \mathbf{0};\\
&\mathbf{x}_{k+2^2N} = P_{+}^{(2N)} \circ P_{+}^{(2N)}\mathbf{x}_k = P_{+}^{(2^2N)}\mathbf{x}_k \text{ and } \mathbf{g}(\mathbf{x}_k, \mathbf{x}_{k+2^2N}, \boldsymbol{\lambda}) = \mathbf{0};\\
&\vdots\\
&\mathbf{x}_{k+2^lN} = P_{+}^{(2^{l-1}N)} \circ P_{+}^{(2^{l-1}N)}\mathbf{x}_k = P_{+}^{(2^lN)}\mathbf{x}_k \text{ and } \mathbf{g}(\mathbf{x}_k, \mathbf{x}_{k+2^lN}, \boldsymbol{\lambda}) = \mathbf{0};
\end{aligned}
\tag{4.17}
$$

for positive mappings and

$$
\begin{aligned}
&\mathbf{x}_k = P_{-}^{(N)} \circ P_{-}^{(N)}\mathbf{x}_{k+2N} = P_{-}^{(2N)}\mathbf{x}_{k+2N} \text{ and } \mathbf{g}(\mathbf{x}_k, \mathbf{x}_{k+2N}, \boldsymbol{\lambda}) = \mathbf{0};\\
&\mathbf{x}_k = P_{-}^{(2N)} \circ P_{-}^{(2N)}\mathbf{x}_{k+2^2N} = P_{-}^{(2^2N)}\mathbf{x}_{k+2^2N} \text{ and } \mathbf{g}(\mathbf{x}_k, \mathbf{x}_{k+2^2N}, \boldsymbol{\lambda}) = \mathbf{0};\\
&\vdots\\
&\mathbf{x}_k = P_{-}^{(2^{l-1}N)} \circ P_{-}^{(2^{l-1}N)}\mathbf{x}_{k+2^lN} = P_{-}^{(2^lN)}\mathbf{x}_{k+2^lN} \text{ and } \mathbf{g}(\mathbf{x}_k, \mathbf{x}_{k+2^lN}, \boldsymbol{\lambda}) = \mathbf{0}
\end{aligned}
\tag{4.18}
$$

for negative mapping, then the following statements hold, i.e.,

(i) *The stable chaos generated by the limit state of the stable* $P_{+}^{(2^lN)}$-1 *solutions* ($l \to \infty$) *in sense of* $\mathbf{g}(\mathbf{x}_k, \mathbf{x}_{k+2^lN}, \boldsymbol{\lambda}) = \mathbf{0}$ *is the unstable chaos generated by the limit state of the unstable stable* $P_{-}^{(2^lN)}$-1 *solution* ($l \to \infty$) *in sense of* $\mathbf{g}(\mathbf{x}_k, \mathbf{x}_{k+2^lN}, \boldsymbol{\lambda}) = \mathbf{0}$ *with all eigenvalue distribution of* $DP_{-}^{(2^lN)}$ *outside unit cycle, vice versa. Such a chaos is the "Yang" chaos in nonlinear discrete dynamical systems.*

(ii) *The unstable chaos generated by the limit state of the unstable* $P_{+}^{(2^lN)}$-1 *solutions* ($l \to \infty$) *in sense of* $\mathbf{g}(\mathbf{x}_k, \mathbf{x}_{k+2^lN}, \boldsymbol{\lambda}) = \mathbf{0}$ *with all eigenvalue distribution of* $DP_{+}^{(2^lN)}$ *outside the unit cycle is the stable chaos generated by the limit state of the stable* $P_{-}^{(2^lN)}$-1 *solution* ($l \to \infty$) *in sense of* $\mathbf{g}(\mathbf{x}_k, \mathbf{x}_{k+2^lN}, \boldsymbol{\lambda}) = \mathbf{0}$, *vice versa. Such a chaos is the "Ying" chaos in nonlinear discrete dynamical systems.*

(iii) *The unstable chaos generated by the limit state of the unstable* $P_{+}^{(2^lN)}$-1 *solutions* ($l \to \infty$) *in sense of* $\mathbf{g}(\mathbf{x}_k, \mathbf{x}_{k+2^lN}, \boldsymbol{\lambda}) = \mathbf{0}$ *with all eigenvalue distribution of* $DP_{+}^{(2^lN)}$ *inside and outside the unit cycle is the unstable chaos generated by the limit state of the unstable* $P_{-}^{(2^lN)}$-1 *solution* ($l \to \infty$) *in sense of* $\mathbf{g}(\mathbf{x}_k, \mathbf{x}_{k+2^lN}, \boldsymbol{\lambda}) = \mathbf{0}$ *with switching all eigenvalue distribution of* $DP_{+}^{(2^lN)}$ *inside and outside the unit cycle, vice versa. Such a chaos is the "Ying–Yang" chaos in nonlinear discrete dynamical systems.*

Proof The proof is similar to the proof of Theorem 4.2, and the chaos is obtained by $l \to \infty$. This theorem is proved. ■

4.2 Discrete Systems with Multiple Maps

Definition 4.6 Consider a set of implicit vector functions $\mathbf{f}^{(j)} : D \to D\,(j = 1, 2, \cdots)$ on an open set $D \subset \mathscr{R}^n$ in an n-dimensional discrete dynamical system. For $\mathbf{x}_k, \mathbf{x}_{k+1} \in D$, there is a discrete relation as

$$\mathbf{f}^{(j)}(\mathbf{x}_k, \mathbf{x}_{k+1}, \mathbf{p}^{(j)}) = \mathbf{0} \text{ for } j = 1, 2, \cdots \tag{4.19}$$

where the vector function is $\mathbf{f}^{(j)} = (f_1^{(j)}, f_2^{(j)}, \cdots, f_n^{(j)})^{\mathrm{T}} \in \mathscr{R}^n$ and discrete variable vector is $\mathbf{x}_k = (x_{k1}, x_{k2}, \cdots, x_{kn})^{\mathrm{T}} \in \Omega$ with a parameter vector $\mathbf{p}^{(j)} = (p_1^{(j)}, p_2^{(j)}, \cdots, p_{m_j}^{(j)})^{\mathrm{T}} \in \mathscr{R}^{m_j}$.

Definition 4.7 Consider a set of implicit vector functions $\mathbf{f}^{(j)} : D \to D\,(j = 1, 2, \cdots)$ on an open set $D \subset \mathscr{R}^n$ in an n-dimensional discrete dynamical system.

(i) A set for discrete relations is defined as

$$\Phi = \{\mathbf{f}^{(j)} | \mathbf{f}^{(j)}(\mathbf{x}_k, \mathbf{x}_{k+1}, \mathbf{p}^{(j)}) = \mathbf{0},\ j \in \mathbb{Z}_+;\ k \in \mathbb{Z}\}. \tag{4.20}$$

(ii) The positive and negative discrete sets are defined as

$$\left.\begin{aligned} \Sigma_+ &= \{\mathbf{x}_{k+i} | \mathbf{x}_{k+i} \in \mathscr{R}^n, i \in \mathbb{Z}_+\} \subset D \text{ and} \\ \Sigma_- &= \{\mathbf{x}_{k-i} | \mathbf{x}_{k-i} \in \mathscr{R}^n, i \in \mathbb{Z}_+\} \subset D \end{aligned}\right\} \tag{4.21}$$

respectively, and the total set of the discrete states is

$$\Sigma = \Sigma_+ \cup \Sigma_-. \tag{4.22}$$

(iii) A positive mapping for $\mathbf{f}^{(j)} \in \Phi$ is defined as

$$P_j^+ : \Sigma \to \Sigma_+ \quad \Rightarrow \quad P_j^+ : \mathbf{x}_k \to \mathbf{x}_{k+1} \tag{4.23}$$

and a negative mapping is defined by

$$P_j^- : \Sigma \to \Sigma_- \quad \Rightarrow \quad P_j^- : \mathbf{x}_k \to \mathbf{x}_{k-1}. \tag{4.24}$$

(iv) Two sets for positive and negative mappings are defined as

$$\begin{aligned} \Theta_+ &= \{P_j^+ | P_j^+ : \mathbf{x}_k \to \mathbf{x}_{k+1} \text{ with } \mathbf{f}^{(j)}(\mathbf{x}_k, \mathbf{x}_{k+1}, \mathbf{p}^{(j)}) = \mathbf{0},\ j \in \mathrm{Z}_+;\ k \in \mathbb{Z}\} \\ \Theta_- &= \{P_j^- | P_j^- : \mathbf{x}_{k+1} \to \mathbf{x}_k \text{ with } \mathbf{f}^{(j)}(\mathbf{x}_k, \mathbf{x}_{k+1}, \mathbf{p}^{(j)}) = \mathbf{0},\ j \in \mathbb{Z}_+;\ k \in \mathbb{Z}\} \end{aligned} \tag{4.25}$$

with the total mapping sets as

$$\Theta = \Theta_+ \cup \Theta_-. \tag{4.26}$$

Definition 4.8 Consider a discrete dynamical system with a set of implicit vector functions $\mathbf{f}^{(j)} : D \to D\,(j = 1, 2, \cdots)$. For a mapping $P_j^+ \in \Theta_+$ with N_j-actions and $P_j^- \in \Theta_-$ with N_j-actions. The resultant mapping is defined as

$$P_{j^N}^+ = \underbrace{P_j^+ \circ P_j^+ \circ \cdots \circ P_j^+}_{N} \text{ and } P_{j^N}^- = \underbrace{P_j^- \circ P_j^- \circ \cdots \circ P_j^-}_{N}. \tag{4.27}$$

Definition 4.9 Consider a discrete dynamical system with a set of implicit vector functions $\mathbf{f}^{(j)} : D \to D\,(j = 1, 2, \cdots)$. For the m- positive mappings of $P_{j_i}^+ \in \Theta_+\,(i = 1, 2, \cdots, m)$ with N_{j_i}-actions ($N_{j_i} \in \{0, \mathbb{Z}_+\}$) and the corresponding m- negative mappings of $P_{j_i}^- \in \Theta_-\,(i = 1, 2, \cdots, m)$ with N_{j_i}-actions, the resultant nonlinear mapping cluster with pure positive or negative mappings is defined as

$$\left.\begin{aligned} P^+_{(N_{jm}\cdots N_{j2}N_{j1})} &= \underbrace{P^+_{j_m^{N_{jm}}} \circ \cdots \circ P^+_{j_2^{N_{j2}}} \circ P^+_{j_1^{N_{j1}}}}_{m-terms};\\ P^-_{(N_{j1}N_{j2}\cdots N_{jm})} &= \underbrace{P^-_{j_1^{N_{j1}}} \circ P^-_{j_2^{N_{j2}}} \circ \cdots \circ P^-_{j_m^{N_{jm}}}}_{m-terms}. \end{aligned}\right\} \tag{4.28}$$

in which at least one of mappings ($P_{j_i}^+$ and $P_{j_i}^-$) with $N_{j_i} \in \mathbb{Z}_+$ possesses a nonlinear iterative relation.

Theorem 4.4 *Consider a discrete dynamical system with a set of implicit vector functions* $\mathbf{f}^{(j)} : D \to D\,(j = 1, 2, \cdots)$. *For the m- positive mappings of* $P_{j_i}^+ \in \Theta_+\,(i = 1, 2, \cdots, m)$ *with* N_{j_i}*-actions* ($N_{j_i} \in \{0, \mathbb{Z}_+\}$) *and the corresponding m- negative mappings of* $P_{j_i}^- \in \Theta_-\,(i = 1, 2, \cdots, m)$ *with* N_{j_i}*-actions, the resultant nonlinear mapping with pure positive and negative mappings*

$$\mathbf{x}_{k+\Sigma_{s=1}^m N_{js}} = P^+_{(N_{jm}\cdots N_{j2}N_{j1})}\mathbf{x}_k \text{ and } \mathbf{x}_k = P^-_{(N_{j1}N_{j2}\cdots N_{jm})}\mathbf{x}_{k+\Sigma_{s=1}^m N_{js}}. \tag{4.29}$$

and $\mathbf{x}_{k+i} = P_{j_s}^+\mathbf{x}_{k+i-1}$ *and* $\mathbf{x}_{k+i-1} = P_{j_s}^-\mathbf{x}_{k+i}$ *can be governed by*

$$\mathbf{f}(\mathbf{x}_{k+i-1}, \mathbf{x}_{k+i}, \mathbf{p}) = \mathbf{0} \text{ for } i = 1, 2, \cdots, \Sigma_{s=1}^m N_{j_t}. \tag{4.30}$$

Suppose a differentiable, vector function $\mathbf{g} \in \mathscr{R}^n$ *possesses* $\mathbf{g}(\mathbf{x}_k, \mathbf{x}_{k+\Sigma_{s=1}^m N_{js}}, \boldsymbol{\lambda}) = \mathbf{0}$ *hold. If the solutions* $(\mathbf{x}_k^*, \cdots, \mathbf{x}_{k+\Sigma_{s=1}^m N_{js}})$ *of Eq. (4.29) with* $\mathbf{g}(\mathbf{x}_k, \mathbf{x}_{k+\Sigma_{s=1}^m N_{js}}, \boldsymbol{\lambda}) = \mathbf{0}$ *exist, then the following conclusions in the sense of* $\mathbf{g}(\mathbf{x}_k, \mathbf{x}_{k+\Sigma_{s=1}^m N_{js}}, \boldsymbol{\lambda}) = \mathbf{0}$ *hold.*

(i) *The stable* $P^+_{(N_{jm}\cdots N_{j2}N_{j1})}$*-1 solution is the unstable* $P^-_{(N_{j1}N_{j2}\cdots N_{jm})}$*-1 solutions with all eigenvalues of* $DP^-_{(N_{j1}N_{j2}\cdots N_{jm})}(\mathbf{x}^*_{k+\Sigma_{s=1}^m N_{js}})$ *outside the unit cycle, vice versa.*
(ii) *The unstable* $P^+_{(N_{jm}\cdots N_{j2}N_{j1})}$*-1 solution with all eigenvalues of* $DP^+_{(N_{jm}\cdots N_{j2}N_{j1})}(\mathbf{x}_k^*)$ *outside the unit cycle is the stable* $P^-_{(N_{j1}N_{j2}\cdots N_{jm})}$*-1 solutions, vice versa.*

(iii) *For the unstable* $P^{+}_{(N_{jm}\cdots N_{j2}N_{j1})}$-1 *solution with eigenvalue distribution of* $DP^{+}_{(N_{jm}\cdots N_{j2}N_{j1})}(\mathbf{x}^{*}_{k})$ *inside and outside the unit cycle, the corresponding* $P^{-}_{(N_{j1}N_{j2}\cdots N_{jm})}$-1 *solution is also unstable with switching eigenvalue distribution of* $DP^{-}_{(N_{j1}N_{j2}\cdots N_{jm})}(\mathbf{x}^{*}_{k+\Sigma_{s=1}^{m}N_{js}})$ *inside and outside the unit cycle, vice versa.*

(iv) *All the bifurcations of the stable and unstable* $P^{+}_{(N_{jm}\cdots N_{j2}N_{j1})}$-1 *solution are all the bifurcations of the unstable and stable* $P^{-}_{(N_{j1}N_{j2}\cdots N_{jm})}$-1 *solution, respectively.*

Proof The proof is similar to the proof of Theorem 4.2. This theorem is proved. ■

The chaos generated by the period-doubling of the $P^{+}_{(N_{jm}\cdots N_{j2}N_{j1})}$-1 and $P^{-}_{(N_{j1}N_{j2}\cdots N_{jm})}$-1 solutions can be described through the following theorem.

Theorem 4.5 *Consider a discrete dynamical system with a set of implicit vector functions* $\mathbf{f}^{(j)}: D \to D\,(j = 1, 2, \cdots)$. *For the* m- *positive mappings of* $P^{+}_{j_i} \in \Theta_{+}\ (i = 1, 2, \cdots, m)$ *with* N_{j_i}*-actions* ($N_{j_i} \in \{0, \mathbb{Z}_{+}\}$) *and the corresponding* m-*negative mappings of* $P^{-}_{j_i} \in \Theta_{-}\ (i = 1, 2, \cdots, m)$ *with* N_{j_i}*-actions, the resultant nonlinear mapping with pure positive and negative mappings*

$$\mathbf{x}_{k+\Sigma_{s=1}^{m}N_{js}} = P^{+}_{(N_{jm}\cdots N_{j2}N_{j1})}\mathbf{x}_{k} \text{ and } \mathbf{x}_{k} = P^{-}_{(N_{j1}N_{j2}\cdots N_{jm})}\mathbf{x}_{k+\Sigma_{s=1}^{m}N_{js}}; \tag{4.31}$$

and $\mathbf{x}_{k+i} = P^{+}_{j_s}\mathbf{x}_{k+i-1}$ *and* $\mathbf{x}_{k+i-1} = P^{-}_{j_s}\mathbf{x}_{k+i}$ *can be governed by*

$$\mathbf{f}^{(j)}(\mathbf{x}_{k+i-1}, \mathbf{x}_{k+i}, \mathbf{p}^{(j)}) = \mathbf{0} \text{ for } i = 1, 2, \cdots, \Sigma_{s=1}^{m}N_{j_t}. \tag{4.32}$$

Suppose a differentiable, vector function $\mathbf{g} \in \mathscr{R}^{n}$ *possesses* $\mathbf{g}(\mathbf{x}_{k}, \mathbf{x}_{k+\Sigma_{s=1}^{m}N_{js}}, \boldsymbol{\lambda}) = \mathbf{0}$ *hold. If the period-doubling cascade of the* $P^{+}_{(N_{jm}\cdots N_{j2}N_{j1})}$-1 *and* $P^{-}_{(N_{j1}N_{j2}\cdots N_{jm})}$-1 *solution occurs, the corresponding mapping structures are given by*

$$\left.\begin{aligned}
\mathbf{x}_{k+2\Sigma_{s=1}^{m}N_{js}} &= P^{+}_{(N_{jm}\cdots N_{j2}N_{j1})} \circ P^{+}_{(N_{jm}\cdots N_{j2}N_{j1})}\mathbf{x}_{k}\\
&= P^{+}_{2(N_{jm}\cdots N_{j2}N_{j1})}\mathbf{x}_{k}\\
\mathbf{g}(\mathbf{x}_{k}, \mathbf{x}_{k+2\Sigma_{s=1}^{m}N_{js}}, \boldsymbol{\lambda}) &= \mathbf{0};
\end{aligned}\right\}$$
$$\left.\begin{aligned}
\mathbf{x}_{k+2^{2}\Sigma_{s=1}^{m}N_{js}} &= P^{+}_{2(N_{jm}\cdots N_{j2}N_{j1})} \circ P^{+}_{2(N_{jm}\cdots N_{j2}N_{j1})}\mathbf{x}_{k}\\
&= P^{+}_{2^{2}(N_{jm}\cdots N_{j2}N_{j1})}\mathbf{x}_{k}\\
\mathbf{g}(\mathbf{x}_{k}, \mathbf{x}_{k+2^{2}\Sigma_{s=1}^{m}N_{js}}, \boldsymbol{\lambda}) &= \mathbf{0};
\end{aligned}\right\} \tag{4.33}$$
$$\vdots$$
$$\left.\begin{aligned}
\mathbf{x}_{k+2^{l}\Sigma_{s=1}^{m}N_{js}} &= P^{+}_{2^{l-1}(N_{jm}\cdots N_{j2}N_{j1})} \circ P^{+}_{2^{l-1}(N_{jm}\cdots N_{j2}N_{j1})}\mathbf{x}_{k}\\
&= P^{+}_{2^{l}(N_{jm}\cdots N_{j2}N_{j1})}\mathbf{x}_{k}\\
\mathbf{g}(\mathbf{x}_{k}, \mathbf{x}_{k+2^{l}\Sigma_{s=1}^{m}N_{js}}, \boldsymbol{\lambda}) &= \mathbf{0};
\end{aligned}\right\}$$

for positive mappings and

$$\left.\begin{aligned}\mathbf{x}_k &= P^-_{(N_{j_1}N_{j_2}\cdots N_{jm})} \circ P^-_{(N_{j_1}N_{j_2}\cdots N_{jm})}\mathbf{x}_{k+2\Sigma_{s=1}^m N_{js}}\\ &= P^-_{2(N_{j_1}N_{j_2}\cdots N_{jm})}\mathbf{x}_{k+2\Sigma_{s=1}^m N_{js}}\\ \mathbf{g}&(\mathbf{x}_k, \mathbf{x}_{k+2\Sigma_{s=1}^m N_{js}}, \boldsymbol{\lambda}) = \mathbf{0};\end{aligned}\right\}$$

$$\left.\begin{aligned}\mathbf{x}_k &= P^-_{2(N_{j_1}N_{j_2}\cdots N_{jm})} \circ P^-_{2(N_{j_1}N_{j_2}\cdots N_{jm})}\mathbf{x}_{k+2^2\Sigma_{s=1}^m N_{js}}\\ &= P^-_{2^2(N_{j_1}N_{j_2}\cdots N_{jm})}\mathbf{x}_{k+2^2\Sigma_{s=1}^m N_{js}}\\ \mathbf{g}&(\mathbf{x}_k, \mathbf{x}_{k+2^2\Sigma_{s=1}^m N_{js}}, \boldsymbol{\lambda}) = \mathbf{0};\end{aligned}\right\} \qquad (4.34)$$

$$\vdots$$

$$\left.\begin{aligned}\mathbf{x}_k &= P^-_{2^{l-1}(N_{j_1}N_{j_2}\cdots N_{jm})} \circ P^-_{2^{l-1}(N_{j_1}N_{j_2}\cdots N_{jm})}\mathbf{x}_{k+2^l\Sigma_{s=1}^m N_{js}}\\ &= P^-_{2^l(N_{j_1}N_{j_2}\cdots N_{jm})}\mathbf{x}_{k+2^l\Sigma_{s=1}^m N_{js}}\\ \mathbf{g}&(\mathbf{x}_k, \mathbf{x}_{k+2^l\Sigma_{s=1}^m N_{js}}, \boldsymbol{\lambda}) = \mathbf{0};\end{aligned}\right\}$$

for negative mapping, then the following statements hold, i.e.,

(i) *The stable chaos generated by the limit state of the stable* $P^+_{2^l(N_{jm}\cdots N_{j_2}N_{j_1})}$*-1 solutions* ($l \to \infty$) *in sense of* $\mathbf{g}(\mathbf{x}_k, \mathbf{x}_{k+2^l\Sigma_{s=1}^m N_{js}}, \boldsymbol{\lambda}) = \mathbf{0}$ *is the unstable chaos generated by the limit state of the unstable stable* $P^-_{2^l(N_{j_1}N_{j_2}\cdots N_{jm})}$*-1 solution* ($l \to \infty$) *in sense of* $\mathbf{g}(\mathbf{x}_k, \mathbf{x}_{k+2^l\Sigma_{s=1}^m N_{js}}, \boldsymbol{\lambda}) = \mathbf{0}$ *with all eigenvalue distribution of* $DP^+_{2^l(N_{jm}\cdots N_{j_2}N_{j_1})}$ *outside unit cycle, vice versa. Such a chaos is the "Yang" chaos in nonlinear discrete dynamical systems.*

(ii) *The unstable chaos generated by the limit state of the unstable* $P^+_{2^l(N_{jm}\cdots N_{j_2}N_{j_1})}$*-1 solutions* ($l \to \infty$) *in sense of* $\mathbf{g}(\mathbf{x}_k, \mathbf{x}_{k+2^l\Sigma_{s=1}^m N_{js}}, \boldsymbol{\lambda}) = \mathbf{0}$ *with all eigenvalue distribution of* $P^+_{2^l(N_{jm}\cdots N_{j_2}N_{j_1})}$*-1 outside the unit cycle is the stable chaos generated by the limit state of the stable* $P^-_{2^l(N_{j_1}N_{j_2}\cdots N_{jm})}$*-1 solution* ($l \to \infty$) *in sense of* $\mathbf{g}(\mathbf{x}_k, \mathbf{x}_{k+2^l\Sigma_{s=1}^m N_{js}}, \boldsymbol{\lambda}) = \mathbf{0}$*, vice versa. Such a chaos is the "Ying" chaos in nonlinear discrete dynamical systems.*

(iii) *The unstable chaos generated by the limit state of the unstable* $P^+_{2^l(N_{jm}\cdots N_{j_2}N_{j_1})}$*-1 solutions* ($l \to \infty$) *in sense of* $\mathbf{g}(\mathbf{x}_k, \mathbf{x}_{k+2^l N}, \boldsymbol{\lambda}) = \mathbf{0}$ *with all eigenvalue distribution of* $DP^+_{2^l(N_{jm}\cdots N_{j_2}N_{j_1})}$ *inside and outside the unit cycle is the unstable chaos generated by the limit state of the unstable* $P^-_{2^l(N_{j_1}N_{j_2}\cdots N_{jm})}$*-1 solution* ($l \to \infty$) *in sense of* $\mathbf{g}(\mathbf{x}_k, \mathbf{x}_{k+2^l N}, \boldsymbol{\lambda}) = \mathbf{0}$ *with switching all eigenvalue distribution of* $DP^-_{2^l(N_{j_1}N_{j_2}\cdots N_{jm})}$ *inside and outside the unit cycle, vice versa. Such a chaos is the "Ying–Yang" chaos in nonlinear discrete dynamical systems.*

Proof The proof is similar to the proof of Theorem 4.2, and the chaos is obtained by $l \to \infty$. This theorem is proved. ■

4.3 Complete Dynamics of a Henon Map System

As in Luo and Guo (2010), consider the Henon map system as

$$\left.\begin{aligned} f_1(\mathbf{x}_k, \mathbf{x}_{k+1}, \mathbf{p}) &= x_{k+1} - y_k - 1 + ax_k^2 = 0, \\ f_2(\mathbf{x}_k, \mathbf{x}_{k+1}, \mathbf{p}) &= y_{k+1} - bx_k = 0 \end{aligned}\right\} \tag{4.35}$$

where $\mathbf{x}_k = (x_k, y_k)^{\mathrm{T}}$, $\mathbf{f} = (f_1, f_2)^{\mathrm{T}}$ and $\mathbf{p} = (a, b)^{\mathrm{T}}$. Consider two positive and negative mapping structures as

$$\begin{aligned} \mathbf{x}_{k+N} &= P_+^{(N)}\mathbf{x}_k = \underbrace{P_+ \circ \cdots P_+ \circ P_+}_{N\text{-terms}} \mathbf{x}_k, \\ \mathbf{x}_k &= P_-^{(N)}\mathbf{x}_{k+N} = \underbrace{P_- \circ \cdots P_- \circ P_-}_{N\text{-terms}} \mathbf{x}_{k+N}. \end{aligned} \tag{4.36}$$

Equations (4.35) and (4.36) give

$$\left.\begin{aligned} &\mathbf{f}(\mathbf{x}_k, \mathbf{x}_{k+1}, \mathbf{p}) = \mathbf{0}, \\ &\mathbf{f}(\mathbf{x}_{k+1}, \mathbf{x}_{k+2}, \mathbf{p}) = \mathbf{0}, \\ &\vdots \\ &\mathbf{f}(\mathbf{x}_{k+N-1}, \mathbf{x}_{k+N}, \mathbf{p}) = \mathbf{0} \end{aligned}\right\} \tag{4.37}$$

and

$$\left.\begin{aligned} &\mathbf{f}(\mathbf{x}_{k+N-1}, \mathbf{x}_{k+N}, \mathbf{p}) = \mathbf{0}, \\ &\mathbf{f}(\mathbf{x}_{k+N-2}, \mathbf{x}_{k+N-1}, \mathbf{p}) = \mathbf{0}, \\ &\vdots \\ &\mathbf{f}(\mathbf{x}_k, \mathbf{x}_{k+1}, \mathbf{p}) = \mathbf{0}. \end{aligned}\right\} \tag{4.38}$$

The switching of equation order in Eq. (4.38) shows Eqs. (4.37) and (4.38) are identical. For periodic solutions of the positive and negative maps, the periodicity of the positive and negative mapping structures of the Henon map requires

$$\mathbf{x}_{k+N} = \mathbf{x}_k \text{ or } \mathbf{x}_k = \mathbf{x}_{k+N}. \tag{4.39}$$

So the periodic solutions $\mathbf{x}_{k+j}^*$ $(j = 0, 1, \cdots, N)$ for the negative and positive mapping structures are the same, which are given by solving Eqs. (4.37) and (4.38) with (4.39). However, the stability and bifurcation are different because $\mathbf{x}_{k+j}$ varies with $\mathbf{x}_{k+j-1}$ for the jth positive mapping and $\mathbf{x}_{k+j-1}$ varies with $\mathbf{x}_{k+j}$ for the jth negative mapping. For a small perturbation, equation (4.37) for the positive mapping gives

$$[\frac{\partial \mathbf{f}}{\partial \mathbf{x}_{k+j-1}}] + [\frac{\partial \mathbf{f}}{\partial \mathbf{x}_{k+j}}] \cdot [\frac{\partial \mathbf{x}_{k+j}}{\partial \mathbf{x}_{k+j-1}}]|_{(\mathbf{x}_{k+j-1}^*, \mathbf{x}_{k+j}^*)} = \mathbf{0}, \tag{4.40}$$

where

$$
\begin{aligned}
\left[\frac{\partial \mathbf{f}}{\partial \mathbf{x}_{k+j-1}}\right]_{(\mathbf{x}^*_{k+j-1},\mathbf{x}^*_{k+j})} &= \begin{bmatrix} \dfrac{\partial f_1}{\partial x_{k+j-1}} & \dfrac{\partial f_1}{\partial y_{k+j-1}} \\ \dfrac{\partial f_2}{\partial x_{k+j-1}} & \dfrac{\partial f_2}{\partial y_{k+j-1}} \end{bmatrix}_{(\mathbf{x}^*_{k+j-1},\mathbf{x}^*_{k+j})} \\
&= \begin{bmatrix} 2ax^*_{k+j-1} & -1 \\ -b & 0 \end{bmatrix},
\end{aligned} \tag{4.41}
$$

$$
\begin{aligned}
\left[\frac{\partial \mathbf{f}}{\partial \mathbf{x}_{k+j}}\right]_{(\mathbf{x}^*_{k+j-1},\mathbf{x}^*_{k+j})} &= \begin{bmatrix} \dfrac{\partial f_1}{\partial x_{k+j}} & \dfrac{\partial f_1}{\partial y_{k+j}} \\ \dfrac{\partial f_2}{\partial x_{k+j}} & \dfrac{\partial f_2}{\partial y_{k+j}} \end{bmatrix}_{(\mathbf{x}^*_{k+j-1},\mathbf{x}^*_{k+j})} \\
&= \begin{bmatrix} 1 & 0 \\ 0 & 1 \end{bmatrix}.
\end{aligned} \tag{4.42}
$$

So

$$
\begin{aligned}
DP_+(\mathbf{x}^*_{k+j-1}) &= \left[\frac{\partial \mathbf{x}_{k+j}}{\partial \mathbf{x}_{k+j-1}}\right]_{\mathbf{x}^*_{k+j-1}} = \left[\frac{\partial \mathbf{f}}{\partial \mathbf{x}_{k+j}}\right]^{-1}\left[\frac{\partial \mathbf{f}}{\partial \mathbf{x}_{k+j-1}}\right]_{\mathbf{x}^*_{k+j-1}} \\
&= \begin{bmatrix} 2ax^*_{k+j-1} & -1 \\ -b & 0 \end{bmatrix}.
\end{aligned} \tag{4.43}
$$

Similarly, for the negative mapping,

$$
\left[\frac{\partial \mathbf{f}}{\partial \mathbf{x}_{k+j}}\right] + \left[\frac{\partial \mathbf{f}}{\partial \mathbf{x}_{k+j-1}}\right] \cdot \left[\frac{\partial \mathbf{x}_{k+j-1}}{\partial \mathbf{x}_{k+j}}\right]\Big|_{(\mathbf{x}^*_{k+j-1},\mathbf{x}^*_{k+j})} = \mathbf{0}. \tag{4.44}
$$

With Eqs. (4.41) and (4.42), the foregoing equation gives

$$
\begin{aligned}
DP_-(\mathbf{x}^*_{k+j}) &= \left[\frac{\partial \mathbf{x}_{k+j-1}}{\partial \mathbf{x}_{k+j}}\right]_{\mathbf{x}^*_{k+j}} = \left[\frac{\partial \mathbf{f}}{\partial \mathbf{x}_{k+j-1}}\right]^{-1}\left[\frac{\partial \mathbf{f}}{\partial \mathbf{x}_{k+j}}\right]_{\mathbf{x}^*_{k+j}} \\
&= -\frac{1}{b}\begin{bmatrix} 0 & 1 \\ b & 2ax^*_{k+j-1} \end{bmatrix}.
\end{aligned} \tag{4.45}
$$

Thus, the resultant perturbation of the mapping structure in Eq. (4.36) gives

$$
\begin{aligned}
\delta \mathbf{x}_{k+N} &= DP_+^{(N)}\delta \mathbf{x}_k = \underbrace{DP_+ \cdot \cdots \cdot DP_+ \cdot DP_+}_{N\text{-terms}}\,\delta \mathbf{x}_k, \\
\delta \mathbf{x}_k &= DP_-^{(N)}\delta \mathbf{x}_{k+N} = \underbrace{DP_- \cdot \cdots \cdot DP_- \cdot DP_-}_{N\text{-terms}}\,\delta \mathbf{x}_{k+N}
\end{aligned} \tag{4.46}
$$

where

$$\left.\begin{aligned} DP_+^{(N)} &= \prod_{j=1}^{N} DP_+(\mathbf{x}_{k+N-j}^*),\\ DP_-^{(N)} &= \prod_{j=N}^{1} DP_-(\mathbf{x}_{k+N-j+1}^*). \end{aligned}\right\} \tag{4.47}$$

Consider the eigenvalues λ^- and λ^+ of $DP_-^{(N)}(\mathbf{x}_{k+N}^*)$ and $DP_+^{(N)}(\mathbf{x}_k^*)$, respectively. The following statements hold.

(i) If $|\lambda_{1,2}^+| < 1$ (or $|\lambda_{1,2}^-| < 1$), the periodic solutions of $P_+^{(N)}(\mathbf{x}_k)$ (or $P_-^{(N)}(\mathbf{x}_{k+N})$) are stable.
(ii) If $|\lambda_{1\,\text{or}2}^+| > 1$ (or $|\lambda_{1\,\text{or}\,2}^-| > 1$), the periodic solutions of $P_+^{(N)}(\mathbf{x}_k)$ (or $P_-^{(N)}(\mathbf{x}_{k+N})$) are unstable.
(iii) If real eigenvalues $\lambda_1^+ = -1$ and $|\lambda_2^+| < 1$ (or $\lambda_1^- = -1$ and $|\lambda_2^-| < 1$), the period-doubling (PD) bifurcation of the periodic solutions of $P_+^{(N)}(\mathbf{x}_k)$ (or $P_-^{(N)}(\mathbf{x}_{k+N})$) occurs.
(iv) If real eigenvalues $|\lambda_1^+| < 1$ and $\lambda_2^+ = 1$ (or $|\lambda_1^-| < 1$ and $\lambda_2^- = 1$), then the saddle-node (SN) bifurcation of the periodic solutions relative to $P_+^{(N)}(\mathbf{x}_k)$ (or $P_-^{(N)}(\mathbf{x}_{k+N})$) occurs.
(v) If two complex eigenvalues of $|\lambda_{1,2}^+| = 1$ (or $|\lambda_{1,2}^-| = 1$), the Neimark bifurcation (NB) of the periodic solutions of $P_+^{(N)}(\mathbf{x}_k)$ (or $P_-^{(N)}(\mathbf{x}_{k+N})$) occurs.

A numerical prediction of periodic solutions of the Henon map is presented with varying parameter b for $a = 1.1$, as shown in Figs. 4.2a–d. The dashed vertical lines give the bifurcation points. The acronyms "PD", "SN" and "NB" are represented the period-doubling bifurcation, saddle-node bifurcation and Neimark bifurcation, respectively. All the stable periodic solutions for positive mapping P_+ lie in $b \in (-1.0, 1.0)$. The stable period-1 solution of P_+ is in $b \in (-1.0, -0.22)$. At $b = -1$, the Neimark bifurcation (NB) of the period-1 solution occurs. At $b \approx -0.22$, the period-doubling bifurcation (PD) of the period-1 solution occurs. This point is the saddle-node bifurcation (SN) for the period-2 solution of P_+ (i.e., $P_+^{(2)}$). The periodic solution of $P_+^{(2)}$ is in $b \in (-0.22, 0.1133)$. At $b \approx 0.1133$, the corresponding period-doubling bifurcation (PD) of $P_+^{(2)}$ occurs, which also corresponds to the saddle-node bifurcation (SN) for the period-4 solution (i.e., $P_+^{(4)}$). The periodic solution of $P_+^{(4)}$ exists in $b \in (0.1133, 0.2084)$. At $b \approx 0.2084$, there is a period-doubling bifurcation (PD) of $P_+^{(4)}$, corresponding to the saddle-node bifurcation (SN) of the period-8 solution (i.e., $P_+^{(8)}$). The periodic solution of $P_+^{(8)}$ exist in $b \in (0.2084, 0.2468)$. Also the coexisting periodic solutions of $P_+^{(3)}$ and $P_+^{(6)}$ exist in $b \in (-1.0, -0.344)$. The periodic solution of $P_+^{(3)}$ exists in $b \in (-1.0, -0.8771)$ and $(-0.4, -0.344)$. At $b \approx -0.344$, the saddle-node bifurcation (SN) of $P_+^{(3)}$ occurs where the periodic solution of $P_+^{(3)}$ disappears. At $b \approx -0.4$, the period-doubling bifurcation (PD) of $P_+^{(3)}$ occurs, and the saddle-node bifurcation (SN) for the periodic solution of $P_+^{(6)}$ appears. The periodic solution of

Table 4.1 Parameter ranges for stable periodic and Chaotic solutions in bifurcation scenario ($a = 1.1$)

	Mapping structure	Parameter b
Positive mapping (stable)	P_+	(−1.0,−0.22)
	$P_+^{(2)}$	(−0.22,0.1133)
	$P_+^{(4)}$	(0.1133,0.2084)
	$P_+^{(8)}$	(0.2084,0.2468)
	$P_+^{(3)}$	(−1.0,−0.8771) and (−0.4,−0.344)
	$P_+^{(6)}$	(−0.482,−0.4) and (−0.8771,−0.8112)
Negative mapping (stable)	P_-	$(-\infty, -1.0)$ and $(2.2184, +\infty)$
	$P_-^{(2)}$	(1.1,2.2184)
	$P_-^{(4)}$	((1.0,1.1))
	$P_-^{(3)}$	(−1.0773,−1.0)
Chaos	–	(0.2468,0.523)

$P_+^{(6)}$ lies in $b \in (-0.4582, -0.4)$ and $(-0.8771, -0.8112)$. At $b = 1$, the Neimark bifurcation (NB) of the periodic solution of $P_+^{(4)}$ occurs, which cannot be obtained through the numerical prediction but is predicted by the analytical prediction. After the Neimark bifurcation, the stable periodic solutions for positive mapping P_+ do not exist any more. Such stable periodic solutions for positive mapping P_+ are shown in Figs. 4.2a, b. The stable solution for negative mapping P_- is in $b \in (-\infty, -1.0)$ and $b \in (-1, +\infty)$. At $b = -1$, the Neimark bifurcation (NB) of the periodic solutions of P_- and $P_-^{(3)}$ coexist. The period-1 solution of P_- is in $b \in (-\infty, -1.0)$ and $b \in (2.2184, +\infty)$. The period-doubling bifurcation (PD) of the period-1 solution of P_- occurs at $b \approx 2.2184$ and the bifurcation point is also a saddle-node bifurcation (SN) for the period-2 solution of P_- (i.e., $P_-^{(2)}$). The stable periodic solution of $P_-^{(2)}$ is in $b \in (1.1, 2.2184)$. After the period-doubling bifurcation, the periodic solution of $P_-^{(4)}$ is in $b \in (1.0, 1.1)$. The coexisting periodic solution of $P_-^{(3)}$ is in $b \in (-1.0773, -1.0)$, which is actually connected with the coexisting periodic solution of $P_+^{(3)}$ through the Neimark bifurcation (NB). At $b = 1$, the Neimark bifurcation (NB) of the periodic solution of $P_-^{(4)}$ occurs. Such stable periodic solutions for negative mapping P_- are shown in Figs. 4.2c, d. Finally the chaotic solutions occurs in $b \in (0.2468, 0.523)$. The parameter range of b can be shown from negative to positive infinity. The values of b for bifurcation and stable periodic solution in bifurcation scenario are tabulated in Table 4.1. From the numerical prediction, some of the stable periodic solutions of the Henon map can be obtained. However, the unstable solution for each mapping structure can only be obtained by the analytical prediction due to the sensitive dependency on initial conditions.

The stable and unstable periodic solutions for both positive and negative mappings of the Henon maps are obtained in Figs. 4.3 and 4.4. The acronyms "PD", "SN" and "NB" represent the stable period-doubling bifurcation, stable saddle-node bifurcation and Neimark bifurcation, respectively. The acronyms "UPD", and "USN" represent

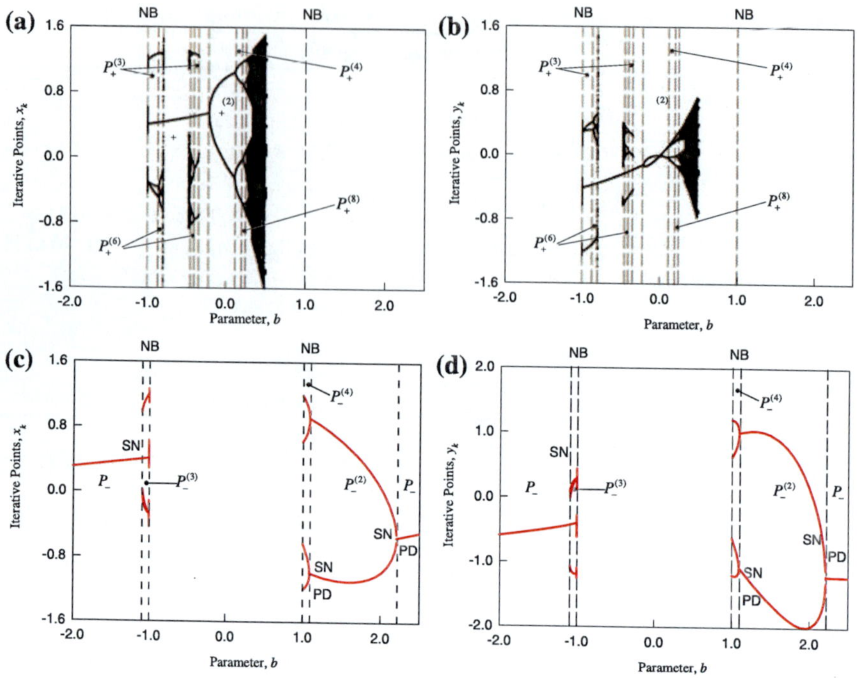

Fig. 4.2 Numerical predictions of periodic solutions of the Henon mapping: (**a**) and (**b**) positive mapping (P_+), (**c**) and (**d**) negative mapping (P_-). ($a=1.1$)

the period-doubling bifurcation from unstable nodes with negative eigenvalues to saddle and the saddle-node bifurcation from unstable nodes with positive eigenvalues to saddle, respectively. The analytical stable and unstable periodic solutions of positive mapping P_+ for $a=1.1$ and $b \in (-\infty, +\infty)$ are presented in Figs. 4.3a–d. The stable period-1 solutions of P_+ lie in $b \in (-1.0, -0.2143)$, matched with the numerical iteration. For $b \in (-0.2143, +\infty)$, the unstable period-1 solution of P_+ is saddle. For $b \in (-\infty, -1.0)$, the unstable period-1 solution of P_+ is an unstable focus. The corresponding bifurcations of the period-1 solution of P_+ are Neimark bifurcation (NB) and period-doubling bifurcation (PD). The stability and bifurcation of the unstable period-1 solution of P_+ is determined by the negative mapping P_-. For $b \in (2.2211, +\infty)$, the period-1 solution of P_+ is an unstable node. The unstable period-1 solution of P_+ is saddle for $b \in (-\infty, 2.2211)$. Thus, the unstable period-doubling bifurcation (UPD) of the period-1 solution ofP_+ occurs at $b \approx 2.2211$. At this point, the unstable periodic solution is from unstable nodes with negative eigenvalues to saddle. Because of the unstable period-doubling bifurcation, the unstable periodic solution of $P_+^{(2)}$ for the unstable nodes with negative eigenvalues is obtained for $b \in (1.0936, 2.2211)$. The unstable periodic solution is from unstable focus to unstable node during the parameter of $b \in (1.0936, 2.2211)$. At $b \approx 2.2211$, the USN bifurcation of the unstable periodic solution of $P_+^{(2)}$ is from unstable nodes with positive eigenvalues to saddle. At such a point, the bifurcation for the mapping of $P_-^{(2)}$ is from the stable node to saddle. At $b \approx 1.0936$, the bifurcation of the unstable periodic solution of $P_+^{(2)}$ is unstable period-doubling bifurcation. The bifurcation of periodic solution of $P_-^{(2)}$ is a period-doubling bifurcation. The stable and unstable periodic solutions of $P_+^{(2)}$ exist in $b \in (-0.2143, 0.1068)$ and in $b \in (0.1068, 1.0932)$ with saddle, respectively. At $b \approx 0.1068$, there is a period-doubling bifurcation of $P_+^{(2)}$ from the stable nodes to saddle. The unstable periodic solution of $P_+^{(2)}$ is saddle. Such a bifurcation point is also a SN bifurcation of $P_+^{(4)}$. The stable periodic solution of $P_+^{(4)}$ is $b \in (0.1068, 0.2083)$ and $(0.9842, 1.0)$. The unstable periodic solution of $P_+^{(4)}$ is saddle in $b \in (0.2083, 0.9842)$. The two points $b = 0.2083$ and 0.9842 are for the period-doubling bifurcation of $P_+^{(4)}$, and for the saddle-node bifurcation of $P_+^{(8)}$, respectively. The unstable periodic solution of $P_+^{(4)}$ are an unstable node in $b \in (1.0, 1.0936)$, determined by the negative mapping of $P_-^{(4)}$. At $b = 1.0$, the Neimark bifurcation (NB) for the period-4 solution of positive mappings (i.e., $P_+^{(4)}$) occurs. The stable period-8 solution lies in $b \in (0.2083, 0.2556)$ and $(0.9451, 0.9842)$. The unstable periodic solution of $P_+^{(8)}$ is saddle in $b \in (0.2556, 0.9451)$. A zoomed view of the periodic solutions of $P_+^{(4)}$ and $P_+^{(8)}$ is given in Fig. 4.3c. The coexisting stable solution of $P_+^{(3)}$ is $b \in (-1.0, -0.8531)$ and $b \in (-0.3981, -0.3408)$. At $b = -1.0$ the Neimark bifurcation (NB) of $P_+^{(3)}$ occurs. At points of $b \approx -0.8531$ and -0.3981, the PD bifurcation of $P_+^{(3)}$ occurs, and the unstable periodic solution of $P_+^{(3)}$ is saddle in $b \in (-0.8531, -0.3981)$. The two points of $b \approx -0.8531$ and -0.3981 are also the SN bifurcation of $P_+^{(6)}$. The unstable periodic solution of $P_+^{(3)}$ in $b \in (-1.0917, -1.0)$ is determined by the negative mapping of $P_-^{(3)}$. At $b \approx -0.3408$, the SN bifurcation of $P_+^{(3)}$ occurs and the corresponding periodic solution disappears. At $b \approx -1.0923$, an USN bifurcation leads to the disappearance of the unstable periodic solution of $P_+^{(3)}$. The stable periodic solution of $P_+^{(6)}$ is in $b \in (-0.8531, -0.8088)$ and $(-0.4384, -0.3947)$, and the unstable periodic solution of $P_+^{(6)}$ is saddle in $b \in (-0.8088, -0.4384)$. A zoomed view of $P_+^{(3)}$ and $P_+^{(6)}$ for coexisting periodic solution of P_+ is presented in Fig. 4.3d. For a clear picture of bifurcation and stability, the bifurcation values and stability ranges of parameter b for the periodic solutions of positive mapping are tabulated in Table 4.2. In a similar fashion, the analytical prediction of stable and unstable periodic solutions for negative mapping P_- for $a = 1.1$ and $b \in (-\infty, +\infty)$ is presented in Figs. 4.4a, b.

The stable periodic solutions for negative mapping P_- lie in $b \in (-\infty, -1.0)$ and $(1.0, +\infty)$, which is the same as in numerical prediction. The stable period-1 solution of P_- is a stable focus in $b \in (-\infty, -1.0)$ and stable nodes in $b \in (2.2211, +\infty)$. For $b \in (-1.0, -0.2143)$, the unstable period-1 solution of P_- is from an unstable focus to unstable node. At $b = -1$, the bifurcation between the stable and unstable

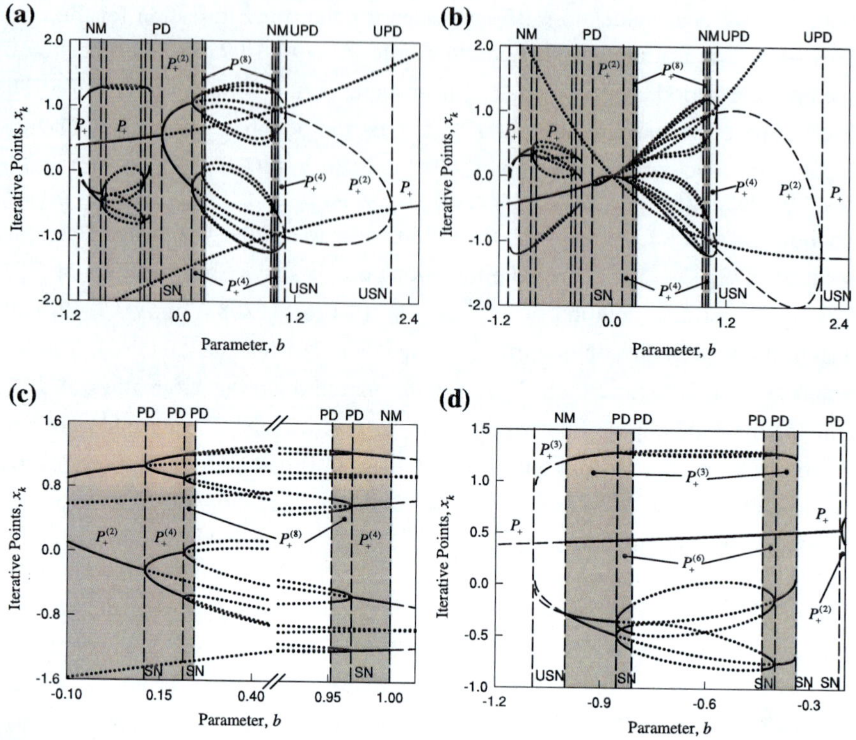

Fig. 4.3 Analytical predictions of *stable* and *unstable* periodic solutions for positive map of the Henon map: (**a**) and (**b**) periodic solutions, (**c**) zoomed period-4 solutions, and (**d**) zoomed period-3 solutions. ($a = 1.1$ and $b \in (-\infty, +\infty)$)

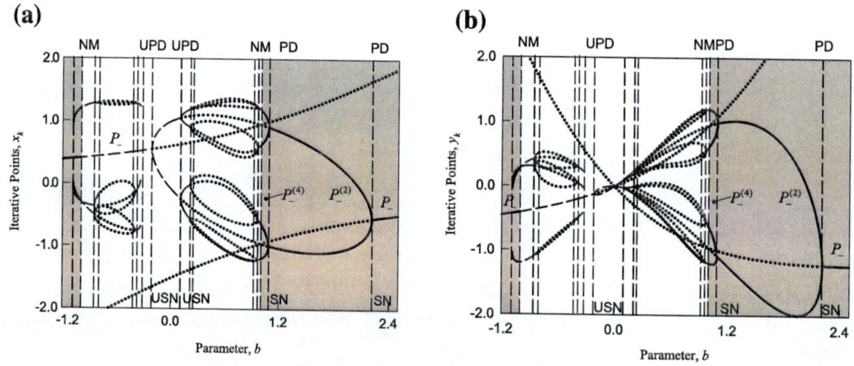

Fig. 4.4 Analytical predictions of *stable* and *unstable* periodic solutions for negative map of the Henon map: (**a**) and (**b**) periodic solutions. ($a = 1.1$ and $b \in (-\infty, +\infty)$)

Table 4.2 Parameter ranges for analytical stable and unstable periodic solutions for positive mappings ($a=1.1$)

	Mapping structure	Parameter b
Stable (nodes & focus)	P_+	(−1.0,−0.2143)
	$P_+^{(2)}$	(−0.2143,0.1068)
	$P_+^{(4)}$	(0.1068,0.2083) and (0.9842,1.0)
	$P_+^{(8)}$	(0.2083,0.2556) and (0.9451,0.9842)
	$P_+^{(3)}$	(−1.0,−0.8531) and (−0.3981,−0.3408)
	$P_+^{(6)}$	(−0.8531,−0.8088) and (−0.4384,−0.3947)
Unstable focus	P_+	$(-\infty, -1.0)$
	$P_+^{(3)}$	(−1.0917,−1.0)
Saddle	P_+	$(-0.2143, +\infty)$ and $(-\infty, 2.2211)$
	$P_+^{(2)}$	(0.1068,1.0932)
	$P_+^{(4)}$	(0.2083,0.9842)
	$P_+^{(8)}$	(0.2556,0.9451)
	$P_+^{(3)}$	(−0.8531,−3981)
	$P_+^{(6)}$	(−0.8088,−0.4384)
Unstable node	P_+	$(2.2211, +\infty)$
	$P_+^{(2)}$	(1.0936,2.2211)
	$P_+^{(4)}$	(1.0,1.0936)

period-1 solution of P_- is the Neimark Bifurcation (NB). For $b \in (-0.2143, +\infty)$, the unstable period-1 solution of P_- is saddle. Thus, the UPD bifurcation of P_- occurs at $b = -0.2143$. For $b \in (-\infty, 2.2211)$, the unstable period-1 solution of P_- is saddle. At $b \approx 2.2211$, the PD bifurcation of the period-1 solution of P_- occurs. The unstable period-2 solution of P_- (i.e., $P_-^{(2)}$) exists in $b \in (-0.2143, 0.1068)$ and$(0.1068, 1.0936)$. For $b \in (1.0936, 2.2211)$, the stable period-2 solution of P_- (i.e., $P_-^{(2)}$) exists from the stable focus to nodes. The point at $b \approx -0.2143$ is the USN bifurcation of the unstable periodic solution of $P_-^{(2)}$. At $b \approx 1.0936$, the PD bifurcation of the stable periodic solution of $P_-^{(2)}$ occurs, and the saddle periodic solution of $P_-^{(2)}$ appears. At $b \approx 2.2211$, the SN bifurcation of $P_-^{(2)}$ takes place. The unstable periodic solution of $P_-^{(4)}$ is in $b \in (0.1068, 0.2083)$ and $(0.9842, 1.0)$. An USN bifurcation for the unstable periodic solution of $P_-^{(4)}$ occurs at $b \approx 0.1068$. At $b \approx 0.2083$ and 0.9842, the UPD bifurcation of the unstable periodic solution of $P_-^{(4)}$ occurs. For the point at $b = 1$, the Neimark bifurcation of the periodic solution of $P_-^{(4)}$ occurs. Thus, the stable periodic solution of $P_-^{(4)}$ is in $b \in (1.0, 1.0936)$. The SN bifurcation of $P_-^{(4)}$ occurs at $b \approx 1.0936$. The unstable periodic solution of $P_-^{(8)}$ is in $b \in (0.2083, 0.2556)$ and $(0.9451, 0.9842)$. The USN bifurcations of $P_-^{(8)}$ occur at $b \approx 0.2083$ and 0.9842. The unstable periodic solution of $P_-^{(8)}$ is saddle in $b \in (0.2556, 0.9451)$, and the UPD bifurcations of $P_-^{(8)}$ are at $b \approx 0.2556$ and 0.9451. The stable periodic solution of $P_-^{(3)}$ coexist with the period-1 motion of

Table 4.3 Parameter ranges for analytical stable and unstable periodic solutions for negative mappings ($a = 1.1$)

	Mapping structure	Parameter b
Stable (nodes & focus)	P_-	$(-\infty, -1.0)$ and $(2.2211, +\infty)$
	$P_-^{(2)}$	(1.0936,2.2211)
	$P_-^{(4)}$	(1.0,1.0936)
	$P_-^{(3)}$	(−1.0917,−1.0)
Unstable focus	P_-	(−1.0,−0.2143)
	$P_-^{(2)}$	(−0.2143,0.1068)
Saddle	P_-	$(-0.2143, +\infty)$ and $(-\infty, 2.2211)$
	$P_-^{(2)}$	(0.1068,1.0932)
	$P_-^{(4)}$	(0.2083,0.9842)
	$P_-^{(8)}$	(0.2556,0.9451)
	$P_-^{(3)}$	(−0.8531,−3981)
	$P_-^{(6)}$	(−0.8088,−0.4384)
Unstable node	$P_-^{(4)}$	(0.1068,0.2083) and (0.9842,1.0)
	$P_-^{(8)}$	(0.2083,0.2556) and (0.9451,0.9842)
	$P_-^{(3)}$	(−1.0,−0.8531) and (−0.3981,−0.3408)
	$P_-^{(6)}$	(−0.8531,−0.8088) and (−0.4384,−0.3947)

P_- in $b \in (-1.0917, -1.0)$, and the unstable periodic solution of $P_-^{(3)}$ is an unstable node in $b \in (-1.0, -0.8531)$ and $(-0.3981, -0.3408)$. The unstable periodic solution of $P_-^{(3)}$ is saddle in $b \in (-0.8531, -0.3981)$. At $b = -1.0$, the Neimark bifurcation of the periodic solutions of $P_-^{(3)}$ and P_- coexist. The unstable periodic solution of $P_-^{(6)}$ is saddle in $b \in (-0.8088, -0.4384)$ and is an unstable node in $b \in (-0.8531, -0.8088)$ and $(-0.4384, -1.0923)$. As in positive mapping, the values of parameter b for stability ranges are listed in Table 4.3.

From the analytical prediction, the following statements are verified.

(i) The stable periodic solution of positive mapping P_+ is the unstable periodic solution of negative mapping P_- with all eigenvalues outside the unit cycle.
(ii) The stable periodic solution of negative mapping P_- is the unstable periodic solution of positive mapping P_+ with all eigenvalues outside the unit cycle.
(iii) The PD and SN bifurcations of the periodic solutions of positive mapping P_+ are the UPD and USN bifurcations of the periodic solutions of negative mapping P_-, vice versa.
(iv) The PD and SN bifurcations of the periodic solutions of negative mapping P_- are the UPD and USN bifurcations of the periodic solutions of positive mapping P_+, vice versa.
(v) If the unstable periodic solutions of positive mapping P_+ are saddle, the corresponding periodic solutions of negative mapping P_- are also saddle.

Table 4.4 Input data for Poincare mappings of period-1 at the Neimark bifurcation ($a = 1.1$ and $b = -1.0$)

(x_k, y_k)	(x_k, y_k)
(0.4083,–0.4083)	(0.4683,–0.4083)
(0.4283,–0.4083)	(0.4783,–0.4083)
(0.4383,–0.4083)	(0.4883,–0.4083)
(0.4483,–0.4083)	(0.4983,–0.4083)
(0.4583,–0.4083)	(0.5131,–0.4083)

Table 4.5 Input data for Poincare mappings of period-3 at the Neimark bifurcation ($a = 1.1$ and $b = -1$)

(x_k, y_k)	(x_k, y_k)
(1.1966,0.2877)	(1.2267,0.2877)
(1.2067,0.2877)	(1.2367,0.2877)
(1.2167,0.2877)	(1.2413,0.2877)

From the analytical prediction, the parameter maps of both the positive and negative mappings are developed. An overall view of the parameter map is given in Fig. 4.5a. The corresponding periodic solutions are labeled by mapping structures. "None" represents no periodic solutions exists, which means the solution goes to infinity. "Chaos" gives the regions for chaotic solutions. The existing theory can only give the periodic solutions relative to the positive mapping. The coexistence of the periodic solutions is observed. The unstable periodic solutions with saddle will not be presented. The positive and negative mappings are separated by the two Neimark bifurcations at $b = \pm 1$. The zoomed views of the parameter map for periodic solutions of $P_+^{(5)}$ and $P_+^{(7)}$ are presented in Figs. 4.5b, c for better illustration, respectively. The Neimark bifurcation of the periodic solution is relative to the unstable and stable focuses, which is presented for a better understanding of the solution switching from positive to negative mappings.

The Poincare mapping relative to the Neimark bifurcation of positive (or negative) mapping at $a = 1.1$ and $b = -1$ is presented in Fig. 4.6. Two Neimark bifurcations coexist with different initial conditions. The Neimark bifurcation of period-1 solution is presented in Fig. 4.6a, and the initial values of (x_k, y_k) are tabulated in Table 4.4. The most inside point $(x_k^*, y_k^*) \approx (0.4083, -0.4083)$ is the point for the period-1 solution of P_+ or P_- relative to the Neimark bifurcation. The most outside curve with the initial condition $(x_k^*, y_k^*) \approx (0.5131, -0.4083)$ is the separatrix for the strange attractors around the period-1 solutions with the Neimark bifurcation. The Neimark bifurcation of period-3 solution is presented in Fig. 4.6b. The initial conditions are listed in Table 4.5. For this case, there are three portions of the strange attractor. The most inside points are $(x_k^*, y_k^*) \approx (-0.2877, -1.1967)$, $(-0.2877, 0.2877)$ and $(1.1966, 0.2877)$ for the period-3 solution of P_+ or P_- relative to the Neimark bifurcation. The initial condition for three portions of the strange attractor is $(x_k^*, y_k^*) \approx (1.2067, 0.2877)$.

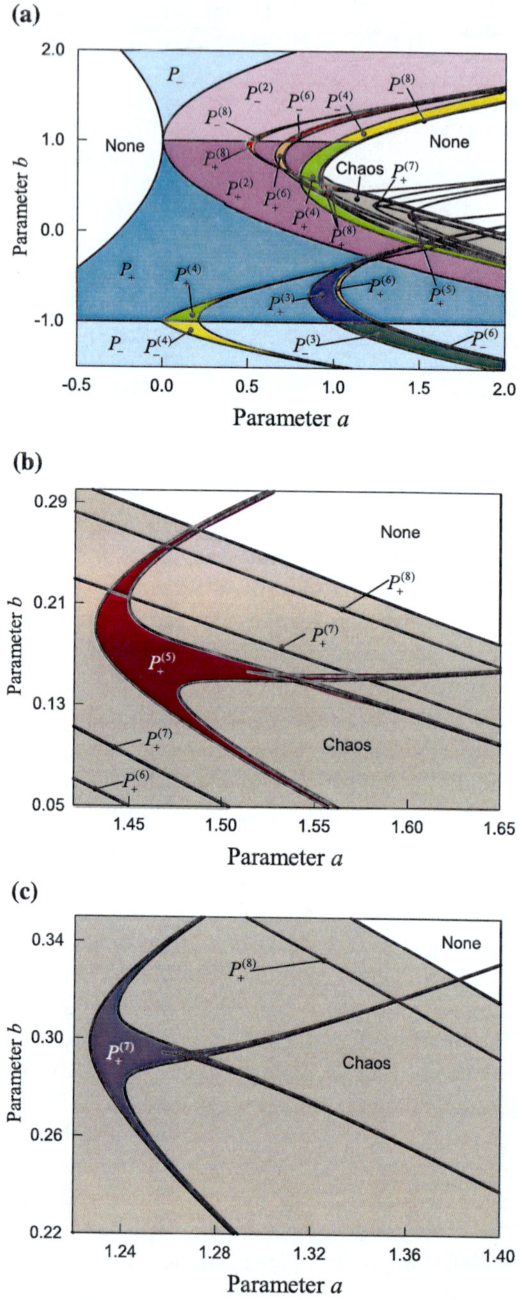

Fig. 4.5 Parameter map of (a,b): (**a**) global view, (**b**) zoomed view for periodic solution of $P_+^{(5)}$ and (**c**) zoomed view for periodic solution of $P_+^{(7)}$

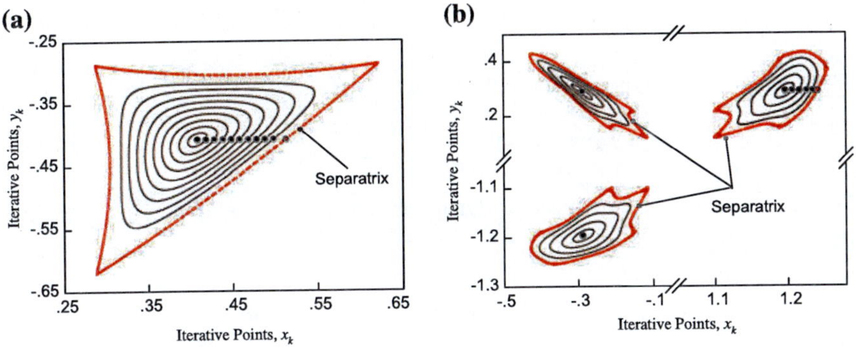

Fig. 4.6 Poincare mappings at the Neimark bifurcation of period-1 and period-3 solution of the Henon map (i.e., $P_+ - 1$ or $P_- - 1$, $P_+^{(3)} - 1$ or $P_-^{(3)} - 1$). (**a**) Neimark bifurcation of period-1 solution, (**b**) Neimark bifurcation of period-3 solution. ($a = 1.1$ and $b = -1$)

4.4 Companion and Synchronization

This section will extend the concepts presented in the previous section. The companion and synchronization of two discrete dynamical systems will be presented.

Definition 4.10 Consider the αth implicit vector function $\mathbf{f}^{(\alpha)} : D \to D\,(\alpha = 1, 2, \cdots, N)$ on an open set $D \subset \mathscr{R}^n$ in an n-dimensional discrete dynamical system. For $\mathbf{x}_k, \mathbf{x}_{k+1} \in D$, there is a discrete relation as

$$\mathbf{f}^{(\alpha)}(\mathbf{x}_k, \mathbf{x}_{k+1}, \mathbf{p}^{(\alpha)}) = \mathbf{0} \tag{4.48}$$

where the vector function is $\mathbf{f}^{(\alpha)} = (f_1^{(\alpha)}, f_2^{(\alpha)}, \cdots, f_n^{(\alpha)})^{\mathrm{T}} \in \mathscr{R}^n$ and discrete variable vector is $\mathbf{x}_k = (x_{k1}, x_{k2}, \cdots, x_{kn})^{\mathrm{T}} \in D$ with the corresponding parameter vector $\mathbf{p}^{(\alpha)} = (p_1^{(\alpha)}, p_2^{(\alpha)}, \cdots, p_{m_\alpha}^{(\alpha)})^{\mathrm{T}} \in \mathscr{R}^{m_\alpha}$.

Similarly, the discrete sets, positive and negative mappings for discrete dynamical system of $\mathbf{f}^{(\alpha)}(\mathbf{x}_k, \mathbf{x}_{k+1}, \mathbf{p}^{(\alpha)}) = \mathbf{0}$ in Eq. (4.48) are defined.

Definition 4.11 For a discrete dynamical system in Eq. (4.48), the positive and negative discrete sets are defined by

$$\left.\begin{array}{l} \Sigma_+^{(\alpha)} = \{\mathbf{x}_{k+i}^{(\alpha)} | \mathbf{x}_{k+i}^{(\alpha)} \in \mathscr{R}^n, i \in \mathbb{Z}_+\} \subset D \text{ and} \\ \Sigma_-^{(\alpha)} = \{\mathbf{x}_{k-i}^{(\alpha)} | \mathbf{x}_{k-i}^{(\alpha)} \in \mathscr{R}^n, i \in \mathbb{Z}_+\} \subset D \end{array}\right\} \tag{4.49}$$

respectively. The corresponding discrete set is

$$\Sigma^{(\alpha)} = \Sigma_+^{(\alpha)} \cup \Sigma_-^{(\alpha)}. \tag{4.50}$$

A positive mapping for discrete dynamical system is defined as

$$P_{\alpha+}:\Sigma^{(\alpha)}\to\Sigma_{+}^{(\alpha)}\quad\Rightarrow\quad P_{\alpha+}:\mathbf{x}_{k}^{(\alpha)}\to\mathbf{x}_{k+1}^{(\alpha)}\tag{4.51}$$

and a negative mapping is defined by

$$P_{\alpha-}:\Sigma^{(\alpha)}\to\Sigma_{-}^{(\alpha)}\quad\Rightarrow\quad P_{\alpha-}:\mathbf{x}_{k}^{(\alpha)}\to\mathbf{x}_{k-1}^{(\alpha)}.\tag{4.52}$$

Definition 4.12 For two discrete dynamical systems in Eq. (4.48), consider two points $\mathbf{x}_k^{(\alpha)},\mathbf{x}_k^{(\beta)}\in D$ and $\mathbf{x}_{k+1}^{(\alpha)},\mathbf{x}_{k+1}^{(\beta)}\in D$, and there is a specific, differentiable, vector function $\boldsymbol{\varphi}=(\varphi_1,\varphi_2,\cdots,\varphi_l)^{\mathrm{T}}\in\mathscr{R}^l$. For a small number $\varepsilon_k>0$, there is a small number $\varepsilon_{k+1}>0$. Suppose there are two subdomains $\mathrm{U}_k^{(\alpha)}\subset D$ and $\mathrm{U}_k^{(\beta)}\subset D$, then for $\mathbf{x}_k^{(\alpha)}\in\mathrm{U}_k^{(\alpha)}$ and $\mathbf{x}_k^{(\beta)}\in\mathrm{U}_k^{(\beta)}$,

$$||\boldsymbol{\varphi}(\mathbf{x}_k^{(\alpha)},\mathbf{x}_k^{(\beta)},\boldsymbol{\lambda})||\le\varepsilon_k.\tag{4.53}$$

(i) For $\varepsilon_{k+1}>0$, there are two subdomains $\mathrm{U}_{k+1}^{(\alpha)}\subset D$ and $\mathrm{U}_{k+1}^{(\beta)}\subset D$. If for $\mathbf{x}_{k+1}^{(\alpha)}\in\mathrm{U}_{k+1}^{(\alpha)}$ and$\mathbf{x}_{k+1}^{(\beta)}\in\mathrm{U}_{k+1}^{(\beta)}$

$$||\boldsymbol{\varphi}(\mathbf{x}_{k+1}^{(\alpha)},\mathbf{x}_{k+1}^{(\beta)},\boldsymbol{\lambda})||\le\varepsilon_{k+1},\tag{4.54}$$

then, the discrete dynamical systems of $\mathbf{f}^{(\alpha)}$ and $\mathbf{f}^{(\beta)}$ are called the companion in sense of $\boldsymbol{\varphi}$ during the kth and $(k+1)$th iteration.

(i_a) The discrete dynamical systems of $\mathbf{f}^{(\alpha)}$ and $\mathbf{f}^{(\beta)}$ are called the finite companion if for $\mathbf{x}_{k+j}^{(\alpha)}\in\mathrm{U}_{k+j}^{(\alpha)}\subset D$ and $\mathbf{x}_{k+j}^{(\beta)}\in\mathrm{U}_{k+j}^{(\beta)}\subset D$

$$||\boldsymbol{\varphi}(\mathbf{x}_{k+j}^{(\alpha)},\mathbf{x}_{k+j}^{(\beta)},\boldsymbol{\lambda})||\le\varepsilon_{k+j}\text{ for }j=1,2,\cdots,N.\tag{4.55}$$

(i_b) The discrete dynamical systems of $\mathbf{f}^{(\alpha)}$ and $\mathbf{f}^{(\beta)}$ are called the absolute permanent companion if $\mathbf{x}_{k+j}^{(\alpha)}\in\mathrm{U}_{k+j}^{(\alpha,\varepsilon)}\subset D$ and $\mathbf{x}_{k+j}^{(\beta)}\in\mathrm{U}_{k+j}^{(\beta,\varepsilon)}\subset D$

$$||\boldsymbol{\varphi}(\mathbf{x}_{k+j}^{(\alpha)},\mathbf{x}_{k+j}^{(\beta)},\boldsymbol{\lambda})||\le\varepsilon_{k+j}\text{ for }j=1,2,\cdots.\tag{4.56}$$

(i_c) The discrete dynamical systems of $\mathbf{f}^{(\alpha)}$ and $\mathbf{f}^{(\beta)}$ are called the repeatable finite companion if $\mathbf{x}_{k+jN(-)}^{(\alpha)}\in\mathrm{U}_{k+jN(-)}^{(\alpha)}\subset D$ and $\mathbf{x}_{k+j(-)}^{(\beta)}\in\mathrm{U}_{k+j(-)}^{(\beta,\varepsilon)}\subset D$

$$\begin{gathered}\Delta\mathbf{I}^{(\alpha)}:\mathbf{x}_{k+jN(-)}^{(\alpha)}\to\mathbf{x}_{k+jN(+)}^{(\alpha)},\text{ and }\Delta\mathbf{I}^{(\beta)}:\mathbf{x}_{k+jN(-)}^{(\beta)}\to\mathbf{x}_{k+jN(+)}^{(\beta)},\\ \mathbf{x}_{k+jN(+)}^{(\alpha)}=\mathbf{x}_{k+jN(-)}^{(\alpha)}+\Delta\mathbf{I}_{jN}^{(\alpha)}\text{ and }\mathbf{x}_{k+jN(+)}^{(\beta)}=\mathbf{x}_{k+jN(-)}^{(\beta)}+\Delta\mathbf{I}_{jN}^{(\beta)};\\ ||\boldsymbol{\varphi}(\mathbf{x}_{k+j(+)}^{(\alpha)},\mathbf{x}_{k+j(+)}^{(\beta)},\boldsymbol{\lambda})||\le\varepsilon_{k+\mathrm{mod(j,N)}}\text{ for }j=1,2,\cdots,\\ \text{with }\mathbf{x}_{k+j(+)}^{(\alpha)}\in\mathrm{U}_{k+\mathrm{mod}(j,N)}^{(\alpha)}\text{ and }\mathbf{x}_{k+j(+)}^{(\beta)}\in\mathrm{U}_{k+\mathrm{mod}(j,N)}^{(\beta)}.\end{gathered}\tag{4.57}$$

(ii) For $\varepsilon_k > 0$, $\varepsilon_{k+(N_\alpha:N_\beta)} > 0$ there are there are two subdomains $\mathrm{U}^{(\alpha)}_{k+N_\alpha} \subset D$ and $\mathrm{U}^{(\beta)}_{k+N_\beta} \subset D$. If for $\mathbf{x}^{(\alpha)}_{k+N_\alpha} \in \mathrm{U}^{(\alpha)}_{k+N_\alpha}$ and $\mathbf{x}^{(\beta)}_{k+N_\beta} \in \mathrm{U}^{(\beta)}_{k+N_\beta}$ if

$$||\boldsymbol{\varphi}(\mathbf{x}^{(\alpha)}_{k+N_\alpha}, \mathbf{x}^{(\beta)}_{k+N_\beta}, \boldsymbol{\lambda})|| \le \varepsilon_{k+(N_\alpha:N_\beta)}, \tag{4.58}$$

then the discrete dynamical systems of $\mathbf{f}^{(\alpha)}$ from the kth to $(k+N_\alpha)$th iteration and $\mathbf{f}^{(\beta)}$ from the kth to $(k+N_\beta)$th iteration are called the $(N_\alpha : N_\beta)$-companion in sense of $\boldsymbol{\varphi}$.

(ii$_a$) The discrete dynamical systems of $\mathbf{f}^{(\alpha)}$ and $\mathbf{f}^{(\beta)}$ are called the finite $(N_\alpha : N_\beta)$ companion if for $\mathbf{x}^{(\alpha)}_{k+jN_\alpha} \in \mathrm{U}^{(\alpha)}_{k+jN_\alpha} \subset D$ and $\mathbf{x}^{(\beta)}_{k+jN_\beta} \in \mathrm{U}^{(\beta)}_{k+jN_\beta} \subset D$

$$||\boldsymbol{\varphi}(\mathbf{x}^{(\alpha)}_{k+jN_\alpha}, \mathbf{x}^{(\beta)}_{k+jN_\beta}, \boldsymbol{\lambda})|| \le \varepsilon_{k+j(N_\alpha:N_\beta)} \text{ for } j = 1, 2, \cdots, N. \tag{4.59}$$

(ii$_b$) The discrete dynamical systems of $\mathbf{f}^{(\alpha)}$ and $\mathbf{f}^{(\beta)}$ are called the absolute permanent $(N_\alpha : N_\beta)$ companion if $\mathbf{x}^{(\alpha)}_{k+jN_\alpha} \in \mathrm{U}^{(\alpha)}_{k+jN_\alpha} \subset D$ and $\mathbf{x}^{(\beta)}_{k+jN_\beta} \in \mathrm{U}^{(\beta)}_{k+jN_\beta} \subset D$

$$||\boldsymbol{\varphi}(\mathbf{x}^{(\alpha)}_{k+jN_\alpha}, \mathbf{x}^{(\beta)}_{k+jN_\beta}, \boldsymbol{\lambda})|| \le \varepsilon_{k+j(N_\alpha:N_\beta)} \text{ for } j = 1, 2, \cdots. \tag{4.60}$$

(ii$_c$) The discrete dynamical systems of $\mathbf{f}^{(\alpha)}$ and $\mathbf{f}^{(\beta)}$ are called the repeatable finite $(N_\alpha : N_\beta)$ companion if $\mathbf{x}^{(\alpha)}_{k+jN_\alpha} \in \mathrm{U}^{(\alpha)}_{k+jN_\alpha} \subset D$ and $\mathbf{x}^{(\beta)}_{k+jN_\beta} \in \mathrm{U}^{(\beta)}_{k+jN_\beta} \subset D$

$$\begin{gathered}
\Delta\mathbf{I}^{(\alpha)} : \mathbf{x}^{(\alpha)}_{k+jN_\alpha(-)} \to \mathbf{x}^{(\alpha)}_{k+jN_\alpha(+)}, \text{ and } \Delta\mathbf{I}^{(\beta)} : \mathbf{x}^{(\beta)}_{k+jN_\beta(-)} \to \mathbf{x}^{(\beta)}_{k+jN_\beta(+)} \\
\mathbf{x}^{(\alpha)}_{k+jN_\alpha(+)} = \mathbf{x}^{(\alpha)}_{k+jN_\alpha(-)} + \Delta\mathbf{I}^{(\alpha)}_{jN_\alpha} \text{ and } \mathbf{x}^{(\beta)}_{k+jN_\beta(+)} = \mathbf{x}^{(\beta)}_{k+jN_\beta(-)} + \Delta\mathbf{I}^{(\beta)}_{jN_\beta} \\
||\boldsymbol{\varphi}(\mathbf{x}^{(\alpha)}_{k+jN_\alpha(+)}, \mathbf{x}^{(\beta)}_{k+jN_\beta(+)}, \boldsymbol{\lambda})|| \le \varepsilon_{k+\bmod(j,N)(N_\alpha:N_\beta)} \text{ for } j = 1, 2, \cdots, \\
\mathbf{x}^{(\alpha)}_{k+jN_\alpha(+)} \in \mathrm{U}^{(\alpha)}_{k+\bmod(j,N)N_\alpha} \text{ and } \mathbf{x}^{(\beta)}_{k+jN_\beta(+)} \in \mathrm{U}^{(\beta)}_{k+\bmod(j,N)N_\beta}.
\end{gathered} \tag{4.61}$$

Definition 4.13 For two discrete dynamical systems in Eq. (4.48), consider two points $\mathbf{x}^{(\alpha)}_k, \mathbf{x}^{(\beta)}_k \in D$ and $\mathbf{x}^{(\alpha)}_{k+1}, \mathbf{x}^{(\beta)}_{k+1} \in D$, and there is a specific, differentiable, vector function $\boldsymbol{\varphi} = (\varphi_1, \varphi_2, \cdots, \varphi_l)^{\mathrm{T}} \in \mathscr{R}^l$. For

$$\boldsymbol{\varphi}(\mathbf{x}^{(\alpha)}_k, \mathbf{x}^{(\beta)}_k, \boldsymbol{\lambda}) = \mathbf{0}. \tag{4.62}$$

(i) If

$$\boldsymbol{\varphi}(\mathbf{x}^{(\alpha)}_{k+1}, \mathbf{x}^{(\beta)}_{k+1}, \boldsymbol{\lambda}) = \mathbf{0}, \tag{4.63}$$

then, discrete dynamical systems of $\mathbf{f}^{(\alpha)}$ and $\mathbf{f}^{(\beta)}$ are called the (1:1) synchronization in sense of $\boldsymbol{\varphi}$;

(ii) If

$$\begin{aligned}
&\boldsymbol{\varphi}(\mathbf{x}_{k+1}^{(\alpha)}, \mathbf{x}_{k+1}^{(\beta)}, \boldsymbol{\lambda}) = \mathbf{0},\\
&\Delta\mathbf{I}^{(\alpha)} : \mathbf{x}_{k+1(-)}^{(\alpha)} \to \mathbf{x}_{k+1(+)}^{(\alpha)} \text{ and } \Delta\mathbf{I}^{(\beta)} : \mathbf{x}_{k+1(-)}^{(\beta)} \to \mathbf{x}_{k+1(+)}^{(\beta)}\\
&\mathbf{x}_{k+1(+)}^{(\alpha)} = \mathbf{x}_{k+1(-)}^{(\alpha)} + \Delta\mathbf{I}^{(\alpha)} \text{ and } \mathbf{x}_{k+1(+)}^{(\beta)} = \mathbf{x}_{k+1(-)}^{(\beta)} + \Delta\mathbf{I}^{(\beta)}\\
&\mathbf{x}_{k+1(+)}^{(\alpha)} = \mathbf{x}_{k}^{(\alpha)} \text{ and } \mathbf{x}_{k+1(+)}^{(\beta)} = \mathbf{x}_{k}^{(\beta)};
\end{aligned} \tag{4.64}$$

then, discrete dynamical systems of $\mathbf{f}^{(\alpha)}$ and $\mathbf{f}^{(\beta)}$ are called the repeatable (1:1) synchronization in sense of $\boldsymbol{\varphi}$.

(iii) If

$$\boldsymbol{\varphi}(\mathbf{x}_{k+N_\alpha}^{(\alpha)}, \mathbf{x}_{k+N_\beta}^{(\beta)}, \boldsymbol{\lambda}) = \mathbf{0} \tag{4.65}$$

then the discrete dynamical systems of $\mathbf{f}^{(\alpha)}$ and $\mathbf{f}^{(\beta)}$ are called the $(N_\alpha : N_\beta)$-synchronization in sense of $\boldsymbol{\varphi}$.

(iv) If

$$\begin{aligned}
&\boldsymbol{\varphi}(\mathbf{x}_{k+N_\alpha}^{(\alpha)}, \mathbf{x}_{k+N_\beta}^{(\beta)}, \boldsymbol{\lambda}) = \mathbf{0} \text{ with}\\
&\Delta\mathbf{I}^{(\alpha)} : \mathbf{x}_{k+N_\alpha}^{(\alpha)} \to \mathbf{x}_{k}^{(\alpha)} \text{ and } \Delta\mathbf{I}^{(\beta)} : \mathbf{x}_{k+N_\beta}^{(\beta)} \to \mathbf{x}_{k}^{(\beta)}\\
&\mathbf{x}_{k+N_\alpha(+)}^{(\alpha)} = \mathbf{x}_{k+N_\alpha(-)}^{(\alpha)} + \Delta\mathbf{I}^{(\alpha)} \text{ and } \mathbf{x}_{k+N_\beta(+)}^{(\beta)} = \mathbf{x}_{k+N_\beta(-)}^{(\beta)} + \Delta\mathbf{I}^{(\beta)}\\
&\mathbf{x}_{k+N_\beta(+)}^{(\alpha)} = \mathbf{x}_{k}^{(\alpha)} \text{ and } \mathbf{x}_{k+N_\beta(+)}^{(\beta)} = \mathbf{x}_{k}^{(\beta)},
\end{aligned} \tag{4.66}$$

then the discrete dynamical systems of $\mathbf{f}^{(\alpha)}$ and $\mathbf{f}^{(\beta)}$ are called the repeatable $(N_\alpha : N_\beta)$-synchronization in sense of $\boldsymbol{\varphi}$.

From the definition, the companions of two discrete dynamical systems are presented in Figs. 4.7 and 4.8. For each step, if the corresponding relation satisfies Eq. (4.62), the companion is called the (1:1)companion, which is presented in Fig. 4.7. The shaded areas are the companion domain which is controlled by ε_k and $\boldsymbol{\varphi}$. For the repeated companion, for each step, the companion with specific impulses will have the same control domains. Such shaded areas can be overlapped or separated. The $(N_\alpha : N_\beta)$ state for $\mathbf{f}^{(\alpha)}$ with N_α-iterations and $\mathbf{f}^{(\beta)}$ with N_β-iterations satisfy Eq. (4.65) is called the $(N_\alpha : N_\beta)$-companion, which is sketched in Fig. 4.8a. This companion does not require each iteration step to do so. The companion states are shaded. For the repeated companion, the companion state with specific impulses will have the same control domains. Similarly, the companion for negative maps can be defined, as shown in Fig. 4.8b.

Consider synchronization of two discrete dynamical systems, as shown in Fig. 4.9, with

$$\mathbf{f}^{(\alpha)}(\mathbf{x}_{k+1}, \mathbf{x}_k, \mathbf{p}^{(\alpha)}) = \mathbf{0} \text{ and } \mathbf{f}^{(\beta)}(\mathbf{y}_{k+1}, \mathbf{y}_k, \mathbf{p}^{(\beta)}) = \mathbf{0}. \tag{4.67}$$

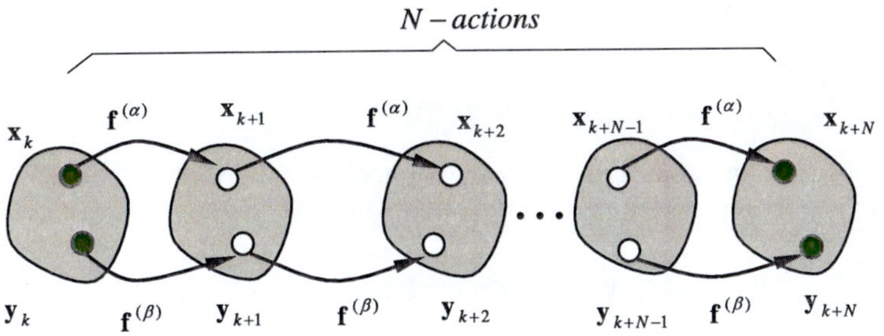

Fig. 4.7 Companion of two discrete dynamical systems

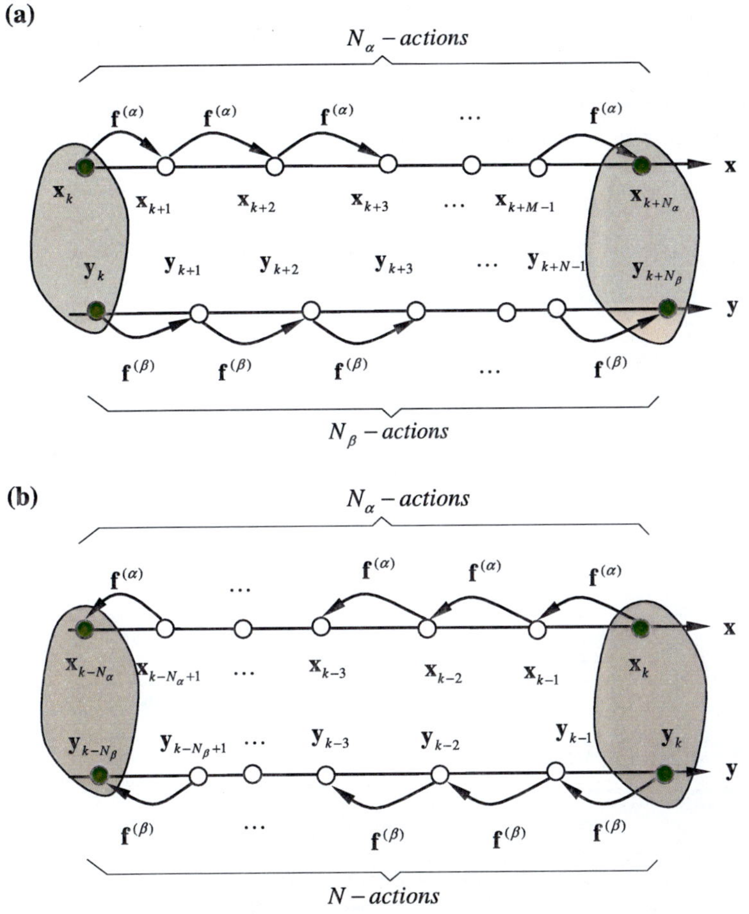

Fig. 4.8 Companion of two discrete nonlinear systems: **a** positive companion and **b** negative companion

For the initial state, there is a relation as

$$\boldsymbol{\varphi}(\mathbf{x}_k, \mathbf{y}_k, \boldsymbol{\lambda}) = \mathbf{0}, \tag{4.68}$$

For the positive synchronization, there are N_α-actions with function $\mathbf{f}^{(\alpha)}$ and mapping $P_{\alpha+}$ and N_β-actions with function $\mathbf{f}^{(\beta)}$ and mapping $P_{\beta+}$

$$\begin{aligned} \mathbf{f}^{(\alpha)}(\mathbf{x}_{k+i}, \mathbf{x}_{k+i-1}, \mathbf{p}^{(\alpha)}) &= \mathbf{0} \text{ for } i = 1, 2, \cdots, N_\alpha \\ \mathbf{f}^{(\beta)}(\mathbf{y}_{k+j}, \mathbf{y}_{k+j-1}, \mathbf{p}^{(\beta)}) &= \mathbf{0} \text{ for } j = 1, 2, \cdots, N_\beta \end{aligned} \tag{4.69}$$

and the synchronization is based on

$$\boldsymbol{\varphi}(\mathbf{x}_{k+N_\alpha}, \mathbf{y}_{k+N_\beta}, \boldsymbol{\lambda}) = \mathbf{0}. \tag{4.70}$$

For the negative synchronization, there are N_α-actions with function $\mathbf{f}^{(\alpha)}$ and mapping $P_{\alpha-}$ and N_β-actions with function $\mathbf{f}^{(\beta)}$ and mapping $P_{\beta-}$

$$\begin{aligned} \mathbf{f}^{(\alpha)}(\mathbf{x}_{k-i+1}, \mathbf{x}_{k-i-1}, \mathbf{p}^{(\alpha)}) &= \mathbf{0} \text{ for } i = 1, 2, \cdots, N_\alpha \\ \mathbf{f}^{(\beta)}(\mathbf{y}_{k-j}, \mathbf{y}_{k-j-1}, \mathbf{p}^{(\beta)}) &= \mathbf{0} \text{ for } j = 1, 2, \cdots, N_\beta \end{aligned} \tag{4.71}$$

and the synchronization is based on

$$\boldsymbol{\varphi}(\mathbf{x}_{k-N_\alpha}, \mathbf{y}_{k-N_\beta}, \boldsymbol{\lambda}) = \mathbf{0}. \tag{4.72}$$

Thus there is a relation

$$\mathbf{x}_k = P_{\varphi-} \circ P_{\beta-}^{(N_\beta)} \circ P_{\varphi+} \circ P_{\alpha+}^{(N_\alpha)} \mathbf{x}_k \tag{4.73}$$

where

$$\begin{aligned} & P_{\varphi-} \circ P_{\beta-}^{(N_\beta)} \circ P_{\varphi+} \circ P_{\alpha+}^{(N_\alpha)} \\ & = P_{\varphi-} \circ \underbrace{P_{\beta-} \circ P_{\beta-} \circ \cdots \circ P_{\beta-}}_{N_\beta-actions} \circ P_{\varphi+} \circ \underbrace{P_{\alpha+} \circ P_{\alpha+} \circ \cdots \circ P_{\alpha+}}_{N_\alpha-actions}. \end{aligned} \tag{4.74}$$

From Eq. (4.73), we have

$$\begin{aligned} & \mathbf{x}_{k+N_\alpha} = P_{\alpha+}^{(N_\alpha)} \mathbf{x}_k \text{ and } \mathbf{y}_{k+N_\beta} = P_\varphi \mathbf{x}_{k+N_\alpha} \\ & \mathbf{y}_k = P_{\beta-}^{(N_\beta)} \mathbf{y}_{k+N_\beta} \text{ and } \mathbf{x}_k = P_{\varphi-} \mathbf{y}_k \end{aligned} \tag{4.75}$$

and

$$\begin{aligned} & \mathbf{x}_{k+N_\alpha} = P_{\alpha+}^{(N_\alpha)} \mathbf{x}_k \text{ and } \mathbf{y}_{k+N_\beta} = P_{\varphi+} \mathbf{x}_{k+N_\alpha} \\ & P_{\varphi+} \mathbf{x}_k = \mathbf{y}_k \text{ and } P_{\beta+}^{(N_\beta)} \mathbf{y}_k = \mathbf{y}_{k+N_\beta}. \end{aligned} \tag{4.76}$$

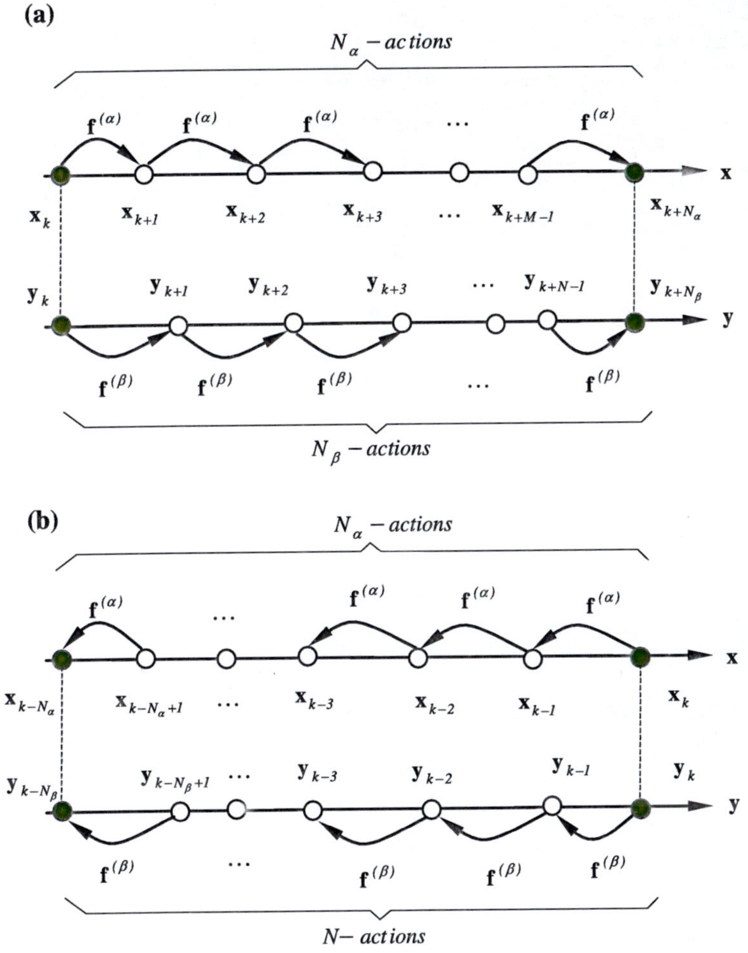

Fig. 4.9 Synchronization of two discrete nonlinear systems: **a** positive synchronization and **b** negative synchronization

The corresponding commutative diagram is given in Fig. 4.10. The solid and dashed arrows give the positive and negative mappings, respectively.

From the above discussion on synchronization of $P_{\alpha+}^{(N_\alpha)}$ and $P_{\beta+}^{(N_\beta)}$ under the constraint $\boldsymbol{\varphi}$, the following relations should exist

$$\begin{aligned} \mathbf{x}'_k &= P_{\boldsymbol{\varphi}-} \circ P_{\beta-}^{(N_\beta)} \circ P_{\boldsymbol{\varphi}+} \circ P_{\alpha+}^{(N_\alpha)} \mathbf{x}_k, \text{ or} \\ \mathbf{x}'_k &= P_{\alpha-}^{(N_\alpha)} \circ P_{\boldsymbol{\varphi}-} \circ P_{\beta+}^{(N_\beta)} \circ P_{\boldsymbol{\varphi}+} \mathbf{x}_k. \end{aligned} \tag{4.77}$$

The above equation forms an iterative mapping. If the fixed point exists, i.e.,

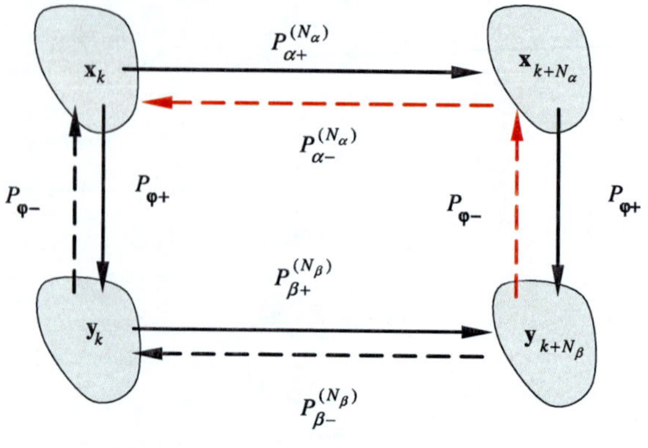

Fig. 4.10 Commutative mapping diagram for synchronization

$$\mathbf{x}'_k = \mathbf{x}_k \tag{4.78}$$

then the synchronization of $P_{\alpha+}^{(N_\alpha)}$ and $P_{\beta-}^{(N_\beta)}$ under the constraint $\boldsymbol{\varphi}$ exists

$$\begin{aligned} &\mathbf{x}_{k+N_\alpha} = P_{\alpha+}^{(N_\alpha)}\mathbf{x}_k, \text{ and } \mathbf{y}_{k+N_\beta} = P_{\beta+}^{(N_\beta)}\mathbf{y}_k; \\ &\mathbf{y}_k = P_{\boldsymbol{\varphi}}\mathbf{x}_k \text{ and } \mathbf{y}_{k+N_\beta} = P_{\boldsymbol{\varphi}}\mathbf{x}_{k+N_\alpha}. \end{aligned} \tag{4.79}$$

Theorem 4.6 *Consider two discrete dynamical systems* $(P_\alpha, \mathbf{f}^{(\alpha)})$ *and* $(P_\beta, \mathbf{f}^{(\beta)})$ *as in Eq.* (4.48) *with*

$$\begin{aligned} &P_{\alpha+} : \mathbf{x}_k \to \mathbf{x}_{k+1} \text{ and } P_{\alpha-} : \mathbf{x}_{k+1} \to \mathbf{x}_k \\ &\mathbf{f}^{(\alpha)}(\mathbf{x}_k, \mathbf{x}_{k+1}, \mathbf{p}^{(\alpha)}) = \mathbf{0} \end{aligned} \tag{4.80}$$

and

$$\begin{aligned} &P_{\beta+} : \mathbf{y}_k \to \mathbf{y}_{k+1} \text{ and } P_{\alpha-} : \mathbf{y}_{k+1} \to \mathbf{y}_k \\ &\mathbf{f}^{(\beta)}(\mathbf{y}_k, \mathbf{y}_{k+1}, \mathbf{p}^{(\beta)}) = \mathbf{0}. \end{aligned} \tag{4.81}$$

For two points $\mathbf{x}_k \in D_\alpha$ *and* $\mathbf{y}_k \in D_\beta$*, there is a specific, differentiable, vector function* $\boldsymbol{\varphi} = (\varphi_1, \varphi_2, \cdots, \varphi_l)^{\mathrm{T}} \in \mathscr{R}^l$. *The synchronization of two discrete dynamical systems* $(P_\alpha, \mathbf{f}^{(\alpha)})$ *and* $(P_\beta, \mathbf{f}^{(\beta)})$ *is under the following constraints*

$$\boldsymbol{\varphi}(\mathbf{x}_k, \mathbf{y}_k, \boldsymbol{\lambda}) = \mathbf{0} \text{ and } \boldsymbol{\varphi}(\mathbf{x}_{k+1}, \mathbf{y}_{k+1}, \boldsymbol{\lambda}) = \mathbf{0}. \tag{4.82}$$

Consider a resultant hybrid mapping relation as

$$\mathbf{x}'_k = P\mathbf{x}_k = P_{\varphi-} \circ P_{\beta-} \circ P_{\varphi+} \circ P_{\alpha+}\mathbf{x}_k \tag{4.83}$$

with

$$
\begin{aligned}
&P_{\alpha+} : \mathbf{x}_k \to \mathbf{x}_{k+1} \text{ with } \mathbf{f}^{(\alpha)}(\mathbf{x}_{k+1}, \mathbf{x}_k, \mathbf{p}^{(\alpha)}) = \mathbf{0}, \\
&P_{\varphi+} : \mathbf{x}_{k+1} \to \mathbf{y}_{k+1} \text{ with } \boldsymbol{\varphi}(\mathbf{x}_{k+1}, \mathbf{y}_{k+1}, \boldsymbol{\lambda}) = \mathbf{0}, \\
&P_{\beta-} : \mathbf{y}_{k+1} \to \mathbf{y}_k \text{ with } \mathbf{f}^{(\beta)}(\mathbf{y}_{k+1}, \mathbf{y}_k, \mathbf{p}^{(\alpha)}) = \mathbf{0}, \\
&P_{\varphi-} : \mathbf{y}_k \to \mathbf{x}'_k \text{ with } \boldsymbol{\varphi}(\mathbf{x}'_k, \mathbf{y}_k, \boldsymbol{\lambda}) = \mathbf{0}; \\
&\mathbf{x}'_k = \mathbf{x}_k
\end{aligned}
\tag{4.84}
$$

and

$$
DP(\mathbf{x}_k^*) = DP_{\varphi-}(\mathbf{y}_k^*) \cdot DP_{\beta-}(\mathbf{y}_{k+1}^*) \cdot DP_{\varphi+}(\mathbf{x}_{k+1}^*) \cdot DP_{\alpha+}(\mathbf{x}_k^*) \tag{4.85}
$$

where

$$
\begin{aligned}
&DP(\mathbf{x}_k^*) = \left[\frac{\partial \mathbf{x}'_k}{\partial \mathbf{x}_k}\right]_{\mathbf{x}_k^*}, \; DP_{\alpha+}(\mathbf{x}_k^*) = \left[\frac{\partial \mathbf{x}_{k+1}}{\partial \mathbf{x}_k}\right]_{\mathbf{x}_k^*}, \; DP_{\varphi+}(\mathbf{x}_{k+1}^*) = \left[\frac{\partial \mathbf{y}_{k+1}}{\partial \mathbf{x}_{k+1}}\right]_{\mathbf{x}_{k+1}^*}, \\
&DP_{\beta-}(\mathbf{y}_{k+1}^*) = \left[\frac{\partial \mathbf{y}_k}{\partial \mathbf{y}_{k+1}}\right]_{\mathbf{y}_{k+1}^*}, \; DP_{\varphi-}(\mathbf{y}_k^*) = \left[\frac{\partial \mathbf{x}'_k}{\partial \mathbf{y}_k}\right]_{\mathbf{y}_k^*}.
\end{aligned}
\tag{4.86}
$$

(i) The (1:1) synchronization of two discrete dynamical systems of $(P_\alpha, \mathbf{f}^{(\alpha)})$ and $(P_\beta, \mathbf{f}^{(\beta)})$ is persistent if and only if all the eigenvalues λ_i $(i = 1, 2, \cdots, n)$ of $DP(\mathbf{x}_k^*)$ lie in the unit circles, i.e.,

$$
|\lambda_i| < 1 \text{ for } i = 1, 2, \cdots, n. \tag{4.87}
$$

(ii) The (1:1) synchronization of two discrete dynamical systems of $(P_\alpha, \mathbf{f}^{(\alpha)})$ and $(P_\beta, \mathbf{f}^{(\beta)})$ is a saddle-node vanishing if and only if at least one of the real eigenvalues λ_i $(i = 1, 2, \cdots, n_1$ and $n_1 \le n)$ of $DP(\mathbf{x}_k^*)$ is positive one (+1) and the other eigenvalues are in the unite circle, i.e.,

$$
\lambda_i = 1 \text{ and } |\lambda_j| < 1 \text{ for } i, j \in \{1, 2, \cdots, n\} \text{ and } j \ne i. \tag{4.88}
$$

(iii) The (1:1) synchronization of two discrete dynamical systems of $(P_\alpha, \mathbf{f}^{(\alpha)})$ and $(P_\beta, \mathbf{f}^{(\beta)})$ is a period-doubling vanishing if and only if at least one of the real eigenvalues λ_i $(i = 1, 2, \cdots, n_1$ and $n_1 \le n)$ of $DP(\mathbf{x}_k^*)$ is negative one (-1) and the other eigenvalues are in the unite circle, i.e.,

$$
\lambda_i = -1 \text{ and } |\lambda_j| < 1 \text{ for } i, j \in \{1, 2, \cdots, n\} \text{ and } j \ne i. \tag{4.89}
$$

(iv) The (1:1) synchronization of two discrete dynamical systems of $(P_\alpha, \mathbf{f}^{(\alpha)})$ and $(P_\beta, \mathbf{f}^{(\beta)})$ is a Neimark vanishing if and only if one pair of all the complex eigenvalues $\lambda_i = \alpha_i \pm \beta_i \mathbf{i}$ $(i = 1, 2, \cdots, n_1$ and $n_1 \le n/2)$ of $DP(\mathbf{x}_k^*)$ are on the unit circle and the other eigenvalues are in the unite circle, i.e.,

$$
|\lambda_i| = \sqrt{\alpha_i^2 + \beta_i^2} = 1 \text{ and } |\lambda_j| < 1 \text{ for } i, j \in \{1, 2, \cdots, n\} \text{ and } j \ne i. \tag{4.90}
$$

(v) The (1:1) synchronization of two discrete dynamical systems of $(P_\alpha, \mathbf{f}^{(\alpha)})$ and $(P_\beta, \mathbf{f}^{(\beta)})$ is a $(l_1 : l_2 : l_3)$ vanishing if and only if l_1 and l_2 real eigenvalues λ_i of $DP(\mathbf{x}_k^*)$ are negative one (-1) and positive one $(+1)$, respectively, and l_3-pairs of complex eigenvalues $\lambda_i = \alpha_i \pm \beta_i \mathbf{i}$ $(i = 1, 2, \cdots, n_1$ and $n_1 \leq n/2)$ of $DP(\mathbf{x}_k^*)$ are on the unit circle and the other eigenvalues are in the unite circle, i.e.,

$$\begin{aligned}
&\lambda_i = -1 \text{ for } i = i_1, i_2, \cdots, i_{l_1} \in \{1, 2, \cdots, n\},\\
&\lambda_j = +1 \text{ for } j = j_1, j_2, \cdots, j_{l_2} \in \{1, 2, \cdots, n\},\\
&|\lambda_r| = \sqrt{\alpha_r^2 + \beta_r^2} = 1 \text{ for } r = r_1, r_2, \cdots, r_{l_3} \in \{1, 2, \cdots, n\},\\
&|\lambda_s| < 1 \text{ for } s \in \{1, 2, \cdots, n\} \text{ and } s \notin \{i, j, r\}.
\end{aligned} \tag{4.91}$$

(vi) The (1:1) synchronization of two discrete dynamical systems of $(P_\alpha, \mathbf{f}^{(\alpha)})$ and $(P_\beta, \mathbf{f}^{(\beta)})$ is instantaneous if and only if at least one of the eigenvalues λ_i $(i = 1, 2, \cdots, n)$ of $DP(\mathbf{x}_k^*)$ lies out of the unit circle, i.e.,

$$|\lambda_i| > 1 \text{ for } i \in \{1, 2, \cdots, n\}. \tag{4.92}$$

Proof From the definition of synchronization, we have

$$\begin{aligned}
&\mathbf{f}^{(\alpha)}(\mathbf{x}_{k+1}, \mathbf{x}_k, \mathbf{p}^{(\alpha)}) = \mathbf{0}, \mathbf{f}^{(\beta)}(\mathbf{y}_{k+1}, \mathbf{y}_k, \mathbf{p}^{(\beta)}) = \mathbf{0};\\
&\boldsymbol{\varphi}(\mathbf{x}_k, \mathbf{y}_k, \boldsymbol{\lambda}) = \mathbf{0}, \boldsymbol{\varphi}(\mathbf{x}_{k+1}, \mathbf{y}_{k+1}, \boldsymbol{\lambda}) = \mathbf{0}.
\end{aligned}$$

Using positive and negative mapping concepts, a resultant mapping in Eq. (4.83) with a relation in Eq.(4.84) can be developed as a hybrid discrete dynamical system with positive and negative mappings, i.e.,

$$\begin{aligned}
&\mathbf{f}^{(\alpha)}(\mathbf{x}_{k+1}, \mathbf{x}_k, \mathbf{p}^{(\alpha)}) = \mathbf{0} \text{ for } P_{\alpha+} : \mathbf{x}_k \to \mathbf{x}_{k+1},\\
&\boldsymbol{\varphi}(\mathbf{x}_{k+1}, \mathbf{y}_{k+1}, \boldsymbol{\lambda}) = \mathbf{0} \text{ for } P_{\varphi+} : \mathbf{x}_{k+1} \to \mathbf{y}_{k+1},\\
&\mathbf{f}^{(\beta)}(\mathbf{y}_{k+1}, \mathbf{y}_k, \mathbf{p}^{(\beta)}) = \mathbf{0} \text{ for } P_{\beta-} : \mathbf{y}_{k+1} \to \mathbf{y}_k,\\
&\boldsymbol{\varphi}(\mathbf{x}'_k, \mathbf{y}_k, \boldsymbol{\lambda}) = \mathbf{0} \text{ for } P_{\varphi-} : \mathbf{y}_k \to \mathbf{x}'_k;\\
&\mathbf{x}'_k = \mathbf{x}_k.
\end{aligned}$$

Thus, two sets of equations are identical to each other. However, the foregoing equation gives an iterative mapping relation as

$$\mathbf{x}'_k = P\mathbf{x}_k = P_{\varphi-} \circ P_{\beta-} \circ P_{\varphi+} \circ P_{\alpha+}\mathbf{x}_k.$$

The fixed point of such a mapping relation yields the solutions of synchronization, and the corresponding stability of the fixed point of the foregoing mapping relation gives the persistence and instant of synchronization. In other words, the eigenvalue analysis of$DP(\mathbf{x}_k^*)$ in Eq. (4.75) gives the stability and bifurcation conditions of the fixed point of the hybrid mapping, which is the persistence, vanishing and instant conditions of the synchronization two discrete dynamical systems of $(P_\alpha, \mathbf{f}^{(\alpha)})$ and $(P_\beta, \mathbf{f}^{(\beta)})$. Thus, statements (i)–(v) can be proved directly. ∎

Theorem 4.7 *Consider two discrete dynamical systems* $(P_\alpha, \mathbf{f}^{(\alpha)})$ *and* $(P_\beta, \mathbf{f}^{(\beta)})$ *as in Eq. (4.48) with*

$$\begin{aligned} &P_{\alpha+} : \mathbf{x}_k \to \mathbf{x}_{k+1} \text{ and } P_{\alpha-} : \mathbf{x}_{k+1} \to \mathbf{x}_k \\ &\mathbf{f}^{(\alpha)}(\mathbf{x}_k, \mathbf{x}_{k+1}, \mathbf{p}^{(\alpha)}) = \mathbf{0} \end{aligned} \tag{4.93}$$

and

$$\begin{aligned} &P_{\beta+} : \mathbf{y}_k \to \mathbf{y}_{k+1} \text{ and } P_{\beta-} : \mathbf{y}_{k+1} \to \mathbf{y}_k \\ &\mathbf{f}^{(\beta)}(\mathbf{y}_k, \mathbf{y}_{k+1}, \mathbf{p}^{(\beta)}) = \mathbf{0} \end{aligned} \tag{4.94}$$

For two points $\mathbf{x}_k \in D_\alpha$ *and* $\mathbf{y}_k \in D_\beta$*, there are two specific, differentiable, vector function* $\boldsymbol{\varphi} = (\varphi_1, \varphi_2, \cdots, \varphi_l)^{\mathrm{T}} \in \mathscr{R}^l$. *The* $(N_\alpha : N_\beta)$*-synchronization of two discrete dynamical systems* $(P_\alpha, \mathbf{f}^{(\alpha)})$ *and* $(P_\beta, \mathbf{f}^{(\beta)})$ *is under the following constraints*

$$\boldsymbol{\varphi}(\mathbf{x}_k, \mathbf{y}_k, \boldsymbol{\lambda}) = \mathbf{0} \text{ and } \boldsymbol{\varphi}(\mathbf{x}_{k+N_\alpha}, \mathbf{y}_{k+N_\beta}, \boldsymbol{\lambda}) = \mathbf{0}. \tag{4.95}$$

Consider a resultant hybrid mapping relation as

$$\mathbf{x}'_k = P^{(N_\alpha:N_\beta)}\mathbf{x}_k = P_{\varphi-} \circ P_{\beta-}^{(N_\beta)} \circ P_\varphi \circ P_{\alpha+}^{(N_\alpha)}\mathbf{x}_k \tag{4.96}$$

with

$$\begin{aligned} &P_{\alpha+}^{(N_\alpha)} : \mathbf{x}_k \to \mathbf{x}_{k+N_\alpha} \text{with} \\ &\left.\begin{aligned} &\mathbf{f}^{(\alpha)}(\mathbf{x}_{k+1}, \mathbf{x}_k, \mathbf{p}^{(\alpha)}) = \mathbf{0} \\ &\mathbf{f}^{(\alpha)}(\mathbf{x}_{k+2}, \mathbf{x}_{k+1}, \mathbf{p}^{(\alpha)}) = \mathbf{0} \\ &\vdots \\ &\mathbf{f}^{(\alpha)}(\mathbf{x}_{k+N_\alpha}, \mathbf{x}_{k+N_\alpha-1}, \mathbf{p}^{(\alpha)}) = \mathbf{0} \end{aligned}\right\}, \\ &P_{\varphi+} : \mathbf{x}_{k+1} \to \mathbf{y}_{k+1} \text{ with } \boldsymbol{\varphi}(\mathbf{x}_{k+N_\alpha}, \mathbf{y}_{k+N_\beta}, \boldsymbol{\lambda}) = \mathbf{0}; \\ &P_{\beta-}^{(N_\beta)} : \mathbf{y}_{k+N_\beta} \to \mathbf{y}_k \text{with} \\ &\left.\begin{aligned} &\mathbf{f}^{(\beta)}(\mathbf{y}_{k+N_\beta}, \mathbf{y}_{k+N_\beta-1}, \mathbf{p}^{(\beta)}) = \mathbf{0} \\ &\vdots \\ &\mathbf{f}^{(\beta)}(\mathbf{y}_{k+2}, \mathbf{y}_{k+1}, \mathbf{p}^{(\beta)}) = \mathbf{0} \\ &\mathbf{f}^{(\beta)}(\mathbf{y}_{k+1}, \mathbf{y}_k, \mathbf{p}^{(\beta)}) = \mathbf{0} \end{aligned}\right\}, \\ &P_{\varphi-} : \mathbf{y}_k \to \mathbf{x}'_k \text{with } \boldsymbol{\varphi}(\mathbf{x}'_k, \mathbf{y}_k, \boldsymbol{\lambda}) = \mathbf{0}; \\ &\mathbf{x}'_k = \mathbf{x}_k \end{aligned} \tag{4.97}$$

and

$$DP^{(N_\alpha:N_\beta)}(\mathbf{x}_k^*) = DP_{\varphi-}(\mathbf{y}_k^*) \cdot DP_{\beta-}^{(N_\beta)}(\mathbf{y}_{k+N_\alpha}^*) \cdot DP_{\varphi+}(\mathbf{x}_{k+N_\alpha}^*) \cdot DP_{\alpha+}^{(N_\alpha)}(\mathbf{x}_k^*) \tag{4.98}$$

where

$$\begin{aligned} DP_{\alpha+}^{(N_\alpha)}(\mathbf{x}_k^*) &= DP_{\alpha+}(\mathbf{x}_{k+N_\alpha-1}^*)\cdot\ldots\cdot DP_{\alpha+}(\mathbf{x}_{k+1}^*)\cdot DP_{\alpha+}(\mathbf{x}_k^*) \\ DP_{\beta-}^{(N_\beta)}(\mathbf{y}_{k+N_\beta}^*) &= DP_{\beta-}(\mathbf{x}_{k+1}^*)\cdot\ldots\cdot DP_{\beta-}(\mathbf{x}_{k+N_\beta-1}^*)\cdot DP_{\beta-}(\mathbf{x}_{k+N_\beta}^*) \end{aligned} \tag{4.99}$$

$$\begin{aligned} &DP^{(N_\alpha:N_\beta)}(\mathbf{x}_k^*) = \left[\frac{\partial \mathbf{x}'_k}{\partial \mathbf{x}_k}\right]_{\mathbf{x}_k^*}, \\ &DP_{\alpha+}^{(N_\alpha)}(\mathbf{x}_k^*) = \prod_{j=N_\alpha}^{1}\left[\frac{\partial \mathbf{x}_{k+j}}{\partial \mathbf{x}_{k+j-1}}\right]_{\mathbf{x}_{k+j-1}^*}, \ DP_{\varphi+}(\mathbf{x}_{k+N_\alpha}^*) = \left[\frac{\partial \mathbf{y}_{k+N_\beta}}{\partial \mathbf{x}_{k+N_\alpha}}\right]_{\mathbf{x}_{k+N_\alpha}^*}, \\ &DP_{\beta-}^{(N_\beta)}(\mathbf{y}_{k+N_\beta}^*) = \prod_{j=1}^{N_\beta}\left[\frac{\partial \mathbf{y}_{k+N_\beta-j}}{\partial \mathbf{y}_{k+N_\beta-j+1}}\right]_{\mathbf{y}_{k+N_\beta-j+1}^*}, \ DP_{\varphi-}(\mathbf{y}_k^*) = \left[\frac{\partial \mathbf{x}'_k}{\partial \mathbf{y}_k}\right]_{\mathbf{y}_k^*}. \end{aligned} \tag{4.100}$$

(i) The $(N_\alpha : N_\beta)$-synchronization of two discrete dynamical systems of $(P_\alpha, \mathbf{f}^{(\alpha)})$ and $(P_\beta, \mathbf{f}^{(\beta)})$ is persistent if and only if all the eigenvalues λ_i $(i = 1, 2, \cdots, n)$ of $DP^{(N_\alpha:N_\beta)}(\mathbf{x}_k^*)$ lie in the unit circles, i.e.,

$$|\lambda_i| < 1 \text{ for } i = 1, 2, \cdots, n. \tag{4.101}$$

(ii) The $(N_\alpha : N_\beta)$-synchronization of two discrete dynamical systems of $(P_\alpha, \mathbf{f}^{(\alpha)})$ and $(P_\beta, \mathbf{f}^{(\beta)})$ is a saddle-node vanishing if and only if at least one of the real eigenvalues λ_i $(i = 1, 2, \cdots, n_1$ and $n_1 \le n)$ of $DP^{(N_\alpha:N_\beta)}(\mathbf{x}_k^*)$ is positive one (+1) and the other eigenvalues are in the unite circle, i.e.,

$$\lambda_i = 1 \text{ and } |\lambda_j| < 1 \text{ for } i, j \in \{1, 2, \cdots, n\} \text{ and } j \neq i. \tag{4.102}$$

(iii) The $(N_\alpha : N_\beta)$ synchronization of two discrete dynamical systems of $(P_\alpha, \mathbf{f}^{(\alpha)})$ and $(P_\beta, \mathbf{f}^{(\beta)})$ is a period-doubling vanishing if and only if at least one of the real eigenvalues λ_i $(i = 1, 2, \cdots, n_1$ and $n_1 \le n)$ of $DP^{(N_\alpha:N_\beta)}(\mathbf{x}_k^*)$ is negative one (-1) and the other eigenavalues are in the unite circle, i.e.,

$$\lambda_i = -1 \text{ and } |\lambda_j| < 1 \text{ for } i, j \in \{1, 2, \cdots, n\} \text{ and } j \neq i. \tag{4.103}$$

(iv) The $(N_\alpha : N_\beta)$-synchronization of two discrete dynamical systems of $(P_\alpha, \mathbf{f}^{(\alpha)})$ and $(P_\beta, \mathbf{f}^{(\beta)})$ is a Neimark vanishing if and only if at least one pair of all the complex eigenvalues $\lambda_i = \alpha_i \pm \beta_i \mathbf{i}$ $(i = 1, 2, \cdots, n_1$ and $n_1 \le n/2)$ of $DP^{(N_\alpha:N_\beta)}(\mathbf{x}_k^*)$ are on the unit circle and the other eigenvalues are in the unite circle, i.e.,

$$|\lambda_i| = \sqrt{\alpha_i^2 + \beta_i^2} = 1 \text{ and } |\lambda_j| < 1 \text{ for } i, j \in \{1, 2, \cdots, n\} \text{ and } j \neq i. \tag{4.104}$$

(v) The $(N_\alpha : N_\beta)$ synchronization of two discrete dynamical systems of $(P_\alpha, \mathbf{f}^{(\alpha)})$ and $(P_\beta, \mathbf{f}^{(\beta)})$ is a $(l_1 : l_2 : l_3)$ vanishing if and only if l_1 and l_2 real eigenvalues λ_i of $DP^{(N_\alpha:N_\beta)}(\mathbf{x}_k^*)$ are negative one (-1) and positive one (+1), respectively, and l_3-pairs of complex eigenvalues $\lambda_i = \alpha_i \pm \beta_i \mathbf{i}$ $(i = 1, 2, \cdots, n_1$ and $n_1 \leq n/2)$ of $DP^{(N_\alpha:N_\beta)}(\mathbf{x}_k^*)$ are on the unit circle and the other eigenvalues are in the unite circle, i.e.,

$$\begin{aligned} &\lambda_i = -1 \text{ for } i = i_1, i_2, \cdots, i_{l_1} \in \{1, 2, \cdots, n\}, \\ &\lambda_j = +1 \text{ for } j = j_1, j_2, \cdots, j_{l_2} \in \{1, 2, \cdots, n\}, \\ &|\lambda_r| = \sqrt{\alpha_r^2 + \beta_r^2} = 1 \text{ for } r = r_1, r_2, \cdots, r_{l_3} \in \{1, 2, \cdots, n\}, \\ &|\lambda_s| < 1 \text{ for } s \in \{1, 2, \cdots, n\} \text{ and } s \notin \{i, j, r\}. \end{aligned} \tag{4.105}$$

(vi) The $(N_\alpha : N_\beta)$ synchronization of two discrete dynamical systems of $(P_\alpha, \mathbf{f}^{(\alpha)})$ and $(P_\beta, \mathbf{f}^{(\beta)})$ is instantaneous if and only if at least one of the eigenvalues λ_i $(i = 1, 2, \cdots, n)$ of $DP^{(N_\alpha:N_\beta)}(\mathbf{x}_k^*)$ lies out of the unit circle, i.e.,

$$|\lambda_i| > 1 \text{ for } i \in \{1, 2, \cdots, n\}. \tag{4.106}$$

Proof The proof is similar to Theorem 4.6. ■

From Chap. 2, fixed points in nonlinear discrete dynamical systems possess many types of unstable states from eigenvalue analysis. From the similar ideas, the instantaneous $(N_\alpha : N_\beta)$ synchronization of two discrete dynamical systems can be classified. Therefore, such instantaneous synchronization classification will not be presented herein. If $N_\alpha \to \infty$ and $N_\beta \to \infty$, the $(N_\alpha : N_\beta)$ synchronization of two discrete dynamical systems should be chaotic. Consider two hybrid maps

$$\begin{aligned} P_+^{(\Sigma_{i=1}^n N_\beta^i \oplus N_\alpha^i)} &= \underbrace{P_{\beta+}^{(N_\beta^n)} \circ P_{\alpha+}^{(N_\alpha^n)} \circ \cdots \circ P_{\beta+}^{(N_\beta^1)} \circ P_{\alpha+}^{(N_\alpha^1)}}_{n-terms}, \\ P_+^{(\Sigma_{j=1}^m M_\beta^j \oplus M_\alpha^j)} &= \underbrace{P_{\beta+}^{(M_\beta^m)} \circ P_{\alpha+}^{(M_\alpha^m)} \circ \cdots \circ P_{\beta+}^{(M_\beta^1)} \circ P_{\alpha+}^{(M_\alpha^1)}}_{m-terms}; \end{aligned} \tag{4.107}$$

$$\begin{aligned} P_-^{(\Sigma_{i=n}^1 N_\alpha^i \oplus N_\beta^i)} &= \underbrace{P_{\alpha-}^{(N_\alpha^1)} \circ P_{\beta-}^{(N_\beta^1)} \circ \cdots \circ P_{\alpha-}^{(N_\alpha^m)} \circ P_{\beta-}^{(N_\beta^m)}}_{n-terms}, \\ P_-^{(\Sigma_{j=m}^1 M_\alpha^j \oplus M_\beta^j)} &= \underbrace{P_{\alpha-}^{(M_\alpha^1)} \circ P_{\beta-}^{(M_\beta^1)} \circ \cdots \circ P_{\alpha-}^{(M_\alpha^m)} \circ P_{\beta-}^{(M_\beta^m)}}_{m-terms}. \end{aligned} \tag{4.108}$$

The $(N_\beta \oplus N_\alpha : M_\beta \oplus M_\alpha)$-hybrid synchronization of two discrete systems with two maps $P_+^{(\Sigma_{i=1}^n N_\beta^i \oplus N_\alpha^i)}$ and $P_+^{(\Sigma_{j=1}^m M_\beta^j \oplus M_\alpha^j)}$ can be investigated via the following map

$$
\begin{aligned}
P^{(N_\beta \oplus N_\alpha : M_\beta \oplus M_\alpha)} \mathbf{x}_k &= P_{\varphi -} \circ P_{-}^{(\Sigma_{j=m}^{1} M_\alpha^j \oplus M_\beta^j)} \circ P_{\varphi +} \circ P_{+}^{(\Sigma_{i=1}^{n} N_\beta^i \oplus N_\alpha^i)} \mathbf{x}_k, \text{ or} \\
P^{(N_\beta \oplus N_\alpha : M_\beta \oplus M_\alpha)} \mathbf{x}_k &= P_{\varphi -} \circ P_{-}^{(\Sigma_{i=n}^{1} N_\alpha^i \oplus N_\beta^i)} \circ P_{\varphi +} \circ P_{+}^{(\Sigma_{j=1}^{m} M_\beta^j \oplus M_\alpha^j)} \mathbf{x}_k.
\end{aligned}
\tag{4.109}
$$

Thus,

$$
\mathbf{x}'_k = P^{(N_\beta \oplus N_\alpha : M_\beta \oplus M_\alpha)} \mathbf{x}_k. \tag{4.110}
$$

Similar to the $(N_\alpha : N_\beta)$-synchronization in Theorem 4.7, the corresponding fixed point and the stability conditions of Eq. (4.110) gives the $(N_\beta \oplus N_\alpha : M_\beta \oplus M_\alpha)$—hybrid synchronization of two discrete systems. This concept can be extended to the discrete dynamical systems with multiple maps

As in discrete dynamical systems with multiple maps in Sect. 4.2, the synchronization for the resultant mappings in multiple different maps can be developed.

Definition 4.14 Consider two sets of discrete dynamical systems $\bigcup_{i=1}(P_{\alpha_i}, \mathbf{f}^{(\alpha_i)})$ and $\bigcup_{j=1}(P_{\beta_j}, \mathbf{f}^{(\beta_j)})$ as in Eq. (4.48) for each discrete system with

$$
\begin{aligned}
&P_{\alpha_i +} : \mathbf{x}_k \to \mathbf{x}_{k+1} \text{ and } P_{\alpha_i -} : \mathbf{x}_{k+1} \to \mathbf{x}_k \\
&\mathbf{f}^{(\alpha_i)}(\mathbf{x}_k, \mathbf{x}_{k+1}, \mathbf{p}^{(\alpha_i)}) = \mathbf{0}
\end{aligned}
\tag{4.111}
$$

and

$$
\begin{aligned}
&P_{\beta_j +} : \mathbf{y}_k \to \mathbf{y}_{k+1} \text{ and } P_{\beta_j -} : \mathbf{y}_{k+1} \to \mathbf{y}_k \\
&\mathbf{f}^{(\beta_j)}(\mathbf{y}_k, \mathbf{y}_{k+1}, \mathbf{p}^{(\beta_j)}) = \mathbf{0}
\end{aligned}
\tag{4.112}
$$

For the two sets of discrete dynamical systems, the resultant mappings are

$$
\begin{aligned}
P^{+}_{(N_{\alpha m} \cdots N_{\alpha_2} N_{\alpha_1})} &= \underbrace{P^{+\,N_{\alpha m}}_{\alpha_m} \circ \cdots \circ P^{+\,N_{\alpha_2}}_{\alpha_2} \circ P^{+\,N_{\alpha_1}}_{\alpha_1}}_{m-terms}; \\
P^{-}_{(N_{\alpha_1} N_{\alpha_2} \cdots N_{\alpha m})} &= \underbrace{P^{-\,N_{\alpha_1}}_{\alpha_1} \circ P^{-\,N_{\alpha_2}}_{\alpha_2} \circ \cdots \circ P^{-\,N_{\alpha m}}_{\alpha_m}}_{m-terms}
\end{aligned}
\tag{4.113}
$$

and

$$
\begin{aligned}
P^{+}_{(N_{\beta n} \cdots N_{\beta_2} N_{\beta_1})} &= \underbrace{P^{+\,N_{\beta n}}_{\beta_n} \circ \cdots \circ P^{+\,N_{\beta_2}}_{\beta_2} \circ P^{+\,N_{\beta_1}}_{\beta_1}}_{n-terms}; \\
P^{-}_{(N_{\beta_1} N_{\beta_2} \cdots N_{\beta n})} &= \underbrace{P^{-\,N_{\beta_1}}_{\beta_n} \circ P^{-\,N_{\beta_2}}_{\beta_2} \circ \cdots \circ P^{-\,N_{\beta n}}_{\beta_n}}_{n-terms}
\end{aligned}
\tag{4.114}
$$

where

$$N_\alpha = \sum_{i=1}^{m} N_{\alpha_i} \text{ and } N_\beta = \sum_{j=1}^{n} N_{\beta_j}. \tag{4.115}$$

For two points $\mathbf{x}_k \in D_{\alpha_1}$ and $\mathbf{y}_k \in D_{\beta_1}$, there is a specific, differentiable, vector function $\boldsymbol{\varphi} = (\varphi_1, \varphi_2, \cdots, \varphi_l)^{\mathrm{T}} \in \mathscr{R}^l$.

(i) If

$$\boldsymbol{\varphi}(\mathbf{x}_{k+N_\alpha}^{(\alpha_m)}, \mathbf{x}_{k+N_\beta}^{(\beta_n)}, \boldsymbol{\lambda}) = \mathbf{0}, \tag{4.116}$$

then the two discrete dynamical systems $\bigcup_{i=1}(P_{\alpha_i}, \mathbf{f}^{(\alpha_i)})$ and $\bigcup_{j=1}(P_{\beta_j}, \mathbf{f}^{(\beta_j)})$ are called the $(N_\alpha : N_\beta)$-synchronization in sense of $\boldsymbol{\varphi}$.

(ii) If

$$\begin{aligned}
&\boldsymbol{\varphi}(\mathbf{x}_{k+N_\alpha}^{(\alpha_m)}, \mathbf{x}_{k+N_\beta}^{(\beta_n)}, \lambda) = \mathbf{0} \text{ with}\\
&\Delta\mathbf{I}^{(\alpha_m\alpha_1)} : \mathbf{x}_{k+N_\alpha(-)}^{(\alpha_m)} \to \mathbf{x}_{k+N_\alpha(+)}^{(\alpha_m)} \text{ and } \Delta\mathbf{I}^{(\beta_n\beta_1)} : \mathbf{x}_{k+N_\beta(-)}^{(\beta_n)} \to \mathbf{x}_{k+N_\beta(+)}^{(\beta_n)}\\
&\mathbf{x}_{k+N_\alpha(+)}^{(\alpha_m)} = \mathbf{x}_{k+N_\alpha(-)}^{(\alpha_m)} + \Delta\mathbf{I}^{(\alpha_m\alpha_1)} \text{ and } \mathbf{x}_{k+N_\beta(+)}^{(\beta_n)} = \mathbf{x}_{k+N_\beta(-)}^{(\beta_n)} + \Delta\mathbf{I}^{(\beta_n\beta_1)}\\
&\mathbf{x}_{k+N_\beta(+)}^{(\alpha_m)} = \mathbf{x}_k^{(\alpha_1)} \text{ and } \mathbf{x}_{k+N_\beta(+)}^{(\beta_n)} = \mathbf{x}_k^{(\beta_1)},
\end{aligned} \tag{4.117}$$

then the two discrete dynamical systems $\bigcup_{i=1}(P_{\alpha_i}, \mathbf{f}^{(\alpha_i)})$ and $\bigcup_{j=1}(P_{\beta_j}, \mathbf{f}^{(\beta_j)})$ are called the repeatable $(N_\alpha : N_\beta)$-synchronization in sense of $\boldsymbol{\varphi}$.

The corresponding theorem can be presented as in Theorem 4.7. For convenience, the statement is given as follows.

Theorem 4.8 *Consider two sets of discrete dynamical systems* $\bigcup_{i=1}(P_{\alpha_i}, \mathbf{f}^{(\alpha_i)})$ *and* $\bigcup_{j=1}(P_{\beta_j}, \mathbf{f}^{(\beta_j)})$ *as in Eq.* (4.48) *for each discrete system with*

$$\begin{aligned}
&P_{\alpha_i+} : \mathbf{x}_k \to \mathbf{x}_{k+1} \text{ and } P_{\alpha_i-} : \mathbf{x}_{k+1} \to \mathbf{x}_k\\
&\mathbf{f}^{(\alpha_i)}(\mathbf{x}_k, \mathbf{x}_{k+1}, \mathbf{p}^{(\alpha_i)}) = \mathbf{0}
\end{aligned} \tag{4.118}$$

and

$$\begin{aligned}
&P_{\beta_j+} : \mathbf{y}_k \to \mathbf{y}_{k+1} \text{ and } P_{\beta_j-} : \mathbf{y}_{k+1} \to \mathbf{y}_k\\
&\mathbf{f}^{(\beta_j)}(\mathbf{y}_k, \mathbf{y}_{k+1}, \mathbf{p}^{(\beta_j)}) = \mathbf{0}.
\end{aligned} \tag{4.119}$$

For two points $\mathbf{x}_k \in D_{\alpha_1}$ *and* $\mathbf{y}_k \in D_{\beta_1}$, *there is a specific, differentiable, vector function* $\boldsymbol{\varphi} = (\varphi_1, \varphi_2, \cdots, \varphi_l)^{\mathrm{T}} \in \mathscr{R}^l$. *The* $(N_\alpha : N_\beta)$*-synchronization of two sets of discrete dynamical systems* $\bigcup_{i=1}(P_{\alpha_i}, \mathbf{f}^{(\alpha_i)})$ *and* $\bigcup_{j=1}(P_{\beta_j}, \mathbf{f}^{(\beta_j)})$ *is under the following constraints*

$$\boldsymbol{\varphi}(\mathbf{x}_k, \mathbf{y}_k, \boldsymbol{\lambda}) = \mathbf{0} \text{ and } \boldsymbol{\varphi}(\mathbf{x}_{k+N_\alpha}, \mathbf{y}_{k+N_\beta}, \boldsymbol{\lambda}) = \mathbf{0}. \tag{4.120}$$

Consider a resultant hybrid mapping relation as

$$\mathbf{x}'_k = P^{(N_\alpha:N_\beta)}\mathbf{x}_k = P_{\varphi-} \circ P_{N_\beta-} \circ P_{\varphi+} \circ P_{N_\alpha+}\mathbf{x}_k \tag{4.121}$$

with

$$P_{N_\alpha+} = P^{+}_{(N_{\alpha_m}\cdots N_{\alpha_2}N_{\alpha_1})} \text{ and } P_{N_\beta-} = P^{-}_{(N_{\beta_1}N_{\beta_2}\cdots N_{\beta_n})} \tag{4.122}$$

$$\begin{aligned}
&P_{N_{\alpha_i}+} : \mathbf{x}_{k+\Sigma_{r=1}^{i-1}N_{\alpha_r}} \to \mathbf{x}_{k+\Sigma_{r=1}^{i}N_{\alpha_r}} \text{ with}\\
&\left.\begin{array}{l}
\mathbf{f}^{(\alpha_i)}(\mathbf{x}_{k+\Sigma_{r=1}^{i-1}N_{\alpha_r}+1}, \mathbf{x}_{k+\Sigma_{r=1}^{i-1}N_{\alpha_r}}, \mathbf{p}^{(\alpha)}) = \mathbf{0}\\
\mathbf{f}^{(\alpha_i)}(\mathbf{x}_{k+\Sigma_{r=1}^{i-1}N_{\alpha_r}+2}, \mathbf{x}_{k+\Sigma_{r=1}^{i-1}N_{\alpha_r}+1}, \mathbf{p}^{(\alpha)}) = \mathbf{0}\\
\vdots\\
\mathbf{f}^{(\alpha_i)}(\mathbf{x}_{k+\Sigma_{r=1}^{i}N_{\alpha_r}}, \mathbf{x}_{k+\Sigma_{r=1}^{i-1}N_{\alpha_r}-1}, \mathbf{p}^{(\alpha_i)}) = \mathbf{0}
\end{array}\right\}\\
&\text{for } i = 1, 2, \cdots, m;\\
&P_{\varphi+} : \mathbf{x}_{k+N_\alpha} \to \mathbf{y}_{k+N_\beta} \text{ with } \boldsymbol{\varphi}(\mathbf{x}_{k+N_\alpha}, \mathbf{y}_{k+N_\beta}, \boldsymbol{\lambda}) = \mathbf{0};\\
&P_{N_{\beta_j}} : \mathbf{y}_{k+N_\beta-\Sigma_{r=j}^{n}N_{\beta_r}} \to \mathbf{y}_{k+N_\beta-\Sigma_{r=j-1}^{j}N_{\beta_r}} \text{ with}\\
&\left.\begin{array}{l}
\mathbf{f}^{(\beta_j)}(\mathbf{y}_{k+N_\beta-\Sigma_{r=j}^{n}N_{\beta_r}}, \mathbf{y}_{k+N_\beta-\Sigma_{r=j}^{n}N_{\beta_r}-1}, \mathbf{p}^{(\beta_j)}) = \mathbf{0}\\
\vdots\\
\mathbf{f}^{(\beta_j)}(\mathbf{y}_{k+N_\beta-\Sigma_{r=j-1}^{j}N_{\beta_r}+2}, \mathbf{y}_{k+N_\beta-\Sigma_{r=j-1}^{j}N_{\beta_r}+1}, \mathbf{p}^{(\beta_j)}) = \mathbf{0}\\
\mathbf{f}^{(\beta_j)}(\mathbf{y}_{k+N_\beta-\Sigma_{r=j-1}^{j}N_{\beta_r}+1}, \mathbf{y}_{k+N_\beta-\Sigma_{r=j-1}^{j}N_{\beta_r}}, \mathbf{p}^{(\beta_j)}) = \mathbf{0}
\end{array}\right\}\\
&\text{for } j = n, n-1, \cdots, 1;\\
&P_{\varphi-} : \mathbf{y}_k \to \mathbf{x}'_k \text{ with } \boldsymbol{\varphi}(\mathbf{x}'_k, \mathbf{y}_k, \boldsymbol{\lambda}) = \mathbf{0};\\
&\mathbf{x}'_k = \mathbf{x}_k
\end{aligned} \tag{4.123}$$

and

$$DP^{(N_\alpha:N_\beta)}(\mathbf{x}_k^*) = DP_{\varphi-}(\mathbf{y}_k^*) \cdot DP^{(N_\beta)}_{\beta-}(\mathbf{y}^*_{k+N_\alpha}) \cdot DP_{\varphi+}(\mathbf{x}^*_{k+N_\alpha}) \cdot DP^{(N_\alpha)}_{\alpha+}(\mathbf{x}_k^*) \tag{4.124}$$

where

$$\begin{aligned}
&DP^{(N_\beta)}_{\beta-}(\mathbf{y}^*_{k+N_\alpha}) = \prod\nolimits_{i=m}^{1} DP^{(N_{\alpha_i})}_{\alpha_i+}(\mathbf{x}_k^*),\\
&DP^{(N_{\alpha_i})}_{\alpha_i+}(\mathbf{x}_k^*)\\
&= DP_{\alpha_i+}(\mathbf{x}^*_{k+\Sigma_{r=1}^{i}N_{\alpha_i}-1})\cdots DP_{\alpha_i+}(\mathbf{x}^*_{k+\Sigma_{r=1}^{i-1}N_{\alpha_i}+1}) \cdot DP_{\alpha_i+}(\mathbf{x}^*_{k+\Sigma_{r=1}^{i-1}N_{\alpha_i}}),
\end{aligned} \tag{4.125}$$

$$DP_{\beta-}^{(N_\beta)}(\mathbf{y}_{k+N_\alpha}^*) = \prod\nolimits_{j=1}^{m} DP_{\beta_j+}^{(N_{\beta_j})}(\mathbf{y}_{k+N_\beta-\Sigma_{r=1}^{j-1}N_{\alpha_i}}^*),$$
$$DP_{\beta_j+}^{(N_{\beta_j})}(\mathbf{y}_{k+N_\beta-\Sigma_{r=1}^{j-1}N_{\alpha_i}}^*)$$
$$= DP_{\beta-}(\mathbf{x}_{k+N_\beta-\Sigma_{r=1}^{j-1}N_{\alpha_i}}^*)\cdots DP_{\beta-}(\mathbf{x}_{k+N_\beta-\Sigma_{r=1}^{j-1}N_{\alpha_i}-1}^*)\cdot DP_{\beta-}(\mathbf{x}_{k+N_\beta-\Sigma_{r=1}^{j}N_{\alpha_i}}^*), \tag{4.126}$$

$$DP^{(N_\alpha:N_\beta)}(\mathbf{x}_k^*) = \left[\frac{\partial \mathbf{x}'_k}{\partial \mathbf{x}_k}\right]_{\mathbf{x}_k^*}, \tag{4.127}$$

$$DP_{\alpha_i+}^{(N_{\alpha_i})}(\mathbf{x}_{k+\Sigma_{r=1}^{i}N_{\alpha_i}}^*) = \prod\nolimits_{s=N_{\alpha_i}}^{1}\left[\frac{\partial \mathbf{x}_{k+\Sigma_{r=1}^{i-1}N_{\alpha_i}+s}}{\partial \mathbf{x}_{k+\Sigma_{r=1}^{i-1}N_{\alpha_i}+s-1}}\right]_{\mathbf{x}_{k+\Sigma_{r=1}^{i-1}N_{\alpha_i}+s-1}^*},$$
$$DP_{\varphi+}(\mathbf{x}_{k+N_\alpha}^*) = \left[\frac{\partial \mathbf{y}_{k+N_\beta}}{\partial \mathbf{x}_{k+N_\alpha}}\right]_{\mathbf{x}_{k+N_\alpha}^*},$$
$$DP_{\beta_j-}^{(N_{\beta_j})}(\mathbf{y}_{k+N_\beta-\Sigma_{r=1}^{j-1}N_{\beta_j}}^*) = \prod\nolimits_{s=1}^{N_{\beta_j}}\left[\frac{\partial \mathbf{y}_{k+N_\beta-\Sigma_{r=1}^{j}N_{\beta_j}-s}}{\partial \mathbf{y}_{k+N_\beta-\Sigma_{r=1}^{j}N_{\beta_j}-s+1}}\right]_{\mathbf{y}_{k+N_\beta-\Sigma_{r=1}^{j}N_{\beta_j}-s+1}^*},$$
$$DP_{\varphi-}(\mathbf{y}_k^*) = \left[\frac{\partial \mathbf{x}'_k}{\partial \mathbf{y}_k}\right]_{\mathbf{y}_k^*}. \tag{4.128}$$

(i) *The $(N_\alpha : N_\beta)$-synchronization of two sets of discrete dynamical systems $\bigcup_{i=1}(P_{\alpha_i}, \mathbf{f}^{(\alpha_i)})$ and $\bigcup_{j=1}(P_{\beta_j}, \mathbf{f}^{(\beta_j)})$ is persistent if and only if all the eigenvalues λ_i $(i = 1, 2, \cdots, n)$ of $DP^{(N_\alpha:N_\beta)}(\mathbf{x}_k^*)$ lie in the unit circles, i.e.,*

$$|\lambda_i| < 1 \textit{ for } i = 1, 2, \cdots, n. \tag{4.129}$$

(ii) *The $(N_\alpha : N_\beta)$-synchronization of two sets of discrete dynamical systems $\bigcup_{i=1}(P_{\alpha_i}, \mathbf{f}^{(\alpha_i)})$ and $\bigcup_{j=1}(P_{\beta_j}, \mathbf{f}^{(\beta_j)})$ is a saddle-node vanishing if and only if at least one of the real eigenvalues λ_i $(i = 1, 2, \cdots, n_1$ and $1 \le n_1 \le n)$ of $DP^{(N_\alpha:N_\beta)}(\mathbf{x}_k^*)$ is positive one (+1) and the other eigenavalues are in the unite circle, i.e.,*

$$\lambda_i = 1 \textit{ and } |\lambda_j| < 1 \textit{ for } i, j \in \{1, 2, \cdots, n\} \textit{ and } j \neq i. \tag{4.130}$$

(iii) *The $(N_\alpha : N_\beta)$ synchronization of two sets of discrete dynamical systems $\bigcup_{i=1}(P_{\alpha_i}, \mathbf{f}^{(\alpha_i)})$ and $\bigcup_{j=1}(P_{\beta_j}, \mathbf{f}^{(\beta_j)})$ is a period-doubling vanishing if and only if at least one of the real eigenvalues λ_i $(i = 1, 2, \cdots, n_1$ and $1 \le n_1 \le n)$*

of $DP^{(N_\alpha:N_\beta)}(\mathbf{x}_k^*)$ *is negative one (-1) and the other eigenavalues are in the unite circle, i.e.,*

$$\lambda_i = -1 \text{ and } |\lambda_j| < 1 \text{ for } i, j \in \{1, 2, \cdots, n\} \text{ and } j \neq i. \tag{4.131}$$

(iv) The $(N_\alpha : N_\beta)$*-synchronization of two sets of discrete dynamical systems* $\bigcup_{i=1}(P_{\alpha_i}, \mathbf{f}^{(\alpha_i)})$ *and* $\bigcup_{j=1}(P_{\beta_j}, \mathbf{f}^{(\beta_j)})$ *is a Neimark vanishing if and only if at least one pair of all the complex eigenvalues* $\lambda_i = \alpha_i \pm \beta_i \mathbf{i}\ (i = 1, 2, \cdots, n_1$ *and* $1 \leq n_1 \leq n/2)$ *of* $DP^{(N_\alpha:N_\beta)}(\mathbf{x}_k^*)$ *are on the unit circle and the other eigenvalues are in the unite circle, i.e.,*

$$|\lambda_i| = \sqrt{\alpha_i^2 + \beta_i^2} = 1 \text{ and } |\lambda_j| < 1 \text{ for } i, j \in \{1, 2, \cdots, n\} \text{ and } j \neq i. \tag{4.132}$$

(v) The $(N_\alpha : N_\beta)$ *synchronization of two sets of discrete dynamical systems* $\bigcup_{i=1}(P_{\alpha_i}, \mathbf{f}^{(\alpha_i)})$ *and* $\bigcup_{j=1}(P_{\beta_j}, \mathbf{f}^{(\beta_j)})$ *is a* $(l_1 : l_2 : l_3)$ *vanishing if and only if* l_1 *and* l_2 *real eigenvalues* λ_i *of* $DP^{(N_\alpha:N_\beta)}(\mathbf{x}_k^*)$ *are* (-1) *and* $(+1)$*, respectively, and* l_3*-pairs of complex eigenvalues* $\lambda_i = \alpha_i \pm \beta_i \mathbf{i}\ (i = 1, 2, \cdots, n_1$ *and* $1 \leq n_1 \leq n/2)$ *of* $DP^{(N_\alpha:N_\beta)}(\mathbf{x}_k^*)$ *are on the unit circle and the other eigenvalues are in the unite circle, i.e.,*

$$\begin{aligned}
&\lambda_i = -1 \text{ for } i = i_1, i_2, \cdots, i_{l_1} \in \{1, 2, \cdots, n\}\\
&\lambda_j = +1 \text{ for } j = j_1, j_2, \cdots, j_{l_2} \in \{1, 2, \cdots, n\}\\
&|\lambda_r| = \sqrt{\alpha_r^2 + \beta_r^2} = 1 \text{ for } r = r_1, r_2, \cdots, r_{l_3} \in \{1, 2, \cdots, n\}\\
&|\lambda_s| < 1 \text{ for } s \in \{1, 2, \cdots, n\} \text{ and } s \notin \{i, j, r\}.
\end{aligned} \tag{4.133}$$

(vi) The $(N_\alpha : N_\beta)$ *synchronization of two sets of discrete dynamical systems* $\bigcup_{i=1}(P_{\alpha_i}, \mathbf{f}^{(\alpha_i)})$ *and* $\bigcup_{j=1}(P_{\beta_j}, \mathbf{f}^{(\beta_j)})$ *is instantaneous if and only if at least one of the eigenvalues* $\lambda_i\ (i = 1, 2, \cdots, n)$ *of* $DP^{(N_\alpha:N_\beta)}(\mathbf{x}_k^*)$ *lies out of the unit circle, i.e.,*

$$|\lambda_i| > 1 \text{ for } i \in \{1, 2, \cdots, n\}. \tag{4.134}$$

Proof The proof is similar to Theorem 4.6. ■

Note that for higher-order singularity, the similar discussion can be done with the stability with the higher-order singularity in Chapter 2.

4.5 Synchronization of Duffing and Henon Maps

As in Luo and Guo (2011), consider an identical synchronization of the Duffing and Henon maps as an example. The Duffing map is

$$x_{1(k+1)} = x_{2(k)} \text{ and } x_{2(k+1)} = -dx_{1(k)} + cx_{2(k)} - x_{2(k)}^3. \tag{4.135}$$

and the Henon map is

$$y_{1(k+1)} = y_{2(k)} + 1 - ay_{1(k)}^2 \text{ and } y_{2(k+1)} = by_{1(k)}. \tag{4.136}$$

Introduce the vectors as

$$\begin{aligned} &\mathbf{x}_k = (x_{1(k)}, x_{2(k)})^{\mathrm{T}} \text{ and } \mathbf{y}_k = (y_{1(k)}, y_{2(k)})^{\mathrm{T}} \\ &\mathbf{f}^{(\alpha)} = (f_1^{(\alpha)}, f_2^{(\alpha)})^{\mathrm{T}} \text{ for } \alpha = 1, 2. \end{aligned} \tag{4.137}$$

Note that $\alpha = 1$ for the Duffing map and $\alpha = 2$ for the Henon map. Thus, the Duffing map is described by

$$P_1 : \mathbf{x}_k \to \mathbf{x}_{k+1} \text{ and } \mathbf{f}^{(1)}(\mathbf{x}_k, \mathbf{x}_{k+1}, \mathbf{p}^{(1)}) = \mathbf{0} \tag{4.138}$$

where

$$\begin{aligned} &f_1^{(1)}(\mathbf{x}_k, \mathbf{x}_{k+1}, \mathbf{p}^{(1)}) = x_{1(k+1)} - x_{2(k)}, \\ &f_2^{(1)}(\mathbf{x}_k, \mathbf{x}_{k+1}, \mathbf{p}^{(1)}) = x_{2(k+1)} + dx_{1(k)} - cx_{2(k)} + x_{2(k)}^3; \\ &\mathbf{p}^{(1)} = (c, d)^{\mathrm{T}}. \end{aligned} \tag{4.139}$$

The Henon map is described by

$$P_2 : \mathbf{y}_k \to \mathbf{y}_{k+1} \text{ and } \mathbf{f}^{(2)}(\mathbf{y}_k, \mathbf{y}_{k+1}, \mathbf{p}^{(2)}) = \mathbf{0} \tag{4.140}$$

where

$$\begin{aligned} &f_1^{(2)}(\mathbf{y}_k, \mathbf{y}_{k+1}, \mathbf{p}^{(2)}) = y_{1(k+1)} - y_{2(k)} - 1 + ay_{1(k)}^2, \\ &f_2^{(2)}(\mathbf{y}_k, \mathbf{y}_{k+1}, \mathbf{p}^{(2)}) = y_{2(k+1)} - by_{1(k)}; \\ &\mathbf{p}^{(2)} = (a, b)^{\mathrm{T}}. \end{aligned} \tag{4.141}$$

Consider the $(N_1 : N_2)$ synchronization of the Duffing and Henon maps with

$$\begin{aligned} &\boldsymbol{\varphi}(\mathbf{x}_k, \mathbf{y}_k, \boldsymbol{\lambda}) = \mathbf{x}_k - \mathbf{y}_k = \mathbf{0}, \\ &\boldsymbol{\varphi}(\mathbf{x}_{k+N_1}, \mathbf{y}_{k+N_2}, \boldsymbol{\lambda}) = \mathbf{x}_{k+N_1} - \mathbf{y}_{k+N_2} = \mathbf{0}. \end{aligned} \tag{4.142}$$

where

$$\begin{aligned} &\mathbf{x}_{k+N_1} = P_1^{(N_1)}\mathbf{x}_k = \underbrace{P_1 \circ P_1 \circ \cdots \circ P_1}_{N_1} \mathbf{x}_k \text{ with} \\ &\left.\begin{aligned} &f_1^{(1)}(\mathbf{x}_{k+i-1}, \mathbf{x}_{k+i}, \mathbf{p}^{(1)}) = x_{1(k+i)} - x_{2(k+i-1)} = 0, \\ &f_2^{(1)}(\mathbf{x}_{k+i-1}, \mathbf{x}_{k+i}, \mathbf{p}^{(1)}) = x_{2(k+i)} + dx_{1(k+i-1)} - cx_{2(k+i-1)} + x_{2(k+i-1)}^3 = 0 \end{aligned}\right\} \\ &\text{for } i = 1, 2, \cdots, N_1 \end{aligned} \tag{4.143}$$

and

$$\mathbf{y}_{k+N_2} = P_2^{(N_2)}\mathbf{y}_k = \underbrace{P_2 \circ P_2 \circ \cdots \circ P_2}_{N_2} \mathbf{y}_k \text{ with}$$

$$\left.\begin{array}{l} f_1^{(2)}(\mathbf{y}_{k+j}, \mathbf{y}_{k+j-1}, \mathbf{p}^{(2)}) = y_{1(k+j)} - y_{2(k+j-1)} - 1 + a y_{1(k+j-1)}^2 = 0, \\ f_2^{(2)}(\mathbf{y}_k, \mathbf{y}_{k+1}, \mathbf{p}^{(2)}) = y_{2(k+j)} - b y_{1(k+j-1)} = 0 \end{array}\right\}$$
$$\text{for } j = 1, 2, \cdots, \mathrm{N}_2 \tag{4.144}$$

For the $(N_1 : N_2)$ synchronization, the equivalent mapping structure is

$$\mathbf{x}'_k = P_\varphi \circ P_{2-}^{(N_2)} \circ P_\varphi \circ P_{1+}^{(N_1)} \mathbf{x}_k. \tag{4.145}$$

If $\mathbf{x}'_k = \mathbf{x}_k$, we have

$$\left.\begin{array}{l} f_1^{(1)}(\mathbf{x}_{k+i-1}, \mathbf{x}_{k+i}, \mathbf{p}^{(1)}) = x_{1(k+i)} - x_{2(k+i-1)} = 0 \\ f_2^{(1)}(\mathbf{x}_{k+i-1}, \mathbf{x}_{k+i}, \mathbf{p}^{(1)}) = x_{2(k+i)} + d x_{1(k+i-1)} - c x_{2(k+i-1)} + x_{2(k+i-1)}^3 = 0 \end{array}\right\}$$
$$\text{for } i = 1, 2, \cdots, N_1;$$
$$\boldsymbol{\varphi}(\mathbf{x}_{k+N_1}, \mathbf{y}_{k+N_2}, \boldsymbol{\lambda}) = \mathbf{x}_{k+N_1} - \mathbf{y}_{k+N_2} = \mathbf{0};$$
$$\left.\begin{array}{l} f_1^{(2)}(\mathbf{y}_{k+j}, \mathbf{y}_{k+j-1}, \mathbf{p}^{(2)}) = y_{1(k+j)} - y_{2(k+j-1)} - 1 + a y_{1(k+j-1)}^2 = 0 \\ f_2^{(2)}(\mathbf{y}_k, \mathbf{y}_{k+1}, \mathbf{p}^{(2)}) = y_{2(k+j)} - b y_{1(k+j-1)} = 0 \end{array}\right\}$$
$$\text{for } j = N_2, \cdots, 2, 1;$$
$$\boldsymbol{\varphi}(\mathbf{x}_k, \mathbf{y}_k, \boldsymbol{\lambda}) = \mathbf{x}_k - \mathbf{y}_k = \mathbf{0}. \tag{4.146}$$

From which the fixed points of Eq. (4.145) can be obtained, $\mathbf{x}_{k+i}^*$ $(i = 1, 2, \cdots, N_1)$ and $\mathbf{y}_{k+j}^*$ $(j = 1, 2, \cdots N_2)$. The corresponding stability boundary of such fixed points is given the eigenvalue analysis, i.e.,

$$\Delta\mathbf{x}'_k = DP_{\varphi-} \cdot DP_{2-}^{(N_2)} \cdot DP_{\varphi+} \cdot DP_{1+}^{(N_1)} \Delta\mathbf{x}_k. \tag{4.147}$$

where

$$\begin{array}{l} DP_{\varphi-}(\mathbf{y}_k^*) = [\dfrac{\partial \mathbf{x}'_k}{\partial \mathbf{y}_k}]_{\mathbf{y}_k^*} = \begin{bmatrix} 1 & 0 \\ 0 & 1 \end{bmatrix}; \\ DP_{2-}^{(N_2)} = \displaystyle\prod_{j=1}^{N_2} DP_{2-}(\mathbf{y}_{k+j}^*), \\ DP_{2-}(\mathbf{y}_{k+j}^*) = [\dfrac{\partial \mathbf{y}_{k+j-1}}{\partial \mathbf{y}_{k+j}}]_{\mathbf{y}_{k+j}^*} = -\dfrac{1}{b}\begin{bmatrix} 0 & 1 \\ b & 2a y_{1(k+j-1)}^* \end{bmatrix}; \\ DP_{\varphi+}(\mathbf{x}_{k+N_1}^*) = [\dfrac{\partial \mathbf{y}_{k+N_2}}{\partial \mathbf{x}_{k+N_1}}]_{\mathbf{x}_{k+N_1}^*} = \begin{bmatrix} 1 & 0 \\ 0 & 1 \end{bmatrix}; \\ DP_{1+}^{(N_1)} = \displaystyle\prod_{i=N_1}^{1} DP_{2-}(\mathbf{x}_{k+j}^*), \\ DP_{1+}(\mathbf{x}_{k+j-1}^*) = [\dfrac{\partial \mathbf{x}_{k+j}}{\partial \mathbf{x}_{k+j-1}}]_{\mathbf{x}_{k+j-1}^*} = \begin{bmatrix} 0 & 1 \\ -d & -c + 3(x_{2(k+j-1)}^*)^2 \end{bmatrix}. \end{array} \tag{4.148}$$

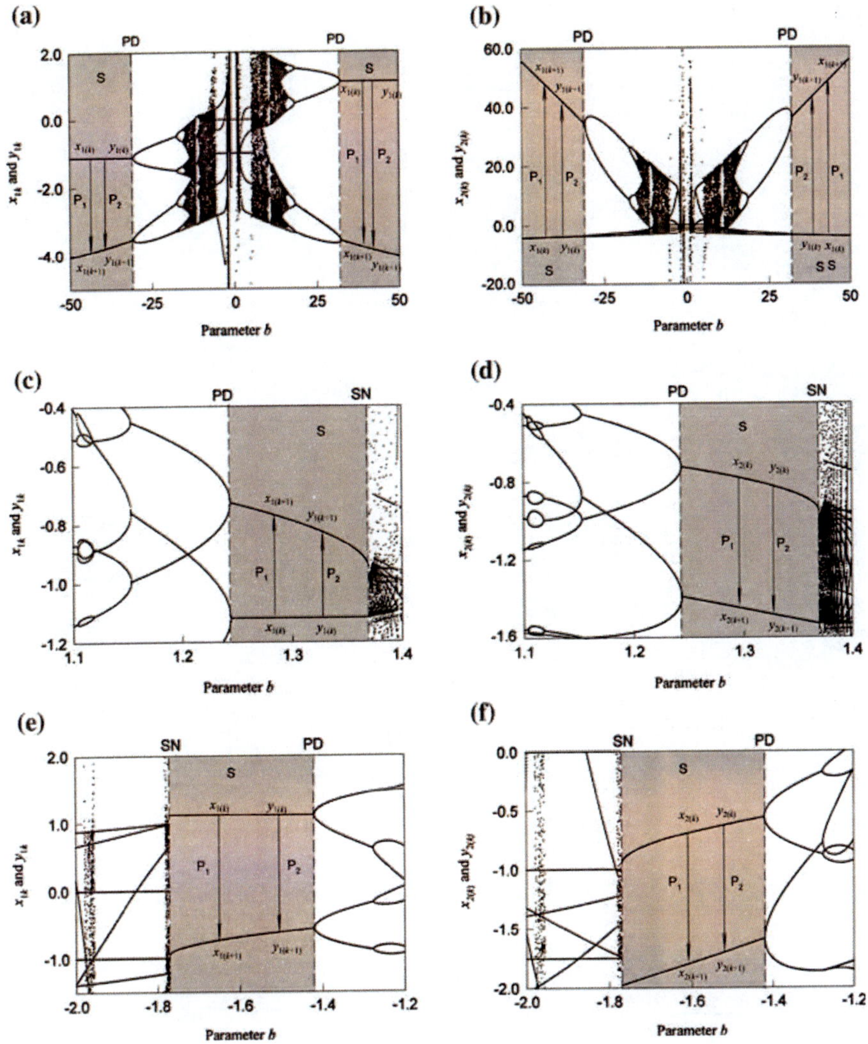

Fig. 4.11 The numerical iteration for the (1:1) synchronization of two discrete dynamical systems with the Duffing and Henon maps. Bifurcation scenario alike plots for $x_{1(k)}$ and $x_{2(k)}$ with $y_{1(k)}$ and$y_{2(k)}$: **a** and **b** for $b \in (-\infty, -30.84)$ and $b \in (33.88, \infty)$; **c** and **d** for $b \in (1.2431, 1.3687)$; **e** and **f** for $b \in (-1.7667, -1.4216)$. The shaded regions are for the (1:1) synchronization. PD and SN represent period-doubling and saddle-node vanishing of the (1:1) synchronization, respectively. ($a = 0.8$, $c = 2.75$ and d=0.2)

Through the above analysis procedure, the $(N_1 : N_2)$ synchronization domains and boundaries can be determined from Theorem 4.7. In Eq. (4.145), we can form a new map iteration

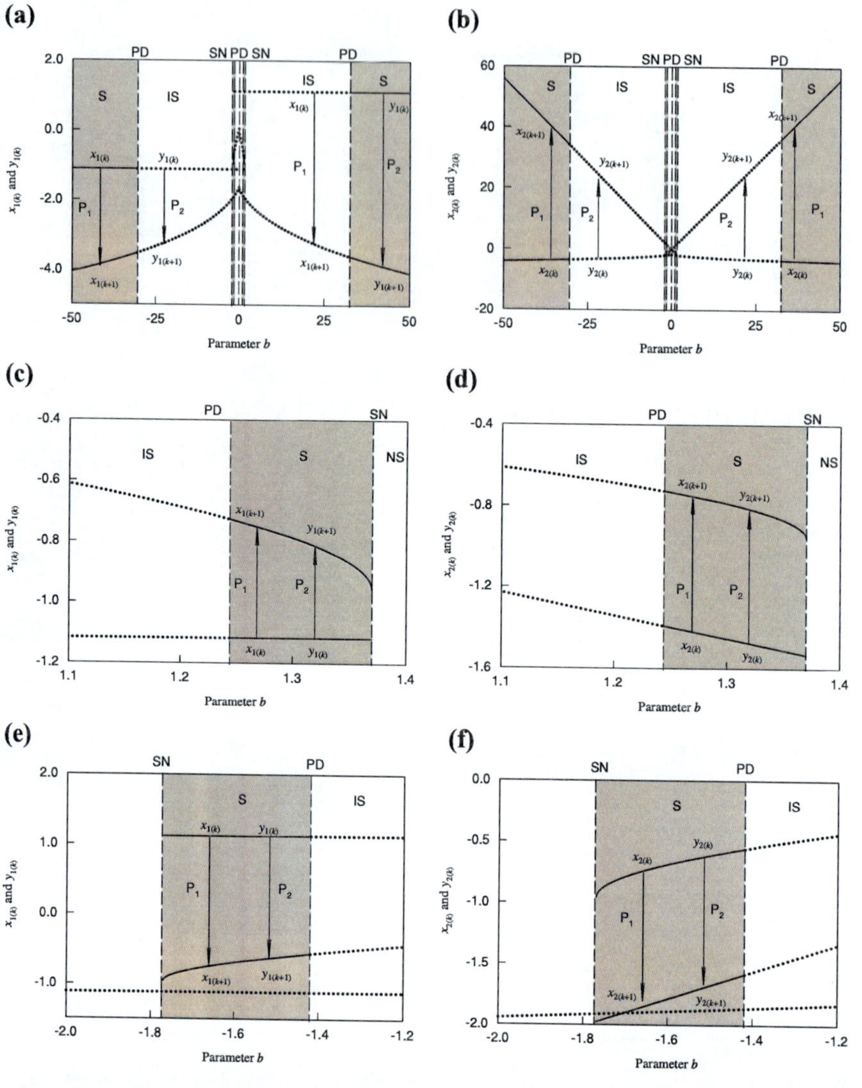

Fig. 4.12 The analytical prediction of the (1:1) synchronization of two discrete dynamical systems with the Duffing and Henon maps. The iterative states $x_{1(k)}$ and $x_{2(k)}$ with $y_{1(k)}$ and $y_{2(k)}$ are presented: **a** and **b** for $b \in (-\infty, -30.84)$ and $b \in (33.88, \infty)$; **c** and **d** for $b \in (1.2431, 1.3687)$; **e** and **f** for $b \in (-1.7667, -1.4216)$. The shaded regions are for the (1:1) synchronization. PD and SN represent period-doubling and saddle-node vanishing of the (1:1) synchronization, respectively. The instantaneous (1:1) synchronizations are represented by the dotted curves. ($a = 0.8,\ c = 2.75$ and $d = 0.2$)

$$\begin{aligned} &\mathbf{x}_{J+1} = P\mathbf{x}_J \text{ with} \\ &\mathbf{x}_J \equiv \mathbf{x}_k \text{ and } P \equiv P_{\varphi-} \circ P_{2-}^{(N_2)} \circ P_{\varphi+} \circ P_{1+}^{(N_1)} \end{aligned} \tag{4.149}$$

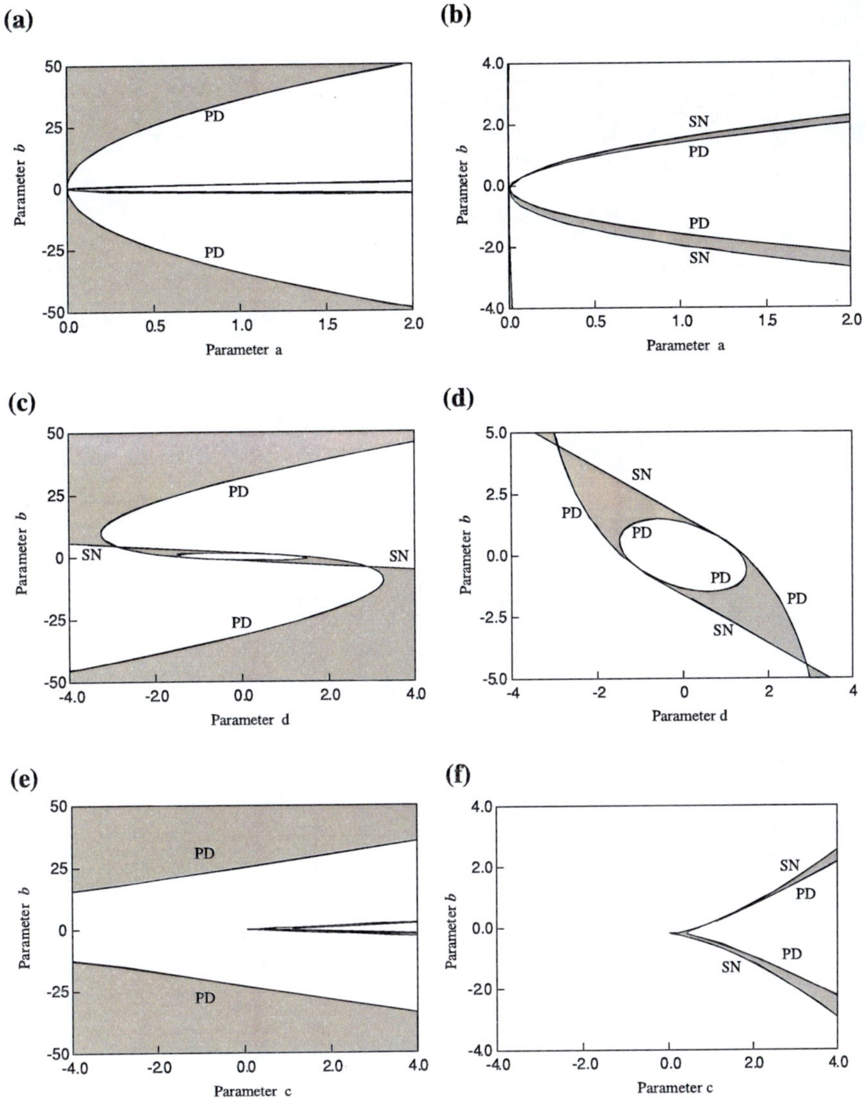

Fig. 4.13 Parameter maps of the (1:1) synchronization of two discrete dynamical systems with the Duffing and Henon maps: **a** and **b** parameter map(a, b) for $c=2.75$ and $d=0.2$; **c** and **d** parameter maps(d, b) for $a=0.8$ and $c=2.75$; **e** and **f** parameter (c, b) for $a=0.8$ and $d=0.2$. The overall views are given on the *left-hand side*, and the zoomed view are given on the *right-hand side*. The shaded regions are for the (1:1) synchronization. PD and SN represent period-doubling and saddle-node vanishing of the (1:1) synchronization, respectively

Using Eq. (4.149), numerical iteration can be done to observe the $(N_1 : N_2)$ identical synchronization of the Duffing and Henon maps.

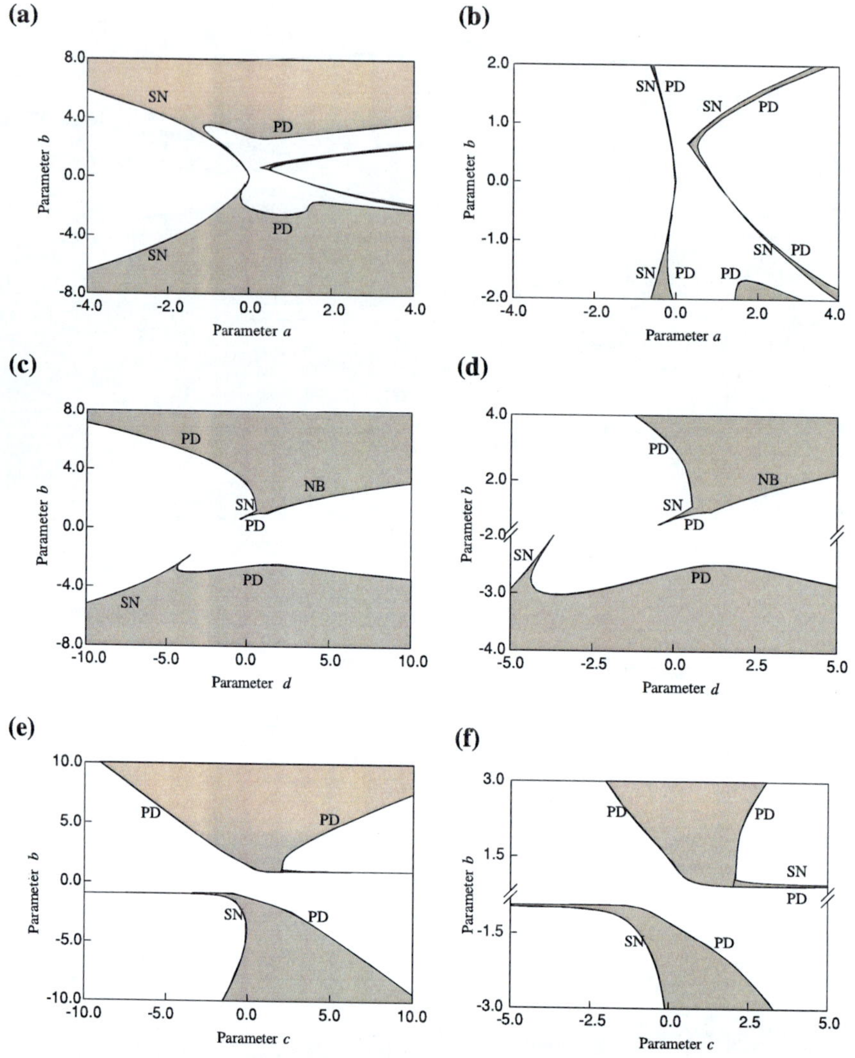

Fig. 4.14 Parameter maps of the (1:2) synchronization of two discrete dynamical systems with the Duffing and Henon maps: **a** and **b** parameter map(a, b) for $c=2.75$ and $d=0.2$; **c** and **d** parameter maps (d, b) for $a=0.8$ and $c=2.75$; **e** and **f** parameter (c, b) for $a=0.8$ and $d=0.2$. The overall views are given on the *left-hand side*, and the zoomed view are given on the *right-hand side*. The shaded regions are for the (1:2) synchronization. PD and SN represent period-doubling and saddle-node vanishing of the (1:2) synchronization, respectively

As in Luo and Guo (2011), consider parameters of $a = 0.8$, $c = 2.75$ and $d=0.2$ From the mapping in Eq. (4.149), the (1:1)-identical synchronization of the Duffing and Henon maps is simulated, as shown in Fig. 4.11. The bifurcation scenario alike plots for $x_{1(k)}$ and $x_{2(k)}$ with $y_{1(k)}$ and $y_{2(k)}$. The shaded regions are for the (1:1)

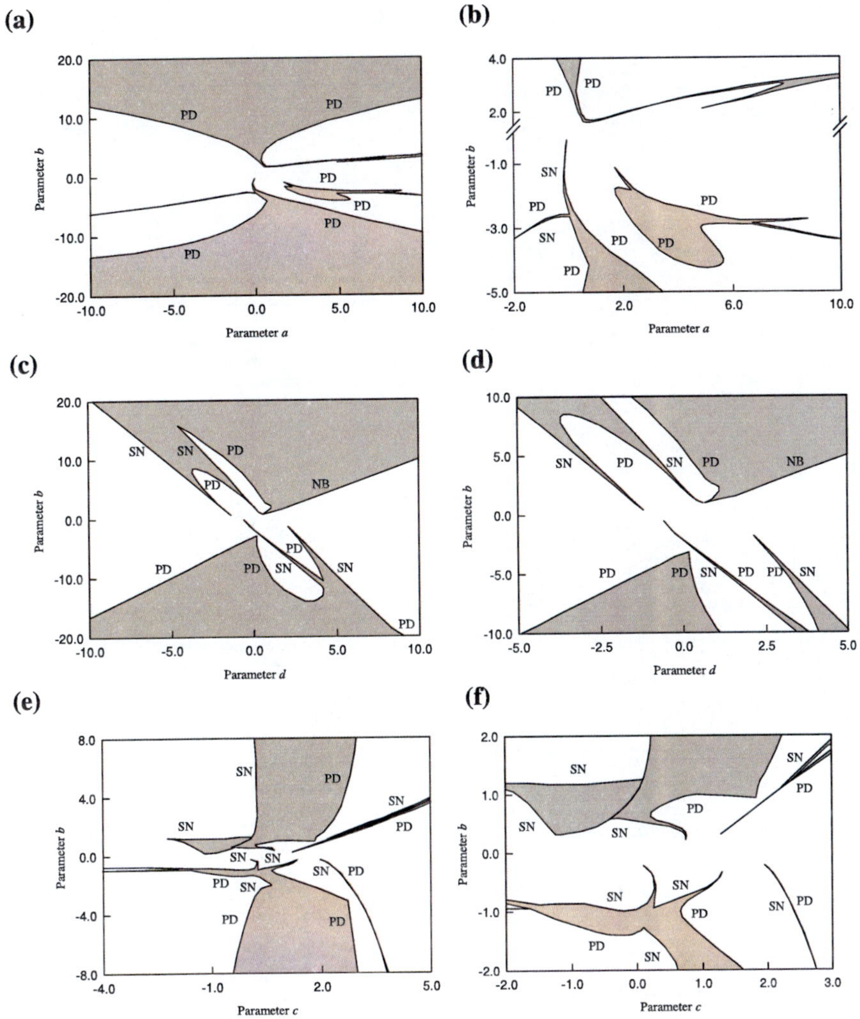

Fig. 4.15 Parameter maps of the (2:2) synchronization of two discrete dynamical systems with the Duffing and Henon maps: **a** and **b** parameter map(a, b) for $c=2.75$ and $d=0.2$; **c** and **d** parameter maps (d, b) for $a=0.8$ and $c=2.75$; **e** and **f** parameter (c, b) for $a=0.8$ and $d=0.2$ The overall views are given on the *left-hand side*, and the zoomed views are given on the *right-hand side*. The shaded regions are for the (2:2) synchronization. PD and SN represent period-doubling and saddle-node vanishing of the (2:2) synchronization, respectively

synchronization. PD and SN represent period-doubling and saddle-node vanishing of the (1:1) synchronization, respectively. The synchronization range is $b\in(-\infty, -30.84)$ and $b\in(33.88, \infty)$ in Figs. 4.11a, b. In Figs. 4.11c–f, the zoomed view for small parameter ranges are presented. The parameter ranges are given by $b\in(1.2431, 1.3687)$ and $b\in(-1.7667, -1.4216)$, respectively. The analytical

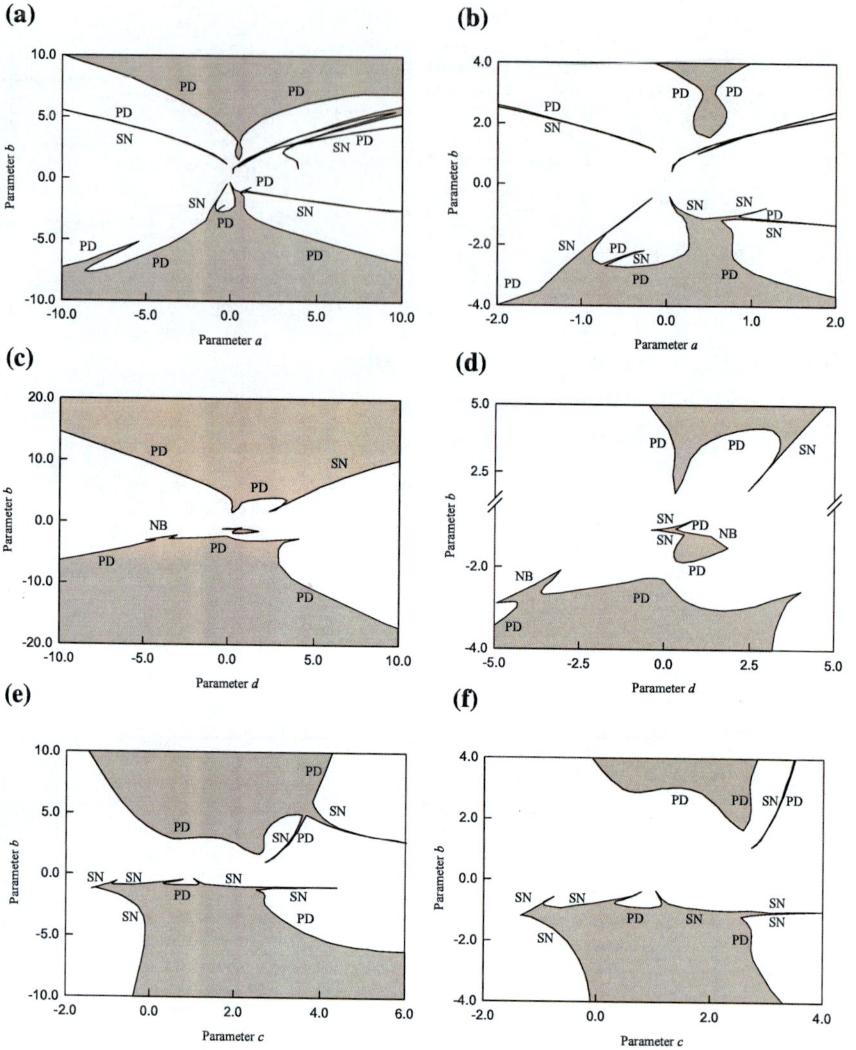

Fig. 4.16 Parameter maps of the (2:3) synchronization of two discrete dynamical systems with the Duffing and Henon maps: **a** and **b** parameter map(a, b) for $c=2.75$ and $d=0.2$; **c** and **d** parameter maps(d, b) for $a=0.8$ and $c=2.75$; **e** and **f** parameter (c, b) for $a=0.8$ and $d=0.2$ The overall views are given on the *left-hand side*, and the zoomed views are given on the *right-hand side*. The shaded regions are for the (2:3) synchronization. PD and SN represent period-doubling and saddle-node vanishing of the (2:3) synchronization, respectively

predictions of the (1:1)-synchronization is presented in Fig. 4.12. The solid curves are the (1:1) synchronizations. PD and SN represent period-doubling and saddle-node vanishing of the (1:1) synchronization, respectively. The instantaneous (1:1) synchronizations are represented by dashed curves. For numerical simulations, the instantaneous synchronization state cannot be achieved. The (1:1)synchronization that is given by the analytical prediction matches with the numerical prediction. The large parameter ranges for the (1:1)synchronization are presented in Figs. 4.12a, b. The small parameter ranges for the (1:1)-synchronization are arranged in Figs. 4.12c–f. The corresponding parameter maps for (1:1)-synchronization are presented in Fig. 4.13. The shaded regions are for the (1:1) synchronization. PD and SN represent period-doubling and saddle-node vanishing of the (1:1) synchronization, respectively. The intersected points of the PD and SN vanishing are (1,1,0)-critical synchronization vanishing with $\lambda_1 = -1$ and $\lambda_2 = 1$. Figures 4.13a, c, e is for overall parameter maps, and Figs. 4.13b, d, f is for the zoomed views of parameter maps. Figures 4.13a, b shows parameter map (a,b) for $c=2.75$ and $d=0.2$. Figures 4.13c, d presents the parameter maps (d,b) for $a=0.8$ and $c=2.75$. Figures 4.13e, f gives the parameter (c,b) for $a=0.8$ and $d=0.2$. For the parameter maps, the (1:1) synchronizations exist in different regions with many cusp points, and such cusp points will be very difficult to be analyzed by the catastrophe analysis.

Similarly, the parameter maps for (1:2), (2:2) and (2:3)-synchronizations are presented in Figs. 4.14, 4.15, 4.16, respectively. The shaded regions are for the $(N_\alpha : N_\beta)$ synchronization. PD, SN and NB represent period-doubling, saddle-node, Neimark bifurcation vanishing of the $(N_\alpha : N_\beta)$ synchronization, respectively. The intersected points of the PD and SN vanishing are $(1, 1, 0)$-critical synchronization vanishing with $\lambda_1 = -1$ and $\lambda_2 = 1$. The intersected points of the PD and NB vanishing are for (1,0,0) or (0,0,1)-critical synchronization vanishing with eigenvalues of $\lambda_1 = -1$ and $|\lambda_{1,2}| = 1$, as observed in Fig. 4.16. For the multiple-step synchronization, the parameter maps become more complicated and many cusps exist. Again, the catastrophe theory to analyze the synchronization is very difficult. Other parameter maps for $(N_\alpha : N_\beta)$ can be developed in the similar fashion. The vanishing boundaries will include all possibility of synchronization vanishing.

References

1. Luo, A.C.J., (2010), A Ying-Yang theory in nonlinear discrete dynamical systems, *International Journal of Bifurcation and Chaos*, **20**:1085–1098.
2. Luo, A.C.J., Guo, Y., (2010), parameter characteristics for stable and unstable solutions in nonlinear discrete dynamical systems, *International Journal of Bifurcation and Chaos*, **20**:3173–3191.
3. Luo, A.C.J., Guo, Y., (2011), Synchronization dynamics of discrete dynamical systems, presented in *The third Conference on Dynamics, Vibration and Control*, August 7–13, 2011. Calgary, Canada.

Chapter 5
Switching Dynamical Systems

In this chapter, dynamics of switching dynamical systems will be presented. A switching system of multiple subsystems with transport laws at switching points will be discussed. The existence and stability of switching dynamical systems will be discussed through equi-measuring functions. The G-function of the equi-measuring functions will be introduced. The local increasing and decreasing of switching systems to equi-measuring functions will be presented. The global increasing and decreasing of the switching systems to equi-measuring functions will be discussed. Based on the global and local properties of the switching dynamical systems to the equi-measuring function, the stability of switching systems can be discussed. To demonstrate flow regularity and complexity of switching systems, the impulsive system is as a special switching system to present, and the quasi-periodic flows and chaotic diffusion of impulsive systems will be presented. A frame work for periodic flows in switching systems will be presented. The periodic flows and stability for linear switching systems will be discussed. This framework can be applied to nonlinear switching systems. The further results on stability and bifurcation of periodic flows in nonlinear switching systems can be discussed.

5.1 Continuous Subsystems

On an open domain $\Omega_i \subset \mathscr{R}^n$, there is a C^{r_i}-continuous system ($r_i > 1$) in the time interval $t \in [t_{k-1}, t_k]$

$$\dot{\mathbf{x}}^{(i)} = \mathbf{F}^{(i)}(\mathbf{x}^{(i)}, t, \mathbf{p}^{(i)}) \in \mathscr{R}^n, \mathbf{x}^{(i)} = (\mathbf{x}_1^{(i)}, \mathbf{x}_2^{(i)}, \cdots, \mathbf{x}_n^{(i)})^{\mathrm{T}} \in \Omega_i. \tag{5.1}$$

The time is t and $\dot{\mathbf{x}}^{(i)} = d\mathbf{x}^{(i)}/dt$. $\mathbf{p}^{(i)} = (p_1^{(i)}, p_2^{(i)}, \cdots, p_{m_i}^{(i)})^{\mathrm{T}} \in \mathscr{R}^{m_i}$ is a parameter vector. On the domain $\Omega_i \subset \mathscr{R}^n$, the vector field $\mathbf{F}^{(i)}(\mathbf{x}^{(i)}, t, \mathbf{p}^{(i)})$ with the parameter vector $\mathbf{p}^{(i)}$ is C^{r_i}-continuous in $\mathbf{x}^{(i)}$ for time interval $t \in [t_{k-1}, t_k]$. With an initial condition $\mathbf{x}^{(i)}(t_{k-1}) = \mathbf{x}_{k-1}^{(i)}$, the dynamical system in Eq. (5.1) possesses a continuous flow as

A. C. J. Luo, *Regularity and Complexity in Dynamical Systems*,
DOI: 10.1007/978-1-4614-1524-4_5, © Springer Science+Business Media, LLC 2012

$$\mathbf{x}^{(i)}(t) = \mathbf{\Phi}^{(i)}(\mathbf{x}_{k-1}^{(i)}, t, \mathbf{p}^{(i)}) \text{ with } \mathbf{x}_{k-1}^{(i)} = \mathbf{\Phi}^{(i)}(\mathbf{x}_{k-1}^{(i)}, t_{k-1}, \mathbf{p}^{(i)}). \tag{5.2}$$

To investigate the switching system consisting of many subsystems, the following assumptions of the ith subsystem should be held.

(A5.1)

$$\left.\begin{array}{l} \mathbf{F}^{(i)}(\mathbf{x}^{(i)}, t, \mathbf{p}^{(i)}) \in C^{r_i}, \\ \mathbf{\Phi}^{(i)}(\mathbf{x}_k^{(i)}, t, \mathbf{p}^{(i)}) \in C^{r_i+1} \end{array}\right\} \text{ on } \Omega_i \text{ for } t \in [t_{k-1}, t_k], \tag{5.3}$$

(A5.2)

$$\left.\begin{array}{l} ||\mathbf{F}^{(i)}|| \leq K_1^{(i)}(\text{const}), \\ ||\mathbf{\Phi}^{(i)}|| \leq K_2^{(i)}(\text{const}) \end{array}\right\} \text{ on } \Omega_i \text{ for } t \in [t_{k-1}, t_k], \tag{5.4}$$

(A5.3)

$$\mathbf{x}^{(i)} = \mathbf{\Phi}^{(i)}(t) \notin \partial\Omega_{ij} \quad \text{for } t \in [t_{k-1}, t_k]. \tag{5.5}$$

(A5.4) The switching of any two subsystems possesses the time-continuity.

From the foregoing assumptions, the subsystem possesses a finite solution in the finite time interval as a candidate to be selected for the resultant switching system. From Assumption (A5.3), any flow in the ith subsystem for $t \in (t_{k-1}, t_k)$ will not arrive to the boundary of the domains before the flow switches to the next subsystem. If the flow of the ith subsystem for $t \in (t_{k-1}, t_k)$ reaches the domain boundary of the systems, it will be discussed in discontinuous switching systems. Suppose the vector $\mathbf{x} \in \mathscr{R}^n$ can be decomposed into two vectors $\mathbf{x}_{n_1} \in \mathscr{R}^{n_1}$ and $\mathbf{x}_{n_2} \in \mathscr{R}^{n_2}$, where $n = n_1 + n_2$ and $\mathbf{x} = (\mathbf{x}_{n_1}, \mathbf{x}_{n_2})^{\mathrm{T}}$. From such concepts, such a finite and bounded solution in phase space is sketched for the time interval $t \in [t_{k-1}, t_k]$ in Fig. 5.1a, and the corresponding time-history of the dynamical flow is presented in Fig. 5.1b.

5.2 Switching Systems

To investigate the switching system, a set of dynamical systems in finite time intervals will be introduced first. From such a set of dynamical systems, the dynamical subsystems in a resultant switching system can be selected.

Definition 5.1 From dynamical systems in Eq. (5.1), a set of dynamical systems on the open domain Ω_i in the time interval $t \in [t_{k-1}, t_k]$ for $i = 1, 2, \cdots, m$ is defined as

$$\mathfrak{S}_{\mho} \equiv \{S_i \,|\, i = 1, 2, \cdots, m\}, \tag{5.6}$$

where

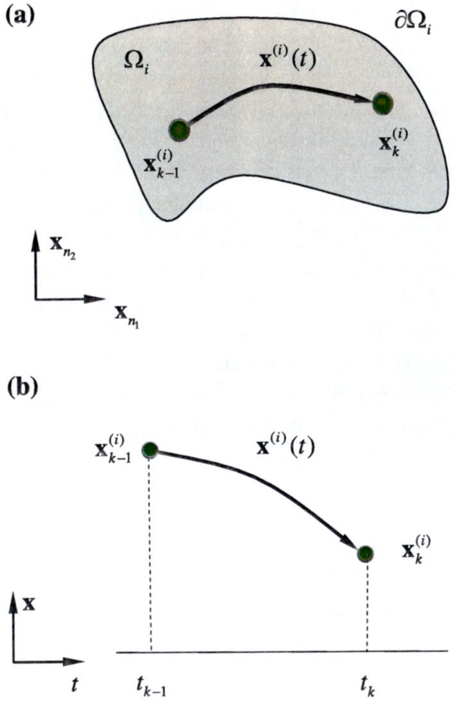

Fig. 5.1 **a** A flow in phase space and **b** a flow in the time-history

$$S_i \equiv \left\{ \dot{\mathbf{x}}^{(i)} = \mathbf{F}^{(i)}(\mathbf{x}^{(i)}, t, \mathbf{p}^{(i)}) \in \mathscr{R}^n \left| \begin{array}{l} \mathbf{x}^{(i)} \in \Omega_i \subset \mathscr{R}^n, \mathbf{p}^{(i)} \in \mathscr{R}^{m_i}; \\ \mathbf{x}^{(i)}(t_{k-1}) = \mathbf{x}^{(i)}_{k-1}; t \in [t_{k-1}, t_k]; \\ k \in \mathbb{N} \end{array} \right. \right\}. \quad (5.7)$$

To discuss the switchability of two subsystems at time t_k, the existence and uniqueness of solutions of the two systems are very important. From Assumptions (A5.1–A5.3), the subsystem possesses a finite solution in the finite time interval and such a solution will not reach the corresponding domain boundary. From the set of subsystems, the corresponding set of solutions for such subsystems can be defined as follows.

Definition 5.2 For the ith dynamical subsystems in Eq. (5.1), with an initial condition $\mathbf{x}^{(i)}_{k-1} \in \Omega_i$ for $k \in \mathbb{N}$, there is a unique solution $\mathbf{x}^{(i)}(t) = \mathbf{\Phi}^{(i)}(\mathbf{x}^{(i)}_{k-1}, t, \mathbf{p}^{(i)})$. For all $i = 1, 2, \cdots, m$, a set of solutions for the ith subsystem in Eq. (5.1) on the open domain D_i in the time interval $t \in [t_{k-1}, t_k]$ is defined as

$$\mathbb{S} = \left\{ \Theta^{(i)} \,|i = 1, 2, \cdots, m \right\}, \quad (5.8)$$

where

$$\Theta^{(i)} \equiv \left\{ \mathbf{x}^{(i)}(t) \left| \begin{array}{l} \mathbf{x}^{(i)}(t) = \mathbf{\Phi}^{(i)}(\mathbf{x}^{(i)}_{k-1}, t, \mathbf{p}^{(i)}) \\ t \in [t_{k-1}, t_k], k \in \mathbb{N} \end{array} \right. \right\}. \quad (5.9)$$

To discuss the switching of two subsystems for time t_k, the overlapping of domains of vector fields should be discussed. For different time intervals, two open domains Ω_i and Ω_j can be overlapped or separated, (i.e., $\Omega_i \cap \Omega_j \neq \varnothing$ or $\Omega_i \cap \Omega_j = \varnothing$). However, in discontinuous dynamical systems, the open domains for subsystems should be $\Omega_i \cap \Omega_j = \varnothing$ (also see, Luo 2006). In other words, the domains for the discontinuous systems cannot be overlapped. If the two domains of the ith and jth subsystems in Eq. (5.1) can be overlapped partially or fully (i.e., $\Omega_i \subseteq \Omega_j$ or $\Omega_i \supseteq \Omega_j$), there are two kinds of switching in the overlapping zone of the two domains. The two switching include (i) the continuous switching and (ii) C^0-discontinuous switching. However, for either the two separated domains or the switching points in the non-overlapping zone, only the C^0-discontinuous switching. Because the time intervals for two systems are different, the domains for the two systems can be overlapped. In other words, on the same domain, many different dynamical systems can be defined. Thus, there is a relation for the ith and jth switchable subsystems

$$\Omega_i \subseteq \Omega_j \ \text{ or } \ \Omega_i \supseteq \Omega_j. \tag{5.10}$$

From the foregoing relation, different systems can be defined on the *full* or *partial*, same domain for different time intervals. To extend such a concept for two subsystems, the domain for the resultant switching system of many subsystems can be defined. Before doing so, as in Luo (2005, 2006, 2008a), the inaccessible and accessible domains in phase space will be defined for discontinuous dynamical systems in phase space.

Definition 5.3 On a domain $\Omega_0 \subset \mathscr{R}^n$ in phase space, if no dynamical system is defined, the domain Ω_0 is called an inaccessible domain.

Definition 5.4 On a domain $\Omega_i \subset \mathscr{R}^n$ in phase space, if the ith-subsystem is defined, the domain Ω_i is called an accessible domain.

Definition 5.5 In the neighborhood of a point $\mathbf{x}_p$, if there is a set of domains Ω_{l_i} ($l_i \in \{1, 2, \cdots, m\}$, $i = 1, 2, \cdots, m_1$) for subsystems and an inaccessible domain Ω_0 in phase space, there is the union of the domains as

$$\mho = \cup_{i=1}^{m_1} \Omega_{l_i} \cup \Omega_0 \text{ and } \Omega_{l_i} \subseteq \mho \tag{5.11}$$

for the subsystems to switch. Such a union is called the resultant domain in the neighborhood of a point $\mathbf{x}_p$ for the switching system of subsystems.

To investigate the responses of the switching system in the resultant domain D in the neighborhood of a point $\mathbf{x}_p$, the switchability of any two systems should be discussed. The concept of the switching from a subsystem to another subsystem is presented.

Definition 5.6 Consider two subsystems $S_i, S_j \in \mathfrak{S}_D$ on the domains Ω_i and Ω_j.

(i) For $\mathbf{x}_k^{(i)}, \mathbf{x}_k^{(j)} \in \Omega_i \cap \Omega_j \neq \varnothing$ at $t = t_k$, if

$$\left.\begin{array}{l}\dfrac{d^s\mathbf{x}_k^{(i)}}{dt^s}=\dfrac{d^s\mathbf{x}_k^{(j)}}{dt^s}\ \text{for } s=0,1,2,\cdots,r;\\ \dfrac{d^{r+1}\mathbf{x}_k^{(i)}}{dt^{r+1}}\neq\dfrac{d^{r+1}\mathbf{x}_k^{(j)}}{dt^{r+1}},\end{array}\right\} \tag{5.12}$$

then the switching of the two subsystems S_i and S_j at time t_k is called a C^r-continuous switching,

(ii) For $\mathbf{x}_k^{(i)} \in \Omega_i$ and $\mathbf{x}_k^{(j)} \in \Omega_j$ at $t = t_k$, if $\mathbf{x}_k^{(i)} \neq \mathbf{x}_k^{(j)}$ and there is an transport law

$$\mathbf{g}^{(ij)}(\mathbf{x}_k^{(i)}, \mathbf{x}_k^{(j)}, \mathbf{p}_{ij}) = \mathbf{0} \tag{5.13}$$

then the switching of the two subsystems S_i and S_j at time t_k is called the C^0-discontinuous switching.

To illustrate the aforementioned concept for the switching of the two subsystems, the switching of the ith and jth subsystems are sketched in Fig. 5.2. A flow from the ith subsystem switching to the jth-subsystem at time t_k is presented. The domain of the ith subsystem is filled with gray color. The flow is depicted by the curves with arrows. The switching points are labeled by circular symbols. The hatched areas are the overlapped domains. In Fig. 5.2a, the two subsystems at time t_k are switched with at least $\mathbf{x}_k^{(i)} = \mathbf{x}_k^{(j)}$. In Fig. 5.2b, the domains of the ith and ith subsystems are separated at time t_k. To complete the switching of the ith and jth subsystems at point t_k, the transport law (i.e., $\mathbf{g}^{(ij)}(\mathbf{x}_k^{(i)}, \mathbf{x}_k^{(j)}, t_k) = 0$) should be used. The transport law is presented with a dashed line with an arrow.

Definition 5.7 For a flow $\mathbf{x}^{(i)} \in \mho$, two positive constants with $0 \leq C_1 < C_2$ exist. If the following relation holds

$$C_1 \leq ||\mathbf{x} - \mathbf{x}_p|| \leq C_2 \tag{5.14}$$

where $||\bullet||$ is a norm, the domain $\mho$ is called the finite domain in the vicinity of point $\mathbf{x}_p$.

From the resultant domain $\mho$ and the switchability conditions between any two subsystems, a resultant switching system can be defined.

Definition 5.8 A switching system on the domain $\mho = \cup_{i=1}^{m_1}\Omega_{l_i} \cup \Omega_0$ is defined as

$$\begin{array}{l}\dot{\mathbf{x}}^{(\alpha_k)} = \mathbf{F}^{(\alpha_k)}(\mathbf{x}^{(\alpha_k)}, t, \mathbf{p}^{(\alpha_k)})\ \text{ for }\ t \in [t_{k-1}, t_k]\\ \alpha_k \in \{l_1, l_2, \cdots, l_{m_1}\} \subseteq \{1, 2, \cdots, m\}, k = 1, 2, \cdots\end{array} \tag{5.15}$$

with an given initial condition $\mathbf{x}^{(\alpha_1)} = \mathbf{x}_0^{(\alpha_1)}$ and the corresponding transport laws at time t_k

$$\mathbf{g}^{(\alpha_k\alpha_{k+1})}(\mathbf{x}_k^{(\alpha_k)}, \mathbf{x}_k^{(\alpha_{k+1})}, \mathbf{p}_{\alpha_k\alpha_{k+1}}) = \mathbf{0} \text{ for } k = 1, 2, \cdots. \tag{5.16}$$

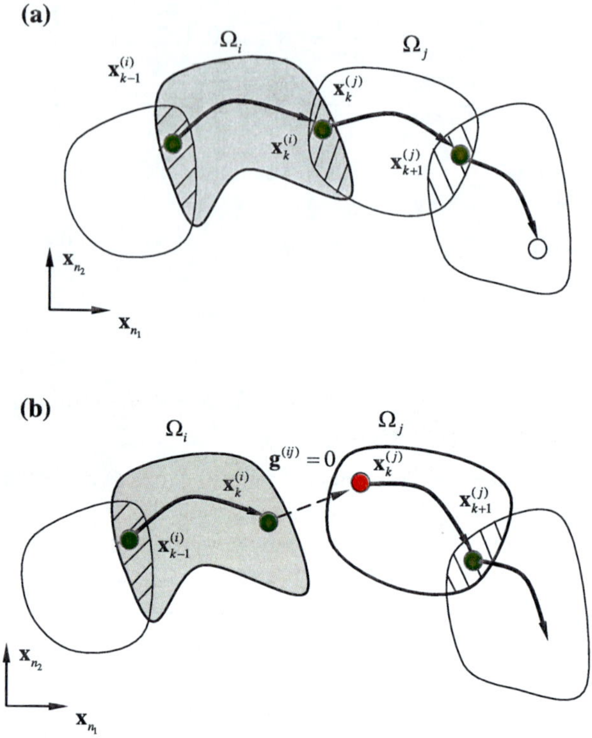

Fig. 5.2 The dynamical system switching: **a** continuous switching, and **b** C^0-discontinuous switching with transport laws $P_0^{(ij)}$ in phase space

At a switching point at time t_k, if the C^r-continuous switching of any the switching system exists, then Eq. (5.16) is expressed by

$$\frac{d^\beta \mathbf{g}^{(\alpha_k \alpha_{k+1})}}{dt^\beta} \equiv \frac{d^\beta \mathbf{x}_k^{(\alpha_{k+1})}}{dt^\beta} - \frac{d^\beta \mathbf{x}_k^{(\alpha_k)}}{dt^\beta} = \mathbf{0} \text{ for } k \in \mathbb{N}, \beta = 0, 1, 2, \cdots, r. \quad (5.17)$$

For $\beta = 0$, one obtains $\mathbf{g}^{(\alpha_k \alpha_{k+1})} \equiv \mathbf{x}_k^{(\alpha_{k+1})} - \mathbf{x}_k^{(\alpha_k)} = \mathbf{0}$. Therefore, no matter how the system is switching, the transport law as a general expression is adopted from now on. Consider a resultant switching system on the domain $\mho = \cup_{i=0}^{m_1} \Omega_{l_i}$ for $l_i \in \{0, 1, 2, \cdots, m\}$ in the vicinity of point $\mathbf{p}_k$ to be formed by

$$\begin{array}{l} S \equiv \cdots \oplus S_{\alpha_k} \oplus \cdots \oplus S_{\alpha_2} \oplus S_{\alpha_1} \\ \text{for } S_{\alpha_k} \in \mathfrak{S}_D, \alpha_k \in \{l_1, l_2, \cdots, l_{m_1}\} \subseteq \{1, 2, \cdots, m\}, k = 1, 2, \cdots \\ t \in \cup_{k=1}^{\infty} [t_k, t_{k-1}] \end{array} \quad (5.18)$$

with the corresponding switching conditions given by the transport law.

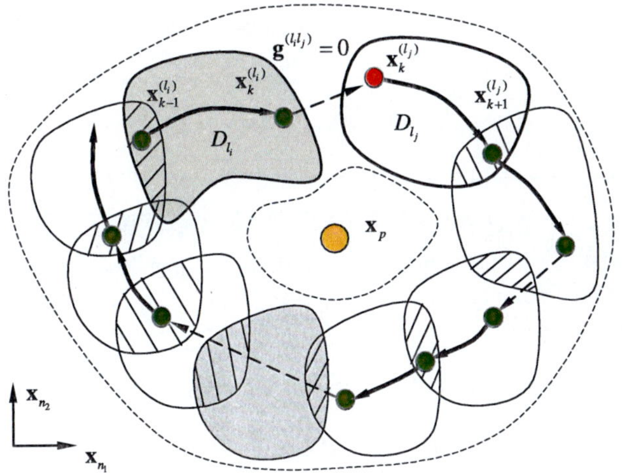

Fig. 5.3 A flow and switching of a switching system on the resultant domain in neighborhood of point $\mathbf{x}_p$

$$\left.\begin{array}{l}\mathbf{x}=\mathbf{x}_0^{(\alpha_1)} \in \Omega_{\alpha_1} \text{ at } t=t_0\\ \mathbf{g}^{(\alpha_k\alpha_{k+1})}(\mathbf{x}_k^{(\alpha_k)},\mathbf{x}_k^{(\alpha_{k+1})},\mathbf{p}_{\alpha_k\alpha_{k+1}})=\mathbf{0} \text{ at } t=t_k,\\ \text{for } k=1,2,\cdots.\end{array}\right\} \tag{5.19}$$

From the set of the solutions in Eq. (5.9) for subsystems, the solution of the switching system is given by

$$\left.\begin{array}{l}\mathbf{x}^{(\alpha_k)}=\mathbf{\Phi}^{(\alpha_k)}(\mathbf{x}_{k-1}^{(\alpha_k)},t,\mathbf{p}^{(\alpha_k)}) \in \Omega_{\alpha_k} \text{ for } t\in[t_{k-1},t_k]\\ \mathbf{g}^{(\alpha_k\alpha_{k+1})}(\mathbf{x}_k^{(\alpha_k)},\mathbf{x}_k^{(\alpha_{k+1})},\mathbf{p}_{\alpha_k\alpha_{k+1}})=\mathbf{0},\\ \alpha_k\in\{l_1,l_2,\cdots,l_{m_1}\}\subseteq\{1,2,\cdots,m\}, k=1,2,\cdots.\end{array}\right\} \tag{5.20}$$

Note that symbol "$\oplus$" means the switching action of two subsystems. To explain the two system switching, on the domain $\mho=\cup_{i=1}^{m_1}\Omega_{l_i}\cup\Omega_0$ a switching dynamical system given by Eq. (5.18) with the switching and initial conditions are sketched in Fig. 5.3. The short-dashed curves are the boundary of the domain D in the neighborhood of point $\mathbf{x}_p$, and the accessible domains are given by closed solid curves. The thick solid curves with arrows are flows. The circular symbols represent the switching points and the dashed lines with arrows are the transport laws for the two systems.

Definition 5.9 For a switching system in Eqs. (5.18) and (5.19), if there is a switching pattern of the subsystems as

$$\begin{array}{l}S\equiv\cdots\oplus S^{(j)}\oplus\cdots\oplus S^{(2)}\oplus S^{(1)} \text{ with } t\in\cup_{j=1}\cup_{i=1}^{m_1}[t_{i-1}^{(j)},t_i^{(j)}]\\ \text{where } S^{(j)}\equiv S_{\alpha_{m_1}}\oplus\cdots\oplus S_{\alpha_2}\oplus S_{\alpha_1} \text{ for } t\in\cup_{i=1}^{m_1}[t_{i-1}^{(j)},t_i^{(j)}]\\ t_0^{(1)}=t_0 \text{ for } j=1,2,\cdots;\end{array} \tag{5.21}$$

then, the switching system S is called *a repeated, switching system of subsystems*. For each repeating pattern of the subsystem, if

$$t_{m_1}^{(j)} - t_0^{(j)} = T = const \text{ for } j = 1, 2, \cdots, \tag{5.22}$$

then, the switching system S in Eq. (5.21) is called *a repeated, switching system with the equi-time interval* (or, an equi-time, repeated, switching system).

Definition 5.10 Under the switching conditions in Eq. (5.19), a switching system of subsystems on the domain $\mho = \cup_{i=1}^{m_1} \Omega_{l_i} \cup \Omega_0$ for $l_i \in \{1, 2, \cdots, m\}$ in the vicinity of point $\mathbf{x}_p$ is defined by

$$S \equiv S_{\alpha_{m_1}} \oplus \cdots \oplus S_{\alpha_2} \oplus S_{\alpha_1} \text{ with } t \in \cup_{i=1}^{m_1} [t_{i-1}, t_i]. \tag{5.23}$$

If the flow of the switching system satisfies the following condition

$$\mathbf{x}_{m_1}^{(\alpha_{m_1})} = \mathbf{x}_0^{(\alpha_1)} \text{ and } t_{m_1} - t_0 \equiv T = const \tag{5.24}$$

then, the switching system S in Eq. (5.23) possesses *a periodic flow on* the domain $\mho$ in the vicinity of point $\mathbf{x}_p$.

Definition 5.11 With the switching conditions in Eq. (5.19), for a switching system S in Eq. (5.23), if there is a new switching system as

$$S' \equiv S \oplus S = \underbrace{S_{\alpha_{m_1}} \oplus \cdots \oplus S_{\alpha_2} \oplus S_{\alpha_1}}_{S} \oplus \underbrace{S_{\alpha_{m_1}} \oplus \cdots \oplus S_{\alpha_2} \oplus S_{\alpha_1}}_{S} \quad \text{for } t \in \cup_{i=1}^{2m_1} [t_{i-1}, t_i] \tag{5.25}$$

with the conditions

$$\mathbf{x}_{2m_1}^{(\alpha_{m_1})} = \mathbf{x}_0^{(\alpha_1)} \text{ and } t_{2m_1} - t_0 \equiv 2T = const \tag{5.26}$$

then, the new switching system S' is called a *period-doubling* system of the switching system S. The corresponding flow is called the *period-doubling* flow of the switching system S.

Definition 5.12 From the switching system S in Eq. (5.23), if there is a new switching system as

$$\begin{aligned} S' &\equiv \underbrace{S \oplus \cdots \oplus S}_{l} \\ &= \underbrace{\underbrace{S_{\alpha_{m_1}} \oplus \cdots \oplus S_{\alpha_2} \oplus S_{\alpha_1}}_{S} \oplus \cdots \oplus \underbrace{S_{\alpha_{m_1}} \oplus \cdots \oplus S_{\alpha_2} \oplus S_{\alpha_1}}_{S}}_{l} \\ &\text{with } t \in \cup_{i=1}^{l \times m_1} [t_{i-1}, t_i] \end{aligned} \tag{5.27}$$

with the conditions

$$\mathbf{x}_{l\times m_1}^{(\alpha_{m_1})} = \mathbf{x}_0^{(\alpha_1)} \text{ and } t_{l\times m_1} - t_0 \equiv l \times T = const \tag{5.28}$$

then, the new switching system S' is called a period-l system of the switching system S. The corresponding flow is called the period-l flow of the switching system S. If $l \to \infty$, the new switching system S' is called a *chaotic* system of the switching system S, and the corresponding flow is called the *chaotic* flow of the switching system S.

5.3 Measuring Functions and Stability

To investigate the stability of the switching systems consisting of many subsystems on the domain in the vicinity of point $\mathbf{x}_p$, a measuring function should be introduced through the relative position vector to point $\mathbf{x}_p$. The relative position vector is given by

$$\mathbf{r} = \mathbf{x} - \mathbf{x}_p. \tag{5.29}$$

From the relative position vector, a distance function of two points $\mathbf{x}_p$ and $\mathbf{x}$ can be defined. Further, using such a distance function, the measuring function can be introduced herein. If $\mathbf{x}_p = \mathbf{0}$, such a position vector is called the absolute position vector.

Definition 5.13 For a given point $\mathbf{x}_p$, consider a flow $\mathbf{x} \in \mho(\mathbf{x}_p)$ in the switching system of Eq. (5.23). A relative distance function for the flow $\mathbf{x}$ to the fixed point $\mathbf{x}_p$ is defined by

$$d(\mathbf{x}, \mathbf{x}_p) = ||\mathbf{x} - \mathbf{x}_p||. \tag{5.30}$$

If $d(\mathbf{x}, \mathbf{x}_p) = C = const$, there is a surface given by

$$||\mathbf{x} - \mathbf{x}_p|| = C \tag{5.31}$$

which is called *the equi-distance surface* of point $\mathbf{x}_p$. Further, if there is a monotonically increasing or decreasing function of the relative distance $d(\mathbf{x}, \mathbf{x}_p)$,

$$E = V(\mathbf{x}, \mathbf{x}_p) \equiv f(d(\mathbf{x}, \mathbf{x}_p)) \tag{5.32}$$

with the following property

$$V(\mathbf{x}, \mathbf{x}_p) = f(d(\mathbf{x}, \mathbf{x}_p)) = \min \text{ (or max) if } d(\mathbf{x}, \mathbf{x}_p) = 0. \tag{5.33}$$

Such a monotonic function $V(\mathbf{x}, \mathbf{x}_p)$ is called a generalized measuring function of switching system in neighborhood of the point $\mathbf{x}_p$. If $E = C =$ const, there is a surface given by

$$V(\mathbf{x}, \mathbf{x}_p) = C \tag{5.34}$$

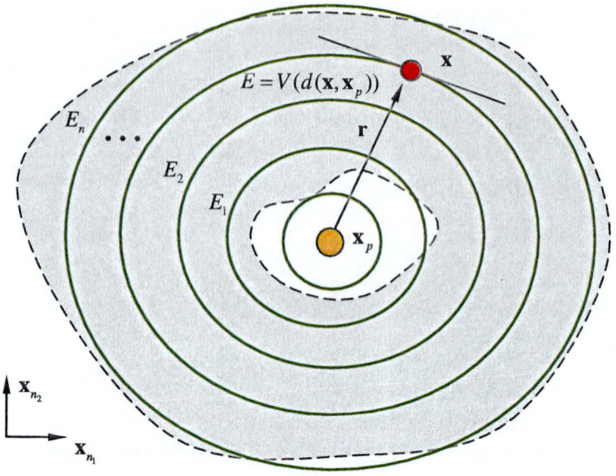

Fig. 5.4 A relative position vector from $\mathbf{x}_p$ to $\mathbf{x}$ and a measuring function with $n_1 + n_2 = n$

which is called the *equi-measuring function* surface.

Consider a distance function $d(\mathbf{x}, \mathbf{x}_p) \equiv [(\mathbf{x} - \mathbf{x}_p) \cdot (\mathbf{x} - \mathbf{x}_p)]^{1/2}$ as an example. From the foregoing definition, consider two monotonically increasing, metric functions to the relative distance to the point $\mathbf{x}_p$ as

$$\begin{aligned} V &= d(\mathbf{x}, \mathbf{x}_p) + C = [(\mathbf{x} - \mathbf{x}_p) \cdot (\mathbf{x} - \mathbf{x}_p)]^{1/2} + C \text{ or} \\ V &= (d(\mathbf{x}, \mathbf{x}_p))^2 + C = (\mathbf{x} - \mathbf{x}_p) \cdot (\mathbf{x} - \mathbf{x}_p) + C \end{aligned} \tag{5.35}$$

where C is an arbitrarily selected constant. Without losing generality, one can choose $C=0$. Similarly, the monotonically decreasing functions can be expressed. Such a selection of the monotonic, measuring functions is dependent on the convenience and efficiency in applications. From the foregoing discussion, the relative distance is a simple measuring function. From Eq. (5.14), the maximum and minimum relative distances in $\mho(\mathbf{x}_p)$ are C_1 and C_2, respectively. The minimum and maximum values of the monotonically increasing metric functions for domain $\mathbf{x} \in \mho(\mathbf{x}_p)$ are $E_{\min} = V(C_1)$ and $E_{\max} = V(C_2)$ from Eq. (5.32). For the two simple metric function in Eq. (5.33), one obtains the minimum and maximum values ($E_{\min} = C_1 + C$ and $E_{\max} = C_2 + C$) or ($E_{\min} = C_1^2 + C$ and $E_{\max} = C_2^2 + C$).

To explain the concept of the measuring function, consider a domain $\mho(\mathbf{x})$ in phase space for a subsystem in the vicinity of the given point $\mathbf{x}_p$ with big circular symbol, as shown in Fig. 5.4. Suppose a point $\mathbf{x}$ with a small circular symbol is the solution of the subsystem. The relative location vector to the given point $\mathbf{x}_p$ is expressed by a vector $\mathbf{r} = \mathbf{x} - \mathbf{x}_p$ and the corresponding relative distance is expressed by $d(\mathbf{x}, \mathbf{x}_p)$. Such a point $\mathbf{x}$ is on the equi-measuring function surface of $E = V(d(\mathbf{x}, \mathbf{x}_p))$. For different values of $E = E_i$ $(i = 1, 2, \cdots)$, a set of the equi-measuring function surfaces will fill the entire domain of $\mho(\mathbf{x})$, which are depicted by the green curves in Fig. 5.4. Based on the equi-measuring function surface, there is a dynamical system.

Definition 5.14 For any *equi-measuring function* surface in Eq. (5.34), there is a dynamical system as

$$\dot{\mathbf{x}}^m = \mathbf{f}^m(\mathbf{x}^m) \tag{5.36}$$

with the initial condition $(\mathbf{x}_0^m, t_0)$ and the equi-measuring function surface can be expressed by

$$V(\mathbf{x}^m, \mathbf{x}_p) = V(\mathbf{x}_0^m, \mathbf{x}_p) = E. \tag{5.37}$$

The dynamical system given in Eq. (5.36) is invariant in sense of the measuring function in Eq. (5.34). The subscript or superscript "m" represents the follow on the "equi-measuring function surface". This system can be designed from the practical applications. To measure the dynamical behaviors of any subsystems to the equi-measuring function surface, from Luo (2008a, b), the following functions are introduced.

Definition 5.15 Consider a flow $\mathbf{x}^{(i)}$ of the ith dynamical subsystem with a vector field $\mathbf{F}^{(i)}(\mathbf{x}^{(i)}, t, \mathbf{p}^{(i)})$ in Eq. (5.1). At time t, if the flow $\mathbf{x}^{(i)}$ arrives to the *equi-measuring function* surface with the corresponding constant ($E = C$) in Eq. (5.33), the kth-order, G-functions at the constant measuring function level are defined as

$$\begin{aligned}
G_m^{(k)}(\mathbf{x}^{(i)}, \mathbf{x}_p, t) &= (k+1)! \lim_{\varepsilon \to 0} \frac{1}{\varepsilon^{k+1}} \Big\{ [\mathbf{n}(\mathbf{x}^m, \mathbf{x}_p, t+\varepsilon)]^{\mathrm{T}} \cdot \mathbf{x}^{(i)}(t+\varepsilon) \\
&\quad - [\mathbf{n}(\mathbf{x}^m, \mathbf{x}_p, t)]^{\mathrm{T}} \cdot \mathbf{x}^{(i)}(t) - \sum_{q=1}^{k} \frac{1}{q!} G_m^{(q-1)}(\mathbf{x}^{(i)}, \mathbf{x}_p, t)\varepsilon^q \Big\} \\
&= \textstyle\sum_{r=1}^{k+1} C_{k+1}^r D^{(k+1-r)} [\mathbf{n}(\mathbf{x}^m, \mathbf{x}_p, t)]^{\mathrm{T}} \cdot [D^{(r-1)} \mathbf{F}^{(i)}(\mathbf{x}^{(i)}, t, \mathbf{p}^{(i)}) \\
&\quad - D^{(r-1)} \mathbf{f}^m(\mathbf{x}^m, \mathbf{x}_p)]\big|_{\mathbf{x}^m = \mathbf{x}^{(i)}}
\end{aligned} \tag{5.38}$$

for $k = 0, 1, 2, \cdots$. The normal vector of the equi-measuring function surface is

$$\mathbf{n}(\mathbf{x}^m, \mathbf{x}_p, t) = \frac{\partial V(\mathbf{x}^m, \mathbf{x}_p)}{\partial \mathbf{x}^m} = \left(\frac{\partial V}{\partial x_1^m}, \frac{\partial V}{\partial x_2^m}, \cdots, \frac{\partial V}{\partial x_n^m}\right)^{\mathrm{T}} \tag{5.39}$$

where the total differential operator is given by

$$\begin{aligned}
&D = \dot{\mathbf{x}} \frac{\partial}{\partial \mathbf{x}} + \frac{\partial}{\partial t}, \\
&D^{(r)}(\cdot) = D \circ D^{(r-1)}(\cdot) = D(D^{(r-1)}(\cdot)), \\
&\text{and } \ D^{(0)} = 1
\end{aligned} \tag{5.40}$$

with

$$C_{k+1}^r = \frac{(k+1)!}{r!(k+1-r)!} \ \text{ and } \ r! = 1 \times 2 \times 3 \cdots \times r. \tag{5.41}$$

From Eq. (5.34), the following relation holds

$$0 = \dot{\mathbf{x}} \cdot \frac{\partial V}{\partial \mathbf{x}} = \mathbf{n}(\mathbf{x}^m, \mathbf{x}_p) \cdot \mathbf{f}^m(\mathbf{x}^m, \mathbf{x}_p). \tag{5.42}$$

For a zero-order G-function ($k = 0$), one obtains

$$\begin{aligned} G_m^{(0)}(\mathbf{x}^{(i)}, \mathbf{x}_p, t) &= [\mathbf{n}(\mathbf{x}^m, \mathbf{x}_p)]^{\mathrm{T}} \cdot [\mathbf{F}^{(i)}(\mathbf{x}^{(i)}, t, \mathbf{p}^{(i)}) - \mathbf{f}^m(\mathbf{x}^m, \mathbf{x}_p)]\Big|_{\mathbf{x}^m = \mathbf{x}^{(i)}} \\ &= [\mathbf{n}(\mathbf{x}^{(i)}, \mathbf{x}_p)]^{\mathrm{T}} \cdot \mathbf{F}^{(i)}(\mathbf{x}^{(i)}, t, \mathbf{p}^{(i)}). \end{aligned} \tag{5.43}$$

The zero-order G-function is the dot product of the vector field $\mathbf{F}^{(i)}(\mathbf{x}^{(i)}, t, \mathbf{p}^{(i)})$ and normal vector $\mathbf{n}(\mathbf{x}^m, \mathbf{x}_p)$ for the ith subsystem. Consider an instantaneous value of the equi-measuring function at time t. In other words, letting $\mathbf{x}^m = \mathbf{x}^i$, equation (5.37) is

$$E^{(i)}(t) = V(\mathbf{x}^{(i)}, \mathbf{x}_p). \tag{5.44}$$

The corresponding time change ratio of the measuring function is

$$\begin{aligned} \frac{dE^{(i)}(t)}{dt} &= \frac{\partial V(\mathbf{x}^{(i)}, \mathbf{x}_p)}{\partial \mathbf{x}^{(i)}} \cdot \dot{\mathbf{x}}^{(i)} \\ &= [\mathbf{n}(\mathbf{x}^{(i)}, \mathbf{x}_p)]^{\mathrm{T}} \cdot \mathbf{F}^{(i)}(\mathbf{x}^{(i)}, t, \mathbf{p}^{(i)}) \\ &= G_m^{(0)}(\mathbf{x}^{(i)}, \mathbf{x}_p, t). \end{aligned} \tag{5.45}$$

From the foregoing equation, the change of the equi-measuring function for the ith dynamical subsystem for time $t \in [t_k, t_{k+1}]$ can be defined from Luo (2008a, b).

Definition 5.16 For a flow $\mathbf{x}^{(i)}$ of the ith dynamical subsystem with a vector field $\mathbf{F}^{(i)}(\mathbf{x}^{(i)}, t, \mathbf{p}^{(i)})$ in Eq. (5.1), consider the equi-measuring function $V(\mathbf{x}^{(i)}, \mathbf{x}_p)$ in Eq. (5.32) to be monotonically increased to a metric function $d(\mathbf{x}, \mathbf{x}_p)$ in Eq. (5.30). The total change of the equi-measuring function for the time interval $[t_k, t]$ is defined as

$$\begin{aligned} L^{(i)}(\mathbf{x}_p, t_k, t) &= \int_{t_k}^{t} \frac{dE^{(i)}(t)}{dt} = \int_{t_k}^{t} G_m^{(0)}(\mathbf{x}^{(i)}, \mathbf{x}_p, t)dt \\ &= \int_{t_k}^{t} [\mathbf{n}(\mathbf{x}^{(i)}, \mathbf{x}_p)]^{\mathrm{T}} \cdot \mathbf{F}^{(i)}(\mathbf{x}^{(i)}, t, \mathbf{p}^{(i)})dt \\ &= V(\mathbf{x}^{(i)}(t), \mathbf{x}_p) - V(\mathbf{x}_k^{(i)}, \mathbf{x}_p) \end{aligned} \tag{5.46}$$

where $\mathbf{x}_k^{(i)} = \mathbf{x}^{(i)}(t_k)$. For a given $t = t_{k+1} > t_k$, the increment of the equi-measuring function to the ith subsystem in Eq. (5.1) for $t \in [t_k, t_{k+1}]$ is

$$\begin{aligned} L^{(i)}(\mathbf{x}_p, t_k, t_{k+1}) &= \int_{t_k}^{t_{k+1}} G_m^{(0)}(\mathbf{x}^{(i)}, \mathbf{x}_p, t)dt \\ &= \int_{t_k}^{t_{k+1}} [\mathbf{n}(\mathbf{x}^{(i)}, \mathbf{x}_p)]^{\mathrm{T}} \cdot \mathbf{F}^{(i)}(\mathbf{x}^{(i)}, t, \mathbf{p}^{(i)})dt \\ &= V(\mathbf{x}_{k+1}^{(i)}, \mathbf{x}_p) - V(\mathbf{x}_k^{(i)}, \mathbf{x}_p). \end{aligned} \tag{5.47}$$

From Eq. (5.47), it is observed that the equi-measuring function quantity is used to measure changes of the ith subsystem. Thus such a function can be used to investigate the stability of dynamical systems. Before the stability theory of the switching system is discussed, the following concepts are defined first.

Definition 5.17 For the ith dynamical subsystem in Eq. (5.1), consider the equi-measuring function $V(\mathbf{x}^{(i)}, \mathbf{x}_p)$ in Eq. (5.32) to be monotonically increased to a metric function $d(\mathbf{x}, \mathbf{x}_p)$ in Eq. (5.30). A flow $\mathbf{x}^{(i)}(t)$ at $\mathbf{x}_k^{(i)}$ for $t = t_k$ in the domain $\Omega_i \subset \mathscr{R}^n$ is:

(i) locally decreasing to the equi-measuring function surface in $\Omega_i \subset \mathscr{R}^n$ if

$$\left.\begin{aligned} V(\mathbf{x}_k^{(i)}, \mathbf{x}_p, t_k) - V(\mathbf{x}_{k-\varepsilon}^{(i)}, \mathbf{x}_p, t_{k-\varepsilon}) < 0, \\ V(\mathbf{x}_{k+\varepsilon}^{(i)}, \mathbf{x}_p, t_{k+\varepsilon}) - V(\mathbf{x}_k^{(i)}, \mathbf{x}_p, t_k) < 0; \end{aligned}\right\} \tag{5.48}$$

(ii) locally increasing to the equi-measuring function surface in $\Omega_i \subset \mathscr{R}^n$ if

$$\left.\begin{aligned} V(\mathbf{x}_k^{(i)}, \mathbf{x}_p, t_k) - V(\mathbf{x}_{k-\varepsilon}^{(i)}, \mathbf{x}_p, t_{k-\varepsilon}) > 0, \\ V(\mathbf{x}_{k+\varepsilon}^{(i)}, \mathbf{x}_p, t_{k+\varepsilon}) - V(\mathbf{x}_k^{(i)}, \mathbf{x}_p, t_k) > 0; \end{aligned}\right\} \tag{5.49}$$

(iii) locally tangential to the equi-measuring function surface in $\Omega_i \subset \mathscr{R}^n$ if

$$\begin{aligned} &\text{either} && \left.\begin{aligned} V(\mathbf{x}_k^{(i)}, \mathbf{x}_p, t_k) - V(\mathbf{x}_{k-\varepsilon}^{(i)}, \mathbf{x}_p, t_{k-\varepsilon}) < 0, \\ V(\mathbf{x}_{k+\varepsilon}^{(i)}, \mathbf{x}_p, t_{k+\varepsilon}) - V(\mathbf{x}_k^{(i)}, \mathbf{x}_p, t_k) > 0; \end{aligned}\right\} \\ &\text{or} && \left.\begin{aligned} V(\mathbf{x}_k^{(i)}, \mathbf{x}_p, t_k) - V(\mathbf{x}_{k-\varepsilon}^{(i)}, \mathbf{x}_p, t_{k-\varepsilon}) > 0, \\ V(\mathbf{x}_{k+\varepsilon}^{(i)}, \mathbf{x}_p, t_{k+\varepsilon}) - V(\mathbf{x}_k^{(i)}, \mathbf{x}_p, t_k) < 0. \end{aligned}\right\} \end{aligned} \tag{5.50}$$

From the previous definitions, the locally increasing and decreasing of a flow $\mathbf{x}^{(i)}(t)$ at $\mathbf{x}_k$ to the measuring function surface can be described in Fig. 5.5 in the vicinity of the point $\mathbf{x}_p$. A flow $\mathbf{x}^{(i)}(t)$ at $\mathbf{x}_k^{(i)}$, locally tangential to the equi-measuring function surface, can be similarly sketched.

Theorem 5.1 *For the ith dynamical subsystem in Eq. (5.1), consider the equi-measuring function $V(\mathbf{x}^{(i)}, \mathbf{x}_p)$ in Eq. (5.32) to be monotonically increased to a metric function $d(\mathbf{x}, \mathbf{x}_p)$ in Eq. (5.30). A flow $\mathbf{x}^{(i)}(t)$ at $\mathbf{x}_k^{(i)}$ for $t = t_k$ in the domain $\Omega_i \subset \mathscr{R}^n$ is:*

(i) *locally decreasing to the equi-measuring function surface in $\Omega_i \subset \mathscr{R}^n$ if and only if*

$$G_m^{(0)}(\mathbf{x}_k^{(i)}, \mathbf{x}_p, t) = [\mathbf{n}(\mathbf{x}_k^{(i)}, \mathbf{x}_p)]^{\mathrm{T}} \cdot \mathbf{F}^{(i)}(\mathbf{x}_k^{(i)}, t_k, \mathbf{p}^{(i)}) < 0; \tag{5.51}$$

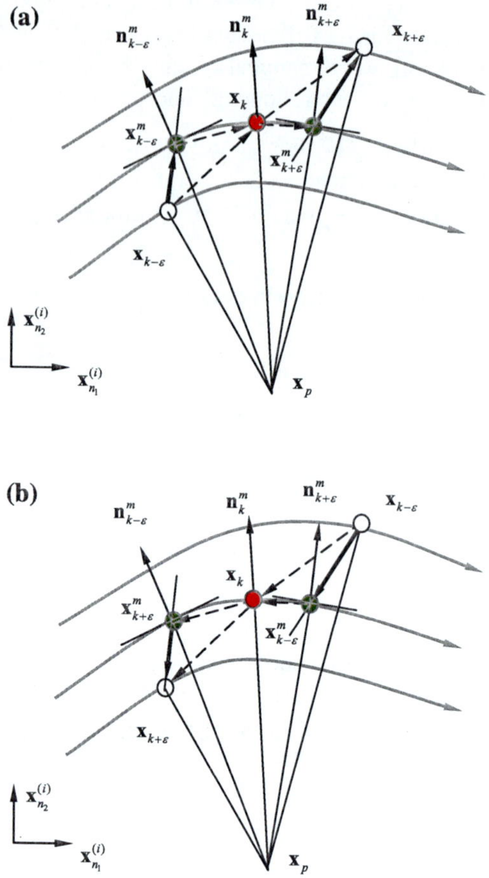

Fig. 5.5 **a** A locally increasing flow, and **b** a locally decreasing flow to a measuring function with $n_1 + n_2 = n$

(ii) *locally increasing to the equi-measuring function surface in* $\Omega_i \subset \mathscr{R}^n$ *if and only if*

$$G_m^{(0)}(\mathbf{x}_k^{(i)}, \mathbf{x}_p, t) = [\mathbf{n}(\mathbf{x}_k^{(i)}, \mathbf{x}_p)]^{\mathrm{T}} \cdot \mathbf{F}^{(i)}(\mathbf{x}_k^{(i)}, t_k, \mathbf{p}^{(i)}) > 0; \tag{5.52}$$

(iii) *locally tangential to the equi-measuring function surface in* $\Omega_i \subset \mathscr{R}^n$ *if and only if*

$$\begin{aligned} &G_m^{(0)}(\mathbf{x}_k^{(i)}, \mathbf{x}_p, t) = [\mathbf{n}(\mathbf{x}_k^{(i)}, \mathbf{x}_p)]^{\mathrm{T}} \cdot \mathbf{F}^{(i)}(\mathbf{x}_k^{(i)}, t_k, \mathbf{p}^{(i)}) = 0; \\ &G_m^{(1)}(\mathbf{x}_k^{(i)}, \mathbf{x}_p, t) \neq 0. \end{aligned} \tag{5.53}$$

Proof Using G-functions and Taylor series expansion, this theorem can be proved directly. ■

Definition 5.18 For the *i*th dynamical subsystem in Eq. (5.1), consider the equi-measuring function $V(\mathbf{x}^{(i)}, \mathbf{x}_p)$ in Eq. (5.32) to be monotonically increased to a metric function $d(\mathbf{x}, \mathbf{x}_p)$ in Eq. (5.30). A flow $\mathbf{x}^{(i)}(t)$ from $\mathbf{x}_k^{(i)}$ to $\mathbf{x}_{k+1}^{(i)}$ for $t = t_s \in [t_k, t_{k+1}]$ in the domain $\Omega_i \subset \mathscr{R}^n$ is:

(i) uniformly decreasing to the equi-measuring function surface in $\Omega_i \subset \mathscr{R}^n$ if

$$\left.\begin{aligned} V(\mathbf{x}_s^{(i)}, \mathbf{x}_p, t_s) - V(\mathbf{x}_{s-\varepsilon}^{(i)}, \mathbf{x}_p, t_{s-\varepsilon}) < 0, \\ V(\mathbf{x}_{s+\varepsilon}^{(i)}, \mathbf{x}_p, t_{s+\varepsilon}) - V(\mathbf{x}_s^{(i)}, \mathbf{x}_p, t_s) < 0; \end{aligned}\right\} \tag{5.54}$$

(ii) uniformly increasing to the equi-measuring function surface in $\Omega_i \subset \mathscr{R}^n$ if

$$\left.\begin{aligned} V(\mathbf{x}_s^{(i)}, \mathbf{x}_p, t_s) - V(\mathbf{x}_{s-\varepsilon}^{(i)}, \mathbf{x}_p, t_{s-\varepsilon}) > 0, \\ V(\mathbf{x}_{s+\varepsilon}^{(i)}, \mathbf{x}_p, t_{s+\varepsilon}) - V(\mathbf{x}_s^{(i)}, \mathbf{x}_p, t_s) > 0; \end{aligned}\right\} \tag{5.55}$$

(iii) uniformly invariant to the equi-measuring function surface in $\Omega_i \subset \mathscr{R}^n$ if

$$V(\mathbf{x}_{s+\varepsilon}^{(i)}, \mathbf{x}_p, t_{s+\varepsilon}) = V(\mathbf{x}_s^{(i)}, \mathbf{x}_p, t_s) = V(\mathbf{x}_{s-\varepsilon}^{(i)}, \mathbf{x}_p, t_{s-\varepsilon}). \tag{5.56}$$

Theorem 5.2 *For the dynamical subsystem in Eq. (5.1), consider the equi-measuring function $V(\mathbf{x}^{(i)}, \mathbf{x}_p)$ in Eq. (5.32) to be monotonically increased to a metric function $d(\mathbf{x}, \mathbf{x}_p)$ in Eq. (5.30). A flow $\mathbf{x}^{(i)}(t)$ from $\mathbf{x}_k^{(i)}$ to $\mathbf{x}_{k+1}^{(i)}$ for $t \in [t_k, t_{k+1}]$ in the domain $\Omega_i \subset \mathscr{R}^n$ is:*

(i) *uniformly decreasing to the equi-measuring function surface in $\Omega_i \subset \mathscr{R}^n$ if and only if all points $\mathbf{x}^{(i)}(t)$ for $t \in [t_k, t_{k+1}]$ on the flow γ satisfy the following condition*

$$G_m^{(0)}(\mathbf{x}^{(i)}, \mathbf{x}_p, t) = [\mathbf{n}(\mathbf{x}^{(i)}, \mathbf{x}_p)]^{\mathrm{T}} \cdot \mathbf{F}^{(i)}(\mathbf{x}^{(i)}, t, \mathbf{p}^{(i)}) < 0; \tag{5.57}$$

(ii) *uniformly increasing to the equi-measuring function surface in $\Omega_i \subset \mathscr{R}^n$ if and only if all points $\mathbf{x}^{(i)}(t)$ for $t \in [t_k, t_{k+1}]$ on the flow γ satisfy the following condition*

$$G_m^{(0)}(\mathbf{x}^{(i)}, \mathbf{x}_p, t) = [\mathbf{n}(\mathbf{x}^{(i)}, \mathbf{x}_p)]^{\mathrm{T}} \cdot \mathbf{F}^{(i)}(\mathbf{x}^{(i)}, t, \mathbf{p}^{(i)}) > 0; \tag{5.58}$$

(iii) *uniformly invariant to the equi-measuring function surface in $\Omega_i \subset \mathscr{R}^n$ if and only if all points $\mathbf{x}^{(i)}(t)$ for $t \in [t_k, t_{k+1}]$ on the flow γ satisfy the following condition*

$$G_m^{(k)}(\mathbf{x}^{(i)}, \mathbf{x}_p, t) = 0 \; k = 0, 1, 2, \cdots . \tag{5.59}$$

Proof Using G-functions and Taylor series expansion, this theorem can be proved directly. ■

Definition 5.19 For the ith dynamical subsystem in Eq. (5.1), consider the equi-measuring function $V(\mathbf{x}^{(i)}, \mathbf{x}_p)$ in Eq. (5.32) to be monotonically increased to a metric function $d(\mathbf{x}, \mathbf{x}_p)$ in Eq. (5.30). A flow $\mathbf{x}^{(i)}(t)$ at $\mathbf{x}_k^{(i)}$ for $t = t_k$ in the domain $\Omega_i \subset \mathscr{R}^n$ is:

(i) locally decreasing with the $(2s)$th-order to the equi-measuring function surface in $\Omega_i \subset \mathscr{R}^n$ if

$$\begin{aligned} &G_m^{(r)}(\mathbf{x}_k^{(i)}, \mathbf{x}_p, t_k) = 0 \text{ for } r = 0, 1, 2, \cdots, 2s-1; \\ &V(\mathbf{x}_k^{(i)}, \mathbf{x}_p, t_k) - V(\mathbf{x}_{k-\varepsilon}^{(i)}, \mathbf{x}_p, t_{k-\varepsilon}) < 0, \\ &V(\mathbf{x}_{k+\varepsilon}^{(i)}, \mathbf{x}_p, t_{k+\varepsilon}) - V(\mathbf{x}_k^{(i)}, \mathbf{x}_p, t_k) < 0; \end{aligned} \tag{5.60}$$

(ii) locally increasing with the $(2s)$th-order to the equi-measuring function surface in $\Omega_i \subset \mathscr{R}^n$ if

$$\begin{aligned} &G_m^{(r)}(\mathbf{x}_k^{(i)}, \mathbf{x}_p, t_k) = 0 \text{ for } r = 0, 1, 2, \cdots, 2s-1; \\ &V(\mathbf{x}_k^{(i)}, \mathbf{x}_p, t_k) - V(\mathbf{x}_{k-\varepsilon}^{(i)}, \mathbf{x}_p, t_{k-\varepsilon}) > 0, \\ &V(\mathbf{x}_{k+\varepsilon}^{(i)}, \mathbf{x}_p, t_{k+\varepsilon}) - V(\mathbf{x}_k^{(i)}, \mathbf{x}_p, t_k) > 0; \end{aligned} \tag{5.61}$$

(iii) locally tangential with the $(2s-1)$th-order to the equi-measuring function surface in $\Omega_i \subset \mathscr{R}^n$ if

$$\begin{aligned} &G_m^{(r)}(\mathbf{x}_k^{(i)}, \mathbf{x}_p, t_k) = 0 \text{ for } r = 0, 1, 2, \cdots, 2s; \\ &\text{either } \left.\begin{aligned} &V(\mathbf{x}_k^{(i)}, \mathbf{x}_p, t_k) - V(\mathbf{x}_{k-\varepsilon}^{(i)}, \mathbf{x}_p, t_{k-\varepsilon}) < 0, \\ &V(\mathbf{x}_{k+\varepsilon}^{(i)}, \mathbf{x}_p, t_{k+\varepsilon}) - V(\mathbf{x}_k^{(i)}, \mathbf{x}_p, t_k) < 0; \end{aligned}\right\} \\ &\text{or } \left.\begin{aligned} &V(\mathbf{x}_k^{(i)}, \mathbf{x}_p, t_k) - V(\mathbf{x}_{k-\varepsilon}^{(i)}, \mathbf{x}_p, t_{k-\varepsilon}) > 0, \\ &V(\mathbf{x}_{k+\varepsilon}^{(i)}, \mathbf{x}_p, t_{k+\varepsilon}) - V(\mathbf{x}_k^{(i)}, \mathbf{x}_p, t_k) > 0. \end{aligned}\right\} \end{aligned} \tag{5.62}$$

Theorem 5.3 *For the ith dynamical subsystem in Eq. (5.1), consider the equi-measuring function $V(\mathbf{x}^{(i)}, \mathbf{x}_p)$ in Eq. (5.32) to be monotonically increased to a metric function $d(\mathbf{x}, \mathbf{x}_p)$ in Eq. (5.30). A flow $\mathbf{x}^{(i)}(t)$ at $\mathbf{x}_k^{(i)}$ for $t = t_k$ in the domain $\Omega_i \subset \mathscr{R}^n$ is:*

(i) *locally decreasing with the $(2s)$th-order to the equi-measuring function surface in $\Omega_i \subset \mathscr{R}^n$ if and only if*

$$\begin{aligned} &G_m^{(r)}(\mathbf{x}_k, \mathbf{x}_p, t_k) = 0, \quad \text{for } r = 0, 1, 2, \cdots, 2s-1 \\ &G_m^{(2s)}(\mathbf{x}_k, \mathbf{x}_p, t_k) < 0; \end{aligned} \tag{5.63}$$

(ii) *locally increasing with the $(2s)$th-order to the equi-measuring function surface in $\Omega_i \subset \mathscr{R}^n$ if and only if*

$$\begin{aligned} &G_m^{(r)}(\mathbf{x}_k, \mathbf{x}_p, t_k) = 0, \quad \text{for } r = 0, 1, 2, \cdots, 2s-1 \\ &G_m^{(2s)}(\mathbf{x}_k, \mathbf{x}_p, t_k) > 0; \end{aligned} \tag{5.64}$$

(iii) *locally tangential with the* $(2s-1)$th-*order to the equi-measuring function surface in* $\Omega_i \subset \mathscr{R}^n$ *if and only if*

$$\begin{aligned} &G_m^{(r)}(\mathbf{x}_k, \mathbf{x}_p, t_k) = 0, \quad \text{for } r = 0, 1, 2, \cdots, 2s \\ &G_m^{(2s+1)}(\mathbf{x}_k, \mathbf{x}_p, t_k) \neq 0. \end{aligned} \tag{5.65}$$

Proof Using G-functions and Taylor series expansion, this theorem can be proved directly. ■

Definition 5.20 For the ith dynamical subsystem in Eq. (5.1), consider the equi-measuring function $V(\mathbf{x}^{(i)}, \mathbf{x}_p)$ in Eq. (5.32) to be monotonically increased to a metric function $d(\mathbf{x}, \mathbf{x}_p)$ in Eq. (5.30). A flow $\mathbf{x}^{(i)}(t)$ from $\mathbf{x}_k^{(i)}$ to $\mathbf{x}_{k+1}^{(i)}$ for $t \in [t_k, t_{k+1}]$ in the domain $\Omega_i \subset \mathscr{R}^n$ is:

(i) globally decreasing to the equi-measuring function surface in $\Omega_i \subset \mathscr{R}^n$ if the equi-measuring function increment for $t \in [t_k, t_{k+1}]$ is less than zero, i.e.,

$$L^{(i)}(\mathbf{x}_p, t_k, t_{k+1}) \equiv \int_{t_k}^{t_{k+1}} [\mathbf{n}(\mathbf{x}^{(i)}, \mathbf{x}_p)]^{\mathrm{T}} \cdot \mathbf{F}^{(i)}(\mathbf{x}^{(i)}, t, \mathbf{p}^{(i)})dt < 0; \tag{5.66}$$

(ii) globally increasing to the equi-measuring function surface in $\Omega_i \subset \mathscr{R}^n$ if the equi-measuring function increment for $t \in [t_k, t_{k+1}]$ is greater than zero, i.e.,

$$L^{(i)}(\mathbf{x}_p, t_k, t_{k+1}) \equiv \int_{t_k}^{t_{k+1}} [\mathbf{n}(\mathbf{x}^{(i)}, \mathbf{x}_p)]^{\mathrm{T}} \cdot \mathbf{F}^{(i)}(\mathbf{x}^{(i)}, t, \mathbf{p}^{(i)})dt > 0; \tag{5.67}$$

(iii) globally invariant to the equi-measuring function surface in $\Omega_i \subset \mathscr{R}^n$ if the equi-measuring function increment for $t \in [t_k, t_{k+1}]$ is equal to zero, i.e.,

$$L^{(i)}(\mathbf{x}_p, t_k, t_{k+1}) \equiv \int_{t_k}^{t_{k+1}} [\mathbf{n}(\mathbf{x}^{(i)}, \mathbf{x}_p)]^{\mathrm{T}} \cdot \mathbf{F}^{(i)}(\mathbf{x}^{(i)}, t, \mathbf{p}^{(i)})dt = 0. \tag{5.68}$$

From the foregoing definition, the global increase, decrease and invariance for a single subsystem are defined. The L-function is pertaining to the averaging of a flow $\mathbf{x}^{(i)}(t)$ to the equi-measuring function surface of $V(\mathbf{x}^{(i)}, \mathbf{x}_p) = C$ for the time period of $[t_k, t_{k+1}]$. To determine the global increase, decrease and invariance for a switching system, the resultant flow of the switching system is the union of all the flows of the subsystems in a certain queue series, and the corresponding L-function can be defined.

Definition 5.21 For a switching system in Eqs. (5.15) and (5.16) on the domain $\mho = \cup_{i=1}^{m_1} \Omega_{l_i} \cup \Omega_0$, there is a resultant flow $\gamma(t_0, t)$, i.e.,

$$\gamma(t_0, t) = \cup_{k=1}^{l-1} \gamma^{(\alpha_k)}(t_{k-1}, t_k) + \gamma^{(\alpha_l)}(t_{l-1}, t), \tag{5.69}$$

where

$$\gamma^{(\alpha_k)}(t_{k-1}, t_k) = \left\{ \mathbf{x}^{(\alpha_k)}(t) \left| \begin{array}{l} S_{\alpha_k} : \dot{\mathbf{x}}^{(\alpha_k)} = \mathbf{F}^{(\alpha_k)}(\mathbf{x}^{(\alpha_k)}, t, \mathbf{p}^{(\alpha_k)}) \\ \text{on } \Omega_{\alpha_k} \text{ with } \mathbf{x}^{(i)}(t_{k-1}) = \mathbf{x}_{k-1}^{(i)} \\ \text{for all } t \in [t_{k-1}, t_k] \end{array} \right. \right\}. \tag{5.70}$$

The corresponding L-function along the resultant flow $\gamma(t_0, t)$ is defined as

$$L(\mathbf{x}_p, t_0, t) = \sum_{k=1}^{l-1} \left[L^{(\alpha_k)}(\mathbf{x}_p, t_{k-1}, t_k) + \Delta^{(\alpha_k \alpha_{k+1})}(t_k) \right] + L^{(\alpha_l)}(\mathbf{x}_p, t_{l-1}, t), \tag{5.71}$$

where $\Delta^{(\alpha_k \alpha_{k+1})}(t_k)$ is the quantity increment of the equi-measuring function surface for the switching from dynamical system S_{α_k} to dynamical systems $S_{\alpha_{k+1}}$ through the transport laws in Eq. (5.16), i.e., the quantity increment is equal to

$$\Delta^{(\alpha_k \alpha_{k+1})}(t_k) = V(\mathbf{x}_k^{(\alpha_{k+1})}, \mathbf{x}_p) - V(\mathbf{x}_k^{(\alpha_k)}, \mathbf{x}_p). \tag{5.72}$$

Note that the resultant L-function for the total flow $\gamma(t_0, t)$ is computed by

$$\begin{aligned} L(\mathbf{x}_p, t_0, t) &= \sum_{k=1}^{l-1} \left[L^{(\alpha_k)}(\mathbf{x}_p, t_{k-1}, t_k) + \Delta^{(\alpha_k \alpha_{k+1})} \right] + L^{(\alpha_l)}(\mathbf{x}_p, t_{l-1}, t) \\ &= \sum_{k=1}^{l-1} [\int_{t_{k-1}}^{t_k} G_m^{(0)}(\mathbf{x}^{(\alpha_k)}, \mathbf{x}_p, t)dt + \Delta^{(\alpha_k \alpha_{k+1})}] \\ &\quad + \int_{t_{l-1}}^{t} G_m^{(0)}(\mathbf{x}^{(\alpha_l)}, \mathbf{x}_p, t)dt \\ &= \sum_{k=1}^{l-1} \left\{ \int_{t_{k-1}}^{t_k} [\mathbf{n}(\mathbf{x}^{(\alpha_k)}, \mathbf{x}_p)]^{\mathrm{T}} \cdot \mathbf{F}^{(\alpha_k)}(\mathbf{x}^{(\alpha_k)}, t, \mathbf{p}^{(\alpha_k)})dt \right. \\ &\quad \left. + \Delta^{(\alpha_k \alpha_{k+1})} \right\} + \int_{t_{l-1}}^{t} [\mathbf{n}(\mathbf{x}^{(\alpha_l)}, \mathbf{x}_p)]^{\mathrm{T}} \cdot \mathbf{F}^{(\alpha_l)}(\mathbf{x}^{(\alpha_l)}, t, \mathbf{p}^{(\alpha_l)})dt \\ &= \sum_{k=1}^{l-1} \left[V(\mathbf{x}_k^{(\alpha_k)}, \mathbf{x}_p) - V(\mathbf{x}_{k-1}^{(\alpha_k)}, \mathbf{x}_p) + \Delta^{(\alpha_k \alpha_{k+1})} \right] \\ &\quad + V(\mathbf{x}^{(\alpha_l)}(t), \mathbf{x}_p) - V(\mathbf{x}_{l-1}^{(\alpha_l)}, \mathbf{x}_p). \end{aligned} \tag{5.73}$$

The resultant L-function in Eq. (5.73) can be sketched through the equi-measuring functions, as shown in Fig. 5.6. The measuring functions for each subsystem may not always increase or decrease with increasing time. The total effect of equi-measuring functions to all the queue series subsystems in the switching system should be considered. The solid curves give the equi-measuring-function values of a flow $\gamma^{(\alpha_k)}(t)$ $(k = 1, 2, \cdots, m, \cdots)$. The dashed lines are jumping changes of the equi-measuring function, and the jumping changes are caused by the transport laws between two adjacent queue subsystems. The circular points represent the switching points for two adjacent queue subsystems. The L-function changes for the switching system is clearly presented.

Since any switching system possesses a queue series of subsystems, the total effect of equi-measuring functions to all the queue series subsystems in the switching system should be considered. If each system is uniformly increasing to the equi-measuring function, with a positive impulse, then the switching system is uniformly

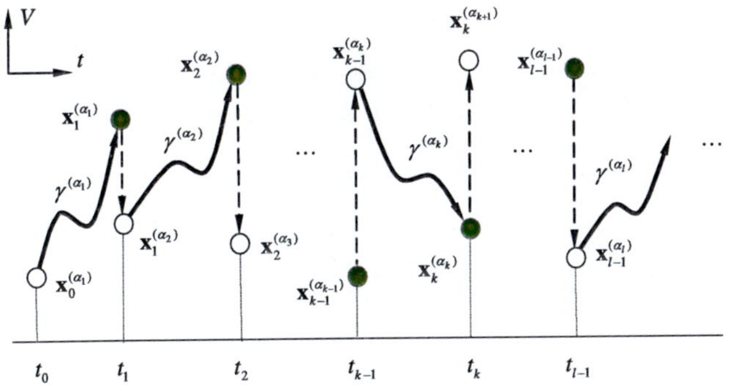

Fig. 5.6 The equi-measuring function varying with flows of sub-systems. The *solid curves* give the equi-measuring function values of a flow $\gamma^{(\alpha_k)}(t)$. The *dashed lines* are jumping changes of the equi-measuring function, and the jumping changes are caused by the transport laws between two adjacent queue subsystems. The circular points represent the switching points for two adjacent queue subsystems

increasing. Otherwise, if each system is uniformly decreasing to the equi-measuring functions, with a negative impulse, then the switching system is uniformly decreasing.

Definition 5.22 For a switching system in Eqs. (5.15) and (5.16) on the domain $\mho = \cup_{i=1}^{m_1} \Omega_{l_i} \cup \Omega_0$, there is an resultant flow $\gamma(t_0, t)$ in Eq. (5.69). For the α_kth dynamical subsystem S_{α_k}, consider the equi-measuring function $V(\mathbf{x}^{(\alpha_k)}, \mathbf{x}_p)$ in Eq. (5.32) to be monotonically increased to a metric function $d(\mathbf{x}, \mathbf{x}_p)$ in Eq. (5.30). A flow $\gamma(t_0, t)$ from $\gamma^{(\alpha_1)}(t_0, t_1)$ to $\gamma^{(\alpha_l)}(t_{l-1}, t)$ for $t \in [t_0, t_l]$ in the domain $\cup_{k=1}^{l} \Omega_{\alpha_k}$ is:

(i) uniformly decreasing to the equi-measuring function surface in $\cup_{k=1}^{l} \Omega_{\alpha_k}$ if

$$\left.\begin{aligned} &V(\mathbf{x}_s^{(\alpha_k)}, \mathbf{x}_p, t_s) - V(\mathbf{x}_{s-\varepsilon}^{(\alpha_k)}, \mathbf{x}_p, t_{s-\varepsilon}) < 0, \\ &V(\mathbf{x}_{s+\varepsilon}^{(\alpha_k)}, \mathbf{x}_p, t_{s+\varepsilon}) - V(\mathbf{x}_s^{(\alpha_k)}, \mathbf{x}_p, t_s) < 0, \\ &\text{for } \mathbf{x}_s^{(\alpha_k)} \in \gamma^{(\alpha_k)}(t_{k-1}, t_k) \text{ with } t_s \in [t_{k-1}, t_k]; \\ &\Delta^{(\alpha_k \alpha_{k+1})}(t_k) < 0 \text{ for } k = 1, 2, \cdots, l; \end{aligned}\right\} \tag{5.74}$$

(ii) uniformly increasing to the equi-measuring function surface in $\cup_{k=1}^{l} \Omega_{\alpha_k}$ if

$$\left.\begin{aligned} &V(\mathbf{x}_s^{(\alpha_k)}, \mathbf{x}_p, t_s) - V(\mathbf{x}_{s-\varepsilon}^{(\alpha_k)}, \mathbf{x}_p, t_{s-\varepsilon}) > 0, \\ &V(\mathbf{x}_{s+\varepsilon}^{(\alpha_k)}, \mathbf{x}_p, t_{s+\varepsilon}) - V(\mathbf{x}_s^{(\alpha_k)}, \mathbf{x}_p, t_s) > 0; \\ &\text{for } \mathbf{x}_s^{(\alpha_k)} \in \gamma^{(\alpha_k)}(t_{k-1}, t_k) \text{ with } t_s \in [t_{k-1}, t_k] \\ &\Delta^{(\alpha_k \alpha_{k+1})}(t_k) > 0 \text{ for } k = 1, 2, \cdots, l; \end{aligned}\right\} \tag{5.75}$$

(iii) uniformly and negatively impulsive to the equi-measuring function surface in $\cup_{k=1}^{l} \Omega_{\alpha_k}$ if

$$\left.\begin{array}{l}V(\mathbf{x}_{s+\varepsilon}^{(\alpha_k)},\mathbf{x}_p,t_{s+\varepsilon})=V(\mathbf{x}_s^{(\alpha_k)},\mathbf{x}_p,t_s)=V(\mathbf{x}_{s-\varepsilon}^{(\alpha_k)},\mathbf{x}_p,t_{s-\varepsilon})\\ \text{for }\ \mathbf{x}_s^{(\alpha_k)}\in\gamma^{(\alpha_k)}(t_{k-1},t_k)\ \text{ with }\ t_s\in[t_{k-1},t_k],\\ \Delta^{(\alpha_k\alpha_{k+1})}(t_k)<0\ \text{ for }\ k=1,2,\cdots,l;\end{array}\right\} \tag{5.76}$$

(iv) uniformly and positively impulsive to the equi-measuring function surface in $\cup_{k=1}^{l}\Omega_{\alpha_k}$ if

$$\left.\begin{array}{l}V(\mathbf{x}_{s+\varepsilon}^{(\alpha_k)},\mathbf{x}_p,t_{s+\varepsilon})=V(\mathbf{x}_s^{(\alpha_k)},\mathbf{x}_p,t_s)=V(\mathbf{x}_{s-\varepsilon}^{(\alpha_k)},\mathbf{x}_p,t_{s-\varepsilon})\\ \text{for }\ \mathbf{x}_s^{(\alpha_k)}\in\gamma^{(\alpha_k)}(t_{k-1},t_k)\ \text{ with }\ t_s\in[t_{k-1},t_k],\\ \Delta^{(\alpha_k\alpha_{k+1})}(t_k)>0\ \text{ for }\ k=1,2,\cdots,l;\end{array}\right\} \tag{5.77}$$

(v) uniformly invariant to the equi-measuring function surface in $\cup_{k=1}^{l}\Omega_{\alpha_k}$ if

$$\left.\begin{array}{l}V(\mathbf{x}_{s+\varepsilon}^{(\alpha_k)},\mathbf{x}_p,t_{s+\varepsilon})=V(\mathbf{x}_s^{(\alpha_k)},\mathbf{x}_p,t_s)=V(\mathbf{x}_{s-\varepsilon}^{(\alpha_k)},\mathbf{x}_p,t_{s-\varepsilon})\\ \text{for }\ \mathbf{x}_s^{(\alpha_k)}\in\gamma^{(\alpha_k)}(t_{k-1},t_k)\ \text{ with }\ t_s\in[t_{k-1},t_k],\\ \Delta^{(\alpha_k\alpha_{k+1})}(t_k)=0\ \text{ for } k=1,2,\cdots,l.\end{array}\right\} \tag{5.78}$$

From the foregoing definition, the uniformly increasing and decreasing of a resultant flow of the switching system to the equi-measuring surface require that each subsystem should be uniformly increasing with a positive impulse and uniformly decreasing with a negative impulse, respectively. For an intuitive illustration, such uniformly increasing and decreasing of a switching system are presented in Fig. 5.7a and b, respectively. The corresponding trajectories in phase space are presented in Fig. 5.8a and b for uniformly increasing and decreasing.

Consider a dynamical system

$$\dot{x}^{(i)}=y^{(i)}\ \text{and}\ \dot{y}^{(i)}=-c^{(i)}x^{(i)}-2d^{(i)}y^{(i)} \tag{5.79}$$

with an impulsive function

$$\begin{array}{l}x_{k+}^{(i)}=x_{k-}^{(i)},\ y_{k+}^{(i)}=y_{k-}^{(i)}+a^{(i)}\mathrm{sgn}(y_{k-}^{(i)})(k=1,2,\cdots)\\ \text{for }\ t_k=kT/m\ \text{ with }\ T=2\pi/\sqrt{c},\ a^{(i)}>0.\end{array} \tag{5.80}$$

For the impulsive system in Eq. (5.79), consider a measuring function

$$V^{(i)}=\frac{1}{2}(y^{(i)})^2+\frac{1}{2}c^{(i)}(x^{(i)})^2. \tag{5.81}$$

The solutions for Eq. (5.79) can be easily obtained. Herein, for $(d^{(i)})^2<c^{(i)}$, the solution is given for $t\in[t_{k+},t_{k+1})$

$$\begin{aligned}x^{(i)}&=e^{-d^{(i)}(t-t_{k+})}[C_1^{(i)}\cos\omega_d^{(i)}(t-t_{k+})+C_2^{(i)}\sin\omega_d^{(i)}(t-t_{k+})],\\ y^{(i)}&=e^{-d^{(i)}(t-t_{k+})}[(C_2^{(i)}\omega_d^{(i)}-C_1^{(i)}d^{(i)})\cos\omega_d^{(i)}(t-t_{k+})\\ &\quad-(C_2^{(i)}d^{(i)}+\omega_d^{(i)}C_1^{(i)})\sin\omega_d^{(i)}(t-t_{k+})],\end{aligned} \tag{5.82}$$

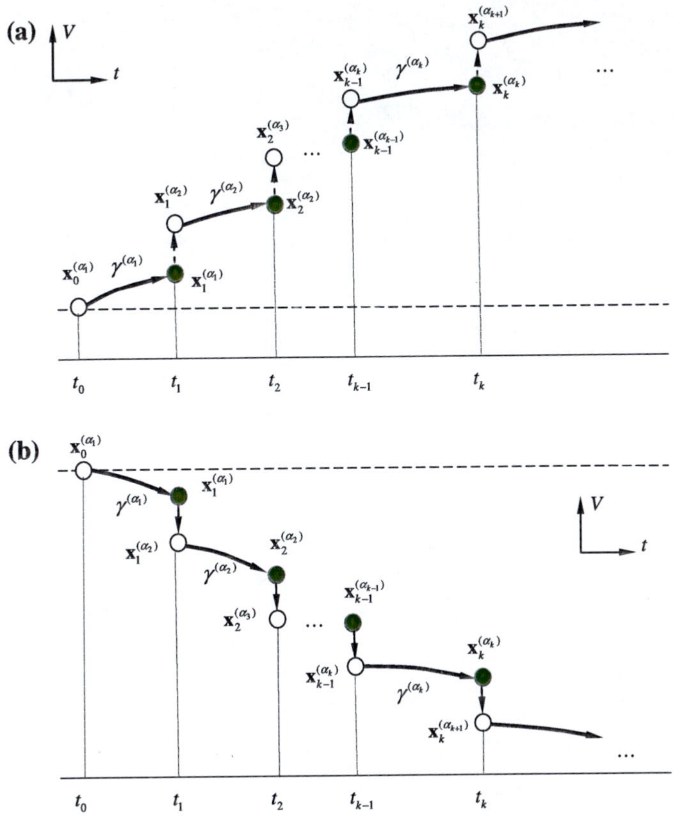

Fig. 5.7 The equi-measuring function varying with time: **a** uniformly increasing, **b** uniformly decreasing. The *solid curves* give the equi-measuring-function values of a flow $\gamma^{(\alpha_k)}(t)$. The *dashed lines* are jumping changes of the equi-measuring function, and the jumping changes are caused by the transport laws between two adjacent queue subsystems. The circular symbols represent the switching points for two adjacent queue subsystems

where

$$C_1^{(i)} = x_{k+}^{(i)}, C_2^{(i)} = (y_{k+}^{(i)} + x_{k+}^{(i)} d^{(i)})/\omega_d^{(i)}; \omega_d^{(i)} = \sqrt{c^{(i)} - (d^{(i)})^2}. \qquad (5.83)$$

For simplicity, consider a system with an impulsive function

$$c^{(i)} = c = 1, d^{(i)} = d = -0.005, a^{(i)} = a = 1.0, m = 2; \qquad (5.84)$$

and the initial condition is

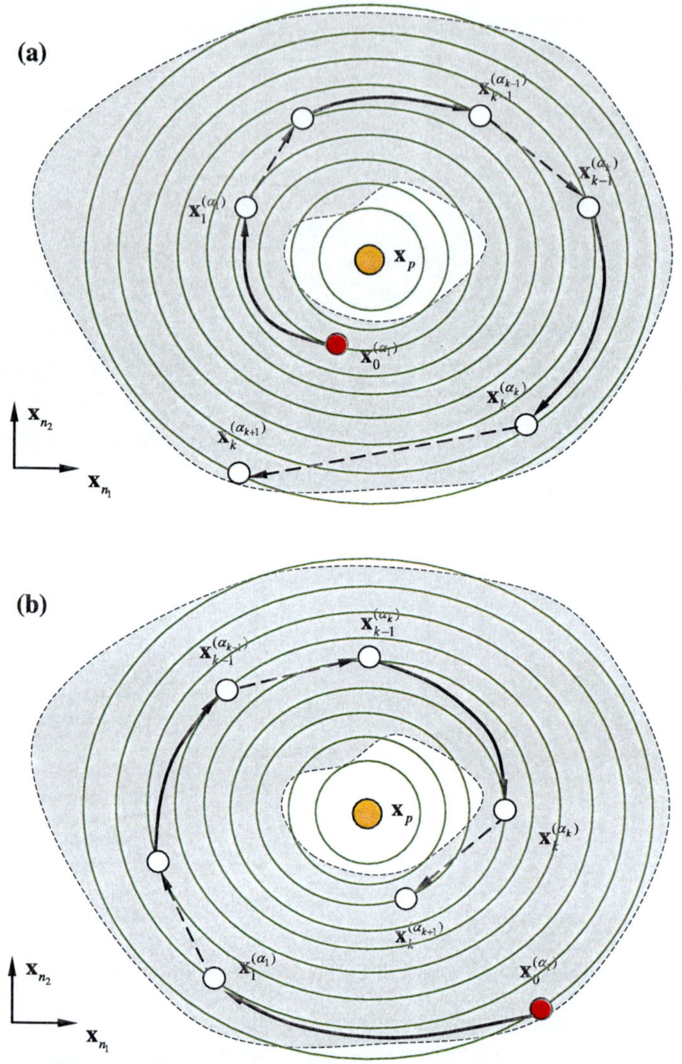

Fig. 5.8 A flow to the equi-measuring function surface in vicinity of point $\mathbf{x}_p$ in phase space: **a** uniform increase, and **b** uniform decrease. The cycles are equi-measuring function surfaces. The *circular symbols* represent the switching points for two adjacent queue subsystems

$$x_0^{(i)} = y_0^{(i)} = 4 \text{ for } t_0 = 0. \tag{5.85}$$

Thus, the dynamical characteristics of the impulsive system are presented in Fig. 5.9. The impulsive system possesses the uniform increases of measuring function with increasing impulses, as presented in Fig. 5.9a. The corresponding trajectory for such an impulsive system is presented in Fig. 5.9b. The time-history of G-function

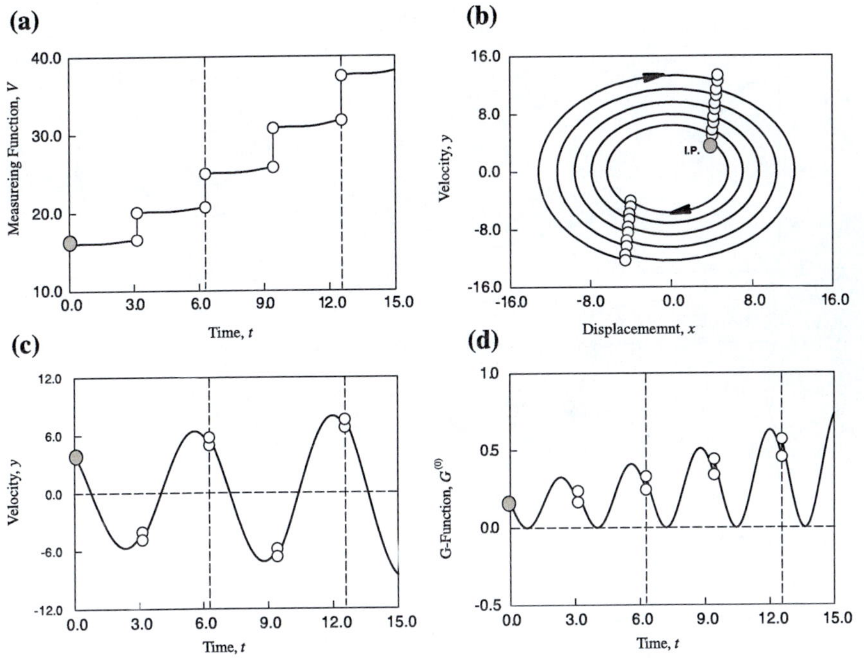

Fig. 5.9 An impulsive system with negative damping: **a** uniformly increasing of measuring function with increasing impulses, **b** trajectory in phase plane, **c** G-function of measuring function, and **d** discontinuous velocity. The *circular symbols* represent the impulsive points for such an impulsive system. Two time impulses for each period are labeled by the *vertical lines*. ($c^{(i)} = c = 1, d^{(i)} = d = -0.01, a^{(i)} = a = 1.0, m = 2$ and $x_0^{(i)} = y_0^{(i)} = 4 \quad \text{for } t_0 = 0$)

of the equi-measuring function is presented in Fig. 5.9c. It is observed that the G-function is less than zero. Since the impulsive effect is exerted, the velocity is C^0-discontinuous, which is presented in Fig. 5.9d. Thus, the corresponding displacement is C^0-continuous, which will not be presented herein. For uniform deceasing of measuring function, the impulsive rule in Eq. (5.80) can be changed as

$$\begin{aligned} &x_{k+}^{(i)} = x_{k-}^{(i)}, y_{k+}^{(i)} = y_{k-}^{(i)} - a^{(i)}\text{sgn}(y_{k-}^{(i)})(k = 1, 2, \cdots) \\ &\text{for } t_k = kT/m \text{ with } T = 2\pi/\sqrt{c}, a^{(i)} > 0. \end{aligned} \tag{5.86}$$

and the impulsive system with an impulsive function has the following parameters to be considered as

$$c^{(i)} = c = 1, d^{(i)} = d = 0.05, a^{(i)} = a = 1.0, m = 2. \tag{5.87}$$

and the initial condition is

$$x_0^{(i)} = y_0^{(i)} = 4 \text{ for } t_0 = 0. \tag{5.88}$$

Thus, such an impulsive system with uniform decrease of measuring function is in Fig. 5.10. The uniform decrease of measuring function with increasing impulses

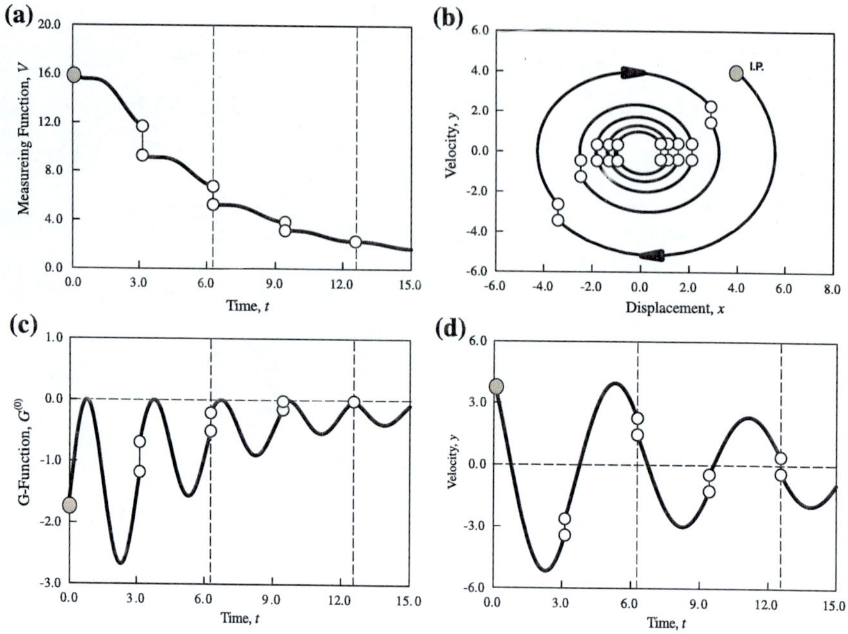

Fig. 5.10 An impulsive system with positive damping: **a** uniformly decreasing of measuring function with decreasing impulses, **b** trajectory in phase plane, **c** G-function of measuring function, and **d** discontinuous velocity. The *circular symbols* represent the impulsive points for such an impulsive system. Two time impulses for each period are labeled by the *vertical lines*. ($c^{(i)} = c = 1$, $d^{(i)} = d = 0.05, a^{(i)} = a = 1.0, m = 2$ and $x_0^{(i)} = y_0^{(i)} = 4$ for $t_0 = 0$)

for the impulsive system is presented in Fig. 5.10a. The correspond trajectory for such an impulsive system is presented in Fig 5.10b. It is observed that the trajectory will approach the equilibrium point ($x = y = 0$). The time-history of G-function of the equi-measuring function is presented in Fig. 5.10c. It is observed that the G-function is greater than zero. Again, since the impulsive effect are exerted, the C^0-discontinuous velocity is presented in Fig. 5.10d.

For the switching system with uniformly increasing (decreasing) of measuring function at impulses, each subsystem in the switching system is uniformly invariant to the equi-measuring function, and only the uniform, positive (or negative) impulses exist in the switching system, as shown in Fig. 5.11a and b. The corresponding trajectories in phase space are also presented in Fig. 5.12a and b. The dashed lines are jumping changes of the equi-measuring function, and the jumping changes are caused by the transport laws between two adjacent queue subsystems. The circular points represent the switching points for two adjacent queue subsystems.

To illustrate this case, consider the dynamical system in Eq. (5.79) without damping ($d^{(i)} = 0$). The corresponding solution is

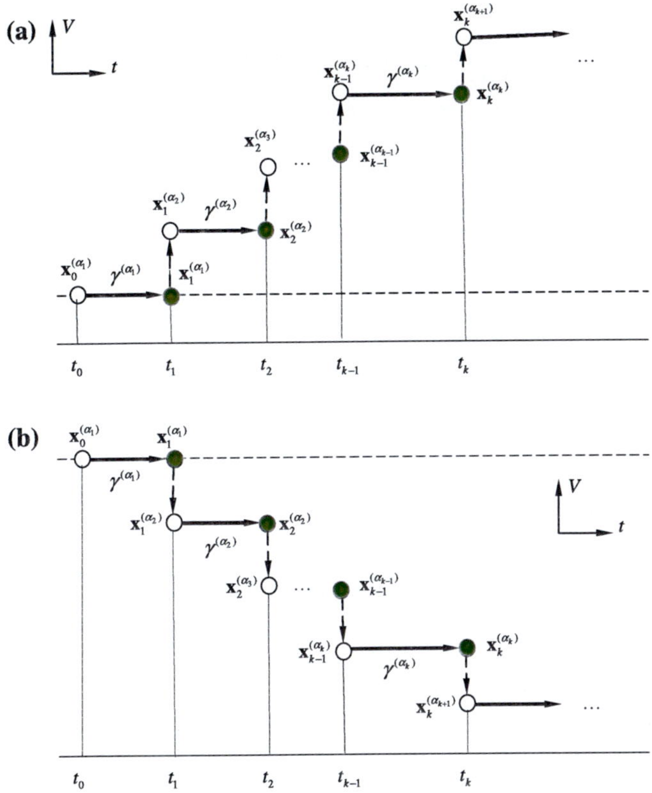

Fig. 5.11 The equi-measuring function varying with time: **a** uniformly positive impulse only, **b** uniformly negative impulse only. The *solid lines* give the invariant equi-measuring-functions of a flow $\gamma^{(\alpha_k)}(t)$. The *dashed lines* are impulsive changes of the equi-measuring function, which are caused by the transport laws between two adjacent queue subsystems. The *circular points* represent the switching points for two adjacent queue subsystems

$$\begin{aligned} x^{(i)} &= [C_1^{(i)} \cos \omega^{(i)}(t - t_{k+}) + C_2^{(i)} \sin \omega^{(i)}(t - t_{k+})], \\ y^{(i)} &= [C_2^{(i)} \omega^{(i)} \cos \omega^{(i)}(t - t_{k+}) - \omega^{(i)} C_1^{(i)} \sin \omega^{(i)}(t - t_{k+})], \end{aligned} \tag{5.89}$$

where

$$C_1^{(i)} = x_{k+}^{(i)}, C_2^{(i)} = y_{k+}^{(i)}/\omega^{(i)}; \omega^{(i)} = \sqrt{c^{(i)}}. \tag{5.90}$$

With the strength of impulsive function ($a = 0.8$) and initial conditions in Eq. (5.87), the corresponding measuring functions and trajectories in phase plane are presented in Figs. 5.13 and 5.14 for the uniformly increasing and decreasing of impulses, respectively.

From the above definition, the corresponding theorem is presented as follows.

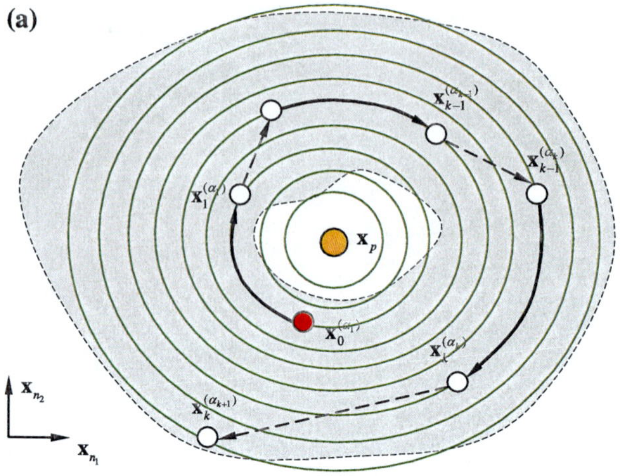

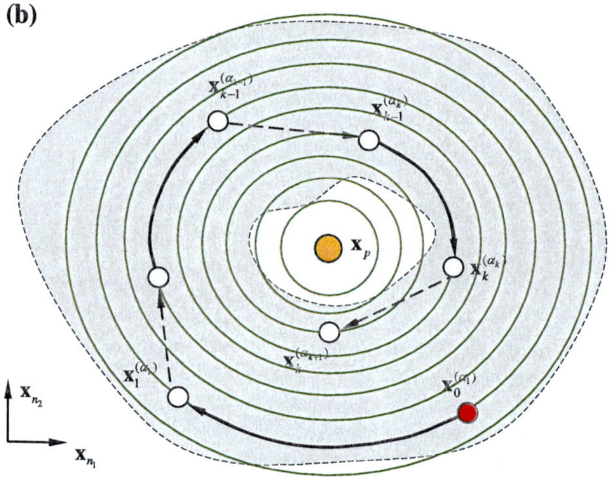

Fig. 5.12 A flow to the equi-measuring function surface in vicinity of point $\mathbf{x}_p$ in phase space: **a** uniformly positive impulse, and **b** uniformly negative impulse. The cycles are equi-measuring function surfaces. The *circular symbols* represent the switching points for two adjacent queue subsystems

Theorem 5.4 *For a switching system in Eqs.* (5.15) *and* (5.16) *on the domain* $\mho = \cup_{i=1}^{m_1}\Omega_{l_i} \cup \Omega_0$, *there is an resultant flow* $\gamma(t_0, t)$ *in Eq.* (5.69). *For the* α_kth *dynamical subsystem* S_{α_k}, consider *the equi-measuring function* $V(\mathbf{x}^{(\alpha_k)}, \mathbf{x}_p)$ *in Eq.* (5.32) *to be monotonically increased to a metric function* $d(\mathbf{x}, \mathbf{x}_p)$ *in Eq.* (5.30). *A flow* $\gamma(t_0, t)$ *from* $\gamma^{(\alpha_1)}(t_0, t_1)$ *to* $\gamma^{(\alpha_l)}(t_{l-1}, t_l)$ *for* $t \in [t_0, t_l]$ *in domain* $\cup_{k=1}^{l}\Omega_{\alpha_k}$ *is:*

(i) *uniformly decreasing to the equi-measuring function surface in* $\cup_{k=1}^{l}\Omega_{\alpha_k}$ *if and only if*

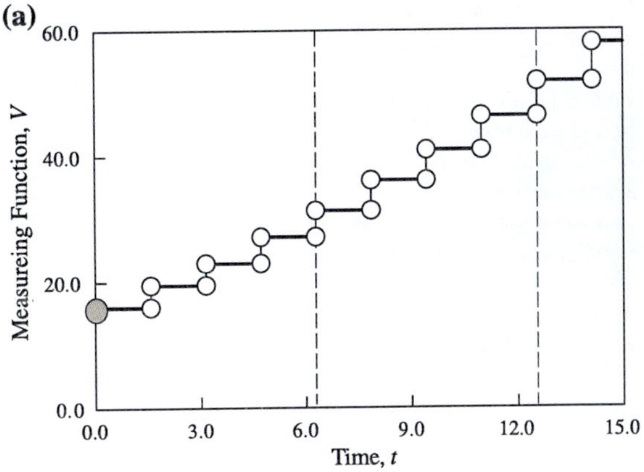

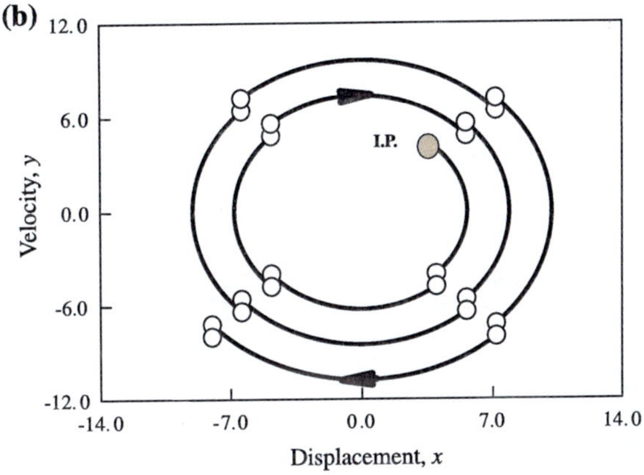

Fig. 5.13 An impulsive system without positive damping: **a** invariant measuring function with increasing impulses, **b** trajectory in phase plane. The *circular symbols* represent the impulsive points for such an impulsive system. Two time impulses for each period are labeled by the *vertical lines*. ($c^{(i)} = c = 1$, $a^{(i)} = a = 0.8, m = 4$ and $x_0^{(i)} = y_0^{(i)} = 4$ for $t_0 = 0$)

$$\begin{aligned} &G_m^{(0)}(\mathbf{x}^{(i)}, \mathbf{x}_p, t) = [\mathbf{n}(\mathbf{x}^{(i)}, \mathbf{x}_p)]^{\mathrm{T}} \cdot \mathbf{F}^{(i)}(\mathbf{x}^{(i)}, t, \mathbf{p}^{(i)}) < 0 \\ &\text{for } \mathbf{x}_s^{(\alpha_k)} \in \gamma^{(\alpha_k)}(t_{k-1}, t_k) \text{ with } t_s \in [t_{k-1}, t_k]; \\ &\Delta^{(\alpha_k \alpha_{k+1})}(t_k) < 0 \text{ for } k = 1, 2, \cdots, l; \end{aligned} \tag{5.91}$$

(ii) *uniformly increasing to the equi-measuring function surface in* $\cup_{k=1}^{l} \Omega_{\alpha_k}$ *if and only if*

$$\begin{aligned} &G_m^{(0)}(\mathbf{x}^{(i)}, \mathbf{x}_p, t) = [\mathbf{n}(\mathbf{x}^{(i)}, \mathbf{x}_p)]^{\mathrm{T}} \cdot \mathbf{F}^{(i)}(\mathbf{x}^{(i)}, t, \mathbf{p}^{(i)}) > 0 \\ &\text{for } \mathbf{x}_s^{(\alpha_k)} \in \gamma^{(\alpha_k)}(t_{k-1}, t_k) \text{ with } t_s \in [t_{k-1}, t_k]; \\ &\Delta^{(\alpha_k \alpha_{k+1})}(t_k) > 0 \text{ for } k = 1, 2, \cdots, l; \end{aligned} \tag{5.92}$$

(iii) uniformly and negatively impulsive (or jumping down, or jumping decreasing) to the equi-measuring function surface in $\cup_{k=1}^{l} \Omega_{\alpha_k}$ *if and only if*

$$\begin{aligned} &G_m^{(r)}(\mathbf{x}^{(\alpha_k)}, \mathbf{x}_p, t) = 0 \text{ for } r = 0, 1, 2, \cdots \\ &\text{for } \mathbf{x}_s^{(\alpha_k)} \in \gamma^{(\alpha_k)}(t_{k-1}, t_k) \text{ with } t_s \in [t_{k-1}, t_k], \\ &\Delta^{(\alpha_k \alpha_{k+1})}(t_k) < 0 \text{ for } k = 1, 2, \cdots, l; \end{aligned} \tag{5.93}$$

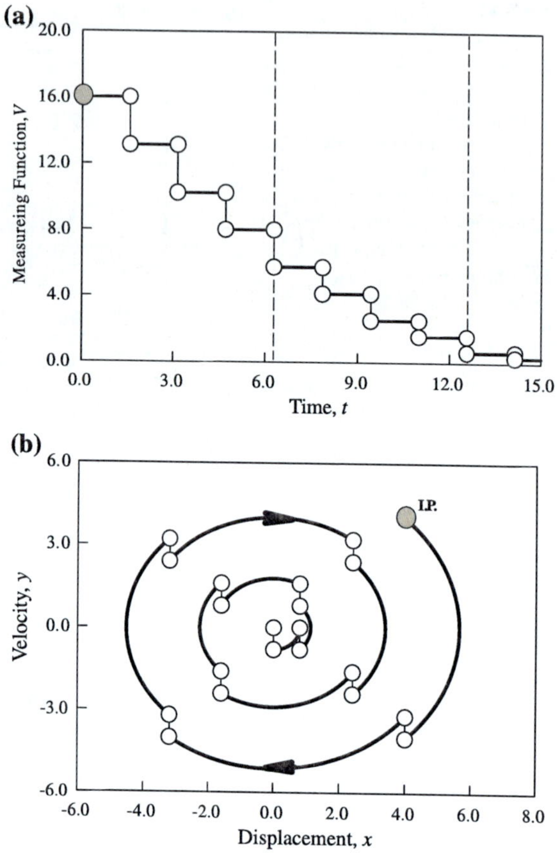

Fig. 5.14 An impulsive system without damping: **a** invariant measuring function with decreasing impulses, **b** trajectory in phase plane. The *circular symbols* represent the impulsive points for such an impulsive system. Two time impulses for each period are labeled by the *vertical lines*. ($c^{(i)} = c = 1, a^{(i)} = a = 0.8, m = 4$ and $x_0^{(i)} = y_0^{(i)} = 4$ for $t_0 = 0$)

(iv) *uniformly and positively impulsive (or jumping up, or jumping increase) to the equi-measuring function surface in* $\cup_{k=1}^{l}\Omega_{\alpha_k}$ *if and only if*

$$
\begin{aligned}
&G_m^{(r)}(\mathbf{x}^{(\alpha_k)}, \mathbf{x}_p, t) = 0 \text{ for } r = 0, 1, 2, \cdots \\
&\text{for } \mathbf{x}_s^{(\alpha_k)} \in \gamma^{(\alpha_k)}(t_{k-1}, t_k) \text{ with } t_s \in [t_{k-1}, t_k], \\
&\Delta^{(\alpha_k\alpha_{k+1})}(t_k) > 0 \text{ for } k = 1, 2, \cdots, l;
\end{aligned}
\tag{5.94}
$$

(v) *uniformly invariant to the equi-measuring function surface in* $\cup_{k=1}^{l}\Omega_{\alpha_k}$ *if and only if*

$$
\begin{aligned}
&G_m^{(r)}(\mathbf{x}^{(\alpha_k)}, \mathbf{x}_p, t) = 0 \text{ for } r = 0, 1, 2, \cdots \\
&\text{for } \mathbf{x}_s^{(\alpha_k)} \in \gamma^{(\alpha_k)}(t_{k-1}, t_k) \text{ with } t_s \in [t_{k-1}, t_k], \\
&\Delta^{(\alpha_k\alpha_{k+1})}(t_k) = 0 \text{ for } k = 1, 2, \cdots, l.
\end{aligned}
\tag{5.95}
$$

Proof Using G-functions and Taylor series expansion, this theorem can be proved directly. ■

For switching dynamical systems, it is very difficult to require all systems are uniformly increasing, or decreasing or invariant. To construct a switching dynamics system, subsystems do not require such strict increase, or decrease and invariance. Thus, as in Definition 5.19, the L-function will be adopted to measure the resultant flow $\gamma(t_0, t)$ with increase, decrease and invariance to the measuring function surface. The increase, decrease and invariance for the entire time interval are called the *global* increase, or decrease, or invariance of the resultant flow to the measuring function surface, which can be used to measure system stability. The corresponding definitions are given as follows.

Definition 5.23 For a switching system in Eqs. (5.15) and (5.16) on the domain $\mho = \cup_{i=1}^{m_1} \Omega_{l_i} \cup \Omega_0$, there is an resultant flow $\gamma(t_0, t)$ in Eq. (5.69). For the α_kth dynamical subsystem S_{α_k}, consider the equi-measuring function $V(\mathbf{x}^{(\alpha_k)}, \mathbf{x}_p)$ in Eq. (5.32) to be monotonically increased to a metric function $d(\mathbf{x}, \mathbf{x}_p)$ in Eq. (5.30). A resultant flow $\gamma(t_0, t_l)$ of the switching system from $\gamma^{(\alpha_1)}(t_0, t_1)$ to $\gamma^{(\alpha_l)}(t_{l-1}, t_l)$ for $t \in [t_0, t_l]$ in the domain $\cup_{k=1}^{l} \Omega_{\alpha_k}$ is:

(i) globally decreasing to the equi-measuring function surface in $\cup_{k=1}^{l} \Omega_{\alpha_k}$ if

$$L(\mathbf{x}_p, t_0, t_l) = \sum\nolimits_{k=1}^{l} \left[L^{(\alpha_k)}(\mathbf{x}_p, t_{k-1}, t_k) + \Delta^{(\alpha_k \alpha_{k+1})} \right] < 0; \qquad (5.96)$$

(ii) globally increasing to the equi-measuring function surface in $\cup_{k=1}^{l} \Omega_{\alpha_k}$ if

$$L(\mathbf{x}_p, t_0, t_l) = \sum\nolimits_{k=1}^{l} \left[L^{(\alpha_k)}(\mathbf{x}_p, t_{k-1}, t_k) + \Delta^{(\alpha_k \alpha_{k+1})} \right] > 0; \qquad (5.97)$$

(iii) globally invariant to the equi-measuring function surface in $\cup_{k=1}^{l} \Omega_{\alpha_k}$ if

$$L(\mathbf{x}_p, t_0, t_l) = \sum\nolimits_{k=1}^{l} \left[L^{(\alpha_k)}(\mathbf{x}_p, t_{k-1}, t_k) + \Delta^{(\alpha_k \alpha_{k+1})} \right] = 0. \qquad (5.98)$$

To illustrate the above definitions, the global increase, global decrease and global invariance of the resultant flow to the equi-measuring function surface can be presented through the equi-measuring function varying with time. The global increase, decrease and invariance of the resultant flow for time interval $t \in [t_0, t_l]$ require the corresponding L-function $L(\mathbf{x}_p, t_0, t_l)$ be *greater than*, *less than* and *equal to* zero. The global increase and decrease are illustrated in Fig. 5.15a and b, respectively. The corresponding trajectories in phase space are presented in Fig. 5.16a and b. The dashed lines are jumping changes of the equi-measuring function, and the jumping changes are caused by the transport laws between two adjacent queue subsystems. The circular points represent the switching points for two adjacent queue subsystems.

Theorem 5.5 *For a switching system on the domain* $\mho = \cup_{i=1}^{m} \Omega_{l_i} \cup \Omega_0$ *in the vicinity of point* $\mathbf{p}_k$ *in Eq.* (5.23), *if there is a periodic flow with periodicity condition in Eq.* (5.24), *then the resultant L-function is zero, i.e.,*

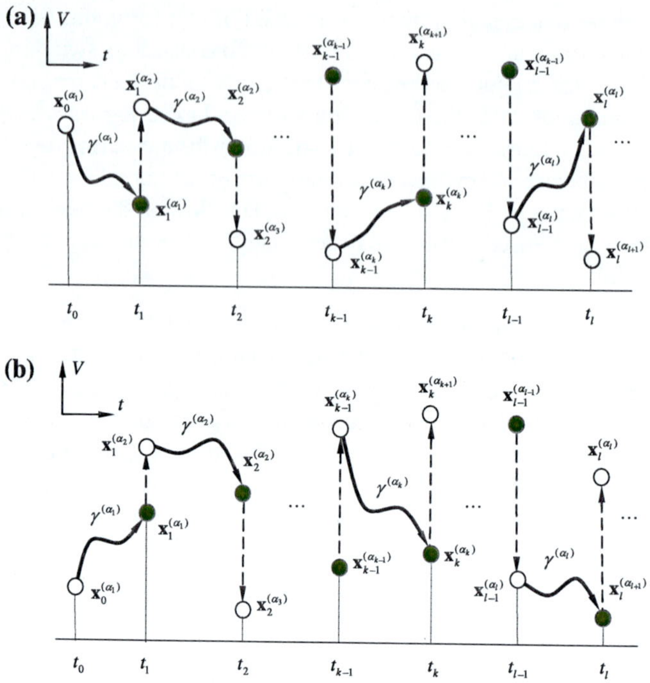

Fig. 5.15 The equi-measuring function varying with time: **a** globally deceasing, **b** globally increasing. The *solid lines* give the invariant equi-measuring-functions of a flow $\gamma^{(\alpha_k)}(t)$. The *dashed lines* are jumping changes of the equi-measuring function, and the jumping changes are caused by the transport laws between two adjacent queue subsystems. The *circular points* represent the switching points for two adjacent queue subsystems

$$\begin{aligned} &L(\mathbf{x}_p, t_0, t_m) = \sum_{k=1}^{m} \left[L^{(\alpha_k)}(\mathbf{x}_p, t_{k-1}, t_k) + \Delta^{(\alpha_k \alpha_{k+1})} \right] = 0 \\ &\mathbf{x}_m^{(\alpha_{m+1})} = \mathbf{x}_0^{(\alpha_1)}, t_m = t_0 + T \text{ and } \alpha_{m+1} = \alpha_1. \end{aligned} \tag{5.99}$$

Proof From Eq. (5.73) with $\Delta^{(\alpha_k \alpha_{k+1})}(t_k) = V(\mathbf{x}_k^{(\alpha_{k+1})}, \mathbf{x}_p) - V(\mathbf{x}_k^{(\alpha_k)}, \mathbf{x}_p)$, one obtains

$$\begin{aligned} L(\mathbf{x}_p, t_0, t_m) &= \sum_{k=1}^{m} \left[L^{(\alpha_k)}(\mathbf{x}_p, t_{k-1}, t_k) + \Delta^{(\alpha_k \alpha_{k+1})} \right] \\ &= \sum_{k=1}^{m} \left[V(\mathbf{x}_k^{(\alpha_k)}, \mathbf{x}_p) - V(\mathbf{x}_{k-1}^{(\alpha_k)}, \mathbf{x}_p) + \Delta^{(\alpha_k \alpha_{k+1})} \right] \\ &= V(\mathbf{x}_m^{(\alpha_{m+1})}, \mathbf{x}_p) - V(\mathbf{x}_0^{(\alpha_1)}, \mathbf{x}_p). \end{aligned}$$

The periodic flow requires $\mathbf{x}_m^{(\alpha_{m+1})} = \mathbf{x}_0^{(\alpha_1)}$, $t_m = t_0 + T$ and $\alpha_{m+1} = \alpha_1$, which implies $V(\mathbf{x}_m^{(\alpha_{m+1})}, \mathbf{x}_p) = V(\mathbf{x}_0^{(\alpha_1)}, \mathbf{x}_p)$. Thus, $L(\mathbf{x}_p, t_0, t_m) = 0$. This theorem is proved. ■

Periodic flows in switching systems with impulses are sketched to help one understand the mechanism, characteristics and construction of periodic flow, which are

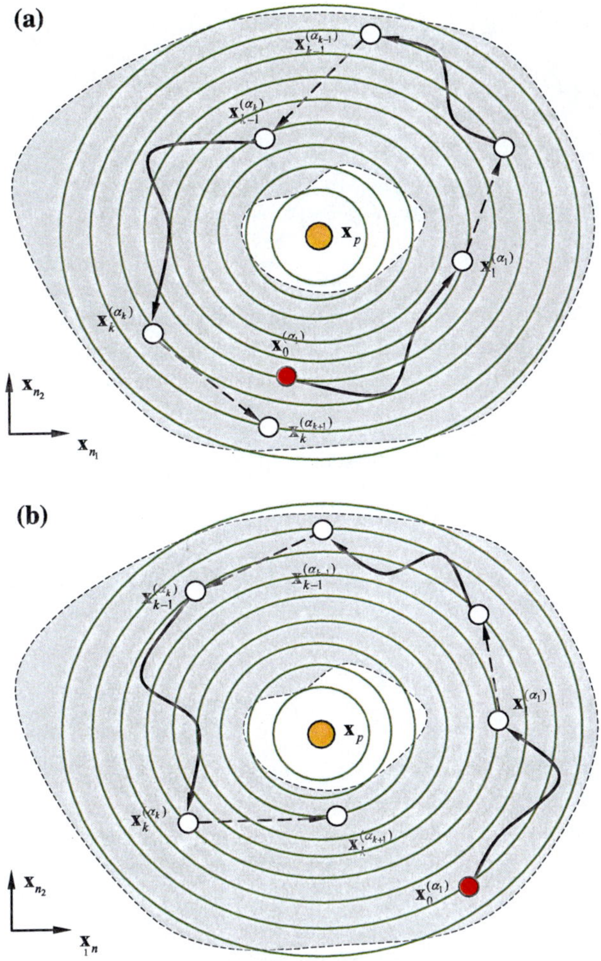

Fig. 5.16 A flow to the equi-measuring function surface in vicinity of point $\mathbf{x}_p$ in phase space: **a** global increase, and **b** global decrease. The cycles are equi-measuring function surfaces. The *circular symbols* represent the switching points for two adjacent queue subsystems

different from the continuous dynamical systems. Herein, the simple cases will be discussed first to build the corresponding concepts. Consider a flow with uniformly invariance to the equi-measuring surface with positive and negative increase with the same amplitude. Such a switching system is a conservative system to the equi-measuring surface for given time intervals with a kind of impulse with a specific transport law at given switching moments. In other words, one has

$$
\begin{aligned}
&G_m^{(r)}(\mathbf{x}^{(\alpha_k)}, \mathbf{x}_p, t) = 0 \ \text{ for } r = 0, 1, 2, \cdots \\
&\text{for } \mathbf{x}^{(\alpha_k)}(t) \in \gamma^{(\alpha_k)}(t_{k-1}, t_k) \ \text{ with } t \in [t_{k-1}, t_k], \\
&\Delta^{(\alpha_k \alpha_{k+1})}(t_k) = -C \ \text{ or } C \ \text{ at switching time } t_k;
\end{aligned}
\tag{5.100}
$$

or

$$\begin{aligned}&V(\mathbf{x}^{(\alpha_k)}(t),\mathbf{x}_p)=V(\mathbf{x}_k^{(\alpha_k)},\mathbf{x}_p)=V(\mathbf{x}_{k-1}^{(\alpha_k)},\mathbf{x}_p)\\&\text{for }\ \mathbf{x}^{(\alpha_k)}(t)\in\gamma^{(\alpha_k)}(t_{k-1},t_k)\ \text{ with }\ t\in[t_{k-1},t_k];\\&\Delta^{(\alpha_k\alpha_{k+1})}(t_k)=C\ \text{ for jumping up},\\&\Delta^{(\alpha_k\alpha_{k+1})}(t_k)=-C\ \text{ for jumping down}\\&\text{with }\ \Delta^{(\alpha_k\alpha_{k+1})}(t_k)=V(\mathbf{x}_k^{(\alpha_{k+1})},\mathbf{x}_p)-V(\mathbf{x}_k^{(\alpha_k)},\mathbf{x}_p).\end{aligned}\tag{5.101}$$

If the switching systems cannot form a periodic flow, then a chaotic flow can be formed or at least a randomly switching flow can be observed. The stochasticity of such a flow can be determined through a random setting of the switching time. However, for $(2m)$ switching with jumping up and down with constant magnitude, if there is a relation as

$$\begin{aligned}&\mathbf{x}_0^{(\alpha_1)}=\mathbf{x}_{2m}^{(\alpha_{2m+1})}\ \text{ and }\ t_{2m}=t_0+T\\&\text{where }\ T\ \text{ is period}\end{aligned}\tag{5.102}$$

then the switching flow can form a periodic flow. In addition, the switching is given by the transport laws

$$\begin{aligned}&\mathbf{g}^{(\alpha_k\alpha_{k+1})}(\mathbf{x}_k^{(\alpha_k)},\mathbf{x}_k^{(\alpha_{k+1})})=0\\&V(\mathbf{x}_k^{(\alpha_{2k+1})},\mathbf{x}_p)=C_1\ \text{ and }\ V(\mathbf{x}_k^{(\alpha_{2k})},\mathbf{x}_p)=C_2.\end{aligned}\tag{5.103}$$

For linear switching systems, such a periodic flow is stable.

For a better understanding of the concepts, the following illustrations are given to form periodic flows. The equi-measuring function varying with time for periodic flows is sketched in Fig. 5.17. In Fig. 5.17a, the periodic motion with positive and negative impulses with the same strength of impulses. In Fig. 5.17b, the periodic flows with positive and negative impulses with different strengths of impulses are presented. The corresponding periodic trajectories for such periodic flows in phase space are presented in Fig. 5.18. For special cases, consider an impulsive system with the measuring functions uniformly increase or decrease with time. However, the impulsive jump can remove such increasing and decreasing of the measuring functions. Thus, the periodic flows can be formed. For intuitive illustration, such periodic flows are presented in Figs. 5.19 and 5.20. The constant increments of measuring function are adopted. In fact, such measuring functions increments are not necessary to be constant. Periodic flows with the uneven increments of measuring functions are sketched in Figs. 5.21 and 5.22. If the total increments of measuring functions can be cancelled by the total increments of impulsive jumps, the periodic flow can formed. In other words, for unstable subsystems in the impulsive system, if impulsive jumps can draw the measuring function to the original level, the resultant impulsive system with unstable subsystems should be stable. For general case, it is not necessary for subsystems to make the measuring function be uniformly increasing or deceasing. The corresponding illustration is given in Fig. 5.23.

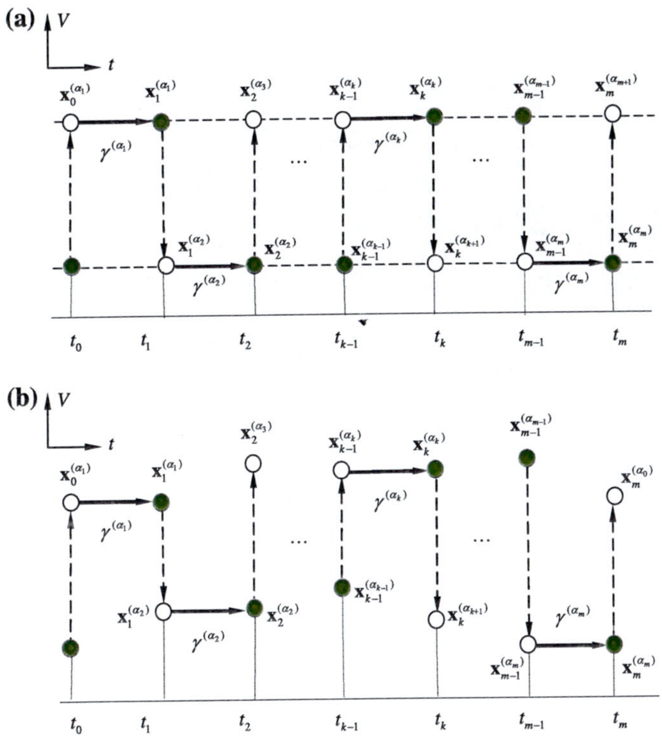

Fig. 5.17 The equi-measuring function varying with time: **a** positive and negative impulses with the same strength, and **b** positive and negative impulses with different strengths. The cycles are equi-measuring function surfaces. The *circular symbols* are for switching points

Consider the dynamical system in Eq. (5.79) without damping ($d^{(i)} = 0$) again. However, the impulsive relation will be changed as

$$\begin{aligned} &x_{k+}^{(i)} = x_{k-}^{(i)},\ y_{k+}^{(i)} = y_{k-}^{(i)} + a^{(i)} (k = 1, 2, \cdots) \\ &\text{for } t_k = kT/m \text{ with } T = 2\pi/\sqrt{c}, a^{(i)} > 0. \end{aligned} \tag{5.104}$$

The corresponding solution in Eq. (5.89) and (5.90) will be used for dynamical system in Eq. (5.79). The measuring function in Eq. (5.81) will be used herein.

With the strength of impulsive function ($a = 1.0$) and initial conditions are

$$x_0^{(i)} = 2.0,\ y_0^{(i)} = 2.0 \text{ for } t = 0.0. \tag{5.105}$$

The periodic motion for the switching system with impulses ($m = 2$) is presented in Fig. 5.24. The measuring functions are two invariants. The impulses possess "jumping up" and "jumping down" with the same increments, which is clearly presented in Fig. 5.24a. The trajectory of the periodic motion in phase plane is presented in

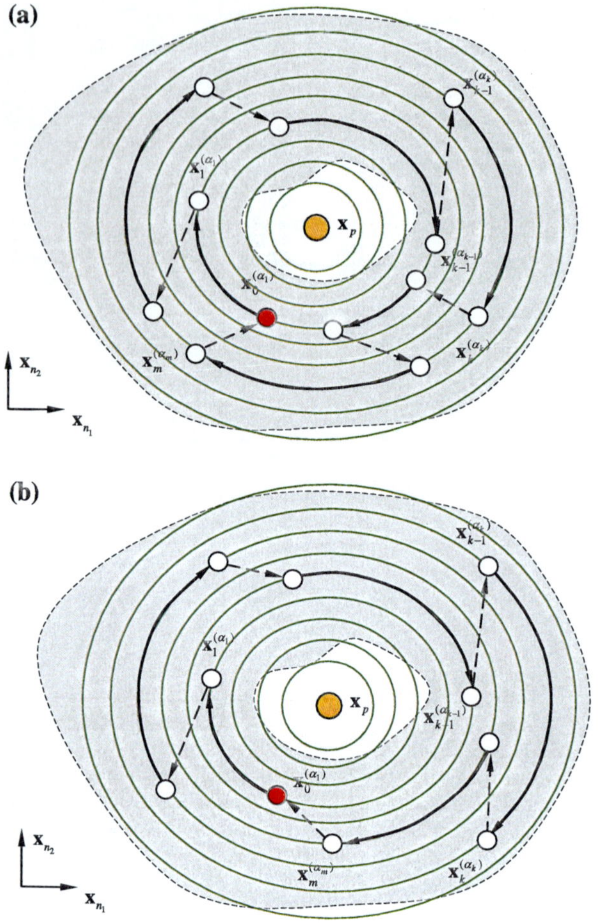

Fig. 5.18 A periodic flow to the equi-measuring function surface in vicinity of point $\mathbf{x}_p$ in phase space: **a** positive and negative impulses with the same strength, and **b** positive and negative impulses with different strengths. The cycles are equi-measuring function surfaces. The *circular symbols* are for switching points

Fig. 5.24b. Since subsystem in Eq. (5.79) with ($d^{(i)} = 0$) is a conservative and the Hamiltonian is selected as a measuring function. Thus, the subsystem possesses the invariant measuring function without impulses. The time-histories of displacement and velocity for such a periodic motion are presented in Fig. 5.24c and d, respectively. Since the impulsive effects are exerted on the velocity. The displacement is C^0-continuous at the impulsive points, but the velocity is C^0-discontinuous. With multiple impulses, the corresponding periodic motions are presented in Fig. 5.25 ($m = 4$). The same initial conditions and system parameters are used. Compared with Fig. 5.24, the impulsive effects on the measuring functions, trajectory in phase plane, displacement and velocity time-histories are observed. If $m \to \infty$, the impul-

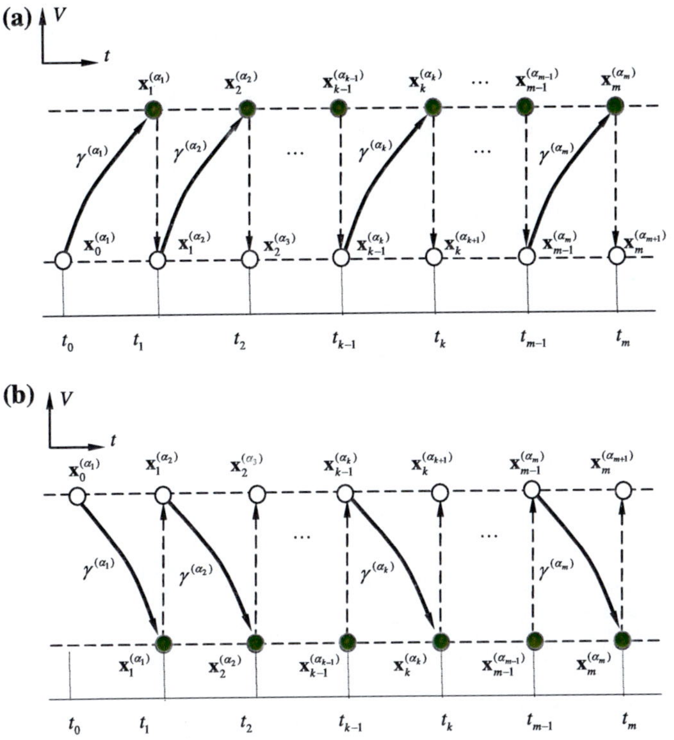

Fig. 5.19 The equi-measuring function varying with time: **a** uniformly increasing with constant jump down only, **b** uniformly increasing with constant jumping up. The *solid lines* give the invariant equi-measuring-functions of a flow $\gamma^{(\alpha_k)}(t)$. The *dashed lines* are jumping changes of the equi-measuring function, and the jumping changes are caused by the transport laws between two adjacent queue subsystems. The *circular points* represent the switching points for two adjacent queue subsystems

sive force will become a constant force exerting on the vibration system. The analytical and numerical solutions can be easily obtained. However, the measuring function is not necessary to be chosen as the Hamiltonian. Consider a measuring function in Eq. (5.81) as

$$V^{(i)} = \frac{1}{2}(y^{(i)})^2 + \frac{1}{4}c^{(i)}(x^{(i)})^2. \tag{5.106}$$

Based on the foregoing measuring functions, the G-function is

$$G^{(0,i)} = y^{(i)}\dot{y}^{(i)} + \frac{1}{2}c^{(i)}x^{(i)}y^{(i)} = -\frac{1}{2}c^{(i)}x^{(i)}y^{(i)}. \tag{5.107}$$

The above measuring function and the corresponding G-functions for the same periodic motion presented in Fig. 5.25 are presented in Fig. 5.26. It is observed that the

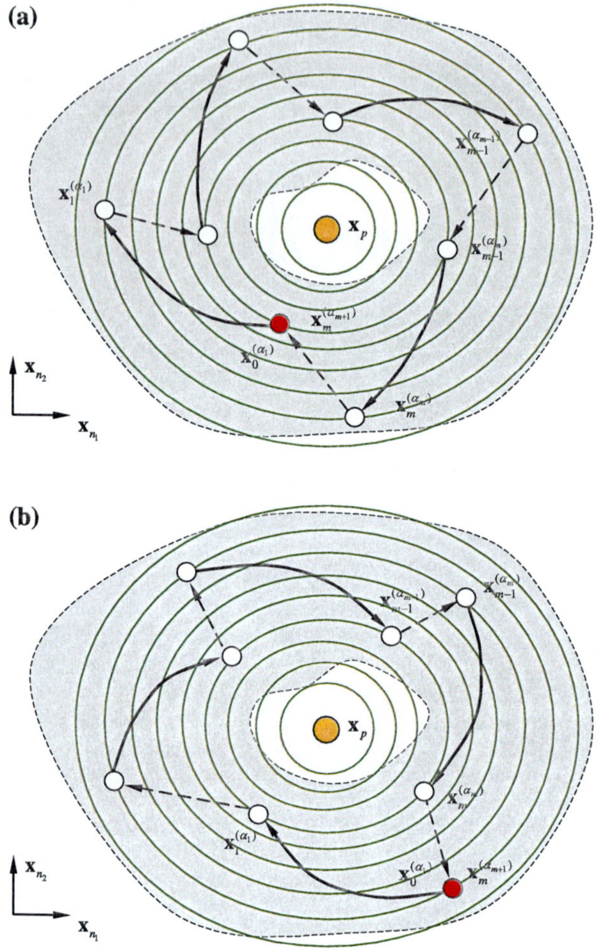

Fig. 5.20 A flow to the equi-measuring function surface in vicinity of point $\mathbf{x}_p$ in phase space: **a** uniform increasing with jump down, and **b** uniform decreasing with jump up. The cycles are equi-measuring function surfaces. The *circular symbols* represent the switching points for two adjacent queue subsystems

measuring function is not uniformly increasing or decreasing. From the traditional Lyapunov method, the stability of this impulsive system cannot be determined. In addition, the switching systems have many subsystems. If one measuring function is used, definitely such a function cannot be the first integral invariant manifolds for all subsystems. Thus, the L-function for measuring function should be adopted.

5.4 Impulsive Systems and Chaotic Diffusions

For further discussion on impulsive systems, consider a dynamical system as

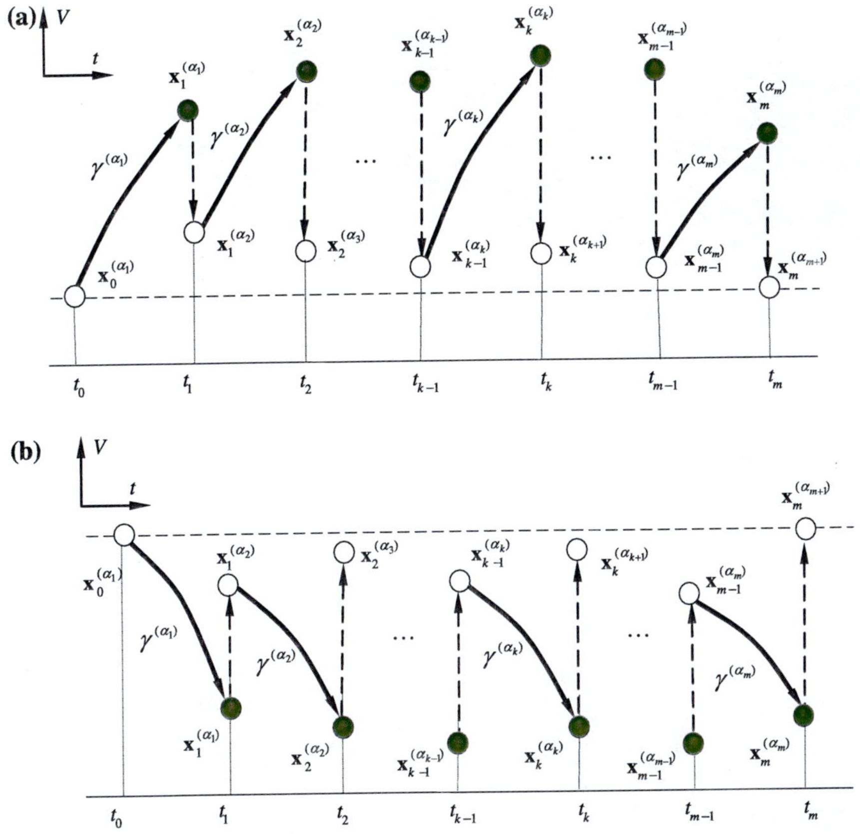

Fig. 5.21 The equi-measuring function varying with time: **a** uniformly invariance with jumps only, **b** uniformly increasing with jumping down. The *solid lines* give the invariant equi-measuring-functions of a flow $\gamma^{(\alpha_k)}(t)$. The *dashed lines* are jumping changes of the equi-measuring function, and the jumping changes are caused by the transport laws between two adjacent queue subsystems. The *circular points* represent the switching points for two adjacent queue subsystems

$$\begin{aligned}&\dot{x}^{(i)} = y^{(i)} \text{ and } \dot{y}^{(i)} = -c^{(i)}x^{(i)} + a^{(i)}f(x,y)\delta(t - knT/m)\\&(k = 1, 2, \cdots; i = 1, 2, \cdots, l).\end{aligned} \tag{5.108}$$

The foregoing equation is equal to subsystems in Eq. (5.79) with $d^{(i)} = 0$ and the impulsive function in Eq. (5.80) becomes

$$\begin{aligned}&x_{k+}^{(i)} = x_{k-}^{(i)}, y_{k+}^{(i)} = y_{k-}^{(i)} + a^{(i)}f^{(i)}(x_{k-}^{(i)}, y_{k-}^{(i)})(k = 1, 2, \cdots)\\&\text{for } t_k = knT/m \text{ with } T = 2\pi/\Omega, a^{(i)} > 0.\end{aligned} \tag{5.109}$$

The solutions in Eqs. (5.89) and (5.90) are adopted. Thus, for $t = t_{(k+1)-}$, we have

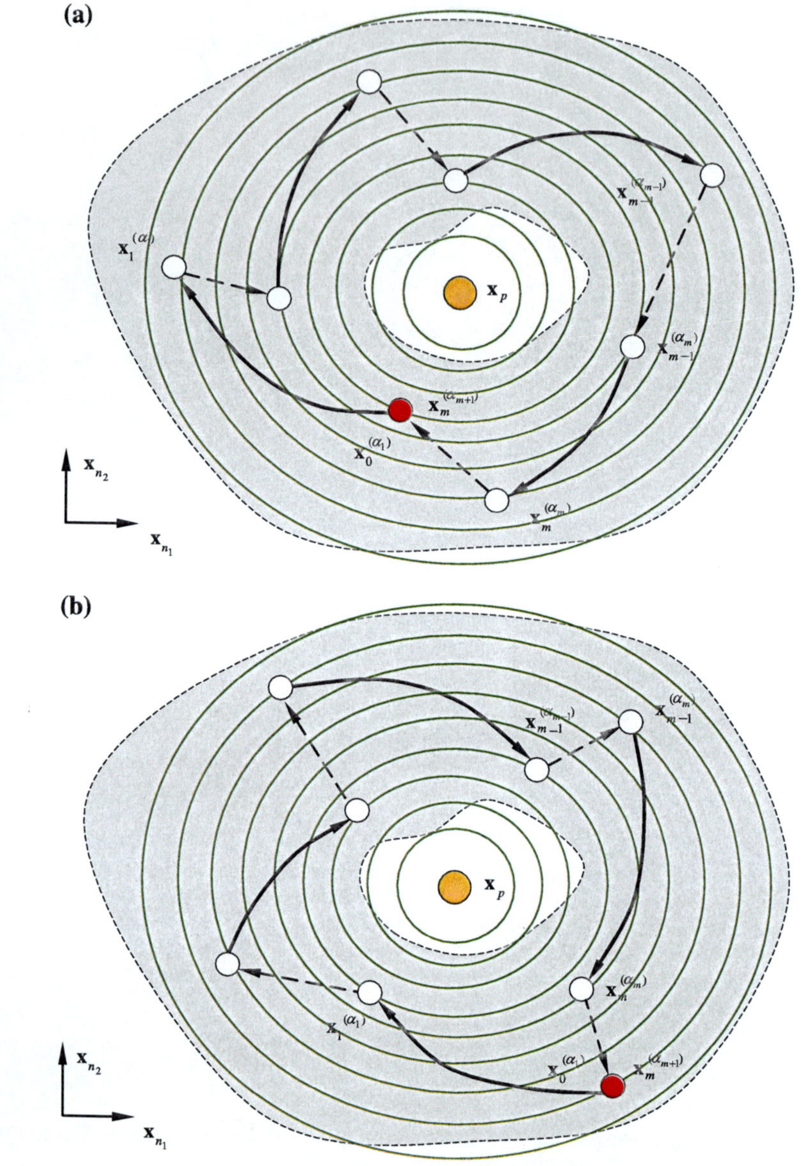

Fig. 5.22 A flow to the equi-measuring function surface in vicinity of point $\mathbf{x}_p$ in phase space: **a** uniform increase, and **b** uniform decrease. The cycles are equi-measuring function surfaces. The *circular symbols* represent the switching points for two adjacent queue subsystems

$$
\begin{aligned}
x_{(k+1)-}^{(i)} &= [x_{k+}^{(i)} \cos \omega^{(i)} (t_{(k+1)-} - t_{k+}) + y_{k+}^{(i)} / \omega^{(i)} \sin \omega^{(i)} (t_{(k+1)-} - t_{k+})], \\
y_{(k+1)-}^{(i)} &= [y_{k+}^{(i)} \cos \omega^{(i)} (t_{(k+1)-} - t_{k+}) - \omega^{(i)} x_{k+}^{(i)} \sin \omega^{(i)} (t_{(k+1)-} - t_{k+})].
\end{aligned}
\tag{5.110}
$$

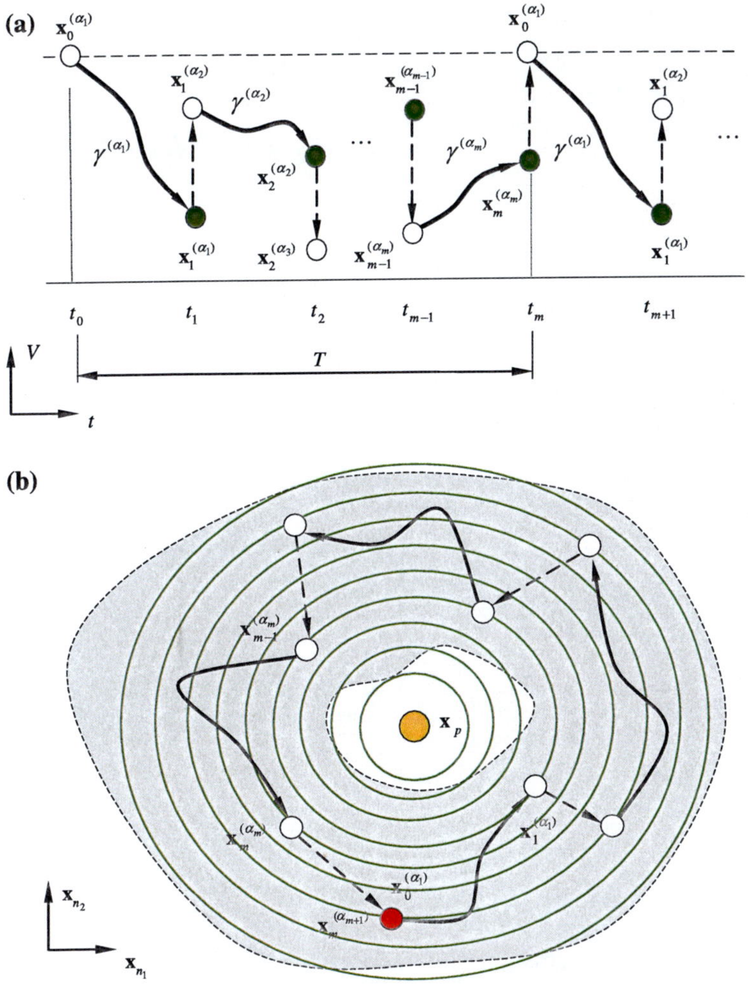

Fig. 5.23 The equi-measuring function varying with time: **a** globally deceasing, **b** globally increasing. The *solid lines* give the invariant equi-measuring-functions of a flow $\gamma^{(\alpha_k)}(t)$. The *dashed lines* are jumping changes of the equi-measuring function, and the jumping changes are caused by the transport laws between two adjacent queue subsystems. The *circular points* represent the switching points for two adjacent queue subsystems

Because

$$\begin{aligned} \cos\omega^{(i)}(t_{(k+1)-}-t_{k+}) &= \cos\frac{2\pi n\omega^{(i)}}{m\Omega} = K_1^{(i)}, \\ \sin\omega^{(i)}(t_{(k+1)-}-t_{k+}) &= \sin\frac{2\pi n\omega^{(i)}}{m\Omega} = K_2^{(i)} \end{aligned} \tag{5.111}$$

with Eq. (5.108), Equation (5.111) becomes

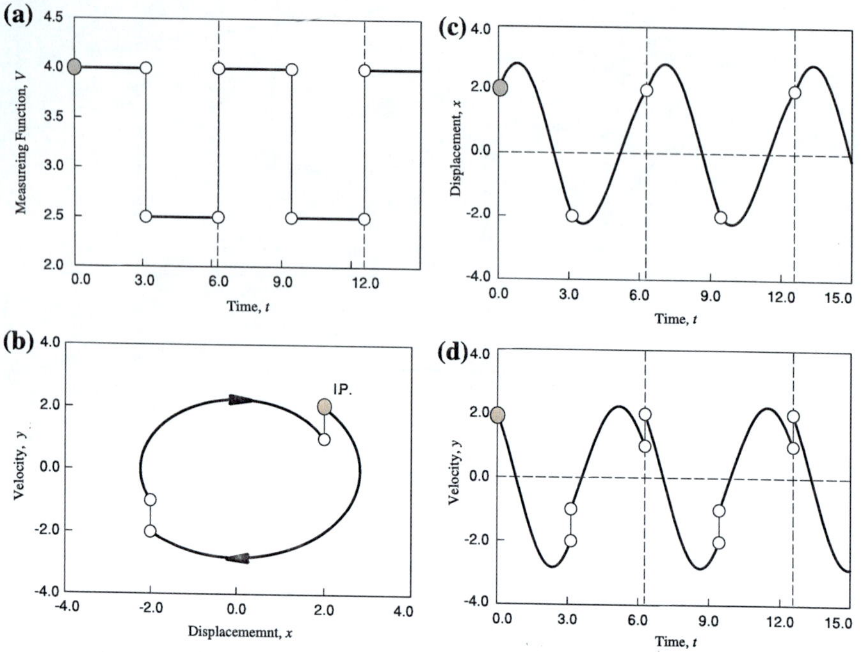

Fig. 5.24 Periodic motions for an impulsive system: **a** Measuring function with impulses, **b** trajectory in phase plane, **c** displacement and **d** discontinuous velocity. The *circular symbols* represent the impulsive points for such an impulsive system. Two time impulses for each period are labeled by the *vertical lines*. ($c^{(i)} = c = 1, a^{(i)} = a = 1.0, m = 2,\ x_0^{(i)} = y_0^{(i)} = 2.0$ for $t_0 = 0$)

$$
\begin{aligned}
x_{(k+1)-}^{(i)} &= K_1^{(i)} x_{k-}^{(i)} + \frac{K_2^{(i)}}{\omega^{(i)}}[y_{k-}^{(i)} + a^{(i)} f^{(i)}(x_{k-}^{(i)}, y_{k-}^{(i)})],\\
y_{(k+1)-}^{(i)} &= K_1^{(i)}[y_{k-}^{(i)} + a^{(i)} f^{(i)}(x_{k-}^{(i)}, y_{k-}^{(i)})] - K_2^{(i)}\omega^{(i)} x_{k-}^{(i)}.
\end{aligned}
\tag{5.112}
$$

Dropping subscript "-", the foregoing equation becomes

$$
\begin{Bmatrix} x_{k+1}^{(i)} \\ y_{k+1}^{(i)} \end{Bmatrix} = \begin{bmatrix} K_1^{(i)} & \dfrac{K_2^{(i)}}{\omega^{(i)}} \\ -K_2^{(i)}\omega^{(i)} & K_1^{(i)} \end{bmatrix} \begin{Bmatrix} x_k^{(i)} \\ y_k^{(i)} \end{Bmatrix} + \begin{Bmatrix} \dfrac{K_2^{(i)}}{\omega^{(i)}} \\ K_1^{(i)} \end{Bmatrix} a^{(i)} f^{(i)}(x_k^{(i)}, y_k^{(i)}). \tag{5.113}
$$

The mapping P_i $(i = 1, 2, \cdots, l)$ of Eq. (5.108) for $t \in [t_{k+1}, t_k]$ is developed. Suppose there are m-subsystems during n-periods (nT). The resultant mapping is

$$
P = P_{l_m} \circ \cdots \circ P_{l_2} \circ P_{l_1}, l_j \in \{1, 2, \cdots, l\} \text{ and } j \in \{1, 2, \cdots, m\}. \tag{5.114}
$$

For $\mathbf{x}_k = (x_k, y_k)^{\mathrm{T}}$, we have

$$
\mathbf{x}_{k+m} = P\mathbf{x}_k = P_{l_m} \circ \cdots \circ P_{l_2} \circ P_{l_1}\mathbf{x}_k. \tag{5.115}
$$

The corresponding relations are

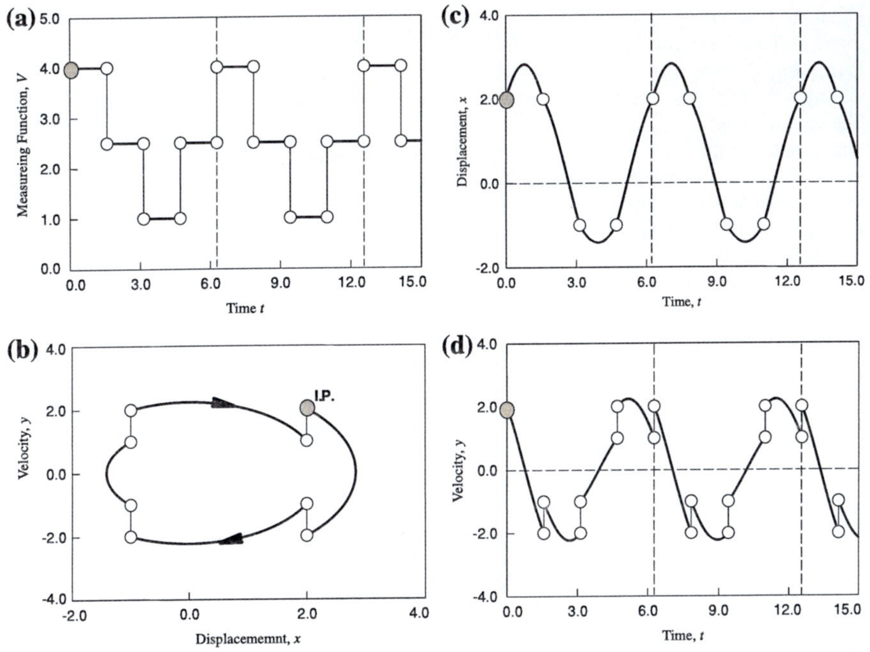

Fig. 5.25 Periodic motions for an impulsive system: **a** Measuring function with impulses, **b** trajectory in phase plane, **c** displacement and **d** discontinuous velocity. The *circular symbols* represent the impulsive points for such an impulsive system. Two time impulses for each period are labeled by the *vertical lines*. ($c^{(i)} = c = 1, a^{(i)} = a = 1.0, m = 4, x_0^{(i)} = y_0^{(i)} = 2.0$ for $t_0 = 0$)

$$\mathbf{x}_{k+j} = \mathbf{g}^{(l_j)}(\mathbf{x}_{k+j-1}) \quad \text{for } j = 1, 2, \cdots, m \tag{5.116}$$

where $\mathbf{g}^{(l_j)}(\mathbf{x}_{k+j-1}) = (g_1^{(l_j)}, g_2^{(l_j)})^{\mathrm{T}}$ with

$$\begin{Bmatrix} g_1^{(l_j)} \\ g_2^{(l_j)} \end{Bmatrix} = \begin{bmatrix} K_1^{(l_j)} & \dfrac{K_2^{(l_j)}}{\omega^{(l_j)}} \\ -K_2^{(l_j)}\omega^{(l_j)} & K_1^{(l_j)} \end{bmatrix} \begin{Bmatrix} x_{k+j-1}^{(l_j)} \\ y_{k+j-1}^{(l_j)} \end{Bmatrix} + \begin{Bmatrix} \dfrac{K_2^{(l_j)}}{\omega^{(l_j)}} \\ K_1^{(l_j)} \end{Bmatrix} a^{(l_j)} f^{(l_j)}(x_{k+j-1}^{(l_j)}, y_{k+j-1}^{(l_j)}). \tag{5.117}$$

For periodic motion during n-periods (nT), the periodicity conditions are

$$\mathbf{x}_{k+m} = \mathbf{x}_k. \tag{5.118}$$

The stability and bifurcation of the periodic solutions can be determined. Herein, it should not be discussed. Similarly problems will be discussed in next section.

To illustrate complex motions in such impulsive system, as in Luo (2011), consider a single system with the following function as example

$$f(x_k, y_k) = \sin x_k. \tag{5.119}$$

Thus, the mapping in Eq. (5.113) becomes

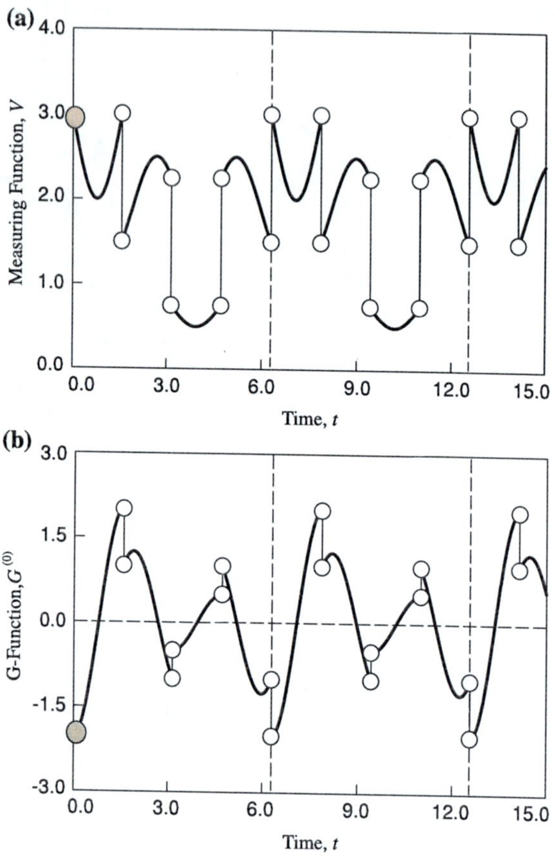

Fig. 5.26 Periodic motions for an impulsive system: **a** Measuring function with impulses, **b** G-functions. The *circular symbols* represent the impulsive points for such an impulsive system. Two time impulses for each period are labeled by the *vertical lines*. ($c^{(i)} = c = 1$, $a^{(i)} = a = 1.0$, $m = 4$, $x_0^{(i)} = y_0^{(i)} = 2.0$ for $t_0 = 0$)

$$\begin{Bmatrix} x_{k+1} \\ y_{k+1} \end{Bmatrix} = \begin{bmatrix} K_1 & \dfrac{K_2}{\omega} \\ -K_2\omega & K_1 \end{bmatrix} \begin{Bmatrix} x_k \\ y_k \end{Bmatrix} + \begin{Bmatrix} \dfrac{K_2}{\omega} \\ K_1 \end{Bmatrix} a \sin x_k. \tag{5.120}$$

During n-periods (nT), the resultant mapping is

$$P^{(m)} = \underbrace{P \circ \cdots \circ P \circ P}_{m}. \tag{5.121}$$

For $\mathbf{x}_k = (x_k, y_k)^{\mathrm{T}}$, we have

$$\mathbf{x}_{k+N} = P^{(N)}\mathbf{x}_k. \tag{5.122}$$

The corresponding relations are

$$\mathbf{x}_{k+j} = \mathbf{g}(\mathbf{x}_{k+j-1}) \quad \text{for } j = 1, 2, \cdots, N \tag{5.123}$$

where $\mathbf{g}(\mathbf{x}_{k+j-1}) = (g_1, g_2)^{\mathrm{T}}$ with

$$\begin{Bmatrix} g_1 \\ g_2 \end{Bmatrix} = \begin{bmatrix} K_1 & \dfrac{K_2}{\omega} \\ -K_2\omega & K_1 \end{bmatrix} \begin{Bmatrix} x_{k+j-1} \\ y_{k+j-1} \end{Bmatrix} + \begin{Bmatrix} \dfrac{K_2}{\omega} \\ K_1 \end{Bmatrix} a \sin x_{k+j-1}. \quad (5.124)$$

From the mapping in Eq. (5.123), the quasi-periodic motions and chaos can be obtained via the impulsive points. The impulsive switching scenario with impulsive strength is presented in Fig. 5.27 for $c^{(i)} = c = 1,\ \Omega = 1.0,\ n = 1,\ m = 3, x_0^{(i)} = y_0^{(i)} = 4.0$ and $a^{(i)} = a$ for $t_0 = 0$. The critical value of impulsive strength $a_{cr} \approx 1.2974$ is for the motion from the quasi-periodic motion to the chaotic motion. If $a > a_{cr}$, the chaotic diffusion will be observed. With such chaotic diffusion, long-range interactions can exist via such an impulsive action. The quasi-periodic motion and chaotic diffusion for $\mathrm{mod}(n\omega, m\Omega) = l$ with $l = 1, 2, \cdots, m-1$ are same. Without loss of generality, we can set $\omega = \Omega$. All possible quasi-periodic motions and chaotic diffusion can be obtained by $\mathrm{mod}(n, m) = l$. The switching sections are same for any specific l with all numbers of n with $\mathrm{mod}(n, m) = l$. For $n > m$, the trajectories in phase plane are same, which are different from the trajectory for $n = l$. However, the time responses are different for different n.

Consider $a = 1.0$ with $n = l = 1$ and m=3 for illustration in Fig. 5.28 with same parameters and same initial conditions. The trajectory in phase plane and impulsive switching sections are presented in Fig. 5.28a and b, respectively. The impulsive switching points before impulse jumps are labeled by circular symbols. To show switching pattern, the impulsive switching points after impulse jumps are not labeled. The impulsive jumps are connected by vertical lines. The switching sections is based on the impulsive switching points before impulsive jumps. The trajectory in phase plane is for $n = l = 1$ and $m = 3$ only. However, the switching section is for the quasi-periodic motion relative to all integers of n with mod $(n, 3) = 1$ for a specific l=1. In other words, for $n = km + 1$ $(k = 0, 1, 2, \cdots)$, the switching section are same. For this problem in Eq. (5.108), the trajectories for $k \neq 0$ and k=0 are different. However, all trajectories for $k \neq 0$ are identical but the corresponding time-histories of responses (e.g., displacement and velocity) are different. In Fig. 5.29a and b, the trajectories in phase plane for n=1,4 are presented, respectively. It is observed that the switching points are same but the trajectories are different. Thus, the time-histories of velocity for n=1,4 are presented in Fig. 5.29c and d, respectively.

As $a > a_{cr}$, the chaotic diffusion generates the chaotic impulsive motion pattern with long-range interactions. The switching sections of chaotic impulsive motion for ($n = 1, m = 3, a = 1.4$ and $a_{cr} \approx 1.2974$) and ($n = 2, m = 3, a = 1.89$ and $a_{cr} \approx 1.8855$) are presented in Fig. 5.30a and b. The 20,000 impulsive switching points are used to generate the switching sections. The chaotic diffusion patterns are very clearly observed for two types of chaotic diffusion. They have three branches for diffusion. The chaotic diffusion patterns for (1:3) and (2:3) resonant impulses are different. For a further view of the chaotic diffusion patterns, the chaotic diffusion patterns for (1:3), (1:4), (1:5) and (1:6) resonant impulses are presented in

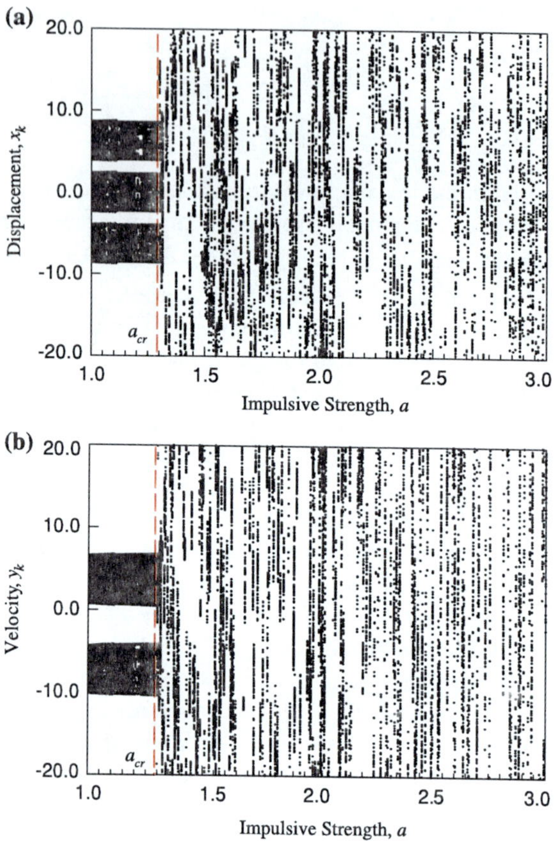

Fig. 5.27 Impulsive switching scenario with impulsive strength: **a** switching displacement, **b** switching velocity. ($c^{(i)} = c = 1, a^{(i)} = a, \quad \Omega = 1., n = 1, m = 3, n = 1, m = 3, a = 1.89$). ($c^{(i)} = c = 1, a^{(i)} = a, \quad \Omega = 1, x_0^{(i)} = y_0^{(i)} = 4.0$ for $t_0 = 0$)

Fig. 5.31a–d, respectively. Each branch of the chaotic diffusion is colored by a different color for a better view of chaotic diffusion pattern. This deterministic system shows random walks properties. The complexity and randomness needs to be further investigated.

5.5 Mappings and Periodic Flows

To describe the switching of sub-systems, consider a switching set for the ith subsystem to be

$$\Sigma^{(i)} = \left\{ \mathbf{x}_k^{(i)} \middle| \mathbf{x}_k^{(i)} = \mathbf{x}^{(i)}(t_k), k \in \{0, 1, 2, \cdots\} \right\}. \tag{5.125}$$

From the solution of the ith subsystem, a mapping P_i for a time interval $[t_{k-1}, t_k]$ is defined as

$$P_i : \Sigma^{(i)} \to \Sigma^{(i)} \quad \text{for } i = 1, 2, \cdots, m \tag{5.126}$$

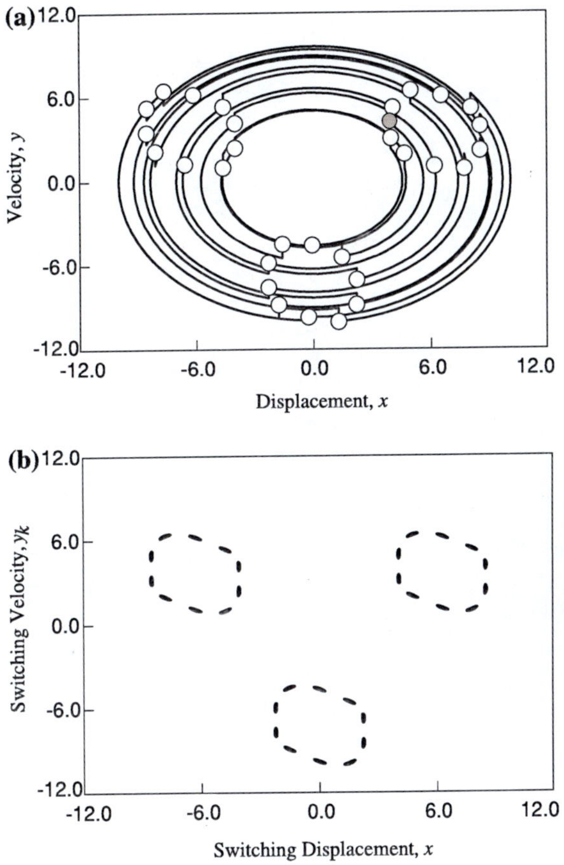

Fig. 5.28 A quasi-periodic impulsive motions: **a** trajectory in phase plane, **b** switching sec-tions. ($c^{(i)} = c = 1, a^{(i)} = a = 1.0, \Omega = 1.n = 1, m = 3,\ x_0^{(i)} = y_0^{(i)} = 4.0$ for $t_0 = 0$)

or

$$P_i : \mathbf{x}_{k-1}^{(i)} \to \mathbf{x}_k^{(i)} \quad \text{for } i = 1, 2, \cdots, m. \tag{5.127}$$

Define a time difference parameter for the ith subsystem for a time interval $[t_{k-1}, t_k]$ which can be set arbitrarily.

$$\alpha_k^{(i)} = t_k - t_{k-1}. \tag{5.128}$$

For simplicity, introduce two vectors herein

$$\mathbf{f}^{(i)} = (f_1^{(i)}, f_2^{(i)}, \cdots f_n^{(i)})^{\mathrm{T}} \text{ and } \mathbf{x}^{(i)} = (x_1^{(i)}, x_2^{(i)}, \cdots x_n^{(i)})^{\mathrm{T}}. \tag{5.129}$$

From the solution in Eq. (5.9) for the ith subsystem, the foregoing equation gives for $(i = 1, 2, \cdots, m)$

$$\mathbf{f}^{(i)}(\mathbf{x}_k^{(i)}, \mathbf{x}_{k-1}^{(i)}) = \mathbf{x}_k^{(i)} - \mathbf{\Phi}^{(i)}(\mathbf{x}_{k-1}^{(i)}, t_k, t_{k-1}, \mathbf{p}^{(i)}) = \mathbf{0}. \tag{5.130}$$

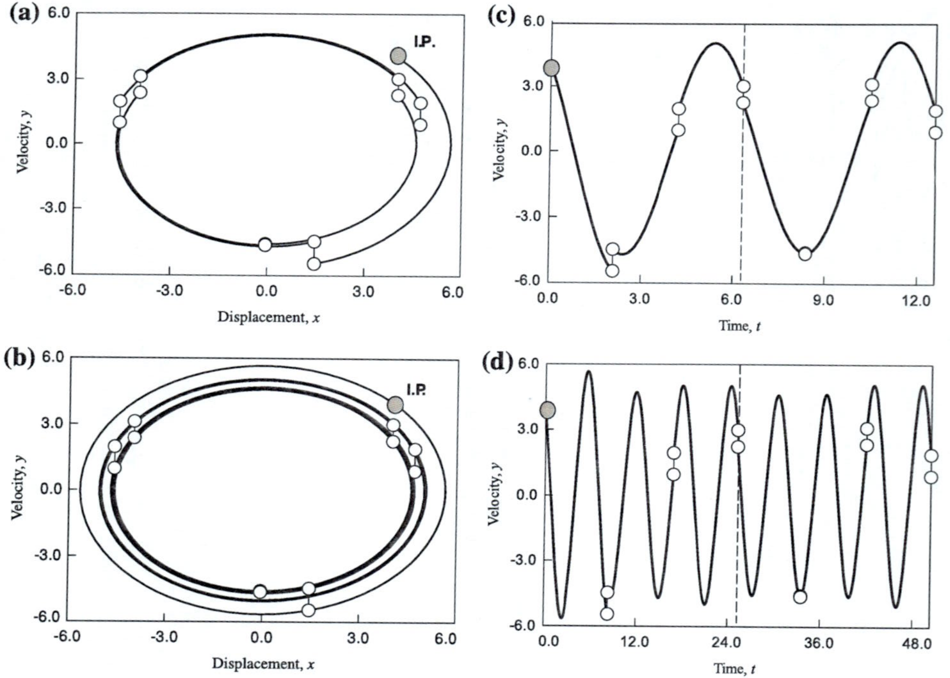

Fig. 5.29 Quasi-periodic impulsive motions ($m = 3$) : **a** trajectory ($n = 1$), **b** trajectory ($n = 4$), **c** velocity ($n = 1$) and **d** velocity ($n = 4$).($c^{(i)} = c = 1$, $a^{(i)} = a = 1.0$, $\Omega = 1$, $x_0^{(i)} = y_0^{(i)} = 4.0$ for $t_0 = 0$)

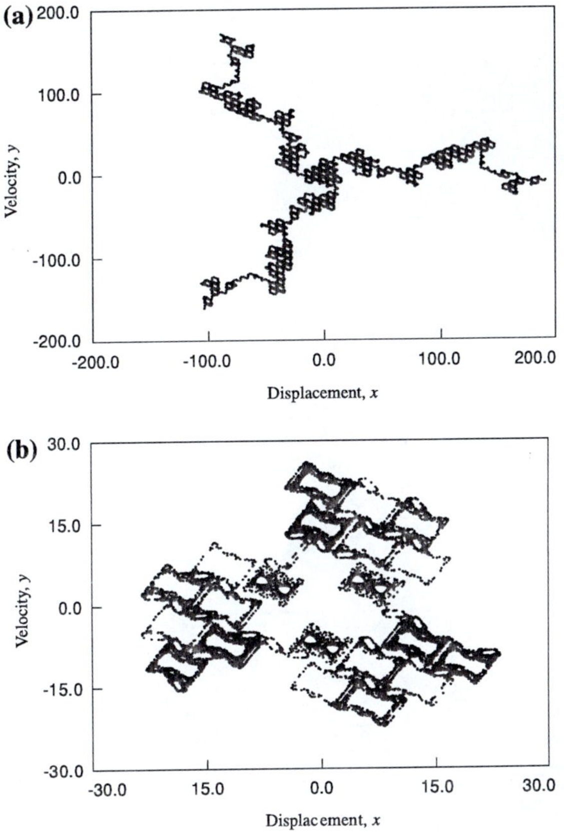

Fig. 5.30 Switching sections for chaotic impulsive motions: **a** $(n = 1,\ m = 3,\ a = 1.4)$ and **b** $(n = 2,\ m = 3,\ a = 1.89)$. $(c^{(i)} = c = 1, a^{(i)} = a, \Omega = 1, x_0^{(i)} = y_0^{(i)} = 4.0$ for $t_0 = 0)$

Suppose the two trajectories of the *i*th and *j*th systems in phase space at the switching time t_k is continuous, i.e.,

$$\mathbf{x}_k^{(j)} = \mathbf{x}_k^{(i)} \text{ at time } t_k \tag{5.131}$$

or

$$(x_{1(k)}^{(j)}, \cdots, x_{n(k)}^{(j)}) = (x_{1(k)}^{(i)}, \cdots, x_{n(k)}^{(i)}) \text{ for } i, j \in \{1, 2, \cdots, m\}. \tag{5.132}$$

If the two solutions of the *i*th and *j*th subsystems at the switching time t_k are discontinuous, for instance, an impulsive switching system needs transport laws. From Luo (2006), a vector for the transport law from the *i*th to *j*th systems is introduced as

$$\mathbf{g}^{(ij)} = (g_1^{(ij)}, g_2^{(ij)}, \cdots g_m^{(ij)})^{\mathrm{T}}. \tag{5.133}$$

So the transport law between the *i*th and *j*th subsystems can be written as

$$\mathbf{g}^{(ij)}(\mathbf{x}_k^{(i)}, \mathbf{x}_k^{(j)}) = \mathbf{0} \text{ at time } t_k. \tag{5.134}$$

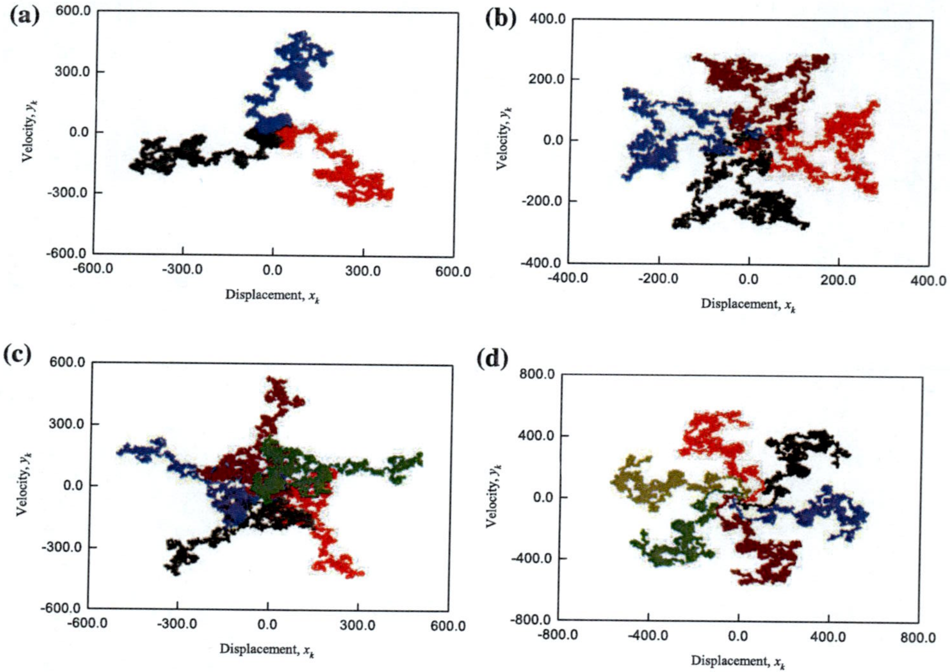

Fig. 5.31 Switching sections for chaotic impulsive motions: **a** ($n = 1, m = 3, a = 3.0$), **b** ($n = 1, m = 4, a = 3.0$), **c** ($n = 1, m = 5, a = 4.0$), **d** ($n = 1, m = 4, a = 4.0$). ($c^{(i)} = c = 1, \quad a^{(i)} = a, \quad \Omega = 1, x_0^{(i)} = y_0^{(i)} = 4.0$ for $t_0 = 0$)

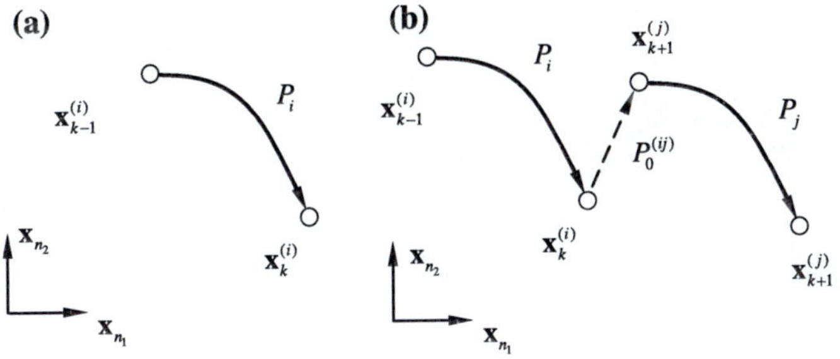

Fig. 5.32 **a** Mapping P_i and **b** transport mapping $P_0^{(ij)}$ in phase plane ($n_1 + n_2 = n$)

In other words, one obtains

$$\left.\begin{aligned} x_{1(k)}^{(j)} &= g_1^{(ij)}(x_{1(k)}^{(i)}, \cdots, x_{n(k)}^{(i)}) \\ x_{2(k)}^{(j)} &= g_2^{(ij)}(x_{1(k)}^{(i)}, \cdots, x_{n(k)}^{(i)}) \\ &\vdots \\ x_{n(k)}^{(j)} &= g_n^{(ij)}(x_{1(k)}^{(i)}, \cdots, x_{n(k)}^{(i)}) \end{aligned}\right\} \text{ for } i, j \in \{1, 2, \cdots, m\}. \tag{5.135}$$

From the transport law, a transport mapping is introduced as for $i, j \in \{1, 2, \cdots, m\}$

$$P_0^{(ij)} : \Sigma^{(i)} \to \Sigma^{(j)}, \tag{5.136}$$

i.e.,

$$P_0^{(ij)} : \mathbf{x}_k^{(i)} \to \mathbf{x}_k^{(j)} \text{ or } P_0^{(ij)} : (x_{1(k)}^{(i)}, \cdots, x_{n(k)}^{(i)}) \to (x_{1(k)}^{(j)}, \cdots, x_{n(k)}^{(j)}). \tag{5.137}$$

The algebraic equations for the transport mapping are given in Eq. (5.135).

The mapping P_i for the ith subsystem for time $t \in [t_{k-1}, t_k]$ and the transport mapping at time $t = t_k$ are sketched in Figs. 5.32 and 5.33. The initial and final points of mapping P_i are $\mathbf{x}_{k-1}^{(i)}$ and $\mathbf{x}_k^{(i)}$. Similarly, the initial and final points of mapping P_j are $\mathbf{x}_k^{(j)}$ and $\mathbf{x}_{k+1}^{(j)}$. The mappings relative to subsystems are sketched by solid curves. The two mappings are connected by a transport mapping at $t = t_k$, which is depicted by the dashed line. In phase space, there is a non-negative distance governed by Eq. (5.134). However, the time-history of flows for the switching system experiences a jump at time $t = t_k$. If the transport law gives a special case to satisfy Eq. (5.131), the solutions of two sub-systems are C^0-continuous at the switching time $t = t_k$. The jump phenomenon will disappear.

Consider a flow of the switching system with a mapping structure for $t \in \cup_{j=1}^{N} [t_{k+j-1}, t_{k+j}]$ as

$$P = P_0^{(l_N l_{N+1})} \circ P_{l_N} \circ \cdots \circ P_{l_2} \circ P_0^{(l_1 l_2)} \circ P_{l_1} \equiv P_{l_N \cdots l_2 l_1}, \tag{5.138}$$

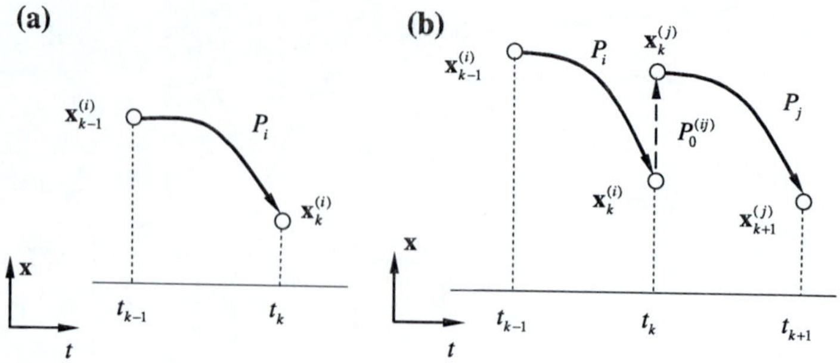

Fig. 5.33 **a** Mapping P_i and **b** transport mapping $P_0^{(ij)}$ in the time-history

where

$$P_{l_j} \in \{ P_i | \, i = 1, 2, \cdots, m\} \;\; \text{for} \;\; j = 1, 2, \cdots, m. \tag{5.139}$$

Consider the initial and final states $\mathbf{x}_k^{(l_1)} = (x_{1(k)}^{(l_1)}, \cdots, x_{n(k)}^{(l_1)})^{\mathrm{T}}$ at $t = t_k$ and $\mathbf{x}_{k+N}^{(l_{N+1})} = (x_{1(k+N)}^{(l_{N+1})}, \cdots, x_{n(k+N)}^{(l_{N+1})})^{\mathrm{T}}$ at $t = t_{k+N}$, respectively.

$$\mathbf{x}_{k+N}^{(l_{N+1})} = P\mathbf{x}_k^{(l_1)} = P_0^{(l_N l_{N+1})} \circ P_{l_N} \circ \cdots \circ P_{l_2} \circ P_0^{(l_1 l_2)} \circ P_{l_1}\mathbf{x}_k^{(l_1)}. \tag{5.140}$$

With each time difference, the total time difference is

$$t_{k+N} - t_k = \sum\nolimits_{j=1}^{N} \alpha_{k+j}^{(l_j)}. \tag{5.141}$$

In addition, equation (5.140) gives the following mapping relations

$$\begin{aligned}
\mathbf{x}_{k+1}^{(l_1)} &= P_{l_1}\mathbf{x}_k^{(l_1)} \Rightarrow P_{l_1} : \mathbf{x}_k^{(l_1)} \to \mathbf{x}_{k+1}^{(l_1)}, \\
\mathbf{x}_{k+1}^{(l_2)} &= P_0^{(l_1 l_2)}\mathbf{x}_{k+1}^{(l_1)} \Rightarrow P_0^{(l_1 l_2)} : \mathbf{x}_{k+1}^{(l_1)} \to \mathbf{x}_{k+1}^{(l_2)}, \\
\mathbf{x}_{k+2}^{(l_2)} &= P_{l_2}\mathbf{x}_{k+1}^{(l_2)} \Rightarrow P_{l_2} : \mathbf{x}_{k+1}^{(l_2)} \to \mathbf{x}_{k+2}^{(l_2)}, \\
\mathbf{x}_{k+1}^{(l_3)} &= P_0^{(l_2 l_3)}\mathbf{x}_{k+2}^{(l_2)} \Rightarrow P_0^{(l_1 l_3)} : \mathbf{x}_{k+2}^{(l_2)} \to \mathbf{x}_{k+2}^{(l_3)}, \\
&\vdots \\
\mathbf{x}_{k+n}^{(l_N)} &= P_{l_N}\mathbf{x}_{k+N-1}^{(l_N)} \Rightarrow P_{l_N} : \mathbf{x}_{k+N-1}^{(l_N)} \to \mathbf{x}_{k+N}^{(l_N)}, \\
\mathbf{x}_{k+N}^{(l_{N+1})} &= P_0^{(l_N l_{N+1})}\mathbf{x}_{k+N}^{(l_N)} \Rightarrow P_0^{(l_N l_{N+1})} : \mathbf{x}_{k+N}^{(l_N)} \to \mathbf{x}_{k+N}^{(l_{N+1})}.
\end{aligned} \tag{5.142}$$

Mapping relations in Eq. (5.137) yields a set of algebraic equations as

$$\left.\begin{aligned}
&\mathbf{f}^{(l_1)}(\mathbf{x}_{k+1}^{(l_1)},\mathbf{x}_k^{(l_1)},\alpha_k^{(l_1)})=\mathbf{0},\\
&\mathbf{g}^{(l_1l_2)}(\mathbf{x}_{k+1}^{(l_1)},\mathbf{x}_{k+1}^{(l_2)})=\mathbf{0},\\
&\vdots\\
&\mathbf{f}^{(l_N)}(\mathbf{x}_{k+N}^{(l_N)},\mathbf{x}_{k+N-1}^{(l_N)},\alpha_{k+N}^{(l_N)})=\mathbf{0},\\
&\mathbf{g}^{(l_Nl_{N+1})}(\mathbf{x}_{k+N}^{(l_N)},\mathbf{x}_{k+N}^{(l_{N+1})})=\mathbf{0}.
\end{aligned}\right\} \tag{5.143}$$

If there is a periodic motion, the periodicity for $t_{k+N}=T+t_k$ is

$$\left.\begin{aligned}
&\mathbf{x}_{k+N}^{(l_{N+1})}=\mathbf{x}_k^{(l_1)}\ \text{ for } l_{N+1}=l_1 \text{ or}\\
&x_{1(k+N)}^{(l_{N+1})}=x_{1(k)}^{(l_1)},\cdots,x_{m(k+N)}^{(l_{N+1})}=\dot{x}_{m(k)}^{(l_1)}
\end{aligned}\right\} \tag{5.144}$$

where T is time period. The resultant periodic solution of the switching system is for $i=1,2,\cdots,m$

$$\left.\begin{aligned}
&\mathbf{x}^{(l_i)}(t)=\left\{\mathbf{x}^{(l_i)}(t)\,\middle|\,t\in[t_{k+Ns+i-1},t_{k+Ns+i}]\ \text{ for } s=0,1,2\cdots\right\},\\
&\mathbf{x}_{k+Ns+i}^{(l_i)}=\mathbf{x}_{k+Ns+i}^{(l\bmod(i,N)+1)}\ \text{ for } s=0,1,2,\cdots.
\end{aligned}\right\} \tag{5.145}$$

From Eqs. (5.138) and (5.139), the corresponding switching points for the periodic motion can be determined. From the time difference, the time interval parameter is defined as

$$q_{k+j}^{(l_j)}=\frac{\alpha_{k+j}^{(l_j)}}{T}. \tag{5.146}$$

Thus, one obtains

$$\sum\nolimits_{j=1}^{N} q_{k+j}^{(l_j)}=1. \tag{5.147}$$

If a set of the time interval parameters for switching subsystems during the next period is the same as during the current period, the periodic flow is called *the equi-time-interval periodic flow*. The pattern of the resultant flow for the switching system during the next period will repeat the pattern of the flow during the current period. If a set of the time interval parameters for the second period is different from the first period, the periodic motion is called the *non-equi-time-interval periodic motion*. For such a flow, the switching pattern during the next period is different from the current one. For a general case, during two periods, only one pattern makes Eq. (5.147) be satisfied. Hence, this switching pattern can be treated as a periodic flow with two periods. To determine the stability of such a periodic motion, the Jacobian matrix can be computed, for $j=1,2,\cdots,n$ i.e.,

$$DP=DP_0^{(l_Nl_1)}\cdot DP_{l_N}\cdot\ldots\cdot DP_{l_2}\cdot DP_0^{(l_1l_2)}\cdot DP_{l_1} \tag{5.148}$$

with

$$DP_{l_j} = \left[\frac{\partial \mathbf{x}_{k+j}^{(l_j)})}{\partial \mathbf{x}_{k+j-1}^{(l_j)}}\right]_{m\times m} = \left[\frac{\partial(x_{1(k+j)}^{(l_j)}, x_{2(k+j)}^{(l_j)}, \cdots x_{m(k+j)}^{(l_j)})}{\partial(x_{1(k+j-1)}^{(l_j)}, x_{2(k+j-1)}^{(l_j)}, \cdots x_{m(k+j-1)}^{(l_j)})}\right]_{m\times m}$$
$$= -\left[\frac{\partial \mathbf{f}^{(l_j)}}{\partial \mathbf{x}_{k+j}^{(l_j)}}\right]_{m\times m}^{-1}\left[\frac{\partial \mathbf{f}^{(l_j)}}{\partial \mathbf{x}_{k+j-1}^{(l_j)}}\right]_{m\times m}. \tag{5.149}$$

Due to Eq. (5.130), one obtains

$$\left[\frac{\partial \mathbf{f}^{(l_j)}}{\partial \mathbf{x}_{k+j-1}^{(l_j)}}\right]_{m\times m} + \left[\frac{\partial \mathbf{f}^{(l_j)}}{\partial \mathbf{x}_{k+j-1}^{(l_j)}}\right]_{m\times m}\left[\frac{\partial \mathbf{x}_{k+j}^{(l_j)}}{\partial \mathbf{x}_{k+j-1}^{(l_j)}}\right]_{m\times m} = \mathbf{0}. \tag{5.150}$$

Similarly, from the transport law, one obtains

$$\left[\frac{\partial \mathbf{g}^{(l_j l_{j+1})}}{\partial \mathbf{x}_{k+j}^{(l_j)}}\right]_{m\times m} + \left[\frac{\partial \mathbf{g}^{(l_j l_{j+1})}}{\partial \mathbf{x}_{k+j}^{(l_{j+1})}}\right]_{m\times m}\left[\frac{\partial \mathbf{x}_{k+j}^{(l_{j+1})}}{\partial \mathbf{x}_{k+j}^{(l_j)}}\right]_{m\times m} = \mathbf{0}, \tag{5.151}$$

$$DP_0^{(l_j l_{j+1})} = \left[\frac{\partial \mathbf{x}_{k+j}^{(l_{j+1})})}{\partial \mathbf{x}_{k+j}^{(l_j)}}\right]_{m\times m} = \left[\frac{\partial(x_{1(k+j)}^{(l_{j+1})}, x_{2(k+j)}^{(l_{j+1})}, \cdots x_{m(k+j)}^{(l_{j+1})})}{\partial(x_{1(k+j)}^{(l_j)}, x_{2(k+j)}^{(l_j)}, \cdots x_{m(k+j)}^{(l_j)})}\right]_{m\times m}$$
$$= -\left[\frac{\partial \mathbf{g}^{(l_j l_{j+1})}}{\partial \mathbf{x}_{k+j}^{(l_{j+1})}}\right]_{m\times m}^{-1}\left[\frac{\partial \mathbf{g}^{(l_j l_{j+1})}}{\partial \mathbf{x}_{k+j}^{(l_j)}}\right]_{m\times m}. \tag{5.152}$$

If the magnitudes of two eigenvalues of the total Jacobian matrix DP are less than 1 (i.e., $|\lambda_\alpha| < 1,\ \alpha = 1, 2, \cdots, m$), the periodic motion is stable. If the magnitude of one of two eigenvalues is greater than 1 ($|\lambda_\alpha| > 1,\ \alpha \in \{1, 2, \cdots, m\}$), the periodic motion is unstable. If one of eigenvalues is a positive one (+1) and the rest of eigenvalues are in the unit cycle, the periodic flow experiences a saddle-node bifurcation. If one of eigenvalues is a negative one (−1) and the rest of eigenvalues are in the unit cycle, the periodic flow possesses a period-doubling bifurcation. If a pair of complex eigenvalues is on the unit cycle and the rest of eigenvalues are in the unit cycle, the Neimark bifurcation of the periodic flow occurs.

Consider a switching system as a combination of two subsystems (i.e., l_1th and l_2th subsystems). A point $\mathbf{x}_k^{(l_1)}$ at $t = t_k$ is selected as the initial point for the l_1th-subsystem. A mapping P_{l_1} maps the initial state to the final state $\mathbf{x}_{k+1}^{(l_1)}$ at $t = t_{k+1}$. For the l_1th subsystem switching to the l_2th subsystem, the transport mapping $P_0^{(l_1 l_2)}$ will be adopted at time $t = t_{k+1}$ and the initial condition of the l_2th subsystem will be obtained (i.e.,$\mathbf{x}_{k+1}^{(l_2)}$). Through the mapping P_{l_2} of the l_2th subsystem, the

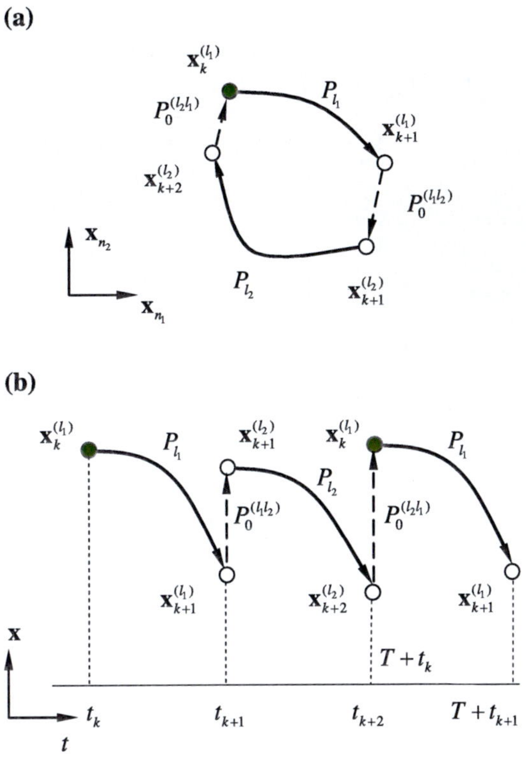

Fig. 5.34 A periodic motion of mapping structure $P_{l_2 l_1}$: **a** phase space and **b** time-history $(n_1 + n_2 = n)$

final state $\mathbf{x}_{k+2}^{(l_2)}$ is obtained at time $t = t_{k+2}$. To form a periodic flow, the transport mapping $P_0^{(l_2 l_1)}$ will map the point $\mathbf{x}_{k+2}^{(l_2)}$ to the starting point $\mathbf{x}_k^{(l_1)}$ with $t_{k+2} = t_k + T$. This process of flow formation is sketched in Fig. 5.34. The solid curves give the mappings for the l_1th and l_2th subsystems, and the dashed lines represent the transport mappings.

5.6 Linear Switching Systems

As in Luo and Wang (2009a, b) consider the ith subsystem in a resultant switching system for a time interval $t \in [t_{k-1}, t_k]$ in a form of

$$\dot{\mathbf{x}} = \mathbf{A}^{(i)}\mathbf{x} + \mathbf{Q}^{(i)}(t) \quad \text{for } i = 1, 2, \cdots, m, \tag{5.153}$$

where $\mathbf{A}^{(i)}$ is a matrix for the ith system matrix in the time interval $t \in [t_{k-1}, t_k]$. $\mathbf{Q}^{(i)}$ is a time-dependent vector function for the ith subsystem.

$$\mathbf{A}^{(i)} = \left(a_{kl}^{(i)}\right)_{n\times n} \text{ and } \mathbf{Q}^{(i)} = \left(Q_{kl}^{(i)}(t)\right)_{n\times 1}. \tag{5.154}$$

Eigenvalues of $\mathbf{A}^{(i)}$, given by $|\lambda^{(i)} I - \mathbf{A}^{(i)}| = 0$, possess p pairs of conjugated complex eigenvalues ($\lambda_{1,2} = u_1 \pm v_1\mathbf{i}$, $\lambda_{3,4} = u_2 \pm v_2\mathbf{i}$, $\cdots$, $\lambda_{2p-1,2p} = u_p \pm v_p\mathbf{i}$)

with $\mathbf{i} = \sqrt{-1}$ and $(n-2p)$ real eigenvalues of $\lambda_{2p+1}, \lambda_{2p+2}, \cdots, \lambda_n$. The closed-form solution of the ith subsystem in Eq. (5.153) will be given herein to demonstrate the methodology, i.e.,

$$\begin{aligned}\mathbf{x}^{(i)} = & \sum_{r=2p+1}^{n} L_r^{(k)} \mathbf{B}_r^{(i)} \exp\,[\lambda_r^{(i)}(t-t_{k-1})] + \sum_{r=1}^{p} \exp\,[u_r^{(i)}(t-t_{k-1})] \\ & \times \left\{ \mathbf{S}_r^{(i)} \cos v_r^{(i)}(t-t_{k-1}) - \mathbf{T}_r^{(i)} \sin v_r^{(i)}(t-t_{k-1}) \right\} + \mathbf{x}_p^{(i)}(t).\end{aligned} \tag{5.155}$$

where

$$\begin{aligned}\mathbf{B}_r^{(i)} &= (1, c_{2(r)}^{(i)}, \cdots, c_{n(r)}^{(i)})^{\mathrm{T}}, \\ \mathbf{S}_j^{(i)} &= (M_j^{(k)}, U_{2(j)}^{(i)} M_j^{(k)} - V_{2(j)}^{(i)} N_j^{(k)}, \cdots, U_{n(j)}^{(i)} M_j^{(k)} - V_{n(j)}^{(i)} N_j^{(k)})^{\mathrm{T}}, \\ \mathbf{T}_j^{(i)} &= (N_j^{(k)}, U_{2(j)}^{(i)} N_j^{(k)} + V_{2(j)}^{(i)} M_j^{(k)}, \cdots, U_{n(j)}^{(i)} N_j^{(k)} + V_{n(j)}^{(i)} M_j^{(k)})^{\mathrm{T}}.\end{aligned} \tag{5.156}$$

and the coefficients can be obtained by

$$\left.\begin{aligned}\mathbf{c}^{(i)} &= \left[\mathbf{C}_{11}^{(i)} - \mathbf{I}\lambda_r^{(i)}\right]^{-1} [-\mathbf{a}^{(i)}], \\ \begin{bmatrix}\mathbf{U}^{(i)} \\ \mathbf{V}^{(i)}\end{bmatrix} &= \begin{bmatrix}\mathbf{C}_{11}^{(i)} - \mathbf{I}a_k^{(i)} & \mathbf{I}b_k^{(i)} \\ -\mathbf{I}b_k^{(i)} & \mathbf{C}_{11}^{(i)} - \mathbf{I}a_k^{(i)}\end{bmatrix}^{-1} \begin{bmatrix}-\mathbf{a}^{(i)} \\ \mathbf{0}\end{bmatrix},\end{aligned}\right\} \tag{5.157}$$

where

$$\begin{aligned}&\mathbf{c}^{(i)} = (c_{2(r)}^{(i)}, c_{3(r)}^{(i)}, \cdots, c_{m(r)}^{(i)})^{\mathrm{T}}, \mathbf{U}^{(i)} = (U_{2(k)}^{(i)}, U_{3(k)}^{(i)}, \cdots, U_{n(k)}^{(i)})^{\mathrm{T}}, \\ &\mathbf{V}^{(i)} = (V_{2(k)}^{(i)}, V_{3(k)}^{(i)}, \cdots, V_{n(k)}^{(i)})^{\mathrm{T}}, \mathbf{a}^{(i)} = [a_{21}^{(i)}, a_{31}^{(i)}, \cdots, a_{n1}^{(i)}]^{\mathrm{T}},\end{aligned} \tag{5.158}$$

and $\mathbf{C}_{11}^{(i)}$ is a matrix relative to the minor of $a_{11}^{(i)}$ in matrix $\mathbf{A}^{(i)}$,

$$\mathbf{C}_{11}^{(i)} = \begin{bmatrix} a_{22}^{(i)} & a_{23}^{(i)} & \cdots & a_{2n}^{(i)} \\ a_{32}^{(i)} & a_{33}^{(i)} & \cdots & a_{3n}^{(i)} \\ \vdots & \vdots & \vdots & \vdots \\ a_{n2}^{(i)} & a_{n3}^{(i)} & \cdots & a_{nn}^{(i)} \end{bmatrix}. \tag{5.159}$$

The corresponding coefficients are

$$\begin{aligned}\mathbf{M}^{(k)} &= [M_1^{(k)}, M_2^{(k)}, \cdots, M_p^{(k)}]^{\mathrm{T}}, \\ \mathbf{N}^{(k)} &= [N_1^{(k)}, N_2^{(k)}, \cdots, N_p^{(k)}]^{\mathrm{T}}, \\ \mathbf{L}^{(k)} &= [L_{2p+1}^{(k)}, L_{2p+2}^{(k)}, \cdots, L_m^{(k)}]^{\mathrm{T}};\end{aligned} \tag{5.160}$$

$$\begin{bmatrix} \mathbf{M}^{(k)} \\ \mathbf{N}^{(k)} \\ \mathbf{L}^{(k)} \end{bmatrix} = \begin{bmatrix} 1 & 0 & \cdots 1 & 0 & 1 & \cdots 1 \\ U_{2(1)}^{(i)} & -V_{2(1)}^{(i)} & \cdots U_{2(p)}^{(i)} & -V_{2(p)}^{(i)} & c_{2(2p+1)}^{(i)} & \cdots c_{2(n)}^{(i)} \\ \vdots & \vdots & \vdots\ \vdots & \vdots & \vdots & \vdots\ \vdots \\ U_{n(1)}^{(i)} & -V_{n(1)}^{(i)} & \cdots U_{n(p)}^{(i)} & -V_{n(p)}^{(i)} & c_{n(2p+1)}^{(i)} & \cdots c_{n(n)}^{(i)} \end{bmatrix}^{-1} \times \left[\mathbf{x}^{(i)}(t_{k-1}) - \mathbf{x}_p^{(i)}(t_{k-1})\right]; \tag{5.161}$$

where $\mathbf{x}_p^{(i)}(t_{k-1})$ is $\mathbf{x}_p^{(i)}(t)$ at time t_{k-1} and $\mathbf{x}_{k-1}^{(i)} = \mathbf{x}^{(i)}(t_{k-1})$.

Consider three simple external forcing as examples herein

$$\begin{aligned} \mathbf{Q}^{(1)}(t) &= \sum_{s=0}^{J_1} \mathbf{a}_s^{(1)}(t - t_{k-1})^s, \\ \mathbf{Q}^{(2)}(t) &= \sum_{s=0}^{J_2} \mathbf{a}_s^{(2)} \cos\left[\Omega_s^{(2)}(t - t_{k-1})\right] + \mathbf{b}_s^{(2)} \sin\left[\Omega_s^{(2)}(t - t_{k-1})\right], \\ \mathbf{Q}^{(3)}(t) &= \sum_{s=0}^{J_3} \mathbf{a}_s^{(3)} \exp\left[\lambda_s^{(3)}(t - t_{k-1})\right] \end{aligned} \tag{5.162}$$

with

$$\begin{aligned} \mathbf{a}_s^{(i)} &= \left[a_{1(s)}^{(i)}\ a_{2(s)}^{(i)} \cdots a_{n(s)}^{(i)}\right]^{\mathrm{T}}, \\ \mathbf{b}_s^{(2)} &= \left[b_{1(s)}^{(2)}\ b_{2(s)}^{(2)} \cdots b_{n(s)}^{(2)}\right]^{\mathrm{T}} \end{aligned} \tag{5.163}$$

where $a_{l(s)}^{(i)}(i = 1, 2, 3; l = 1, \cdots, n)$ and $\Omega_j^{(2)}$, $\varphi_j^{(2)}$ and $\lambda_j^{(3)}$ are coefficients. The corresponding particular solutions are

$$\begin{aligned} \mathbf{x}_p^{(1)}(t) &= \sum_{j=0}^{J_1} \mathbf{A}_j^{(1)}(t - t_{k-1})^j, \\ \mathbf{x}_p^{(2)}(t) &= \sum_{j=0}^{J_2} \mathbf{A}_j^{(2)} \cos\left[\Omega_j^{(2)}(t - t_{k-1})\right] + \mathbf{B}_j^{(2)} \sin\left[\Omega_j^{(2)}(t - t_{k-1})\right], \\ \mathbf{x}_p^{(3)}(t) &= \sum_{j=0}^{J_3} \mathbf{A}_j^{(3)} \exp\left[\lambda_j^{(3)}(t - t_{k-1})\right] \end{aligned} \tag{5.164}$$

where

$$\begin{aligned} &\left.\begin{aligned} \mathbf{A}_s^{(1)} &= \mathbf{A}^{-1}[\mathbf{A}_{s+1}^{(1)}(s+1) - \mathbf{a}_s^{(1)}], \\ \mathbf{A}_{J_1}^{(1)} &= -\mathbf{A}^{-1}\mathbf{a}_{J_1}^{(1)} \end{aligned}\right\} (s = 0, 1, 2, \cdots, J_1 - 1); \\ &\begin{bmatrix} \mathbf{A}_s^{(2)} \\ \mathbf{B}_s^{(2)} \end{bmatrix} = -\begin{bmatrix} \mathbf{A}^{(i)} & -\mathbf{I}\Omega_s^{(2)} \\ \mathbf{I}\Omega_s^{(2)} & \mathbf{A}^{(i)} \end{bmatrix}^{-1} \begin{bmatrix} \mathbf{a}_s^{(2)} \\ \mathbf{b}_s^{(2)} \end{bmatrix}, (s = 0, 1, 2, \cdots, J_2); \\ &\mathbf{A}_s^{(3)} = -\left[\mathbf{A} - I\lambda_s^{(3)}\right]^{-1} \mathbf{a}_s^{(3)}, (s = 0, 1, 2, \cdots, J_3). \end{aligned} \tag{5.165}$$

Consider a periodic flow as in Eq. (5.140). The vector function for mapping P_{l_j} of the l_jth system is

$$\begin{aligned}\mathbf{f}^{(l_j)}(\mathbf{x}_{k+j}^{(l_j)},\mathbf{x}_{k+j-1}^{(l_j)}) &= \mathbf{x}_{k+j}^{(l_j)} - \mathbf{g}^{(l_j)}(\mathbf{x}_{k+j-1}^{(l_j)}) - \mathbf{d}^{(l_j)}(t_{k+j})\\ &= \mathbf{x}_{k+j}^{(l_j)} - \sum\nolimits_{r=2p+1}^{n} L_r^{(k)}\mathbf{B}_r^{(l_j)}(\mathbf{x}_{k+j-1}^{(l_j)})\exp[\lambda_r^{(l_j)}(t_{k+j}-t_{k+j-1})]\\ &\quad - \sum\nolimits_{r=1}^{p} \exp\,[u_r^{(l_j)}(t_{k+j}-t_{k+j-1})]\\ &\quad \times \Big\{\mathbf{S}_r^{(l_j)}(\mathbf{x}_{k+j-1}^{(l_j)})\cos v_r^{(l_j)}(t_{k+j}-t_{k+j-1}) - \mathbf{T}_r^{(l_j)}(\mathbf{x}_{k+j-1}^{(l_j)})\\ &\quad \times \sin v_r^{(l_j)}(t_{k+j}-t_{k+j-1})\Big\} - \mathbf{x}_p^{(l_j)}(t_{k+j}).\end{aligned} \tag{5.166}$$

From $\mathbf{f}^{(l_j)}(\mathbf{x}_{k+j}^{(l_j)},\mathbf{x}_{k+j-1}^{(l_j)}) = 0$, one obtains

$$\mathbf{x}_{k+j}^{(l_j)} = \mathbf{G}^{(l_j)}\mathbf{x}_{k+j-1}^{(l_j)} + \mathbf{d}^{(l_j)} \tag{5.167}$$

where the corresponding mapping and constant vector are given by

$$\mathbf{G}^{(l_j)} = \frac{\partial \mathbf{g}^{(l_j)}(\mathbf{x}_{k+j-1}^{(l_j)})}{\partial \mathbf{x}_{k+j-1}^{(l_j)}} \text{ and } \mathbf{d}^{(l_j)} = \mathbf{x}_p^{(l_j)}. \tag{5.168}$$

Consider the transport law to be an affine transformation

$$\mathbf{g}^{(l_{j+1}l_j)}(\mathbf{x}_{k+j}^{(l_{j+1})},\mathbf{x}_{k+j}^{(l_j)}) = 0 \Rightarrow \mathbf{x}_{k+j}^{(l_{j+1})} = \mathbf{E}^{(l_{j+1}l_j)}\mathbf{x}_{k+j-1}^{(l_j)} + \mathbf{e}^{(l_{j+1}l_j)}. \tag{5.169}$$

For the mapping structure in Eq. (5.140), one obtains

$$\mathbf{x}_{k+j}^{(l_{j+1})} - \mathbf{C}^{(l_{j+1})}\mathbf{x}_k^{(l_1)} = \mathbf{D}^{(l_{j+1})} \tag{5.170}$$

where for $j = 0, 1, 2, \cdots, N$

$$\mathbf{C}^{(l_{j+1})} = \mathbf{E}^{(l_{j+1}l_j)}\mathbf{G}^{(l_{j+1})}\ldots\mathbf{G}^{(l_{j+1})}\mathbf{E}^{(l_{j+1}l_j)}\mathbf{G}^{(l_j)}\ldots\mathbf{E}^{(l_2l_1)}\mathbf{G}^{(l_1)}, \tag{5.171}$$

$$\begin{aligned}\mathbf{D}^{(l_{N+1})} &= (\mathbf{E}^{(l_{N+1}l_N)}\mathbf{G}^{(l_N)})\mathbf{D}^{(l_N)} + \mathbf{e}^{(l_{N+1}l_N)}, \cdots,\\ \mathbf{D}^{(l_{j+1})} &= (\mathbf{E}^{(l_{j+1}l_j)}\mathbf{G}^{(l_j)})\mathbf{D}^{(l_j)} + \mathbf{e}^{(l_{j+1}l_j)}, \cdots,\\ \mathbf{D}^{(l_2)} &= (\mathbf{E}^{(l_2l_1)}\mathbf{G}^{(l_1)})\mathbf{d}^{(l_1)} + \mathbf{e}^{(l_2l_1)}, \mathbf{D}^{(l_1)} = \mathbf{d}^{(l_1)}.\end{aligned} \tag{5.172}$$

The periodic motion requires

$$\mathbf{x}_{k+N}^{(l_{N+1})} = \mathbf{x}_k^{(l_1)}. \tag{5.173}$$

If $\mathbf{D}^{(l_{N+1})} \neq 0$, the existence condition for periodic motion is

$$|\mathbf{I} - \mathbf{C}^{(l_{N+1})}| \neq 0. \tag{5.174}$$

In other words, the solution for the periodic flow does not exist if

$$|\mathbf{I} - \mathbf{C}^{(l_{N+1})}| = 0. \tag{5.175}$$

That is, the existence condition for periodic flow with mapping in Eq. (5.140) is that the eigenvalue of matrix is not one (+1). On the other hand, for the stability of the linear switching system, equation (5.170) for $j = N$ gives

$$\Delta \mathbf{x}_{k+N}^{(l_{N+1})} = DP\Delta \mathbf{x}_k^{(l_1)} = \mathbf{C}^{(l_{N+1})} \Delta \mathbf{x}_k^{(l_1)}. \tag{5.176}$$

Let $\Delta \mathbf{x}_{k+N}^{(l_{N+1})} = \lambda \Delta \mathbf{x}_k^{(l_1)}$, and the foregoing equation gives

$$|DP - \lambda \mathbf{I}| = |\mathbf{C}^{(l_{N+1})} - \lambda \mathbf{I}| = 0. \tag{5.177}$$

If the magnitudes of all eigenvalues λ_j $(j = 1, 2, \cdots, n)$ are less than one (i.e., $|\lambda_j| < 1$), the periodic flow in the switching system is stable. If the magnitude of one of eigenvalues λ_j $(j = 1, 2, \cdots, n)$ is greater than one, the periodic flow in the switching system is unstable. However, the saddle-node bifurcation of such periodic flow occurs if one of eigenvalues λ_j $(j \in \{1, 2, \cdots, n\})$ is one. Meanwhile, the periodic flow with such a mapping structure will disappear.

5.6.1 Vibrations with Piecewise Forces

As in Luo and Wang (2009a), consider the ith oscillator in a resultant switching system for a time interval $t \in [t_{k-1}, t_k]$ as

$$\ddot{x} + 2\delta^{(i)}\dot{x} + (\omega^{(i)})^2 x = Q^{(i)}(t) \text{ for } i = 1, 2, \cdots, m, \tag{5.178}$$

where $\delta^{(i)}$ and $\omega^{(i)}$ are damping coefficient and natural frequency in the ith oscillator in the time interval $t \in [t_{k-1}, t_k]$, respectively. $Q^{(i)}(t)$ is the ith external forcing. For a case $(\delta^{(i)} < |\omega^{(i)}|)$, the closed-form solution of the ith oscillator in Eq. (5.178) will be given herein.

$$\begin{aligned} x^{(i)}(t) &= e^{-\delta^{(i)}(t-t_{k-1})} \left[c_1^{(k)} \cos \omega_d^{(i)}(t - t_{k-1}) + c_2^{(k)} \sin \omega_d^{(i)}(t - t_{k-1}) \right] + x_p^{(i)}(t), \\ \dot{x}^{(i)}(t) &= e^{-\delta^{(i)}(t-t_{k-1})} \Big[(-\delta^{(i)} c_1^{(k)} + \omega_d^{(i)} c_2^{(k)}) \cos \omega_d^{(i)}(t - t_{k-1}) \\ &\quad - (\delta^{(i)} c_2^{(k)} + \omega_d^{(i)} c_1^{(k)}) \sin \omega_d^{(i)}(t - t_{k-1}) \Big] + \dot{x}_p^{(i)}(t); \end{aligned} \tag{5.179}$$

with $\omega_d^{(i)} = \sqrt{(\omega^{(i)})^2 - (\delta^{(i)})^2}$ and the two coefficients $c_1^{(k)}$ and $c_2^{(k)}$ are given by

$$\begin{aligned} c_1^{(k)} &= x_{k-1}^{(i)} - x_p^{(i)}(t_{k-1}), \\ c_2^{(k)} &= \frac{1}{\omega_d^{(i)}}[\dot{x}_{k-1}^{(i)} - \dot{x}_p^{(i)}(t_{k-1}) + \delta^{(i)} c_1^{(k)}], \end{aligned} \tag{5.180}$$

where $x_p^{(i)}(t_{k-1})$ and $\dot{x}_p^{(i)}(t_{k-1})$ are $x_p^{(i)}(t)$ and $\dot{x}_p^{(i)}(t)$ at time t_{k-1}, $x_{k-1}^{(i)} = x^{(i)}(t_{k-1})$ and $\dot{x}_{k-1}^{(i)} = \dot{x}^{(i)}(t_{k-1})$. Consider three simple external forcing as

$$\begin{aligned} Q^{(1)}(t) &= \sum\nolimits_{j=0}^{J_1} a_j^{(1)}(t - t_{k-1})^j, \\ Q^{(2)}(t) &= \sum\nolimits_{j=0}^{J_2} a_j^{(2)} \cos[\Omega_j^{(2)}(t - t_{k-1}) + \varphi_j^{(2)}], \\ Q^{(3)}(t) &= \sum\nolimits_{j=0}^{J_3} a_j^{(3)} \exp[\lambda_j^{(3)}(t - t_{k-1})] \end{aligned} \tag{5.181}$$

where $a_j^{(l)}$ $(l = 1, 2, 3)$ and $\Omega_j^{(2)}, \varphi_j^{(2)}$ and $\lambda_j^{(3)}$ are coefficients. The corresponding particular solutions are

$$\begin{aligned} x_p^{(1)}(t) &= \sum\nolimits_{j=0}^{J_1} A_j^{(1)}(t - t_{k-1})^j, \\ x_p^{(2)}(t) &= \sum\nolimits_{j=1}^{J_2} A_j^{(2)} \cos[\Omega_j^{(2)}(t - t_{k-1}) + \Phi_j^{(2)}], \\ x_p^{(3)}(t) &= \sum\nolimits_{j=0}^{J_3} \frac{a_j^{(3)}}{(\lambda_j^{(3)})^2 + 2\delta^{(i)}\lambda_j^{(3)} + (\omega^{(i)})^2} \exp[\lambda_j^{(3)}(t - t_{k-1})] \end{aligned} \tag{5.182}$$

where

$$\begin{aligned} A_{J_1}^{(1)} &= \frac{a_{J_1}^{(1)}}{(\omega^{(i)})^2}, A_{J_1-1}^{(1)} = \frac{a_{J_1-1}^{(1)} - 2\delta^{(i)} J_1 A_{J_1}^{(1)}}{(\omega^{(i)})^2}, \\ A_J^{(1)} &= \frac{1}{(\omega^{(i)})^2}[a_J^{(1)} - (J+2)(J+1)A_{J+2}^{(1)} - 2\delta^{(i)}(J+1)A_{J+1}^{(1)}] \\ &(J = 0, 1, 2, \cdots, J_1 - 1), \end{aligned} \tag{5.183}$$

$$\begin{aligned} A_j^{(2)} &= \frac{a_j^{(2)}}{\sqrt{[(\omega^{(i)})^2 - (\Omega_j^{(2)})^2]^2 + (2\delta^{(i)}\Omega_j^{(2)})^2}}, \\ \Phi_j^{(2)} &= \varphi_j^{(2)} - \arctan\frac{2\delta^{(i)}\Omega_j^{(2)}}{(\omega^{(i)})^2 - (\Omega_j^{(2)})^2}. \end{aligned} \tag{5.184}$$

Consider a switching set for the ith oscillator to be

$$\Sigma^{(i)} = \left\{ (x_k^{(i)}, \dot{x}_k^{(i)}) \middle| x_k^{(i)} = x^{(i)}(t_k), \dot{x}_k^{(i)} = \dot{x}^{(i)}(t_k), k \in \{0, 1, 2, \cdots\} \right\}. \tag{5.185}$$

From the solution of the ith oscillator, a mapping P_i for a time interval $t \in [t_{k-1}, t_k]$ is defined as

$$P_i : \Sigma^{(i)} \to \Sigma^{(i)} \quad \text{for } i = 1, 2, \cdots, m \tag{5.186}$$

or

$$P_i : (x_{k-1}^{(i)}, \dot{x}_{k-1}^{(i)}) \to (x_k^{(i)}, \dot{x}_k^{(i)}) \quad \text{for } i = 1, 2, \cdots, m. \tag{5.187}$$

For simplicity, introduce two vectors for mapping P_i herein

$$\mathbf{f}^{(i)} = (f_1^{(i)}, f_2^{(i)})^{\mathrm{T}} \text{ and } \mathbf{x}^{(i)} = (x^{(i)}, \dot{x}^{(i)})^{\mathrm{T}}. \tag{5.188}$$

From the displacement and velocity solutions of the ith oscillator for the starting and ending points, the two component functions of the vector function $\mathbf{f}^{(i)}$ can be determined. For instance, from the starting point $(x_{k-1}^{(i)}, \dot{x}_{k-1}^{(i)})$ at time t_{k-1} and the ending points $(x_k^{(i)}, \dot{x}_k^{(i)})$ at time t_k, the two components of the vector function $\mathbf{f}^{(i)}$ can be written for $i = 1, 2, \cdots, m$ by two algebraic equations, i.e.,

$$\left.\begin{aligned} f_1^{(i)}(\mathbf{x}_k^{(i)}, \mathbf{x}_{k-1}^{(i)}) &\equiv f_1^{(i)}(x_k^{(i)}, \dot{x}_k^{(i)}, x_{k-1}^{(i)}, \dot{x}_{k-1}^{(i)},) = 0, \\ f_2^{(i)}(\mathbf{x}_k^{(i)}, \mathbf{x}_{k-1}^{(i)}) &\equiv f_2^{(i)}(x_k^{(i)}, \dot{x}_k^{(i)}, x_{k-1}^{(i)}, \dot{x}_{k-1}^{(i)}) = 0. \end{aligned}\right\} \tag{5.189}$$

Using Eq. (5.179), the algebraic equations of mapping P_i for $t \in [t_{k-1}, t_k]$ are

$$\begin{aligned} f_1^{(i)}(\mathbf{x}_k^{(i)}, \mathbf{x}_{k-1}^{(i)}) = {} & x_k^{(i)} - e^{-\delta^{(i)}(t_k - t_{k-1})}[c_1^{(k-1)} \cos \omega_d^{(i)}(t_k - t_{k-1}) \\ & + c_2^{(k-1)} \sin \omega_d^{(i)}(t_k - t_{k-1})] + x_p^{(i)}(t_k) = 0, \\ f_2^{(i)}(\mathbf{x}_k^{(i)}, \mathbf{x}_{k-1}^{(i)}) = {} & \dot{x}_k^{(i)} - e^{-\delta^{(i)}(t_k - t_{k-1})}[(-\delta^{(i)} c_1^{(k)} + \omega_d^{(i)} c_2^{(k)}) \cos \omega_d^{(i)}(t_k - t_{k-1}) \\ & - (\delta^{(i)} c_2^{(k)} + \omega_d^{(i)} c_1^{(k)}) \sin \omega_d^{(i)}(t_k - t_{k-1})] + \dot{x}_p^{(i)}(t_k) = 0. \end{aligned} \tag{5.190}$$

Suppose the two trajectories of the ith and jth oscillators in phase space at the switching time t_k is continuous, i.e.,

$$(x_k^{(i)}, \dot{x}_k^{(i)}) = (x_k^{(j)}, \dot{x}_k^{(j)}) \quad \text{for } i, j \in \{1, 2, \cdots, m\}. \tag{5.191}$$

If the two responses of the ith and jth oscillators at the switching time t_k are discontinuous, a transport law is needed. The transport law is assumed as

$$\mathbf{x}_k^{(j)} = \mathbf{g}^{(ij)}(\mathbf{x}_k^{(i)}) \text{ where } \mathbf{g}^{(ij)} = (g_1^{(ij)}, g_2^{(ij)})^{\mathrm{T}}. \tag{5.192}$$

In other words, the transport law can be written as

$$\left.\begin{aligned} x_k^{(j)} &= g_1^{(ij)}(x_k^{(i)}, \dot{x}_k^{(i)}) \\ \dot{x}_k^{(j)} &= g_2^{(ij)}(x_k^{(i)}, \dot{x}_k^{(i)}) \end{aligned}\right\} \text{ for } i, j \in \{1, 2, \cdots, m\}. \tag{5.193}$$

The ith and jth dynamical systems are evolving in the time intervals $[t_{k-1}, t_k]$ and $[t_k, t_{k+1}]$, respectively. The switching point for the two dynamical systems is at time t_k. The transport law between the ith and jth dynamical systems is to transport the final state of the ith dynamical system to the initial state of the jth dynamical system. This process makes the responses of the two systems to be continued at time t_k. This phenomenon described by the transport law or the transport mapping extensively exists, such as impact, control logic laws, impulse and other physical laws to make the discontinuity. To describe the entire responses of the resultant systems, the mapping structures are adopted as a kind of symbolic description. Thus, the transport mapping relative to the transport law is used to possess the same function as the transport law. From the transport law, a transport mapping is introduced as

$$P_0^{(ij)} : \Sigma^{(i)} \to \Sigma^{(j)} \text{ for } i, j \in \{1, 2, \cdots, m\}, \tag{5.194}$$

or

$$P_0^{(ij)} : (x_k^{(i)}, \dot{x}_k^{(i)}) \to (x_k^{(j)}, \dot{x}_k^{(j)}) \quad \text{for } i, j \in \{1, 2, \cdots, m\}. \tag{5.195}$$

The algebraic equations for the transport mapping are given by Eq. (5.193). The transport mapping $P_0^{(ij)}$ is to make the final point of the trajectory of the ith system jump to the starting point of the trajectory of the jth system at time t_k.

If the system parameters of the oscillators in Eq. (5.178) will be invariant, consider a switching system governed by a rectangular wave forcing as an example, i.e.,

$$Q^{(i)}(t) = (-1)^{i+1} C^{(i)} \quad \text{for } t \in [t_{k-1}, t_k] \text{ and } i \in \{1, 2\}. \tag{5.196}$$

The corresponding exact solution for each piece of the rectangular wave force is

$$\begin{aligned} x^{(i)}(t) &= e^{-\delta^{(i)}(t-t_{k-1})} \left[c_1^{(k)} \cos \omega_d^{(i)}(t - t_{k-1}) + c_2^{(k)} \sin \omega_d^{(i)}(t - t_{k-1}) \right] \\ &\quad + (-1)^{i+1} \frac{C^{(i)}}{(\omega^{(i)})^2} \\ \dot{x}^{(i)}(t) &= e^{-\delta^{(i)}(t-t_{k-1})} \left[(-\delta^{(i)} c_1^{(k)} + \omega_d^{(i)} c_2^{(k)}) \cos \omega_d^{(i)}(t - t_{k-1}) \right. \\ &\quad \left. - (\delta^{(i)} c_2^{(k)} + \omega_d^{(i)} c_1^{(k)}) \sin \omega_d^{(i)}(t - t_{k-1}) \right] \end{aligned} \tag{5.197}$$

where

$$c_1^{(k)} = x_{k-1}^{(i)} - (-1)^{i+1} \frac{C^{(i)}}{(\omega^{(i)})^2} \text{ and } c_2^{(k)} = \frac{1}{\omega_d^{(i)}} (\dot{x}_{k-1}^{(i)} + \delta^{(i)} c_1^{(k)}). \tag{5.198}$$

For this switching system, the two subsystems switch with the transport law as

$$(x_{k+j}^{(l_1)}, \dot{x}_{k+j}^{(l_1)}) = (x_{k+j}^{(l_2)}, \dot{x}_{k+j}^{(l_2)}) \text{ for } l_1, l_2 \in \{1, 2\} \text{ and } j = 1, 2, \cdots. \tag{5.199}$$

Consider $Q^{(1)}(t) = C^{(1)}$ and $Q^{(2)}(t) = -C^{(2)}$. The total solution of the resultant system is for $i = 1, 2$

$$\left.\begin{aligned}&\mathbf{x}(t) = \left\{ (x^{(i)}(t), \dot{x}^{(i)}(t)) \middle| \, t \in [t_{k+2j+i-1}, t_{k+2j+i}] \text{ and } j = 0, 1, \cdots \right\}\\&(x^{(1)}_{k+j+1}, \dot{x}^{(1)}_{k+j+1}) = (x^{(2)}_{k+j+1}, \dot{x}^{(2)}_{k+j+1}) \text{ for } j = 0, 1, 2, \cdots .\end{aligned}\right\} \quad (5.200)$$

If the piecewise discontinuous forcing in switching systems is periodic, one often uses the Fourier series expansion method to obtain the corresponding responses of the switching systems. However, the Fourier series expansion method has smoothened the discontinuity of piecewise discontinuous forcing. Suppose the infinite-term summation of the Fourier series can approach such piecewise discontinuous forcing. In fact, once the discontinuity of the piecewise forcing is smoothened, the non-smooth responses of the dynamical system cannot be accurately described. For the jump or impulsive discontinuity of the forcing, the Gibbs phenomenon can never be avoided. The method presented herein can treat the switching system through subsystems with switching connections, and the entire exact solutions can be obtained through subsystems in the switching system. Such a method does not use the Dirichlet conditions. If a piecewise forcing in subsystems is non-periodic, one cannot find a traditional method to obtain any approximate solution to approach the exact solution. From the method presented herein, the exact solutions of the switching systems consisting of a group of linear systems can be obtained for any randomly selected piecewise forcing, as well as randomly impulsive forcing (see Luo and Wang, 2009a). One cannot find any Fourier series to model such a random piecewise forcing. The traditional methods based on the Lipschitz condition (e.g., Fourier series and Taylor series expansions) cannot provide adequate solutions for the oscillator with the random piecewise forcing. The method presented herein can easily achieve the exact solution for such a switching linear system. In vibration testing, only random, piecewise, sinusoidal waves can be used in shaking tables. Thus, this method can provide a reasonable wave for actuators to drive the shaking table in random vibration testing.

Consider an excitation of a two-uneven rectangular wave with the time interval spans of $T_1 = 0.75T$ and $T_2 = 0.25T$ (e.g.,$T = 2$) and $C^{(i)} = 1$. The two pieces of the excitation are considered as two different forcing, and the oscillator can be treated as two switching systems to switch at given times. Therefore, from the presented method in the previous section, the responses of the oscillator under such rectangular wave forcing are presented in Fig. 5.35. Such a rectangular excitation forcing is illustrated in Fig. 5.35a. The two pieces of the rectangular uneven wave is illustrated. The time-histories of displacement, velocity and acceleration for such oscillator are presented in Fig. 5.35b–d. It is observed that the exact solution of displacement is smooth. However, the exact velocity response is non-smooth because the rectangular wave forcing is discontinuous. Such a discontinuity causes the acceleration of the oscillator to be discontinuous. However, for the Fourier series solution of this system, the responses of velocity and acceleration are continuous. For a further look at the dynamical behaviors of the oscillator, the trajectory for periodic response is presented in Fig. 5.35e, and the corresponding forcing distribution along the displacement is presented in Fig. 5.35f.

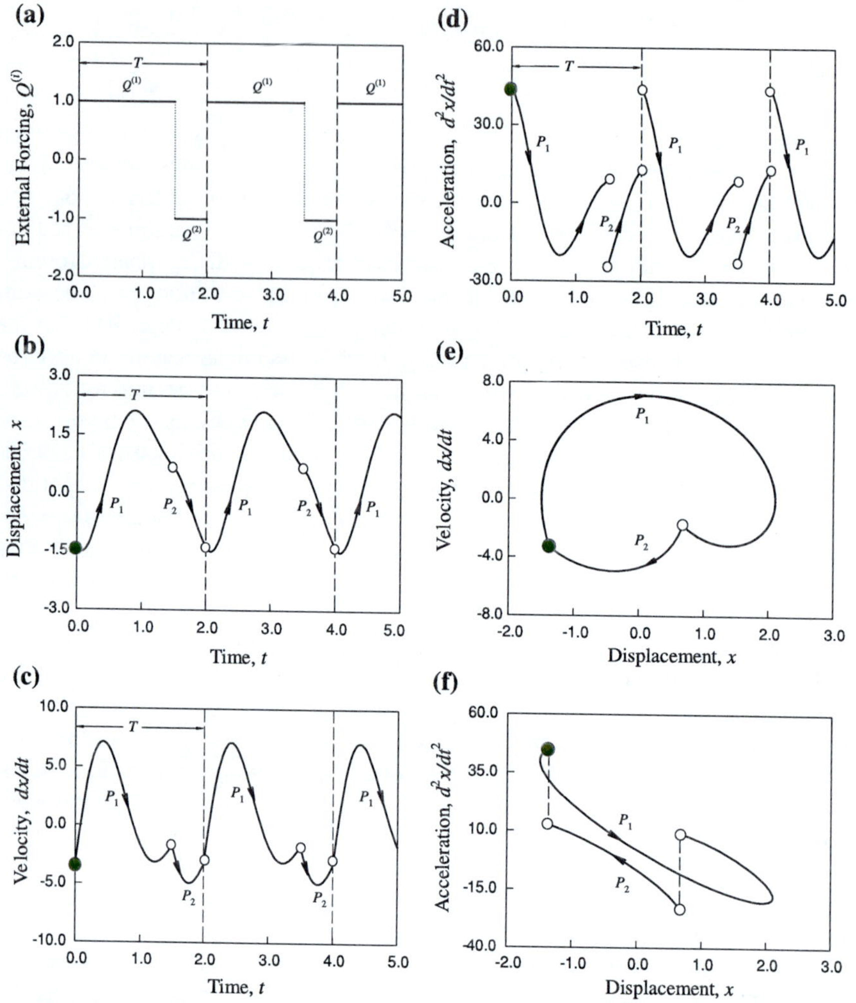

Fig. 5.35 Periodic motion under two pieces of *rectangular waves*: **a** forcing, **b** displacement **c** velocity, **d** acceleration, **e** phase plane and **f** acceleration versus displacement ($\omega^{(i)} = 4.0$, $\delta^{(i)} = 1.0$, $C^{(i)} = 1$, $T_1 = 1.5$, $T_2 = 0.5$, $x(0) = -1.3728$, and $\dot{x}(0) = -3.2947$)

5.6.2 Switching Systems with Impulses

As in Luo and Wang (2009b), consider a switching system with two 2-D subsystems with two matrices for $j = 0, 1, 2, \cdots$

$$\begin{aligned}
\mathbf{A}^{(1)} &= \begin{bmatrix} a_1 & b_1 \\ c_1 & d_1 \end{bmatrix} \quad \text{for } t \in [t_{k+2j}, t_{k+2j+1}], \\
\mathbf{A}^{(2)} &= \begin{bmatrix} a & b \\ c & d \end{bmatrix} \quad \text{for } t \in [t_{k+2j+1}, t_{k+2j+2}], \\
\mathbf{Q}^{(i)} &= (Q_1, Q_2)^{\mathrm{T}} \cos \Omega t \ \text{ for } \ i = 1, 2.
\end{aligned} \tag{5.201}$$

The corresponding switching points are continuous but non-smooth, i.e.,

$$x_1^{(1)}(t_{k+j}) = x_1^{(2)}(t_{k+j}) \ \text{ and } \ x_2^{(1)}(t_{k+j}) = x_2^{(2)}(t_{k+j}). \tag{5.202}$$

Without losing generality, the parameters for a dynamical system are fixed and the parameters for another subsystem are varied. Select parameters as

$$a = a_1 = d_1 = -1, b_1 = 1, c_1 = -2. \tag{5.203}$$

From Eqs. (5.146) and (5.147), the time-interval parameters can be expressed as

$$\begin{aligned}
q_{2k+i-1}^{(i)} &= \frac{t_{2k+i} - t_{2k+i-1}}{T} \ \text{ for } \ i = 1, 2 \ \text{ and} \\
1 &= \sum\nolimits_{i=1}^{2} \sum\nolimits_{k=1}^{N} q_{2k+i-1}^{(i)} \quad \text{for } \ k = 1, 2, \cdots.
\end{aligned} \tag{5.204}$$

If $q_{2k+i-1}^{(i)} = q^{(i)}$, the ith subsystem possesses the equi-time interval. Otherwise, the ith subsystem possesses the non-equi-time interval. Consider a simple periodic motion with a mapping structure of P_{21}, and the corresponding time interval parameter can be set as $q^{(1)}$ and $q^{(2)}$. For $q^{(1)} = 0$ ($q^{(2)} = 1$), the switching system is formed by the second subsystem only. For $q^{(1)} = 1$ ($q^{(2)} = 0$), the switching system is formed by the first subsystem only. For a periodic flow with $P = P_{21}$, consider an initial state $\mathbf{x}_k^{(1)}$ and the final state $\mathbf{x}_{k+1}^{(1)}$ for the first subsystem in the resultant, switching system, and an initial state $\mathbf{x}_{k+1}^{(2)}$ and the final state $\mathbf{x}_{k+2}^{(2)}$ for the second subsystem. The analytical prediction of a periodic flow for such a mapping structure is given in Fig. 5.36 for parameters ($b = -5.0, d = 1.6, q^{(1)} = 0.25$ and $q^{(2)} = 0.75,\ Q_1 = 1.0,\ Q_2 = 1.5,\ \Omega = 1.5$) with other parameters in Eq. (5.202) (i.e., $a = a_1 = d_1 = -1, b_1 = 1, c_1 = -2$). With varying parameter c, the switching points of the periodic flow with P_{21} for the two subsystems are obtained, and the corresponding stability is presented in Fig. 5.37 through the eigenvalue analysis. The stable and unstable ranges of periodic flows of P_{21} for parameter c are clearly observed. With increasing parameter c, the stable range for such a simple periodic flow becomes large. If $c > 10$, all the periodic flow is always stable. The infinity solutions for the switching points are observed when the eigenvalue is equal to positive one (+1), which agreed very well with the afore-mentioned discussion. To further look into the stability of such a periodic flow, the parameter map between the parameters d and c are developed, as shown in Fig. 5.36 for parameters ($b = -5.0,\ \ q^{(1)} = 0.25$ and $q^{(2)} = 0.75\ a = a_1 = d_1 = -1,\ \ b_1 = 1,$

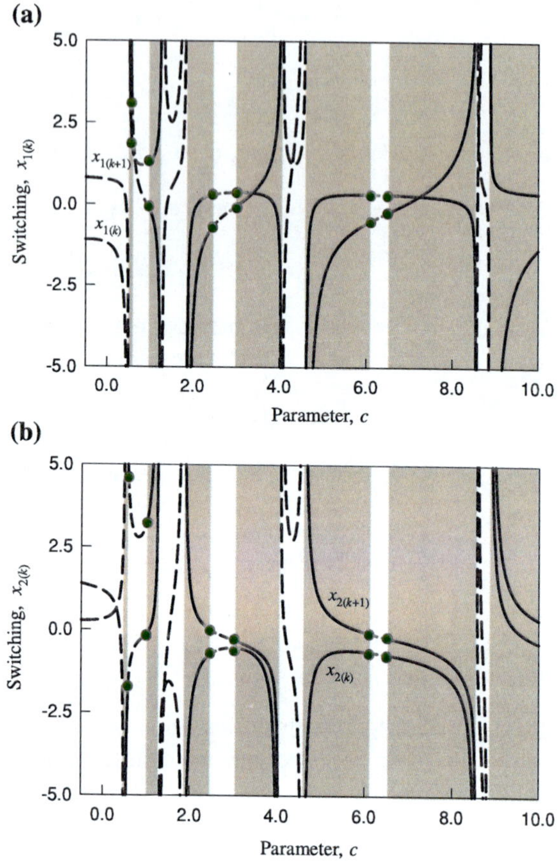

Fig. 5.36 Periodic motion scenario with P_{21} : **a** switching $x_1(k)$ and **b** switching $x_{2(k)}$. ($b = -5.0, d = 1.6,\ q^{(1)} = 0.25$ and $q^{(2)} = 0.75, a = a_1 = d_1 = -1, b_1 = 1, c_1 = -2, Q_1 = 1.0,$ $Q_2 = 1.5,\ \omega = 1.5$). The *solid* and *dashed curves* represent stable and unstable periodic flows with P_{21}, respectively

$c_1 = -2,\quad Q_1 = 1.0,\quad Q_2 = 1.5,\quad \Omega = 1.5$). In Fig. 5.38, the stable and unstable periodic flows lie in the shaded and non-shaded areas, respectively. For the switching system, the time-interval of each sub-system is significant to influence the stability characteristics of the resultant system. The parameter maps between the parameter d and the time interval parameter $q^{(1)}$ for a periodic flow of P_{21} are presented with ($c = -0.5, 0.5, 1.0$ and 2.0) in Fig. 5.39a–d, respectively. For $c = -0.5$, the minimum value of d for the unstable periodic motion is $d \approx -0.271$ with $q^{(1)} \approx 0.09$. The cusp point exists for this parameter map. Only two dents in parameter map exist for the unstable periodic flow. For $c = 0.5$, the minimum value of d for the unstable periodic motion is $d \approx 0.557$ with $q^{(1)} \approx 0.2$. The three dents in parameter map exist for the unstable periodic flow, but the cusp points do not exist. For $c = 1.0$ and 2.0, the number of dents in the parameter maps becomes four and five accordingly. The dents in the stability maps are not uniform.

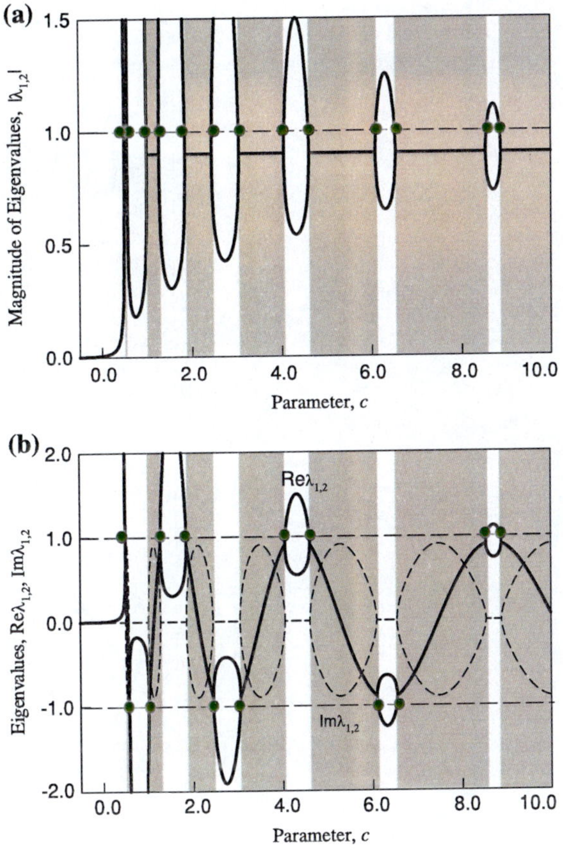

Fig. 5.37 Eigenvalue analysis for periodic motions with P_{21} : **a** magnitudes and **b** real and imaginary parts of eigenvalues ($b = -5.0,\ \ d = 1.6,\ \ q^{(1)} = 0.25,\ \ q^{(2)} = 0.75,\ \ a = a_1 = d_1 = -1,\ \ b_1 = 1,\ \ c_1 = -2,\ \ Q_1 = 1.0,\ \ Q_2 = 1.5,\ \ \Omega = 1.5$)

Table 5.1 Input data for numerical simulations of stable and unstable periodic flows of P_{21} ($b = -5.0, d = 1.6, q^{(1)} = 0.25, q^{(2)} = 0.75,\ \ a = a_1 = d_1 = -1, b_1 = 1, c_1 = -2, Q_1 = 1.0,\ \ Q_2 = 1.5,\ \Omega = 1.5,\ t_0 = 0.0$)

	Parameter c	Initial condition	Stability
Figure 5.41a	$c = 0.45$	$x_1^{(1)} \approx -5.4975, x_2^{(1)} \approx -2.0072$	Unstable
Figure 5.41b	$c = 0.55$	$x_1^{(1)} \approx 5.1645, x_2^{(1)} \approx 6.2506$	stable
Figure 5.41c	$c = 0.80$	$x_1^{(1)} \approx 0.2707, x_2^{(1)} \approx 2.7373$	Unstable
Figure 5.41d	$c = 1.10$	$x_1^{(1)} \approx -0.6420, x_2^{(1)} \approx 4.1622$	Stable

5.6.3 Numerical Illustrations

(A) *A 2-D Switching System*: The matrices and vectors for the 2-dimensional system in Eq. (5.201) will be considered again, and the parameters in Eq. (5.203) will be used. From the analytical prediction, the time histories for state variables for a stable periodic flow of $P_{21} (c = 0.55)$ are plotted in Fig. 5.40 for parameters ($b = -5.0$,

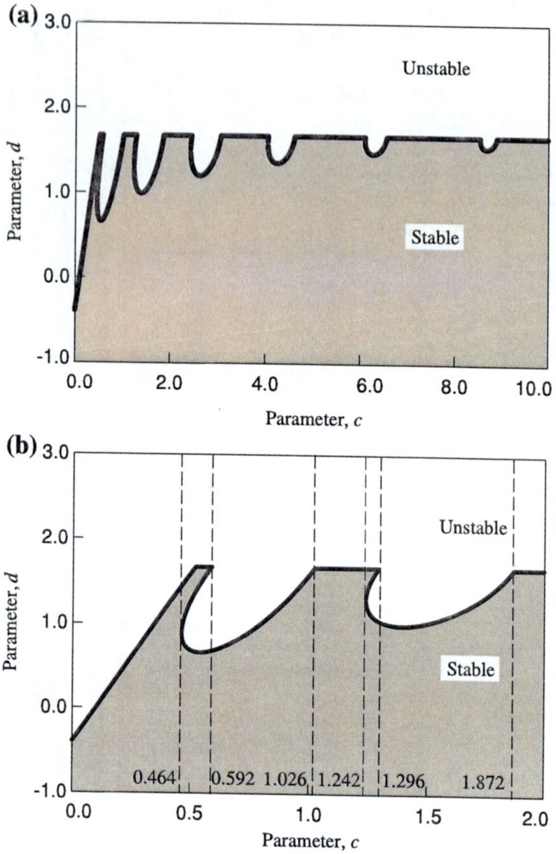

Fig. 5.38 Parameter maps of (d, c): **a** stable and unstable motion regions, **b** zoomed view ($b = -5.0$, $q^{(1)} = 0.25$ and $q^{(2)} = 0.75$, $a = a_1 = d_1 = -1, b_1 = 1, c_1 = -2$, $Q_1 = 1.0$, $Q_2 = 1.5$, $\Omega = 1.5$)

$d = 1.6$, $q^{(1)} = 0.25$, $q^{(2)} = 0.75$, $a = a_1 = d_1 = -1$, $b_1 = 1, c_1 = -2$, $Q_1 = 1.0$, $Q_2 = 1.5$, $\Omega = 1.5$) and the initial conditions ($t_0 = 0.0$, $x_1^{(1)} \approx 5.1645$, $x_2^{(1)} \approx 6.2506$). The non-smooth behavior of the periodic flow for state variables (x_1 and x_2) are observed. For a further observation of non-smooth characteristics of the switching system, the trajectories of stable and unstable periodic flows in phase plane for P_{21} is presented in Fig. 5.41a–d for $d = 1.6$, and the corresponding input data is listed in Table 5.1. The initial conditions are obtained from the analytical prediction. In Fig. 5.41a, the periodic flow for $c = -0.5$ is unstable. The asymptotic instability of the periodic flow in phase plane is observed. With increasing parameter c, a stable periodic flow with $c = 0.5$ is presented in Fig. 5.41b. For $c = 0.8$, another unstable periodic flow can be observed, as plotted in Fig. 5.41c. Compared to the trajectory in Fig. 5.41a, the configuration of trajectory is different. For $c = 1.1$, the stable periodic flow is observed, and the profile of the trajectory in phase plane is changed. The stability of the periodic flows can be obtained through the parameter map in Fig. 5.38.

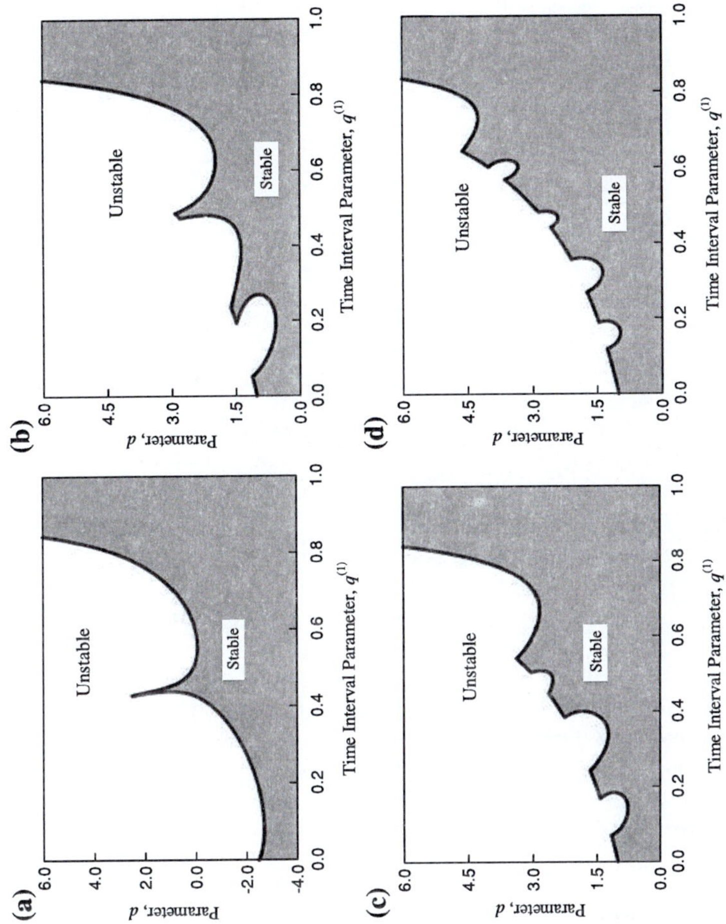

Fig. 5.39 Parameter maps of $(q^{(1)}, d)$: **a** $c = -0.5$, **b** $c = 0.5$, **c** $c = 1.0$ and **d** $c = 2.0$. ($b = -5.0$, $a = a_1 = d_1 = -1$, $b_1 = 1, c_1 = -2$, $Q_1 = 1.0$, $Q_2 = 1.5$, $\Omega = 1.5$)

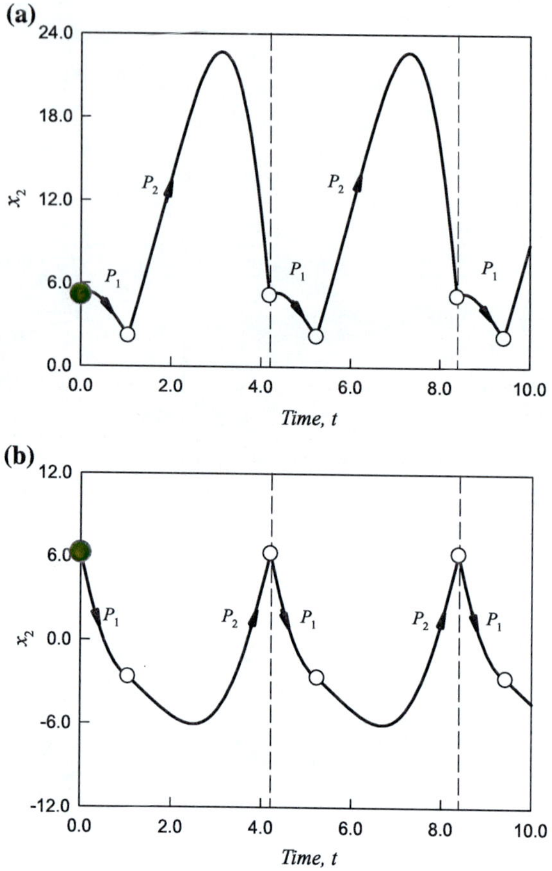

Fig. 5.40 A stable periodic flow of P_{21} ($c = 0.55$) for a 2-D switching system: **a** time-history of x_1 and **b** time-history of x_2 ($b = -5.0, d = 1.6,\ \ q^{(1)} = 0.25,\ q^{(2)} = 0.75, a = a_1 = d_1 = -1,\ \ b_1 = 1, c_1 = -2,\ \ Q_1 = 1.0,\ Q_2 = 1.5,\ \ \Omega = 1.5$) ($t_0$=0.0, $x_1^{(1)} \approx 5.1645$, $x_2^{(1)} \approx 6.2506$)

In the foregoing illustrations, the switching between two subsystems is continuous but non-smooth. In fact, the switching between two systems can be completed through a transport law. To discuss effects of the transport law on periodic motion, the transport law in Eq. (5.169) for the 2-D switching system will be used. Four cases of the linear transformation are:

(i) continuous switching for two sub-systems

$$\mathbf{x}_k^{(1)} = \mathbf{x}_k^{(2)} \quad \text{for } k = 1, 2, \cdots ; \tag{5.205}$$

(ii) translation switching as a switching transport law

$$\left.\begin{aligned} &\mathbf{x}^{(1)}(t_{k+}) = \mathbf{x}^{(2)}(\mathrm{t}_{k-}) + \mathbf{e}^{12)} \\ &\mathbf{x}^{(2)}(t_{k+}) = \mathbf{x}^{(1)}(\mathrm{t}_{k-}) + \mathbf{e}^{(21)} \\ &\mathbf{e}^{(12)}, \mathbf{e}^{(21)} = \text{const} \end{aligned}\right\} \quad \text{for } k = 1, 2, \cdots ; \tag{5.206}$$

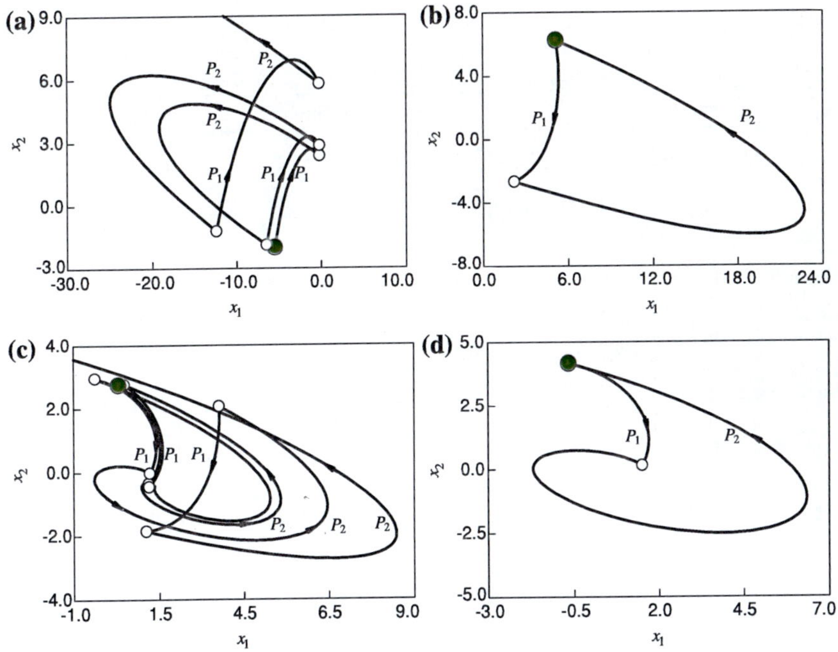

Fig. 5.41 Trajectories of stable and unstable periodic flows in phase plane with P_{21} for a 2-D switching system ($t_0 = 0.0$) : **a** ($c = 0.45, x_1^{(1)} \approx -5.4975, x_2^{(1)} \approx -2.0072$), **b** ($c = 0.55$, $x_1^{(1)} \approx 5.1645$, $x_2^{(1)} \approx 6.2506$), **c** ($c = 0.80$, $x_1^{(1)} \approx 0.2707$, $x_2^{(1)} \approx 2.7373$) and **d** ($c = 1.10$, $x_1^{(1)} \approx -0.6420$, $x_2^{(1)} \approx 4.1622$). ($b = -5.0, d = 1.6$, $q^{(1)} = 0.25, q^{(2)} = 0.75$, $a = a_1 = d_1 = -1$, $b_1 = 1, c_1 = -2$, $Q_1 = 1.0$, $Q_2 = 1.5$, $\Omega = 1.5$)

(iii) scaling switching as a switching transport law

$$\left.\begin{array}{l}\mathbf{x}^{(2)}(t_{k+}) = \mathbf{E}^{(12)} \cdot \mathbf{x}^{(1)}(t_{k-})\\ \mathbf{x}^{(1)}(t_{k+}) = \mathbf{E}^{(21)} \cdot \mathbf{x}^{(2)}(t_{k-})\\ \mathbf{E}^{(12)}, \mathbf{E}^{(21)} \text{ is diaginal matrix}\end{array}\right\} \quad \text{for } k = 1, 2, \cdots ; \tag{5.207}$$

(iv) affine switching as a switching transport law

$$\left.\begin{array}{l}\mathbf{x}^{(2)}(t_{k+}) = \mathbf{E}^{(12)} \cdot \mathbf{x}^{(1)}(t_{k-}) + \mathbf{e}^{(12)}\\ \mathbf{x}^{(1)}(t_{k+}) = \mathbf{E}^{(21)} \cdot \mathbf{x}^{(2)}(t_{k-}) + \mathbf{e}^{(21)}\\ \mathbf{E}^{(12)}, \mathbf{E}^{(21)} \text{ are const matrices}\\ \mathbf{e}^{(12)}, \mathbf{e}^{(21)} \text{ are const vectors}\end{array}\right\} \quad \text{for } k = 1, 2, \cdots . \tag{5.208}$$

Consider the effects of the transport laws to the dynamical behaviors of the resultant system. The same system parameters are used but the switching laws are different. The corresponding input data for numerical simulation is tabulated in Table 5.2 for stable and unstable periodic flows of P_{21} with the transport laws.The trajectories

Table 5.2 Input data for numerical simulations of stable and unstable periodic flows of P_{21} with transportation laws ($a = -1, b = -5,\ c = 1.5,\ d = 0.5,\ q^{(1)} = 0.25$ and $q^{(2)} = 0.75$, $a_1 = d_1 = -1, b_1 = 1, c_1 = -2,\ Q_1 = 1.0,\ Q_2 = 1.5,\ \Omega = 1.5$)

Parameter c			Initial condition (x_{10}, x_{20})	Transport law
Figure 5.42a	$\mathbf{E}^{12} = \begin{bmatrix} 1 & 0 \\ 0 & 1 \end{bmatrix}$	$\mathbf{e}^{12} = \begin{bmatrix} 0 \\ 1 \end{bmatrix}$	$(-8.7597, 8.8674)$	Translation
	$\mathbf{E}^{21} = \begin{bmatrix} 1 & 0 \\ 0 & 1 \end{bmatrix}$	$\mathbf{e}^{21} = \begin{bmatrix} 0 \\ 2 \end{bmatrix}$		
Figure 5.42b	$\mathbf{E}^{12} = \begin{bmatrix} 0.5 & 0 \\ 0 & 0.6 \end{bmatrix}$	$\mathbf{e}^{12} = \begin{bmatrix} 0 \\ 0 \end{bmatrix}$	$(-14.7546, 2.6000)$	Scaling
	$\mathbf{E}^{21} = \begin{bmatrix} 3 & 0 \\ 0 & 4 \end{bmatrix}$	$\mathbf{e}^{21} = \begin{bmatrix} 0 \\ 0 \end{bmatrix}$		
Figure 5.42c	$\mathbf{E}^{12} = \begin{bmatrix} 0.8 & 0.3 \\ 0.2 & 0.6 \end{bmatrix}$	$\mathbf{e}^{12} = \begin{bmatrix} 0 \\ 0 \end{bmatrix}$	$(7.3426, 4.3876)$	Linear transformation
	$\mathbf{E}^{21} = \begin{bmatrix} 8 & 5 \\ 3 & 4 \end{bmatrix}$	$\mathbf{e}^{21} = \begin{bmatrix} 0 \\ 0 \end{bmatrix}$		
Figure 5.42d	$\mathbf{E}^{12} = \begin{bmatrix} 0.5 & 0 \\ 0 & 0.6 \end{bmatrix}$	$\mathbf{e}^{12} = \begin{bmatrix} 0 \\ -1 \end{bmatrix}$	$(3.6392, 3.1714)$	Affine transformation
	$\mathbf{E}^{21} = \begin{bmatrix} 2 & 3 \\ 1.5 & 1 \end{bmatrix}$	$\mathbf{e}^{21} = \begin{bmatrix} 0 \\ 2 \end{bmatrix}$		

of periodic flows with mapping structure P_{21} are plotted in Fig. 5.42. In Fig. 5.42a, the translation switching for the variable x_2 is considered. The jump of variable x_2 is observed in phase plane. For this translation switching, the stability condition is the same as in the continuous switching. Only the switching points are different. In Fig. 5.42b, the scaling transport law is considered. From the first subsystem to the second subsystem, the shrinking transport law is used, but the stretching transport law is adopted from the second subsystem to the first subsystem. Such a stretching and shrinking can be observed very clearly. The scaling transport law will change the stability for the switching of two subsystems. In Fig. 5.42c, a linear transformation is used as a transport law, and the linear transformations with the shrinking and stretching are for the first to second subsystem and for the second to first subsystem, respectively. This linear transformation transport law will also change the stability for the two system switching. Finally, the affine transformations as transport laws are used for the switching of the two sub-systems in Fig. 5.42d. This affine transformation includes the rotation, scaling and translation transformation.

(B) *A 3-D Switching System*: Consider three linear subsystems with matrices as

$$\mathbf{A}^{(1)} = \begin{bmatrix} -1 & 2 & -1 \\ -2 & 1 & -1 \\ 1 & 3 & -3 \end{bmatrix}, \mathbf{A}^{(2)} = \begin{bmatrix} -1 & 0.5 & 2 \\ -1.5 & 1 & -1 \\ -1 & 2 & -1 \end{bmatrix}, \mathbf{A}^{(3)} = \begin{bmatrix} -1 & -1 & 2 \\ 2 & -3 & -1 \\ 1 & -2 & -2 \end{bmatrix} \quad (5.209)$$

and from eigenvalue analysis of the three matrices, $\lambda_1^{(1)} = -1.4735$, $\lambda_{2,3}^{(1)} = -0.7633 \pm 2.0416\mathbf{i}$, $\lambda_1^{(2)} = -3.3672$, $\lambda_{2,3}^{(2)} = 0.1836 \pm 2.4906\mathbf{i}$, and $\lambda_1^{(3)} = -4.4260$, $\lambda_{2,3}^{(3)} = -0.787 \pm 1.1891\mathbf{i}$ where $\mathbf{i} = \sqrt{-1}$. Because the time for

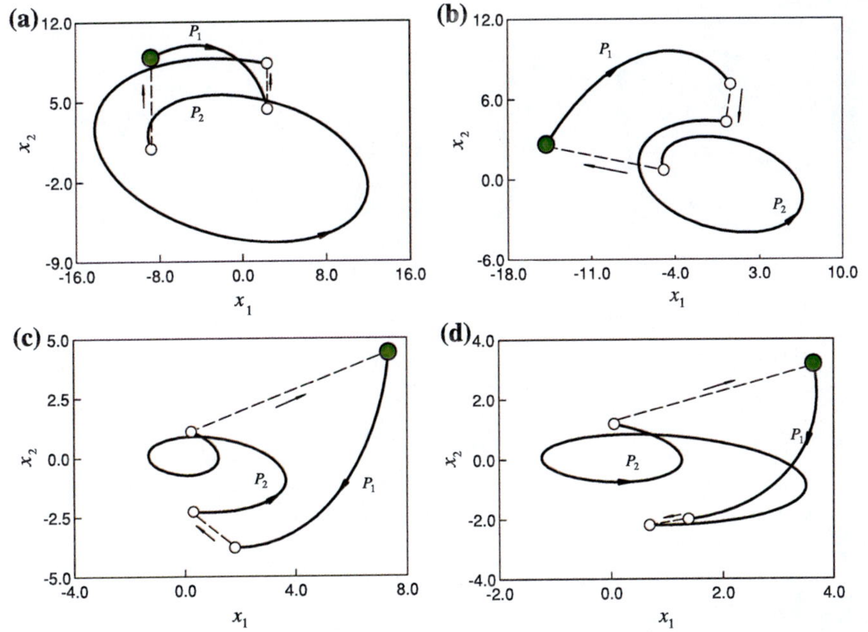

Fig. 5.42 Trajectories of stable periodic flows in phase plane with P_{21} for a 2-D switching system ($t_0 = 0.0$): **a** ($x_1^{(1)} \approx -8.7597, x_2^{(1)} \approx 8.8674$), **b** ($x_1^{(1)} \approx -14.7546, x_2^{(1)} \approx 2.6000$), **c** ($x_1^{(1)} \approx 7.3426, x_2^{(1)} \approx 4.3876$) and **d** ($x_1^{(1)} \approx 3.6392, x_2^{(1)} \approx 3.1714$). ($a = -1, b = -5,\ c = 1.5,\ d = 0.5,\ q^{(1)} = 0.25$ and $q^{(2)} = 0.75,\ a_1 = d_1 = -1, b_1 = 1, c_1 = -2,\ Q_1 = 1.0,\ Q_2 = 1.5,\ \Omega = 1.5$)

subsystems in switching systems is finite, it is not very significant that subsystems are stable or unstable. It is sufficient that the solutions of subsystems exist during the given time interval. The important issue is whether the flow of the resultant switching system is stable or unstable. The time interval for each system in the switching system is a key issue to control the stability of the resultant flow of the switching system. The external excitations for three subsystems in the 3-D switching system are

$$\mathbf{Q}^{(i)} = (A_1^{(i)} e^{t-t_0}, A_2^{(i)} \times (t - t_0), A_3^{(i)})^{\mathrm{T}} \tag{5.210}$$

where $A_1^{(1)} = A_2^{(1)} = A_3^{(1)} = 1, A_1^{(2)} = A_2^{(2)} = A_3^{(3)} = 1, A_3^{(2)} = A_1^{(3)} = A_2^{(3)} = -1$. Select the period of $T = 4.5$ arbitrarily, and consider the time interval parameters as $q^{(1)} = \frac{2}{9},\ \ q^{(2)} = \frac{4}{9}$ and $q^{(3)} = \frac{1}{9}$ for a periodic flow with a mapping structure P_{321} for the 3-D switching system. The continuous and impulsive switching with the same subsystems is considered for illustrations.

For the continuous switching, the analytical prediction gives the initial condition $x_1(0) \approx -2.7348,\ \ x_2(0) \approx -1.2346$ and $x_3(0) \approx -1.7856$ for the periodic flow

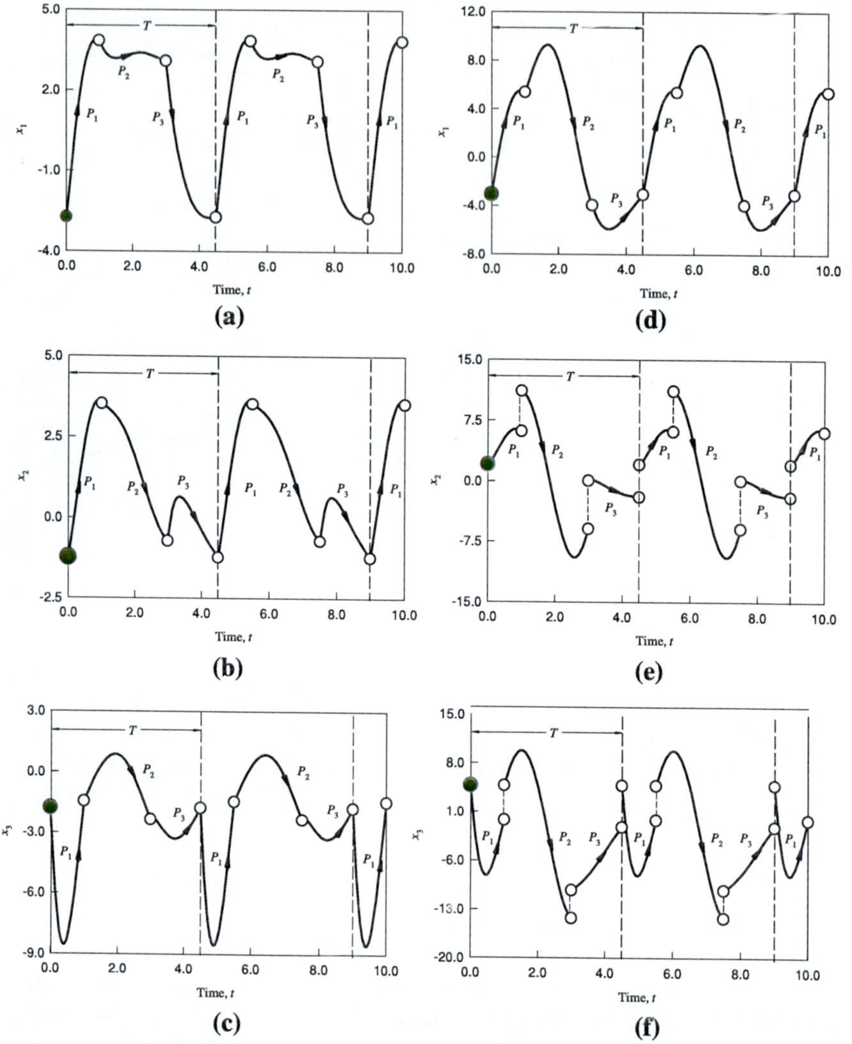

Fig. 5.43 Time-histories of variables ($x_i, i = 1, 2, 3$) for two 3-D switching systems: **a–c** continuous switching ($x_1(0) \approx -3.0672,\ \ x_2(0) \approx 2.0328,\ \ x_3(0) \approx 4.7815$) and **d–f** impulsive switching ($x_1(0) \approx -2.7348;\ \ x_2(0) \approx -1.2346,\ \ x_3(0) \approx -1.7856$). ($A_1^{(1)} = A_2^{(1)} = A_3^{(1)} = 1,\ A_1^{(2)} = A_2^{(2)} = 1,\ A_3^{(2)} = -1,\ A_1^{(3)} = A_2^{(3)} = -1,\ A_3^{(3)} = 1, T = 4.5;\ \ q^{(1)} = \frac{2}{9};\ \ q^{(2)} = \frac{4}{9}; q^{(3)} = \frac{1}{3}$)

of P_{321}. For the impulsive switching, the impulsive vectors between the two subsystems are $\mathbf{e}^{(12)} = (0, 5, 5)^{\mathrm{T}}$, $\mathbf{e}^{(23)} = (0, 6, 4)^{\mathrm{T}}$ and $\mathbf{e}^{(31)} = (0, 4, 6)^{\mathrm{T}}$. The state variable x_1 is continuous, but the stable variables x_2 and x_3 are discontinuous at switching points. The initial condition for the 3-D, impulsive switching system is

Fig. 5.44 Trajectories in phase space ($x_i, i = 1, 2, 3$) for two 3-D switching systems: **a–c** continuous switching ($x_1(0) \approx -3.0672,\ x_2(0) \approx 2.0328,\ x_3(0) \approx 4.7815$) and **d–f** impulsive switching $x_1(0) \approx -2.7348;\ x_2(0) \approx -1.2346,\ x_3(0) \approx -1.7856$. ($A_1^{(1)} = A_2^{(1)} = A_3^{(1)} = 1,\ A_1^{(2)} = A_2^{(2)} = 1,\ A_3^{(2)} = -1,\ A_1^{(3)} = A_2^{(3)} = -1,\ A_3^{(3)} = 1. T = 4.5;\ q^{(1)} = \frac{2}{9};\ q^{(2)} = \frac{4}{9}; q^{(3)} = \frac{1}{3}$)

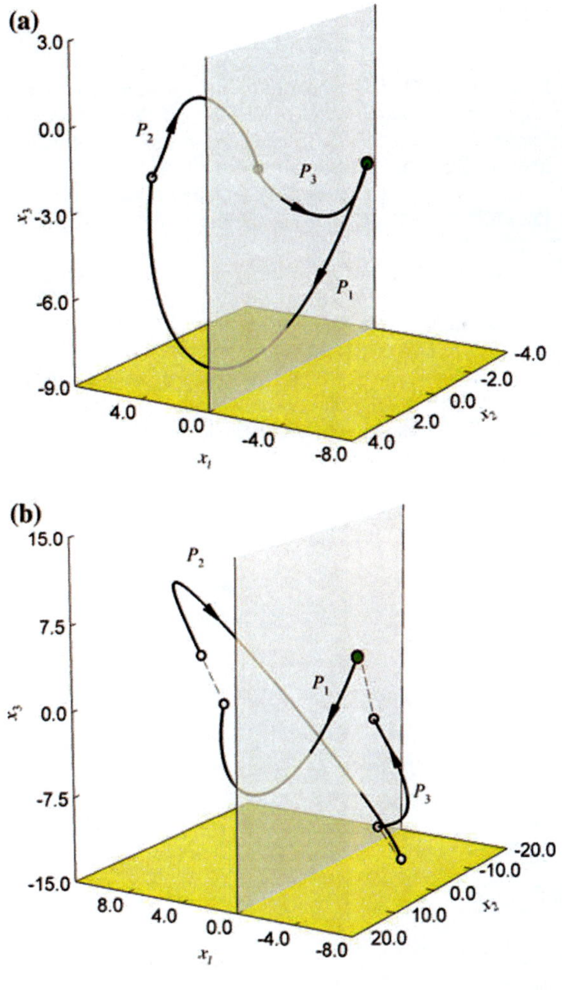

$x_1(0) \approx -3.0672,\ x_2(0) \approx 2.0328,\ x_3(0) \approx 4.7815$. The time-histories of three state variables ($x_i,\ i = 1, 2, 3$) for the periodic flow of the 3-D linear switching system are presented in Fig. 5.43. In Fig. 5.43a–c, the periodic flows for the continuous switching system are presented. It is observed that the periodic flow at all the switching points is continuous but non-smooth. In Fig. 5.43d–f, the periodic flow for the 3-D, impulsive switching system switching are presented. The time-history of the state variable x_1 is continuous, while the time-history of the state variables x_2 and x_3 are discontinuous due to impulsive at switching points. The trajectories in phase space for the continuous and impulsive switching are presented in Fig. 5.44a and b, respectively.

References

Luo, A.C.J., 2005, A theory for non-smooth dynamical systems on connectable domains, *Communication in Nonlinear Science and Numerical Simulation*, **10**, pp. 1–55.

Luo, A.C.J., 2006, *Singularity and Dynamics on Discontinuous Vector Fields*, Amsterdam: Elsevier.

Luo, A.C.J., 2008a, A theory for flow switchability in discontinuous dynamical systems, *Nonlinear Analysis: Hybrid Systems*, **2**(4), pp. 1030–1061.

Luo, A.C.J., 2008b, *Global Transversality, Resonance and Chaotic Dynamics*, Singapore: World Scientific.

Luo, A.C.J., 2011, Chaos and diffusion in dynamical systems with periodic impulses, *Proceedings of the ASME 2011 International Design Engineering Technical Conferences & Computers and Information in Engineering Conference*, IDETC/CIE2011, August 28–31, 2011, Washington, DC, USA, DETC2011-47387.

Luo, A.C.J., Wang, Y., 2009a, Switching dynamics of Multiple oscillators, *Communications in Nonlinear Science and Numerical Simulation*, **44**, pp. 3472–3485.

Luo, A.C.J., Wang, Y., 2009b, Periodic flows and stability of a switching system with multiple subsystems, *Dynamics of Continuous Discrete and Impulsive Systems*, **16**, pp. 825–848.

Chapter 6
Mapping Dynamics and Symmetry

In the previous chapter, dynamics of switching systems were discussed. In this chapter, mapping dynamics and symmetry in discontinuous dynamical systems will be discussed. The G-function of the discontinuous boundary will be presented first. To understand of nonlinear dynamics of a flow from one domain to another domain, mapping dynamics of discontinuous dynamics systems will be presented, which is a generalized symbolic dynamics. Using the mapping dynamics, one can determine periodic and chaotic dynamics of discontinuous dynamical systems, and complex motions can be classified through mapping structure. The nonlinear dynamics of the Chua's circuit system will be presented as an example. The flow grazing property of a flow will be discussed, which is a source to generate the complex motions in discontinuous dynamical systems. The flow symmetry in discontinuous dynamical systems will be discussed through the mapping dynamics and grazing. The strange attractor fragmentation generated by the grazing of flows to the boundary will be presented.

6.1 Discontinuous Dynamical Systems

As in Luo (2005a, 2006a), consider a dynamic system consisting of N sub-dynamic systems in a universal domain $\mho \subset \mathscr{R}^n$. The accessible domain in phase space means that a continuous dynamical system can be defined on such a domain. The inaccessible domain in phase space means that no dynamical system can be defined on such a domain. A universal domain in phase space is divided into N accessible sub-domains Ω_i plus the inaccessible domain Ω_0. The union of all the accessible sub-domains is $\cup_{i=1}^{N}\Omega_i$ and the universal domain is $\mho = \cup_{i=1}^{N}\Omega_i \cup \Omega_0$, which can be expressed through an n_1-dimensional sub-vector $\mathbf{x}_{n_1}$ and an $(n-n_1)$-dimensional sub-vector $\mathbf{x}_{n-n_1}$. Ω_0 is the union of the inaccessible domains. $\Omega_0 = \mho \backslash \cup_{i=1}^{m} \Omega_i$ is the complement of the union of the accessible sub-domain. If all the accessible domains are connected, the universal domain in phase space is called the connectable domain. If the accessible domains are separated by the inaccessible domain, the

A. C. J. Luo, *Regularity and Complexity in Dynamical Systems*,
DOI: 10.1007/978-1-4614-1524-4_6, © Springer Science+Business Media, LLC 2012

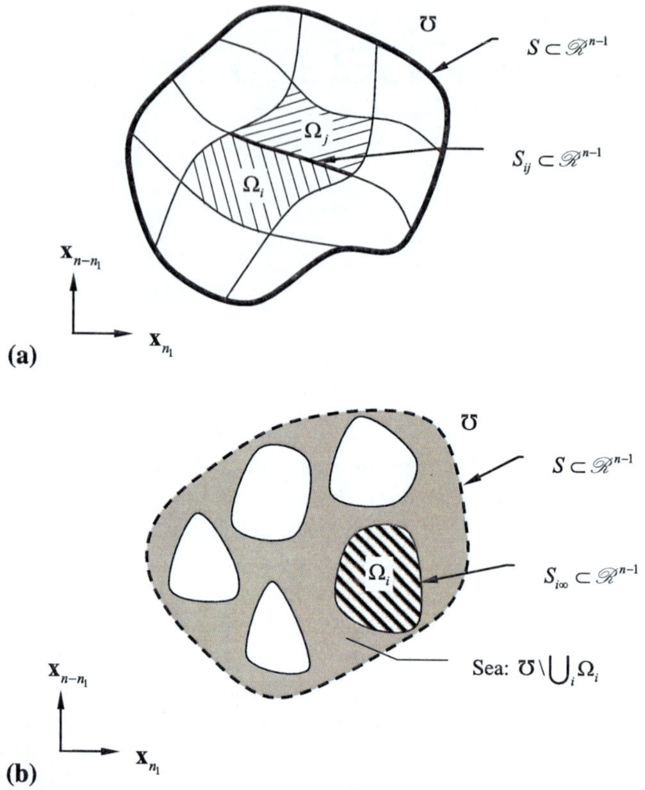

Fig. 6.1 Phase space: **a** connectable and **b** separable domains

universal domain is called the separable domains, as shown in Fig. 6.1. To investigate the relation between two disconnected domains without any common boundary, specific transport laws should be inserted. Such an issue can be referred to in Luo (2006). Herein, the flow switchability in discontinuous dynamical system focuses on dynamics in the two connected domains with a common boundary. For example, the boundary between two domains Ω_i and Ω_j is $\partial\Omega_{ij} = \bar{\Omega}_i \cap \bar{\Omega}_j$, as sketched in Fig. 6.2. This boundary is formed by the intersection of the closed sub-domains.

On the ith open sub-domain Ω_i, there is a C^{r_i}-continuous system ($r_i \geq 1$) in the form of

$$\dot{\mathbf{x}}^{(i)} \equiv \mathbf{F}^{(i)}(\mathbf{x}^{(i)}, t, \mathbf{p}_i) \in \mathscr{R}^n, \quad \mathbf{x}^{(i)} = (\mathrm{x}_1^{(i)}, \quad \mathrm{x}_2^{(i)}, \cdots, \mathrm{x}_n^{(i)})^{\mathrm{T}} \in \Omega_{\mathrm{i}}. \tag{6.1}$$

The time is t and $\dot{\mathbf{x}}^{(i)} = d\mathbf{x}^{(i)}/dt$. In an accessible sub-domain Ω_i, the vector field $\mathbf{F}^{(i)}(\mathbf{x}^{(i)}, t, \mathbf{p}_i)$ with parameter vectors $\mathbf{p} = (p_i^{(1)}, p_i^{(2)}, \cdots, p_i^{(l)})^{\mathrm{T}} \in \mathscr{R}^l$ is C^{r_i}-continuous ($r_i \geq 1$) in a state vector $\mathbf{x}^{(i)}$ and for all time t; and the continuous flow in Eq. (6.1) $\mathbf{x}^{(i)}(t) = \mathbf{\Phi}^{(i)}(\mathbf{x}^{(i)}(t_0), t, \mathbf{p}_i)$ with $\mathbf{x}^{(i)}(t_0) = \mathbf{\Phi}^{(i)}(\mathbf{x}^{(i)}(t_0), t_0, \mathbf{p}_i)$ is C^{r_i+1}-continuous for time t.

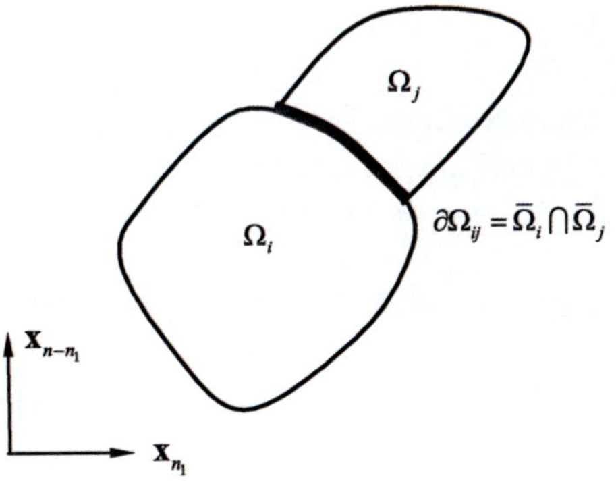

Fig. 6.2 Two adjacent sub-domains Ω_i and Ω_j, the corresponding boundary $\partial\Omega_{ij}$

The discontinuous dynamics theory presented herein holds for the following hypothesis.

H6.1: The switching between two adjacent sub-systems possesses time-continuity.
H6.2: For an unbounded, accessible sub-domain Ω_i, there is a bounded domain $D_i \subset \Omega_i$ and the corresponding vector field and its flow are bounded, i.e.,

$$||\mathbf{F}^{(i)}|| \le K_1(\text{const}) \text{ and } ||\mathbf{\Phi}^{(i)}|| \le K_2(\text{const}) \text{ on } D_i \quad \text{for } t \in [0, \infty). \tag{6.2}$$

H6.3: For a bounded, accessible domain Ω_i, there is a bounded domain $D_i \subset \Omega_i$ and the corresponding vector field is bounded, but the flow may be unbounded, i.e.,

$$||\mathbf{F}^{(i)}|| \le K_1(\text{const}) \text{ and } ||\mathbf{\Phi}^{(i)}|| < \infty \text{ on } D_i \quad \text{for } t \in [0,\infty). \tag{6.3}$$

Because dynamical systems on the different accessible sub-domains are different, a relation between two flows in the two sub-domains should be developed for continuation. For a sub-domain Ω_i, there are k_i-piece boundaries. Consider a boundary set of any two adjacent sub-domains.

Definition 6.1 The boundary in an n-D phase space is defined as

$$\begin{aligned} S_{ij} &\equiv \partial\Omega_{ij} = \bar{\Omega}_i \cap \bar{\Omega}_j \\ &= \left\{\mathbf{x} \,\middle|\, \varphi_{ij}(\mathbf{x}, t, \boldsymbol{\lambda}) = 0,\ \varphi_{ij} \text{ is } C^r\text{-continuous } (r \ge 1)\right\} \subset \mathscr{R}^{n-1}. \end{aligned} \tag{6.4}$$

From the boundary definition, we have $\partial\Omega_{ij} = \partial\Omega_{ji}$. On the separation boundary $\partial\Omega_{ij}$ with $\varphi_{ij}(\mathbf{x}, t, \boldsymbol{\lambda}) = 0$, there is a dynamical system as

$$\dot{\mathbf{x}}^{(0)} = \mathbf{F}^{(0)}(\mathbf{x}^{(0)}, t, \boldsymbol{\lambda}), \tag{6.5}$$

where $\mathbf{x}^{(0)} = (x_1^{(0)}, x_2^{(0)}, \cdots, x_n^{(0)})^{\mathrm{T}}$. The flow of $\mathbf{x}^{(0)}(t) = \mathbf{\Phi}^{(0)}(\mathbf{x}^{(0)}(t_0), t, \boldsymbol{\lambda})$ with $\mathbf{x}^{(0)}(t_0) = \mathbf{\Phi}^{(0)}(\mathbf{x}^{(0)}(t_0), t_0, \boldsymbol{\lambda})$ is C^{r+1}-continuous for time t.

6.2 G-Functions to Boundaries

Consider two infinitesimal time intervals $[t_m - \varepsilon, t_m)$ and $(t_m, t_m + \varepsilon]$. There are two flows in domain $\Omega_\alpha\,(\alpha = i, j)$ and on the boundary $\partial\Omega_{ij}$ in Eqs. (6.1) and (6.5). As in Luo (2008a, b), the vector difference between the two flows for three time instants are given by $\mathbf{x}^{(\alpha)}_{t_m-\varepsilon} - \mathbf{x}^{(0)}_{t_m-\varepsilon}$, $\mathbf{x}^{(\alpha)}_{t_m} - \mathbf{x}^{(0)}_{t_m}$ and $\mathbf{x}^{(\alpha)}_{t_m+\varepsilon} - \mathbf{x}^{(0)}_{t_m+\varepsilon}$. The normal vector of the boundary relative to the corresponding flow $\mathbf{x}^{(0)}(t)$ are expressed by ${}^{t_m-\varepsilon}\mathbf{n}_{\partial\Omega_{ij}}$, ${}^{t_m}\mathbf{n}_{\partial\Omega_{ij}}$ and ${}^{t_m+\varepsilon}\mathbf{n}_{\partial\Omega_{ij}}$, respectively, and the corresponding tangential vectors of the flow $\mathbf{x}^{(0)}(t)$ on the boundary are the tangential vectors expressed by $\mathbf{t}_{\mathbf{x}^{(0)}_{t_m-\varepsilon}}$, $\mathbf{t}_{x^{(0)}_{t_m}}$, and $\mathbf{t}_{\mathbf{x}^{(0)}_{t_m+\varepsilon}}$. From the normal vectors of the boundary $\partial\Omega_{ij}$, the dot products of the normal vectors and the state vector difference between two flows in domain and on the boundary are defined by

$$\left.\begin{aligned} d^{(\alpha)}_{t_m-\varepsilon} &= {}^{t_m-\varepsilon}\mathbf{n}^{\mathrm{T}}_{\partial\Omega_{ij}} \cdot (\mathbf{x}^{(\alpha)}_{t_m-\varepsilon} - \mathbf{x}^{(0)}_{t_m-\varepsilon}), \\ d^{(\alpha)}_{t_m} &= {}^{t_m}\mathbf{n}^{\mathrm{T}}_{\partial\Omega_{ij}} \cdot (\mathbf{x}^{(\alpha)}_{t_m} - \mathbf{x}^{(0)}_{t_m}), \\ d^{(\alpha)}_{t_m+\varepsilon} &= {}^{t_m+\varepsilon}\mathbf{n}^{\mathrm{T}}_{\partial\Omega_{ij}} \cdot (\mathbf{x}^{(\alpha)}_{t_m+\varepsilon} - \mathbf{x}^{(0)}_{t_m+\varepsilon}), \end{aligned}\right\} \tag{6.6}$$

where the normal vector of the boundary surface $\partial\Omega_{ij}$ at point $\mathbf{x}^{(0)}(t)$ is given by

$${}^{t}\mathbf{n}_{\partial\Omega_{ij}}(\mathbf{x}^{(0)}, t, \boldsymbol{\lambda}) = \nabla\varphi_{ij}(\mathbf{x}^{(0)}, t, \boldsymbol{\lambda}) = (\partial_{x_1^{(0)}}\varphi_{ij}, \partial_{x_2^{(0)}}\varphi_{ij}, \cdots, \partial_{x_n^{(0)}}\varphi_{ij})^{\mathrm{T}}_{(t,\mathbf{x}^{(0)})} \tag{6.7}$$

for time t. If the normal vector is a unit vector, the dot product is the normal component, which is a distance of the two points of two flows in the normal direction of the boundary surface.

Definition 6.2 Consider a dynamical system in Eq. (6.1) in domain $\Omega_\alpha\,(\alpha \in \{i, j\})$ which has the flow $\mathbf{x}^{(\alpha)}_t = \boldsymbol{\Phi}(t_0, \mathbf{x}^{(\alpha)}_0, \mathbf{p}, t)$ with the initial condition $(t_0, \mathbf{x}^{(\alpha)}_0)$ and on the boundary $\partial\Omega_{ij}$, there is a flow $\mathbf{x}^{(0)}_t = \boldsymbol{\Phi}(t_0, \mathbf{x}^{(0)}_0, \boldsymbol{\lambda}, t)$ with the initial condition $(t_0, \bar{\mathbf{x}}_0)$. For an arbitrarily small $\varepsilon > 0$, there are two time intervals $[t - \varepsilon, t)$ or $(t, t + \varepsilon]$ for flow $\mathbf{x}^{(\alpha)}_t$ $(\alpha \in \{i, j\})$. The G-functions $(G^{(\alpha)}_{\partial\Omega_{ij}})$ of the flow $\mathbf{x}^{(\alpha)}_t$ to the flow $\mathbf{x}^{(0)}_t$ on the boundary in the normal direction of the boundary $\partial\Omega_{ij}$ are defined as

$$\begin{aligned} &G^{(\alpha)}_{\partial\Omega_{ij}}(\mathbf{x}^{(0)}_t, t_-, \mathbf{x}^{(\alpha)}_{t_-}, \mathbf{p}_\alpha, \boldsymbol{\lambda}) \\ &= \lim_{\varepsilon\to 0}\frac{1}{\varepsilon}\left[{}^{t}\mathbf{n}^{\mathrm{T}}_{\partial\Omega_{ij}} \cdot (\mathbf{x}^{(\alpha)}_{t_-} - \mathbf{x}^{(0)}_t) - {}^{t-\varepsilon}\mathbf{n}^{\mathrm{T}}_{\partial\Omega_{ij}} \cdot (\mathbf{x}^{(\alpha)}_{t-\varepsilon} - \mathbf{x}^{(0)}_{t-\varepsilon})\right], \\ &G^{(\alpha)}_{\partial\Omega_{ij}}(\mathbf{x}^{(0)}_t, t_+, \mathbf{x}^{(\alpha)}_{t_+}, \mathbf{p}_\alpha, \boldsymbol{\lambda}) \\ &= \lim_{\varepsilon\to 0}\frac{1}{\varepsilon}\left[{}^{t+\varepsilon}\mathbf{n}^{\mathrm{T}}_{\partial\Omega_{ij}} \cdot (\mathbf{x}^{(\alpha)}_{t+\varepsilon} - \mathbf{x}^{(0)}_{t+\varepsilon}) - {}^{t}\mathbf{n}^{\mathrm{T}}_{\partial\Omega_{ij}} \cdot (\mathbf{x}^{(\alpha)}_{t_+} - \mathbf{x}^{(0)}_t)\right]. \end{aligned} \tag{6.8}$$

From equation (6.8), since $\mathbf{x}^{(\alpha)}_t$ and $\mathbf{x}^{(0)}_t$ are the solutions of Eqs. (6.1) and (6.5), their derivatives exist. Further, by use of the Taylor series expansion, Eq. (6.8) gives

$$G^{(\alpha)}_{\partial\Omega_{ij}}(\mathbf{x}^{(0)}_t, t_\pm, \mathbf{x}^{(\alpha)}_{t\pm}, \mathbf{p}_\alpha, \boldsymbol{\lambda}) = D_{\mathbf{x}^{(0)}_t}{}^t\mathbf{n}^{\mathrm{T}}_{\partial\Omega_{ij}} \cdot (\mathbf{x}^{(\alpha)}_{t\pm} - \mathbf{x}^{(0)}_t) +{}^t\,\mathbf{n}^{\mathrm{T}}_{\partial\Omega_{ij}} \cdot (\dot{\mathbf{x}}^{(\alpha)}_{t\pm} - \dot{\mathbf{x}}^{(0)}_t) \tag{6.9}$$

where the total derivative $D_{\mathbf{x}^{(0)}_t}(\cdot) = \partial_{\mathbf{x}^{(0)}_t}(\cdot)\dot{\mathbf{x}}^{(0)}_t + \partial_t(\cdot)$. Using Eqs. (6.1) and (6.5), the *G-function* in Eq. (6.9) becomes

$$\begin{aligned} G^{(\alpha)}_{\partial\Omega_{ij}}(\mathbf{x}^{(0)}_t, t, \mathbf{x}^{(\alpha)}_{t\pm}, \mathbf{p}_\alpha, \boldsymbol{\lambda}) &= [\partial_{\mathbf{x}^{(0)}_t}({}^t\mathbf{n}^{\mathrm{T}}_{\partial\Omega_{ij}})\dot{\mathbf{x}}^{(0)}_t + \partial_t({}^t\mathbf{n}^{\mathrm{T}}_{\partial\Omega_{ij}})] \cdot (\mathbf{x}^{(\alpha)}_{t\pm} - \mathbf{x}^{(0)}_t) \\ &\quad +{}^t\mathbf{n}^{\mathrm{T}}_{\partial\Omega_{ij}} \cdot (\mathbf{F}^{(\alpha)}(\mathbf{x}^{(\alpha)}, t_\pm, \mathbf{p}) - \mathbf{F}^{(0)}(\mathbf{x}^{(0)}, t, \boldsymbol{\lambda})). \end{aligned} \tag{6.10}$$

If a flow $\mathbf{x}^{(\alpha)}(t)$ approaches the separation boundary with the zero-order contact at t_m (i.e.,$\mathbf{x}^{(\alpha)}(t_{m-}) = \mathbf{x}_m = \mathbf{x}^{(0)}(t_m)$), the G-function of the zero-order is defined as

$$\begin{aligned} &G^{(\alpha)}_{\partial\Omega_{ij}}(\mathbf{x}_m, t_m, \mathbf{p}_\alpha, \boldsymbol{\lambda}) \\ &\equiv \mathbf{n}^{\mathrm{T}}_{\partial\Omega_{ij}}(\mathbf{x}^{(0)}, t, \boldsymbol{\lambda}) \cdot [\dot{\mathbf{x}}^{(\alpha)}(t) - \dot{\mathbf{x}}^{(0)}(t)]\Big|_{(\mathbf{x}^{(0)}_m, \mathbf{x}^{(\alpha)}_m, t_{m\pm})} \\ &= \mathbf{n}^{\mathrm{T}}_{\partial\Omega_{ij}}(\mathbf{x}^{(0)}, t, \boldsymbol{\lambda}) \cdot \dot{\mathbf{x}}^{(\alpha)}(t) + \partial_t\varphi_{ij}(\mathbf{x}^{(0)}, t, \boldsymbol{\lambda})\Big|_{(\mathbf{x}^{(0)}_m, \mathbf{x}^{(\alpha)}_m, t_{m\pm})} \\ &= \nabla\varphi_{ij}(\mathbf{x}^{(0)}, t, \boldsymbol{\lambda}) \cdot \dot{\mathbf{x}}^{(\alpha)}(t) + \partial_t\varphi_{ij}(\mathbf{x}^{(0)}, t, \boldsymbol{\lambda})\Big|_{(\mathbf{x}^{(0)}_m, \mathbf{x}^{(\alpha)}_m, t_{m\pm})}. \end{aligned} \tag{6.11}$$

With Eqs. (6.1), (6.5), equation (6.11) can be rewritten as

$$\begin{aligned} &G^{(\alpha)}_{\partial\Omega_{ij}}(\mathbf{x}_m, t_m, \mathbf{p}_\alpha, \boldsymbol{\lambda}) \\ &\equiv \mathbf{n}^{\mathrm{T}}_{\partial\Omega_{ij}}(\mathbf{x}^{(0)}, t, \boldsymbol{\lambda}) \cdot [\mathbf{F}(\mathbf{x}^{(\alpha)}, t, \mathbf{p}_\alpha) - \mathbf{F}^{(0)}(\mathbf{x}^{(0)}, t, \boldsymbol{\lambda})]\Big|_{(\mathbf{x}^{(0)}_m, \mathbf{x}^{(\alpha)}_m, t_{m\pm})} \\ &= \mathbf{n}^{\mathrm{T}}_{\partial\Omega_{ij}}(\mathbf{x}^{(0)}, t, \boldsymbol{\lambda}) \cdot \mathbf{F}(\mathbf{x}^{(\alpha)}, t, \mathbf{p}_\alpha) + \partial_t\varphi_{ij}(\mathbf{x}^{(0)}, t, \boldsymbol{\lambda})\Big|_{(\mathbf{x}^{(0)}_m, \mathbf{x}^{(\alpha)}_m, t_{m\pm})} \\ &= \nabla\varphi_{ij}(\mathbf{x}^{(0)}, t, \boldsymbol{\lambda}) \cdot \mathbf{F}(\mathbf{x}^{(\alpha)}, t, \mathbf{p}_\alpha) + \partial_t\varphi_{ij}(\mathbf{x}^{(0)}, t, \boldsymbol{\lambda})\Big|_{(\mathbf{x}^{(0)}_m, \mathbf{x}^{(\alpha)}_m, t_{m\pm})}. \end{aligned} \tag{6.12}$$

Definition 6.3 Consider a dynamical system in Eq. (6.1) in domain $\Omega_\alpha (\alpha \in \{i, j\})$ which has the flow $\mathbf{x}^{(\alpha)}_t = \boldsymbol{\Phi}(t_0, \mathbf{x}^{(\alpha)}_0, \mathbf{p}, t)$ with the initial condition $(t_0, \mathbf{x}^{(\alpha)}_0)$ and on the boundary $\partial\Omega_{ij}$, there is a flow $\mathbf{x}^{(0)}_t = \boldsymbol{\Phi}(t_0, \mathbf{x}^{(0)}_0, \boldsymbol{\lambda}, t)$ with the initial condition $(t_0, \mathbf{x}_0)$. For an arbitrarily small $\varepsilon > 0$, there are two time intervals $[t - \varepsilon, t)$ for flow $\mathbf{x}^{(\alpha)}_t$ $(\alpha \in \{i, j\})$ and $(t, t + \varepsilon]$ for flow $\mathbf{x}^{(\beta)}_t (\beta \in \{i, j\})$. The vector fields $\mathbf{F}^{(\alpha)}(\mathbf{x}^{(\alpha)}, t, \mathbf{p}_\alpha)$ and $\mathbf{F}^{(0)}(\mathbf{x}^{(0)}, t, \boldsymbol{\lambda})$ are $C^r_{[t-\varepsilon,t+\varepsilon]}$-continuous $(r_\alpha \geq k)$ for time t. The flow of $\mathbf{x}^{(\alpha)}_t (\alpha \in \{i, j\})$ and $\mathbf{x}^{(0)}_t$ are $C^{r_\alpha}_{[t-\varepsilon,t)}$ or $C^{r_\alpha}_{(t,t+\varepsilon]}$-continuous $(r_\alpha \geq k + 1)$ for time t, $||d^{r_\alpha+1}\mathbf{x}^{(\alpha)}_t/dt^{r_\alpha+1}|| < \infty$ and $||d^{r_\alpha+1}\mathbf{x}^{(0)}_t/dt^{r_\alpha+1}|| < \infty$. The *G-functions* of *k*th-*order* for a flow $\mathbf{x}_t$ to a boundary flow $\mathbf{x}^{(0)}_t$ in the normal direction of the boundary $\partial\Omega_{ij}$ are defined as

$$\left.\begin{aligned}
&G^{(k,\alpha)}_{\partial\Omega_{ij}}(\mathbf{x}^{(0)}_t, t_-, \mathbf{x}^{(\alpha)}_{t_-}, \mathbf{p}_\alpha, \boldsymbol{\lambda})\\
&= \lim_{\varepsilon\to 0}(k+1)!\frac{(-1)^{k+2}}{\varepsilon^{k+1}}\Big[{}^t\mathbf{n}^{\mathrm{T}}_{\partial\Omega_{ij}}\cdot(\mathbf{x}^{(\alpha)}_{t-}-\mathbf{x}^{(0)}_t) - {}^{t-\varepsilon}\mathbf{n}^{\mathrm{T}}_{\partial\Omega_{ij}}\cdot(\mathbf{x}^{(\alpha)}_{t-\varepsilon}-\mathbf{x}^{(0)}_{t-\varepsilon})\\
&\quad+\sum\nolimits_{s=0}^{k-1}\frac{1}{(s+1)!}G^{(s,\alpha)}_{\partial\Omega_{ij}}(\mathbf{x}^{(0)}_t, t_-, \mathbf{x}^{(\alpha)}_{t-}, \mathbf{p}_\alpha, \boldsymbol{\lambda})(-\varepsilon)^{s+1}\Big],\\
&G^{(k,\alpha)}_{\partial\Omega_{ij}}(\mathbf{x}^{(0)}_t, t_+, \mathbf{x}^{(\alpha)}_{t_+}, \mathbf{p}_\alpha, \boldsymbol{\lambda})\\
&= \lim_{\varepsilon\to 0}(k+1)!\frac{1}{\varepsilon^{k+1}}\Big[{}^{t+\varepsilon}\mathbf{n}^{\mathrm{T}}_{\partial\Omega_{ij}}\cdot(\mathbf{x}^{(\alpha)}_{t+\varepsilon}-\mathbf{x}^{(0)}_{t+\varepsilon}) - {}^{t}\mathbf{n}^{\mathrm{T}}_{\partial\Omega_{ij}}\cdot(\mathbf{x}^{(\alpha)}_{t_+}-\mathbf{x}^{(0)}_{t})\\
&\quad-\sum\nolimits_{s=0}^{k-1}\frac{1}{(s+1)!}G^{(s,\alpha)}_{\partial\Omega_{ij}}(\mathbf{x}^{(0)}_t, t, \mathbf{x}^{(\alpha)}_{t+}, \mathbf{p}_\alpha, \boldsymbol{\lambda})\varepsilon^{s+1}\Big].
\end{aligned}\right\}\tag{6.13}$$

Again, the Taylor series expansion applying to Eq. (6.13) yields

$$\begin{aligned}
&G^{(k,\alpha)}_{\partial\Omega_{ij}}(\mathbf{x}^{(0)}_t, t_\pm, \mathbf{x}^{(\alpha)}_{t\pm}, \mathbf{p}_\alpha, \boldsymbol{\lambda})\\
&= \sum\nolimits_{s=0}^{k+1} C^s_{k+1} D^{k+1-s}_{\mathbf{x}^{(0)}_t}{}^t\mathbf{n}^{\mathrm{T}}_{\partial\Omega_{ij}}\cdot\left(\frac{d^s\mathbf{x}^{(\alpha)}_t}{dt^s}-\frac{d^s\mathbf{x}^{(0)}_t}{dt^s}\right)\Big|_{(\mathbf{x}^{(0)}_t, t, \mathbf{x}^{(\alpha)}_{t\pm})}.
\end{aligned}\tag{6.14}$$

Using Eqs. (6.1) and (6.5), the kth-order G-function becomes

$$\begin{aligned}
&G^{(k,\alpha)}_{\partial\Omega_{ij}}(\mathbf{x}^{(0)}_t, t_\pm, \mathbf{x}^{(\alpha)}_{t\pm}, \mathbf{p}_\alpha, \boldsymbol{\lambda})\\
&= \sum\nolimits_{s=1}^{k+1} C^s_{k+1} D^{k+1-s}_{\mathbf{x}^{(0)}_t}{}^t\mathbf{n}^{\mathrm{T}}_{\partial\Omega_{ij}}\cdot(D^{s-1}_{\mathbf{x}^{(\alpha)}_t}\mathbf{F}(\mathbf{x}^{(\alpha)}_t, t, \mathbf{p})\\
&\quad - D^{s-1}_{\mathbf{x}^{(0)}_t}\mathbf{F}^{(0)}(\mathbf{x}^{(0)}_t, t, \boldsymbol{\mu}))\Big|_{(\mathbf{x}^{(0)}_t, t_\pm, \mathbf{x}^{(\alpha)}_{t\pm})} + D^{k+1}_{\mathbf{x}^{(0)}_t}{}^t\mathbf{n}^{\mathrm{T}}_{\partial\Omega_{ij}}\cdot(\mathbf{x}^{(\alpha)}_{t\pm}-\mathbf{x}^{(0)}_t),
\end{aligned}\tag{6.15}$$

with $C^s_{k+1} = (k+1)k(k-1)\cdots(k-s+2)/s!$ and $C^0_{k+1} = 1$ with $s! = 1\times 2\times\cdots\times s$. The G-function $G^{(k,\alpha)}_{\partial\Omega_{ij}}$ is the time-rate of $G^{(k-1,\alpha)}_{\partial\Omega_{ij}}$. If the flow contacting with the boundary $\partial\Omega_{ij}$ at time t_m (i.e., $\mathbf{x}^{(\alpha)}_{t_m} = \mathbf{x}^{(0)}_{t_m}$) and ${}^t\mathbf{n}^{\mathrm{T}}_{\partial\Omega_{ij}} \equiv \mathbf{n}^{\mathrm{T}}_{\partial\Omega_{ij}}$, the kth-order G-function is computed by

$$\begin{aligned}
&G^{(k,\alpha)}_{\partial\Omega_{ij}}(\mathbf{x}_m, t_{m\pm}, \mathbf{p}_\alpha, \boldsymbol{\lambda})\\
&= \sum\nolimits_{r=1}^{k+1} C^r_{k+1} D^{k+1-s}_{\mathbf{x}^{(0)}}\mathbf{n}^{\mathrm{T}}_{\partial\Omega_{ij}}\cdot\left[\frac{d^s\mathbf{x}}{dt^s}-\frac{d^s\mathbf{x}^{(0)}}{dt^r}\right]\Big|_{(\mathbf{x}^{(0)}_m, \mathbf{x}^{(\alpha)}_{m\pm}, t_{m\pm})}\\
&= \sum\nolimits_{s=1}^{k+1} C^s_{k+1} D^{k+1-s}_{\mathbf{x}^{(0)}}\mathbf{n}^{\mathrm{T}}_{\partial\Omega_{ij}}\cdot\left[D^{s-1}_{\mathbf{x}}\mathbf{F}(\mathbf{x}, t, \mathbf{p}_\alpha) - D^{s-1}_{\mathbf{x}^{(0)}}\mathbf{F}^{(0)}(\mathbf{x}^{(0)}, t, \boldsymbol{\lambda})\right]\Big|_{(\mathbf{x}^{(0)}_m, \mathbf{x}^{(\alpha)}_{m\pm}, t_{m\pm})}.
\end{aligned}\tag{6.16}$$

For $k = 0$, we have $G^{(\alpha,k)}_{\partial\Omega_{ij}}(\mathbf{x}_m, t_{m\pm}, \mathbf{p}_\alpha, \boldsymbol{\lambda}) = G^{(\alpha)}_{\partial\Omega_{ij}}(\mathbf{x}_m, t_{m\pm}, \mathbf{p}_\alpha, \boldsymbol{\lambda})$.

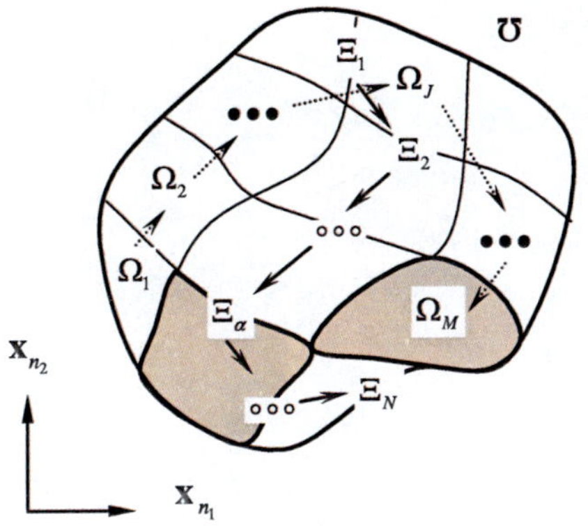

Fig. 6.3 Naming sub-domains and boundaries: the *dotted* route for the order of naming for sub-domains, and the *solid* route for switching surfaces

6.3 Mapping Dynamics

As in Luo (2006a), if a flow of a discontinuous system just in a single sub-domain cannot be intersected with any boundary, such a case will not be discussed because characteristics of such a flow can be determined by the continuous dynamical system theory. The main focus herein is to determine the dynamical properties of global flows intersected with boundaries in discontinuous dynamical systems. To do so, the naming of sub-domains and boundaries in discontinuous dynamic systems is very crucial. Once domains and boundaries in discontinuous dynamical systems are named, the mappings and mapping structures of the global flow in such discontinuous dynamical systems can be developed. Thus, consider a universal domain with M-sub-domains in phase space, and N-boundaries among the M-sub-domains with the universal domains. In the previous discussion, the boundaries are expressed by the neighbored domains. For example, $\partial\Omega_{IJ}$ is the boundary between the sub-domains Ω_I and Ω_J. To define the maps, the switching surfaces relative to boundary should be named, and the domain should be named. The naming of switching surfaces and sub-domains can be independent. In Fig. 6.3, the sub-domains are named through a dotted route, and the switching planes are named by a solid route. In fact, the naming of the sub-domains and the switching planes can be arbitrary.

Consider all the sub-domains and the switching surfaces expressed by Ω_J ($J = 1, 2, \cdots, M$) and Ξ_α ($\alpha = 1, 2, \cdots, N$), respectively. The switching plane Ξ_α ($\alpha = 1, 2, \cdots, N$) defined on the boundary $\partial\Omega_{IJ}$ ($I, J = 1, 2, \cdots, M$) are given by

$$\Xi_\alpha = \{(t_k, \mathbf{x}_k) | \, \varphi_{IJ}(\mathbf{x}_k, t_k) = 0 \text{ for time } t_k\} \subset \partial\Omega_{IJ}. \tag{6.17}$$

The local mappings relative to the switching sets Ξ_α by the dynamical system in the sub-domain Ω_J is

$$P_{J_{\alpha\alpha}} : \Xi_\alpha \to \Xi_\alpha. \tag{6.18}$$

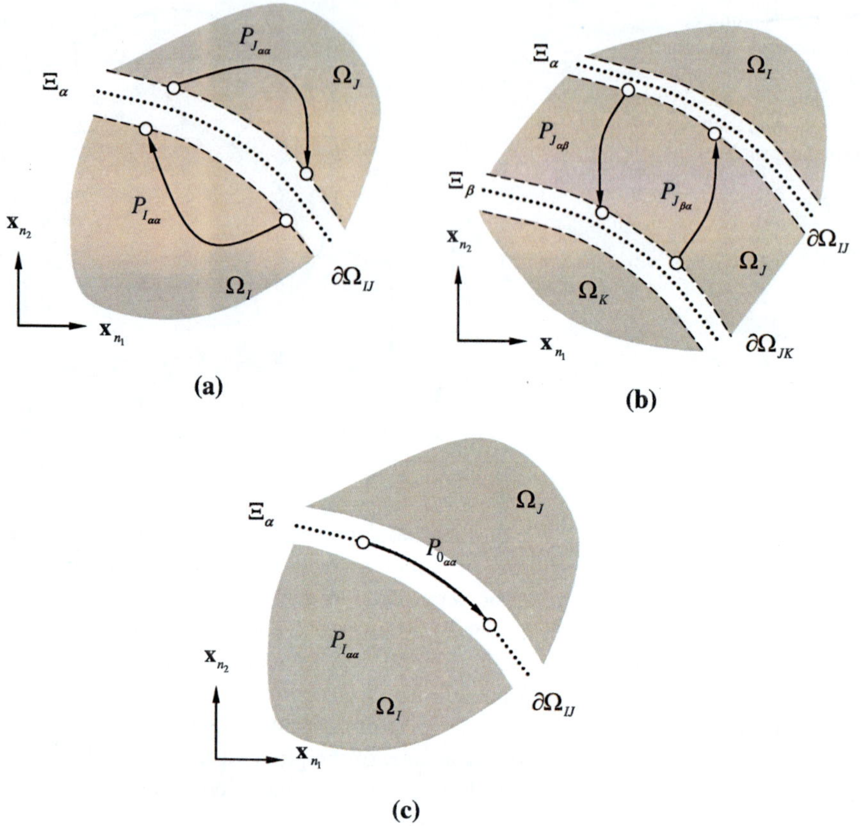

Fig. 6.4 Mappings: **a** local mappings, and **b** global mapping, **c** sliding mapping

The global mapping starting on the switching sets Ξ_α and ending on one of the rest switching sets Ξ_β ($\beta \neq \alpha$) relative to the sub-domain Ω_J is

$$P_{J_{\alpha\beta}} : \Xi_\alpha \to \Xi_\beta. \tag{6.19}$$

The sliding mapping on the switching sets Ξ_α governed by the boundary $\varphi_{IJ}(\mathbf{x}, t) = 0$ is defined by

$$P_{0_{\alpha\alpha}} : \Xi_\alpha \to \Xi_\alpha. \tag{6.20}$$

Notice that no mapping can be defined for the inaccessible domain. The above mappings are described in Fig. 6.4. The local mapping starting and ending at the same switching sets are sketched in Fig. 6.4(a). The global mapping starting and ending different switching sets is presented in Fig. 6.4(b). The sliding mapping is on the switching sets, as shown in Fig. 6.4(c). The special case of the sliding mapping is the sliding mapping on the edge.

For simplicity, the following notation for mapping is introduced as

$$P_{n_j \cdots n_i \cdots n_2 n_1} = P_{n_j} \circ \cdots \circ P_{n_i} \circ \cdots \circ P_{n_2} \circ P_{n_1}, \tag{6.21}$$

where all the local and global mappings are

$$P_{n_i} \in \left\{ P_{J_{\alpha\beta}} | \alpha, \beta \in \{1, 2, \cdots, N\}, J \in \{0, 1, 2, \cdots, M\} \right\} \tag{6.22}$$

for $i \in \{1, 2, \cdots, j, \cdots\}$. The rotation of the mapping of periodic motion in order gives the same motion (i.e., $P_{n_1 n_2 \cdots n_j}, P_{n_2 \cdots n_j n_1}, \cdots, P_{n_j n_1 \cdots n_{k-1}}$), and only the selected Poincare mapping section is different. The flow of the *m-time* repeating of mapping $P_{n_1 n_2 \cdots n_k}$ is defined as

$$\begin{aligned} P^m_{n_j \cdots n_2 n_1} &\equiv P_{(n_j \cdots n_2 n_1)^m} \equiv P_{\underbrace{(n_j \cdots n_2 n_1) \cdots (n_j \cdots n_2 n_1)}_{m}} \\ &\equiv \underbrace{(P_{n_j} \circ \cdots \circ P_{n_2} \circ P_{n_1}) \circ \cdots \circ (P_{n_j} \circ \cdots \circ P_{n_2} \circ P_{n_1})}_{m\text{-sets}}. \end{aligned} \tag{6.23}$$

To extend this concept to the local mapping, define

$$P^m_{n_l \cdots n_j} \equiv P_{(n_l \cdots n_j)^m} \equiv \underbrace{(P_{n_l} \circ \cdots \circ P_{n_j}) \circ \cdots \circ (P_{n_l} \circ \cdots \circ P_{n_j})}_{m\text{-sets}}, \tag{6.24}$$

where the local mappings are

$$P_{n_{j_1}} \in \left\{ P_{J_{\alpha\alpha}} | \alpha \in \{1, 2, \cdots, N\}, J \in \{0, 1, 2, \cdots, M\} \right\} \tag{6.25}$$

for ($j_1 \in \{j, \cdots, l\}$). The Jth-sub-domain for the local mappings should be the neighbored domains of the switching set Ξ_α. For the special combination of global and local mapping, introduce a mapping structure

$$\begin{aligned} &P_{n_k \cdots (n_l \cdots n_j)^m \cdots n_2 n_1} \\ &\equiv P_{n_k} \circ \cdots \circ P^m_{n_l \cdots n_j} \circ \cdots \circ P_{n_2} \circ P_{n_1} \\ &= P_{n_k} \circ \cdots \circ \underbrace{(P_{n_l} \circ \cdots \circ P_{n_j}) \circ \cdots \circ (P_{n_l} \circ \cdots \circ P_{n_j})}_{m-\text{sets}} \circ \cdots \circ P_{n_2} \circ P_{n_1}. \end{aligned} \tag{6.26}$$

From the definition, any global flow of the dynamical systems in Eq. (6.1) can be very easily managed through a certain mapping structures accordingly. Further, all the periodic flow of such a system in Eq. (6.1) can be investigated. For the sub-domain Ω_J, the flow is given by

$$\mathbf{x}^{(J)}(t) = \mathbf{\Phi}^{(J)}(t, \mathbf{x}_0^{(j)}, t_0) \in \mathscr{R}^n. \tag{6.27}$$

Consider the initial condition to be chosen on the discontinuous boundary relative to the switching plane Ξ_α (i.e., $(\mathbf{x}_k, t_k)$). Once the flow in the sub-domain Ω_J for time

$t_{k+1} > t_k$ arrives to the boundary with the switching plane Ξ_β (i.e., $(\mathbf{x}_{k+1}, t_{k+1})$), equation (6.27) becomes

$$\mathbf{x}_{k+1}^{(J)} = \mathbf{\Phi}^{(J)}(t_{k+1}, \mathbf{x}_k^{(J)}, t_k) \in \mathscr{R}^n. \tag{6.28}$$

The foregoing vector equation gives the relationships for mapping $P_{J_{\alpha\beta}}$, which maps the starting point $(\mathbf{x}_k, t_k)$ to the final point $(\mathbf{x}_{k+1}, t_{k+1})$ in the sub-domain Ω_J. For an n-dimensional dynamical system, equation (6.28) gives n-scalar equations. For one-degree-of-freedom systems, two scalar algebraic equations will be given. For the sliding flow on the switching plane Ξ_α, the sliding mapping $P_{0_{\alpha\alpha}}$ with starting and ending points $(\mathbf{x}_k, t_k)$ and $(\mathbf{x}_{k+1}, t_{k+1})$. Consider an $(n-1)$-dimensional boundary $\partial\Omega_{IJ}$, on which the switching set Ξ_α is defined. For the sliding mapping $P_{0_{\alpha\alpha}}$: $\Xi_\alpha \to \Xi_\alpha$, the starting and ending points satisfies $\varphi_{IJ}(\mathbf{x}_k, t_k) = \varphi_{IJ}(\mathbf{x}_{k+1}, t_{k+1}) = 0$. The sliding dynamics on the boundary can be determined by Eq. (6.5), i.e.,

$$\left.\begin{array}{l} \dot{\mathbf{x}}^{(0)} = \mathbf{F}_{IJ}^{(0_{\alpha\alpha})}(\mathbf{x}^{(0)}, t) \in \mathscr{R}^n \\ \varphi_{IJ}(\mathbf{x}^{(0)}, t) = 0 \end{array}\right\} \text{ on } \partial\Omega_{IJ}, \tag{6.29}$$

With the starting point $(\mathbf{x}_k, t_k)$, Eq. (6.29) gives

$$\mathbf{x}^{(0)}(t) = \mathbf{\Psi}^{(0_{\alpha\alpha})}(t, \mathbf{x}_k, t_k) \text{ and } \varphi_{IJ}(\mathbf{x}^{(0)}, t) = 0. \tag{6.30}$$

From the vanishing conditions of the sliding motion on the boundary, at the ending point, equation (6.30) should be satisfied and one of G-functions should be zero. Thus, the sliding mapping $P_{0_{\alpha\alpha}}$ on the boundary with $\varphi_{IJ}(\mathbf{x}_{k+1}, t_{k+1}) = 0$ will be governed by

$$\begin{array}{l} \mathbf{x}_{k+1} = \mathbf{\Psi}^{(0_{\alpha\alpha})}(t_{k+1}, \mathbf{x}_k, t_k) \in \mathscr{R}^n \text{ with } \varphi_{IJ}(\mathbf{x}_k, t_k) = \varphi_{IJ}(\mathbf{x}_{k+1}, t_{k+1}) = 0, \\ G_{\partial\Omega_{IJ}}^{(0,\sigma)}(\mathbf{x}_{k+1}, t_{k+1}) = \mathbf{n}_{\partial\Omega_{IJ}}^{\mathrm{T}} \cdot \mathbf{F}^{(\sigma)}(\mathbf{x}_{k+1}, t_{k+1}) = 0,\ \sigma \in \{I, J\} \\ \left.\begin{array}{l} G_{\partial\Omega_{IJ}}^{(0,\sigma)}(\mathbf{x}_k, t_k) < 0 \text{ and } G_{\partial\Omega_{IJ}}^{(0,\bar{\sigma})}(\mathbf{x}_k, t_k) > 0 \\ G_{\partial\Omega_{IJ}}^{(1,\sigma)}(\mathbf{x}_{k+1}, t_{k+1}) > 0 \end{array}\right\} \text{for } \mathbf{n}_{\partial\Omega_{IJ}} \to \Omega_\sigma; \\ \left.\begin{array}{l} G_{\partial\Omega_{IJ}}^{(0,\sigma)}(\mathbf{x}_k, t_k) > 0 \text{ and } G_{\partial\Omega_{IJ}}^{(0,\bar{\sigma})}(\mathbf{x}_k, t_k) < 0 \\ G_{\partial\Omega_{IJ}}^{(1,\sigma)}(\mathbf{x}_{k+1}, t_{k+1}) < 0 \end{array}\right\} \text{for } \mathbf{n}_{\partial\Omega_{IJ}} \to \Omega_{\bar{\sigma}}, \end{array} \tag{6.31}$$

where $\bar{\sigma} = J$ if $\sigma = I$ or $\bar{\sigma} = I$ if $\sigma = J$. Equation (6.31) is re-written in a general form of

$$\mathbf{x}_{k+1}^{(0_{\alpha\alpha})} = \mathbf{\Phi}^{(0_{\alpha\alpha})}(t_{k+1}, \mathbf{x}_k^{(0_{\alpha\alpha})}, t_k) \in \mathscr{R}^n. \tag{6.32}$$

As in Luo (2011), the transport law can be defined as a transport mapping as $P_{T_{\alpha\beta}}$: $\Xi_\alpha \to \Xi_\beta$. Similarly, using $t_{k+1} - t_k = \Delta$ and the transport law gives

$$\mathbf{x}_{k+1}^{(T_{\alpha\alpha})} = \mathbf{\Phi}^{(T_{\alpha\alpha})}(t_{k+1}, \mathbf{x}_k^{(T_{\alpha\alpha})}, t_k) \in \mathscr{R}^n. \tag{6.33}$$

For practical computations, the transport law is not necessary to be treated as a mapping. Once all the single mappings are determined by the corresponding governing equation. Because of the global flow on the discontinuous boundary in phase space, a new vector $\mathbf{y} = (t, \mathbf{x})^{\mathrm{T}}$ is selected on the boundary. Through the new vectors and boundaries, the global flow based on the mapping structure is expressed by

$$\mathbf{y}_{k+r} = P_{n_k \cdots (n_l \cdots n_j)^m \cdots n_2 n_1} \mathbf{y}_k, \tag{6.34}$$

where r is the total number of mapping actions in the mapping structure. For a global periodic flow, the periodicity conditions are required by

$$(t_{k+r}, \mathbf{x}_{k+r}) = (t_k + NT, \mathbf{x}_k), \tag{6.35}$$

where N is positive integer, and T is period of system in Eq. (6.1). For system without external periodic vector fields, the flow with a certain time difference returns to the selected reference plane, which will be a periodic motion. The governing equations for Eq. (6.34) is

$$\left. \begin{array}{l} \mathbf{x}_{k+\rho}^{(\sigma)} = \mathbf{\Phi}^{(J)}(t_{k+\rho}, \mathbf{x}_{k+\rho-1}^{(\sigma)}, t_{k+\rho-1}), \\ \varphi_{IJ}(\mathbf{x}_{k+\rho-1}^{(\sigma)}, t_{k+\rho-1}) = 0; \end{array} \right\} \tag{6.36}$$

for $\sigma = \{n_1, n_2, \cdots, n_k\}$, $\rho = \{1, 2, \cdots, r\}$ and $I, J = 1, 2, \cdots, M$.

The global periodic flow relative to the mapping structure $P_{n_k \cdots (n_l \cdots n_j)^m \cdots n_2 n_1}$ will be determined by Eqs.(6.35) and (6.36). The global periodic flow may be stable and unstable. The corresponding stability analysis can be completed through the traditional local stability analysis. For the periodic flow with sliding or gazing flows, the local stability analysis may not be useful. The sliding criteria in Luo (2009) should be employed. Although the local stability analysis can be carried out, it cannot provide enough information to check the disappearance of the certain global, periodic flow in discontinuous dynamical systems. For the local stability analysis of the periodic flow, the all switching points are given by $(\mathbf{x}_{k+\rho-1}^{(\sigma)}, t_{k+\rho-1}^{(\sigma)})$ and the corresponding perturbations $\delta \mathbf{y}_{k+\rho-1}^{(\sigma)} = (\delta \mathbf{x}_{k+\rho-1}^{(\sigma)}, \delta t_{k+\rho-1}^{(\sigma)})^{\mathrm{T}}$ are adopted. The perturbed equation for the stable analysis is

$$\delta \mathbf{y}_{k+r} = DP_{n_k \cdots (n_l \cdots n_j)^m \cdots n_2 n_1} \delta \mathbf{y}_k, \tag{6.37}$$

and the Jacobian matrix is

$$DP_{n_k \cdots (n_l \cdots n_j)^m \cdots n_2 n_1} = DP_{n_k} \cdot \ldots \cdot (DP_{n_l} \cdot \ldots \cdot DP_{n_j})^m \cdot \ldots \cdot DP_{n_2} \cdot DP_{n_1}. \tag{6.38}$$

For each single mapping,

$$DP_{\sigma} = \left[\frac{\partial (\mathbf{x}_{k+\rho}^{(\sigma)}, t_{k+\rho})}{\partial (\mathbf{x}_{k+\rho-1}^{(\sigma)}, t_{k+\rho-1})} \right]_{(\mathbf{x}_{k+\rho}^{(\sigma)}, t_{k+\rho})} \tag{6.39}$$

for $\sigma = \{n_1, n_2, \cdots, n_l\}$, $\rho = \{1, 2, \cdots, r\}$ and $I, J = 1, 2, \cdots, M$. The following determinant gives the all eigenvalues to determine the stability, i.e.,

$$|DP_{n_k \cdots (n_l \cdots n_j)^m \cdots n_2 n_1} - \lambda \mathbf{I}| = 0. \tag{6.40}$$

From Chap. 2, for the entire eigenvalues λ_j ($j = 1, \cdots, s$), if the magnitude of all the s-eigenvalues is less than one (i.e., $|\lambda_j| < 1$), the periodic flow determined by Eqs. (6.35) and (6.36) is stable. If at least one of the magnitudes of the s-eigenvalues is greater than one (i.e., $|\lambda_j| < 1$, $j \in \{1, \cdots, s\}$), the periodic flow is unstable. The other stability and bifurcation conditions can be found from Chap. 2.

6.4 Analytical Dynamics of Chua's Circuits

Consider a simple Chua's circuit, as shown in Fig. 6.5. The capacitances of two capacitors are C_1 and C_2. The inductance of inductor is L, and the resistance of linear resistor is R. A nonlinear resistor N_R possesses the V-I characteristic curve in Fig. 6.5. Assume V_1 and V_2 are the voltages of two capacitances, respectively. The current in the inductor is I_3. From the Kirchhoff's law, equations of the Chua's circuit are written as

$$\left.\begin{aligned} C_1 \frac{dV_1}{d\tau} &= \frac{1}{R}(V_2 - V_1) - f(V_1), \\ C_2 \frac{dV_2}{d\tau} &= \frac{1}{R}(V_1 - V_2) + I_3, \\ L \frac{dI_3}{d\tau} &= -V_2, \end{aligned}\right\} \tag{6.41}$$

where

$$f(V_1) = G_b V_1 - \frac{1}{2}(G_a + G_b)(|V_1 + E| - |V_1 - E|). \tag{6.42}$$

Introduce dimensionless variables as

$$\begin{aligned} &x = \frac{V_1}{E},\ \ y = \frac{V_2}{E},\ \ z = \frac{RI_3}{E},\ \ t = \frac{\tau}{RC_2},\ \ \alpha = \frac{C_2}{C_1},\ \ \beta = \frac{C_2 R^2}{L}, \\ &a = RG_a - 1,\ \ b = RG_b + 1. \end{aligned} \tag{6.43}$$

The state variables and vector fields are defined as

$$\mathbf{x} = (x, y, z)^{\mathrm{T}} \text{ and } \mathbf{F}(\mathbf{x}) = (\alpha[y - h(x)], x - y + z, -\beta y)^{\mathrm{T}}, \tag{6.44}$$

where

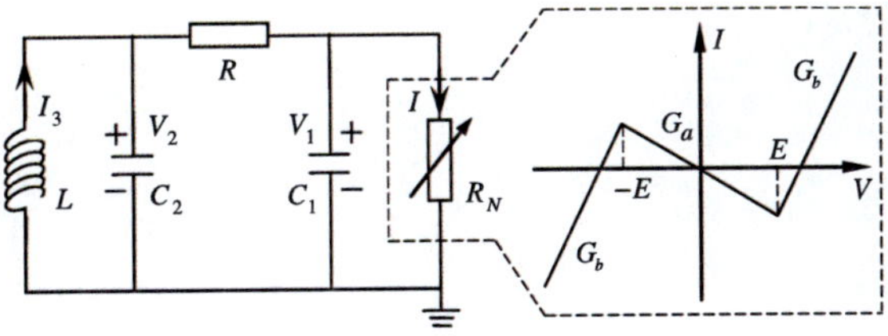

Fig. 6.5 Chua's circuit

$$h(x) = \begin{cases} bx + a + b & x < -1, \\ -ax & |x| \leq 1, \\ bx - a - b & x > 1. \end{cases} \tag{6.45}$$

Equations (6.41) and (6.42) are converted into non-dimensional equations by the dimensionless variables.

$$\dot{\mathbf{x}} = \mathbf{F}(\mathbf{x}, \mathbf{p}), \tag{6.46}$$

where the parameter vector $\mathbf{p}$ includes all parameters (i.e., a, b, α, β). Because the function $h(x)$ is piecewise continuous, there are three vector fields in three regions. Three domains are defined from the three regions as

$$\left.\begin{aligned} \Omega_{-1} &= \{(x, y, z)|x < -1\}, \\ \Omega_0 &= \{(x, y, z)| -1 < x < 1\}, \\ \Omega_1 &= \{(x, y, z)|x > 1\}, \end{aligned}\right\} \tag{6.47}$$

and two corresponding boundaries $\partial\Omega_{(0,-1)}$ and $\partial\Omega_{(0,1)}$ are

$$\left.\begin{aligned} \partial\Omega_{(0,-1)} &= \partial\Omega_{(-1,0)} = \bar{\Omega}_{-1} \cap \bar{\Omega}_0 = \{(x, y, z)|x = -1\}, \\ \partial\Omega_{(0,1)} &= \partial\Omega_{(1,0)} = \bar{\Omega}_1 \cap \bar{\Omega}_0 = \{(x, y, z)|x = 1\}, \end{aligned}\right\} \tag{6.48}$$

where $\bar{\Omega}_i$ $(i = -1, 0, 1)$ means the closure of the three domains Ω_i. The boundaries and domains are sketched in Fig. 6.6.

In three domains, equation (6.46) can be expressed by

$$\dot{\mathbf{x}}^{(i)} = \mathbf{F}^{(i)}(\mathbf{x}^{(i)}, \mathbf{p}^{(i)}) = \mathbf{A}^{(i)}\mathbf{x}^{(i)} + \mathbf{b}^{(i)} \quad (i = -1, 0, 1). \tag{6.49}$$

The linear matrices are for $\sigma = -1, 1$

$$A^{(0)} = \begin{pmatrix} \alpha a & \alpha & 0 \\ 1 & -1 & 1 \\ 0 & -\beta & 0 \end{pmatrix} \text{ and } A^{(\sigma)} = \begin{pmatrix} -\alpha b & \alpha & 0 \\ 1 & -1 & 1 \\ 0 & -\beta & 0 \end{pmatrix} \tag{6.50}$$

with constant vectors

$$\mathbf{b}^{(i)} = (i\alpha(a + b), 0, 0)^{\mathrm{T}}. \tag{6.51}$$

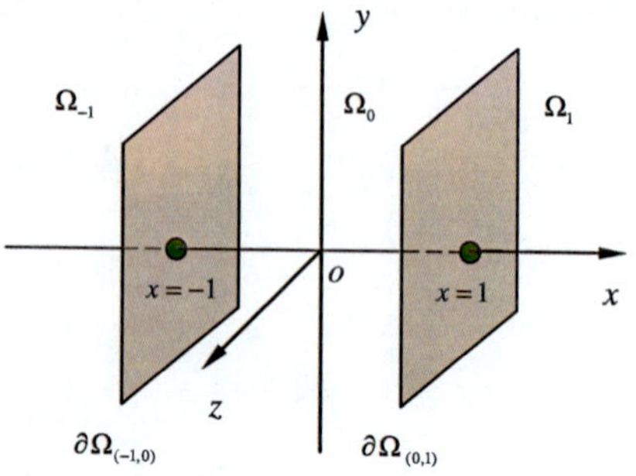

Fig. 6.6 Domains and boundaries

For the homogeneous solution, the eigenvalue $\lambda^{(i)}$ satisfies the equation

$$\left|\mathbf{A}^{(i)} - \lambda^{(i)}\mathbf{I}\right| = \begin{vmatrix} \alpha a^{(i)} - \lambda^{(i)} & \alpha & 0 \\ 1 & -1-\lambda^{(i)} & 1 \\ 0 & -\beta & -\lambda^{(i)} \end{vmatrix} = 0, \tag{6.52}$$

where $a^{(0)} = a$ and $a^{(1)} = a^{(-1)} = a$. Thus,

$$(\lambda^{(i)})^3 - (\alpha a^{(i)} - 1)(\lambda^{(i)})^2 - (\alpha a^{(i)} + \alpha - \beta)\lambda^{(i)} - \alpha\beta a^{(i)} = 0. \tag{6.53}$$

Consider a general form of a cubic equation as

$$f(\lambda^{(i)}) = a_1^{(i)}(\lambda^{(i)})^3 + a_2^{(i)}(\lambda^{(i)})^2 + a_3^{(i)}\lambda^{(i)} + a_4^{(i)} = 0, \quad a_1^{(i)} \neq 0, \tag{6.54}$$

where

$$a_1^{(i)} = 1,\ a_2^{(i)} = -(\alpha a^{(i)} - 1),\ a_3^{(i)} = -(\alpha a^{(i)} + \alpha - \beta),\ a_4^{(i)} = -\alpha\beta a^{(i)}. \tag{6.55}$$

Three cases are distinguished through the discriminant

$$\begin{aligned}\Delta^{(i)} = {} & 4(a_2^{(i)})^3 a_4^{(i)} - (a_2^{(i)} a_3^{(i)})^2 + 4a_1^{(i)}(a_2^{(i)})^3 \\ & - 18a_1^{(i)} a_2^{(i)} a_3^{(i)} a_4^{(i)} + 27(a_1^{(i)} a_4^{(i)})^2.\end{aligned} \tag{6.56}$$

There are three cases:

(i) If $\Delta^{(i)} < 0$, equation (6.56) has three distinct real roots.
(ii) If $\Delta^{(i)} > 0$, equation (6.56) has one real root and a pair of complex conjugate roots.
(iii) If $\Delta^{(i)} = 0$, at least two roots of Eq. (6.56) coincide. For this case, the equation may have a double repeated real root and another distinct single real root, or the coincidence of all three roots yields a triple real root.

Introduce several parameters $p^{(i)}, q^{(i)}, r^{(i)}$ and $s^{(i)}$ as

$$\begin{aligned} q^{(i)} &= \frac{9a_1^{(i)} a_2^{(i)} a_3^{(i)} - 27(a_1^{(i)})^2 a_4^{(i)} - 2(a_2^{(i)})^3}{54(a_1^{(i)})^3}, \\ r^{(i)} &= \sqrt{\left(\frac{3a_1^{(i)} a_3^{(i)} - (a_2^{(i)})^2}{9(a_1^{(i)})^2}\right)^3 + (q^{(i)})^2} = \sqrt{\frac{\Delta^{(i)}}{108(a_1^{(i)})^4}} \end{aligned} \tag{6.57}$$

and

$$s^{(i)} = \sqrt[3]{q^{(i)} + r^{(i)}},\ p^{(i)} = \sqrt[3]{q^{(i)} - r^{(i)}}. \tag{6.58}$$

Thus, the solutions are

$$\left.\begin{aligned} \lambda_1^{(i)} &= s^{(i)} + p^{(i)} - \frac{a_2^{(i)}}{3a_1^{(i)}},\\ \lambda_2^{(i)} &= -\tfrac{1}{2}(s^{(i)} + p^{(i)}) - \frac{a_2^{(i)}}{3a_1^{(i)}} + \frac{\sqrt{3}}{2}(s^{(i)} - p^{(i)})\mathbf{i},\\ \lambda_3^{(i)} &= -\tfrac{1}{2}(s^{(i)} + p^{(i)}) - \frac{a_2^{(i)}}{3a_1^{(i)}} - \frac{\sqrt{3}}{2}(s^{(i)} - p^{(i)})\mathbf{i}, \end{aligned}\right\} \tag{6.59}$$

where $\mathbf{i} = \sqrt{-1}$.

Case 1 Discriminant $\Delta^{(j)} < 0$. There are three different real eigenvalue $\lambda_j^{(i)} (j = 1, 2, 3)$. Modal shape $\mathbf{r}_j^{(i)} = (r_{1i}^{(j)}, r_{2i}^{(j)}, r_{3i}^{(j)})^{\mathrm{T}}$ is computed by

$$\mathbf{r}_j^{(i)} = (r_{1i}^{(j)}, r_{2i}^{(j)}, r_{3i}^{(j)})^{\mathrm{T}} = (1, \frac{\lambda_j^{(i)} - \alpha a^{(i)}}{\alpha}, \frac{a^{(i)}\beta}{\lambda_j^{(i)}}, -\frac{\beta}{\alpha})^{\mathrm{T}}. \tag{6.60}$$

Therefore, the general solution is

$$\begin{aligned} \mathbf{x}^{(i)} &= \mathbf{\Phi}^{(i)}(\mathbf{x}_k^{(i)}, t, t_k) = \sum\nolimits_{i=1}^{3} C_j^{(i)}\mathbf{r}_j^{(i)} e^{\lambda_j^{(i)}(t-t_k)} + \mathbf{x}_p^{(i)}\\ &= C_1^{(i)}\mathbf{r}_1^{(i)} e^{\lambda_1^{(i)}(t-t_k)} + C_2^{(i)}\mathbf{r}_2^{(i)} e^{\lambda_2^{(i)}(t-t_k)} + C_3^{(i)}\mathbf{r}_3^{(i)} e^{\lambda_3^{(i)}(t-t_k)} + \mathbf{x}_p^{(i)}, \end{aligned} \tag{6.61}$$

where

$$\mathbf{x}_p^{(i)} = i(\frac{a+b}{b}, 0, -\frac{a+b}{b})^{\mathrm{T}}. \tag{6.62}$$

This case was discussed in Bartissol and Chua (1988).

Case 2 Discriminant $\Delta^{(j)} > 0$. If $\lambda_1^{(i)} = \lambda^{(i)}$ and $\lambda_{2,3}^{(i)} = \delta^{(i)} \pm \omega^{(i)}\mathbf{i}$, the general solution is

$$\begin{aligned} \mathbf{x}^{(i)}(t) =& \mathbf{\Phi}^{(i)}(\mathbf{x}_k^{(i)}, t, t_k) = \mathbf{x}_p^{(i)} + C_1^{(i)}\mathbf{r}_1^{(i)} e^{\lambda^{(i)}(t-t_k)}\\ &+ (C_2^{(i)}\mathbf{r}_2^{(i)} + C_3^{(i)}\mathbf{r}_3^{(i)}) e^{\delta^{(i)}(t-t_k)} \cos[\omega^{(i)}(t - t_k)]\\ &+ (-C_2^{(i)}\mathbf{r}_3^{(i)} + C_3^{(i)}\mathbf{r}_2^{(i)}) e^{\delta^{(i)}(t-t_k)} \sin[\omega^{(i)}(t - t_k)], \end{aligned} \tag{6.63}$$

where

$$
\left.
\begin{aligned}
\mathbf{r}_1^{(i)} &= (1, \frac{\lambda^{(i)} - \alpha a}{\alpha}, (\frac{\beta a}{\lambda^{(i)}} - \frac{\beta}{\alpha}))^{\mathrm{T}},\\
\mathbf{r}_2^{(i)} &= (1, \frac{\delta^{(i)} - \alpha a}{\alpha}, \frac{\beta a \delta^{(i)}}{(\delta^{(i)})^2 + (\omega^{(i)})^2}, -\frac{\beta}{\alpha})^{\mathrm{T}},\\
\mathbf{r}_3^{(i)} &= (0, \frac{\omega^{(i)}}{\alpha}, -\frac{\beta a \omega^{(i)}}{(\delta^{(i)})^2 + (\omega^{(i)})^2})^{\mathrm{T}}.
\end{aligned}
\right. \tag{6.64}
$$

Case 3 Discriminant $\Delta^{(j)} = 0$. The general solution is

$$
\begin{aligned}
\mathbf{x}^{(i)} =& \boldsymbol{\Phi}^{(i)}(\mathbf{x}_k^{(i)}, t, t_k) = C_1^{(i)} \mathbf{r}_1^{(i)} e^{\lambda_1^{(i)}(t-t_k)} + [C_2^{(i)}(\mathbf{r}_{10}^{(i)}\\
&+ (t - t_k)\mathbf{r}_{11}^{(i)}) + C_3^{(i)}(\mathbf{r}_{20}^{(i)} + (t - t_k)\mathbf{r}_{21}^{(i)})]e^{\lambda_2^{(i)}(t-t_k)} + \mathbf{x}_p^{(i)}
\end{aligned} \tag{6.65}
$$

where

$$
\left.
\begin{aligned}
\mathbf{r}_{10}^{(i)} &= (1, \frac{1}{1 + 2\lambda_2^{(i)}}, 0)^{\mathrm{T}}, \mathbf{r}_{20}^{(i)} = (0, \frac{\beta - (\lambda_2^{(i)})^2}{\beta(1 + 2\lambda_2^{(i)})}, 1)^{\mathrm{T}},\\
\mathbf{r}_{11}^{(i)} &= (\alpha a - \lambda_2^{(i)} + \frac{\alpha}{1 + 2\lambda_2^{(i)}}, 1 - \frac{1 + \lambda_2^{(i)}}{1 + 2\lambda_2^{(i)}}, -\frac{\beta}{1 + 2\lambda_2^{(i)}})^{\mathrm{T}},\\
\mathbf{r}_{21}^{(i)} &= (\frac{\alpha(\beta - (\lambda_2^{(i)})^2)}{\beta(1 + 2\lambda_2^{(i)})}, \frac{(1 + \lambda_2^{(i)})((\lambda_2^{(i)})^2 - \beta)}{\beta(1 + 2\lambda_2^{(i)})} + 1, \frac{(\lambda_2^{(i)})^2 - \beta}{1 + 2\lambda_2^{(i)}} - \lambda_2^{(i)})^{\mathrm{T}}.
\end{aligned}
\right\} \tag{6.66}
$$

The two boundaries are governed by $\varphi\ (x, y, z)\ \equiv x \pm 1 = 0$. The normal vectors of the two boundaries are

$$
\mathbf{n}_{\partial\Omega_{(-1,0)}} = \mathbf{n}_{\partial\Omega_{(0,1)}} = (1, 0, 0)^{\mathrm{T}}. \tag{6.67}
$$

From the theory of discontinuous dynamic systems in Luo (2006, 2008a, b, 2009), the condition for grazing response at the boundary (i.e., $\mathbf{x}_m^{(i)} \in \partial\Omega_{(i,j)}$) at time t_m satisfies

$$
\left.
\begin{aligned}
\mathbf{n}_{\partial\Omega_{(i,j)}}^{\mathrm{T}} \cdot \mathbf{F}^{(i)}(\mathbf{x}_{m\pm}^{(i)}, \mathbf{p}^{(i)}) = 0,\\
\mathbf{n}_{\partial\Omega_{(i,j)}}^{\mathrm{T}} \cdot D\mathbf{F}^{(i)}(\mathbf{x}_{m\pm}^{(i)}, \mathbf{p}^{(i)}) > 0,
\end{aligned}
\right\} \text{ for } (i, j) \in \{(1, 0), (0, -1)\} \tag{6.68}
$$

$$
\left.
\begin{aligned}
\mathbf{n}_{\partial\Omega_{(i,j)}}^{\mathrm{T}} \cdot \mathbf{F}^{(i)}(\mathbf{x}_{m\pm}^{(i)}, \mathbf{p}^{(i)}) = 0,\\
\mathbf{n}_{\partial\Omega_{(i,j)}}^{\mathrm{T}} \cdot D\mathbf{F}^{(i)}(\mathbf{x}_{m\pm}^{(i)}, \mathbf{p}^{(i)}) < 0,
\end{aligned}
\right\} \text{ for } (i, j) \in \{(-1, 0), (0, 1)\}, \tag{6.69}
$$

where

$$
D\mathbf{F}^{(i)}(\mathbf{x}_{m\pm}^{(i)}, \mathbf{p}^{(i)}) = \frac{\partial \mathbf{F}^{(i)}}{\partial \mathbf{x}^{(i)}} \dot{\mathbf{x}}^{(i)}\Big|_{\mathbf{x}_{m\pm}^{(i)}}. \tag{6.70}
$$

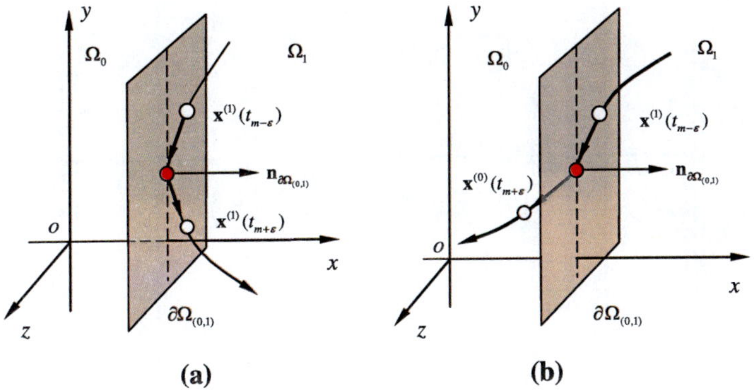

Fig. 6.7 Analytical conditions at $x = 1$: **a** grazing and **b** passable

The condition for passable flows on the boundary (i.e., $\mathbf{x}_m^{(i)} \in \partial\Omega_{(i,j)}$) at time t_m is

$$[\mathbf{n}^{\mathrm{T}}_{\partial\Omega_{(i,j)}} \cdot \mathbf{F}^{(i)}(\mathbf{x}^{(i)}_{m-}, \mathbf{p}^{(i)})] \times [\mathbf{n}^{\mathrm{T}}_{\partial\Omega_{(i,j)}} \cdot \mathbf{F}^{(j)}(\mathbf{x}^{(j)}_{m+}, \mathbf{p}^{(j)})] > 0. \tag{6.71}$$

In other words, one obtains

$$\left.\begin{aligned} \mathbf{n}^{\mathrm{T}}_{\partial\Omega_{(i,j)}} \cdot \mathbf{F}^{(i)}(\mathbf{x}^{(i)}_{m-}, \mathbf{p}^{(i)}) < 0, \\ \mathbf{n}^{\mathrm{T}}_{\partial\Omega_{(i,j)}} \cdot \mathbf{F}^{(j)}(\mathbf{x}^{(j)}_{m+}, \mathbf{p}^{(j)}) < 0 \end{aligned}\right\} \text{ for } (i, j) = (1, 0), (0, -1), \tag{6.72}$$

$$\left.\begin{aligned} \mathbf{n}^{\mathrm{T}}_{\partial\Omega_{(i,j)}} \cdot \mathbf{F}^{(i)}(\mathbf{x}^{(i)}_{m-}, \mathbf{p}^{(i)}) > 0, \\ \mathbf{n}^{\mathrm{T}}_{\partial\Omega_{(i,j)}} \cdot \mathbf{F}^{(j)}(\mathbf{x}^{(j)}_{m+}, \mathbf{p}^{(j)}) > 0 \end{aligned}\right\} \text{ for } (i, j) = (-1, 0), (0, 1). \tag{6.73}$$

For the Chua's system, no sliding responses exist on the two boundaries. The conditions for the sliding responses will not be presented herein. The grazing and passable flows at the boundary $\partial\Omega_{(1,0)}$ are sketched in Fig. 6.7. The normal vector of the boundary points to the domain Ω_1. For a grazing flow in Ω_1, the first equation of Eq. (6.71) is satisfied as a necessary condition. The second equation gives the sufficient condition for the grazing flow, as shown in Fig. 6.7(a). A flow in Ω_1 is passable at the boundary $\partial\Omega_{(1,0)}$, which is sketched in Fig. 6.7(b). It is observed that Eq. (6.71) or (6.72) is satisfied. It means that two normal vector fields have the same direction.

6.4.1 Periodic Flows

For periodic responses, switching sets are introduced from the switching boundaries. The physical meaning of switching sets is a set including the switching points and

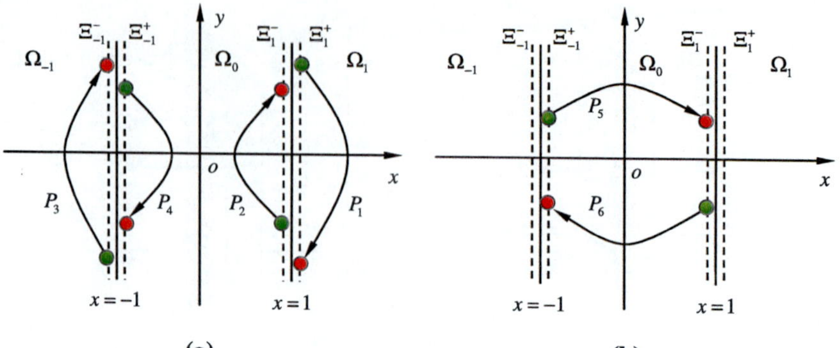

Fig. 6.8 Switching sets and mappings: **a** local mappings and **b** global mappings

switching time. The switching sets are defined as

$$\left.\begin{aligned} \Xi_{-1}^{-} &= \{(x_k, y_k, z_k, t_k)|x_k = -1^{-}\}, \\ \Xi_{-1}^{+} &= \{(x_k, y_k, z_k, t_k)|x_k = -1^{+}\}, \\ \Xi_{1}^{-} &= \{(x_k, y_k, z_k, t_k)|x_k = 1^{-}\}, \\ \Xi_{1}^{+} &= \{(x_k, y_k, z_k, t_k)|x_k = 1^{+}\}. \end{aligned}\right\} \tag{6.74}$$

The superscript "+ " or "−" means that the switching set is close to the *right* or *left* side of the boundary. In Fig. 6.8, the dash lines infinitesimally close to the boundaries represent the switching sets Σ. From the switching sets, six basic mappings are defined as

$$\left.\begin{aligned} &\left.\begin{aligned} P_1 &: \Xi_1^{+} \to \Xi_1^{+},\ P_2 : \Xi_1^{-} \to \Xi_1^{-}, \\ P_3 &: \Xi_{-1}^{-} \to \Xi_{-1}^{-},\ P_4 : \Xi_{-1}^{+} \to \Xi_{-1}^{+} \end{aligned}\right\} \text{for local mappings;} \\ &P_5 : \Xi_1^{-} \to \Xi_{-1}^{+},\ P_6 : \Xi_{-1}^{+} \to \Xi_1^{-} \text{ for global mappings.} \end{aligned}\right\} \tag{6.75}$$

where the mapping definitions do not follow the previous section because this mapping is very simple. The four local mappings, P_l $(l = 1, 2, 3, 4)$, map the initial state to the final state in the same switching sets, as presented in Fig. 6.8(a). The two global mappings, P_5 and P_6, map the initial state from one switching set to another switching set in order to obtain the final state, as in Fig. 6.8(b).

Consider an initial state (x_k, y_k, z_k, t_k) with $x_k \in \{-1, 1\}$ and a final state $(x_{k+1}, y_{k+1}, z_{k+1}, t_{k+1})$ with $x_{k+1} \in \{-1, 1\}$. The four local mappings, P_l $(l = 1, 2, 3, 4)$ are governed by the following algebraic equations

$$\left.\begin{aligned} g_1^{(l)} &\equiv \pm 1 - \Phi_1^{(i)}(\pm 1, y_k, z_k, t_{k+1}, t_k) = 0, \\ g_2^{(l)} &\equiv y_{k+1} - \Phi_2^{(i)}(\pm 1, y_k, z_k, t_{k+1}, t_k) = 0, \\ g_3^{(l)} &\equiv z_{k+1} - \Phi_3^{(i)}(\pm 1, y_k, z_k, t_{k+1}, t_k) = 0 \end{aligned}\right\} \tag{6.76}$$

for $i = -1, 0, 1$. The algebraic equations for global mappings P_l $(l = 5, 6)$ are

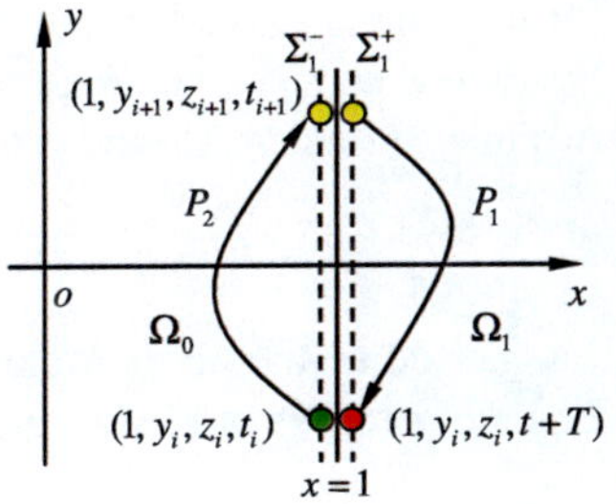

Fig. 6.9 A simple periodic response of P_{12}

$$\left.\begin{aligned} g_1^{(l)} &\equiv \pm 1 - \Phi_1^{(i)}(\mp 1, y_k, z_k, t_{k+1}, t_k) = 0,\\ g_2^{(l)} &\equiv y_{k+1} - \Phi_2^{(i)}(\mp 1, y_k, z_k, t_{k+1}, t_k) = 0,\\ g_3^{(l)} &\equiv z_{k+1} - \Phi_3^{(i)}(\mp 1, y_k, z_k, t_{k+1}, t_k) = 0 \end{aligned}\right\} \tag{6.77}$$

due to $\mathbf{\Phi}^{(i)} = (\Phi_1^{(i)}, \Phi_2^{(i)}, \Phi_3^{(i)})^{\mathrm{T}}$. From such basic mappings, periodic response can be predicted analytically.

Consider a simple periodic flow $P_{12} = P_1 \circ P_2$ in Fig. 6.9 to explain how to predict periodic flows by mapping structures. For a starting point $(1, y_k, z_k, t_k) \in \Xi_1^-$, a flow in domain Ω_0 arrives to the same switching set (i.e., $(1, y_{k+1}, z_{k+1}, t_{k+1}) \in \Xi_1^-$) by mapping P_2 at time t_{k+1}, and the passable condition of the flow is satisfied. The final state $(1, y_{k+1}, z_{k+1}, t_{k+1}) \in \Xi_1^-$ becomes the initial state for mapping P_1 at the switching set Ξ_1^+. One has $(1, y_{k+1}, z_{k+1}, t_{k+1}) \in \Xi_1^+$. The final state of a flow in domain Ω_1 is at the switching set Ξ_1^+ by mapping P_1. In other words, the final state is $(1, y_{k+2}, z_{k+2}, t_{k+2}) \in \Xi_1^+$.

Because of the periodicity of the flow, the following relations are satisfied, i.e.,

$$y_{k+2} = y_k,\ z_{k+2} = z_k,\ t_{k+2} = t_k + T \tag{6.78}$$

where T is period for this periodic flow. Thus, the periodic flow based on mapping structure $P_{12} = P_1 \circ P_2$ is described, as shown in Fig. 6.9. Other periodic responses can be developed in a similar fashion. There are two mappings for the periodic response of P_{12}, and for each mapping, there are three algebraic equations, i.e.,

$$\begin{aligned} &1 - \Phi_1^{(0)}(1, y_k, z_k, t_k, t_{k+1}) = 0,\\ &y_{k+1} - \Phi_2^{(0)}(1, y_k, z_k, t_k, t_{k+1}) = 0,\\ &z_{k+1} - \Phi_3^{(0)}(1, y_k, z_k, t_k, t_{k+1}) = 0 \end{aligned} \tag{6.79}$$

for mapping P_2 and

$$\begin{aligned} &1 - \Phi_1^{(1)}(1, y_{k+1}, z_{k+1}, t_{k+1}, t_{k+2}) = 0,\\ &y_{k+2} - \Phi_2^{(1)}(1, y_{k+1}, z_{k+1}, t_{k+1}, t_{k+2}) = 0,\\ &z_{k+2} - \Phi_3^{(1)}(1, y_{k+1}, z_{k+1}, t_{k+1}, t_{k+2}) = 0 \end{aligned} \tag{6.80}$$

for mapping P_1. Note that t_k and t_{k+1} always appears as $t_{k+1} - t_k$ because in the Eq. (6.79), and t_{k+1} and t_{k+2} always appears as $t_{k+2} - t_{k+1}$ in Eq. (6.80). Introduce two new variables Δt_k and Δt_{k+1} as

$$\Delta t_{k+1} = t_{k+2} - t_{k+1}, \quad \Delta t_k = t_{k+1} - t_k, \quad \Delta t_k + \Delta t_{k+1} = T \tag{6.81}$$

The period of the periodic flow T can be calculated by the summation of all Δt. Equations (6.79) and (6.80) give six equations with six unknowns, i.e.,

$$\begin{aligned}
&f_1^{(2)}(y_{k+1}, z_{k+1}, y_k, z_k, \Delta t_k) = 1 - \Phi_1^{(0)}(1, y_k, z_k, \Delta t_k) = 0,\\
&f_1^{(2)}(y_{k+1}, z_{k+1}, y_k, z_k, \Delta t_k) = y_{k+1} - \Phi_2^{(0)}(1, y_k, z_k, \Delta t_k) = 0,\\
&f_1^{(2)}(y_{k+1}, z_{k+1}, y_k, z_k, \Delta t_k) = z_{k+1} - \Phi_3^{(0)}(1, y_k, z_k, \Delta t_k) = 0,\\
&f_1^{(1)}(y_k, z_k, y_{k+1}, z_{k+1}, \Delta t_{k+1}) = 1 - \Phi_1^{(1)}(1, y_{k+1}, z_{k+1}, \Delta t_{k+1}) = 0,\\
&f_2^{(1)}(y_{k+2}, z_{k+2}, y_{k+1}, z_{k+1}, \Delta t_{k+1}) = y_{k+2} - \Phi_2^{(1)}(1, y_{k+1}, z_{k+1}, \Delta t_{k+1}) = 0,\\
&f_3^{(1)}(y_{k+2}, z_{k+2}, y_{k+1}, z_{k+1}, \Delta t_{k+1}) = z_{k+2} - \Phi_3^{(1)}(1, y_{k+1}, z_{k+1}, \Delta t_{k+1}) = 0.
\end{aligned} \tag{6.82}$$

The switching points of the periodic flow can be obtained by Eqs. (6.82) with (6.78) by any numerical method. Other periodic responses are described in the same way. For instance, a periodic response which contains $2n$ switching points, its algebraic equation set has $3 \times 2n$ unknowns and equations.

To discuss the stability of the periodic response of P_{12}, the switching points $(1,\ y_k^*,\ z_k^*,\ t_k^*)$ of the periodic solution is determined. For the mapping structure of P_{12}, one obtains

$$\mathbf{x}_{k+1} = P_2\mathbf{x}_k \text{ and } \mathbf{x}_{k+2} = P_1\mathbf{x}_{k+1}. \tag{6.83}$$

Plugging in the switching points $\mathbf{x}_k^*$, $\mathbf{x}_{k+1}^*$, and $\mathbf{x}_{k+2}^* = x_k^*$ of the periodic flow yields

$$\mathbf{x}_{k+1}^* = P_2\mathbf{x}_k^* \text{ and } \mathbf{x}_{k+2}^* = P_1\mathbf{x}_{k+1}^*. \tag{6.84}$$

Consider $\mathbf{x}_l$ $(l = k, k+1, k+2)$ to be a disturbance from $\mathbf{x}_l^*$. Using the Taylor series extension and keeping the terms up to $\delta\mathbf{x}_l$, gives

$$\left.\begin{aligned}
\mathbf{x}_{k+1}^* + \delta\mathbf{x}_{k+1} &= P_2\mathbf{x}_k^* + DP_2|_{\mathbf{x}_k^*}\,\delta\mathbf{x}_k + o(\delta\mathbf{x}_k)\\
\mathbf{x}_{k+2}^* + \delta\mathbf{x}_{k+2} &= P_1\mathbf{x}_{k+1}^* + DP_1|_{\mathbf{x}_{k+1}^*}\,\delta\mathbf{x}_{k+1} + o(\delta\mathbf{x}_{k+1})
\end{aligned}\right\}. \tag{6.85}$$

Equations (6.84) and (6.85) yields

$$\delta\mathbf{x}_{k+1} = DP_2|_{\mathbf{x}_k^*}\,\delta\mathbf{x}_k \text{ and } \delta\mathbf{x}_{k+2} = DP_1|_{\mathbf{x}_{k+1}^*}\,\delta\mathbf{x}_{k+1}. \tag{6.86}$$

From the above equation,

$$\delta\mathbf{x}_{k+2} = DP_1|_{\mathbf{x}_{k+1}^*} \cdot DP_2|_{\mathbf{x}_k^*}\,\delta\mathbf{x}_k \equiv DP|_{\mathbf{x}_k^*}\,\delta\mathbf{x}_k, \tag{6.87}$$

where

$$\left.\begin{aligned} DP_1|_{(y_k^*,z_k^*)} &= \left.\frac{\partial(y_{k+1},z_{k+1})}{\partial(y_k,z_k)}\right|_{(y_k^*,z_k^*)} = \left.\begin{bmatrix} \dfrac{\partial y_{k+1}}{\partial y_k} & \dfrac{\partial y_{k+1}}{\partial z_k} \\ \dfrac{\partial z_{k+1}}{\partial y_k} & \dfrac{\partial z_{k+1}}{\partial z_k} \end{bmatrix}\right|_{(y_k^*,z_k^*)} \\ DP_2|_{(y_{k+1}^*,z_{k+1}^*)} &= \left.\frac{\partial(y_{k+2},z_{k+2})}{\partial(y_{k+1},z_{k+1})}\right|_{(y_{k+1}^*,z_{k+1}^*)} = \left.\begin{bmatrix} \dfrac{\partial y_{k+2}}{\partial y_{k+1}} & \dfrac{\partial y_{k+2}}{\partial z_{k+1}} \\ \dfrac{\partial z_{k+2}}{\partial y_{k+1}} & \dfrac{\partial z_{k+2}}{\partial z_{k+1}} \end{bmatrix}\right|_{(y_{k+1}^*,z_{k+1}^*)} \end{aligned}\right\} \tag{6.88}$$

To calculate such DP matrix, equation (6.82) is used. Taking total derivative of each equation with respect to y_σ and z_σ ($\sigma = k, k+1$), respectively, yields,

$$\left.\begin{aligned} \frac{Df_1^{(i)}}{Dy_\sigma} &= \frac{\partial f_1^{(i)}}{\partial y_\sigma} + \frac{\partial f_1^{(i)}}{\partial \Delta t_\sigma}\frac{\partial \Delta t_\sigma}{\partial y_\sigma} = 0, \\ \frac{Df_2^{(i)}}{Dy_\sigma} &= \frac{\partial f_2^{(i)}}{\partial y_\sigma} + \frac{\partial f_2^{(i)}}{\partial y_{\sigma+1}}\frac{\partial y_{\sigma+1}}{\partial y_\sigma} + \frac{\partial f_2^{(i)}}{\partial \Delta t_\sigma}\frac{\partial \Delta t_\sigma}{\partial y_\sigma} = 0, \\ \frac{Df_3^{(i)}}{Dy_\sigma} &= \frac{\partial f_3^{(i)}}{\partial y_\sigma} + \frac{\partial f_3}{\partial z_{\sigma+1}}\frac{\partial z_{\sigma+1}}{\partial y_\sigma} + \frac{\partial f_3}{\partial \Delta t_\sigma}\frac{\partial \Delta t_\sigma}{\partial y_\sigma} = 0, \end{aligned}\right\} \tag{6.89}$$

and

$$\left.\begin{aligned} \frac{Df_1^{(i)}}{Dz_\sigma} &= \frac{\partial f_1^{(i)}}{\partial z_\sigma} + \frac{\partial f_1^{(i)}}{\partial \Delta t_\sigma}\frac{\partial \Delta t_\sigma}{\partial z_\sigma} = 0, \\ \frac{Df_2^{(i)}}{Dz_\sigma} &= \frac{\partial f_2^{(i)}}{\partial z_\sigma} + \frac{\partial f_2^{(i)}}{\partial y_{\sigma+1}}\frac{\partial y_{\sigma+1}}{\partial z_\sigma} + \frac{\partial f_2^{(i)}}{\partial \Delta t_\sigma}\frac{\partial \Delta t_\sigma}{\partial z_\sigma} = 0, \\ \frac{Df_3^{(i)}}{Dz_\sigma} &= \frac{\partial f_3^{(i)}}{\partial z_\sigma} + \frac{\partial f_3^{(i)}}{\partial z_{\sigma+1}}\frac{\partial z_{\sigma+1}}{\partial z_\sigma} + \frac{\partial f_3^{(i)}}{\partial \Delta t_\sigma}\frac{\partial \Delta t_\sigma}{\partial z_\sigma} = 0. \end{aligned}\right\} \tag{6.90}$$

There are three equations with three unknowns in each set of Eqs. (6.89) and (6.90). From the first equation in each set, $\partial\Delta t_\sigma/\partial y_\sigma$ and $\partial\Delta t_\sigma/\partial z_\sigma$ can be solved, then plugging into the other two equations, the components $\partial y_{\sigma+1}/\partial y_\sigma$, $\partial z_{\sigma+1}/\partial y_\sigma$, $\partial y_{\sigma+1}/\partial z_\sigma$ and $\partial z_{\sigma+1}/\partial z_\sigma$ for DP_2 and DP_1 are obtained.

If there exists one eigenvalue of DP matrix with magnitude larger than one, $|\lambda_i| > 1,\ i \in \{1,2\}$, the periodic response based on this Jacobian matrix is unstable. On the other hand, if magnitudes of both eigenvalues are less then one $|\lambda_i| \le 1$, $i = 1, 2$, the periodic response is stable. Especially, if one eigenvalue is positive one (+1), the absolute value of the other eigenvalue is less then one, saddle-node bifurcation happens. Period-doubling bifurcation occurs when one eigenvalue is negative one (-1), the absolute value of the other eigenvalue is less then one. If the two eigenvalues are a pair of complex number with magnitude equals to one, the Neimark bifurcation takes place. Other periodic flows can be discussed in a similar fashion.

6.4.2 Analytical Predictions

As in Luo and Xue (2009), before analytical prediction of periodic flow, a bifurcation scenario based on the switching plane in Eqs. (6.74) and (6.75) is presented first for parameters $a = 0.142857, b = 0.285714$ and $\beta = 9$. The bifurcation scenario is presented via the switching sets versus parameter α. Such a parameter represents the ratio of capacitances, as shown in Fig. 6.10. The acronym "GB" is the grazing bifurcation, depicted by dashed lines, and such a grazing bifurcation occurs at about $\alpha \approx 5.117$. The periodic response of P_{12} lies in $\alpha \in (5.117, 5.890)$. The periodic doubling bifurcation of the periodic flow P_{12} occurs at $\alpha \approx 5.890$. Thus the periodic flow of $P_{(12)^2}$ lies in $\alpha \in (5.890, 6.040)$ and the corresponding period-doubling bifurcation occurs at $\alpha \approx 6.040$. With increasing α, the periodic flow of $P_{(12)^4}$ is obtained. Continuously, the onset of chaos relative to $P_{(12)^{2^n}}$ $(n \to \infty)$ occurs at $\alpha \approx 6.12$. The acronym "PD" represents the period doubling bifurcation. With increasing α, the chaotic responses are observed. However, in the region of chaotic responses, there are many windows of periodic responses. Such periodic responses can be predicted analytically later. In addition, the switching points are plotted through the plane of (y, z) at the boundaries.

For a better understanding of complex responses in the Chua's circuit, the analytical prediction of periodic flows will be presented through the mapping structures as presented in the previous section. From a specific mapping structure, a set of nonlinear algebraic equations will be solved through the Newton-Raphson method. Using the local stability and bifurcation analysis, all the possible stable and unstable responses of the Chua's circuit can be obtained. However, the numerical computation can give only one of possible periodic flows. The disappearance of the periodic responses can be determined by the first saddle-node bifurcation and grazing bifurcation. Once the grazing bifurcation occurs, the mapping structures will be changed, which means the old responses cannot exist any more and the new periodic responses may exist.

The analytical prediction gives the switching sets on the boundaries, as shown in Fig. 6.11. The solid and dotted curves give the stable and unstable periodic responses, respectively. Due to symmetry, switching points in analytical prediction are shown only at the boundary of $x = 1$. The parameter regions for the stable periodic responses are the same as in the bifurcation scenario. The corresponding eigenvalue analysis gives the local stability and bifurcation, as presented in Fig. 6.12. However, grazing bifurcations cannot be determined by such eigenvalue analysis. The analytical conditions in Eqs. (6.68) and (6.69) should be adopted. It is clearly observed that the vanishing of periodic responses is caused by the grazing bifurcation. Only one grazing bifurcation occurs at the stable periodic response, and the other grazing

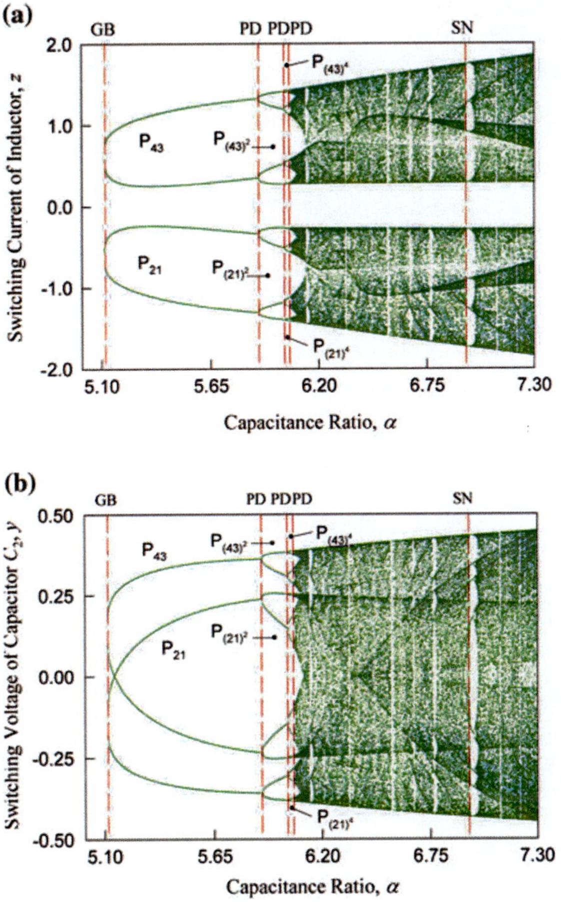

Fig. 6.10 Bifurcation scenario varying with parameter α : **a** switching current in the inductor, **b** switching voltage of the capacitor C_2 and **c** switching current versus switching voltage. ($a = 0.142857$, $b = 0.285714$ and $\beta = 9$)

bifurcations occur at the unstable periodic responses, which cannot be determined through numerical prediction.

To further look into periodic responses embedded in chaotic responses, consider a window of periodic response. The numerical prediction of periodic responses with mapping structure of $P_{(12)^3}$ is given in Fig. 6.13. The analytical prediction of such periodic responses is shown in Fig. 6.14. Two periodic responses with mapping structures of $P_{(12)^3}$ and $P_{(12)^5}$ and their period-doubling responses with $P_{(12)^6}$ and $P_{(12)^{10}}$ are predicted analytically. In Fig. 6.15, the corresponding eigenvalue analysis is given for the local stability and bifurcation. Again, the solid and dotted curves represent the stable and unstable responses, respectively. In Fig. 6.14(a) , the onset of the periodic response of mapping structure $P_{(12)^3}$ is at $\alpha = 8.8784$ owing to the saddle-node-bifurcation. The periodic response of $P_{(12)^3}$ becomes unstable at $\alpha = 8.8894$ and its grazing occurs at $\alpha = 8.8923$. At $\alpha = 8.8894$, the period-doubling bifurcation of the periodic response of $P_{(12)^3}$ occurs and also the saddle-node bifurcation of the

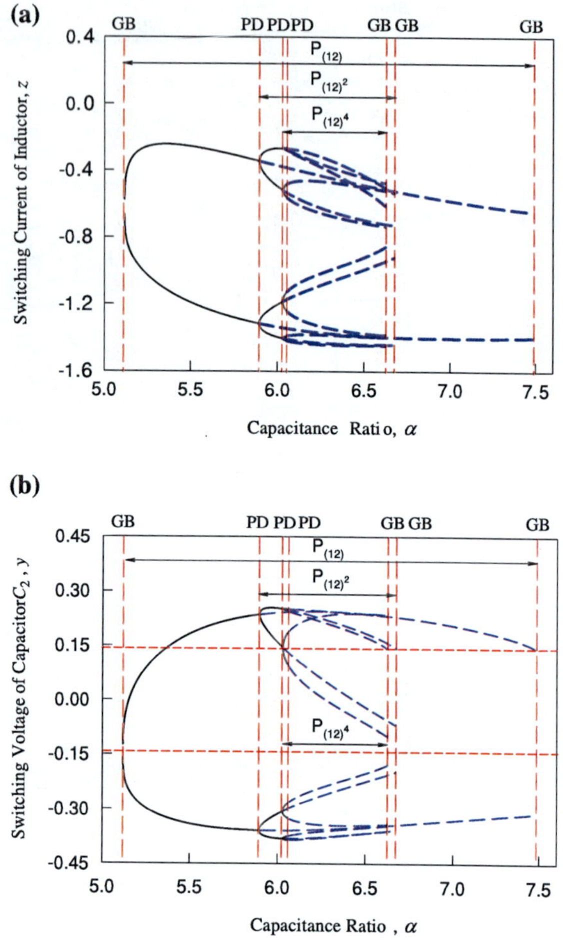

Fig. 6.11 Analytical prediction under different parameter α : **a** switching current in the inductor and **b** switching voltage of the capacitor C_2. ($a = 0.142857, b = 0.285714$ and $\beta = 9$)

periodic response of $P_{(12)^6}$ occurs. The grazing bifurcation for the periodic response of $P_{(12)^6}$ is at $\alpha \approx 8.8898$. Because of the grazing, the new periodic response of $P_{(12)^5}$ appears. With increasing α, the period-doubling bifurcation of such a periodic response is at $\alpha = 8.8948$ and its unstable periodic response grazes at $\alpha = 9.787$. The period-doubling bifurcation of $P_{(12)^5}$ generates the new periodic response of $P_{(12)^{10}}$ for its onset. The corresponding period-doubling bifurcation is at $\alpha = 8.89614$. The grazing bifurcation occurs at the unstable response of $P_{(12)^5}$ with $\alpha = 8.90282$. Without discontinuity, no such grazing bifurcation exist to cause the switching between the periodic responses of $P_{(12)^6}$ and $P_{(12)^5}$. Similarly, the other widows for periodic responses can be carried out.

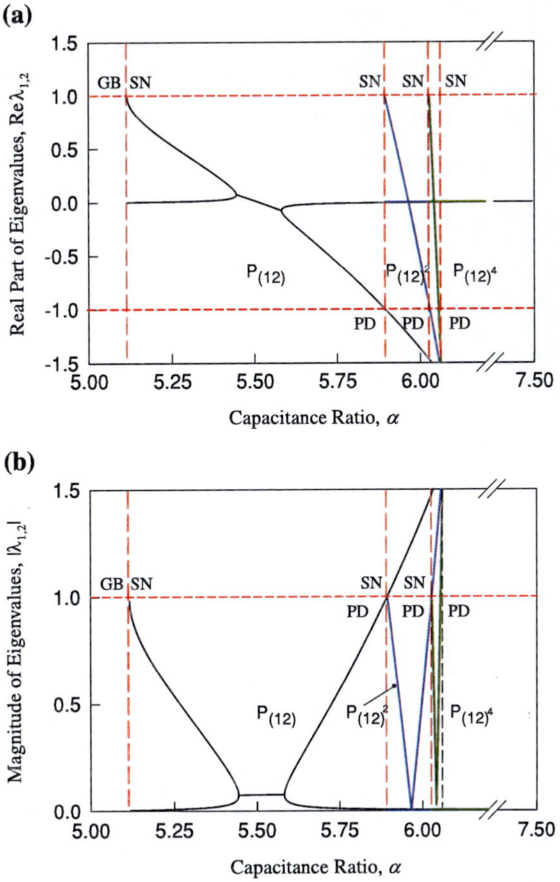

Fig. 6.12 Eigenvalues varying with parameter 2 for periodic responses **a** real part and **b** magnitude (a=0.142857, b=0.285714 and $\alpha = 9$)

6.4.3 Illustrations

From the analytical prediction of periodic responses, the switching sets are obtained. Such switching sets can be used as selected initial conditions, and the periodic responses of the Chua's circuit can be simulated. For example, consider parameters $(\alpha, \beta) = \{(5.86, 9), (5.86, 9), (6.01, 9)$ and $(8.88, 15)\}$ for periodic responses of P_{21}, $P_{(21)^2}$, $P_{21534361}$, and$P_{(21)^3}$, respectively. The initial conditions are given by

$$
\begin{aligned}
&P_{21} : (x_0, y_0, z_0) = (1.0, 0.233577, -0.335582) \text{ for } \alpha = 5.86, \beta = 9,\\
&P_{(21)^2} : (x_0, y_0, z_0) = (1.0, 0.154142, -0.267826) \text{ for } \alpha = 6.01, \beta = 9,\\
&P_{21534361} : (x_0, y_0, z_0) = (1.0, 0.168254, -0.277345) \text{ for } \alpha = 6.34, \beta = 9,\\
&P_{(21)^3} : (x_0, y_0, z_0) = (1.0, 0.179989, -0.083352) \text{ for } \alpha = 8.88, \beta = 15.
\end{aligned}
$$

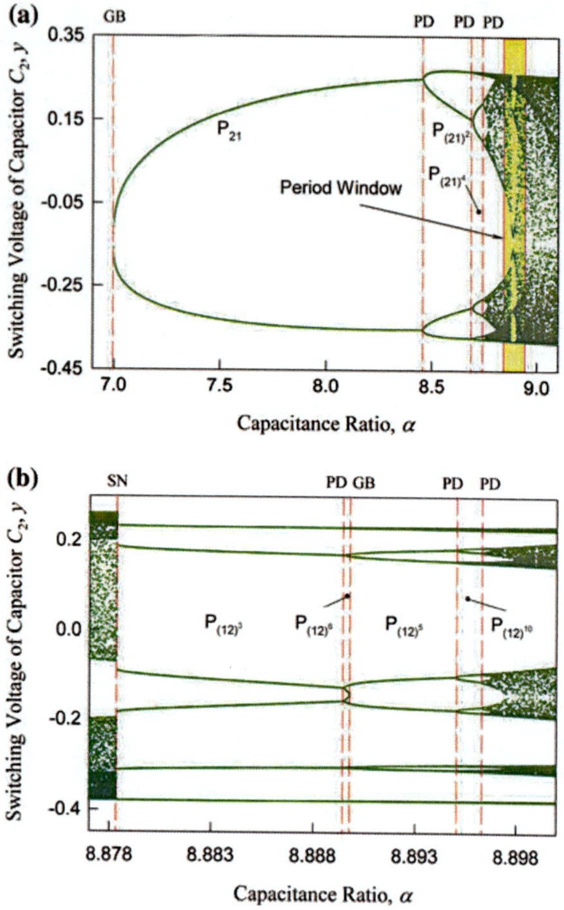

Fig. 6.13 The window of periodic responses of $P_{(12)^3}$: **a** switching voltage of the capacitor C_2 versus parameter α and **b** zoomed area. ($a = 0.142857$, $b = 0.285714$ and $\beta = 15$)

The corresponding trajectories of periodic responses relative to the Chua's circuit in (x, z)-plane are plotted in Fig. 6.16. Dashed lines give the boundaries and the hollow symbols are the switching points for periodic responses. The starting points are given by solid circular symbols. The periodic responses pertaining to mappings P_{21}, $P_{(21)^2}$ and $P_{(21)^3}$ are formed by the local mappings. The three periodic responses are local. However, the periodic response of $P_{21534361}$ connects three domains and such a response is a global response. For a clear illustration of periodic responses, the 3-D view of the periodic responses relative to Fig. 6.16 is given in Fig. 6.17. The starting points are labeled by green circular symbols. The corresponding mappings are labeled. It is clearly observed that how the periodic responses pass through the boundary planes. In other words, the dynamic system will be switched at such boundary planes. If readers are interested in determining the complicated periodic motions in other discontinuous dynamical systems through the mapping structure

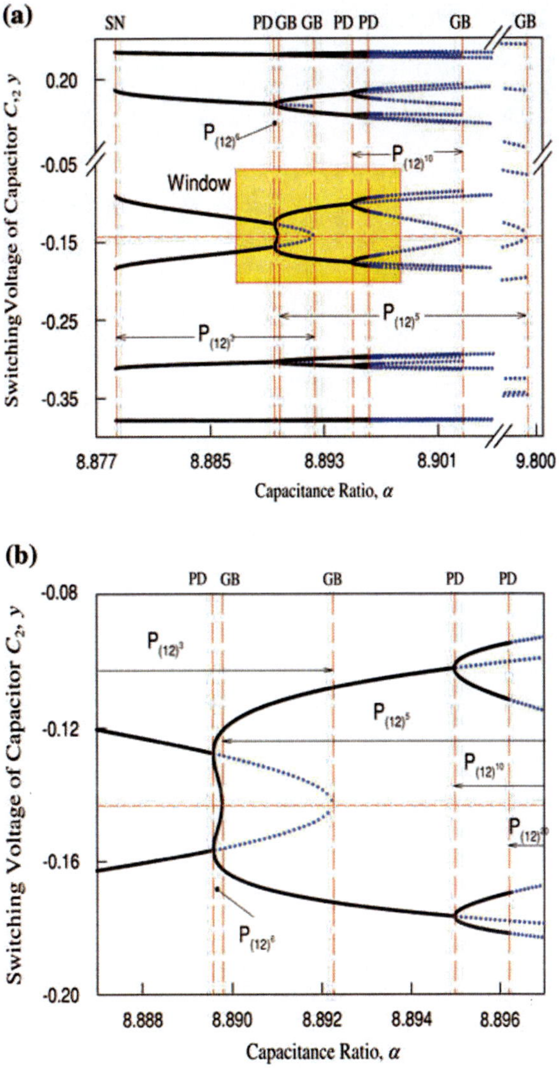

Fig. 6.14 Analytic predictions of periodic responses relative to the $P_{(12)^3}$ window: **a** switching voltage of the capacitor C_2 versus parameter α and **b** zoomed area. ($a = 0.142857$, $b = 0.285714$ and $\beta = 15$)

technique, Luo and coworker's papers (e.g., Han et al 1995; Luo 2002; 2005b; Luo and Chen 2006; Luo and Gegg 2006, 2007; Luo and Zwiegart Jr. 2008; Luo and O'Connor 2009; Luo and Rapp 2009; Luo and Guo 2010) can be referred.

For complicated, discontinuous dynamical systems, the naming system in Sect. 6.1 should be used to identify the possible mappings. The mapping dynamics of periodic motion in any system is to develop the mapping relations from which expected periodic motions can be analytically predicted. The mapping dynamics provides a possibility to obtain all stable and unstable periodic motions in dynamical

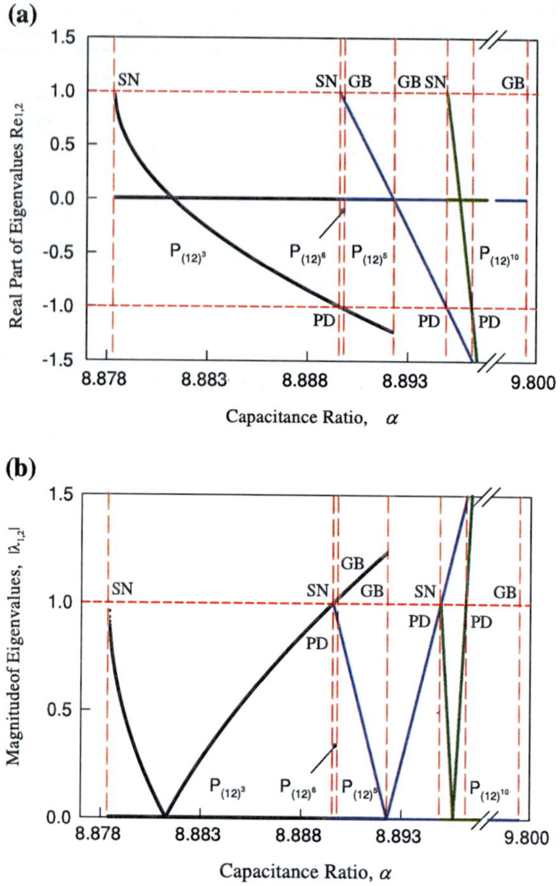

Fig. 6.15 Eigenvalues varying with parameter α for periodic responses in the $P_{(12)^3}$ window: **a** real part and **b** magnitude. ($a = 0.142857$, $b = 0.285714$ and $\beta = 15$)

system rather than only one of stable solutions given by numerical simulations. Based mapping structures, the stochasticity of chaos in discontinuous dynamical systems can be investigated.

6.5 Flow Symmetry

In this section, the symmetry of discontinuous dynamical systems will be discussed. The grazing cluster will be presented as in Luo (2006a). The symmetry of steady-state flows in discontinuous dynamical systems will be discussed in order to determine the co-existing steady-state flows.

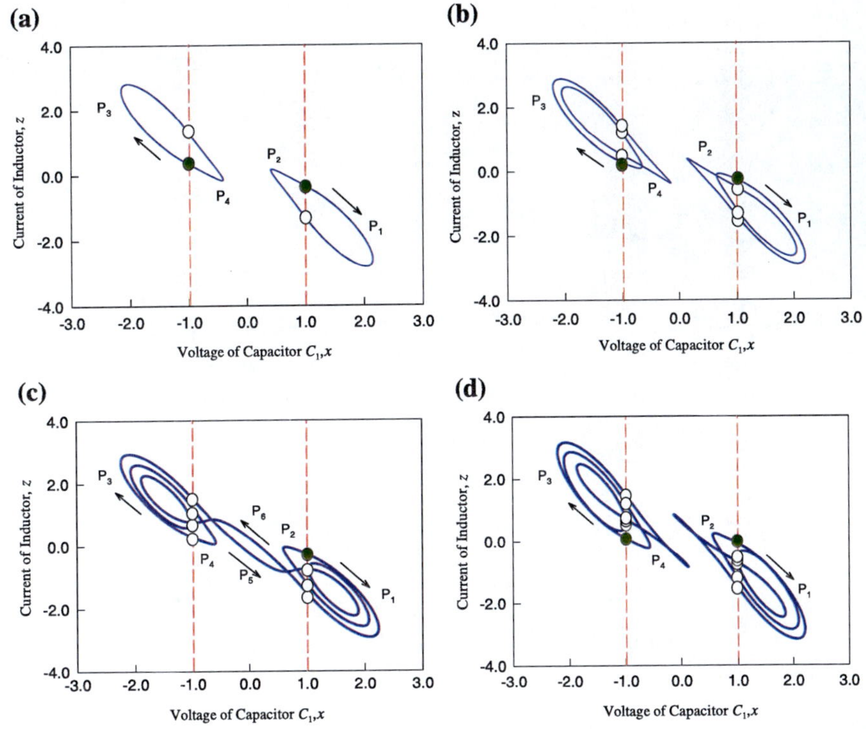

Fig. 6.16 The periodic responses in plane of $(x, z)(a = 0.142857$ and $b = 0.285714)$: **a** $P_{21}(\alpha = 5.86,\ \beta = 9.0, x = 1.0, y = 0.233577, z = -0.335582)$, **b** $P_{(21)^2}(\alpha = 6.01, \beta = 9.0, x = 1.0, y = 0.154142, z = -0.267826)$, **c** $P_{21534361}$ $(\alpha = 6.34, \beta = 9.0, x = 1.0, y = 0.168254, z = -0.277345)$, **d** $P_{(21)^3}$ $(\alpha = 8.88, \beta = 15.0, x = 1.0, y = 0.179989, z = -0.083352)$

6.5.1 Symmetric Discontinuity

As in Sect. 6.1, discontinuous dynamic systems with symmetry can be described. Consider an n-dimensional dynamic system consisting of M -sub-systems on m-accessible, sub-domains Ω_p $(p = 1, 2, \cdots, M)$ in a domain $\mho \subset \mathscr{R}^n$. With the inaccessible domain Ω_0, the universal domain is expressed by $\mho = \cup_{p=1}^{M} \Omega_p \cup \Omega_0$. The M accessible domains and inaccessible domain Ω_0 in phase space are separated by boundary $\partial\Omega_{p_1 p_2} \subset \mathscr{R}^{n-1}$ $(p_1, p_2 \in \{0, 1, 2, \cdots, m\})$, determined by the specific function $\varphi_{p_1 p_2}(\mathbf{x}, t) = 0$. Among all the boundaries, there are some pairs of boundaries $\partial\Omega_{p_1 p_2}$ with symmetry. For the pth accessible domain, there is a continuous system in form of

$$\begin{aligned} \dot{\mathbf{x}} &= \mathbf{F}^{(p)}(\mathbf{x}, t, \mathbf{p}_p) = \mathbf{f}^{(p)}(\mathbf{x}, \boldsymbol{\mu}_p) + \mathbf{g}(\mathbf{x}, t, \boldsymbol{\pi}), \\ \mathbf{x} &= (x_1, x_2, \cdots, x_n)^{\mathrm{T}} \in \Omega_p, \end{aligned} \tag{6.91}$$

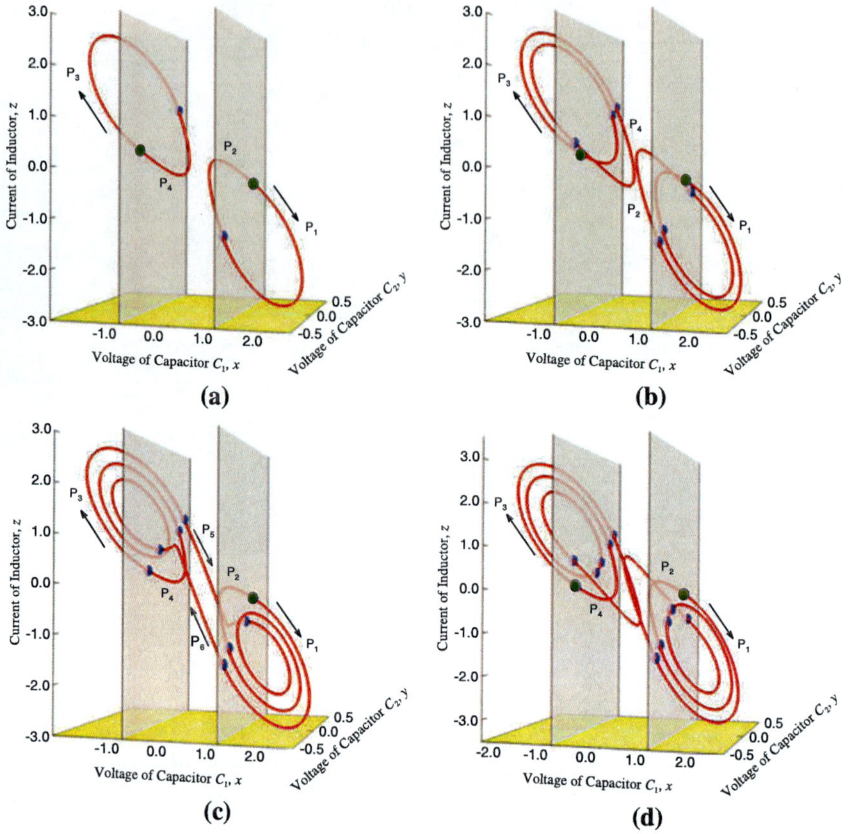

Fig. 6.17 The 3D views of periodic responses $(x, y, z)(a = 0.142857$ and $b = 0.285714)$: **a** $P_{21}(\alpha = 5.86,\ \beta = 9.0, x = 1.0, y = 0.233577, z = -0.335582)$, **b** $P_{(21)^2}(\alpha = 6.01, \beta = 9.0, x = 1.0, y = 0.154142, z = -0.267826)$, **c** $P_{21534361}$ $(\alpha = 6.34, \beta = 9.0, x = 1.0, y = 0.168254, z = -0.277345)$, **d** $P_{(21)^3}$ $(\alpha = 8.88, \beta = 15.0, x = 1.0, y = 0.179989, z = -0.083352)$

where the forcing vector function $\mathbf{g} = (g_1, g_2, \cdots, g_n)^{\mathrm{T}}$ is bounded, periodic functions with period $T = 2\pi/\Omega$. The forcing frequency is Ω. The phase variable is $\varphi = \Omega t$ and a parameter vector $\boldsymbol{\pi} = (\pi_1, \pi_2 \cdots, \pi_{m_1})^{\mathrm{T}}$. The vector function $\mathbf{f}^{(p)} = (f_1^{(p)}, f_2^{(p)}, \cdots f_n^{(p)})^{\mathrm{T}}$ are C^r-continuous $(r \geq 2)$ with system parameter vector $\boldsymbol{\mu}_p = (\mu_1^{(p)}, \mu_2^{(p)}, \cdots, \mu_{m_2}^{(p)})^{\mathrm{T}}$. In all accessible sub-domains Ω_p $(p = 1, 2, \cdots, M)$, the dynamical system in Eq. (6.91) is continuous and there is a continuous flow expressed by

$$\left.\begin{aligned} \mathbf{x}^{(p)}(t) &= \boldsymbol{\Phi}^{(p)}(\mathbf{x}^{(p)}(t_0), t, \boldsymbol{\mu}_p, \boldsymbol{\pi}), \\ \mathbf{x}^{(p)}(t_0) &= \boldsymbol{\Phi}^{(p)}(\mathbf{x}^{(p)}(t_0), t_0, \boldsymbol{\mu}_p, \boldsymbol{\pi}). \end{aligned}\right\} \quad (6.92)$$

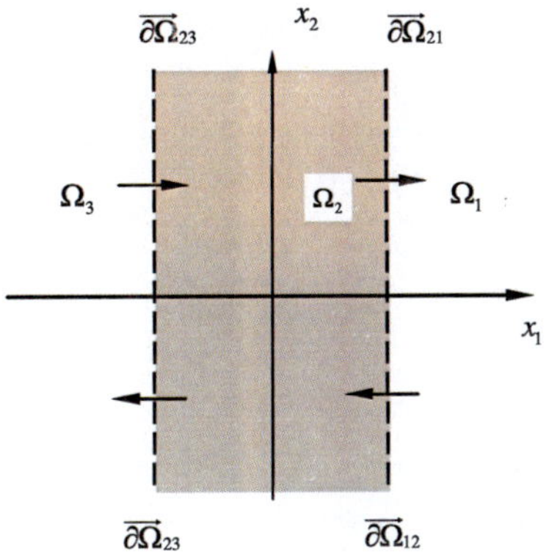

Fig. 6.18 Two symmetric domains in phase plane.

Consider one of the simplest sub-accessible domains in a two-dimensional dynamical system in Fig. 6.18. The universal domain is formed by three accessible sub-domains $\Omega_p (p = 1, 2, 3)$. With oriented directions, two boundaries $\partial\Omega_{p_1 p_2}$ $(p_1, p_2 \in \{1, 2, 3\})$ exist. The accessible sub-domains (Ω_1 and Ω_3) are of the skew-symmetry. The direction-oriented boundaries $(\overrightarrow{\partial\Omega_{21}}, \overrightarrow{\partial\Omega_{12}})$ and $(\overrightarrow{\partial\Omega_{23}}, \overrightarrow{\partial\Omega_{32}})$ are of the skew-symmetry, respectively. On the corresponding symmetric domains, the dynamical systems in Eq. (6.91) are of the skew symmetry. On the boundaries ($\overrightarrow{\partial\Omega_{12}}$ and $\overrightarrow{\partial\Omega_{21}}$), the governing equation is defined through $\varphi_{12}(\mathbf{x}, t) = \varphi_{21}(\mathbf{x}, t) = x_1 - E = 0$, and on the corresponding symmetric boundaries ($\overrightarrow{\partial\Omega_{23}}$ and $\overrightarrow{\partial\Omega_{32}}$), the governing equation is determined by $\varphi_{23}(\mathbf{x}, t) = \varphi_{32}(\mathbf{x}, t) \equiv x_1 + E = 0$. From the above discussion, besides the assumptions in Sect. 6.1, the extra conditions should be added to restrict our discussion. The following assumptions will be considered:
A6.1: The systems in Eq. (6.91) possess time-continuity.
A6.2: For a unbounded domain Ω_p, there is a open domain $D_p \subset \Omega_p$. On the domain D_p, the vector field and the flow are bounded for $t \in [0,\infty)$, i.e.,

$$||\mathbf{f}^{(p)}|| + ||\mathbf{g}|| \leq K_1 \text{ (const) and } ||\mathbf{\Phi}^{(p)}|| \leq K_2\text{(const)}. \tag{6.93}$$

A6.3: For a bounded domain Ω_p, there is a open domain $D_p \subset \Omega_p$. On the domain D_p, the vector field and the flow are bounded for $t \in [0,\infty)$, i.e.,

$$||\mathbf{f}^{(p)}|| + ||\mathbf{g}|| \leq K_1\text{(constant) and } ||\mathbf{\Phi}^{(p)}|| \leq \infty. \tag{6.94}$$

A6.4: The dynamical subsystems possess symmetry at least in two corresponding symmetric domains.
A6.5: The flow on the discontinuous boundaries is C^1-discontinuous or there is a transport law to connect two flows in two different domains.

6.5.2 Switching Sets and Mappings

For convenience, without loss generality, the switching sets are from the boundary $\overrightarrow{\partial\Omega}_{p_1p_2}$ $(q = 0, 1, 2, \cdots, M)$:

$$\Xi_q = \left\{ (\text{mod}(\Omega t_k, 2\pi), \mathbf{x}_k) \middle| \begin{array}{l} \varphi_{p_1p_2}(t_k, \mathbf{x}_k) = 0 \\ p_1, p_2 \in \{0, 1, 2, \cdots, M\} \end{array} \right\} \subseteq \overrightarrow{\partial\Omega}_{p_1p_2}. \tag{6.95}$$

For simplicity in discussion, the following sign function is introduced as

$$\hbar_\alpha = \begin{cases} +1 \text{ for } \mathbf{n}_{\partial\Omega_{ij}} \to \Omega_\beta, \\ -1 \text{ for } \mathbf{n}_{\partial\Omega_{ij}} \to \Omega_\alpha. \end{cases} \tag{6.96}$$

From Luo (2011), the singular sets on the switching sets are stated as follows:

Definition 6.4 For a discontinuous dynamical system of Eq. (6.91), the grazing singular set on the α-side of the boundary $\partial\Omega_{p_1p_2}$ for $\alpha, \beta \in \{p_1, p_2\}$ and $\beta \neq \alpha$ with $p_1, p_2 \in \{1, 2, \cdots, M\}$ is defined as

$$S^{(1,\alpha)}_{p_1p_2} = \left\{ (\text{mod}(\Omega t_m, 2\pi), \mathbf{x}_m) \middle| \begin{array}{l} \hbar_\alpha G^{(\alpha)}_{\partial\Omega_{p_1p_2}}(\mathbf{x}_m, t_{m\pm}) = 0 \text{ and} \\ \hbar_\alpha G^{(1,\alpha)}_{\partial\Omega_{p_1p_2}}(\mathbf{x}_m, t_{m\pm}) < 0 \end{array} \right\} \subset \partial\Omega_{p_1p_2}. \tag{6.97}$$

The double grazing singular set on both sides of the boundary $\partial\Omega_{p_1p_2}$ is defined as

$$\begin{aligned} S^{(1:1)}_{p_1p_2} &= S^{(1,\alpha)}_{p_1p_2} \cup S^{(1,\beta)}_{p_1p_2} \\ &= \left\{ (\text{mod}(\Omega t_m, 2\pi), \mathbf{x}_m) \middle| \begin{array}{l} \hbar_\alpha G^{(\alpha)}_{\partial\Omega_{p_1p_2}}(\mathbf{x}_m, t_{m\pm}) = 0 \text{ and} \\ \hbar_\alpha G^{(1,\alpha)}_{\partial\Omega_{p_1p_2}}(\mathbf{x}_m, t_{m\pm}) < 0, \\ \hbar_\alpha G^{(\beta)}_{\partial\Omega_{p_1p_2}}(\mathbf{x}_m, t_{m\pm}) = 0 \text{ and} \\ \hbar_\alpha G^{(1,\beta)}_{\partial\Omega_{p_1p_2}}(\mathbf{x}_m, t_{m\pm}) > 0 \end{array} \right\} \subset \partial\Omega_{p_1p_2}. \end{aligned} \tag{6.98}$$

Definition 6.5 For a discontinuous dynamical system of Eq. (6.91), the $(2k_\alpha - 1)$th-order, grazing singular set on the α-side of the boundary $\partial\Omega_{p_1p_2}$ for $\alpha, \beta \in \{p_1, p_2\}$ and $\beta \neq \alpha$ with $p_1, p_2 \in \{1, 2, \cdots, M\}$ is defined as

$$S^{(2k_\alpha+1,\alpha)}_{p_1p_2} = \left\{ (\text{mod}(\Omega t_m, 2\pi), \mathbf{x}_m) \middle| \begin{array}{l} \hbar_\alpha G^{(s_\alpha,\alpha)}_{\partial\Omega_{p_1p_2}}(\mathbf{x}_m, t_{m\pm}) = 0 \\ \text{for } s_\alpha = 0, 1, 2, \cdots, 2k_\alpha \\ \text{and } \hbar_\alpha G^{(2k_\alpha+1,\alpha)}_{\partial\Omega_{p_1p_2}}(\mathbf{x}_m, t_{m\pm}) < 0 \end{array} \right\} \subset \partial\Omega_{p_1p_2}. \tag{6.99}$$

The $(2k_\alpha + 1 : 2k_\beta + 1)$-double grazing singular set on both sides of the boundary $\partial\Omega_{p_1p_2}$ is defined as

$$
\begin{aligned}
\mathbf{S}_{p_1p_2}^{(2k_\alpha+1:2k_\beta+1)} &= \mathbf{S}_{p_1p_2}^{(2k_\alpha+1,\alpha)} \cup \mathbf{S}_{p_1p_2}^{(2k_\beta+1,\beta)} \\
&= \left\{ (\mathrm{mod}(\Omega t_m, 2\pi), \mathbf{x}_m) \left| \begin{array}{l} \hbar_\alpha G_{\partial\Omega_{p_1p_2}}^{(s_\beta,\alpha)}(\mathbf{x}_m, t_{m\pm}) = 0 \\ \text{for } s_\alpha = 0, 1, \cdots, 2k_\alpha \\ \text{and } \hbar_\alpha G_{\partial\Omega_{p_1p_2}}^{(2k_\alpha+1,\alpha)}(\mathbf{x}_m, t_{m\pm}) < 0; \\ \hbar_\alpha G_{\partial\Omega_{p_1p_2}}^{(s_\beta,\beta)}(\mathbf{x}_m, t_{m\pm}) = 0 \\ \text{for } s_\beta = 0, 1, \cdots, 2k_\beta \\ \text{and } \hbar_\alpha G_{\partial\Omega_{p_1p_2}}^{(2k_\beta+1,\beta)}(\mathbf{x}_m, t_{m\pm}) > 0 \end{array} \right. \right\} \\
&\subset \partial\Omega_{p_1p_2}.
\end{aligned} \tag{6.100}
$$

A switching set on the discontinuous boundary $\partial\Omega_{p_1p_2}$ is composed of two neighbored boundaries $\Xi_{q_1} \subseteq \overrightarrow{\partial\Omega}_{p_1p_2}$ and $\Xi_{q_2} \subseteq \overleftarrow{\partial\Omega}_{p_1p_2}$ connected by a singular point $\Gamma_{p_1p_2} = \mathbf{S}_{p_1p_2}^{(2k_\alpha+1,\alpha)}$ $(k_\alpha = 0, 1, 2, \cdots)$, i.e.,

$$
\Xi_{q_1q_2} = \Xi_{q_1} \cup \Xi_{q_2} \cup \Gamma_{p_1p_2} \subseteq \partial\Omega_{p_1p_2} \tag{6.101}
$$

for $q_1, q_2 \in \{0, 1, 2, \cdots, M\}$. The local mappings near the switching sets $\Xi_{q_1q_2}$ is defined for specific $J_1, J_2 \in \{1, 2, \cdots, N\}$ as

$$
P_{J_1} : \Xi_{q_1} \to \Xi_{q_2}, P_{J_2} : \Xi_{q_2} \to \Xi_{q_1}. \tag{6.102}
$$

If $\Xi_{q_1} \subseteq \partial\Omega_{p_1p_2}$ and $\Xi_{q_3} \not\subset \partial\Omega_{p_1p_2}$, the global mapping in an accessible domain Ω_α $(\alpha \in \{p_1, p_2\})$ is defined for specific $J_3 \in \{1, 2, \cdots, N\}$ as

$$
P_{J_3} : \Xi_{q_1} \to \Xi_{q_3}. \tag{6.103}
$$

On the switching sets $\Xi_{q_1} \subseteq \partial\Omega_{p_1p_2}$, the sliding mapping is defined for specific $J_4 \in \{1, 2, \cdots, N\}$ as

$$
P_{J_4} : \Xi_{q_1} \to \Xi_{q_1}. \tag{6.104}
$$

For a C^0-discontinuous switching set $\Xi_{q_4} \subseteq \partial\Omega_{p_1p_2}$, there must be a transport law to map it to another switching set, $\Xi_{q_5} \subset \partial\Omega_{p_3p_4}$. Therefore, the transport mapping is defined for specific $J_5 \in \{1, 2, \cdots, N\}$ as

$$
P_{J_5} : \Xi_{q_4} \to \Xi_{q_5}. \tag{6.105}
$$

The switching sets, the local, global, sliding and transport mappings are depicted in Fig. 6.19. The forbidden zone is shown for permanently-non-passable boundary. Consider a discontinuous dynamical system with domains given in Fig. 6.20 to show how to use the above concept. Four switching sets are defined as

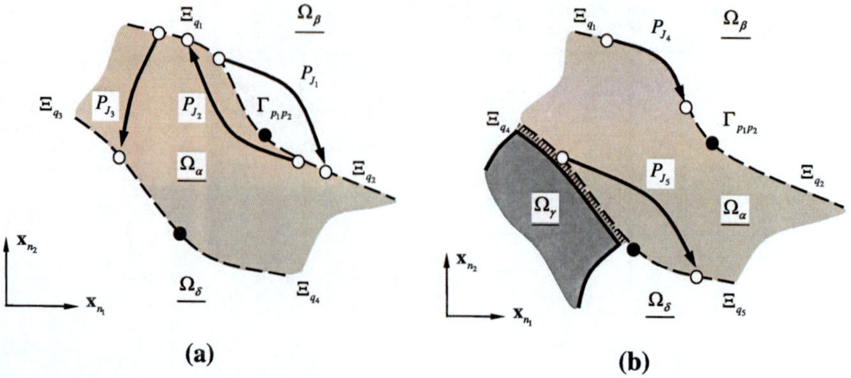

Fig. 6.19 Switching sets and generic mappings: **a** local and global mappings, **b** sliding mapping and transport mapping. The shaded domain Ω_α is accessible, and the inaccessible domain is Ω_γ. The domain Ω_β is accessible and Ω_δ can be accessible or inaccessible

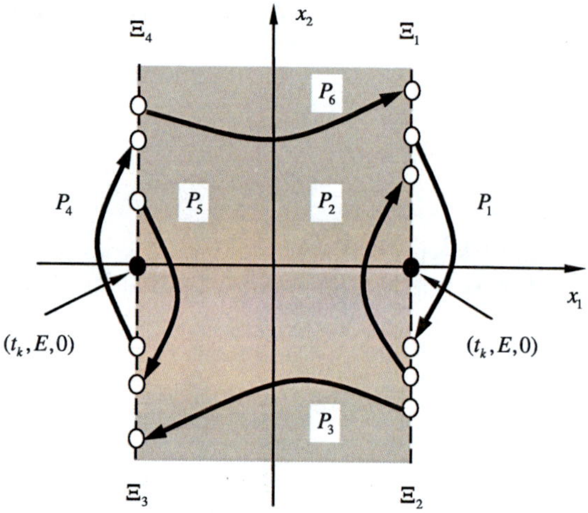

Fig. 6.20 Switching sets and generic mappings for $x_1 = \pm E$.

$$
\begin{aligned}
\Xi_1 &= \{(\mathrm{mod}(\Omega t_m, 2\pi), \mathbf{x}_m) | x_2 > 0, \varphi_{21}(\mathbf{x}_m, t) = x_1 - E = 0\} \subseteq \overrightarrow{\partial\Omega}_{21},\\
\Xi_2 &= \{(\mathrm{mod}(\Omega t_m, 2\pi), \mathbf{x}_m) | x_2 < 0, \varphi_{12}(\mathbf{x}_m, t) = x_1 - E = 0\} \subseteq \overrightarrow{\partial\Omega}_{12},\\
\Xi_3 &= \{(\mathrm{mod}(\Omega t_m, 2\pi), \mathbf{x}_m) | x_2 < 0, \varphi_{23}(\mathbf{x}_m, t) = x_1 + E = 0\} \subseteq \overrightarrow{\partial\Omega}_{23},\\
\Xi_4 &= \{(\mathrm{mod}(\Omega t_m, 2\pi), \mathbf{x}_m) | x_2 > 0, \varphi_{32}(\mathbf{x}_m, t) = x_1 + E = 0\} \subseteq \overrightarrow{\partial\Omega}_{32}
\end{aligned}
\tag{6.106}
$$

and two singular points are

$$
\left.
\begin{aligned}
\Gamma_{\overrightarrow{12}} &= \{(\mathrm{mod}(\Omega t_m, 2\pi), \mathbf{x}_m) | x_2 = 0,\ \varphi_{12} = x_1 - E = 0\} \subset \Gamma_{12}\\
\Gamma_{\overrightarrow{23}} &= \{(\mathrm{mod}(\Omega t_m, 2\pi), \mathbf{x}_m) | x_2 = 0,\ \varphi_{23} = x_1 + E = 0\} \subset \Gamma_{23}
\end{aligned}
\right\}
\tag{6.107}
$$

The two switching sets are

$$\begin{aligned}\Xi_{12} &= \Xi_1 \cup \Xi_2 \cup \Gamma_{12} \subseteq \partial\Omega_{12},\\ \Xi_{34} &= \Xi_3 \cup \Xi_4 \cup \Gamma_{23} \subseteq \partial\Omega_{23}.\end{aligned} \tag{6.108}$$

From four subsets, the local mappings are

$$\begin{aligned}&P_1 : \Xi_1 \to \Xi_2 \text{ and } P_2 : \Xi_1 \to \Xi_2, \text{ on } \partial\Omega_{12},\\ &P_4 : \Xi_3 \to \Xi_4 \text{ and } P_5 : \Xi_4 \to \Xi_3, \text{ on } \partial\Omega_{23}\end{aligned} \tag{6.109}$$

and the global mappings from $\partial\Omega_{12}$ and $\partial\Omega_{23}$ are

$$P_3 : \Xi_2 \to \Xi_3 \text{ and } P_6 : \Xi_4 \to \Xi_1. \tag{6.110}$$

In this discontinuous system, no sliding mapping exists on this C^1-discontinuous boundaries. From the definition of mappings, the mappings P_J $(J = 1, 2, 4, 5)$ relative to one switching section are termed the *local* mapping, and the mappings P_J $(J = 3, 6)$ relative to two switching sections are termed the *global* mapping. The global mapping maps the motion from one switching boundary into another switching boundary. The local mapping is the self-mapping in the corresponding switching section. The six generic mappings are illustrated in Fig. 6.20.

6.5.3 Grazing and Mappings Symmetry

The initial and final times (t_k and t_{k+1}) are used for all the mappings P_J $(J = 1, 2, \cdots, N)$ defined through in Eqs. (6.102)–(8.105), and the corresponding phases are $\varphi_k = \Omega t_k$ and $\varphi_{k+1} = \Omega t_{k+1}$. Equation (6.92) gives

$$\left.\begin{aligned}\mathbf{x}^{(p)}(t_{k+1}) &= \boldsymbol{\Phi}^{(p)}(t_{k+1}, \mathbf{x}^{(p)}(t_k), t_k, \boldsymbol{\mu}_p, \boldsymbol{\pi}), \text{ or}\\ \mathbf{x}^{(p)}(\varphi_{k+1}) &= \boldsymbol{\Phi}^{(p)}(\varphi_{k+1}, \mathbf{x}^{(p)}(\varphi_k), \varphi_k, \boldsymbol{\mu}_p, \boldsymbol{\pi}).\end{aligned}\right\} \tag{6.111}$$

In an accessible domain Ω_p, the foregoing equations with boundary constraint equations give the governing equations for mapping P_J. Consider a notation $\mathbf{y}_k \equiv (\varphi_k, \mathbf{x}_k)^{\mathrm{T}} \in \mathscr{R}^{n+1}$, the governing equations for mapping P_J $(J = 1, 2, \cdots, N)$ with mapping relation $\mathbf{y}_{i+1} = P_J \mathbf{y}_i$ are

$$\mathbf{F}^{(J)}(\varphi_k, \mathbf{x}_k, \varphi_{k+1}, \mathbf{x}_{k+1}, \boldsymbol{\mu}_p, \boldsymbol{\pi}) = \mathbf{0} \tag{6.112}$$

where $\mathbf{F}^{(J)} \in \mathscr{R}^{n+1}$.

For all mappings P_J $(J = 1, 2, \cdots, N)$, there are $2n_1$-symmetric mappings P_q $(q = 1, 2, \cdots, 2n_1)$. The first n_1-mappings P_{q_1} $(q_1 = 1, 2, \cdots, n_1)$ are symmetric with the second n_1-mappings P_{q_2} $(q_2 = n_1 + 1, n_1 + 2, \cdots, 2n_1)$, respectively. From Assumption (A6.4), the subsystems in domain $\Omega_{p_{q_1}}$ and $\Omega_{p_{q_2}} (p_{q_1}, p_{q_2} \in$

$\{1, 2, \cdots, m\}$) are symmetric. Thus, $\mu_{p_{q_1}} = \mu_{p_{q_2}}$. For convenience, the following notation is introduced as

$$\left.\begin{aligned}\hat{q} &= \operatorname{mod}(q + n_1 - 1, 2n_1) + 1,\\ \hat{\varphi}_k^{(q)} &= \operatorname{mod}(\varphi_k^{(q)}, 2(2M_1+1)\pi),\\ \hat{\varphi}_k^{(\hat{q})} &= \operatorname{mod}(\varphi_k^{(\hat{q})}, 2(2M_1+1)\pi).\end{aligned}\right\}, \tag{6.113}$$

Notice that the integer $M_1 \equiv 0, 1, 2, \cdots$ and $\operatorname{mod}(\cdot, \cdot)$ is the modulus function.

Definition 6.6 For a discontinuous dynamical system in Eq. (6.91), under a transformation $T_P : P_q \to P_{\hat{q}}$ during $(2M_1+1)$-periods with

$$\left.\begin{aligned}\hat{\varphi}_k^{(q)} &= \operatorname{mod}((2M_1+1)\pi + \hat{\varphi}_k^{(\hat{q})}, 2(2M_1+1)\pi),\\ \mathbf{x}_k^{(\hat{q})} &= -\mathbf{x}_k^{(q)}\end{aligned}\right\} \tag{6.114}$$

if a relation

$$\mathbf{F}^{(\hat{q})}(\varphi_k^{(\hat{q})}, \mathbf{x}_k^{(\hat{q})}, \varphi_{k+1}^{(\hat{q})}, \mathbf{x}_{k+1}^{(\hat{q})}, \boldsymbol{\mu}_{p_q}, \boldsymbol{\pi}) = -\mathbf{F}^{(q)}(\varphi_k^{(q)}, \mathbf{x}_k^{(q)}, \varphi_{k+1}^{(q)}, \mathbf{x}_{k+1}^{(q)}, \boldsymbol{\mu}_{p_q}, \boldsymbol{\pi}) \tag{6.115}$$

holds, then the mapping pair $(P_q, P_{\hat{q}})$ is of skew-symmetry. If a mapping pair is relative to the *local* (or *global, or sliding or transport*) mapping, such a mapping pair is termed the *local* (or *global, or sliding or transport*) skew-symmetric mapping pair.

Theorem 6.1. *The $2n_1$-mappings P_q ($q = 1, 2, \cdots, 2n_1$) for the dynamical system in Eq. (6.91) are invariant under the two actions of a transformation T_P, i.e., $T_P \circ T_P : P_q \to P_q$.*

Proof Under the transformation T_P, the mapping P_q and $P_{\hat{q}}$ are of skew-symmetry, thus, we have

$$T_P : P_q \to P_{\hat{q}} \text{ and } T_P : P_{\hat{q}} \to P_{\hat{\hat{q}}}.$$

From the foregoing relations, for $\varphi_i^{(q)} \in [0, (2M_1+1)\pi]$, Definition 6.6 gives for a certain given number$N_2 \in \{0, 1, 2, \cdots\}$

$$\left.\begin{aligned}&\varphi_k^{(\hat{q})} = \varphi_k^{(q)} + (2N_2+1)(2M_1+1)\pi, \text{ and } \mathbf{x}_k^{(\hat{q})} = -\mathbf{x}_k^{(q)};\\ &\mathbf{F}^{(\hat{q})}(\varphi_k^{(\hat{q})}, \mathbf{x}_k^{(\hat{q})}, \varphi_{k+1}^{(\hat{q})}, \mathbf{x}_{k+1}^{(\hat{q})}, \boldsymbol{\mu}_{p_q}, \boldsymbol{\pi}) = -\mathbf{F}^{(q)}(\varphi_k^{(q)}, \mathbf{x}_k^{(q)}, \varphi_{k+1}^{(q)}, \mathbf{x}_{k+1}^{(q)}, \boldsymbol{\mu}_{p_q}, \boldsymbol{\pi})\end{aligned}\right\}$$

and

$$\left.\begin{aligned}&\varphi_k^{(\hat{\hat{q}})} = \varphi_k^{(\hat{q})} + (2N_2+1)(2M_1+1)\pi, \text{ and } \mathbf{x}_k^{(\hat{\hat{q}})} = -\mathbf{x}_k^{(\hat{q})};\\ &\mathbf{F}^{(\hat{\hat{q}})}(\varphi_k^{(\hat{\hat{q}})}, \mathbf{x}_k^{(\hat{\hat{q}})}, \varphi_{k+1}^{(\hat{\hat{q}})}, \mathbf{x}_{k+1}^{(\hat{\hat{q}})}, \boldsymbol{\mu}_{p_q}, \boldsymbol{\pi}) = -\mathbf{F}^{(\hat{q})}(\varphi_k^{(\hat{q})}, \mathbf{x}_k^{(\hat{q})}, \varphi_{k+1}^{(\hat{q})}, \mathbf{x}_{k+1}^{(\hat{q})}, \boldsymbol{\mu}_{p_q}, \boldsymbol{\pi}).\end{aligned}\right\}$$

Substitution of the first equation into the second one of the foregoing two equations leads to

$$\left.\begin{array}{l}\varphi_k^{(\hat{\hat{q}})} = \varphi_k^{(q)} + 2(2N_2+1)(2M_1+1)\pi, \text{ and } \mathbf{x}_k^{(\hat{\hat{q}})} = \mathbf{x}_k^{(q)}; \\ \mathbf{F}^{(\hat{\hat{q}})}(\varphi_k^{(\hat{\hat{q}})}, \mathbf{x}_k^{(\hat{\hat{q}})}, \varphi_{k+1}^{(\hat{\hat{q}})}, \mathbf{x}_{k+1}^{(\hat{\hat{q}})}, \boldsymbol{\mu}_{p_q}, \boldsymbol{\pi}) = \mathbf{F}^{(q)}(\varphi_k^{(q)}, \mathbf{x}_k^{(q)}, \varphi_{k+1}^{(q)}, \mathbf{x}_{k+1}^{(q)}, \boldsymbol{\mu}_{p_q}, \boldsymbol{\pi}).\end{array}\right\}$$

It is easily proved that $\hat{\hat{q}} = q$ from the property of the modulus function. Since the period for mapping P_q is $(2M_1+1)$-periods, the phase satisfies a relation as $\mathrm{mod}(\varphi_k^{(q)}, 2(2M_1+1)\pi) = \varphi_k^{(q)}$. In a similar fashion, the case can be proved for $\varphi_k^{(q)} \in [(2M_1+1)\pi, 2(2M_1+1)\pi]$. Thus, the foregoing equation indicates that $T_P \circ T_P : P_q \to P_q$ holds. Namely, the mapping P_q $(q = 1, 2, \cdots, 2n_1)$ for dynamical systems in Eq. (6.91) is invariant under the two actions of transformation T_P.

■

Consider a post-grazing mapping cluster of a specific local or global mapping P_J $(J \in \{1, 2, \cdots, N\})$. For a local mapping P_{J_1} on the boundary $\partial\Omega_{p_1 p_2}$, the pre-grazing, grazing and post-grazing flows are illustrated in Fig. 6.21a–d. There are many clusters of post-grazing mappings, which determines the property of the post-grazing. Two clusters of the post-grazing mappings for mapping P_{J_1} are sketched. After grazing, the relation between the pre-grazing and post-grazing is

$$P_{J_1} \underset{\text{pre-grazing}}{\overset{\text{post-grazing}}{\rightleftharpoons}} \underbrace{P_{J_1} \circ P_{J_{n_2}} \circ \cdots \circ P_{J_3} \circ P_{J_2}}_{\text{local mapping cluster}} = P_{J_1(J_{n_2} \cdots J_3 J_2)} \tag{6.116}$$

for $J_1, J_2, \cdots, J_{n_2} \in \{1, 2, \cdots, N\}$. The index J_2 can be J_1 but the index J_i $(J_i = J_3, J_4, \cdots, J_{n_2})$ should not be J_1. $P_{J_2} \neq P_{J_1}$ can be any mappings on the *same* boundaries. Similarly, consider the global mapping P_{J_1} to map the flow on the boundary $\partial\Omega_{p_1 p_2}$ to another boundary $\partial\Omega_{p_2 p_3}$ in domain Ω_{p_2}. The pre-grazing, grazing and post-grazing flows for the global mapping P_{J_1} are illustrated in Fig. 6.22a–d. The relation between the pre-grazing and post-grazing is given by

$$P_{J_1} \underset{\text{pre-grazing}}{\overset{\text{post-grazing}}{\rightleftharpoons}} \underbrace{P_{J_{n_2}} \circ \cdots \circ P_{J_3}}_{\text{grazing mapping cluster}} \circ P_{J_2} = P_{(J_{n_2} \cdots J_3) J_2} \tag{6.117}$$

for $(J_1, J_2, \cdots, J_{n_2} \in \{1, 2, \cdots, N\})$. The index J_2 can be J_1 but the index J_i $(J_i = J_3, J_4, \cdots, J_{n_2})$ should not be J_1. In post-grazing mapping clusters, the mappings can be local mappings, sliding and transport mappings on the boundary $\partial\Omega_{p_2 p_3}$. Two clusters of post-mapping structure for such a mapping grazing are sketched in Fig. 6.22c and d. For the grazing occurrence of the mapping P_{J_1} on the boundary $\partial\Omega_{p_1 p_2}$, the post-grazing mapping structure is the same as in Eq. (6.116). However, the mapping P_{J_1} is global and the index J_2 cannot be J_1 because P_{J_2} is any mapping rather than P_{J_1} or another global mapping. This mapping can be a local mapping. Without the mapping P_{J_1}, this post-grazing mapping structure is a local

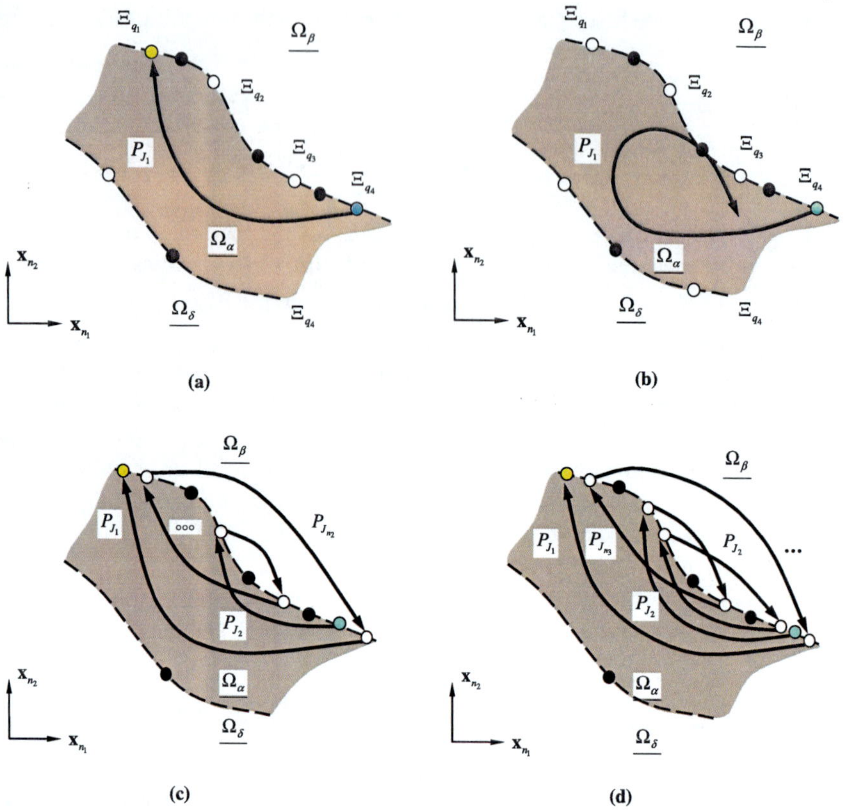

Fig. 6.21 Local mapping grazing switching P_{J_1} $(J_1 \in \{1, 2, \cdots, N\})$: **a** pre-grazing mapping, **b** grazing mapping, **c**, **d** two possible post-grazing mappings. The black solid circular symbols are singular points. The rest circular points are switching points

mapping grazing structure as in Eq. (6.116). This concept is extended to the more generalized case. The post-mapping cluster can include any possible mappings rather than the local mappings.

Consider the generic mappings in Fig. 6.20. Once the grazing occurs, the flow of P_J $(J = 1, 2, \cdots, 6)$ switches from an old flow to a new one, and the corresponding post-grazing mapping structures are from Luo (2005b)

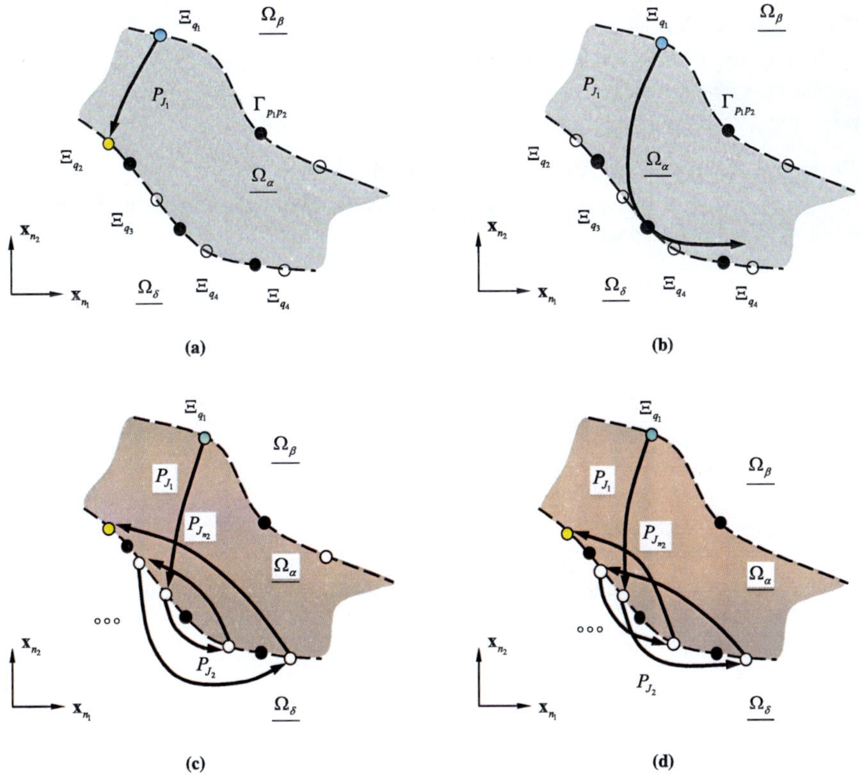

Fig. 6.22 Global mapping grazing switching: **a** pre-grazing mapping, **b** grazing mapping, **c, d** two possible post grazing mappings. The black solid circular symbols are singular points. The rest circular points are switching points

$$\left.\begin{aligned}
&P_J \underset{grazing}{\overset{grazing}{\rightleftharpoons}} P_J \circ P_{J+1} \circ P_J \text{ for } (J = 1, 4);\\
&P_2 \underset{grazing}{\overset{grazing}{\rightleftharpoons}} P_6 \circ P_1 \circ P_3,\ P_2 \underset{grazing}{\overset{grazing}{\rightleftharpoons}} P_2 \circ P_1 \circ P_2;\\
&P_5 \underset{grazing}{\overset{grazing}{\rightleftharpoons}} P_3 \circ P_1 \circ P_6,\ P_5 \underset{grazing}{\overset{grazing}{\rightleftharpoons}} P_5 \circ P_4 \circ P_5;\\
&P_J \underset{grazing}{\overset{grazing}{\rightleftharpoons}} P_J \circ P_{J-2} \circ P_{J-1} \text{ for } (J = 3, 6),\\
&P_J \underset{grazing}{\overset{grazing}{\rightleftharpoons}} P_J \circ P_{\text{mod }(J+1,6)} \circ P_{\text{mod }(J+2,6)} \text{for} (J = 3, 6).
\end{aligned}\right\} \tag{6.118}$$

From the above discussion, the invariance of the post-grazing under the transformation T_P is of great interest.

Theorem 6.2. *For symmetric mappings P_q ($q \in \{1, 2, \cdots, 2n_1\}$, $n_1 \le N/2$) from all N-mappings of the dynamical system in Eq.* (6.91), if the mapping pair $(P_q, P_{\hat{q}})$ *is of skew-symmetry with a symmetric transformation T_P, the post-grazing mapping pair is still of skew-symmetry with the same transformation.*

Proof For the dynamical system in Eq. (6.91), there are n_1 skew-symmetric mapping pairs $(P_q, P_{\hat{q}})$ ($q \in \{1, 2, \cdots, 2n_1\}$) from all the N-mappings. Thus, from Definition 6.6, the skew-symmetry conditions are

$$\varphi_k^{(q_j)} = \varphi_k^{(\hat{q}_j)} + (2M_1 + 1)\pi \text{ and } \mathbf{x}_k^{(q_j)} = -\mathbf{x}_k^{(\hat{q}_j)}$$

for $q_j = \{q_1, q_2 \cdots, q_{n_2}\} \in \{1, 2, \cdots, 2n_1\}$, and

$$\begin{aligned}&\mathbf{F}^{(q_j)}(\varphi_k^{(q_j)}, \mathbf{x}_k^{(q_j)}, \varphi_{k+1}^{(q_j)}, \mathbf{x}_{k+1}^{(q_j)}, \boldsymbol{\mu}_{p_{q_j}}, \boldsymbol{\pi})\\ &= -\mathbf{F}^{(\hat{q}_j)}(\varphi_k^{(\hat{q}_j)}, \mathbf{x}_k^{(\hat{q}_j)}, \varphi_{k+1}^{(\hat{q}_j)}, \mathbf{x}_{k+1}^{(\hat{q}_j)}, \boldsymbol{\mu}_{p_{q_j}}, \boldsymbol{\pi}).\end{aligned}$$

Consider the post-grazing mapping structure of the skew-symmetry mapping pair $(P_{q_1}, P_{\hat{q}_1})$. For a local mapping $P_{\hat{q}_1}$, the post-grazing mapping structure is

$$P_{\hat{q}_1} \underset{\text{pre-grazing}}{\overset{\text{post-grazing}}{\rightleftharpoons}} P_{\hat{q}_1} \circ P_{\hat{q}_{n_2}} \circ \cdots \circ P_{\hat{q}_3} \circ P_{\hat{q}_2}.$$

The corresponding governing equations are

$$\left.\begin{aligned}&\mathbf{F}^{(\hat{q}_j)}(\varphi_{k+j-2}^{(\hat{q}_j)}, \mathbf{x}_{k+j-2}^{(\hat{q}_j)}, \varphi_{k+j-1}^{(\hat{q}_j)}, \mathbf{x}_{k+j-1}^{(\hat{q}_j)}, \boldsymbol{\mu}_{p_{q_j}}, \boldsymbol{\pi}) = \mathbf{0},\\ &\mathbf{F}^{(\hat{q}_1)}(\varphi_{k+n_2-1}^{(\hat{q}_1)}, \mathbf{x}_{k+n_2-1}^{(\hat{q}_1)}, \varphi_{k+n_2}^{(\hat{q}_1)}, \mathbf{x}_{k+n_2}^{(\hat{q}_1)}, \boldsymbol{\mu}_{p_{q_1}}, \boldsymbol{\pi}) = \mathbf{0}.\end{aligned}\right\}$$

for $j = 2, 3, \cdots, n_2$. With Assumption (A6.5), the switching points satisfy

$$\mathbf{x}_{k+j-1}^{(\hat{q}_j)} = \mathbf{x}_{k+j-1}^{(\hat{q}_{j+1})} \text{ and } \mathbf{x}_{k+n_2-1}^{(\hat{q}_{n_2})} = \mathbf{x}_{k+n_2-1}^{(\hat{q}_1)}$$

for $j = 2, 3, \cdots, n_2-1$. From Assumption (A6.1), the system in Eq. (6.91) possesses time-continuity. Therefore the switching phase for $j = 2, 3, \cdots, n_2 - 1$ should be continuous, i.e.,

$$\varphi_{k+j-1}^{(\hat{q}_j)} = \varphi_{k+j-1}^{(\hat{q}_{j+1})} \text{ and } \varphi_{k+n_2-1}^{(\hat{q}_{n_2})} = \varphi_{k+n_2-1}^{(\hat{q}_1)}.$$

Multiplication of negative one (-1) on both sides of the governing equations and the switching point state variable vectors, and simplification with the skew-symmetry conditions gives

$$\left.\begin{aligned}&\mathbf{F}^{(q_j)}(\varphi_{k+j-2}^{(q_j)}, \mathbf{x}_{k+j-2}^{(q_j)}, \varphi_{k+j-1}^{(q_j)}, \mathbf{x}_{k+j-1}^{(q_j)}, \boldsymbol{\mu}_{p_{q_j}}, \boldsymbol{\pi}) = \mathbf{0},\\ &\mathbf{F}^{(q_1)}(\varphi_{k+n_2-1}^{(q_1)}, \mathbf{x}_{k+n_2-1}^{(q_1)}, \varphi_{k+n_2}^{(q_1)}, \mathbf{x}_{k+n_2}^{(q_1)}, \boldsymbol{\mu}_{p_{q_1}}, \boldsymbol{\pi}) = \mathbf{0}\end{aligned}\right\}$$

for $j = 2, 3, \cdots, n_2$. The switching points and switching phases satisfy the following relations ($j = 2, 3, \cdots, n_2$)

$$\left.\begin{aligned} \mathbf{x}_{k+j-1}^{(q_j)} &= \mathbf{x}_{k+j-1}^{(q_{j+1})} \text{ and } \mathbf{x}_{k+n_2-1}^{(q_{n_2})} = \mathbf{x}_{k+n_2-1}^{(q_1)};\\ \varphi_{k+j-1}^{(q_j)} &= \varphi_{k+j-1}^{(q_{j+1})} \text{ and } \varphi_{k+n_2-1}^{(q_{n_2})} = \varphi_{k+n_2-1}^{(q_1)}. \end{aligned}\right\}$$

The foregoing equations form the post-grazing mapping structure of P_{q_1} is

$$P_{q_1} \underset{\text{pre-grazing}}{\overset{\text{post-grazing}}{\rightleftharpoons}} P_{q_1} \circ P_{q_{n_2}} \circ \cdots \circ P_{q_3} \circ P_{q_2}.$$

Thus, the post-grazing mapping structure $(P_{q_1 q_{n_2} \cdots q_3 q_2}, P_{\hat{q}_1 \hat{q}_{n_2} \cdots \hat{q}_3 \hat{q}_2})$ of the skew-symmetric mapping pair $(P_{q_1}, P_{\hat{q}_1})$ is skew-symmetric. Because the local mapping P_{q_1} can be all the local mapping from mapping P_q ($q \in \{1, 2, \cdots, 2n_1\}$), the post-grazing, skew-symmetry for the *local* skew-symmetric mapping pair is proved under the symmetric transformation T_P. Similarly, the post-grazing, skew-symmetry of the global skew-symmetric mapping pair $(P_{q_1}, P_{\hat{q}_1})$ can be proved with a symmetric transformation T_P. Because the skew-symmetry mapping pair $(P_{q_1}, P_{\hat{q}_1})$ are chosen arbitrarily, the post-grazing mapping structure for a mapping pair $(P_q, P_{\hat{q}})$ ($q \in \{1, 2, \cdots, 2n_1\}$) in the skew-symmetry mapping is also skew-symmetric. This theorem is proved. ■

Since the symmetry invariance of the post-grazing of mapping exists, the combination of the symmetric mapping P_q ($q = 1, 2, \cdots, 2n_1$) should possess a symmetry invariance under transformation T_P. For convenience, the following notations for mapping clusters are introduced.

$$\left.\begin{aligned} \hat{P}_{(l_m, n_k)} &= P_{\hat{q}_{n_k} \cdots \hat{q}_{(n_{(k-1)}+1)} (\hat{q}_{n_{(k-1)}} \cdots \hat{q}_{(n_{(k-2)}+1)})^{l_m} \hat{q}_{n_{(k-2)}} \cdots \hat{q}_{(n_2+1)} (\hat{q}_{n_2} \cdots \hat{q}_2)^{l_1} \hat{q}_1},\\ P_{(l_m, n_k)} &= P_{q_{n_k} \cdots q_{(n_{(k-1)}+1)} (q_{n_{(k-1)}} \cdots q_{(n_{(k-2)}+1)})^{l_m} q_{n_{(k-2)}} \cdots q_{(n_2+1)} (q_{n_2} \cdots q_2)^{l_1} q_1}. \end{aligned}\right\} \tag{6.119}$$

To determine such a symmetrical invariance of the skew-symmetry flow, from the above notations, a theorem is stated as follows:

Theorem 6.3. *For mappings P_q ($q = 1, 2, \cdots, 2n_1$) of the dynamical system in Eq. (6.91), if the mapping pair $(P_q, P_{\hat{q}})$ during $(2M_1 + 1)$-periods is of skew-symmetry with a transformation T_P, then the following two mappings*

$$\begin{aligned} &(\underbrace{\hat{P}_{(l_{m_{2L}}, n_{k_{2L}})} \circ P_{(l_{m_{1L}}, n_{k_{1L}})} \circ \cdots \circ \hat{P}_{(l_{m_{21}}, n_{k_{21}})} \circ P_{(l_{m_{11}}, n_{k_{11}})}}_{L\text{-repeating}},\\ &\underbrace{P_{(l_{m_{1L}}, n_{k_{1L}})} \circ \hat{P}_{(l_{m_{2L}}, n_{k_{2L}})} \circ \cdots \circ P_{(l_{m_{11}}, n_{k_{11}})} \circ \hat{P}_{(l_{m_{21}}, n_{k_{21}})}}_{L\text{-repeating}}) \end{aligned} \tag{6.120}$$

form a skew-symmetric mapping pair under the same transformation T_P.

Proof. Consider a motion relative to a mapping

$$\underbrace{\hat{P}_{(l_{m_{2L}},n_{k_{2L}})} \circ P_{(l_{m_{1L}},n_{k_{1L}})} \circ \cdots \circ \hat{P}_{(l_{m_{21}},n_{k_{21}})} \circ P_{(l_{m_{11}},n_{k_{11}})}}_{L-\text{repeating}}$$

with the initial state $\mathbf{y}_k = (\mathbf{x}_k, \varphi_k)$ and the final state $\mathbf{y}_{k+S_L}$ $(S_L = \Sigma_{r=1}^{L} s_r + \hat{s}_r)$, and the mapping equation is

$$\mathbf{y}_{k+S_L} = \underbrace{\hat{P}_{(l_{m_{2L}},n_{k_{2L}})} \circ P_{(l_{m_{1L}},n_{k_{1L}})} \circ \cdots \circ \hat{P}_{(l_{m_{21}},n_{k_{21}})} \circ P_{(l_{m_{11}},n_{k_{11}})}}_{L-\text{repeating}} \mathbf{y}_k.$$

For mapping clusters $P_{(l_{m_{1r}},n_{k_{1r}})}$ and $\hat{P}_{(l_{m_{2r}},n_{k_{2r}})}$ in the rth mapping pair with mapping numbers s_r and $\hat{s}_r$, the governing equations of the foregoing mapping structures for $r = 1, 2, \cdots, S_L$ are

$$\begin{aligned}
&\mathbf{F}^{(q_1)}(\varphi_{k+S_{r-1}}^{(q_1)}, \mathbf{x}_{k+S_{r-1}}^{(q_1)}, \varphi_{k+S_{r-1}+1}^{(q_1)}, \mathbf{x}_{k+S_{r-1}+1}^{(q_1)}, \boldsymbol{\mu}_{p_{q_1}}, \boldsymbol{\pi}) = \mathbf{0},\\
&\vdots\\
&\mathbf{F}^{(q_{n_{k_r}})}(\varphi_{k+S_{r-1}+s_r-1}^{(q_{n_{k_{1r}}})}, \mathbf{x}_{k+S_{r-1}+s_r-1}^{(q_{n_{k_{1r}}})}, \varphi_{k+S_{r-1}+s_r}^{(q_{n_{k_{1r}}})}, \mathbf{x}_{k+S_{r-1}+s_r}^{(q_{n_{k_{1r}}})}, \boldsymbol{\mu}_{p_{q_{n_{k_{1r}}}}}, \boldsymbol{\pi}) = \mathbf{0};\\
&\mathbf{F}^{(\hat{q}_1)}(\varphi_{k+S_{r-1}+s_r}^{(\hat{q}_1)}, \mathbf{x}_{k+S_{r-1}+s_r}^{(\hat{q}_1)}, \varphi_{k+S_{r-1}+s_r+1}^{(\hat{q}_1)}, \mathbf{x}_{k+S_{r-1}+s_r+1}^{(\hat{q}_1)}, \boldsymbol{\mu}_{p_{q_1}}, \boldsymbol{\pi}) = \mathbf{0},\\
&\vdots\\
&\mathbf{F}^{(\hat{q}_{n_k})}(\varphi_{k+S_r-1}^{(\hat{q}_{n_{k_{2r}}})}, \mathbf{x}_{k+S_r-1}^{(\hat{q}_{n_{k_{2r}}})}, \varphi_{k+S_r}^{(\hat{q}_{n_{k_{2r}}})}, \mathbf{x}_{k+S_r}^{(\hat{q}_{n_{k_{2r}}})}, \boldsymbol{\mu}_{p_{q_{n_{k_{2r}}}}}, \boldsymbol{\pi}) = \mathbf{0}.
\end{aligned}$$

where $S_{r-1} = \Sigma_{\sigma=1}^{r-1} s_\sigma + \hat{s}_\sigma$. Application of Eqs. (8.114) and (8.115) to the foregoing equations, and multiplication of negative one (-1) on both sides of the foregoing equations leads to

$$\begin{aligned}
&\mathbf{F}^{(\hat{q}_1)}(\varphi_{k+S_{r-1}}^{(\hat{q}_1)}, \mathbf{x}_{k+S_{r-1}}^{(\hat{q}_1)}, \varphi_{k+S_{r-1}+1}^{(\hat{q}_1)}, \mathbf{x}_{k+S_{r-1}+1}^{(\hat{q}_1)}, \boldsymbol{\mu}_{p_{q_1}}, \boldsymbol{\pi}) = \mathbf{0},\\
&\vdots\\
&\mathbf{F}^{(\hat{q}_{n_k})}(\varphi_{k+S_{r-1}+\hat{s}_r-1}^{(\hat{q}_{n_{k_{2r}}})}, \mathbf{x}_{k+S_{r-1}+\hat{s}_r-1}^{(\hat{q}_{n_{k_{2r}}})}, \varphi_{k+S_{r-1}+\hat{s}_r}^{(\hat{q}_{n_{k_{2r}}})}, \mathbf{x}_{k+S_{r-1}+\hat{s}_r}^{(\hat{q}_{n_{k_{2r}}})}, \boldsymbol{\mu}_{p_{q_{n_{k_{2r}}}}}, \boldsymbol{\pi}) = \mathbf{0},
\end{aligned}$$

$$\begin{aligned}
&\mathbf{F}^{(q_1)}(\varphi_{k+S_{r-1}+\hat{s}_r}^{(q_1)}, \mathbf{x}_{i+S_{r-1}+\hat{s}_r}^{(q_1)}, \varphi_{i+S_{r-1}+\hat{s}_r+1}^{(q_1)}, \mathbf{x}_{i+S_{r-1}+\hat{s}_r+1}^{(q_1)}, \boldsymbol{\mu}_{p_{q_1}}, \boldsymbol{\pi}) = \mathbf{0},\\
&\vdots\\
&\mathbf{F}^{(q_{n_{k_r}})}(\varphi_{k+S_r-1}^{(q_{n_{k_{1r}}})}, \mathbf{x}_{k+S_r-1}^{(q_{n_{k_{1r}}})}, \varphi_{i+S_r}^{(q_{n_{k_{1r}}})}, \mathbf{x}_{i+S_r}^{(q_{n_{k_{1r}}})}, \boldsymbol{\mu}_{p_{q_{n_{k_{1r}}}}}, \boldsymbol{\pi}) = \mathbf{0}
\end{aligned}$$

being governing equations for

$$\mathbf{y}_{k+S_L} = \underbrace{P_{(l_{m1L}, n_{k1L})} \circ \hat{P}_{(l_{m2L}, n_{k2L})} \circ \cdots \circ P_{(l_{m11}, n_{k11})} \circ \hat{P}_{(l_{m21}, n_{k21})}}_{L-\text{repeating}} \mathbf{y}_k.$$

Therefore, the following mapping pair

$$(\underbrace{\hat{P}_{(l_{m2L}, n_{k2L})} \circ P_{(l_{m1L}, n_{k1L})} \circ \cdots \circ \hat{P}_{(l_{m21}, n_{k21})} \circ P_{(l_{m11}, n_{k11})}}_{L-\text{repeating}},$$

$$\underbrace{P_{(l_{m1l}, n_{k1L})} \circ \hat{P}_{(l_{m2L}, n_{k2L})} \circ \cdots \circ P_{(l_{m11}, n_{k11})} \circ \hat{P}_{(l_{m21}, n_{k21})}}_{L-\text{repeating}})$$

is skew-symmetric under a transformation T_P. ■

6.5.4 Symmetry for Steady-State Flows and Chaos

The symmetry invariance of combined mapping structures has been discussed. The flow symmetry in discontinuous dynamical systems will be presented in this section. Consider a skew-symmetry mapping cluster pair$(P_{(l_m,n_k)}, \hat{P}_{(l_m,n_k)})$. The mapping numbers are s for two mapping clusters in the mapping pair. The corresponding flows should be of the skew-symmetry. The theorem is presented herein.

Theorem 6.4. *For mappings P_q $(q = 1, 2, \cdots, 2n_1)$ of the dynamical system in Eq. (6.91), if the mapping pair $(P_q, P_{\hat{q}})$ under $(2M_1+1)$-periods is of skew-symmetry with a transformation T_P, then the asymmetric flows respectively relative to two mappings $P_{(l_m,n_k)}\mathbf{y} = \mathbf{y}$ and $\hat{P}_{(l_m,n_k)}\mathbf{y} = \mathbf{y}$ with the same mapping number s under N_1-periods with a periodicity*

$$\mathbf{y}_{k+s} = \mathbf{y}_k \text{ or } (\varphi_{k+s}, \mathbf{x}_{k+s})^{\mathrm{T}} \equiv (\varphi_k + 2N_1\pi, \mathbf{x}_k)^{\mathrm{T}} \tag{6.121}$$

have the corresponding solutions $(\varphi^{\mathrm{I}}_{k+j}, \mathbf{x}^{\mathrm{I}}_{k+j})$ and $(\varphi^{\mathrm{II}}_{k+j}, \mathbf{x}^{\mathrm{II}}_{k+j})$ for $j = \{0, 1, \cdots, s\}$ on the discontinuous boundaries to satisfy the following conditions

$$\begin{aligned} \hat{\varphi}^{\mathrm{I}}_{k+j} &= \operatorname{mod}((2M_1 + 1)\pi + \hat{\varphi}^{\mathrm{II}}_{k+j}, 2(2M_1 + 1)\pi), \\ \mathbf{x}^{\mathrm{I}}_{k+j} &= -\mathbf{x}^{\mathrm{II}}_{k+j}. \end{aligned} \tag{6.122}$$

Superscripts I and II denote solutions of $P_{(l_m,n_k)}\mathbf{y} = \mathbf{y}$ and $\hat{P}_{(l_m,n_k)}\mathbf{y} = \mathbf{y}$, respectively.

Proof From Definition 6.6, the solutions $(\varphi^{\mathrm{I}}_{i+j}, \mathbf{x}^{\mathrm{I}}_{i+j})$ and $(\varphi^{\mathrm{II}}_{i+j}, \mathbf{x}^{\mathrm{II}}_{i+j})$ for the mapping $P_{(l_m,n_k)}\mathbf{y} = \mathbf{y}$ and the skew-symmetry mapping $\hat{P}_{(l_m,n_k)}\mathbf{y} = \mathbf{y}$ satisfy Eq. (8.115). ■

If the local mappings in the mapping cluster exist only on a special boundary, the two skew-symmetric, local flows relative to $P_{(l_m,n_k)}\mathbf{y} = \mathbf{y}$ and $\hat{P}_{(l_m,n_k)}\mathbf{y} = \mathbf{y}$ are separated, as in Sect. 6.4. From the above theorem, once the solutions for one of the two mapping clusters are determined, the solutions for the other mapping clusters can be obtained right away. If the two mapping clusters are combined with global mappings, the symmetric flows relative to a global symmetric mapping structure $\hat{P}_{(l_m,n_k)} \circ P_{(l_m,n_k)}$ are discussed first, and then the asymmetrical flows for such a global symmetric mapping structure are presented. The theorem for the symmetrical flow is stated as follows.

Theorem 6.5. *For mappings P_q ($q = 1, 2, \cdots, 2n_1$) of the dynamical system in Eq. (6.91), if the mapping pair (P_q, $P_{\hat{q}}$) under $(2M_1+1)$-periods is of skew-symmetry with a transformation T_P, then the symmetric flow relative to a mapping $\hat{P}_{(l_m,n_k)} \circ P_{(l_m,n_k)}\mathbf{y} = \mathbf{y}$ with the same mapping number s of two mapping clusters under N_1-periods with a periodicity condition*

$$\mathbf{y}_{k+2s} = \mathbf{y}_k \text{ or } (\varphi_{k+2s}, \mathbf{x}_{k+2s})^{\mathrm{T}} \equiv (\varphi_k + 2N_1\pi, \mathbf{x}_k)^{\mathrm{T}} \tag{6.123}$$

have the corresponding solutions $(\varphi_{i+j}, \mathbf{x}_{i+j})$ for $j = 0, 1, \cdots, 2s$ on the discontinuous boundaries to satisfy the following conditions

$$\begin{aligned} \hat{\varphi}_{k+j} &= \operatorname{mod}((2M_1 + 1)\pi + \hat{\varphi}_{k+\operatorname{mod}(s+j,2s)}, 2(2M_1 + 1)\pi), \\ \mathbf{x}_{k+j} &= -\mathbf{x}_{k+\operatorname{mod}(s+j,2s)}. \end{aligned} \tag{6.124}$$

Proof For the symmetrical flow of a mapping $\hat{P}_{(l_m,n_k)} \circ P_{(l_m,n_k)}\mathbf{y} = \mathbf{y}$ with Eq. (8.123), the governing equations are

$$\begin{aligned} &\mathbf{F}^{(q_1)}(\varphi_k^{(q_1)}, \mathbf{x}_k^{(q_1)}, \varphi_{k+1}^{(q_1)}, \mathbf{x}_{k+1}^{(q_1)}, \boldsymbol{\mu}_{p_{q_1}}, \boldsymbol{\pi}) = \mathbf{0}, \\ &\vdots \\ &\mathbf{F}^{(q_{n_k})}(\varphi_{k+s-1}^{(q_{n_k})}, \mathbf{x}_{k+s-1}^{(q_{n_k})}, \varphi_{k+s-1}^{(q_{n_k})}, \mathbf{x}_{k+s}^{(q_{n_k})}, \boldsymbol{\mu}_{p_{q_{n_k}}}, \boldsymbol{\pi}) = \mathbf{0}; \end{aligned}$$

$$\begin{aligned} &\mathbf{F}^{(\hat{q}_1)}(\varphi_{k+s}^{(\hat{q}_1)}, \mathbf{x}_{k+s}^{(\hat{q}_1)}, \varphi_{k+s+1}^{(\hat{q}_1)}, \mathbf{x}_{k+s+1}^{(\hat{q}_1)}, \boldsymbol{\mu}_{p_{q_1}}, \boldsymbol{\pi}) = \mathbf{0}, \\ &\vdots \\ &\mathbf{F}^{(\hat{q}_{n_k})}(\varphi_{k+2s-1}^{(\hat{q}_{n_k})}, \mathbf{x}_{k+2s-1}^{(\hat{q}_{n_k})}, \varphi_{k+2s}^{(\hat{q}_{n_k})}, \mathbf{x}_{k+2s}^{(\hat{q}_{n_k})}, \boldsymbol{\mu}_{p_{q_{n_k}}}, \boldsymbol{\pi}) = \mathbf{0}. \end{aligned}$$

Substitution of Eq. (6.124) into the foregoing equations and using modulus $\operatorname{mod}(s + j, 2s)$ from (8.124) gives for a given number $N_2 \in \{0, 1, 2, \cdots\}$

$$\mathbf{F}^{(q_1)}(\varphi_{k+s}^{(q_1)} + (2N_2+1)(2M_1+1)\pi, -\mathbf{x}_{k+s}^{(q_1)},$$
$$\varphi_{k+s+1}^{(q_1)} + (2N_2+1)(2M_1+1)\pi, -\mathbf{x}_{k+s+1}^{(q_1)}, \boldsymbol{\mu}_{p_{q_1}}, \boldsymbol{\pi}) = \mathbf{0},$$
$$\vdots$$
$$\mathbf{F}^{(q_{n_k})}(\varphi_{i+2s-1}^{(q_{n_k})} + (2N_2+1)(2M_1+1)\pi, -\mathbf{x}_{i+2s-1}^{(q_{n_k})},$$
$$\varphi_{i+2s}^{(q_{n_k})} + (2N_2+1)(2M_1+1)\pi, -\mathbf{x}_{i+2s}^{(q_{n_k})}, \boldsymbol{\mu}_{p_{q_{n_k}}}, \boldsymbol{\pi}) = \mathbf{0};$$

$$\mathbf{F}^{(\hat{q}_1)}(\varphi_{k+2s}^{(\hat{q}_1)} + (2N_2+1)(2M_1+1)\pi, -\mathbf{x}_{k+2s}^{(\hat{q}_1)},$$
$$\varphi_{k+2s+1}^{(\hat{q}_1)} + (2N_2+1)(2M_1+1)\pi, -\mathbf{x}_{k+2s+1}^{(\hat{q}_1)}, \boldsymbol{\mu}_{p_{q_1}}, \boldsymbol{\pi}) = \mathbf{0},$$
$$\vdots$$
$$\mathbf{F}^{(\hat{q}_{n_k})}(\varphi_{k+3s-1}^{(\hat{q}_{n_k})} + (2N_2+1)(2M_1+1)\pi, -\mathbf{x}_{k+3s-1}^{(\hat{q}_{n_k})},$$
$$\varphi_{k+3s}^{(\hat{q}_{n_k})} + (2N_2+1)(2M_1+1)\pi, -\mathbf{x}_{k+3s}^{(\hat{q}_{n_k})}, \boldsymbol{\mu}_{p_{q_{n_k}}}, \boldsymbol{\pi}) = \mathbf{0}.$$

Using Eqs. (6.114) and (6.115) in Definition 6.6 into the foregoing equations yields

$$\mathbf{F}^{(\hat{q}_1)}(\varphi_{k+s}^{(\hat{q}_1)}, \mathbf{x}_{k+s}^{(\hat{q}_1)}, \varphi_{k+s+1}^{(\hat{q}_1)}, \mathbf{x}_{k+s+1}^{(\hat{q}_1)}, \boldsymbol{\mu}_{p_{q_1}}, \boldsymbol{\pi}) = \mathbf{0},$$
$$\vdots$$
$$\mathbf{F}^{(\hat{q}_{n_k})}(\varphi_{k+2s-1}^{(\hat{q}_{n_k})}, \mathbf{x}_{k+2s-1}^{(\hat{q}_{n_k})}, \varphi_{k+2s}^{(\hat{q}_{n_k})}, \mathbf{x}_{k+2s}^{(\hat{q}_{n_k})}, \boldsymbol{\mu}_{p_{q_{n_k}}}, \boldsymbol{\pi}) = \mathbf{0};$$

$$\mathbf{F}^{(q_1)}(\varphi_{k}^{(q_1)}, \mathbf{x}_{k}^{(q_1)}, \varphi_{k+1}^{(q_1)}, \mathbf{x}_{k+1}^{(q_1)}, \boldsymbol{\mu}_{p_{q_1}}, \boldsymbol{\pi}) = \mathbf{0},$$
$$\vdots$$
$$\mathbf{F}^{(q_{n_k})}(\varphi_{k+s-1}^{(q_{n_k})}, \mathbf{x}_{k+s-1}^{(q_{n_k})}, \varphi_{k+s}^{(q_{n_k})}, \mathbf{x}_{k+s}^{(q_{n_k})}, \boldsymbol{\mu}_{p_{q_{n_k}}}, \boldsymbol{\pi}) = \mathbf{0}.$$

After exchanging the order of the above two equations, they are identical to the mapping structure of $\hat{P}_{(l_m,n_k)} \circ P_{(l_m,n_k)}\mathbf{y} = \mathbf{y}$. If $(\varphi_{k+j}, \mathbf{x}_{k+j})^{\mathrm{T}}$ is the solutions of periodic flow, then $(\varphi_{k+\bmod(s+j,2s)}, \mathbf{x}_{k+\bmod(s+j,2s)})^{\mathrm{T}}$ is also a skew-symmetry solution satisfying Eq. (6.124). ■

The foregoing theorem discussed about the symmetrical solutions of period-1 motion associated with mapping $\hat{P}_{(l_m,n_k)} \circ P_{(l_m,n_k)}$. This structure is quite stable. For instance, the symmetrical period-1 motion of impacting oscillators can be referred in references (e.g., Luo 2002; Luo and Chen 2005b). One thought this motion may have period-doubling bifurcation. In fact, no period-doubling bifurcation exists. The symmetrical motion will be converted into the asymmetrical period-1 motion with the same mapping structures through the saddle-node bifurcation of the first kind and an unstable region. The flow symmetry for such an asymmetric period-$2^L(L = 0, 1, 2 \cdots \infty)$ motion is presented in the following theorem.

Theorem 6.6 *For mappings P_q $(q = 1, 2, \cdots, 2n_1)$ of the dynamical system in Eq. (8.1), if the mapping pair $(P_q, P_{\hat{q}})$ under $(2M_1+1)$-periods is of skew-symmetry with a transformation T_P, then the two asymmetric flows relative to a mapping*

$$\underbrace{\hat{P}_{(l_m,n_k)} \circ P_{(l_m,n_k)} \circ \cdots \circ \hat{P}_{(l_m,n_k)} \circ P_{(l_m,n_k)}}_{2^L-\text{repeating}} \mathbf{y} = \mathbf{y} \tag{6.125}$$

for $L = 0, 1, 2, \cdots, \infty$ with the same mapping number (i.e., s) in the two mapping clusters under N_1-periods with a periodicity condition

$$\mathbf{y}_{k+2^{L+1}s} = \mathbf{y}_k \text{ or } (\varphi_{k+2^{L+1}s}, \mathbf{x}_{k+2^{L+1}s})^{\mathrm{T}} \equiv (\varphi_k + 2N_1\pi, \mathbf{x}_k)^{\mathrm{T}} \tag{6.126}$$

possess the following solution properties

$$\begin{aligned} \hat{\varphi}^{\mathrm{I}}_{k+2(r-1)s+j} &= \mathrm{mod}((2M_1+1)\pi + \hat{\varphi}^{\mathrm{II}}_{k+2(r-1)s+\mathrm{mod}\,(s+j,2s)}, 2(2M_1+1)\pi), \\ \mathbf{x}^{\mathrm{I}}_{k+2(r-1)s+j} &= -\mathbf{x}^{\mathrm{II}}_{k+2(r-1)s+\mathrm{mod}\,(s+j,2s)}; \end{aligned} \tag{6.127}$$

for $r = 1, 2, \cdots, 2^L$ and $j = 0, 1, \cdots, 2s$. Superscripts I and II denote the two asymmetrical solutions.

Proof Suppose the mapping

$$\underbrace{\hat{P}_{(l_m,n_k)} \circ P_{(l_m,n_k)} \circ \cdots \circ \hat{P}_{(l_m,n_k)} \circ P_{(l_m,n_k)}}_{2^L-\text{repeating}} \mathbf{y} = \mathbf{y}$$

has a set of solutions $(\varphi^{\mathrm{I}}_{k+2(r-1)s+j}, \mathbf{x}^{\mathrm{I}}_{k+2(r-1)s+j})$ for $j = 0, 1, \cdots, 2s$ with $r = 1, 2, \cdots, 2^N$, the governing equations are

$$\begin{aligned} &\mathbf{F}^{(q_1)}(\varphi^{\mathrm{I}(q_1)}_{k+2(r-1)s}, \mathbf{x}^{\mathrm{I}(q_1)}_{k+2(r-1)s}, \varphi^{\mathrm{I}(q_1)}_{k+2(r-1)s+1}, \mathbf{x}^{\mathrm{I}(q_1)}_{k+2(r-1)s+1}, \boldsymbol{\mu}_{p_{q_1}}, \boldsymbol{\pi}) = \mathbf{0}, \\ &\vdots \\ &\mathbf{F}^{(q_{n_k})}(\varphi^{\mathrm{I}(q_{n_k})}_{k+(2r-1)s-1}, \mathbf{x}^{\mathrm{I}(q_{n_k})}_{k+(2r-1)s-1}, \varphi^{\mathrm{I}(q_{n_k})}_{k+(2r-1)s}, \mathbf{x}^{\mathrm{I}(q_{n_k})}_{k+(2r-1)s}, \boldsymbol{\mu}_{p_{q_{n_k}}}, \boldsymbol{\pi}) = \mathbf{0}; \end{aligned}$$

$$\begin{aligned} &\mathbf{F}^{(\hat{q}_1)}(\varphi^{\mathrm{I}(\hat{q}_1)}_{k+(2r-1)s}, \mathbf{x}^{\mathrm{I}(\hat{q}_1)}_{k+(2r-1)s}, \varphi^{\mathrm{I}(\hat{q}_1)}_{k+(2r-1)s+1}, \mathbf{x}^{\mathrm{I}(\hat{q}_1)}_{k+(2r-1)s+1}, \boldsymbol{\mu}_{p_{q_1}}, \boldsymbol{\pi}) = \mathbf{0}, \\ &\vdots \\ &\mathbf{F}^{(\hat{q}_{n_k})}(\varphi^{\mathrm{I}(\hat{q}_{n_k})}_{k+2rs-1}, \mathbf{x}^{\mathrm{I}(\hat{q}_{n_k})}_{k+2rs-1}, \varphi^{\mathrm{I}(\hat{q}_{n_k})}_{k+2rs}, \mathbf{x}^{\mathrm{I}(\hat{q}_{n_k})}_{k+2rs}, \boldsymbol{\mu}_{p_{q_{n_k}}}, \boldsymbol{\pi}) = \mathbf{0}. \end{aligned}$$

Substitution of Eq. (6.126) into the foregoing equation and using modulus function of mod $(s+j, 2s)$ from Eq. (6.125) gives for a given number $N_2 \in \{0, 1, 2, \cdots\}$

$$\begin{aligned}
&\mathbf{F}^{(q_1)}(\varphi^{\mathrm{II}(q_1)}_{k+(2r-1)s} + (2N_2+1)(2M_1+1)\pi, -\mathbf{x}^{\mathrm{II}(q_1)}_{k+(2r-1)s},\\
&\quad \varphi^{\mathrm{II}(q_1)}_{k+(2r-1)s+1} + (2N_2+1)(2M_1+1)\pi, -\mathbf{x}^{\mathrm{II}(q_1)}_{k+(2r-1)s} + 1, \boldsymbol{\mu}_{p_{q_1}}, \boldsymbol{\pi}) = \mathbf{0},\\
&\vdots\\
&\mathbf{F}^{(q_{n_k})}(\varphi^{\mathrm{II}(q_{n_k})}_{k+2rs-1} + (2N_2+1)(2M_1+1)\pi, -\mathbf{x}^{\mathrm{II}(q_{n_k})}_{k+2rs-1},\\
&\quad \varphi^{\mathrm{II}(q_{n_k})}_{k+2rs} + (2N_2+1)(2M_1+1)\pi, -\mathbf{x}^{\mathrm{II}(q_{n_k})}_{k+2rs}, \boldsymbol{\mu}_{p_{q_{n_k}}}, \boldsymbol{\pi}) = \mathbf{0};
\end{aligned}$$

$$\begin{aligned}
&\mathbf{F}^{(\hat{q}_1)}(\varphi^{\mathrm{II}(\hat{q}_1)}_{k+2rs} + (2N_2+1)(2M_1+1)\pi, -\mathbf{x}^{\mathrm{II}(\hat{q}_1)}_{k+2rs},\\
&\quad \varphi^{\mathrm{II}(\hat{q}_1)}_{k+2rs+1} + (2N_2+1)(2M_1+1)\pi, -\mathbf{x}^{\mathrm{II}(\hat{q}_1)}_{k+2rs+1}, \boldsymbol{\mu}_{p_{q_1}}, \boldsymbol{\pi}) = \mathbf{0},\\
&\vdots\\
&\mathbf{F}^{(\hat{q}_{n_k})}(\varphi^{\mathrm{II}(\hat{q}_{n_k})}_{k+(2r+1)s-1} + (2N_2+1)(2M_1+1)\pi, -\mathbf{x}^{\mathrm{II}(\hat{q}_{n_k})}_{k+(2r+1)s-1},\\
&\quad \varphi^{\mathrm{II}(\hat{q}_{n_k})}_{k+(2r+1)s} + (2N_2+1)(2M_1+1)\pi, -\mathbf{x}^{\mathrm{II}(\hat{q}_{n_k})}_{k+(2r+1)s}, \boldsymbol{\mu}_{p_{q_{n_k}}}, \boldsymbol{\pi}) = \mathbf{0}.
\end{aligned}$$

Using Eqs. (6.114) and (6.115) in Definition 6.6 into the foregoing equations and taking modulus of the index yields

$$\begin{aligned}
&\mathbf{F}^{(\hat{q}_1)}(\varphi^{\mathrm{II}(\hat{q}_1)}_{k+(2r-1)s}, \mathbf{x}^{\mathrm{II}(\hat{q}_1)}_{k+(2r-1)s}, \varphi^{\mathrm{II}(\hat{q}_1)}_{k+(2r-1)s+1}, \mathbf{x}^{\mathrm{II}(\hat{q}_1)}_{k+(2r-1)s+}, \boldsymbol{\mu}_{p_{q_1}}, \boldsymbol{\pi}) = \mathbf{0},\\
&\vdots\\
&\mathbf{F}^{(\hat{q}_{n_k})}(\varphi^{\mathrm{II}(\hat{q}_{n_k})}_{k+2rs-1}, \mathbf{x}^{\mathrm{II}(\hat{q}_{n_k})}_{k+2rs-1}, \varphi^{\mathrm{II}(\hat{q}_{n_k})}_{k+2rs}, \mathbf{x}^{\mathrm{II}(\hat{q}_{n_k})}_{k+2rs}, \boldsymbol{\mu}_{p_{q_{n_k}}}, \boldsymbol{\pi}) = \mathbf{0};
\end{aligned}$$

$$\begin{aligned}
&\mathbf{F}^{(q_1)}(\varphi^{\mathrm{II}(q_1)}_{k+2(r-1)s}, \mathbf{x}^{\mathrm{II}(q_1)}_{k+2(r-1)s}, \varphi^{\mathrm{II}(q_1)}_{k+2(r-1)s+1}, \mathbf{x}^{\mathrm{II}(q_1)}_{k+2(r-1)s+1}, \boldsymbol{\mu}_{p_{q_1}}, \boldsymbol{\pi}) = \mathbf{0},\\
&\vdots\\
&\mathbf{F}^{(q_{n_k})}(\varphi^{\mathrm{II}(q_{n_k})}_{k+(2r-1)s-1}, \mathbf{x}^{\mathrm{II}(q_{n_k})}_{k+(2r-1)s-1}, \varphi^{\mathrm{II}(q_{n_k})}_{k+(2r-1)s}, \mathbf{x}^{\mathrm{II}(q_{n_k})}_{k+(2r-1)s}, \boldsymbol{\mu}_{p_{q_{n_k}}}, \boldsymbol{\pi}) = \mathbf{0}.
\end{aligned}$$

After exchanging the order of the above two sets of equations, it is clear that $(\varphi^{\mathrm{II}}_{i+2(r-1)s+j}, \mathbf{x}^{\mathrm{II}}_{i+2(r-1)s+j})$ for $j = 0, 1, \cdots, 2s$ with $r = 1, 2, \cdots, 2^N$ is another solution for mapping structures

$$\underbrace{\hat{P}_{(l_m,n_k)} \circ P_{(l_m,n_k)} \circ \cdots \circ \hat{P}_{(l_m,n_k)} \circ P_{(l_m,n_k)}}_{2^L-\text{repeating}} \mathbf{y} = \mathbf{y}.$$

This theorem is proved. ■

The above theorem discussed about the asymmetrical solutions of periodic and chaotic flows induced by period-doubling bifurcation of $\hat{P}_{(l_m,n_k)} \circ P_{(l_m,n_k)}\mathbf{y} = \mathbf{y}$. Similarly, the motions pertaining to the asymmetric mapping $\hat{P}_{(l_{m2},n_{k2})} \circ P_{(l_{m1},n_{k1})}\mathbf{y} = \mathbf{y}$ are presented in the following theorem.

Theorem 6.7. *For mappings P_q $(q = 1, 2, \cdots, 2n_1)$ of the dynamical system in Eq.(6.91), if the mapping pair $(P_q, P_{\hat{q}})$ under $(2M_1+1)$-periods is of skew-symmetry with a transformation T_P, then the asymmetric flows respectively relative to two mappings*

$$\underbrace{\hat{P}_{(l_{m2},n_{k2})} \circ P_{(l_{m1},n_{k1})} \circ \cdots \circ \hat{P}_{(l_{m2},n_{k2})} \circ P_{(l_{m1},n_{k1})}}_{2^L-\text{repeating}} \mathbf{y} = \mathbf{y} \text{ and}$$
$$\underbrace{\hat{P}_{(l_{m1},n_{k1})} \circ P_{(l_{m2},n_{k2})} \circ \cdots \circ \hat{P}_{(l_{m1},n_{k1})} \circ P_{(l_{m2},n_{k2})}}_{2^L-\text{repeating}} \mathbf{y} = \mathbf{y} \tag{6.128}$$

for $L = 0, 1, 2, \cdots, \infty$ with different mapping numbers (i.e.,s_1 and s_2) in two mapping clusters under N_1-periods with a periodicity condition

$$\mathbf{y}_{k+2^L(s_1+s_2)} = \mathbf{y}_k \text{ or } (\varphi_{k+2^L(s_1+s_2)}, \mathbf{x}_{k+2^L(s_1+s_2)})^{\mathrm{T}} = (\varphi_k + 2N_1\pi, \mathbf{x}_k)^{\mathrm{T}} \tag{6.129}$$

are $(\varphi^{\mathrm{I}}_{k+(r-1)(s_1+s_2)+j}, \mathbf{x}^{\mathrm{I}}_{k+(r-1)(s_1+s_2)+j})$ and$(\varphi^{\mathrm{II}}_{k+(r-1)(s_1+s_2)+j}, \mathbf{x}^{\mathrm{II}}_{k+(r-1)(s_1+s_2)+j})$ *with*

$$\begin{aligned}\hat{\varphi}^{\mathrm{I}}_{k+(r-1)(s_1+s_2)+j} &= \operatorname{mod}((2M_1+1)\pi \\ &\quad + \hat{\varphi}^{\mathrm{II}}_{k+(r-1)(s_1+s_2)+\operatorname{mod}(s_2+j,s_1+s_2)}, 2(2M_1+1)\pi), \\ \mathbf{x}^{\mathrm{I}}_{k+(r-1)(s_1+s_2)+j} &= -\mathbf{x}^{\mathrm{II}}_{k+(r-1)(s_1+s_2)+\operatorname{mod}(s_2+j,s_1+s_2)};\end{aligned} \tag{6.130}$$

or

$$\begin{aligned}\hat{\varphi}^{\mathrm{II}}_{i+(r-1)(s_1+s_2)+j} &= \operatorname{mod}((2M_1+1)\pi \\ &\quad + \hat{\varphi}^{\mathrm{I}}_{i+(r-1)(s_1+s_2)+\operatorname{mod}(s_1+j,s_1+s_2)}, 2(2M_1+1)\pi), \\ \mathbf{x}^{\mathrm{II}}_{i+(r-1)(s_1+s_2)+j} &= -\mathbf{x}^{\mathrm{I}}_{i+(r-1)(s_1+s_2)+\operatorname{mod}(s_1+j,s_1+s_2)}.\end{aligned} \tag{6.131}$$

For $r = 1, 2, \cdots, 2^L$ *and* $j = 0, 1, \cdots, (s_1 + s_2)$. *Superscripts I and II denote two asymmetrical flows.*

Proof Follow the same proof procedure of Theorem 6.5. Suppose

$$\underbrace{\hat{P}_{(l_{m2},n_{k2})} \circ P_{(l_{m1},n_{k1})} \circ \cdots \circ \hat{P}_{(l_{m2},n_{k2})} \circ P_{(l_{m1},n_{k1})}}_{2^L-\text{repeating}} \mathbf{y} = \mathbf{y}$$

has a solution $(\varphi^{\mathrm{I}}_{k+(r-1)(s_1+s_2)+j}, \mathbf{x}^{\mathrm{I}}_{k+(r-1)(s_1+s_2)+j})$ for $r = 1, 2, \cdots, 2^L$ and $j = 0, 1, \cdots, 2s$, the governing equations are

$$\mathbf{F}^{(q_1)}(\varphi^{\mathrm{I}(q_1)}_{k+(r-1)(s_1+s_2)}, \mathbf{x}^{\mathrm{I}(q_1)}_{k+(r-1)(s_1+s_2)}, \varphi^{\mathrm{I}(q_1)}_{k+(r-1)(s_1+s_2)+1}, \mathbf{x}^{\mathrm{I}(q_1)}_{k+(r-1)(s_1+s_2)+1}, \boldsymbol{\mu}_{p_{q_1}}, \boldsymbol{\pi}) = \mathbf{0},$$

$$\vdots$$

$$\mathbf{F}^{(q_{n_k})}(\varphi^{\mathrm{I}(q_{n_k})}_{k+(r-1)s_2+rs_1-1}, \mathbf{x}^{\mathrm{I}(q_{n_k})}_{k+(r-1)s_2+rs_1-1}, \varphi^{\mathrm{I}(q_{n_k})}_{k+(r-1)s_2+rs_1}, \mathbf{x}^{\mathrm{I}(q_{n_k})}_{k+(r-1)s_2+rs_1}, \boldsymbol{\mu}_{p_{q_{n_k}}}, \boldsymbol{\pi}) = \mathbf{0};$$

$$\mathbf{F}^{(\hat{q}_1)}(\varphi^{\mathrm{I}(\hat{q}_1)}_{k+(r-1)s_2+rs_1}, \mathbf{x}^{\mathrm{I}(\hat{q}_1)}_{k+(r-1)s_2+rs_1}, \varphi^{\mathrm{I}(\hat{q}_1)}_{k+(r-1)s_2+rs_1+1}, \mathbf{x}^{\mathrm{I}(\hat{q}_1)}_{k+(r-1)s_2+rs_1+1}, \boldsymbol{\mu}_{p_{q_1}}, \boldsymbol{\pi}) = \mathbf{0},$$

$$\vdots$$

$$\mathbf{F}^{(\hat{q}_{n_k})}(\varphi^{\mathrm{I}(\hat{q}_{n_k})}_{k+r(s_1+s_2)-1}, \mathbf{x}^{\mathrm{I}(\hat{q}_{n_k})}_{k+r(s_1+s_2)-1}, \varphi^{\mathrm{I}(\hat{q}_{n_k})}_{k+r(s_1+s_2)}, \mathbf{x}^{\mathrm{I}(\hat{q}_{n_k})}_{k+r(s_1+s_2)}, \boldsymbol{\mu}_{p_{q_{n_k}}}, \boldsymbol{\pi}) = \mathbf{0}.$$

Substitution of Eq. (6.130) into the foregoing equations and using modulus mod $(s+j, 2s)$ from Eq. (6.129) gives for a given number $N_2 \in \{0, 1, 2, \cdots\}$

$$\begin{aligned}\mathbf{F}^{(q_1)}(&\varphi^{\mathrm{II}(q_1)}_{k+(r-1)s_1+rs_2} + (2N_2+1)(2M_1+1)\pi), -\mathbf{x}^{\mathrm{II}(q_1)}_{k+(r-1)s_1+rs_2},\\ &\varphi^{\mathrm{II}(q_1)}_{k+(r-1)s_1+rs_2+1} + (2N_2+1)(2M_1+1)\pi, -\mathbf{x}^{\mathrm{II}(q_1)}_{k+(r-1)s_1+rs_2}, \boldsymbol{\mu}_{p_{q_1}}, \boldsymbol{\pi}) = \mathbf{0},\end{aligned}$$

$$\vdots$$

$$\begin{aligned}\mathbf{F}^{(q_{n_k})}(&\varphi^{\mathrm{II}(q_{n_k})}_{k+r(s_1+s_2)-1} + (2N_2+1)(2M_1+1)\pi, -\mathbf{x}^{\mathrm{II}(q_{n_k})}_{k+r(s_1+s_2)-1},\\ &\varphi^{\mathrm{II}(q_{n_k})}_{k+r(s_1+s_2)} + (2N_2+1)(2M_1+1)\pi, -\mathbf{x}^{\mathrm{II}(q_{n_k})}_{k+r(s_1+s_2)}, \boldsymbol{\mu}_{p_{q_{n_k}}}, \boldsymbol{\pi}) = \mathbf{0};\end{aligned}$$

$$\begin{aligned}\mathbf{F}^{(\hat{q}_1)}(&\varphi^{\mathrm{II}(\hat{q}_1)}_{k+r(s_1+s_2)} + (2N_2+1)(2M_1+1)\pi, -\mathbf{x}^{\mathrm{II}(\hat{q}_1)}_{k+r(s_1+s_2)},\\ &\varphi^{\mathrm{II}(\hat{q}_1)}_{k+r(s_1+s_2)+1} + (2N_2+1)(2M_1+1)\pi, -\mathbf{x}^{\mathrm{II}(\hat{q}_1)}_{k+r(s_1+s_2)+1}, \boldsymbol{\mu}_{p_{q_1}}, \boldsymbol{\pi}) = \mathbf{0},\end{aligned}$$

$$\vdots$$

$$\begin{aligned}\mathbf{F}^{(\hat{q}_{n_k})}(&\varphi^{\mathrm{II}(\hat{q}_{n_k})}_{k+r(s_1+s_2)+s_2-1} + (2N_2+1)(2M_1+1)\pi, -\mathbf{x}^{\mathrm{II}(\hat{q}_{n_k})}_{k+r(s_1+s_2)+s_2-1},\\ &\varphi^{\mathrm{II}(\hat{q}_{n_k})}_{k+r(s_1+s_2)+s_2} + (2N_2+1)(2M_1+1)\pi, -\mathbf{x}^{\mathrm{II}(\hat{q}_{n_k})}_{k+r(s_1+s_2)+s_2}, \boldsymbol{\mu}_{p_{q_{n_k}}}, \boldsymbol{\pi}) = \mathbf{0}.\end{aligned}$$

Using Eqs. (6.114) and (6.115) in Definition 6.6 into the foregoing equations and taking modulus of the index yields

$$\mathbf{F}^{(\hat{q}_1)}(\varphi^{\mathrm{II}(\hat{q}_1)}_{k+rs_2+(r-1)s_1}, \mathbf{x}^{\mathrm{II}(\hat{q}_1)}_{k+rs_2+(r-1)s_1}, \varphi^{\mathrm{II}(\hat{q}_1)}_{k+rs_2+(r-1)s_1+1}, \mathbf{x}^{\mathrm{II}(\hat{q}_1)}_{k+rs_2+(r-1)s_1+1}, \boldsymbol{\mu}_{p_{q_1}}, \boldsymbol{\pi}) = \mathbf{0},$$

$$\vdots$$

$$\mathbf{F}^{(\hat{q}_{n_k})}(\varphi^{\mathrm{II}(\hat{q}_{n_k})}_{k+r(s_1+s_2)-1}, \mathbf{x}^{\mathrm{II}(\hat{q}_{n_k})}_{k+r(s_1+s_2)-1}, \varphi^{\mathrm{II}(\hat{q}_{n_k})}_{k+r(s_1+s_2)}, \mathbf{x}^{\mathrm{II}(\hat{q}_{n_k})}_{k+r(s_1+s_2)}, \boldsymbol{\mu}_{p_{q_{n_k}}}, \boldsymbol{\pi}) = \mathbf{0};$$

$$\mathbf{F}^{(q_1)}(\varphi^{\mathrm{II}(q_1)}_{k+(r-1)(s_1+s_2)}, \mathbf{x}^{\mathrm{II}(q_1)}_{k+(r-1)(s_1+s_2)}, \varphi^{\mathrm{II}(q_1)}_{k+(r-1)(s_1+s_2)+1}, \mathbf{x}^{\mathrm{II}(q_1)}_{k+(r-1)(s_1+s_2)+1}, \boldsymbol{\mu}_{p_{q_1}}, \boldsymbol{\pi}) = \mathbf{0},$$

$$\vdots$$

$$\mathbf{F}^{(q_{n_k})}(\varphi^{\mathrm{II}(q_{n_k})}_{k+(r-1)s_1+rs_2-1}, \mathbf{x}^{\mathrm{II}(q_{n_k})}_{k+(r-1)s_1+rs_2-1}, \varphi^{\mathrm{II}(q_{n_k})}_{k+(r-1)s_1+rs_2}, \mathbf{x}^{\mathrm{II}(q_{n_k})}_{k+(r-1)s_1+rs_2}, \boldsymbol{\mu}_{p_{q_{n_k}}}, \boldsymbol{\pi}) = \mathbf{0}.$$

After exchanging the order of the above two sets of equations, it is clear that $(\varphi^{\mathrm{II}}_{k+(r-1)(s_1+s_2)+j}, \mathbf{x}^{\mathrm{II}}_{k+(r-1)(s_1+s_2)+j})$ for $j = 0, 1, \cdots, 2s$ with $r = 1, 2, \cdots, 2^N$ is a set of solutions for the mapping structure

$$\underbrace{\hat{P}_{(l_{m1},n_{k1})} \circ P_{(l_{m2},n_{k2})} \circ \cdots \circ \hat{P}_{(l_{m1},n_{k1})} \circ P_{(l_{m2},n_{k2})}}_{2^N-\text{repeating}} \mathbf{y} = \mathbf{y}.$$

This theorem is proved. ■

The above results can be generalized in the following theorem.

Theorem 6.8 *For mappings P_q ($q = 1, 2, \cdots, 2n_1$) of the dynamical system in Eq. (6.91), if the mapping pair $(P_q, P_{\hat{q}})$ under $(2M_1+1)$-periods is of skew-symmetry with a transformation T_P, then the asymmetric flows respectively relative to two mappings*

$$\begin{gathered}\underbrace{\hat{P}_{(l_{m2L},n_{k2L})} \circ P_{(l_{m1L},n_{k1L})} \circ \cdots \circ \hat{P}_{(l_{m21},n_{k21})} \circ P_{(l_{m11},n_{k11})}}_{L-\text{repeating}} \mathbf{y} = \mathbf{y} \text{ and} \\ \underbrace{P_{(l_{m1L},n_{k1L})} \circ \hat{P}_{(l_{m2N},n_{kL})} \circ \cdots \circ P_{(l_{m11},n_{k11})} \circ \hat{P}_{(l_{m21},n_{k21})}}_{L-\text{repeating}} \mathbf{y} = \mathbf{y}\end{gathered} \tag{6.132}$$

for $N = 1, 2, \cdots, \infty$ with different mapping numbers (i.e., s_{1r} and s_{2r}) of the two mapping clusters $P_{(l_{m1r},n_{k1r})}$ and $\hat{P}_{(l_{m2r},n_{k2r})}(r = 1, 2, \cdots, N)$ under N_1-periods with a periodicity condition

$$\mathbf{y}_{k+\Sigma_{r=1}^{N}s_{1r}+s_{2r}} = \mathbf{y}_k \text{ or } (\varphi_{k+\Sigma_{r=1}^{N}s_{1r}+s_{2r}}, \mathbf{x}_{k+\Sigma_{r=1}^{N}s_{1r}+s_{2r}})^{\mathrm{T}} \equiv (\varphi_k + 2N_1\pi, \mathbf{x}_k)^{\mathrm{T}} \tag{6.133}$$

are $(\varphi^{\mathrm{I}}_{k+\Sigma_{\rho=1}^{N}s_{1\rho}+s_{2\rho}+j}, \mathbf{x}_{k+\Sigma_{\rho=1}^{N}s_{1\rho}+s_{2\rho}+j}\mathrm{I})$ and $(\varphi^{\mathrm{II}}_{k+\Sigma_{\rho=1}^{N}s_{1\rho}+s_{2\rho}+j}, \mathbf{x}^{\mathrm{II}}_{k+\Sigma_{\rho=1}^{N}s_{1\rho}+s_{2\rho}+j})$ with a solution structure

$$\begin{aligned}\hat{\varphi}_{k+\Sigma_{\rho=1}^{r-1}s_{1\rho}+s_{2\rho}+j}\mathrm{I} &= \mathrm{mod}((2M_1+1)\pi + \hat{\varphi}^{\mathrm{II}}_{k+\Sigma_{\rho=1}^{r-1}s_{1\rho}+s_{2\rho}+j+\,\mathrm{mod}\,(s_{2r}+j,s_{1r}+s_{2r})}, \\ &\qquad 2(2M_1+1)\pi), \\ \mathbf{x}^{\mathrm{I}}_{k+\Sigma_{\rho=1}^{r-1}s_{1\rho}+s_{2\rho}+j} &= -\mathbf{x}^{\mathrm{II}}_{k+\Sigma_{\rho=1}^{r-1}s_{1\rho}+s_{2\rho}+j+\,\mathrm{mod}\,(s_{2r}+j,s_{1r}+s_{2r})};\end{aligned} \tag{6.134}$$

or

$$\begin{aligned}\hat{\varphi}^{\mathrm{II}}_{k+\Sigma_{\rho=1}^{r-1}s_{1\rho}+s_{2\rho}+j} &= \mathrm{mod}((2M_1+1)\pi + \hat{\varphi}^{\mathrm{I}}_{k+\Sigma_{\rho=1}^{r-1}s_{1\rho}+s_{2\rho}+j+\,\mathrm{mod}\,(s_{1r}+j,s_{1r}+s_{2r})},\\ &\quad 2(2M_1+1)\pi),\\ \mathbf{x}^{\mathrm{II}}_{k+\Sigma_{\rho=1}^{r-1}s_{1\rho}+s_{2\rho}+j} &= -\mathbf{x}^{\mathrm{I}}_{k+\Sigma_{\rho=1}^{r-1}s_{1\rho}+s_{2\rho}+j+\,\mathrm{mod}\,(s_{1r}+j,s_{1r}+s_{2r})}.\end{aligned} \tag{6.135}$$

for $j = \{0, 1, \cdots, (s_{1r} + s_{2r})\}$. *Superscripts I and II denote two skew-symmetric solutions of flows.*

Proof From Theorem 6.3, the two mapping structures are skew-symmetric. Thus, the two solutions of the two mapping structures are skew-symmetric, which implies that Eqs. (6.132)–(6.135) hold. The alternative proof can be completed through the procedure used in proof of Theorem 6.7. ∎

The symmetry of flow in discontinuous dynamical systems with mapping clusters on many discontinuous boundaries is discussed. The grazing does not change the symmetry invariance of mapping structures in such dynamical systems, and the periodic and chaotic motions in such a dynamical system possess the symmetry invariance as same as the basic mappings. From the discussion, the group structure of mapping combination exists. Thus the further investigation on such an issue should be carried out. The illustrations for the symmetry of periodic motions in symmetric, discontinuous dynamic systems can be found. The detailed presentations can be also seen in Luo (2005c).

6.6 Strange Attractor Fragmentation

Before the discussion of the fragmentation mechanism, the initial and final sets of grazing mapping will be presented. Because the grazing is strongly dependent on the singular sets in Eqs.(6.97)–(6.100), the definitions are given as follows.

Definition 6.7 For a discontinuous dynamical system in Eq. (6.91), consider an initial switching point $(\mathrm{mod}(\varphi_k, 2\pi), \mathbf{x}_k)$

$$\hbar_{p_1} G^{(p_1)}_{\partial\Omega_{p_1p_2}}(\mathbf{x}_k, t_{k+}) = \hbar_{p_1}\mathbf{n}^{\mathrm{T}}_{\partial\Omega_{p_1p_2}} \cdot \mathbf{F}^{(p_1)}(\varphi_k, \mathbf{x}_{k+}) < 0. \tag{6.136}$$

If $\Xi_{q_1} \subseteq \partial\Omega_{p_1p_2}$ and $\Xi_{q_2} \subseteq \partial\Omega_{p_1\alpha}$ ($\alpha \in \{p_2, p_3\}$, $p_i \in \{1, 2, \cdots, m\}$ for $i = 1, 2, 3$), the following subset ${}^{(\mathrm{i})}\Gamma_J$ for mapping $P_J : \Xi_{q_1} \to \Xi_{q_2}$ ($J \in \{1, 2, \cdots, N\}$ and $q_1, q_2 \in \{0, 1, 2, \cdots, M\}$) is called *the initial set of grazing mapping*,

$${}^{(\mathrm{i})}\Gamma_J \equiv \left\{ (\mathrm{mod}(\varphi_k, 2\pi), \mathbf{x}_k) \left| \begin{array}{l} P_J\mathbf{y}_k = \mathbf{y}_{k+1}, \\ G^{(p_1)}_{\partial\Omega_{p_1\alpha}}(\mathbf{x}_{k+1}, t_{(k+1)\pm}) = 0 \\ \hbar_{p_1} G^{(1,p_1)}_{\partial\Omega_{p_1\alpha}}(\mathbf{x}_{k+1}, t_{(k+1)\pm}) < 0 \end{array} \right. \right\} \subset \Xi_{q_1} \tag{6.137}$$

where $\mathbf{y}_k = (\varphi_k, \mathbf{x}_k)^{\mathrm{T}}$. The corresponding grazing set is defined as

$$
{}^{(\mathrm{f})}\mathrm{G_J} \equiv \left\{ (\mathrm{mod}(\varphi_{k+1}, 2\pi), \mathbf{x}_{k+1}) \left| \begin{array}{l} P_J \mathbf{y}_k = \mathbf{y}_{k+1}, \\ G^{(p_1)}_{\partial\Omega_{p_1\alpha}}(\mathbf{x}_{k+1}, t_{(k+1)\pm}) = 0 \\ \hbar_{p_1} G^{(1,p_1)}_{\partial\Omega_{p_1\alpha}}(\mathbf{x}_{k+1}, t_{(k+1)\pm}) < 0 \end{array} \right. \right\} \subset \Xi_{q_2}. \tag{6.138}
$$

Definition 6.8 For a discontinuous dynamical system in Eq. (6.91), consider a final switching point $(\mathrm{mod}(\varphi_{k+1}, 2\pi), \mathbf{x}_{k+1})$ with

$$
\hbar_{p_1} G^{(p_1)}_{\partial\Omega_{p_1\alpha}}(\mathbf{x}_{k+1}, t_{(k+1)-}) = \hbar_{p_1} \mathbf{n}^{\mathrm{T}}_{\partial\Omega_{p_1p_2}} \cdot \mathbf{F}^{(p_1)}(\varphi_{k+1}, \mathbf{x}_{(k+1)-}) < 0. \tag{6.139}
$$

If $\Xi_{q_1} \subseteq \partial\Omega_{p_1p_2}$ and $\Xi_{q_2} \subseteq \partial\Omega_{p_1\alpha}(\alpha \in \{p_2, p_3\}, p_i \in \{1, 2, \cdots, m\}$ for $i = 1, 2, 3)$, the following subset ${}^{(\mathrm{f})}\Gamma_J$ for mapping $P_J : \Xi_{q_1} \to \Xi_{q_2}$ $(J \in \{1, 2, \cdots, N\}$ and $q_1, q_2 \in \{0, 1, 2, \cdots, M\})$ is called the final set of grazing post-mapping:

$$
{}^{(\mathrm{f})}\Gamma_J \equiv \left\{ (\mathrm{mod}(\varphi_{k+1}, 2\pi), \mathbf{x}_{k+1}) \left| \begin{array}{l} P_J \mathbf{y}_k = \mathbf{y}_{k+1}, \\ G^{(p_1)}_{\partial\Omega_{p_1p_2}}(\mathbf{x}_k, t_{k\pm}) = 0 \\ \hbar_{p_1} G^{(1,p_1)}_{\partial\Omega_{p_1p_2}}(\mathbf{x}_k, t_{k\pm}) < 0 \end{array} \right. \right\} \subset \Xi_{q_2} \tag{6.140}
$$

where $\mathbf{y}_k = (\Omega t_k, \mathbf{x}_k)^{\mathrm{T}}$. The corresponding grazing set of the post-grazing is defined as

$$
{}^{(\mathrm{i})}\mathrm{G}_J \equiv \left\{ (\mathrm{mod}(\varphi_k, 2\pi), \mathbf{x}_k) \left| \begin{array}{l} P_J \mathbf{y}_k = \mathbf{y}_{k+1}, \\ G^{(p_1)}_{\partial\Omega_{p_1p_2}}(\mathbf{x}_{k+1}, t_{k\pm}) = 0 \\ \hbar_{p_1} G^{(1,p_1)}_{\partial\Omega_{p_1p_2}}(\mathbf{x}_{k+1}, t_{k\pm}) < 0 \end{array} \right. \right\} \subset \Xi_{q_1}. \tag{6.141}
$$

Definition 6.9 For a discontinuous dynamical system in Eq. (6.91), consider an initial switching point $(\mathrm{mod}(\varphi_k, 2\pi), \mathbf{x}_k)$

$$
\begin{aligned}
&G^{(s_{p_1},p_1)}_{\partial\Omega_{p_1p_2}}(\mathbf{x}_k, t_{k+}) = 0 \text{ for } s_{p_1} = 0, 1, \cdots, 2k_{p_1} - 1; \\
&\hbar_{p_1} G^{(2k_{p_1},p_1)}_{\partial\Omega_{p_1p_2}}(\mathbf{x}_k, t_{k+}) < 0.
\end{aligned} \tag{6.142}
$$

If $\Xi_{q_1} \subseteq \partial\Omega_{p_1p_2}$ and $\Xi_{q_2} \subseteq \partial\Omega_{p_1\alpha}(\alpha \in \{p_2, p_3\}, p_i \in \{1, 2, \cdots, m\}$ for $i = 1, 2, 3)$, the following subset ${}^{(\mathrm{i})}\Gamma_J$ for mapping $P_J : \Xi_{q_1} \to \Xi_{q_2}$ $(J \in \{1, 2, \cdots, N\}$ and $q_1, q_2 \in \{0, 1, 2, \cdots, M\})$ is called *the initial set of grazing mapping*,

$$
{}^{(\mathrm{i})}\Gamma_J \equiv \left\{ (\mathrm{mod}(\varphi_k, 2\pi), \mathbf{x}_k) \left| \begin{array}{l} P_J \mathbf{y}_k = \mathbf{y}_{k+1}, \\ G^{(s_{p_1},p_1)}_{\partial\Omega_{p_1\alpha}}(\mathbf{x}_{k+1}, t_{(k+1)\pm}) = 0 \\ \text{for } s_{p_1} = 0, 1, \cdots, 2k_{p_1} \text{ and} \\ \hbar_{p_1} G^{(2k_{p_1}+1,p_1)}_{\partial\Omega_{p_1\alpha}}(\mathbf{x}_{k+1}, t_{(k+1)\pm}) < 0 \end{array} \right. \right\} \subset \Xi_{q_1} \tag{6.143}
$$

where $\mathbf{y}_k = (\varphi_k, \mathbf{x}_k)^{\mathrm{T}}$. The corresponding grazing set is defined as

$$
{}^{(\mathrm{f})}\mathrm{G}_J \equiv \left\{ (\mathrm{mod}(\varphi_{k+1}, 2\pi), \mathbf{x}_{k+1}) \left| \begin{array}{l} P_J\mathbf{y}_k = \mathbf{y}_{k+1}, \\ G^{(s_{p_1}, p_1)}_{\partial\Omega_{p_1\alpha}}(\mathbf{x}_{k+1}, t_{(k+1)\pm}) = 0 \\ \text{for } s_{p_1} = 0, 1, \cdots, 2k_{p_1} \text{ and} \\ \hbar_{p_1} G^{(2k_{p_1}+1, p_1)}_{\partial\Omega_{p_1\alpha}}(\mathbf{x}_{k+1}, t_{(k+1)\pm}) < 0 \end{array} \right. \right\} \subset \Xi_{q_2}. \tag{6.144}
$$

Definition 6.10 For a discontinuous dynamical system in Eq. (6.91), consider a final switching point $(\mathrm{mod}(\varphi_{k+1}, 2\pi), \mathbf{x}_{k+1})$

$$
\begin{aligned} & G^{(s_{p_1}, p_1)}_{\partial\Omega_{p_1\alpha}}(\mathbf{x}_{k+1}, t_{(k+1)-}) = 0 \text{ for } s_{p_1} = 0, 1, \cdots, 2k_{p_1} - 1; \\ & \hbar_{p_1} G^{(2k_{p_1}, p_1)}_{\partial\Omega_{p_1\alpha}}(\mathbf{x}_{k+1}, t_{(k+1)-}) < 0. \end{aligned} \tag{6.145}
$$

If $\Xi_{q_1} \subseteq \partial\Omega_{p_1 p_2}$ and $\Xi_{q_2} \subseteq \partial\Omega_{p_1\alpha}(\alpha \in \{p_2, p_3\}$ and $p_i \in \{1, 2, \cdots, m\}$ for $i = 1, 2, 3)$, the following subset ${}^{(\mathrm{f})}\Gamma_J$ for mapping $P_J : \Xi_{q_1} \to \Xi_{q_2} (J \in \{1, 2, \cdots, N\}$ and $q_1, q_2 \in \{0, 1, 2, \cdots, M\})$ is called the final set of grazing post-mapping:

$$
{}^{(\mathrm{f})}\Gamma_J \equiv \left\{ (\mathrm{mod}(\varphi_{k+1}, 2\pi), \mathbf{x}_{k+1}) \left| \begin{array}{l} P_J\mathbf{y}_k = \mathbf{y}_{k+1}, \\ G^{(s_\alpha, \alpha)}_{\partial\Omega_{p_1 p_2}}(\mathbf{x}_k, t_{k\pm}) = 0 \\ s_\alpha = 0, 1, \cdots, 2k_\alpha \text{ and} \\ \hbar_\alpha G^{(2k_\alpha+1, \alpha)}_{\partial\Omega_{p_1 p_2}}(\mathbf{x}_k, t_{k\pm}) < 0 \end{array} \right. \right\} \subset \Xi_{q_2}, \tag{6.146}
$$

where $\mathbf{y}_k = (\Omega t_k, \mathbf{x}_k)^{\mathrm{T}}$. The corresponding grazing set of the post-grazing is defined as

$$
{}^{(\mathrm{i})}\mathrm{G}_J \equiv \left\{ (\mathrm{mod}(\varphi_k, 2\pi), \mathbf{x}_k) \left| \begin{array}{l} P_J\mathbf{y}_k = \mathbf{y}_{k+1}, \\ \mathrm{G}^{(s_\alpha, \alpha)}_{\partial\Omega_{p_1 p_2}}(\mathbf{x}_k, t_{k\pm}) = 0 \\ s_\alpha = 0, 1, \cdots, 2k_\alpha \text{ and} \\ \hbar_\alpha G^{(2k_\alpha+1, \alpha)}_{\partial\Omega_{p_1 p_2}}(\mathbf{x}_k, t_{k\pm}) < 0 \end{array} \right. \right\} \subset \Xi_{q_1}. \tag{6.147}
$$

For global and local grazing mappings, the grazing and post-grazing mapping are sketched in Figs. 6.23 and 6.24 through mapping P_J $(J \in \{J_1, \cdots, J_m\})$ on the boundary $\partial\Omega_{p_1 p_2}$. The grazing in domain Ω_{p_1} occurs at the final points of the grazing mapping P_J on the boundary $\partial\Omega_{p_1\alpha}(\alpha \in \{p_1, p_3\})$. The above definitions for both the initial grazing sets of grazing mapping and the final sets of grazing post-mapping are illustrated. The hollow symbol is the initial point for grazing mapping or the final point of grazing post-mapping. The circular symbolis the grazing point of the grazing or post grazing mappings. The governing equations of mapping P_J with the final point on the boundary $\partial\Omega_{p_1\alpha}$ in Ω_α ($\alpha \in \{p_2, p_3\}$) are expressed by

$$\left.\begin{array}{l}\mathbf{F}^{(J)}(\varphi_k, \mathbf{x}_k, \varphi_{k+1}, \mathbf{x}_{k+1}) = \mathbf{0};\\ \varphi_{p_1p_2}(\mathbf{x}_k, \varphi_k) = 0, \varphi_{p_1\alpha}(\mathbf{x}_{k+1}, \varphi_{k+1}) = 0.\end{array}\right\} \tag{6.148}$$

From Luo (2009), the grazing necessary condition for mapping P_J at the singular set $\Gamma^{(0)}_{p_1\alpha}$ in the sub-domain Ω_α ($\alpha, \beta = \{p_1, p_2\}, \alpha \neq \beta$) is:

$$G^{(p_1)}_{\partial\Omega_{p_1p_2}}(\mathbf{x}_{k+1}, t_{(k+1)\pm}) = \mathbf{n}^{\mathrm{T}}_{\partial\Omega_{p_1p_2}} \cdot \mathbf{F}^{(p_1)}(\varphi_{k+1}, \mathbf{x}_{k+1}) = 0. \tag{6.149}$$

To guarantee the occurrence of the grazing flow at the singular points, the sufficient condition is

$$\hbar_{p_1} G^{(1,p_1)}_{\partial\Omega_{p_1p_2}}(\mathbf{x}_{k+1}, t_{(k+1)\pm}) < 0. \tag{6.150}$$

From Eqs. (6.148) and (6.149), the initial set of grazing mapping is on an $(n-1)$-dimensional surface because of the $(n+3)$-equations with $2(n+1)$-unknowns. Equation (6.150) is the sufficient condition for the initial set. The $(n-1)$-dimensional surface in phase space (mod $(\varphi_k, 2\pi)$, $\mathbf{x}_k \cap \partial\Omega_{p_1p_2}$) is called the *initial grazing manifold*. Such a manifold will be used in discussion of the strange attractor fragmentation. Owing to $\varphi_k = \Omega t_k$, in computation, the switching time conditions $t_{k+1} > t_k$ should be inserted. The boundary $\partial\Omega_{p_1\alpha}$ is given by $\varphi_{p_1\alpha}(\mathbf{x}_k, \varphi_k) = 0$. In addition, equation (6.136) should be satisfied. Similarly, when φ_k and φ_{k+1} in Eqs. (6.148)–(6.150) are exchanged, *the final grazing manifold* can be determined through the final set of grazing post-mapping under the condition in Eq. (6.139).

For higher order singularity, the grazing necessary condition for mapping P_J at the singular set $\Gamma^{(0)}_{p_1\alpha}$ in domain Ω_α ($\alpha, \beta = \{p_1, p_2\}, \alpha \neq \beta$) is:

$$G^{(2k_{p_1}, p_1)}_{\partial\Omega_{p_1p_2}}(\mathbf{x}_{k+1}, t_{(k+1)\pm}) = 0. \tag{6.151}$$

To guarantee the occurrence of the grazing flow at the singular point, the sufficient condition is

$$\hbar_{p_1} G^{(2k_{p_1}+1, p_1)}_{\partial\Omega_{p_1p_2}}(\mathbf{x}_{k+1}, t_{(k+1)\pm}) < 0. \tag{6.152}$$

To investigate the strange attractor fragmentation of chaos through the switching sets, consider each switching set Ξ_q consisting of a set of finite, independent subsets, (i.e., $\Xi_q = \cup_{\kappa=1}^{K} S_q^{(\kappa)}$ and $K < \infty$) and the subsets possesses the following properties

$$S_q^{(\kappa)} \subset \Xi_q, S_q^{(\kappa)} \cap S_q^{(\lambda)} = \varnothing; \kappa, \lambda \in \{1, 2, \cdots, K\} \text{ but } \kappa \neq \lambda. \tag{6.153}$$

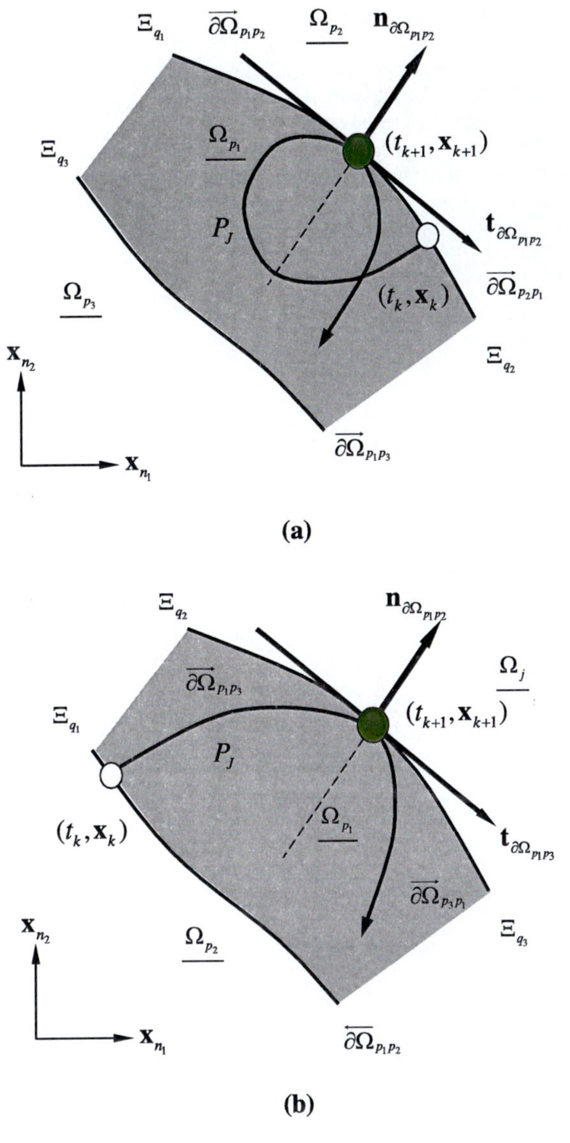

Fig. 6.23 **a** Local and **b** global grazing mappings. The filled solid circular symbols are grazing points. The hollow circular symbols are initial switching points. $(n_1 + n_2 = n)$

For periodic flows, fixed points in the subset $S_{ij}^{(\kappa)}$ for specific κ are countable and finite, and both the measure and Hausdorff dimension of all the subsets are zero. However, from the definition of strange attractors for chaotic flows, the fixed points in such bounded subsets are infinite and countable, and the corresponding Hausdorff dimension is nonzero. In addition, the compact subsets are not hyperbolic. Based on the mapping P_J on the compact subsets of the strange attractor, the corresponding

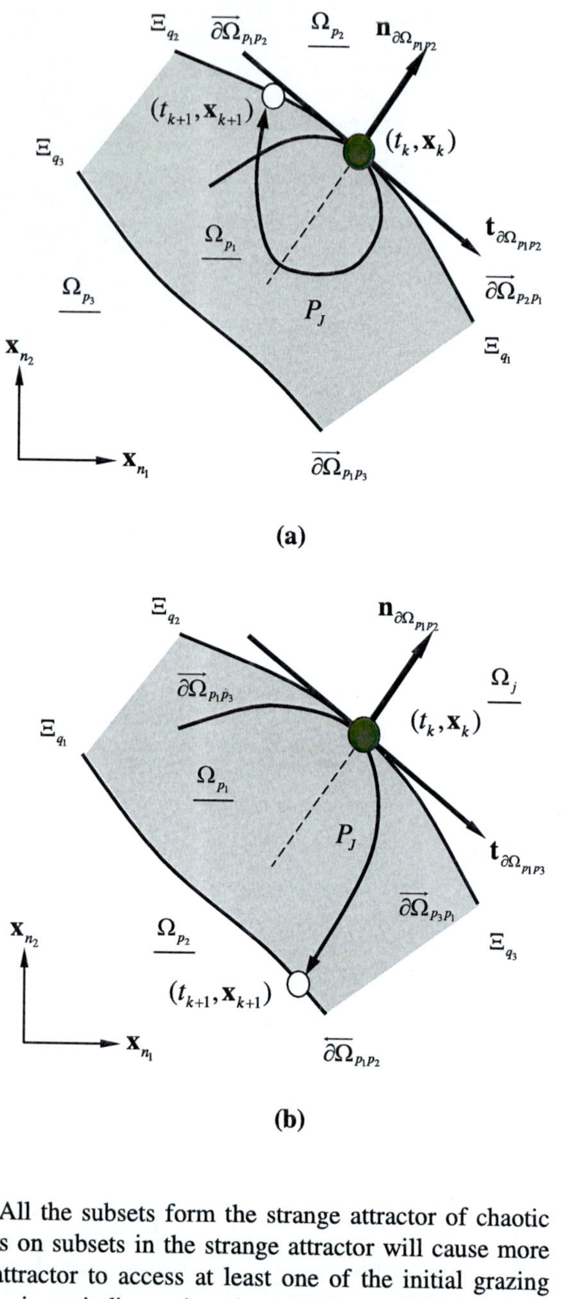

Fig. 6.24 **a** Local and **b** global grazing post-mappings mappings. The filled solid circular symbols are grazing points. The hollow circular symbols are initial switching points. $(n_1 + n_2 = n)$

flows in domains are dense. All the subsets form the strange attractor of chaotic flows. The denseness of flows on subsets in the strange attractor will cause more possibilities for the strange attractor to access at least one of the initial grazing manifolds. However, the flows in periodic motions have less possibility to access the initial grazing manifolds. Without chaotic flows, the transition between the pre- and post-grazing periodic flows can be carried out by a grazing catastrophe. The grazing of periodic flows will be further discussed. To describe the fragmentation of the strange attractors in chaotic flows in discontinuous dynamical systems, the bounded subsets of the strange attractors are used as the initial and final sets of the local and global mappings.

Definition 6.11 For a discontinuous dynamical system in Eq. (6.91), the subsets $S_{q1}^{(\kappa)} \subset \Xi_{q1}$ ($\kappa = 1, 2, \cdots, K$) and $S_{q2}^{(\lambda)} \subset \Xi_{q2}$ ($\lambda = 1, 2, \cdots, L$) are termed the *initial* and *final* subsets of the mapping P_J ($J \in \{1, 2, \cdots, N\}$) if there is a mapping $P_J : S_{q1}^{(\kappa)} \to S_{q2}^{(\lambda)}$ ($q_1, q_2 \in \{0, 1, 2, \cdots, M\}$, $q_1 \neq q_2$).

For convenience, after grazing, the post-grazing mapping structure is

$$^{G}P_J \equiv P_{J_{n+1}} \circ \underbrace{P_{J_n} \circ \cdots \circ P_{J_2} \circ P_{J_1}}_{\text{post-grazing mapping cluster}} = P_{J_{n+1}} \circ {}^{C}P_{(J_n \cdots J_1)}. \tag{6.154}$$

Once the intersection exists between the invariant subsets of strange attractor and one of the *initial* grazing manifolds of generic mappings, the strange attractor fragmentation will occur. For a subset $S_{q1}^{(\kappa)}$ and an initial, grazing manifold $^{(\mathrm{i})}\Gamma_J$ relative to a mapping P_J, if $S_{q1}^{(\kappa)} \cap {}^{(\mathrm{i})}\Gamma_J \neq \varnothing$, then the final subset of the mapping P_J will be fragmentized. If one of the all initial, grazing manifolds is tangential to one of the strange attractor subsets, the strange attractor fragmentation may appear or vanish. Thus, a mathematical definition of strange attractor fragmentation is given as follows.

Definition 6.12 For a discontinuous dynamical system in Eq. (6.91), there is a mapping cluster $P_{(l_m, n_k)}$ with s mappings to generate the strange attractor of chaotic flows on the switching set Ξ_q ($q \in \{0, 1, 2, \cdots, M\}$). For a mapping $P_J : S_{q1}^{(\kappa)} \to S_{q2}^{(\lambda)}$ ($J \in \{1, 2, \cdots, N\}$), if $^{(i)}\Pi_{q1}^{(\kappa)} \equiv S_{q1}^{(\kappa)} \cap {}^{(\mathrm{i})}\Gamma_J \neq \varnothing$ and $S_{q1}^{(\kappa)} = {}^{\mathrm{F}}S_{q1}^{(\kappa)} \cup {}^{\mathrm{U}}S_{q1}^{(\kappa)} \cup {}^{(\mathrm{i})}\Pi_{q1}^{(\kappa)}$, then $^{(\mathrm{f})}\Pi_{q2}^{(\lambda)} \equiv S_{q2}^{(\lambda)} \cap {}^{(\mathrm{f})}\Gamma_J \neq \varnothing$ with $S_{q2}^{(\lambda)} = {}^{\mathrm{F}}S_{q2}^{(\lambda)} \cup {}^{\mathrm{U}}S_{q2}^{(\lambda)} \cup {}^{(\mathrm{f})}\Pi_{q2}^{(\lambda)}$ exist to make the following mappings hold

$$P_J : {}^{(\mathrm{i})}\Pi_{q1}^{(\kappa)} \to {}^{(\mathrm{f})}G_J^{(\kappa)}, \text{ for } {}^{(\mathrm{f})}G_J^{(\kappa)} \subset {}^{(\mathrm{f})}G_J, \tag{6.155}$$

$$P_J : {}^{\mathrm{U}}S_{q1}^{(\kappa)} \to {}^{\mathrm{U}}S_{q2}^{(\lambda)} \text{ and } {}^{G}P_J : {}^{\mathrm{F}}S_{q1}^{(\kappa)} \to {}^{\mathrm{F}}S_{q2}^{(\lambda)}. \tag{6.156}$$

For the post-grazing mapping cluster $^{G}P_J = P_{J_{n+1}} \circ P_{J_n} \circ \cdots \circ P_{J_2} \circ P_{J_1}$, if there is a mapping chain as

$$^{\mathrm{F}}S_{q1}^{(\kappa)} \xrightarrow{P_{J_1}} {}^{\mathrm{F}}\mathfrak{A}_{\rho_1}^{(\kappa_1)} \ldots {}^{\mathrm{F}}\mathfrak{A}_{\rho_{n-1}}^{(\kappa_{n-1})} \xrightarrow{P_{J_n}} {}^{\mathrm{F}}\mathfrak{A}_{\rho_n}^{(\kappa_n)} \xrightarrow{P_{J_{n+1}}} {}^{\mathrm{F}}S_{q2}^{(\lambda)}. \tag{6.157}$$

The union of all the switching sets generated by the mapping cluster of $^{G}P_J$, $\mathfrak{A}_\rho = \cup_{i=1}^{n} \mathfrak{A}_{\rho_i}^{(\kappa_i)}$ ($\mathfrak{A}_{\rho_i}^{(\kappa_i)} \subset \Xi_{\rho_i}$ for $\rho_i \in \{0, 1, 2, \cdots, M\}$) is termed *the fragmentation set* of the invariant set $S_{q1}^{(\kappa)} \subset \Xi_{q1}$ under the mapping cluster $P_{(l_m, n_k)}$.

To intuitively demonstrate the concept introduced above, consider the invariant subsets of a strange attractor on the boundary $\partial\Omega_{p_1 p_2}$ for the following mapping structure

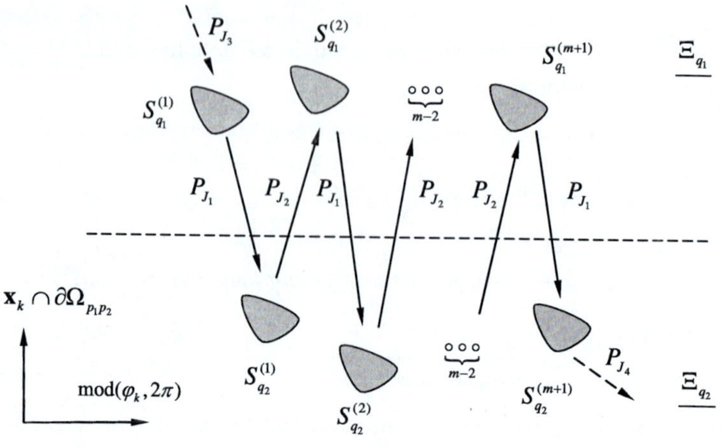

(a)

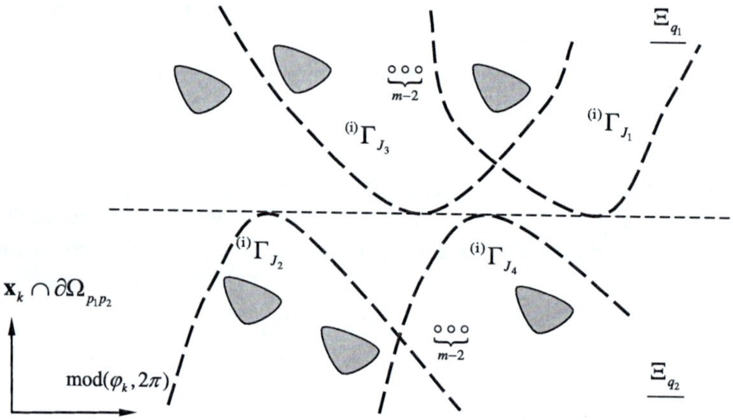

(b)

Fig. 6.25 **a** Invariant subsets and **b** invariant initial manifolds of grazing mapping on the boundary $\partial\Omega_{p_1 p_2}$

$$P_{\cdots J_4 (J_1 J_2)^m J_1 J_3 \cdots} = \circ \cdots \circ P_{J_4} \circ \underbrace{(P_{J_1} \circ P_{J_2})}_{\text{m-sets}} \circ P_{J_1} \circ P_{J_3} \circ \cdots \circ . \tag{6.158}$$

The invariant sets are generated by

$$P^{(l)}_{\cdots J_4 (J_1 J_2)^m J_1 J_3 \cdots} = \underbrace{P_{\cdots J_4 (J_1 J_2)^m J_1 J_3 \cdots} \circ \cdots \circ P_{\circ \cdots \circ J_4 (J_1 J_2)^m J_1 J_3 \cdots}}_{l \to \infty} \tag{6.159}$$

and the corresponding invariant sets on the boundary $\partial\Omega_{p_1 p_2}$ are

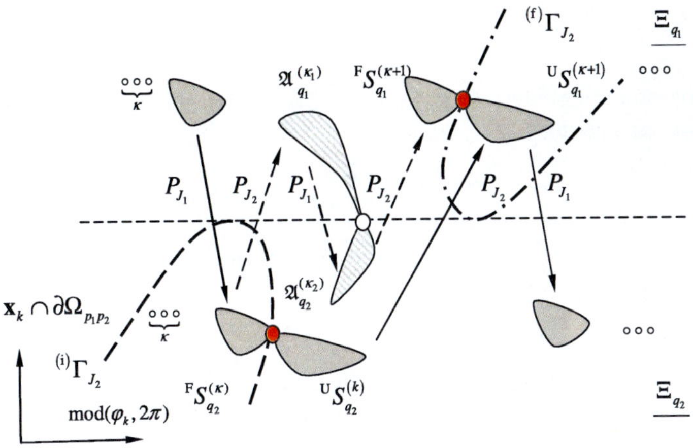

Fig. 6.26 Invariant sets fragmentation on the boundary $\partial\Omega_{p_1 p_2}$

$$\Xi_{q1} = \cup_{\kappa=1}^{m+1} S_{q1}^{(\kappa)} \text{ and } \Xi_{q2} = \cup_{\kappa=1}^{m+1} S_{q2}^{(\kappa)}. \tag{6.160}$$

The foregoing invariant subsets are illustrated in Fig. 6.25a through the filled areas on the switching plane $(\mathrm{mod}(\varphi_i, 2\pi), \mathbf{x}_i) \cap \partial\Omega_{p_1 p_2}$. The mapping relation for the mapping structure in Eq.(6.148) is also presented. The initial grazing manifolds ${}^{(i)}\Gamma_J$ $(J \in \{J_1, J_2, J_3, J_4\})$ are sketched by the dashed curves in Fig. 6.25b in the switching planes. For this case, $\Xi_{q1} \cap {}^{(i)}\Gamma_J = \varnothing$ and $\Xi_{q2} \cap {}^{(i)}\Gamma_J = \varnothing$, no fragmentation of the strange attractor occurs for the aforementioned mapping structure. If $\Xi_{q1} \cap {}^{(i)}\Gamma_J \neq \varnothing$ and/or $\Xi_{q2} \cap {}^{(i)}\Gamma_J \neq \varnothing$, the fragmentation will occur. The fragmentation of strange attractors on $\partial\Omega_{p_1 p_2}$ is sketched in Fig. 6.26. Suppose ${}^{(i)}\Pi_{q2}^{(\kappa)} = S_{q2}^{(\kappa)} \cap {}^{(i)}\Gamma_{J_2} \neq \varnothing$ be represented by solid symbols, $S_{q2}^{(\kappa)} = {}^{\mathrm{F}}S_{q2}^{(\kappa)} \cup {}^{\mathrm{U}}S_{q2}^{(\kappa)} \cup {}^{(i)}\Pi_{J_2}^{(\kappa)}$. The initial grazing manifold is depicted by the dashed curve. After fragmentation, two new invariant sets $\mathfrak{A}_{q2}^{(\kappa_2)} \subset \Xi_{q2}$ and $\mathfrak{A}_{q1}^{(\kappa_1)} \subset \Xi_{q1}$ exist, which are showed by hatched areas. The non-fragmentized mapping are: $P_{J_1} : {}^{\mathrm{U}}S_{q2}^{(\kappa)} \to {}^{\mathrm{U}}S_{q1}^{(\kappa+1)}$ and the mappings relative to the fragmentation are: $P_{J_2} : {}^{\mathrm{F}}S_{q2}^{(\kappa)} \to \mathfrak{A}_{q1}^{(\kappa_1)}$, $P_{J_1} : \mathfrak{A}_{q1}^{(\kappa_1)} \to \mathfrak{A}_{q2}^{(\kappa_2)}$ and $P_{J_2} : \mathfrak{A}_{q2}^{(\kappa_1)} \to {}^{\mathrm{F}}S_{q1}^{(\kappa+1)}$. From the new subsets to the non-fragmentized subsets, the final manifold of the post-grazing ${}^{(f)}\Gamma_{J_2}$, expressed by the dotted dash curve, separates the invariant subset into two parts ${}^{\mathrm{U}}S_{q1}^{(\kappa+1)} \cup {}^{\mathrm{F}}S_{q1}^{(\kappa+1)}$ plus the intersected set ${}^{(f)}\Pi_{q1}^{(\kappa+1)} = S_{q1}^{(\kappa+1)} \cap {}^{(f)}\Gamma_{J_1} \neq \varnothing$. Note that if ${}^{(i)}\Pi_{q2}^{(\kappa)}$ possesses n-values, there are n-pieces of non-intersected, invariant subsets in $\mathfrak{A}_{q2}^{(1)}$. As $n \to \infty$, countable, infinite pieces of non-intersected invariant subsets are obtained by such an attractor fragmentation. The initial sets of grazing mapping are presented and the corresponding, initial grazing manifolds are presented. Finally, the grazing-induced fragmentation of strange attractors of chaotic motions in discontinuous dynamical systems is discussed. The theory for such a fragmentation of strange attractors should be further developed.

6.7 Fragmentized Strange Attractors

To demonstrate the fragmentized strange attractor, consider a periodically excited, piecewise linear system as

$$\ddot{x} + 2d\dot{x} + k(x) = a\cos\Omega t, \tag{6.161}$$

where $\dot{x} = dx/dt$. The parameters (Ω and a) are excitation frequency and amplitude, respectively. The restoring force is

$$k(x) = \begin{cases} cx - e, & \text{for } x \in [E, \infty); \\ 0, & \text{for } x \in [-E, E]; \\ cx + e, & \text{for } x \in (-\infty, -E]; \end{cases} \tag{6.162}$$

with $E = e/c$. The foregoing system possesses three linear regions of the restoring force (Region I: $x \geq E$, Region II: $-E \leq x \leq E$ and Region III: $x \leq -E$). From Eqs. (6.161) and (6.162), the dynamical systems in Regions I and III do not have any singularity. The motions of dynamical systems in Region I and III are finite. In Region II, the motion of dynamical systems is unstable. However, the displacement of the domain on which the dynamical system is defined is bounded. Since the velocity is the derivative of displacement with respect to time, the flows of the system in the displacement-bounded, Region II is bounded.

The phase space in Eq. (6.161) is divided into three sub-domains, and the three sub-domains are defined by

$$\left.\begin{aligned} \Omega_1 &= \{(x, y) | \, x \in [E, \infty), \, y \in (-\infty, \infty)\}, \\ \Omega_2 &= \{(x, y) | \, x \in [-E, E], \, y \in (-\infty, \infty)\}, \\ \Omega_3 &= \{(x, y) | \, x \in (-\infty, -E], \, y \in (-\infty, \infty)\}. \end{aligned}\right\} \tag{6.163}$$

The entire phase space is given by

$$\Omega = \cup_{\alpha=1}^{3} \Omega_\alpha. \tag{6.164}$$

The corresponding boundaries are

$$\left.\begin{aligned} \partial\Omega_{12} &= \Omega_1 \cap \Omega_2 = \{(x, y) | \, \varphi_{12}(x, y) \equiv x - E = 0\}, \\ \partial\Omega_{23} &= \Omega_2 \cap \Omega_3 = \{(x, y) | \, \varphi_{23}(x, y) \equiv x + E = 0\}. \end{aligned}\right\} \tag{6.165}$$

Such domains and the boundary are sketched in Fig. 6.27. From the above definitions, Eqs. (6.161) and (6.162) give

$$\mathbf{x}^{(\alpha)} = \mathbf{F}^{(\alpha)}(\mathbf{x}^{(\alpha)}, t, \boldsymbol{\mu}_\alpha, \boldsymbol{\pi}) \text{for } \alpha = 1, 2, 3, \tag{6.166}$$

where

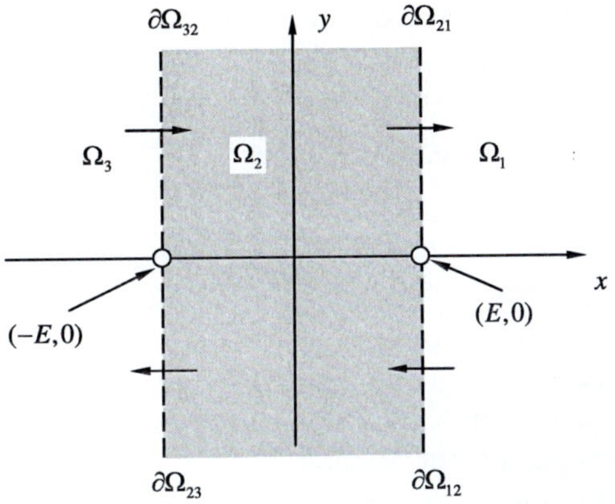

Fig. 6.27 Sub-domains, boundaries and equilibriums $(\pm E, 0)$

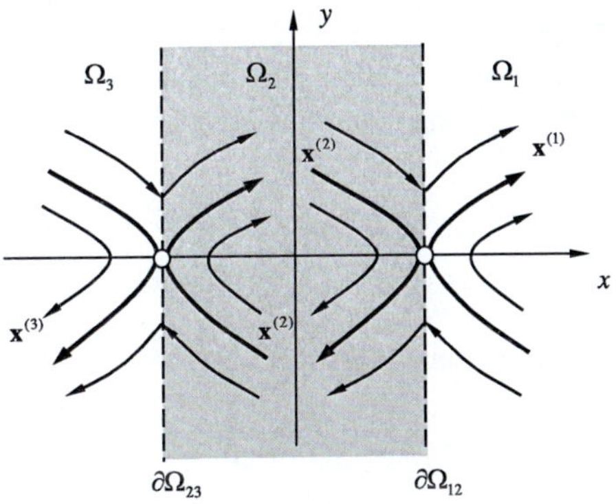

Fig. 6.28 Phase portraits near equilibriums $(\pm E, 0)$ on the boundaries

$$
\begin{aligned}
&\mathbf{F}^{(1)}(\mathbf{x}^{(1)}, t, \mu_1, \boldsymbol{\pi}) = (y^{(1)}, -2dy^{(1)} - cx^{(1)} + a\cos\omega t)^{\mathrm{T}}, && \text{for } \mathbf{x}^{(1)} \in \Omega_1;\\
&\mathbf{F}^{(2)}(\mathbf{x}^{(2)}, t, \mu_2, \boldsymbol{\pi}) = (y^{(2)}, -2dy^{(2)} + a\cos\omega t)^{\mathrm{T}}, && \text{for } \mathbf{x}^{(2)} \in \Omega_2;\\
&\mathbf{F}^{(3)}(\mathbf{x}^{(3)}, t, \mu_3, \boldsymbol{\pi}) = (y^{(3)}, -2dy^{(3)} - cx^{(3)} + a\cos\omega t)^{\mathrm{T}}, && \text{for } \mathbf{x}^{(3)} \in \Omega_3.
\end{aligned}
\tag{6.167}
$$

Note that $\mu_1 = \mu_3 = (c, d)^{\mathrm{T}}, \mu_2 = (0, d)^{\mathrm{T}}$ and $\boldsymbol{\pi} = (\Omega, a)^{\mathrm{T}}$. To investigate the global dynamics of Eq. (6.161), an understanding of the local singularity of the flow the boundary is very important. From Eq. (6.165), the normal vectors of the boundaries (i.e.,$\mathbf{n}_{\partial\Omega_{ij}} = \nabla\varphi_{ij}$) is given by

$$
\mathbf{n}_{\partial\Omega_{12}} = \mathbf{n}_{\partial\Omega_{23}} = (1, 0)^{\mathrm{T}}. \tag{6.168}
$$

From Eqs. (6.167) and (6.168), one gets

$$
\begin{aligned}
&\mathbf{n}^{\mathrm{T}}_{\partial\Omega_{12}} \cdot \mathbf{F}^{(1)}(\mathbf{x}^{(1)}, t, \mu_1, \boldsymbol{\pi}) = y^{(1)}, \mathbf{n}^{\mathrm{T}}_{\partial\Omega_{12}} \cdot \mathbf{F}^{(2)}(\mathbf{x}^{(2)}, \mathrm{t}, \mu_2, \boldsymbol{\pi}) = \mathrm{y}^{(2)};\\
&\mathbf{n}^{\mathrm{T}}_{\partial\Omega_{23}} \cdot \mathbf{F}^{(2)}(\mathbf{x}^{(2)}, t, \mu_2, \boldsymbol{\pi}) = y^{(2)}, \mathbf{n}^{\mathrm{T}}_{\partial\Omega_{23}} \cdot \mathbf{F}^{(3)}(\mathbf{x}^{(3)}, \mathrm{t}, \mu_3, \boldsymbol{\pi}) = \mathrm{y}^{(3)}.
\end{aligned}
\tag{6.169}
$$

From Eqs. (6.168) and (6.169), points $(\pm E, 0)$ are critical for a flow to be tangential to the boundary. From Luo (2006a, 2009, 2011), the grazing bifurcation of a flow to the boundary in this discontinuous system are

$$\begin{aligned} y^{(1)} &= 0,\ \dot{y}^{(1)} > 0 \text{ at } x^{(1)} = E,\\ y^{(2)} &= 0,\ \dot{y}^{(2)} > 0 \text{ at } x^{(2)} = -E,\\ y^{(2)} &= 0,\ \dot{y}^{(2)} < 0 \text{ at } x^{(2)} = E,\\ y^{(3)} &= 0,\ \dot{y}^{(3)} < 0 \text{ at } x^{(3)} = -E. \end{aligned} \tag{6.170}$$

Therefore, in the neighborhoods of the two equilibrium points, the local topological structures of flows in the system given in Eq. (6.161) are sketched in Fig. 6.28. The detailed discussion can be refereed to Luo (2006b). The switching sets and mappings can be defined in Eqs.(6.106)–(6.110). The mappings are switched in Fig. 6.20. Consider the initial and final states of $(t, x, \dot{x})$ to be (t_k, x_k, y_k) and $(t_{k+1}, x_{k+1}, y_{k+1})$ in the sub-domain Ω_α $(\alpha = 1, 2, 3)$, respectively. The local mappings are $\{P_1, P_3, P_5, P_6\}$ and the global mappings are $\{P_2, P_4\}$. The displacement and velocity equations in the linear system in Eq. (6.161) with initial conditions give the governing equations for mapping P_j $(j = 1, 2, \cdots, 6)$, i.e.,

$$P_j : \begin{cases} f_1^{(j)}(x_k, y_k, t_k, x_{k+1}, y_{k+1}, t_{k+1}) = 0,\\ f_2^{(j)}(x_k, y_k, t_k, x_{k+1}, y_{k+1}, t_{k+1}) = 0. \end{cases} \tag{6.171}$$

The necessary and sufficient conditions for all the six generic mappings are:

$$\left.\begin{aligned} y_{k+1}^{(j)} &= 0;\\ \dot{y}_{k+1}^{(j)} &= a\cos\Omega t_{k+1} > 0 \text{ for } P_j\ (j = 1, 2, 6),\\ \dot{y}_{k+1}^{(j)} &= a\cos\Omega t_{k+1} < 0 \text{ for } P_j\ (j = 3, 4, 5). \end{aligned}\right\} \tag{6.172}$$

With the foregoing equation, once one of the initial time and velocity is selected, the grazing bifurcation can be determined.

The fragmentation of strange attractors in chaotic motion is illustrated by the Poincare mapping sections for four sub-switching planes. As in Fig. 6.21, the subsets (i.e., Ξ_2 and Ξ_1) of the switching plane Ξ_{12} for strange attractors are on the upper and lower dashed line. Similarly, the subsets (i.e., Ξ_3 and Ξ_4) of the switching plane Ξ_{34} separated by a dashed line are presented as well. The dashed curves are the *initial* grazing, switching manifolds computed by Eq. (6.172) for specific parameters. The location of grazing points are labeled by "Grazing". The parameters ($a = 20, c = 100, E = 1, d = 0.5$, and $x_i = 1$) are used. Consider a motion with a single grazing first for $\Omega = 2.10$ and the initial condition $(\text{mod}(\Omega t_k, 2\pi), y_k) \approx (6.1117, 6.5251)$ at $x_i = 1$. The Poincare mapping sections are plotted by the switching planes in Fig. 6.29 for a strange attractor of chaotic motion relative to mapping structures ($P_{643(12)1}$, P_{6431} and$P_{6(45)431}$). The initial, grazing switching manifold of P_2 has three intersected points with the strange attractor in Ξ_1 Based on three intersection points, the grazing points

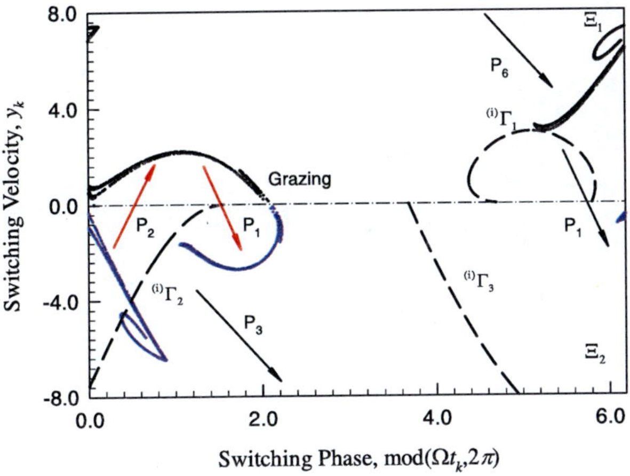

(a)

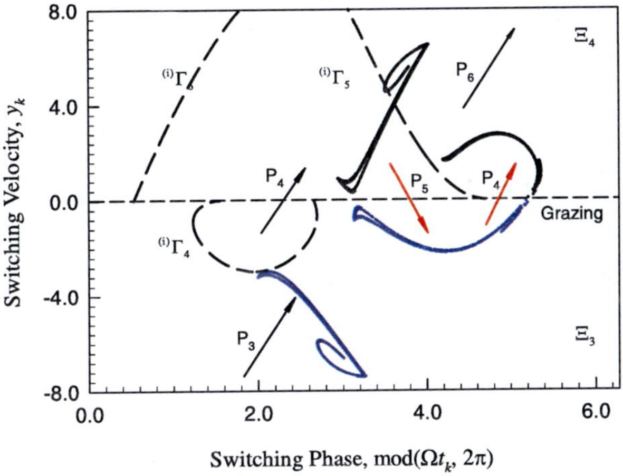

(b)

Fig. 6.29 Chaotic motion associated with mappings ($P_{6(45)431}$, P_{6431} and $P_{643(12)1}$) : **a** subsets of switching plane Ξ_{12} ($\Xi_1$andΞ_2) and **b** subsets of switching plane Ξ_{34} ($\Xi_3$andΞ_4). ($a = 20, c = 100, E = 1, d = 0.5, \Omega t_k \approx 6.1117, x_k = 1, y_k \approx 6.5251$ and $\Omega = 2.1$)

are (mod($\Omega t_k, 2\pi$), y_k) $\approx$ (2.056, 0.0), (2.087, 0.0) and (2.094, 0.0), labeled by "Grazing". The new invariant set of the strange attractor has two branches in both Ξ_1 and Ξ_2 compared to the strange attractor relative to P_{6431}. Hence, there are two mappings of $P_2 : \Xi_2 \to \Xi_1$ and one mapping of $P_1 : \Xi_1 \to \Xi_2$. The chaotic motion is relative to the mapping structure $P_{643(12)1}$ and P_{6431}. However, for the switching section of Ξ_3, the initial, grazing switching manifold of mapping P_5 is similar to mapping P_2 which possesses three intersected points with the strange attractor. The

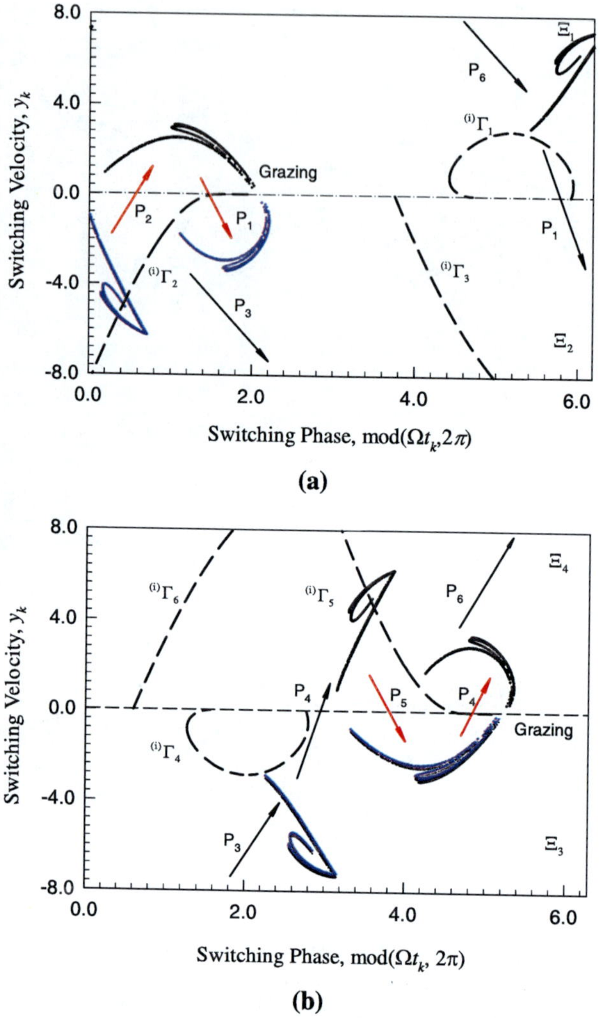

Fig. 6.30 Fragmented strange attractor for chaotic motion associated with mappings ($P_{6(45)431}$, P_{6431} and $P_{643(12)1}$) : **a** subsets of switching plane Ξ_{12} ($\Xi_1$andΞ_2) and **b** subsets of switching plane Ξ_{34} ($\Xi_3$andΞ_4). ($a = 20, c = 100, E = 1, d = 0.5, \Omega t_k \approx 5.7257, x_k = 1, y_k \approx 6.2363$ and $\Omega = 1.89$)

grazing locations are close to $(\text{mod}(\Omega t_k, 2\pi), y_k) \approx (5.200, 0.0)$, $(5.229, 0.0)$ and $(5.235, 0.0)$ accordingly. The initial grazing manifold of mapping P_1 is almost tangential to the strange attractor at point $(\text{mod}(\Omega t_k, 2\pi), y_k) \approx (5.24, 2.93)$ in Ξ_1. Some points in the strange attractor in Ξ_1 and Ξ_2 are close to the grazing point $(\text{mod}(\Omega t_k, 2\pi), y_k) \approx (0.0193, 0.0)$, as shown in Fig. 6.29. Similarly, the initial grazing manifold of mapping P_3 almost tangential to the strange attractor is observed.

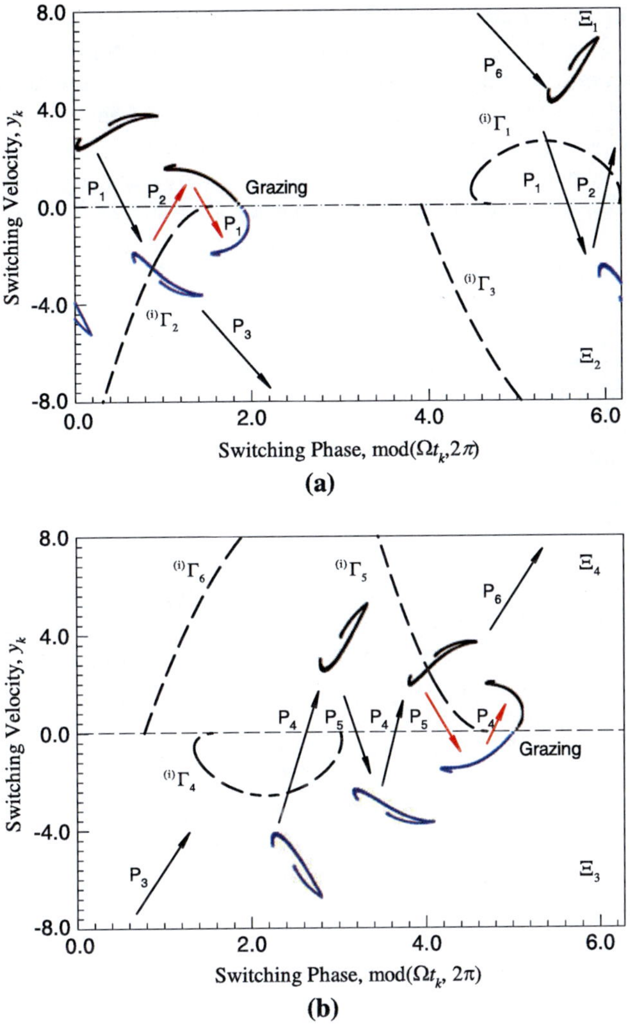

Fig. 6.31 Chaotic motion relative to mappings ($P_{6(45)43(12)1}$, $P_{6(45)43(12)^21}$ and $P_{6(45)^243(12)1}$) : **a** subsets of switching plane Ξ_{12} (Ξ_1 and Ξ_2) and **b** subsets of switching plane Ξ_{34} (Ξ_3 and Ξ_4). ($a = 20, c = 100, E = 1, d = 0.5, \Omega t_k \approx 0.1062, x_k = 1, y_k \approx 2.4511$ and $\Omega = 1.4$)

Further illustrations of the fragmentized attractor relative to mapping P_{6431} are given for a better understanding of strange attractor fragmentation. For $\Omega = 1.89$ with the same other system parameters, the fragmentized strange attractor of chaotic motion at $x_k = E$ is presented in Fig. 6.30 with $(\Omega t_k, y_k) \approx (5.7257, 6.2363)$. The shape of the strange attractor is distinct from the one in Fig. 6.29. The grazing locations on the switching set Ξ_{12} are close to $(\Omega t_k, y_k) \approx (2.018, 0.0)$, $(1.991, 0.0)$, $(1.956, 0.0)$ and $(1.929, 0.0)$. The grazing locations on the switching set Ξ_{34} are near by $(\Omega t_k, y_k) \approx (5.159, 0.0)$, $(5.133, 0.0)$, $(5.098, 0.0)$ and $(5.071, 0.0)$. It is observed that the two fragmentized attractors are distinguishing. With varying system parameters, such fragmentized strange attractors of chaotic motions will disappear. For instance, decreasing excitation frequency yields the symmetric and asymmetric, periodic motion relative to mapping $P_{6(45)43(12)1}$. If the grazing of the asymmetric periodic motion occurs, the fragmentized strange attractor will exist. Consider another excitation frequency $\Omega = 1.4$ with the initial condition $(\Omega t_k, y_k) \approx (0.1062, 2.4511)$ at $x_k = E$. The Poincare mapping sections of the strange attractor of chaotic motion are shown in Fig. 6.31. There are three branches of the strange attractor. The initial, grazing manifolds of the mappings P_2 in Ξ_2 and P_5 in Ξ_4 have one intersected point with one of three branches of the strange attractor, and the corresponding grazing points are $(\Omega t_k, y_k) \approx (1.876, 0.0)$ and $(5.0375, 0.0)$ at Ξ_{12} and Ξ_{34}, respectively. Owing to grazing, one of three branches of the strange attractor is produced by such a grazing. Thus, the strange attractor of the chaotic motion possesses the mapping structures of $P_{6(45)43(12)1}$, $P_{6(45)^2 43(12)1}$ and $P_{6(45)43(12)^2 1}$. The other fragmentized strange attractors of the chaotic motion can be illustrated in the similar fashion.

From the foregoing discussion, the initial and final grazing, switching manifolds are invariant for given system parameters, which is an important clue to investigate the mechanism of strange attractor fragmentation. The criteria and topological structure for the fragmentation of the strange attractor need to be further developed as in hyperbolic strange attractors. The fragmentation of the strange attractors extensively exists in discontinuous dynamical systems, which will help us better understand motion complexity in discontinuous dynamic systems.

From this investigation, the dynamical behaviors in dynamical systems may not need the Axioms in Smale (1967). The corresponding spectral decomposition of diffeomorphisms can be directly obtained from the switching sets of discontinuous boundary. The strange attractor fragmentation caused by grazing singularity is a key to open the door for the topological structures of chaos in dynamical systems. The hyperbolic set needed in Smale (1967) may be for separatrix as a hidden discontinuity. No matter how, the continuous flow should be employed to discuss the topological structures of chaos, rather than mappings only.

References

Bartissol, P. and Chua, L.O. 1988, The Double Hook *Nonlinear Chaotic Circuits, IEEE Transactions on Circuits and Systems* **35**:1512–1522.

Han, R.P.S., Luo, A..C.J. and Deng, W. 1995, Chaotic motion of a horizontal impact pair, *Journal of Sound and Vibration*, **181**:231–250.

Luo, A.C.J., 2002, An unsymmetrical motion in a horizontal impact oscillator, *ASME Journal of Vibration and Acoustics* **124**:420–426.

Luo, A.C.J., 2005a, A theory for non-smooth dynamical systems on connectable domains, *Communication in Nonlinear Science and Numerical Simulation* **10**:1–55.

Luo, A.C.J., 2005b, The mapping dynamics of periodic motions for a three-piecewise linear system under a periodic excitation, *Journal of Sound and Vibration* **283**:723–748.

Luo, A.C.J., 2005c, The symmetry of steady-state solutions in non-smooth dynamical systems with two constraints, *IMechE Part K Journal of Multi-body Dynamics* **219**:09–124.

Luo, A.C.J., 2006, *Singularity and Dynamics on Discontinuous Vector Fields*, Amsterdam: Elsevier.

Luo, A.C.J., 2006b, Grazing and chaos in a periodically forced, piecewise linear system, *ASME Journal of Vibration and Acoustics* **128**:28–34.

Luo, A.C.J., 2008a, A theory for flow switchability in discontinuous dynamical systems, *Nonlinear Analysis: Hybrid Systems*, **2**(4):1030–1061.

Luo, A.C.J., 2008b, *Global Transversality, Resonance and Chaotic Dynamics*, World Scientific: Singapore.

Luo, A.C.J., 2009, *Discontinuous Dynamical Systems on Time varying Domains*, HEP-Springer: Heidelberg.

Luo, A.C.J., 2011, *Discontinuous Dynamical Systems*, HEP-Springer: Heidelberg.

Luo, A.C.J. and Chen, L.D., 2006, Grazing phenomena and fragmented strange attractors in a harmonically forced, piecewise, linear system with impacts, IMeChe Part K:*Journal of Multi-body Dynamics*, **220**:35–51.

Luo, A.C.J. and Gegg, B.C., 2006, Stick and non-stick, periodic motions of a periodically forced, linear oscillator with dry friction, *Journal of Sound and Vibration* **291**:132–168

Luo, A.C.J. and Gegg, B.C., 2007, An analytical prediction of sliding motions along discontinuous boundary in non-smooth dynamical systems, *Nonlinear Dynamics* **40**:401–424.

Luo, A.C.J. and Guo, Y., 2010, Switching mechanism and complex motion in a generalized Fermi accelleration oscillator, *ASME Journal of Computational and Nonlinear Dynamics* **5**(4):041007 (1–14).

Luo, A.C.J. and O'Connor, D., 2009, Impact chatter in a gear transmission system with two oscillators, *IMrChe Part K: Journal of Multi-body Dynamics* **223**:159–188.

Luo, A.C.J., Rapp, B.M., 2009 Flow switchability and periodic motions in a periodically forced, discontinuous dynamical system, *Nonlinear Analysis: Real World Applications* **10**:3028–3044.

Luo, A.C.J. and Zwiegart Jr., P., 2008, Existence and analytical prediction of periodic motions in a periodically forced, nonlinear friction oscillator, *Journal of Sound and Vibration* **309**:129–149.

Luo, A.C.J., Xue, B., 2009, An analytical prediction of periodic flows in the Chua's Circuit system, *International Journal of Bifurcation and Chaos* **19**:2165–2180.

Smale, S., 1967, Differential dynamical systems, *Bulletin of the American Mathematical Society*, **73**, pp.747–817.

Appendix A
Linear Continuous Dynamical Systems

In this Appendix, the theory of linear systems will be presented to review the traditional linear dynamical systems. Separated linear systems and diagonalization of square matrix will be discussed first. The linear operator exponentials will be presented. The fundamental solutions of autonomous linear systems will be given with the matrix possessing real eigenvalues, complex eigenvalues and repeated eigenvalues. The stability theory for autonomous linear systems will be discussed. The solutions of non-autonomous linear systems will be discussed and steady state solutions will be presented. A generalized "resonance" concept will be introduced, and the resonant solutions will be presented. Lower-dimensional linear systems will be discussed in detail for solutions and stability.

A.1 Basic Solutions

Definition A.1 Consider a linear dynamical system

$$\dot{\mathbf{x}} = \mathbf{A}\mathbf{x} + \mathbf{Q}(t) \text{ for } t \in \mathscr{R} \text{ and } \mathbf{x} = (x_1, x_2, \ldots, x_n)^{\mathrm{T}} \in \mathscr{R}^n \tag{A.1}$$

where $\dot{\mathbf{x}} = d\mathbf{x}/dt$ is differentiation with respect to time t. $\mathbf{A}$ is an $n \times n$ matrix and $\mathbf{Q}(t)$ is a continuous vector function. If $\mathbf{Q}(t) = \mathbf{0}$, the linear dynamical system in Eq. (A.1) is autonomous. Equation (A.1) becomes

$$\dot{\mathbf{x}} = \mathbf{A}\mathbf{x} \text{ for } t \in \mathscr{R} \text{ and } \mathbf{x} \in \mathscr{R}^n \tag{A.2}$$

which is called an autonomous linear system or a homogenous linear system. With an initial condition of $\mathbf{x}(t_0) = \mathbf{x}_0$, the solution of Eq. (A.2) is given by

$$\mathbf{x}(t) = e^{\mathbf{A}(t-t_0)}\mathbf{x}_0. \tag{A.3}$$

If $\mathbf{Q}(t) \neq \mathbf{0}$, the linear dynamical system in Eq. (A.1) is non-autonomous, and such a non-autonomous system is also called a nonhomogenous linear system. With an initial condition of $\mathbf{x}(t_0) = \mathbf{x}_0$, the solution of Eq. (A.1) is given by

A. C. J. Luo, *Regularity and Complexity in Dynamical Systems*,
DOI: 10.1007/978-1-4614-1524-4, © Springer Science+Business Media, LLC 2012

$$\mathbf{x}(t) = \Phi(t)\Phi^{-1}(t_0)\mathbf{x}_0 + \int_{t_0}^{t} \Phi(t)\Phi^{-1}(\tau)\mathbf{Q}(\tau)d\tau. \tag{A.4}$$

where $\Phi(t)$ is a fundamental matrix of the homogenous linear system in Eq. (A.2) with

$$\dot{\Phi}(t) = \mathbf{A}\Phi(t) \text{ for all } t \in I \subseteq \mathscr{R}. \tag{A.5}$$

Definition A.2 For a linear dynamical system in Eq. (A.2), if the linear matrix $\mathbf{A} = diag(\lambda_1, \lambda_2, \ldots, \lambda_n)$ is a diagonal matrix, then the linear dynamical system in Eq. (A.2) is called an uncoupled linear homogenous system. With an initial condition of $\mathbf{x}(t_0) = \mathbf{x}_0$, the solution of the uncoupled linear homogenous solution is

$$\mathbf{x}(t) = diag[e^{\lambda_1(t-t_0)}, e^{\lambda_2(t-t_0)}, \ldots, e^{\lambda_n(t-t_0)}]\mathbf{x}_0. \tag{A.6}$$

Theorem A.1 *Consider a linear dynamical system* $\dot{\mathbf{x}} = \mathbf{A}\mathbf{x}$ *in Eq.* (A.2) *with the initial condition of* $\mathbf{x}(t_0) = \mathbf{x}_0$. *If the real and distinct eigenvalues of the* $n \times n$ *matrix* $\mathbf{A}$ *are* $\lambda_1, \lambda_2, \ldots, \lambda_n$, *then a set of corresponding eigenvectors* $\{\mathbf{v}_1, \mathbf{v}_2, \ldots, \mathbf{v}_n\}$ *is determined by*

$$(\mathbf{A} - \lambda_i \mathbf{I})\mathbf{v}_i = \mathbf{0} \tag{A.7}$$

which forms a basis in $\Omega \subseteq \mathscr{R}^n$. *The eigenvector matrix of* $\mathbf{P} = [\mathbf{v}_1, \mathbf{v}_2, \ldots, \mathbf{v}_n]$ *is invertible and*

$$\mathbf{P}^{-1}\mathbf{A}\mathbf{P} = diag[\lambda_1, \lambda_2, \ldots, \lambda_n]. \tag{A.8}$$

Thus, with an initial condition of $\mathbf{x}(t_0) = \mathbf{x}_0$, *the solution of linear dynamical system in Eq.* (A.2) *is*

$$\begin{aligned} \mathbf{x}(t) &= \mathbf{P}diag[e^{\lambda_1(t-t_0)}, e^{\lambda_2(t-t_0)}, \ldots, e^{\lambda_n(t-t_0)}]\mathbf{P}^{-1}\mathbf{x}_0 \\ &= \mathbf{P}\mathbf{E}(t-t_0)\mathbf{P}^{-1}\mathbf{x}_0 \end{aligned} \tag{A.9}$$

where the diagonal matrix $\mathbf{E}(t)$ *is given by*

$$\mathbf{E}(t-t_0) = diag[e^{\lambda_1(t-t_0)}, e^{\lambda_2(t-t_0)}, \ldots, e^{\lambda_n(t-t_0)}]. \tag{A.10}$$

Proof Assuming $\mathbf{x}(t) = \mathbf{C}e^{\lambda t} = C\mathbf{v}e^{\lambda t}$, equation (A.2) gives $(\mathbf{A} - \lambda\mathbf{I})\mathbf{v} = \mathbf{0}$. Since $\det(\mathbf{A} - \lambda\mathbf{I}) = 0$ gives real and distinct eigenvalues λ_i $(i = 1, 2, \ldots, n)$, one gets $(\mathbf{A} - \lambda_i\mathbf{I})\mathbf{v}_i = \mathbf{0}$.

$$[\mathbf{A}\mathbf{v}_1, \mathbf{A}\mathbf{v}_2, \ldots, \mathbf{A}\mathbf{v}_n] = [\lambda_1\mathbf{v}_1, \lambda_2\mathbf{v}_2, \ldots \lambda_n\mathbf{v}_n].$$

Deformation of the foregoing equation gives

$$\mathbf{A}[\mathbf{v}_1, \mathbf{v}_2, \ldots, \mathbf{v}_n] = [\mathbf{v}_1, \mathbf{v}_2, \ldots, \mathbf{v}_n]\, diag[\lambda_1, \lambda_2, \ldots, \lambda_n].$$

Further

$$\mathbf{AP} = \mathbf{P}\, diag[\lambda_1, \lambda_2, \ldots, \lambda_n].$$

The left multiplication of $\mathbf{P}^{-1}$ on both sides of equation yields

$$\mathbf{P}^{-1}\mathbf{AP} = \mathbf{P}^{-1}\mathbf{P}\, diag[\lambda_1, \lambda_2, \ldots, \lambda_n] = diag[\lambda_1, \lambda_2, \ldots, \lambda_n].$$

Consider a new variable $\mathbf{y} = \mathbf{P}^{-1}\mathbf{x}$. Thus, application of $\mathbf{x} = \mathbf{Py}$ to Eq. (A.2) yields

$$\dot{\mathbf{y}} = \mathbf{P}^{-1}\dot{\mathbf{x}} = \mathbf{P}^{-1}\mathbf{Ax} = \mathbf{P}^{-1}\mathbf{APy} = diag[\lambda_1, \lambda_2, \ldots, \lambda_n]\mathbf{y}.$$

With initial conditions $\mathbf{y}_0 = \mathbf{P}^{-1}\mathbf{x}_0$, the uncoupled linear system has a solution as

$$\mathbf{y}(t) = diag[e^{\lambda_1(t-t_0)}, e^{\lambda_2(t-t_0)}, \ldots, e^{\lambda_n(t-t_0)}]\mathbf{y}_0.$$

Using $\mathbf{x} = \mathbf{Py}$ and $\mathbf{y}_0 = \mathbf{P}^{-1}\mathbf{x}_0$, we have

$$\mathbf{x}(t) = \mathbf{P}diag[e^{\lambda_1(t-t_0)}, e^{\lambda_2(t-t_0)}, \ldots, e^{\lambda_n(t-t_0)}]\mathbf{P}^{-1}\mathbf{x}_0 = \mathbf{PE}(t - t_0)\mathbf{P}^{-1}\mathbf{x}_0$$

where

$$\mathbf{E}(t - t_0) = diag[e^{\lambda_1(t-t_0)}, e^{\lambda_2(t-t_0)}, \ldots, e^{\lambda_n(t-t_0)}].$$

This theorem is proved. ∎

It is very important to compute the eigenvector, which is a key to obtain the general solution of linear dynamical systems. The eigenvector of $\mathbf{v}_i$ is assumed as

$$\mathbf{v}_i = \begin{Bmatrix} 1 \\ \mathbf{r}_i \end{Bmatrix} v_i. \tag{A.11}$$

From Eq. (A.7), we have

$$\begin{bmatrix} a_{11} - \lambda_i & \mathbf{b}_{1\times(n-1)} \\ \mathbf{c}_{(n-1)\times 1} & \mathbf{A}_{11} - \lambda_i\mathbf{I}_{(n-1)\times(n-1)} \end{bmatrix} \begin{Bmatrix} 1 \\ \mathbf{r}_i \end{Bmatrix} v_i = \mathbf{0}, \tag{A.12}$$

where the minor of matrix $\mathbf{A}$ is $\mathbf{A}_{11}$, and other vectors are defined by

$$\begin{aligned} \mathbf{c}_{(n-1)\times 1} &= (a_{i1})_{(n-1)\times 1}(i = 2, 3, \ldots, n) \\ \mathbf{b}_{1\times(n-1)} &= (a_{1j})_{1\times(n-1)}(j = 2, 3, \ldots, n) \\ \mathbf{A}_{11} &= (a_{ij})_{(n-1)\times(n-1)}(i, j = 2, 3, \ldots, n) \end{aligned} \tag{A.13}$$

Thus,

$$\mathbf{r}_i = (\mathbf{A}_{11} - \lambda_i \mathbf{I}_{(n-1)\times(n-1)})^{-1}\mathbf{c}_{n\times 1}. \tag{A.14}$$

The solution of linear dynamical system in Eq. (A.2) is

$$\begin{aligned}\mathbf{x}(t) &= \sum\nolimits_{i=1}^{n} C_i \mathbf{v}_i e^{\lambda_i(t-t_0)}\\ &= [\mathbf{v}_1, \mathbf{v}_2, \ldots, \mathbf{v}_n] diag[e^{\lambda_1(t-t_0)}, e^{\lambda_2(t-t_0)}, \ldots, e^{\lambda_n(t-t_0)}]\mathbf{C}\\ &= \mathbf{P} diag[e^{\lambda_1(t-t_0)}, e^{\lambda_2(t-t_0)}, \ldots, e^{\lambda_n(t-t_0)}]\mathbf{C}\end{aligned} \tag{A.15}$$

where

$$\mathbf{C} = (C_1, C_2, \ldots, C_n)^{\mathrm{T}}. \tag{A.16}$$

For $t = t_0$, the initial conditions is $\mathbf{x}(t) = \mathbf{x}_0$. Thus,

$$\mathbf{C} = \mathbf{P}^{-1}\mathbf{x}_0. \tag{A.17}$$

Therefore, the solution is expressed by

$$\mathbf{x}(t) = \mathbf{P} diag[e^{\lambda_1(t-t_0)}, e^{\lambda_2(t-t_0)}, \ldots, e^{\lambda_n(t-t_0)}]\mathbf{P}^{-1}\mathbf{x}_0 = \mathbf{PE}(t-t_0)\mathbf{P}^{-1}\mathbf{x}_0. \tag{A.18}$$

The two methods give the same expression.

Theorem A.2 *Consider a linear dynamical system $\dot{\mathbf{x}} = \mathbf{Ax}$ in Eq. (A.2) with the initial condition of $\mathbf{x}(t_0) = \mathbf{x}_0$. If the distinct complex eigenvalues of the $2n \times 2n$ matrix $\mathbf{A}$ are $\lambda_j = \alpha_j + \mathbf{i}\beta_j$ and $\bar{\lambda}_j = \alpha_j - \mathbf{i}\beta_j$ with corresponding eigenvectors $\mathbf{w}_j = \mathbf{u}_j + \mathbf{i}\mathbf{v}_j$ and $\bar{\mathbf{w}}_j = \mathbf{u}_j - \mathbf{i}\mathbf{v}_j$ ($j = 1, 2, \ldots, n$ and $\mathbf{i} = \sqrt{-1}$), then the corresponding eigenvectors $\mathbf{u}_j$ and $\mathbf{v}_j$ ($j = 1, 2, \ldots, n$) are determined by*

$$\begin{aligned}&(\mathbf{A} - (\alpha_j + \mathbf{i}\beta_j)\mathbf{I})(\mathbf{u}_j + \mathbf{i}\mathbf{v}_j) = \mathbf{0}, \text{ or}\\ &(\mathbf{A} - (\alpha_j - \mathbf{i}\beta_j)\mathbf{I})(\mathbf{u}_j - \mathbf{i}\mathbf{v}_j) = \mathbf{0}.\end{aligned} \tag{A.19}$$

which forms a basis in $\Omega \subseteq \mathscr{R}^{2n}$. The corresponding eigenvector matrix of $\mathbf{P} = [\mathbf{u}_1, \mathbf{v}_1, \mathbf{u}_2, \mathbf{v}_2, \ldots, \mathbf{u}_n, \mathbf{v}_n]$ is invertible and

$$\mathbf{P}^{-1}\mathbf{AP} = diag(\mathbf{B}_1, \mathbf{B}_2, \ldots, \mathbf{B}_n). \tag{A.20}$$

where

$$\mathbf{B}_j = \begin{bmatrix} \alpha_j & \beta_j \\ -\beta_j & \alpha_j \end{bmatrix} (j = 1, 2, \ldots, n). \tag{A.21}$$

Thus, with an initial condition of $\mathbf{x}(t_0) = \mathbf{x}_0$, the solution of the linear dynamical system in Eq. (A.2) is

$$\mathbf{x}(t) = \mathbf{PE}(t-t_0)\mathbf{P}^{-1}\mathbf{x}_0 \tag{A.22}$$

where the diagonal matrix $\mathbf{E}(t - t_0)$ *is given by*

$$\begin{aligned}\mathbf{E}(t) &= diag[\mathbf{E}_1(t-t_0), \mathbf{E}_2(t-t_0), \ldots, \mathbf{E}_n(t-t_0)],\\ \mathbf{E}_j(t-t_0) &= e^{\alpha_j(t-t_0)}\begin{bmatrix} \cos\beta_j(t-t_0) & \sin\beta_j(t-t_0)\\ -\sin\beta_j(t-t_0) & \cos\beta_j(t-t_0)\end{bmatrix}.\end{aligned} \tag{A.23}$$

Proof Assuming $\mathbf{x}(t) = \mathbf{C}e^{\lambda t} = C\mathbf{w}e^{\lambda t}$, equation (A.2) gives $(\mathbf{A} - \lambda\mathbf{I})\mathbf{w} = \mathbf{0}$. Since $\det(\mathbf{A} - \lambda\mathbf{I}) = 0$ gives n distinct pairs of complex eigenvalues $\lambda_j = \alpha_j + \mathbf{i}\beta_j$ and $\bar{\lambda}_j = \alpha_j - \mathbf{i}\beta_j$ $(j = 1, 2, \ldots, n)$, with conjugate vectors $\mathbf{w}_j = \mathbf{u}_j + \mathbf{i}\mathbf{v}_j$ and $\bar{\mathbf{w}}_j = \mathbf{u}_j - \mathbf{i}\mathbf{v}_j$, we have

$$\begin{aligned}(\mathbf{A} - \alpha_j\mathbf{I})\mathbf{u}_j + \beta_j\mathbf{I}\mathbf{v}_j &= \mathbf{0},\\ -\beta_j\mathbf{I}\mathbf{u}_j + (\mathbf{A} - \alpha_j\mathbf{I})\mathbf{v}_j &= \mathbf{0}.\end{aligned}$$

$$\mathbf{A}\mathbf{u}_j = (\mathbf{u}_j, \mathbf{v}_j)\begin{Bmatrix}\alpha_j\\ -\beta_j\end{Bmatrix} \quad \text{and} \quad \mathbf{A}\mathbf{v}_j = (\mathbf{u}_j, \mathbf{v}_j)\begin{Bmatrix}\beta_j\\ \alpha_j\end{Bmatrix}.$$

$$\mathbf{A}(\mathbf{u}_j, \mathbf{v}_j) = (\mathbf{u}_j, \mathbf{v}_j)\begin{bmatrix}\alpha_j & \beta_j\\ -\beta_j & \alpha_j\end{bmatrix}.$$

Assembling $\mathbf{A}(\mathbf{u}_j, \mathbf{v}_j)$ for $(j = 1, 2, \ldots, n)$ gives

$$\mathbf{A}\mathbf{P} = \mathbf{P}diag(\begin{bmatrix}\alpha_1 & \beta_1\\ -\beta_1 & \alpha_1\end{bmatrix}, \begin{bmatrix}\alpha_2 & \beta_2\\ -\beta_2 & \alpha_2\end{bmatrix}, \ldots, \begin{bmatrix}\alpha_n & \beta_n\\ -\beta_n & \alpha_n\end{bmatrix})$$

where

$$\mathbf{P} = (\mathbf{u}_1, \mathbf{v}_1, \mathbf{u}_2, \mathbf{v}_2, \ldots, \mathbf{u}_n, \mathbf{v}_n).$$

The left multiplication of $\mathbf{P}^{-1}$ on both sides of equation yields

$$\begin{aligned}\mathbf{P}^{-1}\mathbf{A}\mathbf{P} &= \mathbf{P}^{-1}\mathbf{P}diag(\begin{bmatrix}\alpha_1 & \beta_1\\ -\beta_1 & \alpha_1\end{bmatrix}, \begin{bmatrix}\alpha_2 & \beta_2\\ -\beta_2 & \alpha_2\end{bmatrix}, \ldots, \begin{bmatrix}\alpha_n & \beta_n\\ -\beta_n & \alpha_n\end{bmatrix})\\ &= diag(\begin{bmatrix}\alpha_1 & \beta_1\\ -\beta_1 & \alpha_1\end{bmatrix}, \begin{bmatrix}\alpha_2 & \beta_2\\ -\beta_2 & \alpha_2\end{bmatrix}, \ldots, \begin{bmatrix}\alpha_n & \beta_n\\ -\beta_n & \alpha_n\end{bmatrix}).\end{aligned}$$

Consider a new variable $\mathbf{y} = \mathbf{P}^{-1}\mathbf{x}$. Thus, application of $\mathbf{x} = \mathbf{P}\mathbf{y}$ to Eq. (A.2) yields

$$\begin{aligned}\dot{\mathbf{y}} &= \mathbf{P}^{-1}\dot{\mathbf{x}} = \mathbf{P}^{-1}\mathbf{A}\mathbf{x} = \mathbf{P}^{-1}\mathbf{A}\mathbf{P}\mathbf{y}\\ &= diag(\begin{bmatrix}\alpha_1 & \beta_1\\ -\beta_1 & \alpha_1\end{bmatrix}, \begin{bmatrix}\alpha_2 & \beta_2\\ -\beta_2 & \alpha_2\end{bmatrix}, \ldots, \begin{bmatrix}\alpha_n & \beta_n\\ -\beta_n & \alpha_n\end{bmatrix})\mathbf{y}.\end{aligned}$$

With the initial condition, $\mathbf{y}_0 = \mathbf{P}^{-1}\mathbf{x}_0$, the uncoupled linear system has a solution as

$$\begin{aligned}\mathbf{y}(t) &= diag[\mathbf{E}_1(t-t_0), \mathbf{E}_2(t-t_0), \ldots, \mathbf{E}_n(t-t_0)]\mathbf{y}_0 \\ &= \mathbf{E}(t-t_0)\mathbf{y}_0\end{aligned}$$

where

$$\mathbf{E}(t-t_0) = diag[\mathbf{E}_1(t-t_0), \mathbf{E}_2(t-t_0), \ldots, \mathbf{E}_n(t-t_0)],$$

$$\mathbf{E}_j(t-t_0) = e^{\alpha_j(t-t_0)}\begin{bmatrix} \cos\beta_j(t-t_0) & \sin\beta_j(t-t_0) \\ -\sin\beta_j(t-t_0) & \cos\beta_j(t-t_0)\end{bmatrix}.$$

Using $\mathbf{x} = \mathbf{P}\mathbf{y}$ and $\mathbf{y}_0 = \mathbf{P}^{-1}\mathbf{x}_0$, we have

$$\begin{aligned}\mathbf{x}(t) &= \mathbf{P}diag[\mathbf{E}_1(t-t_0), \mathbf{E}_2(t-t_0), \ldots, \mathbf{E}_n(t-t_0)]\mathbf{P}^{-1}\mathbf{x}_0 \\ &= \mathbf{P}\mathbf{E}(t-t_0)\mathbf{P}^{-1}\mathbf{x}_0.\end{aligned}$$

This theorem is proved. ■

Compared to the real eigenvectors, the computation of the complex eigenvectors is much complicated, and the corresponding, detailed procedure is presented herein. The conjugate complex eigenvectors are assumed as

$$\mathbf{u}_i + \mathbf{i}\mathbf{v}_i = C_i\begin{Bmatrix} 1 \\ \mathbf{r}_i \end{Bmatrix} \quad \text{and} \quad \mathbf{u}_i - \mathbf{i}\mathbf{v}_i = \bar{C}_i\begin{Bmatrix} 1 \\ \bar{\mathbf{r}}_i \end{Bmatrix} \tag{A.24}$$

where the conjugate complex constants are assumed as

$$\begin{aligned} &C_i = \frac{1}{2}(M_i - \mathbf{i}N_i) \text{ and } \bar{C}_i = \frac{1}{2}(M_i + \mathbf{i}N_i), \\ &\mathbf{r}_i = \mathbf{U}_i + \mathbf{i}\mathbf{V}_i \text{ and } \bar{\mathbf{r}}_i = \mathbf{U}_i - \mathbf{i}\mathbf{V}_i. \end{aligned} \tag{A.25}$$

From Eq. (A.19), we have

$$\begin{bmatrix} a_{11} - \alpha_i - \mathbf{i}\beta_i & \mathbf{b}_{1\times(n-1)} \\ \mathbf{c}_{(n-1)\times 1} & \mathbf{A}_{11} - (\alpha_i + \mathbf{i}\beta_i)\mathbf{I}_{(n-1)\times(n-1)} \end{bmatrix}\begin{Bmatrix} 1 \\ \mathbf{r}_i \end{Bmatrix} C_i = \mathbf{0}, \tag{A.26}$$

Thus, the foregoing equation gives

$$\mathbf{c} + [(\mathbf{A}_{11} - \alpha_i\mathbf{I}) - \mathbf{i}\beta_i\mathbf{I}]\mathbf{r}_i = \mathbf{0}, \tag{A.27}$$

$$\mathbf{r}_i = \left[(\mathbf{A}_{11} - \alpha_i\mathbf{I})^2 + \beta_i^2\mathbf{I}\right]^{-1}\left[(\mathbf{A}_{11} - \alpha_i\mathbf{I}) + \mathbf{i}\beta_i\mathbf{I}\right]\mathbf{c} = \mathbf{U}_i + \mathbf{i}\mathbf{V}_i, \tag{A.28}$$

where

$$\mathbf{U}_i = \left[(\mathbf{A}_{11} - \alpha_i\mathbf{I})^2 + \beta_i^2\mathbf{I}\right]^{-1}(\mathbf{A}_{11} - \alpha_i\mathbf{I})\mathbf{c},$$
$$\mathbf{V}_i = \left[(\mathbf{A}_{11} - \alpha_i\mathbf{I})^2 + \beta_i^2\mathbf{I}\right]^{-1}\beta_i\mathbf{c}. \tag{A.29}$$

The solution of the linear dynamical system in Eq. (A.2) is

$$\begin{aligned}
\mathbf{x}(t) &= \sum\nolimits_{i=1}^{n} \frac{1}{2}(M_i - \mathbf{i}N_i)\begin{Bmatrix} 1 \\ \mathbf{U}_i + \mathbf{i}\mathbf{V}_i \end{Bmatrix} e^{(\alpha_i + \mathbf{i}\beta_i)(t-t_0)} \\
&\quad + \frac{1}{2}(M_i + \mathbf{i}N_i)\begin{Bmatrix} 1 \\ \mathbf{U}_i - i\mathbf{V}_i \end{Bmatrix} e^{(\alpha_i - \mathbf{i}\beta_i)(t-t_0)} \\
&= \sum\nolimits_{i=1}^{n} e^{\alpha_i(t-t_0)}\left[\begin{Bmatrix} M_i \\ M_i\mathbf{U}_i + N_i\mathbf{V}_i \end{Bmatrix}\cos\beta_i(t-t_0)\right. \\
&\quad \left. + \begin{Bmatrix} N_i \\ N_i\mathbf{U}_i - M_i\mathbf{V}_i \end{Bmatrix}\sin\beta_i(t-t_0)\right] \\
&= \sum\nolimits_{i=1}^{n} e^{\alpha_i(t-t_0)}\left[(M_i\begin{Bmatrix} 1 \\ \mathbf{U}_i \end{Bmatrix} + N_i\begin{Bmatrix} 0 \\ \mathbf{V}_i \end{Bmatrix})\cos\beta_i(t-t_0)\right. \\
&\quad \left. +(N_i\begin{Bmatrix} 1 \\ \mathbf{U}_i \end{Bmatrix} - M_i\begin{Bmatrix} 0 \\ \mathbf{V}_i \end{Bmatrix})\sin\beta_i(t-t_0)\right] \\
&= \sum\nolimits_{i=1}^{n} e^{\alpha_i(t-t_0)}(\mathbf{u}_i, \mathbf{v}_i)\begin{bmatrix} \cos\beta_i(t-t_0) & \sin\beta_i(t-t_0) \\ -\sin\beta_i(t-t_0) & \cos\beta_i(t-t_0) \end{bmatrix}\begin{Bmatrix} M_i \\ N_i \end{Bmatrix} \\
&= \mathbf{PE}(t-t_0)\mathbf{C}
\end{aligned} \tag{A.30}$$

where

$$\begin{aligned}
\mathbf{P} &= [\mathbf{u}_1, \mathbf{v}_1, \ldots, \mathbf{u}_n, \mathbf{v}_n], \\
\mathbf{E}(t-t_0) &= diag[\mathbf{E}_1(t-t_0), \mathbf{E}_2(t-t_0), \ldots, \mathbf{E}_n(t-t_0)], \\
\mathbf{C} &= (M_1, N_1, \ldots, M_n, N_n)^{\mathrm{T}}, \\
\mathbf{E}_i(t-t_0) &= e^{\alpha_i(t-t_0)}\begin{bmatrix} \cos\beta_i(t-t_0) & \sin\beta_i(t-t_0) \\ -\sin\beta_i(t-t_0) & \cos\beta_i(t-t_0) \end{bmatrix}, \\
\mathbf{u}_i &= \begin{Bmatrix} 1 \\ \mathbf{U}_i \end{Bmatrix} \quad \text{and} \quad \mathbf{v}_i = \begin{Bmatrix} 0 \\ \mathbf{V}_i \end{Bmatrix}.
\end{aligned} \tag{A.31}$$

For $t = t_0$, the initial conditions is $\mathbf{x}(t) = \mathbf{x}_0$. Thus,

$$\mathbf{C} = \mathbf{P}^{-1}\mathbf{x}_0. \tag{A.32}$$

Therefore, the solution is expressed by

$$\begin{aligned}\mathbf{x}(t) &= \mathbf{P}diag[\mathbf{E}_1(t-t_0), \mathbf{E}_2(t-t_0), \ldots, \mathbf{E}_n(t-t_0)]\mathbf{P}^{-1}\mathbf{x}_0 \\ &= \mathbf{PE}(t-t_0)\mathbf{P}^{-1}\mathbf{x}_0.\end{aligned} \tag{A.33}$$

The two methods give the same expression.

Theorem A.3 *Consider a linear dynamical system $\dot{\mathbf{x}} = \mathbf{Ax}$ in Eq. (A.2) with the initial condition of $\mathbf{x}(t_0) = \mathbf{x}_0$. If the eigenvalues of the $n \times n$ matrix $\mathbf{A}$ possesses p-pairs of distinct complex eigenvalues with $\lambda_j = \alpha_j + \mathbf{i}\beta_j$ and $\bar{\lambda}_j = \alpha_j - \mathbf{i}\beta_j$ with corresponding eigenvectors $\mathbf{w}_j = \mathbf{u}_j + \mathbf{i}\mathbf{v}_j$ and $\mathbf{w}_j = \mathbf{u}_j - \mathbf{i}\mathbf{v}_j$ ($j = 1, 2, \ldots, p$ and $\mathbf{i} = \sqrt{-1}$), and $(n-2p)$ distinct real eigenvalues of $\lambda_{2p+1}, \lambda_{2p+2}, \ldots, \lambda_n$, then the corresponding eigenvectors $\mathbf{u}_j$ and $\mathbf{v}_j$ for complex eigenvalues $(\lambda_j, \bar{\lambda}_j)$ ($j = 1, 2, \ldots, p$) are determined by*

$$\begin{aligned}&(\mathbf{A} - (\alpha_j + \mathbf{i}\beta_j)\mathbf{I})(\mathbf{u}_j + \mathbf{i}\mathbf{v}_j) = \mathbf{0}, \text{ or} \\ &(\mathbf{A} - (\alpha_j - \mathbf{i}\beta_j)\mathbf{I})(\mathbf{u}_j - \mathbf{i}\mathbf{v}_j) = \mathbf{0}\end{aligned} \tag{A.34}$$

and the eigenvectors $\{\mathbf{v}_{2p+1}, \mathbf{v}_{2p+2}, \ldots, \mathbf{v}_n\}$ for real eigenvalues are determined by

$$(\mathbf{A} - \lambda_i\mathbf{I})\mathbf{v}_i = \mathbf{0} \tag{A.35}$$

which forms a basis in $\Omega \subseteq \mathscr{R}^n$. The eigenvector matrix of

$$\mathbf{P} = [\mathbf{u}_1, \mathbf{v}_1, \mathbf{u}_2, \mathbf{v}_2, \ldots, \mathbf{u}_p, \mathbf{v}_p, \mathbf{v}_{2p+1}, \mathbf{v}_{2p+2}, \ldots, \mathbf{v}_n] \tag{A.36}$$

is invertible and

$$\mathbf{P}^{-1}\mathbf{AP} = diag(\mathbf{B}_1, \mathbf{B}_2, \ldots, \mathbf{B}_p, \lambda_{2p+1}, \lambda_{2p+2}, \ldots, \lambda_n) \tag{A.37}$$

where

$$\mathbf{B}_j = \begin{bmatrix} \alpha_j & \beta_j \\ -\beta_j & \alpha_j \end{bmatrix} (j = 1, 2, \ldots, n). \tag{A.38}$$

Thus, with an initial condition of $\mathbf{x}(t_0) = \mathbf{x}_0$, the solution of linear dynamical system in Eq. (A.2) is

$$\begin{aligned}\mathbf{x}(t) &= \mathbf{P}diag[\mathbf{E}_1(t-t_0), \mathbf{E}_2(t-t_0), \ldots, \mathbf{E}_p(t-t_0), \\ &\qquad e^{\lambda_{2p+1}(t-t_0)}, e^{\lambda_{2p+2}(t-t_0)}, \ldots, e^{\lambda_n(t-t_0)}]\mathbf{P}^{-1}\mathbf{x}_0 \\ &= \mathbf{PE}(t-t_0)\mathbf{P}^{-1}\mathbf{x}_0\end{aligned} \tag{A.39}$$

where the diagonal matrix $\mathbf{E}(t-t_0)$ *is given by*

$$\begin{aligned}\mathbf{E}(t-t_0) &= diag[\mathbf{E}_1(t-t_0), \mathbf{E}_2(t-t_0), \ldots, \mathbf{E}_p(t-t_0),\\ &\quad e^{\lambda_{2p+1}(t-t_0)}, e^{\lambda_{2p+2}(t-t_0)}, \ldots, e^{\lambda_n(t-t_0)}];\\ \mathbf{E}_j(t-t_0) &= e^{\alpha_j(t-t_0)}\begin{bmatrix}\cos\beta_j(t-t_0) & \sin\beta_j(t-t_0)\\ -\sin\beta_j(t-t_0) & \cos\beta_j(t-t_0)\end{bmatrix} (j=1,2,\ldots,p).\end{aligned} \tag{A.40}$$

Proof The proof of the theorem is from the proof of Theorems A.1 and A.2. ∎

A.2 Operator Exponentials

Definition A.3 Consider a linear operator $\mathbf{A}: \mathscr{R}^n \to \mathscr{R}^n$ in linear operator space (i.e., $\mathbf{A} \in L(R^n)$). The operator norm of $\mathbf{A}$ is defined by

$$||\mathbf{A}|| = \max_{||\mathbf{x}|| \le 1} ||\mathbf{A}(\mathbf{x})|| \tag{A.41}$$

where $||\mathbf{x}||$ is the Euclidean norm of $\mathbf{x} \in \mathscr{R}^n$. The operator norm has the following properties for $\mathbf{A}, \mathbf{B} \in L(\mathscr{R}^n)$:

(i) $||\mathbf{A}|| \ge 0$ and $||\mathbf{A}|| = 0$ if and only if $\mathbf{A} = \mathbf{0}$.
(ii) $||k\mathbf{A}|| = k||\mathbf{A}||$ for $k \in \mathscr{R}$.
(iii) $||\mathbf{A}+\mathbf{B}|| \le ||\mathbf{A}|| + ||\mathbf{B}||$.

Definition A.4 Consider a sequence of linear operator $\mathbf{A}_k \in L(\mathscr{R}^n)$ and the linear operator $\mathbf{A} \in L(\mathscr{R}^n)$. For any $\varepsilon > 0$, there exists an N such that for $k \ge N$,

$$||\mathbf{A} - \mathbf{A}_k|| < \varepsilon. \tag{A.42}$$

Thus, the sequence of linear operator $\mathbf{A}_k$ is called to be convergent to a linear operator $\mathbf{A}$ as $k \to \infty$, i.e.,

$$\lim_{k\to\infty} \mathbf{A}_k = \mathbf{A}. \tag{A.43}$$

Theorem A.4 *For* $\mathbf{A}, \mathbf{B} \in L(\mathscr{R}^n)$ *and* $\mathbf{x} \in \mathscr{R}^n$,

(i) $||\mathbf{A}\mathbf{x}|| \le ||\mathbf{A}|| \times ||\mathbf{x}||$,
(ii) $||\mathbf{A}\mathbf{B}|| \le ||\mathbf{A}|| \times ||\mathbf{B}||$, *and*
(iii) $||\mathbf{A}^k|| \le ||\mathbf{A}||^k$.

Proof

(i) Consider $\mathbf{y} = \mathbf{x}/||\mathbf{x}||$. The definition of the norm of linear operator gives

$$||\mathbf{A}|| \geq ||\mathbf{Ay}|| = \frac{||\mathbf{Ax}||}{||\mathbf{x}||} \Rightarrow ||\mathbf{Ax}|| \leq ||\mathbf{A}|| \times ||\mathbf{x}||.$$

(ii) For $||\mathbf{x}|| \leq 1$, the foregoing relation give

$$||\mathbf{ABx}|| \leq ||\mathbf{A}|| \times ||\mathbf{Bx}|| \leq ||\mathbf{A}|| \times ||\mathbf{B}|| \times ||\mathbf{x}|| \leq ||\mathbf{A}|| \times ||\mathbf{B}||.$$

Thus

$$||\mathbf{AB}|| = \max_{||\mathbf{x}|| \leq 1} ||\mathbf{ABx}|| \leq ||\mathbf{A}|| \times ||\mathbf{B}||.$$

(iii) For $\mathbf{B} = \mathbf{A}$, the foregoing equation gives $||\mathbf{A}^2|| \leq ||\mathbf{A}||^2$, and continuously $||\mathbf{A}^k|| \leq ||\mathbf{A}^{k-1}|| \times ||\mathbf{A}|| \leq ||\mathbf{A}||^k$.

Theorem A.5 *For* $\mathbf{A} \in L(\mathscr{R}^n)$ *and* $t \in \mathscr{R}$ *with* $t_0 > 0$, *a series* $\mathbf{I} + \sum_{k=1}^{\infty} \frac{1}{k!} \mathbf{A}^k t^k$ *is absolutely and uniformly convergent for* $|t| \leq t_0$.

Proof For $|t| \leq t_0$,

$$||\frac{\mathbf{A}^k t^k}{k!}|| \leq \frac{||\mathbf{A}^k|| \times |t|^k}{k!} \leq \frac{||\mathbf{A}||^k t_0^k}{k!}.$$

Using the Taylor series gives

$$\sum_{k=0}^{\infty} \frac{||\mathbf{A}||^k t_0^k}{k!} = e^{||\mathbf{A}|| t_0}.$$

The triangle inequality gives

$$||\mathbf{I} + \sum_{k=1}^{\infty} \frac{\mathbf{A}^k t^k}{k!}|| \leq \sum_{k=0}^{\infty} ||\frac{\mathbf{A}^k t^k}{k!}|| \leq \sum_{k=0}^{\infty} \frac{||\mathbf{A}||^k t_0^k}{k!}.$$

Therefore, the series $\mathbf{I} + \sum_{k=1}^{\infty} \frac{1}{k!} \mathbf{A}^k t^k$ is absolutely and uniformly convergent for $|t| \leq t_0$. ■

Definition A.5 The exponential of a linear operator $\mathbf{A} \in L(\mathscr{R}^n)$ is defined by

$$e^{\mathbf{A}} = \mathbf{I} + \sum_{k=1}^{\infty} \frac{1}{k!} \mathbf{A}^k. \tag{A.44}$$

If $\mathbf{A}$ is an $n \times n$ matrix, for $t \in \mathscr{R}$,

$$e^{\mathbf{A}t} = \mathbf{I} + \sum_{k=1}^{\infty} \frac{1}{k!} \mathbf{A}^k t^k. \tag{A.45}$$

Theorem A.6 *For* $\mathbf{A}, \mathbf{B}, \mathbf{P} \in L(\mathscr{R}^n)$,

(i) *if* $\mathbf{P}$ *is nonsingular, then* $e^{\mathbf{P}^{-1}\mathbf{A}\mathbf{P}} = \mathbf{P}^{-1}e^{\mathbf{A}}\mathbf{P}$;
(ii) *if* $\mathbf{AB} = \mathbf{BA}$, *then* $e^{\mathbf{A}+\mathbf{B}} = e^{\mathbf{A}}e^{\mathbf{B}}$;
(iii) $e^{-\mathbf{A}} = (e^{\mathbf{A}})^{-1}$.
(iv) *If* $\mathbf{A} = \lambda\mathbf{I}$, *then* $e^{\mathbf{A}} = e^{\lambda}\mathbf{I}$.

Proof

(i) Since the following relations exist

$$\mathbf{P}^{-1}(\mathbf{A}+\mathbf{B})\mathbf{P} = \mathbf{P}^{-1}\mathbf{A}\mathbf{P} + \mathbf{P}^{-1}\mathbf{B}\mathbf{P} \text{ and}$$
$$(\mathbf{P}^{-1}\mathbf{A}\mathbf{P})^k = (\mathbf{P}^{-1}\mathbf{A}\mathbf{P})(\mathbf{P}^{-1}\mathbf{A}\mathbf{P})\ldots(\mathbf{P}^{-1}\mathbf{A}\mathbf{P}) = \mathbf{P}^{-1}\mathbf{A}^k\mathbf{P},$$

Definition A.5 gives

$$e^{\mathbf{P}^{-1}\mathbf{A}\mathbf{P}} = \sum\nolimits_{k=0}^{\infty} \frac{1}{k!}(\mathbf{P}^{-1}\mathbf{A}\mathbf{P})^k = \mathbf{P}^{-1}\sum\nolimits_{k=0}^{\infty} \frac{1}{k!}\mathbf{A}^k\mathbf{P} = \mathbf{P}^{-1}e^{\mathbf{A}}\mathbf{P}.$$

(ii) Because

$$(\mathbf{A}+\mathbf{B})^n = n! \sum_{\substack{j+k=n \\ (j=0,1,\ldots,n)}} \frac{\mathbf{A}^j}{j!}\frac{\mathbf{B}^k}{k!},$$

we have

$$e^{\mathbf{A}+\mathbf{B}} = \sum\nolimits_{n=0}^{\infty} \frac{1}{n!}(\mathbf{A}+\mathbf{B})^n = \sum\nolimits_{n=0}^{\infty} \frac{1}{n!}\Big(n! \sum_{\substack{j+k=n \\ (j=0,1,\ldots,n)}} \frac{\mathbf{A}^j}{j!}\frac{\mathbf{B}^k}{k!}\Big)$$
$$= \sum\nolimits_{j=0}^{n} \frac{\mathbf{A}^j}{j!}\sum\nolimits_{k=0}^{n} \frac{\mathbf{B}^k}{k!} = e^{\mathbf{A}}e^{\mathbf{B}}.$$

(iii) If $\mathbf{B} = -\mathbf{A}$, then the case of (ii) gives

$$\mathbf{I} = e^{\mathbf{A}-\mathbf{A}} = e^{\mathbf{A}}e^{-\mathbf{A}}.$$

Thus

$$(e^{\mathbf{A}})^{-1} = e^{-\mathbf{A}}.$$

(iv) If $\mathbf{A} = \lambda\mathbf{I}$, from definition, one obtains

$$e^{\lambda\mathbf{I}} = \mathbf{I} + \sum\nolimits_{k=1}^{\infty} \frac{\lambda^k}{k!}\mathbf{I}^k = \Big(1 + \sum\nolimits_{k=1}^{\infty} \frac{\lambda^k}{k!}\Big)\mathbf{I} = e^{\lambda}\mathbf{I}.$$

This theorem is proved. ∎

Lemma A.1 *For an $n \times n$ matrix* $\mathbf{A}$,

$$\frac{d}{dt}e^{\mathbf{A}t} = \mathbf{A}e^{\mathbf{A}t}. \tag{A.46}$$

Proof From the derivative definition,

$$\begin{aligned}\frac{d}{dt}e^{\mathbf{A}t} &= \lim_{\Delta t \to 0}\frac{e^{\mathbf{A}(t+\Delta t)} - e^{\mathbf{A}t}}{\Delta t} = \lim_{\Delta t \to 0} e^{\mathbf{A}t}\frac{e^{\mathbf{A}\Delta t} - \mathbf{I}}{\Delta t}\\ &= e^{\mathbf{A}t}\lim_{\Delta t \to 0}\lim_{m \to \infty}(\mathbf{A} + \sum\nolimits_{k=2}^{m}\frac{\mathbf{A}(\Delta t)^{k-1}}{k!})\\ &= \mathbf{A}e^{\mathbf{A}t}.\end{aligned}$$

This lemma is proved. ■

Theorem A.7 *Consider a linear dynamical system $\dot{\mathbf{x}} = \mathbf{A}\mathbf{x}$ in Eq. (A.2) with the initial condition of $\mathbf{x}(0) = \mathbf{x}_0$. The solution of the linear dynamical system is unique, which is given by*

$$\mathbf{x} = e^{\mathbf{A}t}\mathbf{x}_0. \tag{A.47}$$

Proof From Lemma A.1, if $\mathbf{x}(t) = e^{\mathbf{A}t}\mathbf{x}_0$ exist, then for $t \in I \subseteq \mathscr{R}$, we have

$$\dot{\mathbf{x}} = \frac{d}{dt}(e^{\mathbf{A}t}\mathbf{x}_0) = \mathbf{A}e^{\mathbf{A}t}\mathbf{x}_0 = \mathbf{A}\mathbf{x}.$$

In addition, $\mathbf{x}(0) = \mathbf{I}\mathbf{x}_0 = \mathbf{x}_0$ is an initial condition. Therefore, $\mathbf{x}(t) = e^{\mathbf{A}t}\mathbf{x}_0$ is a solution. If there is another solution $\mathbf{x}_1(t) = e^{-\mathbf{A}t}\mathbf{x}(t)$, then for $t \in I \subseteq \mathscr{R}$,

$$\begin{aligned}\dot{\mathbf{x}}_1(t) &= e^{-\mathbf{A}t}\dot{\mathbf{x}}(t) - \mathbf{A}e^{-\mathbf{A}t}\mathbf{x}(t)\\ &= e^{-\mathbf{A}t}\mathbf{A}\mathbf{x}(t) - \mathbf{A}e^{-\mathbf{A}t}\mathbf{x}(t)\\ &= \mathbf{A}e^{-\mathbf{A}t}\mathbf{x}(t) - \mathbf{A}e^{-\mathbf{A}t}\mathbf{x}(t)\\ &= \mathbf{0}.\end{aligned}$$

Thus, $\mathbf{x}_1(t) = \mathbf{C}$ (constant). Let $t = 0$, $\mathbf{x}_1(0) = \mathbf{I}\mathbf{x}(0) = \mathbf{x}_0$. So the deformation of $\mathbf{x}_0 = e^{-\mathbf{A}t}\mathbf{x}(t)$ gives $\mathbf{x}(t) = e^{\mathbf{A}t}\mathbf{x}_0$. This theorem is proved. ■

A.3 Linear Systems with Repeated Eigenvalues

Definition A.6 Consider a linear dynamical system $\dot{\mathbf{x}} = \mathbf{A}\mathbf{x}$ in Eq. (A.2) with the initial condition of $\mathbf{x}(0) = \mathbf{x}_0$. If the $n \times n$ matrix $\mathbf{A}$ has an m-repeated real eigenvalue of λ with $(m \leq n)$, then any nonzero eigenvector of

$$(\mathbf{A} - \lambda\mathbf{I})^m\mathbf{v} = \mathbf{0} \tag{A.48}$$

is called a generalized eigenvector of $\mathbf{A}$.

Definition A.7 An $n \times n$ matrix $\mathbf{N}$ is called a nilpotent matrix of order k if $\mathbf{N}^{k-1} \neq \mathbf{0}$ and $\mathbf{N}^{k} = \mathbf{0}$.

Theorem A.8 *Consider a linear dynamical system* $\dot{\mathbf{x}} = \mathbf{A}\mathbf{x}$ *in Eq.* (A.2) *with the initial condition of* $\mathbf{x}(0) = \mathbf{x}_0$. *There is a repeated eigenvalue* λ_j *with m-times among the real eigenvalues* $\lambda_1, \lambda_2, \ldots, \lambda_n$ *of the* $n \times n$ *matrix* $\mathbf{A}$. *If a set of generalized eigenvectors* $\{\mathbf{v}_1, \mathbf{v}_2, \ldots, \mathbf{v}_n\}$ *forms a basis in* $\Omega \subseteq \mathcal{R}^n$. *The eigenvector matrix of* $\mathbf{P} = [\mathbf{v}_1, \mathbf{v}_2, \ldots, \mathbf{v}_n]$ *is invertible. For the repeated eigenvalue* λ_j, *the matrix* $\mathbf{A}$ *can be decomposed by*

$$\mathbf{A} = \mathbf{S} + \mathbf{N} \tag{A.49}$$

where

$$\mathbf{P}^{-1}\mathbf{S}\mathbf{P} = diag[\lambda_j]_{n\times n}, \tag{A.50}$$

and the matrix $\mathbf{N} = \mathbf{A} - \mathbf{S}$ *is nilpotent of order* $m \leq n$ $(\mathbf{N}^m = \mathbf{0})$*with* $\mathbf{S}\mathbf{N} = \mathbf{N}\mathbf{S}$.

$$\mathbf{P}^{-1}\mathbf{A}\mathbf{P} = diag[\lambda_1, \ldots \lambda_{j-1}, \underbrace{\lambda_j, \ldots, \lambda_j}_{m}, \lambda_{j+m}, \ldots, \lambda_n]. \tag{A.51}$$

Thus, with an initial condition of $\mathbf{x}(t_0) = \mathbf{x}_0$, *the solution of linear dynamical system in Eq.* (A.2) *is*

$$\mathbf{x}(t) = \mathbf{P}\mathbf{E}(t)\mathbf{P}^{-1}\left[\mathbf{I} + \sum\nolimits_{k=1}^{m-1} \frac{\mathbf{N}^k t^k}{k!}\right]\mathbf{x}_0 \tag{A.52}$$

where

$$\mathbf{E}(t) = diag[e^{\lambda_1 t}, \ldots, e^{\lambda_{j-1} t}, \underbrace{e^{\lambda_j t}, \ldots, e^{\lambda_j t}}_{m}, e^{\lambda_{j+m} t}, \ldots, e^{\lambda_n t}]. \tag{A.53}$$

Proof Consider the repeated real eigenvalue λ_j of the matrix $\mathbf{A}$. The method of coefficient variation is adopted, and the corresponding solution is assumed as

$$\mathbf{x}^{(j)} = \mathbf{C}^{(j)}(t)e^{\lambda_j t},$$

$$\dot{\mathbf{x}}^{(j)} = \dot{\mathbf{C}}^{(j)}e^{\lambda_j t} + \lambda_j \mathbf{C}^{(j)}e^{\lambda_j t} = \mathbf{A}\mathbf{C}^{(j)}e^{\lambda_j t}.$$

Therefore,

$$\dot{\mathbf{C}}^{(j)} = (\mathbf{A} - \lambda_j \mathbf{I})\mathbf{C}^{(j)}.$$

Let

$$\mathbf{C}^{(j)} = V^{(j)}\mathbf{v}_j \text{ and } \mathbf{C}_0^{(j)} = V_0^{(j)}\mathbf{v}_j.$$

Consider the constant vector and eigenvector matrix as

$$\mathbf{V}^{(j)} = (0,\ldots,0,V^{(j)},0,\ldots,0)^{\mathrm{T}}$$
$$\mathbf{P} = (\mathbf{v}_1,\ldots,\mathbf{v}_{j-1},\mathbf{v}_j,\mathbf{v}_{j+1},\ldots,\mathbf{v}_n).$$

Thus

$$\mathbf{P}\dot{\mathbf{V}}^{(j)} = (\mathbf{A}-\lambda_j\mathbf{I})\mathbf{P}\mathbf{V}^{(j)} \Rightarrow \dot{\mathbf{V}}^{(j)} = \mathbf{P}^{-1}(\mathbf{A}-\lambda_j\mathbf{I})\mathbf{P}\mathbf{V}^{(j)}.$$

Let $\mathbf{A} = \mathbf{S}+\mathbf{N}$, thus

$$\begin{aligned}\dot{\mathbf{V}}^{(j)} &= \mathbf{P}^{-1}(\mathbf{S}+\mathbf{N}-\lambda_j\mathbf{I})\mathbf{P}\mathbf{V}^{(j)}\\ &= (\mathbf{P}^{-1}\mathbf{S}\mathbf{P}-\lambda_j\mathbf{I}+\mathbf{P}^{-1}\mathbf{N}\mathbf{P})\mathbf{V}^{(j)}.\end{aligned}$$

Because of $\mathbf{P}^{-1}\mathbf{S}\mathbf{P} = diag[\lambda_j]$, the solution of the foregoing equation is

$$\begin{aligned}\mathbf{V}^{(j)} &= (\mathbf{I}+\sum\nolimits_{k=1}^{m-1}\frac{(\mathbf{P}^{-1}\mathbf{N}\mathbf{P})^k t^k}{k!})\mathbf{V}_0^{(j)} = (\mathbf{I}+\sum\nolimits_{k=1}^{m-1}\frac{\mathbf{P}^{-1}\mathbf{N}^k\mathbf{P}t^k}{k!})\mathbf{V}_0^{(j)}\\ &= \mathbf{P}^{-1}(\mathbf{I}+\sum\nolimits_{k=1}^{m-1}\frac{\mathbf{N}^k t^k}{k!})\mathbf{P}\mathbf{V}_0^{(j)}.\end{aligned}$$

Therefore, the coefficient for the repeated eigenvalue λ_j

$$\mathbf{C}^{(j)} = (\mathbf{I}+\sum\nolimits_{k=1}^{m-1}\frac{\mathbf{N}^k t^k}{k!})\mathbf{C}_0^{(j)}.$$

Further

$$\mathbf{x}^{(j)} = e^{\lambda_j t}(\mathbf{I}+\sum\nolimits_{k=1}^{m-1}\frac{\mathbf{N}^k t^k}{k!})\mathbf{C}_0^{(j)}.$$

Assuming

$$\mathbf{x}^{(j)} = \sum\nolimits_{k=0}^{m-1}\mathbf{C}_k^{(j)}e^{\lambda_j t}t^k,$$

one obtains

$$k!\mathbf{C}_k^{(j)} = \mathbf{N}^k\mathbf{C}_0^{(j)} \text{ or } (k+1)\mathbf{C}_{k+1}^{(j)} = \mathbf{N}\mathbf{C}_k^{(j)}.$$

If $V_k^{(j)} = V_0^{(j)}$, then

$$k!\mathbf{v}_k^{(j)} = \mathbf{N}^k\mathbf{v}_0^{(j)} \quad \text{or} \quad (k+1)\mathbf{v}_{k+1}^{(j)} = \mathbf{N}\mathbf{v}_k^{(j)}.$$

Let

$$\mathbf{C} = (V_1,\ldots,V_{j-1},\underbrace{V_0^{(j)},\ldots,V_0^{(j)}}_{m},V_{j+m},\ldots,V_n)^{\mathrm{T}},$$

$$\mathbf{P} = (\mathbf{v}_1,\ldots,\mathbf{v}_{j-1},\underbrace{\mathbf{v}_j,\ldots,\mathbf{v}_{j+m-1}}_{m},\mathbf{v}_{j+m},\ldots,\mathbf{v}_n).$$

Thus, there is a relation

$$\mathbf{P}^{-1}\mathbf{A}\mathbf{P} = diag[\lambda_1, \cdots, \lambda_{j-1}, \underbrace{\lambda_j, \ldots, \lambda_j}_{m}, \lambda_{j+m}, \ldots, \lambda_n],$$

and the resultant solution is

$$\begin{aligned}\mathbf{x} &= \mathbf{x}^{(1)} + \ldots + \mathbf{x}^{(j-1)} + \underbrace{\mathbf{x}_0^{(j)} + \mathbf{x}_1^{(j)} + \cdots + \mathbf{x}_{m-1}^{(j)}}_{m} + \mathbf{x}^{(j+m)} + \cdots + \mathbf{x}^{(n)} \\ &= \mathbf{P}diag(e^{\lambda_1 t}, \ldots, e^{\lambda_{j-1} t}\underbrace{e^{\lambda_j t}, \ldots, e^{\lambda_j t}}_{m}, e^{\lambda_{j+m} t}, \ldots, e^{\lambda_n t})(\mathbf{I} + \sum_{k=1}^{m-1} \frac{\mathbf{N}^k t^k}{k!})\mathbf{C}.\end{aligned}$$

For $t = 0$, using $\mathbf{x} = \mathbf{x}_0$, the foregoing equation give $\mathbf{PC} = \mathbf{x}_0$, so $\mathbf{C} = \mathbf{P}^{-1}\mathbf{x}_0$. Because of

$$(\mathbf{P}^{-1}\mathbf{N}\mathbf{P})^k = \mathbf{P}^{-1}\mathbf{N}^k\mathbf{P}$$

One obtains

$$\begin{aligned}\mathbf{x} &= \mathbf{P}diag(e^{\lambda_1 t}, \ldots, e^{\lambda_{j-1} t}, \underbrace{e^{\lambda_j t}, \ldots, e^{\lambda_j t}}_{m}, e^{\lambda_{j+m} t}, \ldots, e^{\lambda_n t})(\mathbf{I} + \sum_{k=1}^{m-1} \frac{\mathbf{N}^k t^k}{k!})\mathbf{P}^{-1}\mathbf{x}_0 \\ &= \mathbf{P}diag(e^{\lambda_1 t}, \ldots, e^{\lambda_{j-1} t}, \underbrace{e^{\lambda_j t}, \ldots, e^{\lambda_j t}}_{m}, e^{\lambda_{j+m} t}, \ldots, e^{\lambda_n t})\mathbf{P}^{-1}(\mathbf{I} + \sum_{k=1}^{m-1} \frac{\mathbf{N}^k t^k}{k!})\mathbf{x}_0.\end{aligned}$$

Therefore,

$$\mathbf{x} = \mathbf{P}\mathbf{E}(t)\mathbf{P}^{-1}(\mathbf{I} + \sum\nolimits_{k=1}^{m-1} \frac{\mathbf{N}^k t^k}{k!})\mathbf{x}_0,$$

where

$$\mathbf{E}(t) = diag(e^{\lambda_1 t}, \ldots, e^{\lambda_{j-1} t} e^{\lambda_j t}, \ldots, e^{\lambda_j t}, e^{\lambda_{j+m} t}, \ldots, e^{\lambda_n t}).$$

This theorem is proved. ■

Consider the solution for repeated eigenvalues of a linear dynamical system as

$$\mathbf{x}^{(j)}(t) = \sum\nolimits_{k=0}^{m-1} e^{\lambda_j t}(C_k^{(j)}\mathbf{v}_k^{(j)} t^k), \tag{A.54}$$

$$\dot{\mathbf{x}}^{(j)}(t) = \sum\nolimits_{k=0}^{m-1} \lambda_j e^{\lambda_j t}(C_k^{(j)}\mathbf{v}_k^{(j)} t^k) + e^{\lambda_j t}(kC_k^{(j)}\mathbf{v}_k^{(j)} t^{k-1}). \tag{A.55}$$

Submission of Eqs. (A.54) and (A.55) into $\dot{\mathbf{x}} = \mathbf{A}\mathbf{x}$ in Eq. (A.2) gives

$$\sum\nolimits_{k=0}^{m-1} (\mathbf{A} - \lambda_j\mathbf{I})e^{\lambda_j t}(C_k^{(j)}\mathbf{v}_k^{(j)} t^k) - e^{\lambda_j t}(kC_k^{(j)}\mathbf{v}_k^{(j)} t^{k-1}) = \mathbf{0}. \tag{A.56}$$

Thus for $k = 0, 1, 2, \ldots, m-2$

$$(\mathbf{A} - \lambda_j \mathbf{I}) C_k^{(j)} \mathbf{v}_k^{(j)} - (k+1) C_{k+1}^{(j)} \mathbf{v}_{k+1}^{(j)} = \mathbf{0}. \tag{A.57}$$

With $(\mathbf{A} - \lambda_j \mathbf{I}) \mathbf{v}_{m-1}^{(j)} = \mathbf{0}$, once eigenvectors are determined, the constants $C_k^{(j)}$ are obtained. On the other hand, let

$$C_k^{(j)} = (k+1) C_{k+1}^{(j)}. \tag{A.58}$$

Thus, one obtains

$$\begin{aligned} &(\mathbf{A} - \lambda_j \mathbf{I}) \mathbf{v}_{m-1}^{(j)} = \mathbf{0}, \\ &(\mathbf{A} - \lambda_j \mathbf{I}) \mathbf{v}_k^{(j)} = \mathbf{v}_{k+1}^{(j)} \quad (k = 0, 1, 2, \ldots, m-2). \end{aligned} \tag{A.59}$$

Deformation of the second equation of Eq. (A.59) gives

$$\begin{aligned} &\mathbf{A}(\underbrace{\mathbf{0}, \ldots, \mathbf{0}}_{j-1}, \underbrace{\mathbf{v}_0^{(j)}, \mathbf{v}_1^{(j)}, \ldots, \mathbf{v}_{m-1}^{(j)}}_{m}, \underbrace{\mathbf{0}, \ldots, \mathbf{0}}_{n-m-j+1}) \\ &= (\underbrace{\mathbf{0}, \ldots, \mathbf{0}}_{j-1}, \underbrace{\mathbf{v}_0^{(j)}, \mathbf{v}_1^{(j)}, \ldots, \mathbf{v}_{m-1}^{(j)}}_{m}, \underbrace{\mathbf{0}, \ldots, \mathbf{0}}_{n-m-j+1}) \mathbf{B}; \end{aligned} \tag{A.60}$$

where the Jordan matrix is

$$\mathbf{B}^{(j)} = \begin{bmatrix} \lambda_j & 0 & 0 & \ldots & 0 & 0 \\ 1 & \lambda_j & 0 & \ldots & 0 & 0 \\ 0 & 1 & \lambda_j & \ldots & 0 & 0 \\ \vdots & \vdots & \vdots & \ldots & \vdots & \vdots \\ 0 & 0 & 0 & \ldots & \lambda_j & 0 \\ 0 & 0 & 0 & \ldots & 1 & \lambda_j \end{bmatrix}_{m \times m}, \tag{A.61}$$

$$\mathbf{B} = diag(\mathbf{0}_{(j-1)\times(j-1)}, \mathbf{B}^{(j)}, \mathbf{0}_{(n-m-j+1)\times(n-m-j+1)}). \tag{A.62}$$

Thus

$$\begin{aligned} \mathbf{AP} &= \mathbf{P} diag(\lambda_1, \ldots, \lambda_{j-1}, \mathbf{B}^{(j)}|_{m\times m}, \lambda_{j+m}, \ldots, \lambda_n), \\ \mathbf{P}^{-1}\mathbf{AP} &= diag(\lambda_1, \ldots, \lambda_{j-1}, \mathbf{B}^{(j)}|_{m\times m}, \lambda_{j+m}, \ldots, \lambda_n), \end{aligned} \tag{A.63}$$

where

$$\begin{aligned} \mathbf{P} &= (\mathbf{v}^{(1)}, \ldots, \mathbf{v}^{(j-1)}, \mathbf{v}_0^{(j)}, \mathbf{v}_1^{(j)}, \ldots, \mathbf{v}_{m-1}^{(j)}, \mathbf{v}^{(j+m)}, \ldots, \mathbf{v}^{(n)}) \\ &= (\mathbf{v}_1, \mathbf{v}_2, \ldots, \mathbf{v}_n). \end{aligned} \tag{A.64}$$

With Eqs. (A.58), equation (A.54) becomes

$$\begin{aligned}\mathbf{x}^{(j)}(t) &= \sum_{k=0}^{m-1} e^{\lambda_j t}(C_k^{(j)}\mathbf{v}_k^{(j)}t^k)\\ &= (\mathbf{0},\ldots,\mathbf{0},\mathbf{v}_0^{(j)},\mathbf{v}_1^{(j)},\ldots,\mathbf{v}_{m-1}^{(j)},\mathbf{0},\ldots,\mathbf{0})e^{\mathbf{B}t}\mathbf{C}^{(j)}\end{aligned} \tag{A.65}$$

where

$$e^{\mathbf{B}t} = diag(\mathbf{0}_{(j-1)\times(j-1)}, e^{\mathbf{B}_j t}, \mathbf{0}_{(n-m-j+1)\times(n-m-j+1)}),$$

$$e^{\mathbf{B}_j t} = e^{\lambda_j t}\begin{bmatrix} 1 & 0 & 0 & \ldots & 0 & 0\\ t & 1 & 0 & \ldots & 0 & 0\\ t^2/2! & t & 1 & \ldots & 0 & 0\\ \vdots & \vdots & \vdots & \ldots & \vdots & \vdots\\ t^{m-2}/(m-2)! & t^{m-3}/(m-3)! & t^{m-4}/(m-4)! & \ldots & 1 & 0\\ t^{m-1}/(m-1)! & t^{m-2}/(m-2)! & t^{m-3}/(m-3)! & \ldots & t & 1\end{bmatrix}, \tag{A.66}$$

$$\mathbf{C}_0^{(j)} = (\underbrace{0,\ldots,0}_{j-1},\underbrace{C_0^{(j)},\ldots,C_{m-1}^{(j)}}_{m}\underbrace{0,\ldots,0}_{n-m-j+1})^{\mathrm{T}}. \tag{A.67}$$

Therefore,

$$\begin{aligned}\mathbf{x}(t) &= \mathbf{P}diag(e^{\lambda_1 t},\ldots,e^{\lambda_{j-1}t},e^{\mathbf{B}_j t},e^{\lambda_{j+m}t},\ldots,e^{\lambda_n t})\mathbf{C}\\ &= \mathbf{P}\bar{\mathbf{E}}(t)\mathbf{C}\end{aligned} \tag{A.68}$$

where

$$\begin{aligned}\mathbf{C} &= (C_1,\ldots,C_{j-1},C_0^{(j)},\ldots,C_{m-1}^{(j)},C_{j+m},\ldots,C_n)^{\mathrm{T}},\\ \bar{\mathbf{E}}(t) &= diag(e^{\lambda_1 t},\ldots,e^{\lambda_{j-1}t},e^{\mathbf{B}_j t},e^{\lambda_{j+m}t},\ldots,e^{\lambda_n t}).\end{aligned} \tag{A.69}$$

From initial condition, we have $\mathbf{x}_0 = \mathbf{PC}$. So, $\mathbf{C} = \mathbf{P}^{-1}\mathbf{x}_0$. Further,

$$\mathbf{x}(t) = \mathbf{P}\bar{\mathbf{E}}(t)\mathbf{P}^{-1}\mathbf{x}_0 = \mathbf{P}\bar{\mathbf{E}}(t)\mathbf{P}^{-1}\mathbf{x}_0. \tag{A.70}$$

Deformation of $\bar{\mathbf{E}}(t)$ gives

$$\begin{aligned}\bar{\mathbf{E}}(t) &= diag(e^{\lambda_1 t},\ldots,e^{\lambda_{j-1}t},\underbrace{0,\ldots0}_{m},e^{\lambda_{j+m}t},\ldots,e^{\lambda_n t})\\ &\quad + diag(0,\ldots,0,(e^{\mathbf{B}_j t})_{m\times m},0,\ldots,0),\end{aligned} \tag{A.71}$$

$$e^{\mathbf{B}_j t} = e^{\lambda_j t}(\mathbf{I} + t\bar{\mathbf{N}} + \ldots + \frac{\bar{\mathbf{N}}^{m-1}t^{m-1}}{(m-1)!})_{m\times m} = e^{\lambda_j t}(\mathbf{I} + \sum_{k=1}^{m-1}\frac{\bar{\mathbf{N}}^k t^k}{k!})_{m\times m}. \tag{A.72}$$

The $m \times m$ nilpotent matrix of order m is

$$\bar{\mathbf{N}} = \begin{bmatrix} 0 & 0 & 0 & \cdots & 0 & 0 \\ 1 & 0 & 0 & \cdots & 0 & 0 \\ \vdots & \vdots & \vdots & \cdots & \vdots & \vdots \\ 0 & 0 & 0 & \cdots & 0 & 0 \\ 0 & 0 & 0 & \cdots & 1 & 0 \end{bmatrix}_{m\times m}, \tag{A.73}$$

and

$$\bar{\mathbf{N}}^2 = \begin{bmatrix} 0 & 0 & 0 & \cdots & 0 & 0 \\ 0 & 0 & 0 & \cdots & 0 & 0 \\ 1 & 0 & 0 & \cdots & 0 & 0 \\ 0 & 1 & 0 & \cdots & 0 & 0 \\ \vdots & \vdots & \vdots & \vdots & \vdots & \vdots \\ 0 & 0 & 0 & \cdots & 0 & 0 \end{bmatrix}_{m\times m}, \cdots, \bar{\mathbf{N}}^{m-1} = \begin{bmatrix} 0 & 0 & 0 & \cdots & 0 & 0 \\ 0 & 0 & 0 & \cdots & 0 & 0 \\ 0 & 0 & 0 & \cdots & 0 & 0 \\ \vdots & \vdots & \vdots & \cdots & \vdots & \vdots \\ 0 & 0 & 0 & \cdots & 0 & 0 \\ 1 & 0 & 0 & \cdots & 0 & 0 \end{bmatrix}_{m\times m} \tag{A.74}$$

where

$$\begin{aligned} &\mathbf{N} = diag(\mathbf{0}_{(j-1)\times(j-1)}, \bar{\mathbf{N}}|_{m\times m}, \mathbf{0}_{(n-j-m+1)\times(n-j-m+1)}), \\ &\mathbf{N}^2 = diag(\mathbf{0}_{(j-1)\times(j-1)}, \bar{\mathbf{N}}^2|_{m\times m}, \mathbf{0}_{(n-j-m+1)\times(n-j-m+1)}), \\ &\vdots \\ &\mathbf{N}^{m-1} = diag(\mathbf{0}_{(j-1)\times(j-1)}, \bar{\mathbf{N}}^{m-1}|_{m\times m}, \mathbf{0}_{(n-j-m+1)\times(n-j-m+1)}). \end{aligned} \tag{A.75}$$

Finally,

$$\mathbf{x}(t) = \mathbf{PE}(t)(\mathbf{I} + \sum\nolimits_{k=1}^{m-1} \frac{\mathbf{N}^k t^k}{k!})\mathbf{P}^{-1}\mathbf{x}_0 = \mathbf{PE}(t)\mathbf{P}^{-1}(\mathbf{I} + \sum\nolimits_{k=1}^{m-1} \frac{\mathbf{N}^k t^k}{k!})\mathbf{x}_0 \tag{A.76}$$

where

$$\mathbf{E}(t) = diag(e^{\lambda_1 t}, \ldots, e^{\lambda_{j-1} t}, \underbrace{e^{\lambda_j t}, \ldots, e^{\lambda_j t}}_{m}, e^{\lambda_{j+m} t}, \ldots, e^{\lambda_n t}). \tag{A.77}$$

The foregoing discussion shows how to determine the Jordan matrix and the corresponding vectors for the repeated real eigenvalues. If one does not choose the relation in Eq. (A.58), one can get the general form as stated in Theorem A.8.

Theorem A.9 *Consider a linear dynamical system* $\dot{\mathbf{x}} = \mathbf{A}\mathbf{x}$ *in Eq.* (A.2) *with the initial condition of* $\mathbf{x}(0) = \mathbf{x}_0$. *A pair of repeated complex eigenvalue with m-times among the n-pairs of complex eigenvalues of the* $2n \times 2n$ *matrix* $\mathbf{A}$ *is* $\lambda_j = \alpha_j + \mathbf{i}\beta_j$ *and* $\bar{\lambda}_j = \alpha_j - \mathbf{i}\beta_j$ $(j = 1, 2, \ldots, n$ *and* $\mathbf{i} = \sqrt{-1})$. *The corresponding eigenvectors are* $\mathbf{w}_j = \mathbf{u}_j + \mathbf{i}\mathbf{v}_j$ *and* $\bar{\mathbf{w}}_j = \mathbf{u}_j - \mathbf{i}\mathbf{v}_j$. *If the corresponding eigenvector matrix of*

$\mathbf{P} = [\mathbf{u}_1, \mathbf{v}_1, \mathbf{u}_2, \mathbf{v}_2, \ldots, \mathbf{u}_n, \mathbf{v}_n]$ *is invertible as a basis in* $\Omega \subseteq \mathscr{R}^{2n}$. *For the repeated complex eigenvalue* λ_j, *the matrix* $\mathbf{A}$ *can be decomposed by*

$$\mathbf{A} = \mathbf{S} + \mathbf{N} \tag{A.78}$$

where

$$\mathbf{P}^{-1}\mathbf{S}\mathbf{P} = diag\left(\begin{bmatrix} \alpha_j & \beta_j \\ -\beta_j & \alpha_j \end{bmatrix}\right)_{n\times n}, \tag{A.79}$$

the matrix $\mathbf{N} = \mathbf{A} - \mathbf{S}$ *is nilpotent of order* $m \leq n$ *(i.e.,*$\mathbf{N}^m = \mathbf{0}$)*with* $\mathbf{SN} = \mathbf{NS}$.

$$\mathbf{P}^{-1}\mathbf{A}\mathbf{P} = diag[\mathbf{B}_1, \ldots, \mathbf{B}_{j-1}, \underbrace{\mathbf{B}_j, \ldots, \mathbf{B}_j}_{m}, \mathbf{B}_{j+m}, \ldots, \mathbf{B}_n], \tag{A.80}$$

where

$$\mathbf{B}_k = \begin{bmatrix} \alpha_k & \beta_k \\ -\beta_k & \alpha_k \end{bmatrix} (k = 1, 2, \ldots, n). \tag{A.81}$$

Thus, with an initial condition of $\mathbf{x}(t_0) = \mathbf{x}_0$, *the solution of linear dynamical system in Eq.* (A.2) *is*

$$\begin{aligned} \mathbf{x}(t) &= \mathbf{P}diag[\mathbf{E}_1(t), \ldots, \mathbf{E}_{j-1}(t), \underbrace{\mathbf{E}_j(t), \ldots, \mathbf{E}_j(t)}_{m}, \mathbf{E}_{j+m}(t), \ldots, \mathbf{E}_n(t)]\mathbf{P}^{-1} \\ &\quad \times [\mathbf{I} + \mathbf{N}t + \cdots + \frac{\mathbf{N}^{m-1}t^{m-1}}{(m-1)!}]\mathbf{x}_0 \\ &= \mathbf{P}\mathbf{E}(t)\mathbf{P}^{-1}[\mathbf{I} + \mathbf{N}t + \cdots + \frac{\mathbf{N}^{m-1}t^{m-1}}{(m-1)!}]\mathbf{x}_0, \end{aligned} \tag{A.82}$$

where the diagonal matrix $\mathbf{E}(t)$ *is given by*

$$\begin{aligned} \mathbf{E}(t) &= diag[\mathbf{E}_1(t), \ldots, \mathbf{E}_{j-1}(t), \underbrace{\mathbf{E}_j(t), \ldots, \mathbf{E}_j(t)}_{m}, \mathbf{E}_{j+m}(t), \ldots, \mathbf{E}_n(t)], \\ \mathbf{E}_k(t) &= e^{\alpha_k t}\mathbf{R}^{(k)} = e^{\alpha_k t}\begin{bmatrix} \cos\beta_k t & \sin\beta_k t \\ -\sin\beta_k t & \cos\beta_k t \end{bmatrix} \quad (k = 1, 2, \ldots, n). \end{aligned} \tag{A.83}$$

Proof Consider a pair of repeated complex eigenvalues with $\lambda_j = \alpha_i + \mathbf{i}\beta_j$ and $\bar{\lambda}_j = \alpha_i - \mathbf{i}\beta_j$ of matrix $\mathbf{A}$. The method of coefficient variation is adopted, and a pair of solutions relative to the two conjugate complex eigenvalue is

$$\mathbf{x}_+^{(j)} = \mathbf{C}^{(j)}(t)e^{(\alpha_j + \mathbf{i}\beta_j)t} \text{ and } \mathbf{x}_-^{(j)} = \bar{\mathbf{C}}^{(j)}(t)e^{(\alpha_j - \mathbf{i}\beta_j)t}.$$

Assume the coefficient vectors for complex eigenvalues as

$$\begin{aligned}\mathbf{C}^{(j)} &= \mathbf{U}^{(j)} + \mathbf{i}\mathbf{V}^{(j)} = \frac{1}{2}(U^{(j)} - \mathbf{i}V^{(j)})(\mathbf{u}^{(j)} + \mathbf{i}\mathbf{v}^{(j)})\\ &= \frac{1}{2}(U^{(j)}\mathbf{u}^{(j)} + V^{(j)}\mathbf{v}^{(j)}) - \frac{1}{2}\mathbf{i}(V^{(j)}\mathbf{u}^{(j)} - U^{(j)}\mathbf{v}^{(j)}),\end{aligned}$$

and

$$\begin{aligned}\bar{\mathbf{C}}^{(j)} &= \mathbf{U}^{(j)} - \mathbf{i}\mathbf{V}^{(j)} = \frac{1}{2}(U^{(j)} + \mathbf{i}V^{(j)})(\mathbf{u}^{(j)} - \mathbf{i}\mathbf{v}^{(j)})\\ &= \frac{1}{2}(U^{(j)}\mathbf{u}^{(j)} + V^{(j)}\mathbf{v}^{(j)}) + \frac{1}{2}\mathbf{i}(V^{(j)}\mathbf{u}^{(j)} - U^{(j)}\mathbf{v}^{(j)}).\end{aligned}$$

Thus,

$$\begin{aligned}\mathbf{x}_{+}^{(j)} + \mathbf{x}_{-}^{(j)} &= \mathbf{C}^{(j)}(t)e^{(\alpha_j+\mathbf{i}\beta_j)t} + \bar{\mathbf{C}}^{(j)}(t)e^{(\alpha_j-\mathbf{i}\beta_j)t}\\ &= (\mathbf{U}^{(j)} + \mathbf{i}\mathbf{V}^{(j)})e^{(\alpha_j+\mathbf{i}\beta_j)t} + (\mathbf{U}^{(j)} - \mathbf{i}\mathbf{V}^{(j)})e^{(\alpha_j-\mathbf{i}\beta_j)t}\\ &= (\mathbf{u}^{(j)}, \mathbf{v}^{(j)})e^{\alpha_j t}\begin{bmatrix} \cos\beta_j t & \sin\beta_j t\\ -\sin\beta_j t & \cos\beta_j t\end{bmatrix}\begin{Bmatrix} U^{(j)}\\ V^{(j)}\end{Bmatrix}.\end{aligned}$$

Further,

$$\begin{aligned}\dot{\mathbf{x}}_{+}^{(j)} + \dot{\mathbf{x}}_{-}^{(j)} = {} & \alpha_j e^{\alpha_j t}(\mathbf{u}^{(j)}, \mathbf{v}^{(j)})\begin{bmatrix} \cos\beta_j t & \sin\beta_j t\\ -\sin\beta_j t & \cos\beta_j t\end{bmatrix}\begin{Bmatrix} U^{(j)}\\ V^{(j)}\end{Bmatrix}\\ & + \beta_j e^{\alpha_j t}(\mathbf{u}^{(j)}, \mathbf{v}^{(j)})\begin{bmatrix} -\sin\beta_j t & \cos\beta_j t\\ -\cos\beta_j t & -\sin\beta_j t\end{bmatrix}\begin{Bmatrix} U^{(j)}\\ V^{(j)}\end{Bmatrix}\\ & + e^{\alpha_j t}(\mathbf{u}^{(j)}, \mathbf{v}^{(j)})\begin{bmatrix} \cos\beta_j t & \sin\beta_j t\\ -\sin\beta_j t & \cos\beta_j t\end{bmatrix}\begin{Bmatrix} \dot{U}^{(j)}\\ \dot{V}^{(j)}\end{Bmatrix},\\ \mathbf{A}(\mathbf{x}_{+}^{(j)} + \mathbf{x}_{-}^{(j)}) = {} & \mathbf{A}(\mathbf{u}^{(j)}, \mathbf{v}^{(j)})e^{\alpha_j t}\begin{bmatrix} \cos\beta_j t & \sin\beta_j t\\ -\sin\beta_j t & \cos\beta_j t\end{bmatrix}\begin{Bmatrix} U^{(j)}\\ V^{(j)}\end{Bmatrix}.\end{aligned}$$

The equation of $\dot{\mathbf{x}}_{+}^{(j)} + \dot{\mathbf{x}}_{-}^{(j)} = \mathbf{A}(\mathbf{x}_{+}^{(j)} + \mathbf{x}_{-}^{(j)})$ gives

$$\begin{aligned}&(\mathbf{u}^{(j)}, \mathbf{v}^{(j)})\mathbf{R}^{(j)}\begin{Bmatrix} \dot{U}^{(j)}\\ \dot{V}^{(j)}\end{Bmatrix}\\ &\quad = \left\{\mathbf{A} - \alpha_j\begin{bmatrix} 1 & 0\\ 0 & 1\end{bmatrix}\mathbf{I}_{2n\times 2n} + \beta_j\begin{bmatrix} 0 & 1\\ -1 & 0\end{bmatrix}\mathbf{I}_{2n\times 2n}\right\}(\mathbf{u}^{(j)}, \mathbf{v}^{(j)})\mathbf{R}^{(j)}\begin{Bmatrix} U^{(j)}\\ V^{(j)}\end{Bmatrix}\end{aligned}$$

where

$$\mathbf{R}^{(j)} = \begin{bmatrix} \cos\beta_j t & \sin\beta_j t\\ -\sin\beta_j t & \cos\beta_j t\end{bmatrix}.$$

Consider the constant vector and eigenvector matrix as

$$\begin{aligned}\mathbf{D}^{(j)} &= (0,0,\ldots,0,0,U^{(j)},V^{(j)},0,0,\ldots,0,0)^{\mathbf{T}}_{1\times 2n},\\ \mathbf{T} &= diag(\mathbf{0},\ldots,\mathbf{0},\mathbf{R}^{(j)},\mathbf{0},\ldots,\mathbf{0})^{\mathbf{T}}_{n\times n},\\ \mathbf{P} &= (\mathbf{u}_1,\mathbf{v}_1,\ldots,\mathbf{u}_{j-1},\mathbf{v}_{j-1},\mathbf{u}_j,\mathbf{v}_j,\mathbf{u}_{j+1},\mathbf{v}_{j+1},\ldots,\mathbf{u}_n,\mathbf{v}_n).\end{aligned}$$

Thus

$$\mathbf{P}(\mathbf{T}\dot{\mathbf{D}}^{(j)}) = (\mathbf{A} - \begin{bmatrix} \alpha_j & \beta_j \\ -\beta_j & \alpha_j \end{bmatrix}\mathbf{I}_{2n\times 2n})\mathbf{P}(\mathbf{T}\mathbf{D}^{(j)}).$$

Let $\mathbf{A} = \mathbf{S} + \mathbf{N}$, thus

$$\dot{\mathbf{D}}^{(j)} = (\mathbf{T})^{-1}(\mathbf{P}^{-1}\mathbf{S}\mathbf{P} - \mathbf{B}_j\mathbf{I}_{2n\times 2n} + \mathbf{P}^{-1}\mathbf{N}\mathbf{P})\mathbf{T}\mathbf{D}^{(j)}$$

where

$$\mathbf{B}_j = \begin{bmatrix} \alpha_j & \beta_j \\ -\beta_j & \alpha_j \end{bmatrix}, \mathbf{I}_{2\times 2} = \begin{bmatrix} 1 & 0 \\ 0 & 1 \end{bmatrix} \text{ and } \mathbf{I}_{2n\times 2n} = diag(\mathbf{I}_{2\times 2},\mathbf{I}_{2\times 2},\ldots,\mathbf{I}_{2\times 2})_{n\times n}.$$

Because of

$$\mathbf{P}^{-1}\mathbf{S}\mathbf{P} = diag(\begin{bmatrix} \alpha_j & -\beta_j \\ \beta_j & \alpha_j \end{bmatrix})_{n\times n} = diag(\mathbf{B}_j)_{n\times n} = \mathbf{B}_j\mathbf{I}_{2n\times 2n}$$

the solution of the foregoing equation is

$$\begin{aligned}\mathbf{D}^{(j)} &= (\mathbf{I}_{2n\times 2n} + \sum\nolimits_{k=1}^{m-1}\frac{[(\mathbf{T}^{-1}\mathbf{P}^{-1})\mathbf{N}(\mathbf{P}\mathbf{T})]^k t^k}{k!})\mathbf{D}_0^{(j)}\\ &= (\mathbf{I}_{2n\times 2n} + \sum\nolimits_{k=1}^{m-1}\frac{(\mathbf{T}^{-1}\mathbf{P}^{-1})\mathbf{N}^k(\mathbf{P}\mathbf{T})t^k}{k!})\mathbf{D}_0^{(j)}\\ &= (\mathbf{T}^{-1}\mathbf{P}^{-1})(\mathbf{I}_{2n\times 2n} + \sum\nolimits_{k=1}^{m-1}\frac{\mathbf{N}^k t^k}{k!})(\mathbf{P}\mathbf{T})\mathbf{D}_0^{(j)}.\end{aligned}$$

Therefore, the coefficient for the repeated complex eigenvalue of λ_j

$$\mathbf{R}^{(j)}\begin{Bmatrix} U^{(j)} \\ V^{(j)} \end{Bmatrix} = \mathbf{P}^{-1}(\mathbf{I}_{2n\times 2n} + \sum\nolimits_{k=1}^{m-1}\frac{\mathbf{N}^k t^k}{k!})\mathbf{P}\mathbf{R}^{(j)}\begin{Bmatrix} U_0^{(j)} \\ V_0^{(j)} \end{Bmatrix}.$$

Further

$$\mathbf{x}_+^{(j)} + \mathbf{x}_-^{(j)} = \mathbf{P}e^{\alpha_j t}\begin{bmatrix} \cos\beta_j t & \sin\beta_j t \\ -\sin\beta_j t & \cos\beta_j t \end{bmatrix}(\mathbf{I}_{2n\times 2n} + \sum\nolimits_{k=1}^{m-1}\frac{\mathbf{N}^k t^k}{k!})\mathbf{D}_k^{(j)}.$$

On the other hand, assuming

$$\mathbf{x}_{+}^{(j)}+\mathbf{x}_{-}^{(j)}=\sum\nolimits_{k=0}^{m-1}\mathbf{D}_{k}^{(j)}e^{\alpha_{j}t}\begin{bmatrix}\cos\beta_{j}t & \sin\beta_{j}t\\ -\sin\beta_{j}t & \cos\beta_{j}t\end{bmatrix}t^{k}\mathbf{D}_{k}^{(j)},$$

one obtains

$$k!\mathbf{D}_{k}^{(j)}=\mathbf{N}^{k}\mathbf{D}_{0}^{(j)}\text{ or }(k+1)\mathbf{D}_{k+1}^{(j)}=\mathbf{N}\mathbf{D}_{k}^{(j)}.$$

If $U_{k}^{(j)}=U_{0}^{(j)}$ and $V_{k}^{(j)}=V_{0}^{(j)}$, then

$$k!\mathbf{u}_{k}^{(j)}=\mathbf{N}^{k}\mathbf{u}_{0}^{(j)}\text{ or }(k+1)\mathbf{u}_{k+1}^{(j)}=\mathbf{N}\mathbf{u}_{k}^{(j)};$$
$$k!\mathbf{v}_{k}^{(j)}=\mathbf{N}^{k}\mathbf{v}_{0}^{(j)}\text{ or }(k+1)\mathbf{v}_{k+1}^{(j)}=\mathbf{N}\mathbf{v}_{k}^{(j)}.$$

Consider the total solution of the complex eigenvalues. Let

$$\mathbf{D}=(U_{1},V_{1},\ldots,U_{j-1},V_{j-1},\underbrace{U_{0}^{(j)},V_{0}^{(j)},\ldots,U_{0}^{(j)},V_{0}^{(j)}}_{m},U_{j+m},V_{j+m},\cdots,U_{n},V_{n})^{\mathrm{T}},$$

$$\mathbf{P}=(\mathbf{u}_{1},\mathbf{v}_{1},\ldots,\mathbf{u}_{j-1},\mathbf{v}_{j-1},\mathbf{u}_{j},\mathbf{v}_{j},\ldots,\mathbf{u}_{j+m-1},\mathbf{v}_{j+m-1},\mathbf{u}_{j+m},\mathbf{v}_{j+m},\ldots,\mathbf{u}_{n},\mathbf{v}_{n}).$$

Thus, there is a relation

$$\mathbf{P}^{-1}\mathbf{A}\mathbf{P}=diag[\mathbf{B}_{1},\ldots,\mathbf{B}_{j-1},\underbrace{\mathbf{B}_{j},\ldots,\mathbf{B}_{j}}_{m},\mathbf{B}_{j+m},\ldots,\mathbf{B}_{n}]$$

and, the resultant solution for the repeated eigenvalues is

$$\begin{aligned}\mathbf{x}(t)&=\sum\nolimits_{l=0}^{n}(\mathbf{x}_{+}^{(l)}+\mathbf{x}_{-}^{(l)})(1-\delta_{l}^{j})+\sum\nolimits_{k=0}^{m-1}(\mathbf{x}_{k+}^{(j)}+\mathbf{x}_{k-}^{(j)})\\ &=\mathbf{P}\mathbf{E}(t)\left[\mathbf{I}_{2n\times 2n}+\sum\nolimits_{k=1}^{m-1}\frac{\mathbf{N}^{k}t^{k}}{k!}\right]\mathbf{D}.\end{aligned}$$

where the diagonal matrix $\mathbf{E}(t)$ is given by

$$\mathbf{E}(t)=diag\left[\mathbf{E}_{1}(t),\ldots,\mathbf{E}_{j-1}(t),\underbrace{\mathbf{E}_{j}(t),\ldots,\mathbf{E}_{j}(t)}_{m},\mathbf{E}_{j+m}(t),\ldots,\mathbf{E}_{n}(t)\right],$$

$$\mathbf{E}_{k}(t)=e^{\alpha_{k}t}\begin{bmatrix}\cos\beta_{k}t & \sin\beta_{k}t\\ -\sin\beta_{k}t & \cos\beta_{k}t\end{bmatrix}(k=1,2,\ldots,n).$$

For $t=0$, using $\mathbf{x}=\mathbf{x}_{0}$, one obtains $\mathbf{PD}=\mathbf{x}_{0}$, so $\mathbf{D}=\mathbf{P}^{-1}\mathbf{x}_{0}$. Thus,

$$\mathbf{x}(t)=\mathbf{P}\mathbf{E}(t)\mathbf{P}^{-1}\left[\mathbf{I}_{2n\times 2n}+\sum\nolimits_{k=1}^{m-1}\frac{\mathbf{N}^{k}t^{k}}{k!}\right]\mathbf{x}_{0}.$$

This theorem is proved. ■

Consider the solution for repeated eigenvalues of a linear dynamical system as

$$\mathbf{x}_{+}^{(j)}=\sum\nolimits_{k=0}^{m-1}C_{k}^{(j)}\mathbf{c}_{k}^{(j)}e^{(\alpha_j+\mathbf{i}\beta_j)t}t^{k}\text{ and }\mathbf{x}_{-}^{(j)}=\sum\nolimits_{k=0}^{m-1}\bar{C}_{k}^{(j)}\bar{\mathbf{c}}_{k}^{(j)}e^{(\alpha_j-\mathbf{i}\beta_j)t}t^{k}. \tag{A.84}$$

$$\begin{aligned}\dot{\mathbf{x}}_{+}^{(j)}&=\sum\nolimits_{k=0}^{m-1}(\alpha_j+\mathbf{i}\beta_j)C_{k}^{(j)}\mathbf{c}_{k}^{(j)}e^{(\alpha_j+\mathbf{i}\beta_j)t}]t^{k}+kC_{k}^{(j)}\mathbf{c}_{k}^{(j)}e^{(\alpha_j+\mathbf{i}\beta_j)t}t^{k-1},\\ \dot{\mathbf{x}}_{-}^{(j)}&=\sum\nolimits_{k=0}^{m-1}(\alpha_j-\mathbf{i}\beta_j)\bar{C}_{k}^{(j)}\bar{\mathbf{c}}_{k}^{(j)}e^{(\alpha_j-\mathbf{i}\beta_j)t}t^{k}+k\bar{C}_{k}^{(j)}\bar{\mathbf{c}}_{k}^{(j)}e^{(\alpha_j-\mathbf{i}\beta_j)t}t^{k-1}.\end{aligned} \tag{A.85}$$

Submission of Eqs. (A.84) and (A.85) into $\dot{\mathbf{x}}_{+}^{(j)}=\mathbf{A}\mathbf{x}_{+}^{(j)}$ and $\dot{\mathbf{x}}_{-}^{(j)}=\mathbf{A}\mathbf{x}_{-}^{(j)}$ gives

$$\begin{aligned}&\sum\nolimits_{k=0}^{m-1}[\mathbf{A}-(\alpha_j+\mathbf{i}\beta_j)\mathbf{I}_{2n\times 2n}]C_{k}^{(j)}\mathbf{c}_{k}^{(j)}e^{(\alpha_j+\mathbf{i}\beta_j)t}t^{k}-kC_{k}^{(j)}\mathbf{c}_{k}^{(j)}e^{(\alpha_j+\mathbf{i}\beta_j)t}t^{k-1}=\mathbf{0},\\ &\sum\nolimits_{k=0}^{m-1}[\mathbf{A}-(\alpha_j-\mathbf{i}\beta_j)\mathbf{I}_{2n\times 2n}]\bar{C}_{k}^{(j)}\bar{\mathbf{c}}_{k}^{(j)}e^{(\alpha_j-\mathbf{i}\beta_j)t}t^{k}-k\bar{C}_{k}^{(j)}\bar{\mathbf{c}}_{k}^{(j)}e^{(\alpha_j+\mathbf{i}\beta_j)t}t^{k-1}=\mathbf{0}.\end{aligned} \tag{A.86}$$

Thus

$$\begin{aligned}&[\mathbf{A}-(\alpha_j+\mathbf{i}\beta_j)\mathbf{I}_{2n\times 2n}]C_{k}^{(j)}\mathbf{c}_{k}^{(j)}-(k+1)C_{k+1}^{(j)}\mathbf{c}_{k+1}^{(j)}=\mathbf{0},\\ &[\mathbf{A}-(\alpha_j+\mathbf{i}\beta_j)\mathbf{I}_{2n\times 2n}]\bar{C}_{k}^{(j)}\bar{\mathbf{c}}_{k}^{(j)}-(k+1)\bar{C}_{k+1}^{(j)}\bar{\mathbf{c}}_{k+1}^{(j)}=\mathbf{0}\\ &(k=0,1,2,\ldots m-2).\end{aligned} \tag{A.87}$$

With $[\mathbf{A}-(\alpha_j+\mathbf{i}\beta_j)\mathbf{I}_{2n\times 2n}]\mathbf{c}_{m-1}^{(j)}=\mathbf{0}$ and $[\mathbf{A}-(\alpha_j-\mathbf{i}\beta_j)\mathbf{I}_{2n\times 2n}]\bar{\mathbf{c}}_{m-1}^{(j)}=\mathbf{0}$, once eigenvectors are determined, the constants $C_{k}^{(j)}$ are obtained. On the other hand, let

$$C_{k}^{(j)}=(k+1)C_{k+1}^{(j)}\text{ and }\bar{C}_{k}^{(j)}=(k+1)\bar{C}_{k+1}^{(j)}. \tag{A.88}$$

Thus, one obtains

$$\begin{aligned}&[\mathbf{A}-(\alpha_j+\mathbf{i}\beta_j)\mathbf{I}_{2n\times 2n}]\mathbf{c}_{m-1}^{(j)}=\mathbf{0},\\ &[\mathbf{A}-(\alpha_j+\mathbf{i}\beta_j)\mathbf{I}_{2n\times 2n}]\mathbf{c}_{k}^{(j)}-\mathbf{c}_{k+1}^{(j)}=\mathbf{0}\\ &[\mathbf{A}-(\alpha_j-\mathbf{i}\beta_j)\mathbf{I}_{2n\times 2n}]\bar{\mathbf{c}}_{m-1}^{(j)}=\mathbf{0},\\ &[\mathbf{A}-(\alpha_j-\mathbf{i}\beta_j)\mathbf{I}_{2n\times 2n}]\bar{\mathbf{c}}_{k}^{(j)}-\bar{\mathbf{c}}_{k+1}^{(j)}=\mathbf{0}\\ &(k=0,1,2,\ldots m-2).\end{aligned} \tag{A.89}$$

Assuming

$$\mathbf{c}_{k}^{(j)}=\mathbf{u}_{k}^{(j)}+\mathbf{i}\mathbf{v}_{k}^{(j)}\text{ and }\bar{\mathbf{c}}_{k}^{(j)}=\mathbf{u}_{k}^{(j)}-\mathbf{i}\mathbf{v}_{k}^{(j)}, \tag{A.90}$$

deformation of Eq. (A.89) gives

$$\begin{aligned}&\mathbf{A}(\mathbf{u}_{m-1}^{(j)},\mathbf{v}_{m-1}^{(j)})=(\mathbf{u}_{m-1}^{(j)},\mathbf{v}_{m-1}^{(j)})\begin{bmatrix}\alpha_j & \beta_j\\ -\beta_j & \alpha_j\end{bmatrix},\\ &\mathbf{A}(\mathbf{u}_{k}^{(j)},\mathbf{v}_{k}^{(j)})=(\mathbf{u}_{k}^{(j)},\mathbf{v}_{k}^{(j)})\begin{bmatrix}\alpha_j & \beta_j\\ -\beta_j & \alpha_j\end{bmatrix}+(\mathbf{u}_{k+1}^{(j)},\mathbf{v}_{k+1}^{(j)})\\ &(k=0,1,2,\ldots m-2);\end{aligned} \tag{A.91}$$

$$\begin{aligned}
&\mathbf{A}(\underbrace{\mathbf{0},\ldots,\mathbf{0}}_{2(j-1)},\underbrace{\mathbf{u}_0^{(j)},\mathbf{v}_0^{(j)},\mathbf{u}_1^{(j)},\mathbf{v}_1^{(j)},\ldots,\mathbf{u}_{m-1}^{(j)},\mathbf{v}_{m-1}^{(j)}}_{2m},\underbrace{\mathbf{0},\ldots,\mathbf{0}}_{2(n-(j+m)+1)})\\
&=(\underbrace{\mathbf{0},\ldots,\mathbf{0}}_{2(j-1)},\underbrace{\mathbf{u}_0^{(j)},\mathbf{v}_0^{(j)},\mathbf{u}_1^{(j)},\mathbf{v}_1^{(j)},\ldots,\mathbf{u}_{m-1}^{(j)},\mathbf{v}_{m-1}^{(j)}}_{2m},\underbrace{\mathbf{0},\ldots,\mathbf{0}}_{2(n-(j+m)+1)})\mathbf{B},
\end{aligned} \tag{A.92}$$

where the Jordan matrix is

$$\begin{aligned}
&\mathbf{D}^{(j)}=\begin{bmatrix}\alpha_j & \beta_j\\ -\beta_j & \alpha_j\end{bmatrix}\text{and}\\
&\mathbf{B}=diag(\mathbf{0}_{2(j-1)\times 2(j-1)},\mathbf{B}^{(j)},\mathbf{0}_{2(n-m-j+1)\times 2(n-m-j+1)})\\
&\mathbf{B}^{(j)}=\begin{bmatrix}
\mathbf{D}^{(j)} & \mathbf{0} & \mathbf{0} & \ldots & \mathbf{0} & \mathbf{0}\\
\mathbf{I}_{2\times 2} & \mathbf{D}^{(j)} & \mathbf{0} & \ldots & \mathbf{0} & \mathbf{0}\\
\mathbf{0} & \mathbf{I}_{2\times 2} & \mathbf{D}^{(j)} & \ldots & \mathbf{0} & \mathbf{0}\\
\vdots & \vdots & \vdots & \ldots & \vdots & \vdots\\
\mathbf{0} & \mathbf{0} & \mathbf{0} & \ldots & \mathbf{D}^{(j)} & \mathbf{0}\\
\mathbf{0} & \mathbf{0} & \mathbf{0} & \ldots & \mathbf{I}_{2\times 2} & \mathbf{D}^{(j)}
\end{bmatrix}_{2m\times 2m}.
\end{aligned} \tag{A.93}$$

Thus

$$\begin{aligned}
\mathbf{AP}&=\mathbf{P}daig(\mathbf{D}^{(1)},\ldots\mathbf{D}^{(j-1)},\mathbf{B}^{(j)},\mathbf{D}^{(j+m)},\ldots\mathbf{D}^{(n)}),\\
\mathbf{P}^{-1}\mathbf{AP}&=daig(\mathbf{D}^{(1)},\ldots\mathbf{D}^{(j-1)},\mathbf{B}^{(j)},\mathbf{D}^{(j+m)},\ldots\mathbf{D}^{(n)}),
\end{aligned} \tag{A.94}$$

where

$$\begin{aligned}
\mathbf{P}&=(\mathbf{u}^{(1)},\mathbf{v}^{(1)},\ldots,\mathbf{u}^{(j-1)},\mathbf{v}^{(j-1)},\mathbf{u}_0^{(j)},\mathbf{v}_0^{(j)},\mathbf{u}_1^{(j)},\mathbf{v}_1^{(j)},\\
&\quad\ldots,\mathbf{u}_{m-1}^{(j)},\mathbf{v}_{m-1}^{(j)},\mathbf{u}^{(j+m)},\mathbf{v}^{(j+m)},\ldots,\mathbf{u}^{(n)},\mathbf{v}^{(n)})\\
&=(\mathbf{u}_1,\mathbf{v}_1,\mathbf{u}_2,\mathbf{v}_2,\ldots,\mathbf{u}_n,\mathbf{v}_n).
\end{aligned} \tag{A.95}$$

Suppose two conjugate constants are

$$C_k^{(j)}=\frac{1}{2}(U_k^{(j)}-\mathbf{i}V_k^{(j)})\text{ and }\bar{C}_k^{(j)}=\frac{1}{2}(U_k^{(j)}+\mathbf{i}V_k^{(j)}). \tag{A.96}$$

With Eq. (A.88), Eq. (A.83) becomes

$$\begin{aligned}
\mathbf{x}^{(j)}(t)&=\mathbf{x}_+^{(j)}+\mathbf{x}_-^{(j)}\\
&=\sum\nolimits_{k=0}^{m-1}C_k^{(j)}\mathbf{c}_k^{(j)}e^{(\alpha_j+\mathbf{i}\beta_j)t}t^k+\sum\nolimits_{k=0}^{m-1}\bar{C}_k^{(j)}\bar{\mathbf{c}}_k^{(j)}e^{(\alpha_j-\mathbf{i}\beta_j)t}t^k\\
&=(\mathbf{u}_0^{(j)},\mathbf{v}_0^{(j)},\mathbf{u}_1^{(j)},\mathbf{v}_1^{(j)},\ldots,\mathbf{u}_{m-1}^{(j)},\mathbf{v}_{m-1}^{(j)})e^{\mathbf{B}_j t}\mathbf{C}^{(j)}
\end{aligned} \tag{A.97}$$

where

$$e^{\mathbf{B}_j t} = e^{\alpha_j t}\begin{bmatrix} \mathbf{R}^{(j)} & \mathbf{0} & \cdots & \mathbf{0} & \mathbf{0} \\ \mathbf{R}^{(j)}t & \mathbf{R}^{(j)} & \cdots & \mathbf{0} & \mathbf{0} \\ \mathbf{R}^{(j)}t^2/2! & \mathbf{R}^{(j)}t & \cdots & \mathbf{0} & \mathbf{0} \\ \vdots & \vdots & \vdots & \vdots & \vdots \\ \mathbf{R}^{(j)}t^{m-2}/(m-2)! & \mathbf{R}^{(j)}t^{m-3}/(m-3)! & \cdots & \mathbf{R}^{(j)} & \mathbf{0} \\ \mathbf{R}^{(j)}t^{m-1}/(m-1)! & \mathbf{R}^{(j)}t^{m-2}/(m-2)! & \cdots & \mathbf{R}^{(j)}t & \mathbf{R}^{(j)} \end{bmatrix}, \tag{A.98}$$

$$\mathbf{C}^{(j)} = (\underbrace{0,\ldots 0}_{2(j-1)}, \underbrace{U_0^{(j)}, V_0^{(j)}\ldots, U_{m-1}^{(j)}, V_{m-1}^{(j)}}_{2m}, \underbrace{0,\ldots,0}_{2(n-m-j+1)})^{\mathrm{T}}. \tag{A.99}$$

Therefore,

$$\begin{aligned} \mathbf{x}(t) &= \mathbf{P}diag(\mathbf{E}_1(t),\ldots,\mathbf{E}_{j-1}(t), e^{\mathbf{B}_j t}, \mathbf{E}_{j+m}(t),\ldots,\mathbf{E}_n(t))\mathbf{C} \\ &= \mathbf{P}\bar{E}(t)\mathbf{C} \end{aligned} \tag{A.100}$$

where

$$\begin{aligned} \mathbf{C} &= (U^{(1)}, V^{(1)},\ldots, U^{(j-1)}, V^{(j-1)}, \underbrace{U_0^{(j)}, V_0^{(j)},\ldots, U_{m-1}^{(j)}, V_{m-1}^{(j)}}_{m}, \\ &\quad U^{(j+m)}, V^{(j+m)},\ldots, U^{(n)}, V^{(n)})^{\mathrm{T}}; \\ \bar{\mathbf{E}}(t) &= \mathbf{P}diag(\mathbf{E}_1(t),\ldots,\mathbf{E}_{j-1}(t), e^{\mathbf{B}_j t}, \mathbf{E}_{j+m}(t),\ldots,\mathbf{E}_n(t)). \end{aligned} \tag{A.101}$$

For initial conditions, we have $\mathbf{x}_0 = \mathbf{PC}$. So, $\mathbf{C} = \mathbf{P}^{-1}\mathbf{x}_0$. Further,

$$\mathbf{x}(t) = \mathbf{PE}(t)\mathbf{P}^{-1}\mathbf{x}_0 = \mathbf{PE}(t)\mathbf{P}^{-1}\mathbf{x}_0 \tag{A.102}$$

where

$$\begin{aligned} \bar{\mathbf{E}}(t) &= diag(\mathbf{E}_1(t),\ldots,\mathbf{E}_{j-1}(t), e^{\mathbf{B}_j t}, \mathbf{E}_{j+m}(t),\ldots,\mathbf{E}_n(t)) \\ &= diag(\mathbf{E}_1(t),\ldots,\mathbf{E}_{j-1}(t), \underbrace{\mathbf{0},\ldots,\mathbf{0}}_{m}, \mathbf{E}_{j+m}(t),\ldots,\mathbf{E}_n(t)) \\ &\quad + diag(\mathbf{0},\ldots,\mathbf{0}, (e^{\mathbf{B}_j t})_{2m\times 2m}, \mathbf{0},\ldots,\mathbf{0}) \end{aligned} \tag{A.103}$$

and

$$e^{\mathbf{B}_j t} = e^{\alpha_j}\mathbf{R}^{(j)}(\mathbf{I} + \sum_{k=1}^{m-1}\frac{\bar{\mathbf{N}}^k t^k}{k!})_{2m\times 2m}. \tag{A.104}$$

The $2m \times 2m$ nilpotent matrix of order m is

$$\bar{\mathbf{N}} = \begin{bmatrix} \mathbf{0} & \mathbf{0} & \mathbf{0} & \dots & \mathbf{0} & \mathbf{0} \\ \mathbf{I}_{2\times2} & \mathbf{0} & \mathbf{0} & \dots & \mathbf{0} & \mathbf{0} \\ \mathbf{0} & \mathbf{I}_{2\times2} & \mathbf{0} & \dots & \mathbf{0} & \mathbf{0} \\ \vdots & \vdots & \vdots & \vdots & \vdots & \vdots \\ \mathbf{0} & \mathbf{0} & \mathbf{0} & \dots & \mathbf{0} & \mathbf{0} \\ \mathbf{0} & \mathbf{0} & \mathbf{0} & \dots & \mathbf{I}_{2\times2} & \mathbf{0} \end{bmatrix}_{2m\times2m}, \tag{A.105}$$

and

$$\begin{aligned} \bar{\mathbf{N}}^2 &= \begin{bmatrix} \mathbf{0} & \mathbf{0} & \mathbf{0} & \dots & \mathbf{0} \\ \mathbf{0} & \mathbf{0} & \mathbf{0} & \dots & \mathbf{0} \\ \mathbf{I}_{2\times2} & \mathbf{0} & \mathbf{0} & \dots & \mathbf{0} \\ \mathbf{0} & \mathbf{I}_{2\times2} & \mathbf{0} & \dots & \mathbf{0} \\ \vdots & \vdots & \vdots & \dots & \vdots \\ \mathbf{0} & \mathbf{0} & \mathbf{0} & \dots & \mathbf{0} \end{bmatrix}_{2m\times2m}, \\ &\vdots \\ \bar{\mathbf{N}}^{m-1} &= \begin{bmatrix} \mathbf{0} & \mathbf{0} & \mathbf{0} & \dots & \mathbf{0} & \mathbf{0} \\ \mathbf{0} & \mathbf{0} & \mathbf{0} & \dots & \mathbf{0} & \mathbf{0} \\ \mathbf{0} & \mathbf{0} & \mathbf{0} & \dots & \mathbf{0} & \mathbf{0} \\ \vdots & \vdots & \vdots & \vdots & \vdots & \vdots \\ \mathbf{0} & \mathbf{0} & \mathbf{0} & \dots & \mathbf{0} & \mathbf{0} \\ \mathbf{I}_{2\times2} & \mathbf{0} & \mathbf{0} & \dots & \mathbf{0} & \mathbf{0} \end{bmatrix}_{2m\times2m}. \end{aligned} \tag{A.106}$$

where

$$\begin{aligned} \mathbf{N} &= diag(\mathbf{0}_{2(j-1)\times2(j-1)}, \bar{\mathbf{N}}|_{2m\times2m}, \mathbf{0}_{2(n-j-m+1)\times2(n-j-m+1)}), \\ \mathbf{N}^2 &= diag(\mathbf{0}_{2(j-1)\times2(j-1)}, \bar{\mathbf{N}}^2|_{2m\times2m}, \mathbf{0}_{2(n-j-m+1)\times2(n-j-m+1)}), \\ &\vdots \\ \mathbf{N}^{m-1} &= diag(\mathbf{0}_{2(j-1)\times2(j-1)}, \bar{\mathbf{N}}^{m-1}|_{2m\times2m}, \mathbf{0}_{2(n-j-m+1)\times2(n-j-m+1)}). \end{aligned} \tag{A.107}$$

Finally,

$$\begin{aligned} \mathbf{x}(t) &= \mathbf{PE}(t)(\mathbf{I} + \sum\nolimits_{k=1}^{m-1} \frac{\mathbf{N}^k t^k}{k!})\mathbf{P}^{-1}\mathbf{x}_0 \\ &= \mathbf{PE}(t)\mathbf{P}^{-1}(\mathbf{I} + \sum\nolimits_{k=1}^{m-1} \frac{\mathbf{N}^k t^k}{k!})\mathbf{x}_0. \end{aligned} \tag{A.108}$$

where

$$\mathbf{E}(t) = diag(\mathbf{E}^{(1)}, \dots, \mathbf{E}^{(j-1)}, \underbrace{\mathbf{E}^{(j)}, \dots, \mathbf{E}^{(j)}}_{m}, \mathbf{E}^{(j+m)}, \dots, \mathbf{E}^{(n)}). \tag{A.109}$$

The foregoing discussion shows how to determine the Jordan matrix and the corresponding vectors for the repeated complex eigenvalues. If one does not choose the relation in Eq. (A.88), one can get the general form as stated in Theorem A.9.

A.4 Nonhomogeneous Linear Systems

Definition A.8 For a linear dynamical system $\dot{\mathbf{x}} = \mathbf{A}\mathbf{x}$ in Eq. (A.2), there is a fundamental matrix solution $\mathbf{\Phi}(t)$ to satisfy

$$\dot{\mathbf{\Phi}}(t) = \mathbf{A}\mathbf{\Phi}(t) \text{ for all } t \in \mathscr{R} \tag{A.110}$$

where $\mathbf{\Phi}(t)$ is an $n \times n$ nonsingular matrix function.

Theorem A.10 *For a linear dynamical system* $\dot{\mathbf{x}} = \mathbf{A}\mathbf{x} + \mathbf{Q}(t)$*in Eq.* (A.1) *with the initial condition of* $\mathbf{x}(t_0) = \mathbf{x}_0$, *if* $\mathbf{\Phi}(t)$ *is a fundamental matrix solution of* $\dot{\mathbf{x}} = \mathbf{A}\mathbf{x}$ *in Eq.* (A.2) , *then the solution of Eq.* (A.1) *is given by*

$$\mathbf{x}(t) = \mathbf{\Phi}(t)\mathbf{\Phi}^{-1}(t_0)\mathbf{x}_0 + \int_{t_0}^{t} \mathbf{\Phi}(t)\mathbf{\Phi}^{-1}(\tau)\mathbf{Q}(\tau)d\tau. \tag{A.111}$$

Proof Using the variation of coefficient, assume the solution as $\mathbf{x}(t) = \mathbf{\Phi}(t)\mathbf{C}(t)$. Thus, one obtains

$$\dot{\mathbf{x}}(t) = \dot{\mathbf{\Phi}}(t)\mathbf{C}(t) + \mathbf{\Phi}(t)\dot{\mathbf{C}}(t),$$
$$\mathbf{\Phi}(t)\dot{\mathbf{C}}(t) = \mathbf{Q}(t) \Rightarrow \dot{\mathbf{C}}(t) = \mathbf{\Phi}^{-1}(t)\mathbf{Q}(t).$$

So

$$\mathbf{C}(t) = \int_{t_0}^{t} \mathbf{\Phi}^{-1}(\tau)\mathbf{Q}(\tau)d\tau + \mathbf{C}_0.$$
$$\mathbf{x}(t) = \mathbf{\Phi}(t)\mathbf{C}(t) = \mathbf{\Phi}(t)\int_{t_0}^{t} \mathbf{\Phi}^{-1}(\tau)\mathbf{Q}(\tau)d\tau + \mathbf{\Phi}(t)\mathbf{C}_0.$$

With initial conditions,

$$\mathbf{x}(t) = \mathbf{\Phi}(t)\int_{t_0}^{t} \mathbf{\Phi}^{-1}(\tau)\mathbf{Q}(\tau)d\tau + \mathbf{\Phi}(t)\mathbf{\Phi}^{-1}(t_0)\mathbf{x}_0.$$

The derivative of $\mathbf{x}(t)$ from the foregoing equation gives

$$\begin{aligned}\dot{\mathbf{x}}(t) &= \dot{\mathbf{\Phi}}(t)\mathbf{\Phi}^{-1}(t_0)\mathbf{x}_0 + \int_{t_0}^{t} \dot{\mathbf{\Phi}}(t)\mathbf{\Phi}^{-1}(\tau)\mathbf{Q}(\tau)d\tau + \mathbf{\Phi}(t)\mathbf{\Phi}^{-1}(t)\mathbf{Q}(t)\\ &= \mathbf{A}\left[\mathbf{\Phi}(t)\mathbf{\Phi}^{-1}(t_0)\mathbf{x}_0 + \int_{t_0}^{t} \mathbf{\Phi}(t)\mathbf{\Phi}^{-1}(\tau)\mathbf{Q}(\tau)d\tau\right] + \mathbf{Q}(t)\\ &= \mathbf{A}\mathbf{x}(t) + \mathbf{Q}(t).\end{aligned}$$

This theorem is proved. ■

Notice that the fundamental solution matrix is expressed by

$$\boldsymbol{\Phi}(t) = e^{\mathbf{A}(t-t_0)}. \tag{A.112}$$

Using Eqs. (A.112), (A.111) becomes

$$\mathbf{x}(t) = e^{\mathbf{A}(t-t_0)}\mathbf{x}_0 + \int_{t_0}^{t} e^{\mathbf{A}(t-\tau)}\mathbf{Q}(\tau)d\tau. \tag{A.113}$$

For distinct real eigenvalues, the expression of the homogeneous solution for $t_0 = 0$ is

$$\begin{aligned} \boldsymbol{\Phi}(t) &= \mathbf{P}diag(e^{\lambda_1 t}, e^{\lambda_2 t}, \ldots, e^{\lambda_n t}), \\ \mathbf{P} &= (\mathbf{v}_1, \mathbf{v}_2, \ldots, \mathbf{v}_n)_{n\times n}. \end{aligned} \tag{A.114}$$

For distinct complex eigenvalues, the expression of the homogeneous solution for $t_0 = 0$ is

$$\begin{aligned} &\boldsymbol{\Phi}(t) = \mathbf{P}diag(e^{\alpha_1 t}\mathbf{B}_1, e^{\alpha_2 t}\mathbf{B}_2, \ldots, e^{\alpha_n t}\mathbf{B}_n), \\ &\mathbf{P} = (\mathbf{u}_1, \mathbf{v}_1, \mathbf{u}_2, \mathbf{v}_2, \ldots, \mathbf{u}_n, \mathbf{v}_n)_{2n\times 2n} \text{ and } \mathbf{B}_j = \begin{bmatrix} \cos\beta_j t & \sin\beta_j t \\ -\sin\beta_j t & \cos\beta_j t \end{bmatrix}. \end{aligned} \tag{A.115}$$

For repeated eigenvalues, the fundamental solution matrix can be obtained from the previous discussion.

Definition A.9 For a linear dynamical system $\dot{\mathbf{x}} = \mathbf{A}\mathbf{x} + \mathbf{Q}(t)$ in Eq. (A.1) with

$$\mathbf{Q}(t) = \mathbf{Q}_0^{(k)} e^{\lambda_k t}. \tag{A.116}$$

The system in Eq. (A.1) is called an excited nonhomogeneous system in the direction of eigenvector $\mathbf{v}_k$ if λ_k is one of eigenvalues of $\mathbf{A}$, or

$$\det(\mathbf{A} - \lambda_k \mathbf{I}) = 0. \tag{A.117}$$

Definition A.10 For a linear dynamical system $\dot{\mathbf{x}} = \mathbf{A}\mathbf{x} + \mathbf{Q}(t)$ in Eq. (A.1) with

$$\mathbf{Q}(t) = \mathbf{Q}_1^{(k)} e^{\alpha_k t}\cos\beta_k t + \mathbf{Q}_2^{(k)} e^{\alpha_k t}\sin\beta_k t. \tag{A.118}$$

(i) The system in Eq. (A.1) is called an excited nonhomogeneous system in the spiral oscillation if $\alpha_k \pm \mathbf{i}\beta_k$ is a pair of eigenvalues of $\mathbf{A}$.
(ii) The system in Eq. (A.1) is called a *resonant* nonhomogeneous system (simply say, *resonance*) in such oscillation if $\alpha_k = 0$ and $\pm\mathbf{i}\beta_k$ is a pair of eigenvalues of $\mathbf{A}$.

For $\det(\mathbf{A} - \lambda_k \mathbf{I}) \neq 0$, consider a few special cases herein. For the first case,

$$\mathbf{Q}(t) = \sum\nolimits_{k=1}^{m} \mathbf{Q}_0^{(k)} e^{\lambda_k t}, \tag{A.119}$$

Assuming

$$\mathbf{x}_p(t) = \sum\nolimits_{k=1}^{m} \mathbf{D}^{(k)} e^{\lambda_k t}, \tag{A.120}$$

$$\dot{\mathbf{x}}_p(t) = \sum_{k=1}^{m} \lambda_k \mathbf{D}^{(k)} e^{\lambda_k t}, \tag{A.121}$$

Thus

$$\sum_{k=1}^{m} \lambda_k \mathbf{D}^{(k)} e^{\lambda_k t} = \mathbf{A} \sum_{k=1}^{m} \mathbf{D}^{(k)} e^{\lambda_k t} + \sum_{k=1}^{m} \mathbf{Q}_0^{(k)} e^{\lambda_k t}, \tag{A.122}$$

Therefore, if $|\mathbf{A} - \lambda_k \mathbf{I}| \neq 0$,

$$\mathbf{D}^{(k)} = d^{(k)} \mathbf{v}_k = -(\mathbf{A} - \lambda_k \mathbf{I})^{-1} \mathbf{Q}_0^{(k)}. \tag{A.123}$$

The particular solution of the nonhomogeneous solution is

$$\mathbf{x}_p(t) = -\sum_{k=1}^{m} (\mathbf{A} - \lambda_k \mathbf{I})^{-1} \mathbf{Q}_0^{(k)} e^{\lambda_k t}, \tag{A.124}$$

and the total solution is expressed by

$$\begin{aligned} \mathbf{x}(t) &= \mathbf{x}_h(t) + \mathbf{x}_p(t) \\ &= e^{\mathbf{A}(t-t_0)} \mathbf{C} - \sum_{k=1}^{m} (\mathbf{A} - \lambda_k \mathbf{I})^{-1} \mathbf{Q}_0^{(k)} e^{\lambda_k t}. \end{aligned} \tag{A.125}$$

With an initial condition, the total solution is

$$\begin{aligned} \mathbf{x}(t) &= e^{\mathbf{A}(t-t_0)} (\mathbf{x}_0 - \mathbf{x}_p^0) - \sum_{k=1}^{m} (\mathbf{A} - \lambda_k \mathbf{I})^{-1} \mathbf{Q}_0^{(k)} e^{\lambda_k t}, \\ \mathbf{x}_p^0 &= -\sum_{k=1}^{m} (\mathbf{A} - \lambda_k \mathbf{I})^{-1} \mathbf{Q}_0^{(k)} e^{\lambda_k t_0}. \end{aligned} \tag{A.126}$$

Consider the second example as

$$\mathbf{Q}(t) = \sum_{k=1}^{m} \mathbf{Q}_1^{(k)} e^{\alpha_k t} \cos \beta_k t + \mathbf{Q}_1^{(k)} e^{\alpha_k t} \sin \beta_k t, \tag{A.127}$$

The particular solutions is assumed as

$$\begin{aligned} \mathbf{x}_p(t) &= \sum_{k=1}^{m} \mathbf{C}_1^{(k)} e^{(\alpha_k + \mathbf{i}\beta_k)t} + \mathbf{C}_2^{(k)} e^{(\alpha_k - \mathbf{i}\beta_k)t} \\ &= \sum_{k=1}^{m} \frac{1}{2} (\mathbf{c}_1^{(k)} - \mathbf{i}\mathbf{c}_2^{(k)}) e^{(\alpha_k + \mathbf{i}\beta_k)t} + \frac{1}{2} (\mathbf{c}_1^{(k)} + \mathbf{i}\mathbf{c}_2^{(k)}) e^{(\alpha_k - \mathbf{i}\beta_k)t} \\ &= \sum_{k=1}^{m} \mathbf{c}_1^{(k)} e^{\alpha_k t} \cos \beta_k t + \mathbf{c}_2^{(k)} e^{\alpha_k t} \sin \beta_k t, \end{aligned} \tag{A.128}$$

$$\begin{aligned} \dot{\mathbf{x}}_p(t) &= \sum_{k=1}^{m} (\alpha_k + \mathbf{i}\beta_k) \mathbf{C}_1^{(k)} e^{(\alpha_k + \mathbf{i}\beta_k)t} + (\alpha_k - \mathbf{i}\beta_k) \mathbf{C}_2^{(k)} e^{(\alpha_k - \mathbf{i}\beta_k)t} \\ &= \frac{1}{2} \sum_{k=1}^{m} \Big[(\alpha_k + \mathbf{i}\beta_k)(\mathbf{c}_1^{(k)} - \mathbf{i}\mathbf{c}_2^{(k)}) e^{(\alpha_k + \mathbf{i}\beta_k)t} \\ &\quad + (\alpha_k - \mathbf{i}\beta_k)(\mathbf{c}_1^{(k)} + \mathbf{i}\mathbf{c}_2^{(k)}) e^{(\alpha_k - \mathbf{i}\beta_k)t} \Big] \\ &= \sum_{k=1}^{m} (\alpha_k \mathbf{c}_1^{(k)} + \beta_k \mathbf{c}_2^{(k)}) e^{\alpha_k t} \cos \beta_k t + (\alpha_k \mathbf{c}_2^{(k)} - \beta_k \mathbf{c}_1^{(k)}) e^{\alpha_k t} \sin \beta_k t. \end{aligned} \tag{A.129}$$

Thus

$$\begin{bmatrix} (\mathbf{A}-\alpha_k\mathbf{I}) & \beta_k\mathbf{I} \\ -\beta_k\mathbf{I} & (\mathbf{A}-\alpha_k\mathbf{I}) \end{bmatrix} \begin{Bmatrix} \mathbf{c}_1^{(k)} \\ \mathbf{c}_2^{(k)} \end{Bmatrix} = -\begin{Bmatrix} \mathbf{Q}_1^{(k)} \\ \mathbf{Q}_2^{(k)} \end{Bmatrix}$$
$$\begin{Bmatrix} \mathbf{c}_1^{(k)} \\ \mathbf{c}_2^{(k)} \end{Bmatrix} = -\begin{bmatrix} (\mathbf{A}-\alpha_k\mathbf{I}) & \beta_k\mathbf{I} \\ -\beta_k\mathbf{I} & (\mathbf{A}-\alpha_k\mathbf{I}) \end{bmatrix}^{-1} \begin{Bmatrix} \mathbf{Q}_1^{(k)} \\ \mathbf{Q}_2^{(k)} \end{Bmatrix} \tag{A.130}$$

Consider the third example as

$$\mathbf{Q}(t) = \sum\nolimits_{k=1}^{m} \mathbf{Q}_k t^k. \tag{A.131}$$

The corresponding particular solution is

$$\begin{aligned} \mathbf{x}_p(t) &= \sum\nolimits_{k=0}^{m} \mathbf{C}_k t^k, \\ \dot{\mathbf{x}}_p(t) &= \sum\nolimits_{k=1}^{m} k\mathbf{C}_k t^{k-1} = \sum\nolimits_{k=0}^{m-1} (k+1)\mathbf{C}_{k+1} t^k, \end{aligned} \tag{A.132}$$

and

$$\sum\nolimits_{k=0}^{m-1} (k+1)\mathbf{C}_{k+1} t^k = \mathbf{A}\sum\nolimits_{k=0}^{m} \mathbf{C}_k t^k + \sum\nolimits_{k=0}^{m-1} \mathbf{Q}_k t^k. \tag{A.133}$$

Thus, if $|\mathbf{A}| \neq 0$,

$$\begin{aligned} &\mathbf{C}_m = \mathbf{A}^{-1}\mathbf{Q}_m, \\ &\mathbf{C}_k = \mathbf{A}^{-1}[(k+1)\mathbf{C}_{k+1} - \mathbf{Q}_k] \\ &(k = 0, 1, \ldots, m-1). \end{aligned} \tag{A.134}$$

A.5 Linear Systems with Periodic Coefficients

Theorem A.10 (Floquet) *Consider a linear dynamical system*

$$\dot{\mathbf{x}} = \mathbf{A}(t)\mathbf{x} \tag{A.135}$$

with $\mathbf{A}(t)$ *is a continuous T-periodic,* $n \times n$ *matrix. The fundamental matrix* $\mathbf{\Phi}(t)$ *is determined by*

$$\mathbf{\Phi}(t) = \mathbf{P}(t)e^{\mathbf{B}t} \tag{A.136}$$

where $\mathbf{P}(t)$ *is T-periodic and* $\mathbf{B}$ *is a constant* $n \times n$ *matrix.*

Proof If $\mathbf{\Phi}(t)$ is the fundamental matrix of $\dot{\mathbf{x}} = \mathbf{A}(t)\mathbf{x}$, then it consists of n - independent solutions. Letting $\tau = t + T$,

$$\frac{d\mathbf{x}}{dt} = \mathbf{A}(t)\mathbf{x} \Rightarrow \frac{d\mathbf{x}}{d\tau} = \mathbf{A}(\tau - T)\mathbf{x} = \mathbf{A}(\tau)\mathbf{x}.$$

Thus, $\mathbf{\Phi}(\tau)$ is also the fundamental matrix. Further, $\mathbf{\Phi}(t)$ and $\mathbf{\Phi}(\tau)$ are linearly dependent with a constant nonsingular matrix $\mathbf{C}$, i.e.,

$$\mathbf{\Phi}(\tau) = \mathbf{\Phi}(t)\mathbf{C} \Rightarrow \mathbf{\Phi}(t+T) = \mathbf{\Phi}(t)\mathbf{C}.$$

Consider the time-varying eigenvector matrix as a constant matrix,

$$\mathbf{P}(t) = \mathbf{\Phi}(t)e^{-\mathbf{B}t}$$

For time $\tau = t + T$, the time-varying eigenvector matrix is

$$\mathbf{P}(t+T) = \mathbf{\Phi}(t+T)e^{-\mathbf{B}(t+T)} = \mathbf{\Phi}(t)\mathbf{C}e^{-\mathbf{B}T}e^{-\mathbf{B}t}.$$

If there is a constant matrix $\mathbf{B}$ with

$$\mathbf{C} = e^{\mathbf{B}T},$$

Then $\mathbf{P}(t)$ is T-periodic, i.e.,

$$\mathbf{P}(t+T) = \mathbf{\Phi}(t)e^{-\mathbf{B}t} = \mathbf{P}(t).$$

Therefore, the fundamental matrix is determined by

$$\mathbf{P}(t) = \mathbf{\Phi}(t)e^{-\mathbf{B}t}.$$

where $\mathbf{P}(t)$ is T-periodic, and $\mathbf{B}$ is a constant $n \times n$ matrix. ■

Using the transformation

$$\mathbf{x} = \mathbf{P}(t)\mathbf{y}.$$

Taking derivatives gives

$$\begin{aligned}\dot{\mathbf{x}} &= \dot{\mathbf{P}}(t)\mathbf{y} + \mathbf{P}(t)\dot{\mathbf{y}} = \mathbf{A}(t)\mathbf{P}(t)\mathbf{y},\\ \dot{\mathbf{y}} &= \mathbf{P}^{-1}(\mathbf{AP} - \dot{\mathbf{P}})\mathbf{y}.\end{aligned}$$

However,

$$\begin{aligned}\dot{\mathbf{P}}(t) &= \dot{\mathbf{\Phi}}(t)e^{-\mathbf{B}t} + \mathbf{\Phi}(t)e^{-\mathbf{B}t}(-\mathbf{B})\\ &= \mathbf{A\Phi}e^{-\mathbf{B}t} - \mathbf{\Phi}e^{-\mathbf{B}t}\mathbf{B}\\ &= \mathbf{AP} - \mathbf{PB}.\end{aligned}$$

Therefore,

$$\dot{\mathbf{y}} = \mathbf{P}^{-1}\mathbf{PBy} = \mathbf{By}.$$

Note that the periodic solution of Eq. (A.135) and corresponding stability are determined by the eigenvalues of the matrix $\mathbf{B}$. The necessary condition of periodic solution is that there is at least a pair of purely imaginary eigenvalue of the matrix $\mathbf{B}$.

Theorem A.11 *Consider a linear dynamical system of* $\dot{\mathbf{x}} = \mathbf{A}(t)\mathbf{x}$ *in Eq.* (A.135) *with characteristic multiplies*

$$\rho_k = e^{\lambda_k T} \tag{A.137}$$

where $\lambda_k (k = 1, 2, \ldots, n)$ *are exponents of matrix* $\mathbf{A}(t)$. *The characteristic multiplies and exponents have following characteristic.*

$$\begin{aligned} &\prod\nolimits_{k=1}^{n} \rho_k = e^{\int_0^T \mathrm{tr}\mathbf{A}(t)dt}, \\ &\sum\nolimits_{k=1}^{n} \lambda_k = \frac{1}{T}\int_0^T \mathrm{tr}\mathbf{A}(t)dt. \end{aligned} \tag{A.138}$$

Proof If $\mathbf{\Phi}(t)$ is the fundamental matrix of $\dot{\mathbf{x}} = \mathbf{A}(t)\mathbf{x}$ with $\mathbf{\Phi}(0) = \mathbf{I}_{n\times n}$, then $\dot{\mathbf{\Phi}}(t) = \mathbf{A}(t)\mathbf{\Phi}(t)$.

$$\begin{aligned} \mathbf{\Phi}(t+\varepsilon) &= \mathbf{\Phi}(t) + \dot{\mathbf{\Phi}}\varepsilon + o(\varepsilon) \\ &= \mathbf{\Phi}(t) + \mathbf{A}(t)\mathbf{\Phi}\varepsilon + o(\varepsilon). \end{aligned}$$

The determinant of the fundamental matrix is

$$\begin{aligned} \det\mathbf{\Phi}(t+\varepsilon) &= \det(\mathbf{\Phi}(t) + \mathbf{A}(t)\mathbf{\Phi}\varepsilon + o(\varepsilon)) \\ &= \det(\mathbf{I} + \mathbf{A}(t)\varepsilon + o(\varepsilon))\det\mathbf{\Phi}(t) \\ &= (1 + \mathrm{tr}\mathbf{A}(t)\varepsilon + o(\varepsilon))\det\mathbf{\Phi}(t). \end{aligned}$$

Further, one obtains

$$\begin{aligned} &\frac{d}{dt}\det\mathbf{\Phi}(t) = \mathrm{tr}\mathbf{A}(t)\det\mathbf{\Phi}(t), \\ &\det\mathbf{\Phi}(t) = \mathbf{C}e^{\int_0^t \mathrm{tr}\mathbf{A}(t)dt} \Rightarrow \mathbf{C} = \det\mathbf{\Phi}(0) = \mathbf{I}. \end{aligned}$$

So

$$\det\mathbf{\Phi}(t) = e^{\int_0^t \mathrm{tr}\mathbf{A}(t)dt} \Rightarrow \det\mathbf{\Phi}(T) = e^{\int_0^T \mathrm{tr}\mathbf{A}(t)dt}.$$

Since

$$\begin{aligned} &\mathbf{\Phi}(t+T) = \mathbf{P}(t+T)e^{\mathbf{B}(t+T)} = \mathbf{P}(t)e^{\mathbf{B}t}e^{\mathbf{B}T} = \mathbf{\Phi}(t)e^{\mathbf{B}T}, \\ &\mathbf{\Phi}(T) = \mathbf{\Phi}(0)e^{\mathbf{B}T} = e^{\mathbf{B}T} \end{aligned}$$

we have

$$\det\mathbf{\Phi}(T) = \det(e^{\mathbf{B}T}) = e^{\int_0^T \mathrm{tr}\mathbf{A}(t)dt}.$$

Since

$$e^{\mathbf{B}T} = diag(e^{\lambda_1 T}, e^{\lambda_2 T}, \ldots, e^{\lambda_n T}),$$

the determinant of the foregoing equation gives

$$\det e^{\mathbf{B}T} = \det e^{\mathbf{B}T} = \prod_{k=1}^{n} e^{\lambda_k T} = \prod_{k=1}^{n} \rho_k \Rightarrow \prod_{k=1}^{n} \rho_k = e^{\int_0^T tr\mathbf{A}(t)dt}.$$

So we have

$$\sum_{k=1}^{n} \lambda_k = \frac{1}{T}\int_0^T \mathrm{tr}\mathbf{A}(t)dt.$$

This theorem is proved. ■

A.6 Stability Theory of Linear Systems

In this section, the stability of linear dynamical systems will be presented.

Definition A.11 For a linear dynamical system $\dot{\mathbf{x}} = \mathbf{A}\mathbf{x}$ in Eq. (A.2), consider a real eigenvalue λ_k of matrix $\mathbf{A}(k \in N = \{1, 2, \ldots, n\})$ and there is a corresponding eigenvector $\mathbf{v}_k$. For $\mathbf{x}^{(k)} = c^{(k)}\mathbf{v}_k$, $\dot{\mathbf{x}}^{(k)} = \dot{c}^{(k)}\mathbf{v}_k = \lambda_k c^{(k)}\mathbf{v}_k$, thus $\dot{c}^{(k)} = \lambda_k c^{(k)}$.

(i) $\mathbf{x}^{(k)}$ on the direction $\mathbf{v}_k$ is stable if

$$\lim_{t\to\infty} c^{(k)} = \lim_{t\to\infty} c_0^{(k)} e^{\lambda_k t} = 0 \text{ for } \lambda_k < 0. \tag{A.139}$$

(ii) $\mathbf{x}^{(k)}$ on the direction $\mathbf{v}_k$ is unstable if

$$\lim_{t\to\infty} |c^{(k)}| = \lim_{t\to\infty} |c_0^{(k)} e^{\lambda_k t}| = \infty \text{ for } \lambda_k > 0. \tag{A.140}$$

(iii) $\mathbf{x}^{(i)}$ on the direction $\mathbf{v}_i$ is invariant if

$$\lim_{t\to\infty} c^{(k)} = \lim_{t\to\infty} e^{\lambda_k t} c_0^{(k)} = c_0^{(k)} \text{ for } \lambda_k = 0. \tag{A.141}$$

Definition A.12 For a linear dynamical system $\dot{\mathbf{x}} = \mathbf{A}\mathbf{x}$ in Eq. (A.2), consider a pair of complex eigenvalue $\alpha_k \pm \mathbf{i}\beta_k$ of matrix $\mathbf{A}(k \in N = \{1, 2, \ldots, n\}, \ \mathbf{i} = \sqrt{-1})$ and there is a pair of eigenvectors $\mathbf{u}_k \pm \mathbf{i}\mathbf{v}_k$. On the invariant plane of $(\mathbf{u}_k, \mathbf{v}_k)$, consider $\mathbf{x}^{(k)} = \mathbf{x}_+^{(k)} + \mathbf{x}_-^{(k)}$ with

$$\mathbf{x}^{(k)} = c^{(k)}\mathbf{u}_k + d^{(k)}\mathbf{v}_k, \ \dot{\mathbf{x}}^{(k)} = \dot{c}^{(k)}\mathbf{u}_k + \dot{d}^{(k)}\mathbf{v}_k. \tag{A.142}$$

Thus, $\mathbf{c}^{(k)} = (c^{(k)}, d^{(k)})^{\mathrm{T}}$ with

$$\dot{\mathbf{c}}^{(k)} = \mathbf{E}_k \mathbf{c}^{(k)} \Rightarrow \mathbf{c}^{(k)} = e^{\alpha_k t} \mathbf{B}_k \mathbf{c}_0^{(k)} \tag{A.143}$$

where

$$\mathbf{E}_i = \begin{bmatrix} \alpha_k & \beta_k \\ -\beta_k & \alpha_k \end{bmatrix} \text{ and } \mathbf{B}_k = \begin{bmatrix} \cos\beta_k t & \sin\beta_k t \\ -\sin\beta_k t & \cos\beta_k t \end{bmatrix}. \tag{A.144}$$

(i) $\mathbf{x}^{(k)}$ on the plane of $(\mathbf{u}_k, \mathbf{v}_k)$ is spirally stable if

$$\lim_{t\to\infty} ||\mathbf{c}^{(k)}|| = \lim_{t\to\infty} e^{\alpha_k t} ||\mathbf{B}_k|| \times ||\mathbf{c}_0^{(k)}|| = 0 \text{ for } \mathrm{Re}\lambda_k = \alpha_k < 0. \tag{A.145}$$

(ii) $\mathbf{x}^{(k)}$ on the plane of $(\mathbf{u}_k, \mathbf{v}_k)$ is spirally unstable if

$$\lim_{t\to\infty} ||\mathbf{c}^{(k)}|| = \lim_{t\to\infty} e^{\alpha_k t} ||\mathbf{B}_k|| \times ||\mathbf{c}_0^{(k)}|| = \infty \text{ for } \mathrm{Re}\lambda_k = \alpha_k > 0. \tag{A.146}$$

(iii) $\mathbf{x}^{(k)}$ on the plane of $(\mathbf{u}_k, \mathbf{v}_k)$ is on the invariant circle if

$$\lim_{t\to\infty} ||\mathbf{c}^{(k)}|| = \lim_{t\to\infty} e^{\alpha_k t} ||\mathbf{B}_k|| \times ||\mathbf{c}_0^{(k)}|| = ||\mathbf{c}_0^{(k)}|| \text{ for } \mathrm{Re}\lambda_k = \alpha_k = 0. \tag{A.147}$$

(iv) $\mathbf{x}^{(k)}$ on the plane of $(\mathbf{u}_k, \mathbf{v}_k)$ is degenerate in the direction of $\mathbf{u}_k$ if $\mathrm{Im}\lambda_k = 0$.

Definition A.13 For a linear dynamical system $\dot{\mathbf{x}} = \mathbf{A}\mathbf{x}$ in Eq. (A.2), the matrix $\mathbf{A}$ has n_1 real eigenvalues $\lambda_k < 0 (k \in N_1)$, n_2 real eigenvalues $\lambda_k > 0$ $(k \in N_2)$ and n_3 real eigenvalues $\lambda_k = 0$ $(k \in N_3)$. The corresponding vectors for the negative, positive and zero eigenvalues of $D\mathbf{f}(\mathbf{x}^*, \mathbf{p})$ are $\{\mathbf{u}_k\}$ $(k \in N_i,\ i = 1,2,3)$, respectively. Set $N_i = \{i_1, i_2, \ldots, i_{n_i}\} \cup \varnothing \subseteq N \cup \varnothing$ and $N = \{1,2,\ldots,n\}$ with $l_j \in N$ $(j = 1,2,\ldots,n_i,\ i = 1,2,3)$. $\cup_{i=1}^3 N_i = N, N_i \cap N_l = \varnothing (l \neq i)$ and $\Sigma_{i=1}^3 n_i = n$. The stable, unstable, and invariant subspaces of $\dot{\mathbf{x}} = \mathbf{A}\mathbf{x}$ in Eq. (A.2) are linear subspaces spanned by $\{\mathbf{u}_{k_i}\} (k_i \in N_i, i = 1,2,3)$, respectively; i.e.,

$$\begin{aligned} \mathscr{E}^{\mathrm{s}} &= span\{\mathbf{u}_k | (\mathbf{A} - \lambda_k \mathbf{I})\mathbf{u}_k = \mathbf{0}, \lambda_k < 0, k \in N_1 \subseteq N \cup \varnothing\}; \\ \mathscr{E}^{\mathrm{u}} &= span\{\mathbf{u}_k | (\mathbf{A} - \lambda_k \mathbf{I})\mathbf{u}_k = \mathbf{0}, \lambda_k > 0, k \in N_2 \subseteq N \cup \varnothing\}; \\ \mathscr{E}^{\mathrm{i}} &= span\{\mathbf{u}_k | (\mathbf{A} - \lambda_k \mathbf{I})\mathbf{u}_k = \mathbf{0}, \lambda_k = 0, k \in N_3 \subseteq N \cup \varnothing\}. \end{aligned} \tag{A.148}$$

Definition A.14 For a linear dynamical system $\dot{\mathbf{x}} = \mathbf{A}\mathbf{x}$ in Eq. (A.2), the $2n \times 2n$ matrix $\mathbf{A}$ has complex eigenvalues $\alpha_k \pm \mathbf{i}\beta_k$ with eigenvectors $\mathbf{u}_k \pm \mathbf{i}\mathbf{v}_k$ $(k \in \{1,2,\ldots,n\})$ and the base of vector is

$$\mathbf{B} = \{\mathbf{u}_1, \mathbf{v}_1, \ldots, \mathbf{u}_k, \mathbf{v}_k, \ldots, \mathbf{u}_n, \mathbf{v}_n\}. \tag{A.149}$$

The stable, unstable, center subspaces of $\dot{\mathbf{x}} = \mathbf{A}\mathbf{x}$ in Eq. (A.2) are linear subspaces spanned by $\{\mathbf{u}_k, \mathbf{v}_k\}$ $(k \in N_i,\ i = 1, 2, 3)$, respectively. Set $N_i = \{i_1, i_2, \ldots, i_{n_i}\} \cup \varnothing \subseteq N \cup \varnothing$ and $N = \{1, 2, \ldots, n\}$ with $l_j \in N\ (j = 1, 2, \ldots, n_i, i = 1, 2, 3)$ and $\sum_{i=1}^{3} n_i = n$. $\cup_{i=1}^{3} N_i = N$ and $N_i \cap N_l = \varnothing\ (l \neq i)$. $N_i = \varnothing$ if $n_i = 0$. The stable, unstable, center subspaces of $\dot{\mathbf{x}} = \mathbf{A}\mathbf{x}$ in Eq. (A.2) are defined by

$$\begin{aligned}
\mathscr{E}^{\mathrm{s}} &= span\left\{ (\mathbf{u}_k, \mathbf{v}_k) \left| \begin{array}{c} \alpha_k < 0, \beta_k \neq 0, \\ (\mathbf{A} - (\alpha_k \pm \mathbf{i}\beta_k)\mathbf{I})(\mathbf{u}_k \pm \mathbf{i}\mathbf{v}_k) = \mathbf{0}, \\ k \in N_1 \subseteq \{1, 2, \ldots, n\} \cup \varnothing \end{array} \right. \right\}; \\
\mathscr{E}^{\mathrm{u}} &= span\left\{ (\mathbf{u}_k, \mathbf{v}_k) \left| \begin{array}{c} \alpha_k > 0, \beta_k \neq 0, \\ (\mathbf{A} - (\alpha_k \pm \mathbf{i}\beta_k)\mathbf{I})(\mathbf{u}_k \pm \mathbf{i}\mathbf{v}_k) = \mathbf{0}, \\ k \in N_2 \subseteq \{1, 2, \ldots, n\} \cup \varnothing \end{array} \right. \right\}; \\
\mathscr{E}^{\mathrm{c}} &= span\left\{ (\mathbf{u}_k, \mathbf{v}_k) \left| \begin{array}{c} \alpha_k = 0, \beta_k \neq 0, \\ (\mathbf{A} - (\alpha_k \pm \mathbf{i}\beta_k)\mathbf{I})(\mathbf{u}_k \pm \mathbf{i}\mathbf{v}_k) = \mathbf{0}, \\ k \in N_3 \subseteq \{1, 2, \ldots, n\} \cup \varnothing \end{array} \right. \right\}.
\end{aligned} \tag{A.150}$$

Definition A.15 For a linear dynamical system $\dot{\mathbf{x}} = \mathbf{A}\mathbf{x} \in \mathscr{R}^n$ in Eq. (A.2), set $N = \{1, 2, \ldots, m, m+1, \ldots, (n-m)/2\}$, $N_i = \{i_1, i_2, \ldots, i_{n_i}\} \cup \varnothing$ with $i_j \in N\ (j = 1, 2, \ldots, n_i, i = 1, 2, \ldots, 6)$, $\Sigma_{i=1}^{3} n_i = m$ and $2\Sigma_{i=4}^{6} n_i = n - m$. $\cup_{i=1}^{6} N_i = N$ and $N_i \cap N_l = \varnothing\ (l \neq i)$. $N_i = \varnothing$ if $n_i = 0$. If the matrix $\mathbf{A}$ possesses n_1-stable, n_2-unstable and n_3-invariant real eigenvectors plus n_4-stable, n_5-unstable and n_6-center pairs of complex eigenvectors, a flow $\Phi(t)$ of $\dot{\mathbf{x}} = \mathbf{A}\mathbf{x}$ is called an $(n_1 : n_2 : n_3 | n_4 : n_5 : n_6)$ flow.

Definition A.16 For a linear dynamical system $\dot{\mathbf{x}} = \mathbf{A}\mathbf{x}$ in Eq. (A.2), a subspace of $\mathscr{E} \subset \mathscr{R}^n$ is termed to be invariant with respect to flow $\Phi(t) = e^{\mathbf{A}t} : \mathscr{R}^n \to \mathscr{R}^n$ if $e^{\mathbf{A}t}\mathscr{E} \subset \mathscr{E}$ for all $t \in \mathscr{R}$.

Lemma A.2 *For a linear dynamical system $\dot{\mathbf{x}} = \mathbf{A}\mathbf{x}$ in Eq. (A.2), if a generalized eigenspace of $\mathbf{A}$ corresponding to λ is $E \subset \mathscr{R}^n$, then $\mathbf{A}\mathscr{E} \subset \mathscr{E}$.*

Proof For a generalized eigenvector $\{\mathbf{v}_k\}$ $(k = 1, 2, \ldots, n)$ for $\mathscr{E} \subset \mathscr{R}^n$, consider a new vector $\mathbf{v} \in \mathscr{E}$, we have

$$\mathbf{v} = \sum_{k=1}^{n} c_k \mathbf{v}_k \Rightarrow \mathbf{A}\mathbf{v} = \sum_{k=1}^{n} c_k \mathbf{A}\mathbf{v}_k.$$

For each $\mathbf{v}_k$, the following relation exists

$$(\mathbf{A} - \lambda\mathbf{I})\mathbf{v}_k = \mathbf{0}$$

with a minimal n_k. Thus,

$$(\mathbf{A} - \lambda\mathbf{I})\mathbf{v}_k = \mathbf{V}_k$$

where $\mathbf{V}_k \in \ker(\mathbf{A} - \lambda\mathbf{I})|_{n_k} \subset \mathscr{E}$ with dimension n_k. Thus $\mathbf{A}\mathbf{v}_k = \lambda\mathbf{v}_k + \mathbf{V}_k \in \mathscr{E}$, i.e.

$$\mathbf{A}\mathbf{v} = \sum\nolimits_{k=1}^{n} c_k \mathbf{A}\mathbf{v}_k \in \mathscr{E}.$$

So, one achieves $\mathbf{A}\mathscr{E} \subset \mathscr{E}$. ■

Theorem A.12 *For a linear dynamical system* $\dot{\mathbf{x}} = \mathbf{A}\mathbf{x}$ *in Eq.* (A.2), *the eigenspace of* $\mathbf{A}$ *(i.e.,* $\mathscr{E} \subseteq \mathscr{R}^n$*) is expressed by direct sum of three subspaces*

$$\mathscr{E} = \mathscr{E}^s \oplus E^u \oplus \mathscr{E}^c. \tag{A.151}$$

where $\mathscr{E}^s$, $\mathscr{E}^u$ and $\mathscr{E}^c$ *are the stable, unstable and center spaces, respectively. They are invariant with respect to the flow* $\Phi(t) = e^{\mathbf{A}t}$ *of* $\dot{\mathbf{x}} = \mathbf{A}\mathbf{x}$.

Proof For $\mathbf{x}_0 \in \mathscr{E}^s$ with n_s -dimensions, one gets

$$\mathbf{x}_0 = \sum\nolimits_{k=1}^{n_s} c_k \mathbf{V}_k \text{ and}$$
$$\mathbf{V}_k \in \{\mathbf{V}_k\}_{k=1}^{n_s} \subset \mathbf{B} = \{\mathbf{u}_1, \mathbf{v}_1, \ldots, \mathbf{u}_j, \mathbf{v}_j, \mathbf{u}_{j+1}, \ldots, \mathbf{u}_n\}.$$

The linearity of $e^{\mathbf{A}t}$ gives

$$\mathbf{x} = e^{\mathbf{A}t}\mathbf{x}_0 = \sum\nolimits_{k=1}^{n_s} c_k e^{\mathbf{A}t}\mathbf{V}_k,$$
$$e^{\mathbf{A}t}\mathbf{V}_k = \lim_{m\to\infty} (\mathbf{I} + \sum\nolimits_{j=1}^{m} \frac{\mathbf{A}^j t^j}{j!})\mathbf{V}_k \in \mathscr{E}^s,$$
$$\mathbf{x} = \sum\nolimits_{k=1}^{n_s} c_k e^{\mathbf{A}t}\mathbf{V}_k \in \mathscr{E}^s \subset \mathscr{E} \subseteq \mathscr{R}^n.$$

Therefore, $e^{\mathbf{A}t}\mathscr{E}^s \subset \mathscr{E}^s$. That is, $\mathscr{E}^s$ is invariant under the flow $e^{\mathbf{A}t}$. Similarly, $\mathscr{E}^u$ and $\mathscr{E}^c$ are invariant under the flow $e^{\mathbf{A}t}$. ■

Definition A.17 For a linear dynamical system $\dot{\mathbf{x}} = \mathbf{A}\mathbf{x}$ in Eq. (A.2),

(i) the linear system is asymptotically stable to the origin if

$$\left.\begin{aligned} &\lim_{t\to\infty} \mathbf{x}(t) = \lim_{t\to\infty} e^{\mathbf{A}t}\mathbf{x}_0 = \mathbf{0} \text{ for } \mathbf{x}_0 \in \Omega \subseteq \mathscr{R}^n, \text{or} \\ &\lim_{t\to-\infty} ||\mathbf{x}(t)|| = \lim_{t\to-\infty} ||e^{\mathbf{A}t}\mathbf{x}_0|| = \infty \text{ for } \mathbf{x}_0 \in \Omega \text{ but } \mathbf{x}_0 \neq \mathbf{0}; \end{aligned}\right. \tag{A.152}$$

(ii) the linear system is unstable if

$$\left.\begin{aligned} &\lim_{t\to\infty} ||\mathbf{x}(t)|| = \lim_{t\to\infty} ||e^{\mathbf{A}t}\mathbf{x}_0|| = \infty \text{ for } \mathbf{x}_0 \in \Omega \subseteq \mathscr{R}^n \text{ but } \mathbf{x}_0 \neq \mathbf{0}, \text{or} \\ &\lim_{t\to-\infty} \mathbf{x}(t) = \lim_{t\to-\infty} e^{\mathbf{A}t}\mathbf{x}_0 = \mathbf{0} \text{ for } \mathbf{x}_0 \in \Omega \subseteq \mathscr{R}^n; \end{aligned}\right. \tag{A.153}$$

(iii) the origin of the linear system is a center if

$$||\mathbf{x}(t)|| \leq C||\mathbf{x}_0|| \text{ for positve constant } C > 0. \tag{A.154}$$

Theorem A.13 *Consider a linear dynamical system* $\dot{\mathbf{x}} = \mathbf{A}\mathbf{x}$ *in Eq. (A.2) and the matrix* $\mathbf{A}$ *possesses n eigenvalues* λ_k $(k = 1, 2, \ldots n)$. *Let* $N = \{1, 2, \ldots, n\}$, *and* $N = N_1 \cup N_2$ *with* $N_1 \cap N_2 = \varnothing$.

(i) *If Re* $\lambda_k > 0$ *for* $k \in N$, *the linear system is unstable.*
(ii) *If Re* $\lambda_k < 0$ *for all* $k \in N$, *the linear system is asymptotically stable to the origin.*
(iii) *If Re* $\lambda_k < 0$ *for all* $k \in N_1 \neq N$ *and* $\mathrm{Re}\,\lambda_j = 0$ *for all* $j \in N_2 \subseteq N$ *with different eigenvalues, the linear system is stable. The linear system is also said to be Lyapunov-stable to the origin, and the origin is a center for this system.*
(iv) *If Re* $\lambda_k < 0$ *for all* $k \in N_1 \neq N$ *and* $\mathrm{Re}\,\lambda_j = 0$ *for all* $j \in N_2 \subseteq N$ *with repeated eigenvalues with* $\mathbf{N}^m = \mathbf{0}$ $(1 < m \le n)$, *the linear system is unstable.*
(v) *If Re* $\lambda_k < 0$ *for all* $k \in N_1 \neq N$ *and* $\mathrm{Re}\,\lambda_j = 0$ *for all* $j \in N_2 \subseteq N$ *with repeated eigenvalues with* $\mathbf{N} = \mathbf{0}$, *the linear system is stable.*

Proof Consider one of eigenvalue of $\mathbf{A}$ as $\lambda_i = \alpha_i + \mathbf{i}\beta_i$ which is an m-repeated eigenvalue $(m \le n)$. The corresponding solution of $\dot{\mathbf{x}} = \mathbf{A}\mathbf{x}$ in Eq. (A.2) is a linear combination of functions of the form $t^k e^{\alpha_i t} \cos \beta_i t$ and /or $t^k e^{\alpha_i t} \sin \beta_i t$ $(0 \le k \le m - 1)$, so that $||e^{At}\mathbf{x}_0|| \ge e^{\alpha_i t}||\mathbf{x}_0||$.

(I) For $\alpha_i > 0$

$$\lim_{t\to\infty} |t^k e^{\alpha_i t} \cos \beta_i t| = \infty \text{ and } \lim_{t\to\infty} |t^k e^{\alpha_i t} \sin \beta_i t| = \infty,$$

Thus $||e^{At}\mathbf{x}_0|| \to \infty$ as $t \to \infty$. In other words,

$$\lim_{t\to\infty} ||e^{At}\mathbf{x}_0|| = \infty \text{ or } \lim_{t\to\infty} e^{At}\mathbf{x}_0 \neq \mathbf{0}.$$

Therefore, if $\mathrm{Re}\,(\lambda_i) > 0$ $(i \in \{1, 2, \ldots, n\})$, the origin of the linear system is unstable.

(II) For $\alpha_i = 0$ and $k \neq 0$, the eigenvalues with $\mathrm{Re}\,(\lambda_i) = 0$ are repeated. At least for one eigenvalue, one gets

$$\lim_{t\to\infty} |t^k \cos \beta_i t| = \infty \text{ and } \lim_{t\to\infty} |t^k \sin \beta_i t| = \infty.$$

Thus

$$\lim_{t\to\infty} ||e^{At}\mathbf{x}_0|| = \infty \text{ or } \lim_{t\to\infty} e^{At}\mathbf{x}_0 \neq \mathbf{0}.$$

(III) For all $\alpha_i < 0$ $(i = 1, 2, \ldots, n)$

$$\lim_{t\to\infty} |t^k e^{\alpha_k t} \cos \beta_k t| = 0 \text{ and } \lim_{t\to\infty} |t^k e^{\alpha_k t} \sin \beta_k t| = 0.$$

Thus

$$
\begin{aligned}
&||e^{\mathbf{A}t}\mathbf{x}_0|| \\
&\quad \le ||\sum\nolimits_i c_i\mathbf{u}_i t^k e^{\alpha_i t}\cos\beta_i t|| + ||\sum\nolimits_i d_i\mathbf{v}_i t^k e^{\alpha_i t}\sin\beta_i t|| \\
&\quad \le \sum\nolimits_i |c_i| \times ||\mathbf{u}_i|| \times |t^k e^{\alpha_i t}\cos\beta_i t| + \sum\nolimits_i |d_i| \times ||\mathbf{v}_i|| \times |t^k e^{\alpha_i t}\sin\beta_i t|.
\end{aligned}
$$

Since c_i and d_i are constant and the norms of eigenvector $||\mathbf{u}_i||$ and $||\mathbf{v}_i||$ are finite, one obtains

$$
\begin{aligned}
&\lim_{t\to\infty} ||\mathbf{x}(t)|| \\
&\quad \le \sum\nolimits_i |c_i| \times ||\mathbf{u}_i|| \lim_{t\to\infty} |t^k e^{\alpha_i t}\cos\beta_i t| + \sum\nolimits_i |d_i| \times ||\mathbf{v}_i|| \lim_{t\to\infty} |t^k e^{\alpha_i t}\sin\beta_i t| \\
&\quad = 0.
\end{aligned}
$$

Therefore

$$\lim_{t\to\infty} e^{At}\mathbf{x}_0 = \mathbf{0}.$$

(IV) For $\alpha_j = 0$ ($j \in \{1, 2, \ldots, n\}$) and k=0

$$|\cos\beta_j t| \le 1 \text{and} |\sin\beta_j t| \le 1.$$

If $\alpha_i \ne 0$, then $\alpha_i < 0$. From the case (III), one obtains

$$
\begin{aligned}
||e^{\mathbf{A}t}\mathbf{x}_0|| \le\ & ||\sum\nolimits_{i\ne j} c_i\mathbf{u}_i t^k e^{\alpha_i t}\cos\beta_i t|| + ||\sum\nolimits_{i\ne j} d_i\mathbf{v}_i t^k e^{\alpha_i t}\sin\beta_i t|| \\
& + ||\sum\nolimits_j c_j\mathbf{u}_j\cos\beta_j t|| + ||\sum\nolimits_j d_j\mathbf{v}_j\sin\beta_j t||.
\end{aligned}
$$

Therefore,

$$\lim_{t\to\infty} ||\mathbf{x}(t)|| = \lim_{t\to\infty} ||e^{At}\mathbf{x}_0|| = C||\mathbf{x}_0||.$$

This theorem is proved. ■

Definition A.18 For an n-dimensional, linear dynamical system $\dot{\mathbf{x}} = \mathbf{A}\mathbf{x}$ in Eq. (A.2), the matrix **A** possesses n eigenvalues λ_k ($k = 1, 2, \ldots, n$).

(i) The origin is called a sink for the linear system if Re $\lambda_k < 0$ ($k = 1, 2, \ldots, n$).
(ii) The origin is called a source for the linear system if Re $\lambda_k > 0$ ($k = 1, 2, \ldots, n$).
(iii) The origin is called a center for the linear system if Re $\lambda_k = 0$ ($k = 1, 2, \ldots, n$) with distinct eigenvalues.
(iv) The origin is called a source for the linear system if Re$\lambda_k = 0$ ($k \in \{1, 2, \ldots, n\}$) with at least one-time repeated eigenvalues with $\mathbf{N}^m = \mathbf{0}$ ($1 < m < n$).

Definition A.19 For an n-dimensional, linear dynamical system $\dot{\mathbf{x}} = \mathbf{A}\mathbf{x}$ in Eq. (A.2), the matrix **A** possesses n real eigenvalues λ_k ($k = 1, 2, \ldots, n$).

(i) The origin is called a stable node for the linear system if $\lambda_k < 0$ ($k = 1, 2, \ldots, n$).
(ii) The origin is called an unstable node for the linear system if $\lambda_k > 0$ ($k = 1, 2, \ldots, n$).

(iii) The origin is called a saddle for the linear system if $\lambda_k > 0$ and $\lambda_j < 0$ ($j,k \in \{1,2,\ldots,n\}$ and $j \neq k$).
(iv) The origin is called a degenerate case for the linear system if $\lambda_k = 0$ ($k = 1,2,\ldots,n$).

Definition A.20 For a $2n$-dimensional, linear dynamical system $\dot{\mathbf{x}} = \mathbf{A}\mathbf{x}$ in Eq. (A.2), the matrix $\mathbf{A}$ possesses n-pairs of eigenvalues λ_k ($k = 1,2,\ldots,n$).

(i) The origin is called a spiral sink for the linear system if $\mathrm{Re}\lambda_k < 0$ ($k = 1, 2,\ldots,n$) and $\mathrm{Im}\lambda_j \neq 0$ ($j \in \{1,2,\ldots,n\}$).
(ii) The origin is called a spiral source for the linear system if $\mathrm{Re}\lambda_k > 0$ ($k \in \{1,2,\ldots,n\}$) with $\mathrm{Im}\lambda_j \neq 0$ ($j \in \{1,2,\ldots,n\}$).

The above classification of stability is very rough. Thus, the refined classification should be discussed. The generalized structures of stability characteristics of flows in linear dynamical systems in Eq. (A.2) will be given first.

Definition A.21 For an n-dimensional, linear dynamical system $\dot{\mathbf{x}} = \mathbf{A}\mathbf{x}$ in Eq. (A.2), the matrix $\mathbf{A}$ possesses n eigenvalues λ_k ($k = 1,2,\ldots,n$). Set $N = \{1,2,\ldots,m,m+1,\ldots,(n-m)/2\}$, $N_i = \{i_1,i_2,\ldots,i_{n_i}\} \cup \varnothing$ with $i_j \in N$ ($j = 1,2,\ldots,n_i$, $i = 1,2,\ldots,6$), $\Sigma_{i=1}^{3} n_i = m$ and $2\Sigma_{i=4}^{6} n_i = n - m$. $\cup_{i=1}^{6} N_i = N$ and $N_i \cap N_l = \varnothing\,(l \neq i)$. $N_i = \varnothing$ if $n_i = 0$. The matrix $\mathbf{A}$ possesses n_1 -stable, n_2 -unstable and n_3 -invariant real eigenvectors plus n_4 -stable, n_5 -unstable and n_6 -center pairs of complex eigenvectors. Without repeated complex eigenvalues of $\mathrm{Re}\lambda_k = 0$ ($k \in N_3 \cup N_6$), the flow $\boldsymbol{\Phi}(t)$ of $\dot{\mathbf{x}} = \mathbf{A}\mathbf{x}$ is an $(n_1 : n_2 : [n_3; m_3] | n_4 : n_5 : n_6)$ flow. However, with repeated complex eigenvalues of $\mathrm{Re}\lambda_k = 0$ ($k \in N_3 \cup N_6$), the flow $\boldsymbol{\Phi}(t)$ of $\dot{\mathbf{x}} = \mathbf{A}\mathbf{x}$ is an $(n_1 : n_2 : [n_3; m_3] | n_4 : n_5 : [n_6, l; m_6])$ flow $\boldsymbol{\Phi}(t)$of $\dot{\mathbf{x}} = \mathbf{A}\mathbf{x}$ is an$(n_1 : n_2 : [n_3; m_3] | n_4 : n_5 : [n_6, l; \kappa_6])$ flow $\kappa_3 \in \{\varnothing, m_3\}$, $\kappa_6 = (\kappa_{61}, \kappa_{62}, \cdots, \kappa_{6l})^{\mathrm{T}}$ with $\kappa_{6i} \in \{\varnothing, m_{6i}\}$ ($i = 1,2,\cdots,l$) $\mathbf{m}_6 = (m_{61}, m_{62}, \cdots, m_{6l})^{\mathrm{T}}$. The meanings of notations in the aforementioned structures are defined as follows:

(i) n_1 represents exponential sinks on n_1 -directions of $\mathbf{v}_k$ ($k \in N_1$) if $\lambda_k < 0$ ($k \in N_1$ and $1 \leq n_1 \leq m$) with distinct or repeated eigenvalues.
(ii) n_2 represents exponential sources on n_2-directions of $\mathbf{v}_k$ ($k \in N_2$) if $\lambda_k > 0$ ($k \in N_2$ and $1 \leq n_2 \leq m$) with distinct or repeated eigenvalues.
(iii) $n_3 = 1$ represents an invariant center on 1-direction of $\mathbf{v}_k$ ($k \in N_3$) if $\lambda_k = 0$ ($k \in N_3$ and $n_3 = 1$).
(iv) n_4 represents spiral sinks on n_4 -pairs of $(\mathbf{u}_k, \mathbf{v}_k)(k \in N_4)$ if $\mathrm{Re}\lambda_k < 0$ and $\mathrm{Im}\lambda_k \neq 0$ ($k \in N_4$ and $1 \leq n_4 \leq (n-m)/2$) with distinct or repeated eigenvalues.
(v) n_5 represents spiral sources on n_5-pairs of $(\mathbf{u}_k, \mathbf{v}_k)$ ($k \in N_5$) if $\mathrm{Re}\lambda_k > 0$ and $\mathrm{Im}\lambda_k \neq 0$ ($k \in N_5$ and $1 \leq n_5 \leq (n-m)/2$) with distinct or repeated eigenvalues.
(vi) n_6 represents invariant centers on n_6 -pairs of $(\mathbf{u}_k, \mathbf{v}_k)$ ($k \in N_6$) if $\mathrm{Re}\lambda_k = 0$ and $\mathrm{Im}\lambda_k \neq 0$ ($k \in N_6$ and $1 \leq n_6 \leq (n-m)/2$) with distinct eigenvalues

(vii) $\varnothing$ represents none or empty if $n_i = 0$ $(i \in \{1, 2, \ldots, 6\})$.

(viii) $[n_3; m_3]$ represents invariant centers on $(n_3 - m_3)$ -directions of $\mathbf{v}_{k_3}$ $(k_3 \in N_3)$ and sources in m_3-directions of $\mathbf{v}_{j_3}$ $(j_3 \in N_3$ and $j_3 \neq k_3)$ if $\lambda_k = 0$ $(k \in N_3$ and $n_3 \leq n)$ with the $(m_3 + 1)$th -order nilpotent matrix $\mathbf{N}_3^{m_3+1} = \mathbf{0}$ $(0 < m_3 \leq n_3 - 1)$.

(ix) $[n_3; \varnothing]$ represents invariant centers on n_3 -directions of $\mathbf{v}_k$ $(k \in N_3)$ if$\lambda_k = 0$ $(k \in N_3$ and $1 < n_3 \leq m)$ with a nilpotent matrix $\mathbf{N}_3 = \mathbf{0}$.

(x) $[n_6, l; m_6]$ represents invariant centers on $(n_6 - \Sigma_{s=1}^{l} m_{6s})$ -pairs of $(\mathbf{u}_{k_6}, \mathbf{v}_{k_6})$ $(k_6 \in N_6)$ and sources in $\Sigma_{s=1}^{l} m_{6s}$ -pairs of $(\mathbf{u}_{j_6}, \mathbf{v}_{j_6})$ $(j_6 \in N_6$ and $j_6 \neq k_6)$ if $\mathrm{Re}\lambda_k = 0$ and $\mathrm{Im}\lambda_k \neq 0$ $(k \in N_6$ and $n_6 \leq (n - m)/2)$ for $\Sigma_{s=1}^{l} m_{6s}$ -pairs of repeated eigenvalues with the $(m_{6s} + 1)$th -order nilpotent matrix $\mathbf{N}_6^{m_{6s}+1} = \mathbf{0}(0 < m_{6s} \leq l,\ s = 1, 2, \cdots, l)$.

(xi) $[n_6, l; \varnothing]$ represents invariant centers on n_6 -pairs of $(\mathbf{u}_k, \mathbf{v}_k)$ $(k \in N_6)$ if $\mathrm{Re}\lambda_k = 0$ and $\mathrm{Im}\lambda_k \neq 0$ $(k \in N_6$ and $1 \leq n_6 \leq (n - m)/2)$ for $\Sigma_{s=1}^{l} m_{6s}$ pairs of repeated eigenvalues with a nilpotent matrix $\mathbf{N}_6 = \mathbf{0}$.

Definition A.22 For an n-dimensional, linear dynamical system $\dot{\mathbf{x}} = \mathbf{A}\mathbf{x}$ in Eq. (A.2), the matrix $\mathbf{A}$ possesses n eigenvalues λ_k $(k = 1, 2, \ldots, n)$. Set $N = \{1, 2, \ldots, m, m + 1, \ldots, (n - m)/2\}$, $N_i = \{i_1, i_2, \ldots, i_{n_i}\} \cup \varnothing$ with $i_j \in N$ $(j = 1, 2, \ldots, n_i,\ i = 1, 2, \ldots, 6)$, $\Sigma_{i=1}^{3} n_i = m$ and $2\Sigma_{i=4}^{6} n_i = n - m$. $\cup_{i=1}^{6} N_i = N$ and $N_i \cap N_l = \varnothing (l \neq i)$. $N_i = \varnothing$ if $n_i = 0$. The matrix $\mathbf{A}$ possesses n_1-stable, n_2-unstable and n_3-invariant real eigenvectors plus n_4-stable, n_5-unstable and n_6-center pairs of complex eigenvectors. $\kappa_i \in \{\varnothing, m_i\} (i = 3,\ 6)$. The flow $\mathbf{\Phi}(t)$ of $\underline{\mathbf{x}} = \mathbf{A}\mathbf{x}$ is an$(n_1 : n_2 : [n_3; m_3] \| n_4 : n_5 : [n_6, l; \kappa_6])$ flow $\kappa_3 \in \{\varnothing, m_3\}$, $\kappa_6 = (\kappa_{61}, \kappa_{62}, \cdots, \kappa_{6l})^{\mathrm{T}}$ with $\kappa_{6i} \in \{\varnothing, m_{6i}\}$ $(i = 1, 2, \cdots, l)$. $\mathbf{m}_6 = (m_{61}, m_{62}, \cdots, m_{6l})^{\mathrm{T}}$. The flow $\Phi(t)$ of $\dot{\mathbf{x}} = \mathbf{A}\mathbf{x}$ is an $(n_1 : n_2 : [n_3; m_3] \| n_4 : n_5 : [n_6, l; \kappa_6])$ is an $\kappa_3 \in \{\varnothing, m_3\}$, $\kappa_6 = (\kappa_{61}, \kappa_{62}, \cdots, \kappa_{6l})^{\mathrm{T}}$ with $\kappa_{6i} \in \{\varnothing, m_{6i}\} (i = 1, 2, \cdots, l)$ $\mathbf{m}_6 = (m_{61}, m_{62}, \cdots, m_{6l})^{\mathrm{T}}$.

I. *Non-degenerate cases*

(i) The origin is an $(n_1 : n_2 : \varnothing | n_4 : n_5 : \varnothing)$ hyperbolic point (or saddle) for the linear system.

(ii) The origin is an $(n_1 : \varnothing : \varnothing | n_4 : \varnothing : \varnothing)$ -sink for the linear system.

(iii) The origin is an $(\varnothing : n_2 : \varnothing | \varnothing : n_5 : \varnothing)$ -source for the linear system.

(iv) The origin is an $(\varnothing : \varnothing : \varnothing | \varnothing : \varnothing : n/2)$ -circular center for the linear system.

(v) The origin is an $(\varnothing : \varnothing : \varnothing | \varnothing : \varnothing : [n/2, l; \varnothing])$ -circular center for the linear system.

(vi) The origin is an $(\varnothing : \varnothing : \varnothing | \varnothing : \varnothing : [n/2, l; \mathbf{m}_6])$ point for the linear system.

(vii) The origin is an $(n_1 : \varnothing : \varnothing | n_4 : \varnothing : n_6)$ -point for the linear system.

(viii) The origin is an $(\varnothing : n_2 : \varnothing | \varnothing : n_5 : n_6)$ -point for the linear system.

(ix) The origin is an $(n_1 : n_2 : \varnothing | n_4 : n_5 : n_6)$ -point for the linear system.

II. *Simple degenerate cases*

(i) The origin is an $(\varnothing : \varnothing : [n;\varnothing]|\varnothing : \varnothing : \varnothing)$ -invariant (or static) center for the linear system.
(ii) The origin is an $(\varnothing : \varnothing : [n;m]|\varnothing : \varnothing : \varnothing)$ point for the linear system.
(iii) The origin is an $(\varnothing : \varnothing : [n_3;\varnothing]|\varnothing : \varnothing : n_6)$ point for the linear system.
(iv) The origin is an $(\varnothing : \varnothing : [n_3;m]|\varnothing : \varnothing : n_6)$ point for the linear system
(v) The origin is an $(\varnothing : \varnothing : [n_3;\varnothing]|\varnothing : \varnothing : [n_6, l;\varnothing])$ point for the linear system.
(vi) The origin is an $(\varnothing : \varnothing : [n_3;m_3]|\varnothing : \varnothing : [n_6, l;\varnothing])$ point for the linear system.
(vii) The origin is an $(\varnothing : \varnothing : [n_3;\varnothing]|\varnothing : \varnothing : [n_6, l;\mathbf{m}_6])$ point for the linear system.
(viii) The origin is an $(\varnothing : \varnothing : [n_3;m_3]|\varnothing : \varnothing : [n_6, l;\mathbf{m}_6])$ point for the linear system.

III. *Complex degenerate cases*

(i) The origin is an $(n_1 : \varnothing : [n_3;\varnothing]|n_4 : \varnothing : \varnothing)$ point for the linear system.
(ii) The origin is an $(n_1 : \varnothing : [n_3;m]|n_4 : \varnothing : \varnothing)$ point for the linear system.
(iii) The origin is an $(\varnothing : n_2 : [n_3;\varnothing]|\varnothing : n_5 : \varnothing)$ point for the linear system.
(iv) The origin is an $(\varnothing : n_2 : [n_3;m]|\varnothing : n_5 : \varnothing)$ point for the linear system.
(v) The origin is an $(n_1 : \varnothing : [n_3;\varnothing]|n_4 : \varnothing : n_6)$ point for the linear system.
(vi) The origin is an $(n_1 : \varnothing : [n_3;m]|n_4 : \varnothing : n_6)$ point for the linear system.
(vii) The origin is an $(\varnothing : n_2 : [n_3;\varnothing]|\varnothing : n_5 : n_6)$ point for the linear system.
(viii) The origin is an $(\varnothing : n_2 : [n_3;m]|\varnothing : n_5 : n_6)$ point for the linear system.

IV. *Simple critical cases*

(i) An $(n_1 : n_2 : 1|n_4 : n_5 : \varnothing)$ state of the origin for the linear system is a boundary of its $(n_1 : n_2 + 1 : \varnothing|n_4 : n_5 : \varnothing)$ -spiral saddle and $(n_1 + 1 : n_2 : \varnothing|n_4 : n_5 : \varnothing)$ - spiral saddle.
(ii) An $(n_1 - 1 : \varnothing : 1|n_4 : \varnothing : \varnothing)$ state of the origin for the linear system is a boundary of its $(n_1 - 1 : 1 : \varnothing|n_4 : \varnothing : \varnothing)$ -spiral saddle and $(n_1 : \varnothing : \varnothing|n_4 : \varnothing : \varnothing)$ -spiral sink.
(iii) An $(\varnothing : n_2 - 1 : 1|\varnothing : n_5 : \varnothing)$ state of the origin for the linear system is a boundary of its$(1 : n_2 - 1 : \varnothing|\varnothing : n_5 : \varnothing)$ -spiral saddle and $(\varnothing : n_2 : \varnothing|\varnothing : n_5 : \varnothing)$ -spiral source.
(iv) An $(n_1 : n_2 : \varnothing|n_4 : n_5 : 1)$ state of the origin for the linear system is a boundary of its $(n_1 : n_2 : \varnothing|n_4 + 1 : n_5 : \varnothing)$ -spiral saddle and $(n_1 : n_2 : \varnothing|n_4 : n_5 + 1 : \varnothing)$ - spiral saddle.
(v) An $(n_1 : \varnothing : \varnothing|n_4 - 1 : \varnothing : 1)$ state of the origin for the linear system is a boundary of its $(n_1 : \varnothing : \varnothing|n_4 - 1 : 1 : \varnothing)$ -spiral saddle and $(n_1 : \varnothing : \varnothing|n_4 : \varnothing : \varnothing)$ -spiral sink.
(vi) An $(\varnothing : n_2 : \varnothing|\varnothing : n_5 - 1 : 1)$ state of the origin for the linear system is a boundary of its $(\varnothing : n_2 : \varnothing|1 : n_5 - 1 : \varnothing)$ -spiral saddle and $(\varnothing : n_2 : \varnothing|\varnothing : n_5 : \varnothing)$ -spiral source.

(vii) An $(n_1 : n_2 : 1|n_4 : n_5 : n_6)$ state of the origin for the linear system is a boundary of its $(n_1 + 1 : n_2 : \varnothing|n_4 : n_5 : n_6)$ state and $(n_1 : n_2 + 1 : \varnothing|n_4 : n_5 : n_6)$ state.

(viii) An $(n_1 : n_2 : 1|n_4 : n_5 : [n_6, l; \boldsymbol{\kappa}_6])$ state of the origin for the linear system is a boundary of its $(n_1 + 1 : n_2 : \varnothing|n_4 : n_5 : [n_6, l; \boldsymbol{\kappa}_6])$ state and $(n_1 : n_2 + 1 : \varnothing|n_4 : n_5 : [n_6, l; [n_6, l; \boldsymbol{\kappa}_6])$ state.

(ix) An $(n_1 : n_2 : \varnothing|n_4 : n_5 : n_6 + 1)$ state of the origin for the linear system is a boundary of its $(n_1 : n_2 : \varnothing|n_4 + 1 : n_5 : n_6)$ state and $(n_1 : n_2 : \varnothing|n_4 : n_5 + 1 : n_6)$ state.

(x) An $(n_1 : n_2 : [n_3; \varnothing]|n_4 : n_5 : n_6 + 1)$ state of the origin for the linear system is a boundary of its $(n_1 : n_2 : [n_3; \varnothing]|n_4 + 1 : n_5 : n_6)$ state and $(n_1 : n_2 : [n_3; \varnothing]|n_4 : n_5 + 1 : n_6)$ state.

V. *Complex critical cases*

(i) An $(n_1 : n_2 : [n_3; \kappa_3]|n_4 : n_5 : n_6)$ -state of the origin for the linear system is a boundary of its $(n_1 + n_3 : n_2 : \varnothing|n_4 : n_5 : n_6)$ state and $(n_1 : n_2 + n_3 : \varnothing|n_4 : n_5 : n_6)$ state.

(ii) An $(n_1 - m : n_2 : [n_3 + m; \kappa_3]|n_4 : n_5 : n_6)$ -state of the origin for the linear system is a boundary of its $(n_1 : n_2 : [n_3; \kappa_3']|n_4 : n_5 : n_6)$ state and $(n_1 - m : n_2 + m : [n_3; \kappa_3']|n_4 : n_5 : n_6)$ state.

(iii) An $(n_1 : n_2 - m : [n_3 + m; \kappa_3]|n_4 : n_5 : n_6)$ -state of the origin for the linear system is a boundary of its $(n_1 : n_2 : [n_3; \kappa_3']|n_4 : n_5 : n_6)$ -state and $(n_1 + m : n_2 - m : [n_3; \kappa_3']|n_4 : n_5 : n_6)$ -state.

(iv) An $(n_1 - m_1 : n_2 : [n_3 + m_1; \kappa_3]|n_4 : n_5 : n_6)$ -state of the origin for the linear system is a boundary of $(n_1 : n_2 : [n_3; \kappa_3']|n_4 : n_5 : n_6)$ -state and $(n_1 - m_1 : n_2 + m_2 : [n_3 + m_1 - m_2; \kappa_3'']|n_4 : n_5 : n_6)$ -state.

(v) An $(n_1 : n_2 - m_2 : [n_3 + m_2; \kappa_3]|n_4 : n_5 : n_6)$ -state of the origin for the linear system is a boundary of $(n_1 : n_2 : [n_3; \kappa_3']|n_4 : n_5 : n_6)$ -state and $(n_1 + m_1 : n_2 - m_2 : [n_3 + m_2 - m_1; \kappa_3'']|n_4 : n_5 : n_6)$ -state

(vi) An $(n_1 : n_2 : [n_3; \kappa_3]|n_4 : n_5 : [n_6, l; \boldsymbol{\kappa}_6])$ state of the origin for the linear system is a boundary of its $(n_1 : n_2 : [n_3; \kappa_3]|n_6 + n_4 : n_5 : \varnothing)$ state and $(n_1 : n_2 : [n_3; \kappa_3]|n_4 : n_5 + n_6 : \varnothing)$ state.

(vii) An $(n_1 : n_2 : [n_3; \kappa_3]|n_4 - m : n_5 : [n_6 + m, l_1; \boldsymbol{\kappa}_6])$ state of the origin for the linear system is a boundary of its $(n_1 : n_2 : [n_3; \kappa_3]|n_4 : n_5 : [n_6, l_2; \boldsymbol{\kappa}_6])$ -state and $(n_1 : n_2 : [n_3; \kappa_3]|n_4 - m : n_5 + m : [n_6, l_2; \boldsymbol{\kappa}_6])$ -state.

(viii) An $(n_1 : n_2 : [n_3; \kappa_3]|n_4 : n_5 - m : [n_6 + m_4, l_1; \boldsymbol{\kappa}_6])$ of the origin for the linear system is a boundary of its $(n_1 : n_2 : [n_3; \kappa_3]|n_4 : n_5 : [n_6, l_2; \boldsymbol{\kappa}_6'])$ state and $(n_1 : n_2 : [n_3; \kappa_3]|n_4 + m : n_5 - m : [n_6, l_2; \boldsymbol{\kappa}_6''])$ state of equilibrium point $(\mathbf{x}^*, \mathbf{p})$.

(ix) An $(n_1 : n_2 : [n_3; \kappa_3]|n_4 - m_4 : n_5 : [n_6 + m_4, l_1; \boldsymbol{\kappa}_6])$ state of the origin for the linear system is a boundary of its $(n_1 : n_2 : [n_3; \kappa_3]|n_4 : n_5 : [n_6, l_2; \boldsymbol{\kappa}_6'])$ -state and $(n_1 : n_2 : [n_3; \kappa_3]|n_4 - m_4 : n_5 + m_5 : [n_6 + m_4 - m_5, l_3; \boldsymbol{\kappa}_6''])$ -state of equilibrium point $(\mathbf{x}^*, \mathbf{p})$.

(x) An $(n_1 : n_2 : [n_3; \kappa_3]|n_4 : n_5 - m_5 : [n_6 + m_5, l_1; \boldsymbol{\kappa}_6])$ state of the origin for the linear system is a boundary of $(n_1 : n_2 : [n_3; \kappa_3]|n_4 : n_5 : [n_6, l_2; \boldsymbol{\kappa}'_6])$ - state and $(n_1 : n_2 : [n_3; \kappa_3]|n_4 + m_4 : n_5 - m_5 : [n_6 + m_5 - m_4, l_3; \boldsymbol{\kappa}''_6])$ - state of equilibrium point $(\mathbf{x}^*, \mathbf{p})$.

Definition A.23 For an n-dimensional, linear dynamical system $\dot{\mathbf{x}} = \mathbf{A}\mathbf{x}$ in Eq. (A.2), the matrix $\mathbf{A}$ possesses n real eigenvalues λ_k $(k = 1, 2, \ldots, n)$. Set $N = \{1, 2, \ldots, n\}$, $N_i = \{i_1, i_2, \ldots, i_{n_i}\} \cup \varnothing$ with $i_j \in N$ $(j = 1, 2, \ldots, n_i, i = 1, 2, 3)$ and $\Sigma_{i=1}^3 n_i = n$. $\cup_{i=1}^3 N_i = N$ and $N_i \cap N_l = \varnothing$ $(l \neq i)$. $N_i = \varnothing$ if $n_i = 0$. The matrix $\mathbf{A}$ possesses n_1 -stable, n_2 -unstable, and n_3 -invariant real eigenvectors. Without repeated eigenvalues of $\lambda_k = 0$ $(k \in N_3)$, the flow $\Phi(t)$ of $\dot{\mathbf{x}} = \mathbf{A}\mathbf{x}$ is an $(n_1 : n_2 : \varnothing|$ or $(n_1 : n_2 : 1|$ flow. However, with repeated eigenvalues of $\lambda_k = 0$ $(k \in N_3)$, the flow $\Phi(t)$ of $\dot{\mathbf{x}} = \mathbf{A}\mathbf{x}$ is an $(n_1 : n_2 : [n_3; m_3]|$ flow $\kappa_3 \in \{\varnothing, m_3\}$.

I. *Non-degenerate cases*

(i) The origin is an $(n : \varnothing : \varnothing|$ -stable node for the linear system.
(ii) The origin is an $(\varnothing : n : \varnothing|$ -unstable node for the linear system.
(iii) The origin is an $(n_1 : n_2 : \varnothing|$ -saddle for the linear system.

II. *Degenerate cases*

(i) The origin is in an $(n_1 : n_2 : 1|$ -critical point for the linear system.
(ii) The origin is an $(n_1 : n_2 : [n_3; \varnothing]|$ -point for the linear system.
(iii) The origin is an $(n_1 : n_2 : [n_3; m_3]|$ -point for the linear system.

III. *Simple critical cases*

(i) An $(n_1 : n_2 : 1|$ state of the origin for the linear system is a boundary of its $(n_1 + 1 : n_2 : \varnothing|$ -saddle and $(n_1 : n_2 + 1 : \varnothing|$ -saddle.
(ii) An $(n - 1 : \varnothing : 1|$ state of the origin for the linear system is a boundary of its $(n - 1 : 1 : \varnothing|$ -saddle and $(n : \varnothing : \varnothing|$ -stable node.
(iii) An $(\varnothing : n - 1 : 1|$ state of the origin for the linear system is a boundary of $(1 : n - 1 : \varnothing|$ -saddle and $(\varnothing : n : \varnothing|$ -unstable node.
(iv) An $(n_1 - 1 : n_2 : [n_3 + 1; \kappa_3]|$ state of the origin for the linear system is a boundary of $(n_1 : n_2 : [n_3; \kappa_3]|$ -degenerate saddle and $(n_1 - 1 : n_2 + 1 : [n_3; \kappa_3]|$ -degenerate saddle.

IV. *Complex critical cases*

(i) An $(n_1 : n_2 : [n_3; \kappa_3]|$ -state of the origin for the linear system is a boundary of $(n_1 + n_3 : n_2 : \varnothing|$ -saddle and $(n_1 : n_2 + n_3 : \varnothing|$ -saddle.
(ii) An $(n_1 - m : n_2 : [n_3 + m; \kappa_3]|$ -state of the origin for the linear system is a boundary of its $(n_1 : n_2 : [n_3; \kappa'_3]|$ -state and $(n_1 - m : n_2 + m : [n_3; \kappa'_3]|$ -state.
(iii) An$(n_1 : n_2 - m : [n_3 + m; \kappa_3]|$ -state of the origin for the linear system is a boundary of its $(n_1 : n_2 : [n_3; \kappa'_3]|$ -state and $(n_1 + m : n_2 - m : [n_3; \kappa'_3]|$ -state.
(iv) An$(n_1 - m_1 : n_2 : [n_3 + m_1; \kappa_3]|$ -state of the origin for the linear system is a boundary of $(n_1 : n_2 : [n_3; \kappa'_3]|$ -state and $(n_1 - m_1 : n_2 + m_2 : [n_3 + m_1 - m_2; \kappa''_3]|$ -state.

(v) An $(n_1 : n_2 - m_2 : [n_3 + m_2; \kappa_3]|$ -state of the origin for the linear system is a boundary of $(n_1 : n_2 : [n_3; \kappa_3']|$ -state and $(n_1 + m_1 : n_2 - m_2 : [n_3 + m_2 - m_1; \kappa_3'']|$ -state.

Definition A.24 For a $2n$-dimensional, linear dynamical system $\dot{\mathbf{x}} = \mathbf{A}\mathbf{x}$ in Eq. (A.2), the matrix **A** possesses n-pairs of complex eigenvalues λ_k $(k = 1, 2, \ldots, n)$. Set $N = \{1, 2, \ldots, n\}$, $N_i = \{i_1, i_2, \ldots, i_{n_i}\} \cup \varnothing$ with $l_j \in N$ $(j = 1, 2, \ldots, n_i;\ i = 4, 5, 6)$ and $\Sigma_{i=4}^{6} n_i = n$. $\cup_{i=4}^{6} N_i = N$ and $N_i \cap N_l = \varnothing$ $(l \neq i)$. $N_i = \varnothing$ if $n_i = 0$. The matrix **A** possesses n_4 -stable, n_5 -unstable and n_6 -center pairs of complex eigenvectors. Without repeated eigenvalues of $\mathrm{Re}\lambda_k = 0$ $(k \in N_6)$, the flow $\Phi(t)$ of $\dot{\mathbf{x}} = \mathbf{A}\mathbf{x}$ is an $|n_4 : n_5 : n_6, l; \boldsymbol{\kappa}_6)$ flow. However, with repeated eigenvalues of $\mathrm{Re}\lambda_k = 0$ $(k \in N_6)$, the flow $\Phi(t)$ of $\dot{\mathbf{x}} = \mathbf{A}\mathbf{x}$ is an $|n_4 : n_5 : [n_6; l; \boldsymbol{\kappa}_6])$ flow. $\boldsymbol{\kappa}_6 = (\kappa_{61}, \kappa_{62}, \cdots, \kappa_{6l})^{\mathrm{T}}$ with $\kappa_{6i} \in \{\varnothing, m_{6i}\} (i = 1, 2, \cdots, l)$.

I. *Non-degenerate cases*

(i) The origin is an $|n : \varnothing : \varnothing)$ -spiral sink for the linear system.
(ii) The origin is an $|\varnothing : n : \varnothing)$ -spiral source for the linear system.
(iii) The origin is an $|\varnothing : \varnothing : n)$ -circular center for the linear system.
(iv) The origin is an $|n_4 : n_5 : \varnothing)$ -spiral saddle for the linear system.

II. *Quasi-degenerate cases*

(i) The origin is an $|n_4 : \varnothing : n_6)$ -point for the linear system.
(ii) The origin is an $|\varnothing : n_5 : n_6)$ -point for the linear system.
(iii) The origin is an $|n_4 : \varnothing : [n_6, l; \varnothing])$ -point for the linear system.
(iv) The origin is an $|n_4 : \varnothing : [n_6, l; \boldsymbol{\kappa}_6])$ -point for the linear system.
(v) The origin is an $|\varnothing : n_5 : [n_6, l; \varnothing])$ -point for the linear system.
(vi) The origin is an $|\varnothing : n_5 : [n_6, l; \boldsymbol{\kappa}_6])$ -point for the linear system.

III. *Simple critical cases*

(i) An $|n_4 : n_5 : 1)$ -state of the origin for the linear system is a boundary of its $|n_4 + 1 : n_5 : \varnothing)$ -spiral saddle and $|n_4 : n_5 + 1 : \varnothing)$ -spiral saddle.
(ii) An $|n_4 : \varnothing : 1)$ - state of the origin for the linear system is a boundary of its $|n_4 + 1 : \varnothing : \varnothing)$ -spiral sink and $|n_4 : 1 : \varnothing)$ -spiral saddle.
(iii) An $|\varnothing : n_5 : 1)$ -state of the origin for the linear system is a boundary of $|\varnothing : n_5 + 1 : \varnothing)$ -purely spiral source and $|1 : n_5 : \varnothing)$ -spiral saddle.
(iv) An $|n_4 : n_5 : n_6 + 1)$ -state of the origin for the linear system is a boundary of its $|n_4 + 1 : n_5 : n_6)$ -state and $|n_4 : n_5 + 1 : n_6)$ -state.
(v) An $|n_4 : \varnothing : n_6 + 1)$ -state of the origin for the linear system is a boundary of its $|n_4 + 1 : \varnothing : n_6)$ -state and $|n_4 : 1 : n_6)$ -state.
(vi) An $|\varnothing : n_5 : n_6 + 1)$ -state of the origin for the linear system is a boundary of $|\varnothing : n_5 + 1 : n_6)$ -state source and $|1 : n_5 : n_6)$ -state.

IV. *Complex critical cases*

(i) An $|n_4 : n_5 : [n_6, l; \boldsymbol{\kappa}_6])$ -state of the origin for the linear system is a boundary of its $|n_6 + n_4 : n_5 : \varnothing)$ -spiral saddle and $|n_4 : n_5 + n_6 : \varnothing)$ -spiral saddle.

(ii) An $|n_4 - m : n_5 : [n_6 + m, l_2; \boldsymbol{\kappa}_6])$ -state of the origin for the linear system is a boundary of its $|n_4 : n_5 : [n_6, l_1; \boldsymbol{\kappa}'_6])$ -state and $|n_4 - m : n_5 + m : [n_6, l_3; \boldsymbol{\kappa}''_6])$ -state.

(iii) An $|n_4 : n_5 - m : [n_6 + m, l_2; \boldsymbol{\kappa}_6])$ - state of the origin for the linear system is a boundary of its $|n_4 : n_5 : [n_6, l_1; \boldsymbol{\kappa}'_6])$ -state and $|n_4 + m : n_5 - m : [n_6, l_3; \boldsymbol{\kappa}''_6])$ -state.

(iv) An $|n_4 - m_4 : n_5 : [n_6 + m_4, l_2; \boldsymbol{\kappa}_6])$ state of the origin for the linear system is a boundary of its $|n_4 : n_5 : [n_6; l_1; \boldsymbol{\kappa}'_6])$ -state and $|n_4 - m_4 : n_5 + m_5 : [n_6 + m_4 - m_5, l_3; \boldsymbol{\kappa}''_6])$ -state.

(v) An $|n_4 : n_5 - m_5 : [n_6 + m_5; l_2; \boldsymbol{\kappa}_6])$ -state of the origin for the linear system is a boundary of $|n_4 : n_5 : [n_6, l_1; \boldsymbol{\kappa}'_6])$ -state and $|n_4 + m_4 : n_5 - m_5 : [n_6 + m_5 - m_4, l_3; \boldsymbol{\kappa}''_6])$ -state.

For the linear system in Eq. (A.135) with periodic coefficients, the sufficient and necessary conditions of stability to the origin are $\mathrm{Re}\lambda_k < 0$ $(k = 1, 2, \ldots n)$ of matrix **B**. However, the transformation matrix $\mathbf{P}(t)$ is very difficult to be determined. Thus one uses the sum of characteristic exponents in Eq. (A.138) to determine stability. The conclusion is that the linear system to the origin is unstable if $\Sigma_{k=1}^{n}\lambda_k > 0$. However, if $\Sigma_{k=1}^{n}\lambda_k \leq 0$, one cannot conclude the linear system to the origin is stable. Therefore, for each problem, it should be treated specially.

A.7 Lower-Dimensional Dynamical Systems

Consider a one-dimensional linear system as

$$\dot{x} = \lambda x \tag{A.155}$$

with initial condition $x(t_0) = x_0$. The solution is

$$x = x_0 e^{\lambda(t - t_0)}. \tag{A.156}$$

The following properties of the solution exist.

(i) $\lim_{t\to\infty} |x(t)| = 0$, and the system to the origin is stable if $\lambda < 0$;
(ii) $\lim_{t\to\infty} |x(t)| = \infty$, and the system to the origin is unstable if $\lambda > 0$;
(iii) $x(t) = x_0$, and the origin to the system is center if $\lambda = 0$.

The above solutions are illustrated in Fig. A.1 . The solutions and phase lines for the unstable, stable and invariant linear systems are presented in Fig. A.1a–c, respectively. The gray points are the values of λ.

Consider a one-dimensional linear system with external excitation

$$\dot{x} = \lambda x + f(t) \tag{A.157}$$

with initial condition $x(t_0) = x_0$. The solution is

$$x = x_0 e^{\lambda(t - t_0)} + e^{\lambda t} \int_{t_0}^{t} e^{-\lambda\tau} f(\tau) d\tau. \tag{A.158}$$

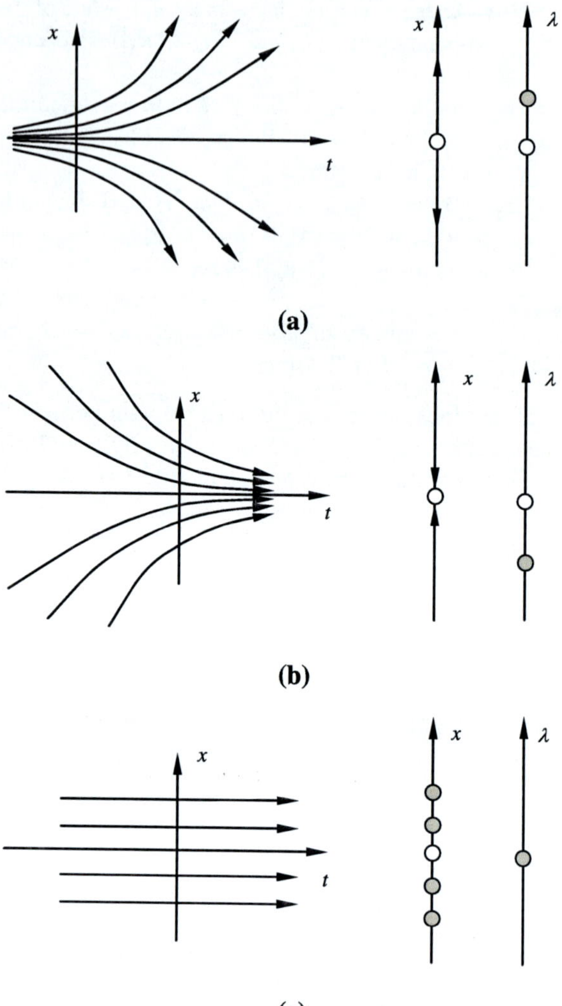

(c)

Fig. A.1 Solution and phase line of $\dot{x} = \lambda x$: **a** an $(\varnothing : 1 : \varnothing|$ -unstable node $(\lambda > 0)$, **b** a $(1 : \varnothing : \varnothing|$ -stable node $(\lambda < 0)$ and **c** an $(\varnothing : \varnothing : 1|$ -static invariance $(\lambda = 0)$.

A.7.1 Planar Dynamical Systems

Consider a two-dimensional linear system as

$$\dot{\mathbf{x}} = \mathbf{A}\mathbf{x} \tag{A.159}$$

with initial condition $\mathbf{x}(t_0) = \mathbf{x}_0$, and

$$\mathbf{A} = \begin{bmatrix} a_{11} & a_{12} \\ a_{21} & a_{22} \end{bmatrix}. \tag{A.160}$$

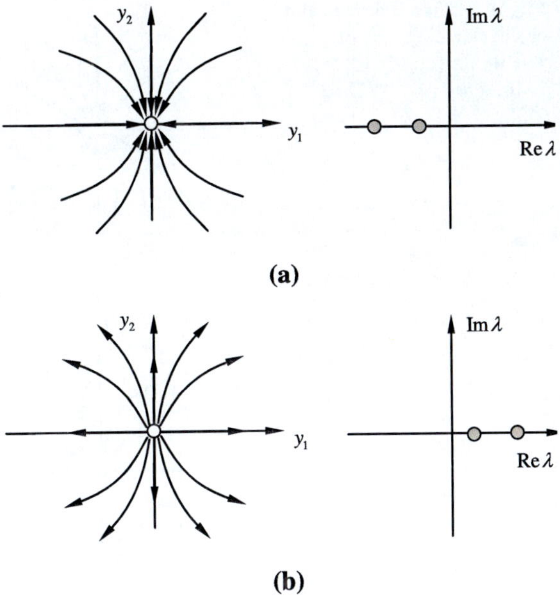

Fig. A.2 Phase portraits and eigenvalue diagrams of $\dot{\mathbf{y}} = \mathbf{B}\mathbf{y}$: **a** a $(2 : \varnothing : \varnothing|$ -stable node ($\lambda_k < 0,\ k = 1, 2$), **b** an $(\varnothing : 2 : \varnothing|$ -unstable node $\lambda_k > 0$ ($k = 1, 2$).

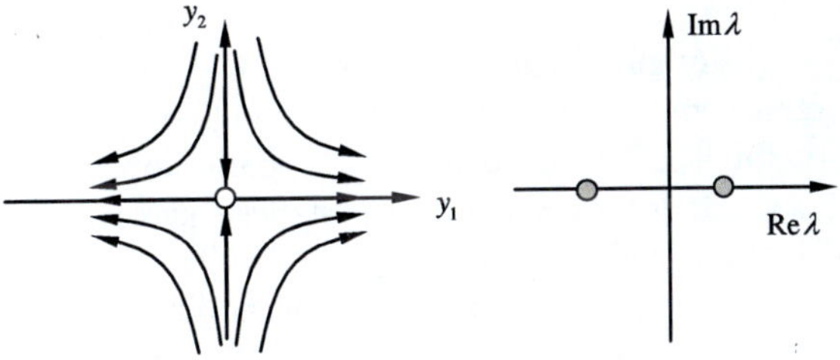

Fig. A.3 Phase portraits and eigenvalue diagrams of a $(1 : 1 : \varnothing|$ -saddle for $\dot{\mathbf{y}} = \mathbf{B}\mathbf{y}$ with$\lambda_1 > 0$ and $\lambda_2 < 0$.

If $\det \mathbf{A} \neq 0$, $\mathbf{x} = \mathbf{0}$ is a unique equilibrium point. With a nonsingular transform matrix $\mathbf{P}$, $\mathbf{B} = \mathbf{P}^{-1}\mathbf{A}\mathbf{P}$. With $\mathbf{x} = \mathbf{P}\mathbf{y}$,

$$\dot{\mathbf{y}} = \mathbf{B}\mathbf{y} \tag{A.161}$$

where

$$\mathbf{B} = \begin{bmatrix} \lambda_1 & 0 \\ 0 & \lambda_2 \end{bmatrix}. \tag{A.162}$$

There are four cases:

(A) For two real distinct eigenvalues ($\lambda_1 \neq \lambda_2$), the solution is expressed by

$$\mathbf{B} = \begin{bmatrix} \lambda_1 & 0 \\ 0 & \lambda_2 \end{bmatrix} \text{ and } \mathbf{y}(t) = \begin{bmatrix} e^{\lambda_1(t-t_0)} & 0 \\ 0 & e^{\lambda_2(t-t_0)} \end{bmatrix}\mathbf{y}_0. \tag{A.163}$$

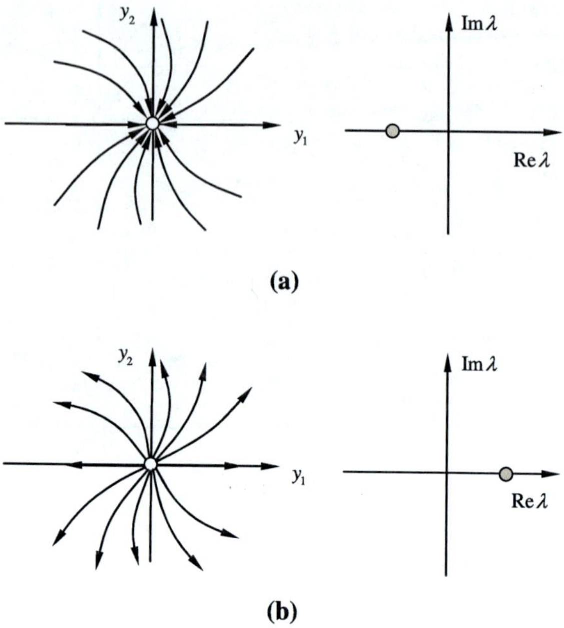

Fig. A.4 Phase portraits and eigenvalue diagrams of $\dot{\mathbf{y}} = \mathbf{B}\mathbf{y}$: **a** $(2:\varnothing:\varnothing|\ (\lambda_k = \lambda<0,\ k=1,2)$, **b** $(\varnothing:2:\varnothing|\ (\lambda_k = \lambda > 0, k=1,2)$.

The origin is called a node of the linear system if two real eigenvalues have the same sign. If $\lambda_k < 0$ $(k = 1,2)$, the origin is a stable node. If $\lambda_k > 0$ $(k = 1,2)$, the origin is an unstable node. The corresponding phase portraits and eigenvalue diagrams for the stable and unstable nodes of the linear systems are sketched in Fig. A.2a and b, respectively.

The origin is called a saddle of the linear system if two real eigenvalues have different signs $(\lambda_1 > 0$ and $\lambda_2 < 0)$. The linear system is unstable. The corresponding phase portraits and eigenvalue diagram are presented in Fig. A.3. On the eigenvector direction, the flows will come to or leave the origin.

(B) For two real repeated eigenvalues $(\lambda_1 = \lambda_2 = \lambda)$, the solution is given by

$$\begin{aligned} \mathbf{B} &= \begin{bmatrix} \lambda & 1 \\ 0 & \lambda \end{bmatrix} \text{ and } \mathbf{y}(t) = e^{\lambda(t-t_0)} \begin{bmatrix} 1 & t \\ 0 & 1 \end{bmatrix} \mathbf{y}_0. \\ \mathbf{B} &= \begin{bmatrix} \lambda & 0 \\ 0 & \lambda \end{bmatrix} \text{ and } \mathbf{y}(t) = e^{\lambda(t-t_0)} \begin{bmatrix} 1 & 0 \\ 0 & 1 \end{bmatrix} \mathbf{y}_0. \end{aligned} \tag{A.164}$$

For repeated eigenvalues $\lambda_k = \lambda < 0$ $(k = 1,2)$, the origin is a stable node. If repeated eigenvalues $\lambda_k > 0$ $(k = 1,2)$, the origin is also an unstable node. The corresponding phase portraits and eigenvalue diagram for the stable and unstable nodes are shown in Fig. A.4a and b. For the second equation of Eq. (A.164), the line exist in phase portrait. If $\lambda = 0$, then $y_2 = c$ and $y_1 = c_0 + ct$. This is the

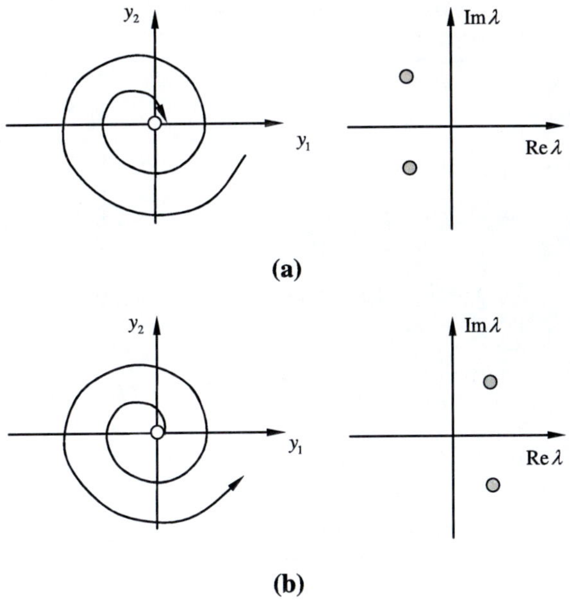

Fig. A.5 Phase portraits and eigenvalue diagrams of $\dot{\mathbf{y}} = \mathbf{B}\mathbf{y}$ ($\mathrm{Im}\lambda_k = \pm\beta \neq 0, k = 1,2$) : **a** for a $|1:\varnothing:\varnothing)$ -stable focus ($\mathrm{Re}\lambda_k = \alpha < 0$), **b** an $|\varnothing:1:\varnothing)$ -unstable focus ($\mathrm{Re}\lambda_k = \alpha > 0$).

constant velocity case. If $c = 0$, the dynamical system is in static state forever. For the second case with $\lambda = 0$, it gives stationary points in phase portrait.

(C) For $\lambda_1 = \alpha + \mathbf{i}\beta$ and $\lambda_2 = \alpha - \mathbf{i}\beta$, the solution is given by

$$\mathbf{B} = \begin{bmatrix} \alpha & \beta \\ -\beta & \alpha \end{bmatrix} \text{ and } \mathbf{y}(t) = e^{\alpha(t-t_0)} \begin{bmatrix} \cos\beta(t-t_0) & \sin\beta(t-t_0) \\ \sin\beta(t-t_0) & \cos\beta(t-t_0) \end{bmatrix} \mathbf{y}_0. \quad \text{(A.165)}$$

The origin is called a focus of the linear system if the real part of two complex eigenvalues are nonzero ($\mathrm{Re}\lambda_k = \alpha \neq 0$ for $k = 1,2$). The origin is called a stable focus if $\mathrm{Re}\lambda_k = \alpha < 0$. The origin is called an unstable focus if $\mathrm{Re}\lambda_k = \alpha > 0$. From the solutions, the phase portraits and eigenvalue diagram for stable and unstable focuses are presented in Fig. A.5a and b, respectively. The eigenvalues are a pair of complex eigenvalues. The initial point for the unstable focus cannot be selected at the origin. For the stable focus, the solution of the linear system will approach the origin as $t \to \infty$.

The origin is called the sink of the linear system in Eq. (A.159) if the real parts of all eigenvalues are less than zero ($\mathrm{Re}\lambda_k < 0$ for $k = 1,2$). The origin is called the source of the linear system in Eq. (A.159) if the real parts of all eigenvalues are greater than zero ($\mathrm{Re}\lambda_k > 0$ for $k = 1,2$). Compared to the node and saddle-nodes, the stable and unstable focuses make a flow spirally come to the origin or spirally leave for infinity, respectively.

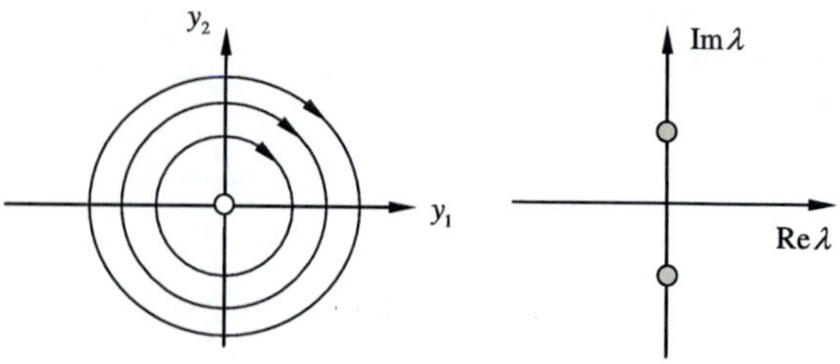

Fig. A.6 Phase portrait and eigenvalue diagram for an $|\varnothing : \varnothing : 1)$ -center for $\dot{\mathbf{y}} = \mathbf{B}\mathbf{y}$

(D) For $\lambda_1 = \mathbf{i}\beta$ and $\lambda_2 = -\mathbf{i}\beta$, the solution is given by

$$\mathbf{B} = \begin{bmatrix} 0 & \beta \\ -\beta & 0 \end{bmatrix} \text{and } \mathbf{y}(t) = \begin{bmatrix} \cos\beta(t-t_0) & \sin\beta(t-t_0) \\ -\sin\beta(t-t_0) & \cos\beta(t-t_0) \end{bmatrix} \mathbf{y}_0. \quad \text{(A.166)}$$

The origin is called a center of the linear system if the real part of two complex eigenvalues are zero ($\text{Re}\lambda_k = \alpha \neq 0$ and $\text{Im}\lambda_k = \pm\beta \neq 0$ for $k = 1, 2$). For this case, the phase portrait is a family of circles, and the eigenvalues lie on the imaginary axes, as sketched in Fig. A.6.

The eigenvalues of $\mathbf{A}$ are determined by $\det(\mathbf{A} - \lambda\mathbf{I}) = 0$, i.e.

$$\lambda^2 - \text{tr}(\mathbf{A})\lambda + \det(\mathbf{A}) = 0. \quad \text{(A.167)}$$

where

$$\text{tr}(\mathbf{A}) = a_{11} + a_{22} \text{ and } \det(\mathbf{A}) = \begin{vmatrix} a_{11} & a_{12} \\ a_{21} & a_{22} \end{vmatrix}. \quad \text{(A.168)}$$

The corresponding eigenvalues are

$$\lambda_{1,2} = \frac{\text{tr}(\mathbf{A}) \pm \sqrt{\Delta}}{2} \text{ and } \Delta = (\text{tr}(\mathbf{A}))^2 - 4\det(\mathbf{A}). \quad \text{(A.169)}$$

The linear system in Eq. (A.159) possesses

(i) a saddle at the origin for $\det(\mathbf{A}) < 0$ with $\lambda_1 < 0$ and $\lambda_2 > 0$;
(ii) a stable node at the origin for $\det(\mathbf{A}) > 0$, $\text{tr}(\mathbf{A}) < 0$ and $\Delta > 0$ with $\lambda_1 < 0$ and $\lambda_2 < 0$;
(iii) an unstable node at the origin for $\det(\mathbf{A}) > 0$, $\text{tr}(\mathbf{A}) > 0$ and $\Delta > 0$ with $\lambda_1 > 0$ and $\lambda_2 > 0$;
(iv) a stable focus at the origin for $\det(\mathbf{A}) > 0$, $\text{tr}(\mathbf{A}) < 0$ and $\Delta < 0$ with $\lambda_{1,2} = \text{tr}(\mathbf{A}) \pm \mathbf{i}\sqrt{|\Delta|}$;
(v) an unstable focus at the origin for $\det(\mathbf{A}) > 0$, $\text{tr}(\mathbf{A}) > 0$ and $\Delta < 0$ with $\lambda_{1,2} = \text{tr}(\mathbf{A}) \pm \mathbf{i}\sqrt{|\Delta|}$;
(vi) a center at the origin for $\det(\mathbf{A}) > 0$ and $\text{tr}(\mathbf{A}) = 0$ with $\lambda_{1,2} = \pm\mathbf{i}\sqrt{|\Delta|}$;
(vii) a degenerate equilibrium point at the origin for $\det(\mathbf{A}) = 0$.

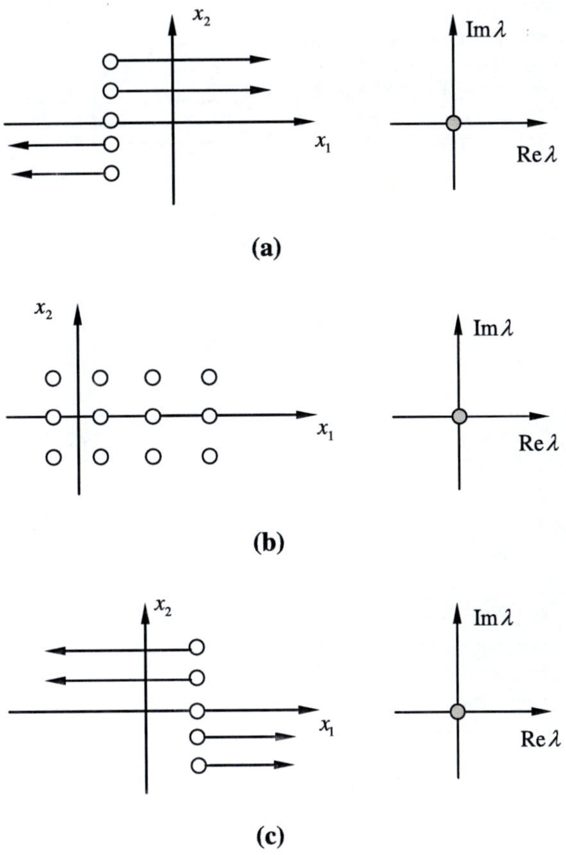

Fig. A.7 Phase portraits and eigenvalue diagram of an $(\varnothing:\varnothing:2|$ -critical case for $\det(\mathbf{A})=0$ and $\operatorname{tr}(\mathbf{A})=0$: **a** one-dimensional source $a_{12}>0$, **b** invariance ($a_{12}=0$) and **c** one dimensional source ($a_{12}<0$)

For the degenerate case, there are three cases

$$\mathbf{A}=\begin{bmatrix} a_{11} & 0 \\ 0 & 0 \end{bmatrix}, \mathbf{A}=\begin{bmatrix} 0 & a_{12} \\ 0 & 0 \end{bmatrix} \text{ and } \mathbf{A}=\begin{bmatrix} 0 & 0 \\ 0 & 0 \end{bmatrix} \tag{A.170}$$

and the corresponding solutions are

$$\mathbf{x}(t)=\begin{bmatrix} e^{a_{11}t} & 0 \\ 0 & 1 \end{bmatrix}\mathbf{x}_0, \mathbf{x}(t)=\begin{bmatrix} 1 & a_{12}t \\ 0 & 1 \end{bmatrix}\mathbf{x}_0 \text{ and } \mathbf{x}(t)=\begin{bmatrix} 1 & 0 \\ 0 & 1 \end{bmatrix}\mathbf{x}_0. \tag{A.171}$$

The phase portraits and eigenvalue diagrams for degenerate cases are presented in Figs. A.7 and A.8.

The summarization of stability and its boundary for the linear system in Eq. (A.159) are intuitively illustrated in Fig. A.9. through the complex plane of eigenvalue. The shaded area is for focus and center. The area above the shaded area is for unstable node, and the area below the shaded area is for stable node. The left area of the axis $\operatorname{tr}(\mathbf{A})$ is for saddle. The center is on the positive axis of $\det(\mathbf{A})$. The phase portrait is based on the transformed system in Eq. (A.161).

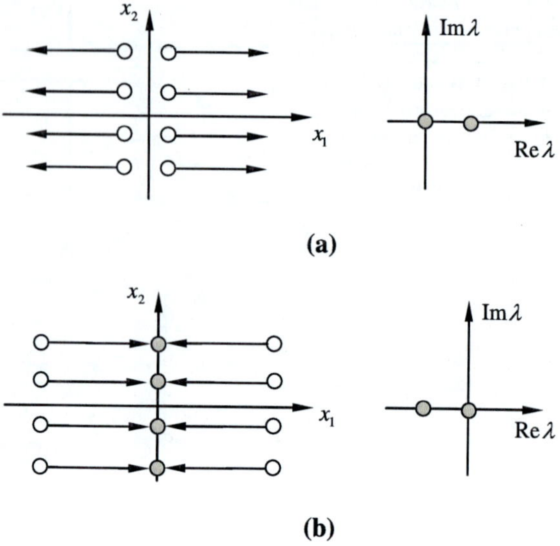

Fig. A.8 Solution and phase portraits for $\det(\mathbf{A}) = 0$: **a** $(\varnothing : 1 : 1|$ -one-dimensional source ($\mathrm{tr}(\mathbf{A}) > 0$), and **b** a $(1 : \varnothing : 1|$ -one-dimensional sink ($\mathrm{tr}(\mathbf{A}) < 0$)

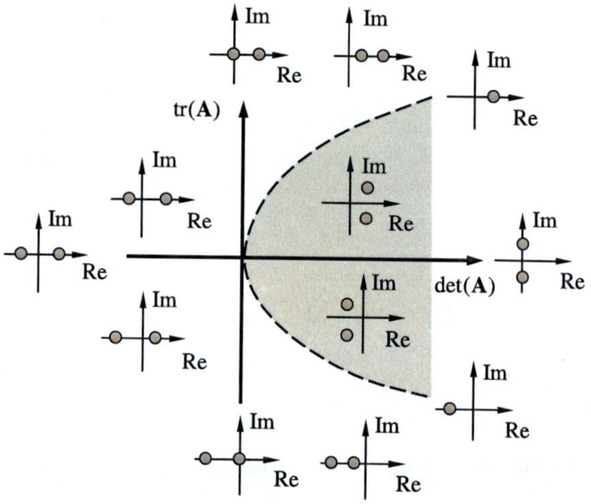

Fig. A.9 Stability and its boundary diagram through the complex plane of eigenvalues

The solutions of $\dot{\mathbf{x}} = \mathbf{A}\mathbf{x}$ in Eq. (A.159) is given by $\mathbf{x} = \mathbf{P}\mathbf{y}$. So the phase portrait of $\dot{\mathbf{x}} = \mathbf{A}\mathbf{x}$ can be obtained by the transform of $\mathbf{x} = \mathbf{P}\mathbf{y}$.

A.7.2 Three-Dimensional Dynamical Systems

Consider a three-dimensional linear system as

$$\dot{\mathbf{x}} = \mathbf{A}\mathbf{x} \tag{A.172}$$

with initial condition $\mathbf{x}(t_0) = \mathbf{x}_0$, and

$$\mathbf{A} = \begin{bmatrix} a_{11} & a_{12} & a_{13} \\ a_{21} & a_{22} & a_{23} \\ a_{31} & a_{32} & a_{33} \end{bmatrix}. \tag{A.173}$$

If $\det \mathbf{A} \neq 0$, $\mathbf{x} = \mathbf{0}$ is a unique equilibrium point. With a nonsingular transform matrix $\mathbf{P}$, $\mathbf{B} = \mathbf{P}^{-1}\mathbf{A}\mathbf{P}$. With $\mathbf{x} = \mathbf{P}\mathbf{y}$,

$$\dot{\mathbf{y}} = \mathbf{B}\mathbf{y} \tag{A.174}$$

where

$$\mathbf{B} = \begin{bmatrix} \lambda_1 & 0 & 0 \\ 0 & \lambda_2 & 0 \\ 0 & 0 & \lambda_3 \end{bmatrix}. \tag{A.175}$$

(A) If three real eigenvalues are different ($\lambda_1 \neq \lambda_2 \neq \lambda_3$), the solution is

$$\mathbf{B} = \begin{bmatrix} \lambda_1 & 0 & 0 \\ 0 & \lambda_2 & 0 \\ 0 & 0 & \lambda_3 \end{bmatrix} \text{and}\, \mathbf{y}(t) = \begin{bmatrix} e^{\lambda_1(t-t_0)} & 0 & 0 \\ 0 & e^{\lambda_2(t-t_0)} & 0 \\ 0 & 0 & e^{\lambda_3(t-t_0)} \end{bmatrix} \mathbf{y}_0. \tag{A.176}$$

The origin is called a node of the linear system if three real eigenvalues have the same sign. If $\lambda_k < 0$ $(k = 1, 2, 3)$, the origin is a stable node. If $\lambda_k > 0$ $(k = 1, 2, 3)$, the origin is an unstable node. The phase portraits and eigenvalue diagrams for the linear system with stable and unstable nodes at the origin are sketched in Fig. A.10a and b with one eighth view. All flows will come to the origin as the stable node. However, the flows in a linear system with an unstable node at the origin will leave away from the origin.

The origin is called a saddle-node of the linear system if three real eigenvalues have the different signs. If $\lambda_k < 0 (k = 1, 2)$ with $\lambda_3 > 0$, the origin is a saddle-node with two-directional attraction and one-directional expansion. If $\lambda_k > 0$ $(k = 1, 2)$ with $\lambda_3 < 0$, the origin is a saddle-node with one-directional attraction and two-directional expansion. The phase portraits and eigenvalue diagrams for the linear system with two saddle-nodes at the origin are sketched in Fig. A.11a and b with one-eighth view. The flows in the linear systems with saddle-nodes shrink in the attraction direction(s) and stretch in the expansion direction(s).

(B) For two repeated real eigenvalues ($\lambda_1 = \lambda_2 = \lambda$ and λ_3), the solutions are

$$\mathbf{B} = \begin{bmatrix} \lambda & 0 & 0 \\ 0 & \lambda & 0 \\ 0 & 0 & \lambda_3 \end{bmatrix} \text{and}\, \mathbf{y}(t) = \begin{bmatrix} e^{\lambda(t-t_0)} & 0 & 0 \\ 0 & e^{\lambda(t-t_0)} & 0 \\ 0 & 0 & e^{\lambda_3(t-t_0)} \end{bmatrix} \mathbf{y}_0. \tag{A.177}$$

Fig. A.10 One-eighth phase portrait and eigenvalue diagram: **a** a $(3:\varnothing:\varnothing|$ stable node (or sink) $\lambda_k<0$ $(k=1,2,3)$ and **b** a $(\varnothing:3:\varnothing|$ - unstable node (or a source) $\lambda_k>0$ $(k=1,2,3)$.

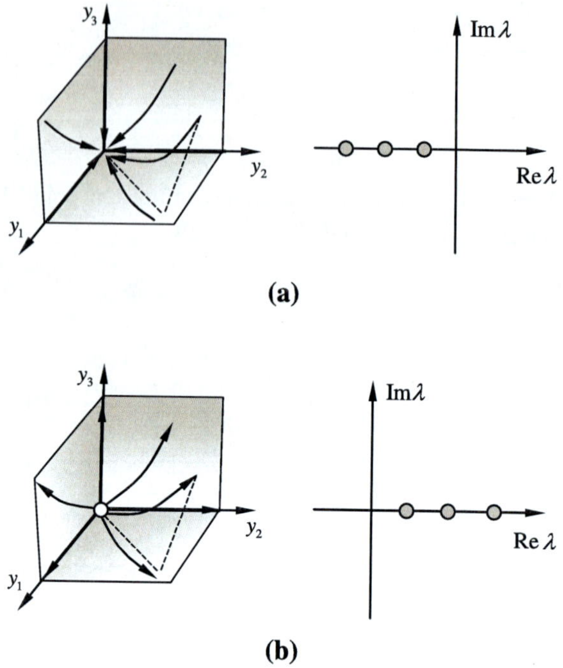

Fig. A.11 One-eighth phase portrait and eigenvalue diagrams:**a** a $(1:2:\varnothing|$ -saddle $(\lambda_k>0,\ k=1,2$ with $\lambda_3<0)$, and **b** a $(2:1:\varnothing|$ - saddle $(\lambda_k<0,\ k=1,2$ with$\lambda_3>0)$.

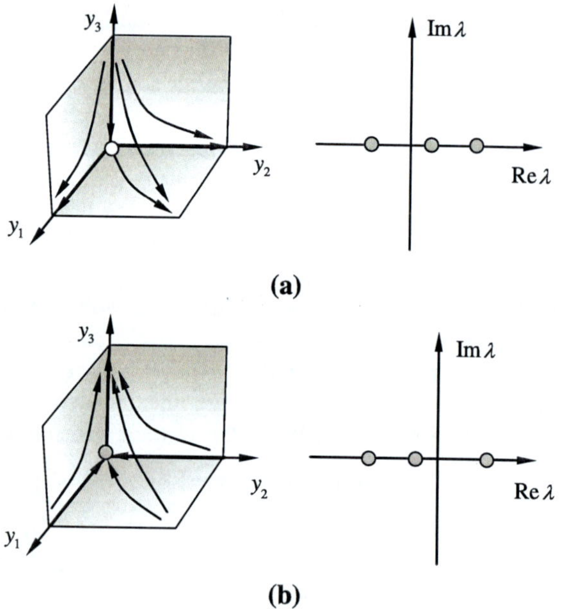

$$\mathbf{B} = \begin{bmatrix} \lambda & 1 & 0 \\ 0 & \lambda & 0 \\ 0 & 0 & \lambda_3 \end{bmatrix} \text{and}\, \mathbf{y}(t) = \begin{bmatrix} e^{\lambda(t-t_0)} & (t-t_0)e^{\lambda(t-t_0)} & 0 \\ 0 & e^{\lambda(t-t_0)} & 0 \\ 0 & 0 & e^{\lambda_3(t-t_0)} \end{bmatrix} \mathbf{y}_0. \quad (A.178)$$

The stability characteristics of Eq. (A.172) with two repeated real eigenvalues are similar to the case of three real distinct eigenvalues. The origin is a stable node (sink) with ($\lambda<0$ and $\lambda_3<0$), an unstable node (source) with ($\lambda>0$ and $\lambda_3>0$), and a saddle-node ($\lambda<0$ and $\lambda_3>0$ or $\lambda>0$ and $\lambda_3<0$) for the linear system. The phase portraits and eigenvalue diagram will not be presented.

(C) For two repeated real eigenvalues ($\lambda_1=\lambda_2=\lambda_3=\lambda$), the solutions are

$$\mathbf{B} = \begin{bmatrix} \lambda & 0 & 0 \\ 0 & \lambda & 0 \\ 0 & 0 & \lambda \end{bmatrix} \text{and}\, \mathbf{y}(t) = \begin{bmatrix} e^{\lambda(t-t_0)} & 0 & 0 \\ 0 & e^{\lambda(t-t_0)} & 0 \\ 0 & 0 & e^{\lambda(t-t_0)} \end{bmatrix} \mathbf{y}_0. \quad (A.179)$$

$$\mathbf{B} = \begin{bmatrix} \lambda & 1 & 0 \\ 0 & \lambda & 0 \\ 0 & 0 & \lambda \end{bmatrix} \text{and}\, \mathbf{y}(t) = \begin{bmatrix} e^{\lambda(t-t_0)} & (t-t_0)e^{\lambda(t-t_0)} & 0 \\ 0 & e^{\lambda(t-t_0)} & 0 \\ 0 & 0 & e^{\lambda(t-t_0)} \end{bmatrix} \mathbf{y}_0. \quad (A.180)$$

$$\mathbf{B} = \begin{bmatrix} \lambda & 1 & 0 \\ 0 & \lambda & 1 \\ 0 & 0 & \lambda \end{bmatrix} \text{ and } \mathbf{y}(t) = \begin{bmatrix} e^{\lambda(t-t_0)} & (t-t_0)e^{\lambda(t-t_0)} & \frac{1}{2}(t-t_0)^2 e^{\lambda(t-t_0)} \\ 0 & e^{\lambda(t-t_0)} & (t-t_0)e^{\lambda(t-t_0)} \\ 0 & 0 & e^{\lambda(t-t_0)} \end{bmatrix} \mathbf{y}_0. \quad (A.181)$$

The stability characteristics of Eq. (A.172) with three repeated real eigenvalues are similar to the case of three real distinct eigenvalues. The origin is a stable node (sink) with $\lambda<0$, an unstable node (source) with $\lambda>0$ for the linear system. The phase portraits and eigenvalue diagram will not be presented.

(D) For ($\lambda_{1,2}=\alpha\pm\mathbf{i}\beta$) and $\mathrm{Im}\lambda_3=0$, the solution is

$$\mathbf{B} = \begin{bmatrix} \alpha & \beta & 0 \\ -\beta & \alpha & 0 \\ 0 & 0 & \lambda_3 \end{bmatrix}, \text{and}$$

$$\mathbf{y}(t) = \begin{bmatrix} e^{\alpha(t-t_0)}\cos\beta(t-t_0) & e^{\alpha(t-t_0)}\sin\beta(t-t_0) & 0 \\ -e^{\alpha(t-t_0)}\sin\beta(t-t_0) & e^{\alpha(t-t_0)}\cos\beta(t-t_0) & 0 \\ 0 & 0 & e^{\lambda_3(t-t_0)} \end{bmatrix} \mathbf{y}_0. \quad (A.182)$$

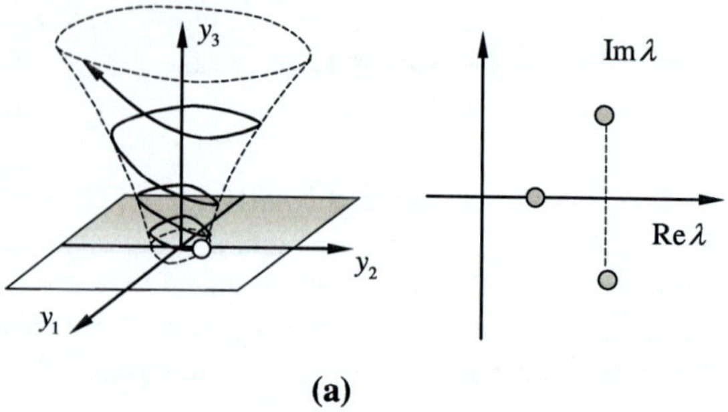

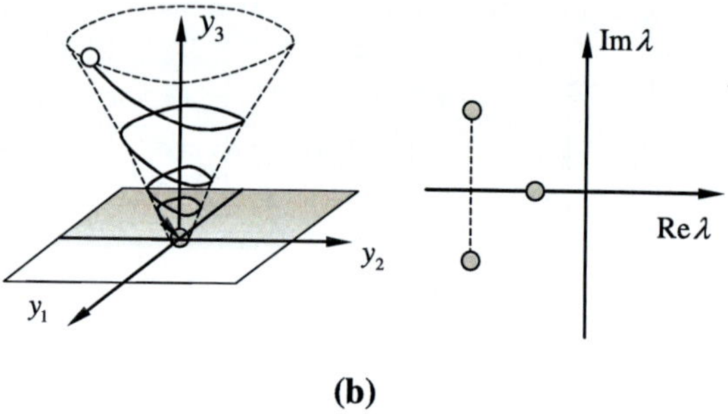

Fig. A.12 Positive half spiral flows and eigenvalue diagrams: **a** $(\varnothing : 1 : \varnothing|\varnothing : 1 : \varnothing)$ -spiral source $\mathrm{Re}\lambda_k > 0$ $(k = 1, 2, 3)$, and **b** $(1 : \varnothing : \varnothing|1 : \varnothing : \varnothing)$ -spiral sink $\mathrm{Re}\lambda_k < 0$ $(k = 1, 2, 3)$

The origin is called a spiral focus of the linear system if the real parts of three eigenvalues have the same sign. If $\mathrm{Re}\lambda_k < 0$ $(k = 1, 2, 3)$, the origin is called a stable spiral focus (or a spiral sink). If $\mathrm{Re}\lambda_k > 0$ $(k = 1, 2, 3)$, the origin is called an unstable spiral focus (or a spiral source or a tornado). The linear system with stable and unstable spiral focuses at the origin is sketched in Fig. A.12a and b with a half space view. The spiral flows and eigenvalue diagrams are presented. All flows with a spiral sink spirally come to the origin. However, the flows in linear system with a spiral source at the origin will spirally leave away from the origin like a tornado. The origin is called a spiral saddle with a spiral-exponential attraction and expansion of the linear system if the real parts of three eigenvalues have different signs. If $\mathrm{Re}\lambda_k = \alpha < 0$ $(k = 1, 2)$ with $\lambda_3 > 0$, the origin is a saddle of the first kind which has an $(\varnothing : 1 : \varnothing|1 : \varnothing : \varnothing)$ spiral attraction and an exponential expansion. If $\mathrm{Re}\lambda_k = \alpha > 0$ $(k = 1, 2)$ with $\lambda_3 < 0$, the origin is a saddle of the second kind which has a $(1 : \varnothing : \varnothing|\varnothing : 1 : \varnothing)$ spiral expansion with an exponential attraction and two-directional expansion. The flows and eigenvalue diagrams for the two cases of the linear system are sketched in Fig. A.13a and b, respectively.

(v) For $(\lambda_{1,2} = \pm \mathbf{i}\beta)$ and $\mathrm{Im}\,\lambda_3 = 0$, the solution is given by

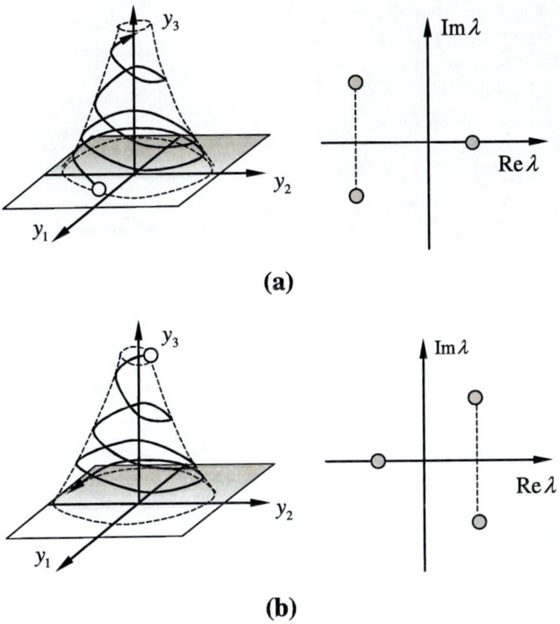

Fig. A.13 Positive half spiral saddle flows and eigenvalue diagrams: **a** an $(\varnothing:1:\varnothing|1:\varnothing:\varnothing)$ -spiral attraction and exponential expansion ($\mathrm{Re}\lambda_k = \alpha<0$ $(k=1,2)$ with $\lambda_3>0$) and **b** a $(1:\varnothing:\varnothing|\varnothing:1:\varnothing)$ -spiral expansion and exponential attraction ($\mathrm{Re}\lambda_k = \alpha<0$ $(k=1,2)$ with $\lambda_3<0$).

$$\mathbf{B} = \begin{bmatrix} 0 & \beta & 0 \\ -\beta & 0 & 0 \\ 0 & 0 & \lambda_3 \end{bmatrix}, \text{ and}$$
$$\mathbf{y}(t) = \begin{bmatrix} \cos\beta(t-t_0) & \sin\beta(t-t_0) & 0 \\ -\sin\beta(t-t_0) & \cos\beta(t-t_0) & 0 \\ 0 & 0 & e^{\lambda_3(t-t_0)} \end{bmatrix} \mathbf{y}_0. \tag{A.183}$$

The origin is called a cylindrical spiral of the linear system if $\mathrm{Re}\lambda_k = \alpha = 0$. If $\lambda_k = \pm\mathbf{i}\beta$ $(k=1,2)$ with $\lambda_3>0$, the origin is a center of an unstable cylindrical spiral. If $\lambda_k = \pm\mathbf{i}\beta$ $(k=1,2)$ with $\lambda_3<0$, the origin is a center of a stable cylindrical spiral. The flows and eigenvalue diagrams for the two special cases of the linear system are sketched in Fig. A.14**a** and **b**.

The eigenvalues of **A** in Eq. (A.172) are determined by $\det(\mathbf{A}-\lambda\mathbf{I})=0$, i.e.,

$$\lambda^3 + I_1\lambda^2 + I_2\lambda + I_3 = 0 \tag{A.184}$$

where

$$\begin{aligned} I_1 &= \mathrm{tr}(\mathbf{A}) = a_{11}+a_{22}+a_{33}, \\ I_2 &= a_{11}a_{22}+a_{22}a_{33}+a_{33}a_{11}-a_{12}a_{21}-a_{23}a_{32}-a_{13}a_{31}, \\ I_3 &= \det(\mathbf{A}) \\ &= a_{11}a_{22}a_{33}+a_{12}a_{23}a_{31}+a_{13}a_{32}a_{21}-a_{11}a_{32}a_{23}-a_{22}a_{13}a_{31}-a_{33}a_{12}a_{21}. \end{aligned} \tag{A.185}$$

Fig. A.14 Positive cylindrically spiral flows and eigenvalue diagram($\lambda_k = \pm \mathbf{i}\beta$, $k = 1, 2$) : **a** a $(1 : \varnothing : \varnothing | \varnothing : \varnothing : 1)$ -cylindrically sink flow ($\lambda_3 > 0$), and **b** a $(\varnothing : 1 : \varnothing | \varnothing : \varnothing : 1)$ cylindrically spiral source flow ($\lambda_3 < 0$).

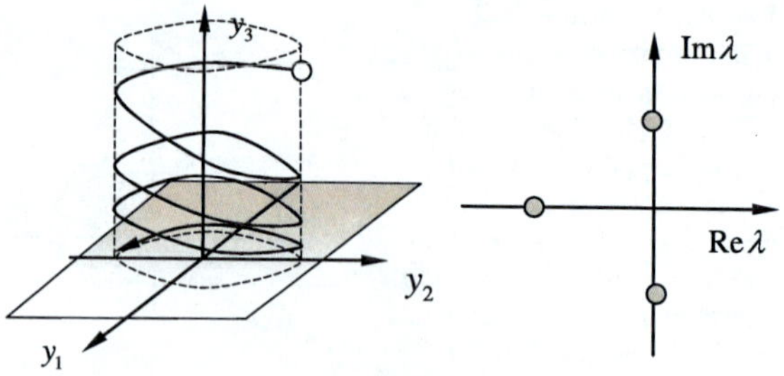

(a)

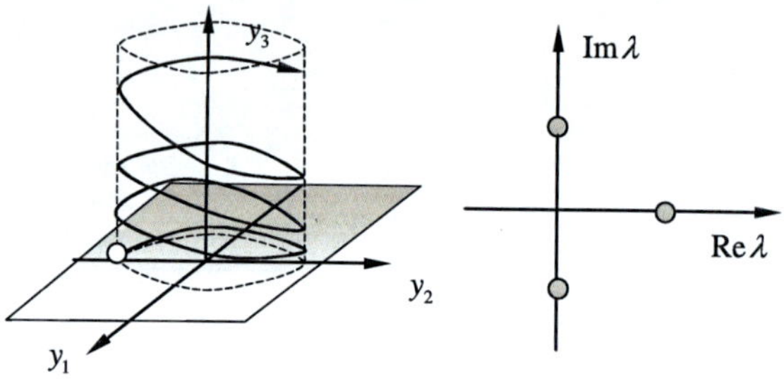

(b)

The corresponding eigenvalues are

$$
\begin{aligned}
\lambda_1 &= \sqrt[3]{\Delta_1} + \sqrt[3]{\Delta_2},\\
\lambda_2 &= \omega_1 \sqrt[3]{\Delta_1} + \omega_2 \sqrt[3]{\Delta_2} \text{ and}\\
\lambda_3 &= \omega_2 \sqrt[3]{\Delta_1} + \omega_1 \sqrt[3]{\Delta_2}
\end{aligned}
\tag{A.186}
$$

where

$$
\begin{aligned}
\omega_1 &= \frac{-1 + \mathbf{i}\sqrt{3}}{2} \text{ and } \omega_2 = \frac{-1 - \mathbf{i}\sqrt{3}}{2};\\
\Delta_{1,2} &= -\frac{q}{2} \pm \sqrt{\Delta} \text{ and } \Delta = (\frac{q}{2})^2 + (\frac{p}{3})^3;\\
p &= I_2 - \frac{1}{3} I_1^2 \text{ and } q = I_3 - \frac{1}{3} I_1 I_2 + \frac{2}{27} I_1^3.
\end{aligned}
\tag{A.187}
$$

The linear system in Eq. (A.172) possesses the following characteristics

(i) For $\Delta > 0$, the matrix $\mathbf{A}$ has one real eigenvalue and a pair of complex eigenvalues. The spiral sink, spiral source, and spiral saddle exist at the origin, i.e., $(1 : \varnothing : \varnothing | 1 : \varnothing : \varnothing)$, $(\varnothing : 1 : \varnothing | \varnothing : 1 : \varnothing)$, $(\varnothing : 1 : \varnothing | 1 : \varnothing : \varnothing)$ and $(1 : \varnothing : \varnothing | \varnothing : 1 : \varnothing)$.

(ii) For $\Delta = 0$ and $p = q = 0$, the matrix $\mathbf{A}$ has three repeated eigenvalues. Stable and unstable nodes exist at the origin.

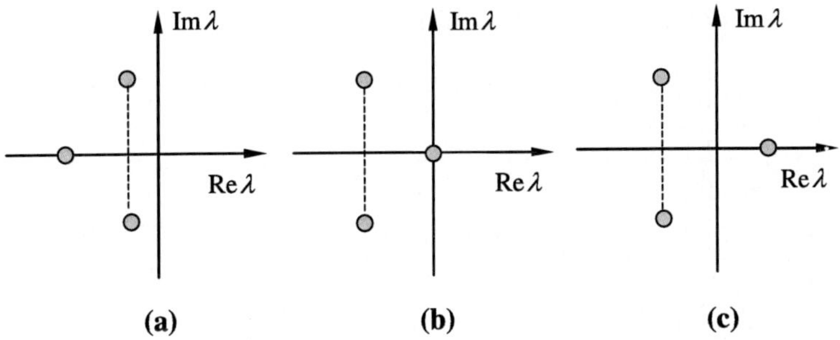

Fig. A.15 Eigenvalue diagrams ($\text{Re}\lambda_k < 0$ and $\text{Im}\lambda_k \neq 0, k = 1, 2$) : **a** an $(1 : \varnothing : \varnothing | 1 : \varnothing : \varnothing)$ -stable spiral sink ($\lambda_3 < 0$), **b** an $(\varnothing : \varnothing : 1 | 1 : \varnothing : \varnothing)$ stable focus in $(\mathbf{u}_1, \mathbf{v}_1)$ -plane (or a boundary of spiral sink and spiral saddle) ($\lambda_3 = 0$), and **c** an $(\varnothing : 1 : \varnothing | 1 : \varnothing : \varnothing)$ -spiral saddle (a spiral saddle of the first kind) ($\lambda_3 > 0$)

(iii) For $\Delta = 0$ and $q^2 = -4p^3/27 \neq 0$, the matrix $\mathbf{A}$ has two repeated real eigenvalues. Sink, source and saddle-node exist at the origin.

(iv) For $\Delta < 0$, the matrix $\mathbf{A}$ has three different real eigenvales. There are six cases at the origin: $(3 : \varnothing : \varnothing|$ -stable node (sink); $(\varnothing : 3 : \varnothing|$ -unstable node (source); $(2 : 1 : \varnothing|$ and $(1 : 2 : \varnothing|$ -saddles; the $(2 : \varnothing : 1|$ boundary of $(2 : 1 : \varnothing|$ -saddle with $(3 : \varnothing : \varnothing|$ -stable node; the $(\varnothing : 2 : 1|$ boundary of $(1 : 2 : \varnothing|$ -saddle with $(\varnothing : 3 : \varnothing|$ -unstable node; and the $(1 : 1 : 1|$ boundary $(2 : 1 : \varnothing|$ -saddle with $(1 : 2 : \varnothing|$ -saddle.

(v) A degenerate equilibrium point is at the origin for $\det(\mathbf{A}) = 0$.

One of eigenvalues is zero, which is a degenerated case. The total degenerate cases are given as follows. For the 1-dimensional degenerate case, there are four basic cases.

$$\mathbf{B} = \begin{bmatrix} \alpha & \beta & 0 \\ -\beta & \alpha & 0 \\ 0 & 0 & 0 \end{bmatrix}, \text{ and } \mathbf{B} = \begin{bmatrix} \lambda_1 & 0 & 0 \\ 0 & \lambda_2 & 0 \\ 0 & 0 & 0 \end{bmatrix}. \tag{A.188}$$

$$\mathbf{B} = \begin{bmatrix} \lambda & 0 & 0 \\ 0 & \lambda & 0 \\ 0 & 0 & 0 \end{bmatrix}, \text{ and } \mathbf{B} = \begin{bmatrix} \lambda & b_{12} & 0 \\ 0 & \lambda & 0 \\ 0 & 0 & 0 \end{bmatrix}. \tag{A.189}$$

For the two-dimensional degenerate case, there are six basic cases.

$$\mathbf{A} = \begin{bmatrix} a_{11} & 0 & 0 \\ 0 & 0 & 0 \\ 0 & 0 & 0 \end{bmatrix}, \mathbf{A} = \begin{bmatrix} a_{11} & a_{12} & 0 \\ 0 & 0 & 0 \\ 0 & 0 & 0 \end{bmatrix} \text{ and } \mathbf{A} = \begin{bmatrix} a_{11} & 0 & 0 \\ 0 & 0 & a_{23} \\ 0 & 0 & 0 \end{bmatrix}. \tag{A.190}$$

$$\mathbf{A} = \begin{bmatrix} a_{11} & a_{12} & 0 \\ 0 & 0 & a_{23} \\ 0 & 0 & 0 \end{bmatrix}, \mathbf{A} = \begin{bmatrix} a_{11} & 0 & a_{13} \\ 0 & 0 & a_{23} \\ 0 & 0 & 0 \end{bmatrix} \text{ and } \mathbf{A} = \begin{bmatrix} a_{11} & a_{12} & a_{13} \\ 0 & 0 & a_{23} \\ 0 & 0 & 0 \end{bmatrix}. \tag{A.191}$$

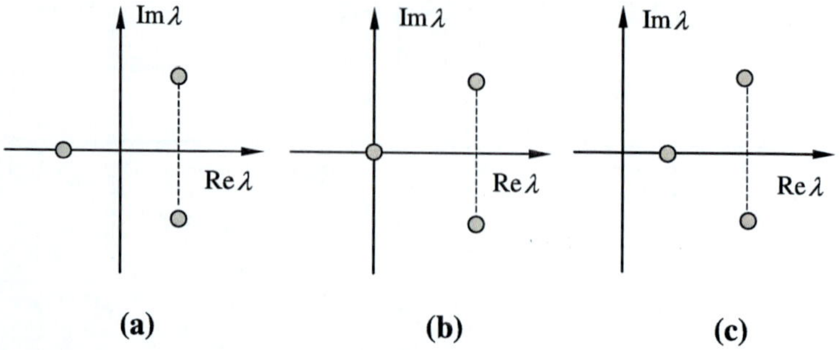

Fig. A.16 Eigenvalue diagrams ($\mathrm{Re}\lambda_k > 0$ and $\mathrm{Im}\lambda_k \neq 0, k = 1,2$) : **a** $(1:\varnothing:\varnothing|\varnothing:1:\varnothing)$ - spiral saddle (or spiral saddle of the second) ($\lambda_3 < 0$), **b** an $(\varnothing:\varnothing:1|\varnothing:1:\varnothing)$ - unstable focus in $(\mathbf{u}_1, \mathbf{v}_1)$ -plane (or a boundary of spiral source and spiral saddle) ($\lambda_3 = 0$) and **c** $(\varnothing:1:\varnothing|\varnothing:1:\varnothing)$ -unstable node (or spiral source) ($\lambda_3 > 0$)

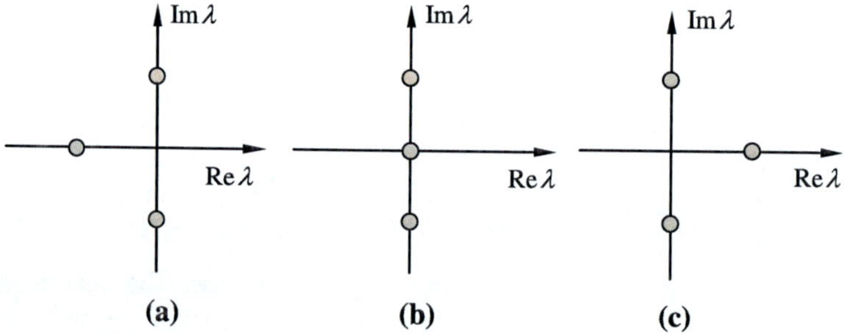

Fig. A.17 Eigenvalue diagrams ($\mathrm{Re}\lambda_k = 0$ and$\mathrm{Im}\lambda_k \neq 0, k = 1,2$) : **a** an $(1:\varnothing:\varnothing|\varnothing:\varnothing:1)$ stable cylindrical spiral ($\lambda_3 < 0$), **b** an $(\varnothing:\varnothing:1|\varnothing:\varnothing:1)$ -center in$(\mathbf{u}_1, \mathbf{v}_1)$ -plane ($\lambda_3 = 0$), and **c** an $(\varnothing:1:\varnothing|\varnothing:\varnothing:1)$ unstable cylindrical spiral ($\lambda_3 > 0$)

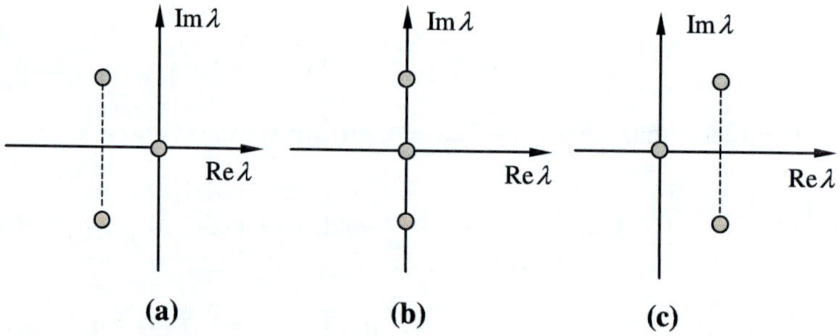

Fig. A.18 Eigenvalue diagrams for degenerate cases ($\lambda_3 = 0$ and $\mathrm{Im}\lambda_k \neq 0, k = 1,2$): **a** $(\varnothing:\varnothing:1|1:\varnothing:\varnothing)$ stable focus (or spiral sink) in plane $(\mathbf{u}_1, \mathbf{v}_1)$ ($\mathrm{Re}\lambda_k < 0$), **b** a$(\varnothing:\varnothing:1|\varnothing:\varnothing:1)$ boundary of sink and source in plane $(\mathbf{u}_1, \mathbf{v}_1)$) ($\mathrm{Re}\lambda_k = 0$) and **c** $(\varnothing:\varnothing:1|\varnothing:1:\varnothing)$ -spiral source in plane $(\mathbf{u}_1, \mathbf{v}_1)$ ($\mathrm{Re}\lambda_k > 0$)

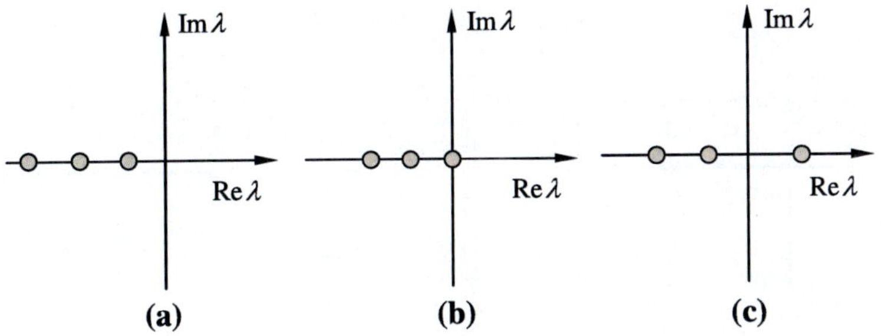

Fig. A.19 Eigenvalue diagrams: **a** $(3:\varnothing:\varnothing|$ -node (or stable node or sink) $(\lambda_k<0,\ k=1,2,3)$, **b** $(2:\varnothing:1|$ saddle-node (or a boundary of saddle-stable node) $(\lambda_k<0,\ k=1,2$ and $\lambda_3=0)$ and **c** $(2:1:\varnothing|$ -saddle (or a saddle of the first kind) $(\lambda_k<0,\ k=1,2$ and $\lambda_3>0)$.

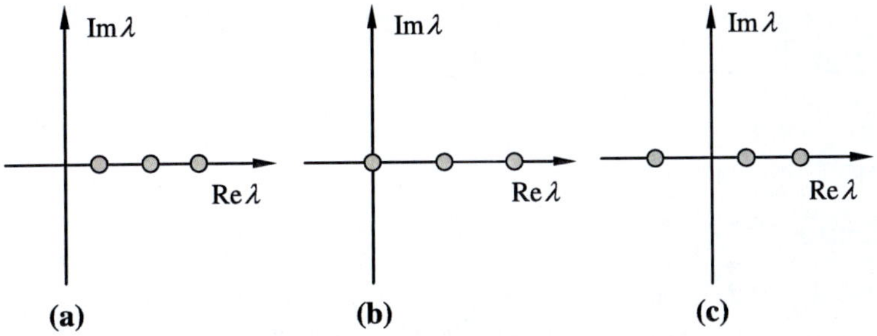

Fig. A.20 Eigenvalue diagrams: **a** $(\varnothing:3:\varnothing|$ -unstable node (source) $(\lambda_k>0,\ k=1,2,3)$, **b** $(\varnothing:2:1|$ boundary of $(1:2:\varnothing|$ saddle-$(\varnothing:3:\varnothing|$ unstable node $(\lambda_k>0,\ k=1,2$ and $\lambda_3=0)$ and **c** $(1:2:\varnothing|$ -saddle (or a saddle of the second kind) $(\lambda_k>0,\ k=1,2$ and $\lambda_3<0)$

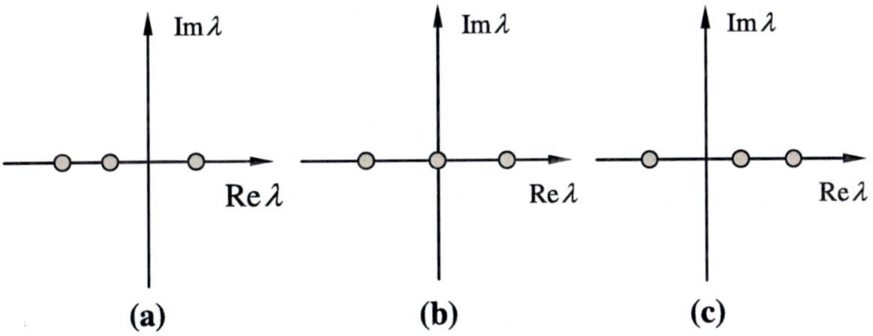

Fig. A.21 Eigenvalue diagram: **a** a $(2:1:\varnothing|$ -saddle (or a saddle of the first kind) $(\lambda_k<0,\ k=1,2$ and $\lambda_3>0)$. **b** a $(1:1:1|$ boundary of the $(2:1:\varnothing|$ saddle to $(1:2:\varnothing|$ saddle $(\lambda_1>0, \lambda_2=0$ and $\lambda_3<0)$, and **c** a $(1:2:\varnothing|$ -saddle (or saddle of the second kind) $(\lambda_k>0,\ k=1,2$ and $\lambda_3<0)$

For the three-dimensional degenerate case, there are six basic cases.

$$\mathbf{A}=\begin{bmatrix}0&0&0\\0&0&0\\0&0&0\end{bmatrix}, \mathbf{A}=\begin{bmatrix}0&0&0\\0&0&a_{23}\\0&0&0\end{bmatrix}, \text{ and } \mathbf{A}=\begin{bmatrix}0&a_{12}&0\\0&0&a_{23}\\0&0&0\end{bmatrix}. \tag{A.192}$$

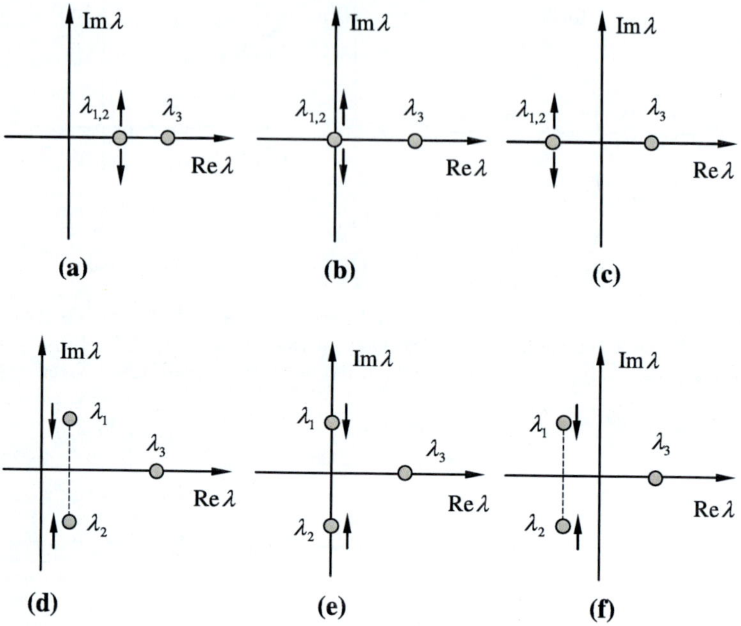

Fig. A.22 Eigenvalue diagrams ($\lambda_3 > 0$): **a** $(\varnothing : 3 : \varnothing|$ -unstable node (source) ($\lambda_k = \lambda > 0,\ k = 1,2$), **b** an $(\varnothing : 1 : 2|$ boundary of the $(2 : 1 : \varnothing|$ -saddle to $(\varnothing : 3 : \varnothing|$ -unstable node ($\lambda_k = \lambda = 0,\ k = 1,2$) and **c** $(2 : 1 : \varnothing|$ -saddle (or a saddle of the first kind) ($\lambda_k = \lambda < 0,\ k = 1,2$); The reduction froms: **d** an $(\varnothing : 1 : \varnothing|\varnothing : 1 : \varnothing)$ -unstable spiral node (spiral source) ($\lambda_k = \lambda > 0, k = 1,2$), **e** an $(\varnothing : 1 : \varnothing|\ \ \varnothing : \varnothing : 1)$ boundary of the $(1 : \varnothing : \varnothing|\varnothing : 1 : \varnothing)$ -spiral saddle to $(\varnothing : 1 : \varnothing|\varnothing : 1 : \varnothing)$ -spiral node ($\lambda_k = \lambda = 0,\ k = 1,2$), and **f** an $(\varnothing : 1 : \varnothing|1 : \varnothing : \varnothing)$ -spiral saddle ($\lambda_k = \lambda < 0,\ k = 1,2$)

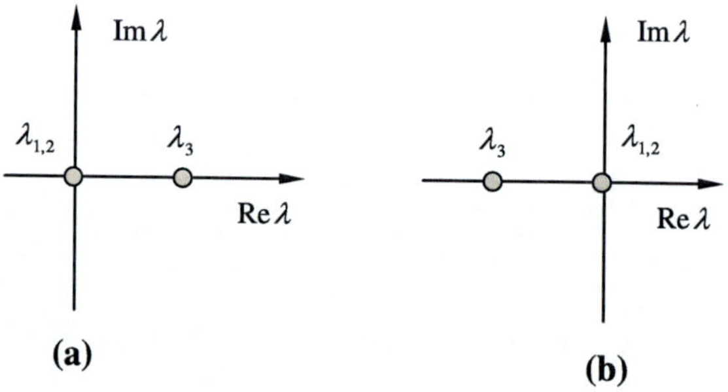

Fig. A.23 Eigenvalue diagrams ($\lambda_k = \lambda = 0, k = 1,2$): **a** an $(\varnothing : 1 : [2;1]|$ -one-dimensional unstable source flows ($\lambda_3 > 0$), **b** a $(1 : \varnothing : [2;1]|$ -one-dimensional stable sink flows ($\lambda_3 < 0$)

$$\mathbf{A} = \begin{bmatrix} 0 & a_{12} & a_{13} \\ 0 & 0 & a_{23} \\ 0 & 0 & 0 \end{bmatrix}, \mathbf{A} = \begin{bmatrix} 0 & 0 & a_{13} \\ 0 & 0 & a_{23} \\ 0 & 0 & 0 \end{bmatrix}, \text{ and } \mathbf{A} = \begin{bmatrix} 0 & a_{12} & a_{23} \\ 0 & 0 & 0 \\ 0 & 0 & 0 \end{bmatrix}. \quad \text{(A.193)}$$

The eigenvalue diagrams are presented in Figs. A.15–A.26 for all possible flows and boundaries. The distinct eigenvalue diagrams are sketched in Figs. A.15–A.21. The repeated eigenvalues are presented in Figs. A.22–A.26.

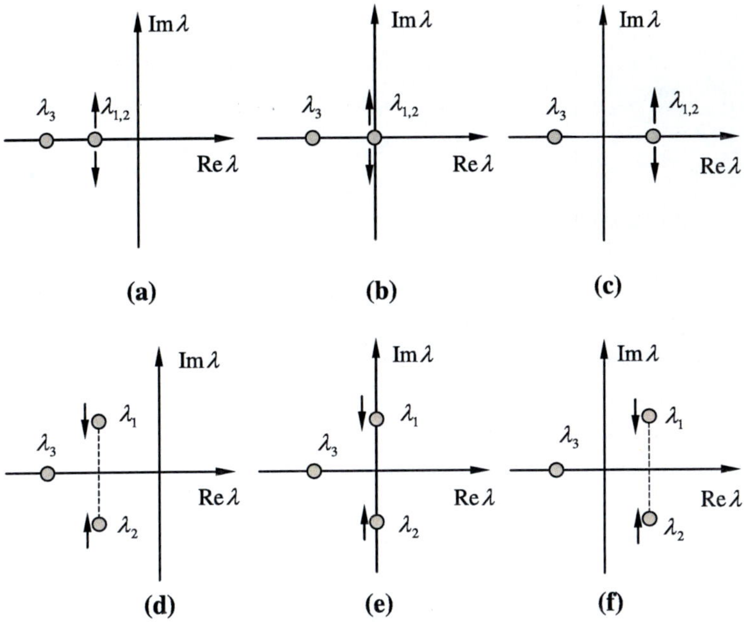

Fig. A.24 Eigenvalue diagrams ($\lambda_3 < 0$): **a** a $(3:\varnothing:\varnothing|$ -stable node (sink) ($\lambda_k = \lambda < 0$, $k = 1,2$), **b** a $(1:\varnothing:2|$ boundary of $(1:2:\varnothing|$ -saddle to $(3:\varnothing:\varnothing|$ -node , or the saddle-node of the second kind) ($\lambda_k = \lambda = 0$, $k = 1,2$) and **c** $(1:2:\varnothing|$ -saddle (or a saddle of the second kind) ($\lambda_k = \lambda > 0$, $k = 1,2$); **d** a $(1:\varnothing:\varnothing|1:\varnothing:\varnothing)$ -node (stable spiral node or spiral sink) ($\mathrm{Re}\lambda_k < 0, k = 1,2$), **e** a $(1:\varnothing:\varnothing|\varnothing:\varnothing;1)$ -boundary of $(1:\varnothing:\varnothing|1:\varnothing:\varnothing)$ -saddle to $(1:\varnothing:\varnothing|\varnothing:\varnothing:1)$ -saddle, or the spiral saddle-spiral node of the second kind) ($\mathrm{Re}\lambda_k = 0$, $k = 1,2$), and **f** a $(1:\varnothing:\varnothing|\varnothing:1:\varnothing)$ -saddle (or a spiral saddle of the second kind) ($\mathrm{Re}\lambda_k > 0$, $k = 1,2$).

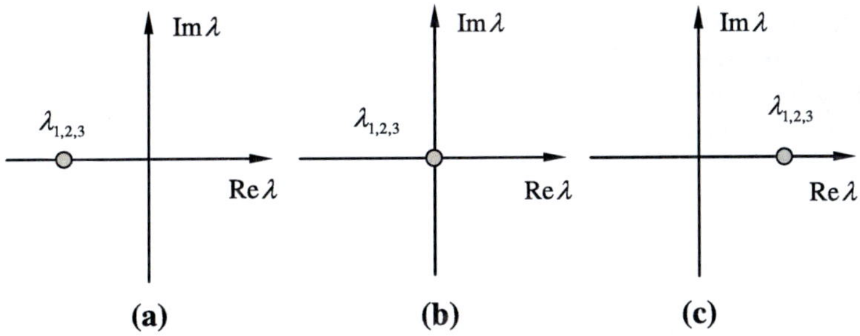

Fig. A.25 Three repeated eigenvalue diagrams: **a** a $(3:\varnothing:\varnothing|$ -stable node ($\lambda_{1,2,3} = \lambda < 0$), **b** an $(\varnothing:\varnothing:[3;2]|$ boundary of the $(3:\varnothing:\varnothing|$ -stable node to $(\varnothing:3:\varnothing|$ - unstable node ($\lambda_{1,2,3} = \lambda = 0$) and **c** an $(\varnothing:3:\varnothing|$ -unstable node ($\lambda_{1,2,3} = \lambda > 0$)

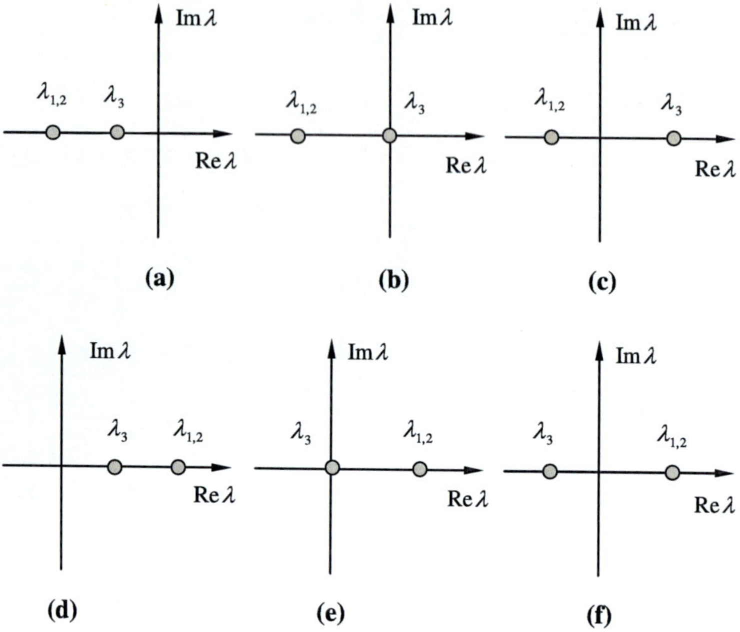

Fig. A.26 Eigenvalue diagram ($\lambda_k = \lambda < 0,\ k = 1, 2$): **a** a $(3 : \varnothing : \varnothing|$ -stable node ($\lambda_3 > 0$), **b** a $(2 : \varnothing : 1|$ boundary of $(2 : 1 : \varnothing|$ -saddle to $(3 : \varnothing : \varnothing|$ -node ($\lambda_3 = 0$), and **c** a $(2 : 1 : \varnothing|$ -saddle ($\lambda_3 < 0$). Eigenvalue diagram ($\lambda_k = \lambda > 0,\ k = 1, 2$) : **d** an $(\varnothing : 3 : \varnothing|$ -node ($\lambda_3 > 0$), **e** an $(\varnothing : 2 : 1|$ boundary of $(1 : 2 : \varnothing|$ -saddle to $(\varnothing : 3 : \varnothing|$ -node ($\lambda_3 = 0$), and **f** a $(1 : 2 : \varnothing|$ -saddle ($\lambda_3 < 0$)

It is very difficult to illustrate multiplicity of eigenvalues for nilpotent matrix $\mathbf{N}^{m+1} = 0\ (m > 1)$. $\varnothing$ can be used "0", and herein $\varnothing$ represents "no existing". For degenerate cases, most of cases cannot be illustrated through eigenvalue diagrams. For more detail discussion, the textbook on linear dynamical systems can be referred to Coddington and Levinson (1955), Hirsch et al (2004), Perko (1991) and Verhulst (1996).

References

Coddington E.A., Levinson N. (1955) Theory of Ordinary Differential Equations. New York: McGraw-Hill.

Hirsch M.W., Smale S., Devaney R.L. (2004) Differential Equations, Dynamical Systems, and An Introduction to Chaos. Amsterdam: Elsevier.

Perko L. (1991) Differential Equations and Dynamical Systems. New York: Springer-Verlag.

Verhulst F. (1996) Nonlinear Differential Equations and Dynamical Systems. Berlin: Springer.

Appendix B
Linear Discrete Dynamical Systems

In this Appendix, the theory of linear discrete dynamical systems will be presented. The basic iterative solution for linear discrete systems will be presented first. The iterative solutions based on distinct and repeated eigenvalues will be discussed. The stability theory for linear discrete dynamical systems will be presented. Compared to linear continuous systems, the stability of linear discrete dynamical systems will be discussed from the oscillatory, monotonic and spiral convergence and divergence. The invariant, flip and circular critical boundaries of the stability in the specific eigenvector will be classified. The lower-dimensional linear discrete systems will be discussed.

B.1 Basic Iterative Solutions

Basic concepts of discrete dynamical systems will be presented herein before further discussion on the iterative solutions and stability.

Definition B.1 Consider a discrete linear dynamical system based on a linear map $P : \mathbf{x}_k \to \mathbf{x}_{k+1}$ with the corresponding relation

$$\mathbf{x}_{k+1} = \mathbf{A}\mathbf{x}_k + \mathbf{B} \text{ for } k \in \mathbb{Z} \text{ and } \mathbf{x}_k = (x_{1k}, x_{2k}, \ldots, x_{nk})^{\mathrm{T}} \in \mathscr{R}^n \tag{B.1}$$

where $\mathbf{A}$ is an $n \times n$ matrix and $\mathbf{B}$ is a constant vector function. If $\mathbf{B} = \mathbf{0}$, the discrete linear dynamical system in Eq. (B.1) is homogeneous. Equation (B.1) becomes

$$\mathbf{x}_{k+1} = \mathbf{A}\mathbf{x}_k \text{ for } k \in \mathrm{Z} \text{ and } \mathbf{x}_k \in \mathscr{R}^n \tag{B.2}$$

which is called a homogeneous linear discrete system. If $\mathbf{B} \neq \mathbf{0}$, the linear discrete dynamical system in Eq. (B.1) is nonhomogeneous, and the corresponding linear discrete system is an nonhomogeneous, linear, discrete system.

A. C. J. Luo, *Regularity and Complexity in Dynamical Systems*,
DOI: 10.1007/978-1-4614-1524-4,

Consider $P^j : \mathbf{x}_k \to \mathbf{x}_{k+j}$ with $P^j = P \circ P^{j-1}$ and $P^0 = \mathbf{I}$. With Eq. (B.2), the final state $\mathbf{x}_{k+j}$ of mapping P^j is given by

$$\mathbf{x}_{k+j} = \mathbf{A}\mathbf{x}_{k+j-1} = \ldots = \mathbf{A}^j\mathbf{x}_k. \tag{B.3}$$

For Eq. (B.1), the final state $\mathbf{x}_{k+j}$ of mapping P^j is given by

$$\mathbf{x}_{k+j} = \mathbf{A}\mathbf{x}_{k+j-1} + \mathbf{B} = \mathbf{A}(\mathbf{A}\mathbf{x}_{k+j-2} + \mathbf{B}) + \mathbf{B} = \mathbf{A}^j\mathbf{x}_0 + \sum\nolimits_{m=0}^{j-1}\mathbf{A}^m\mathbf{B}. \tag{B.4}$$

If $\det(\mathbf{I} - \mathbf{A}) \neq 0$, one obtains

$$\sum\nolimits_{m=0}^{j-1}\mathbf{A}^m(\mathbf{I} - \mathbf{A}) = \mathbf{I} - \mathbf{A}^j \Rightarrow \sum\nolimits_{j=0}^{j-1}\mathbf{A}^m = (\mathbf{I} - \mathbf{A}^j)(\mathbf{I} - \mathbf{A})^{-1}. \tag{B.5}$$

Thus, the final state $\mathbf{x}_{k+j}$ of mapping P^j is given by

$$\mathbf{x}_{k+j} = \mathbf{A}^j\mathbf{x}_k + (\mathbf{I} - \mathbf{A}^j)(\mathbf{I} - \mathbf{A})^{-1}\mathbf{B}. \tag{B.6}$$

Definition B.2 Consider a discrete linear dynamical system based on a linear map $P : \mathbf{x}_k \to \mathbf{x}_{k+1}$ with $\mathbf{x}_{k+1} = \mathbf{A}\mathbf{x}_k + \mathbf{B}$ in Eq. (B.1). If $\mathbf{x}_{k+1} = \mathbf{x}_k = \mathbf{x}_k^*$, then the point $\mathbf{x}_k^*$ is called the fixed point (or period-1 solution) which is determined by

$$\mathbf{x}_k^* = \mathbf{A}\mathbf{x}_k^* + \mathbf{B} \tag{B.7}$$

For map $P^j : \mathbf{x}_k \to \mathbf{x}_{k+j}$, if $\mathbf{x}_{k+j} = \mathbf{x}_k = \mathbf{x}_k^*$, then the point $\mathbf{x}_k^*$ is called the period-j solution which is determined by

$$\begin{aligned} \mathbf{x}_{k+1}^* &= \mathbf{A}\mathbf{x}_k^* + \mathbf{B}, \\ \mathbf{x}_{k+2}^* &= \mathbf{A}\mathbf{x}_{k+1}^* + \mathbf{B}, \\ &\vdots \\ \mathbf{x}_{k+j}^* &= \mathbf{A}\mathbf{x}_{k+j-1}^* + \mathbf{B} = \mathbf{x}_k^*. \end{aligned} \tag{B.8}$$

From the definition, the unique fixed point in Eq. (B.7) is given by

$$\mathbf{x}_k^* = (\mathbf{I} - \mathbf{A})^{-1}\mathbf{B} \text{ if } \det(\mathbf{I} - \mathbf{A}) \neq 0. \tag{B.9}$$

(i) If $\mathbf{B} \neq \mathbf{0}$ and $\det(\mathbf{I} - \mathbf{A}) = 0$, then the fixed point $\mathbf{x}_k^*$ is no solution.
(ii) If $\mathbf{B} = \mathbf{0}$ and $\det(\mathbf{I} - \mathbf{A}) \neq 0$, then the fixed point $\mathbf{x}_k^* = \mathbf{0}$ is unique.
(iii) If $\mathbf{B} = \mathbf{0}$ and $\det(\mathbf{I} - \mathbf{A}) = 0$, then the fixed point $\mathbf{x}_k^*$ is uncertain.

Equation (B.6) with $\mathbf{x}_{k+j} = \mathbf{x}_k = \mathbf{x}_k^*$ gives

$$\mathbf{x}_k^* = (\mathbf{I} - \mathbf{A})^{-1}\mathbf{B} \text{ if } \det(\mathbf{I} - \mathbf{A}^j) \neq 0. \tag{B.10}$$

B.2 Linear Discrete Systems with Distinct Eigenvalues

In this section, the solutions for linear discrete dynamical systems with distinct eigenvalues will be presented.

Definition B.3 For a linear dynamical system of $\mathbf{x}_{k+1} = \mathbf{A}\mathbf{x}_k$ in Eq. (B.2), if the linear matrix $\mathbf{A} = diag[\lambda_1, \lambda_2, \ldots, \lambda_n]$ is a diagonal matrix, then the linear discrete dynamical system in Eq. (B.2) is called an uncoupled, linear, discrete homogenous system. With an initial state of $\mathbf{x}_0$, the solution of the uncoupled linear homogenous system is

$$\mathbf{x}_k = diag[\lambda_1^k, \lambda_2^k, \ldots, \lambda_n^k]\mathbf{x}_0. \tag{B.11}$$

Theorem B.1 *Consider a linear dynamical system $\mathbf{x}_{k+1} = \mathbf{A}\mathbf{x}_k$ in Eq. (B.2) with the initial state of $\mathbf{x}_k$. If the real distinct eigenvalues of the $n \times n$ matrix $\mathbf{A}$ are $\lambda_1, \lambda_2, \ldots, \lambda_n$, then a set of eigenvectors $\{\mathbf{v}_k^{(1)}, \mathbf{v}_k^{(2)}, \ldots, \mathbf{v}_k^{(n)}\}$ is determined by*

$$(\mathbf{A} - \lambda_i \mathbf{I})\mathbf{v}_k^{(i)} = \mathbf{0}. \tag{B.12}$$

which forms a basis in $\Omega \subseteq \mathscr{R}^n$. The eigenvector matrix of $\mathbf{P} = [\mathbf{v}_k^{(1)}, \mathbf{v}_k^{(2)}, \ldots, \mathbf{v}_k^{(n)}]$ is invertible and

$$\mathbf{P}^{-1}\mathbf{A}\mathbf{P} = diag[\lambda_1, \lambda_2, \ldots, \lambda_n]. \tag{B.13}$$

Thus, with an initial state of $\mathbf{x}_k$, the solution of discrete linear dynamical system in Eq. (B.2) is

$$\mathbf{x}_{k+1} = \mathbf{P}diag[\lambda_1, \lambda_2, \ldots, \lambda_n]\mathbf{P}^{-1}\mathbf{x}_k = \mathbf{P}\mathbf{E}\mathbf{P}^{-1}\mathbf{x}_k \tag{B.14}$$

where the diagonal matrix $\mathbf{E}$ is given by

$$\mathbf{E} = diag[\lambda_1, \lambda_2, \ldots, \lambda_n]. \tag{B.15}$$

The iteration solution of discrete linear dynamical system in Eq. (B.2) is

$$\mathbf{x}_k = \mathbf{P}diag[\lambda_1^k, \lambda_2^k, \ldots, \lambda_n^k]\mathbf{P}^{-1}\mathbf{x}_0 = \mathbf{P}\mathbf{E}^k\mathbf{P}^{-1}\mathbf{x}_0. \tag{B.16}$$

Proof Assuming $\mathbf{x}_{k+1} = \lambda\mathbf{x}_k$ and $\mathbf{x}_k = c\mathbf{v}_k$, equation (B.2) gives $c(\mathbf{A} - \lambda\mathbf{I})\mathbf{v}_k = \mathbf{0}$. Thus, $(\mathbf{A} - \lambda_i\mathbf{I})\mathbf{v}_k^{(i)} = \mathbf{0}$ yields

$$[\mathbf{A}\mathbf{v}_k^{(1)}, \mathbf{A}\mathbf{v}_k^{(2)}, \ldots, \mathbf{A}\mathbf{v}_k^{(n)}] = [\lambda_1\mathbf{v}_k^{(1)}, \lambda_2\mathbf{v}_k^{(2)}, \ldots, \lambda_n\mathbf{v}_k^{(n)}].$$

Deformation of the foregoing equation gives

$$\mathbf{A}[\mathbf{v}_k^{(1)}, \mathbf{v}_k^{(2)}, \ldots, \mathbf{v}_k^{(n)}] = [\mathbf{v}_k^{(1)}, \mathbf{v}_k^{(2)}, \ldots, \mathbf{v}_k^{(n)}]diag[\lambda_1, \lambda_2, \ldots, \lambda_n].$$

Further

$$\mathbf{A}\mathbf{P} = \mathbf{P}diag[\lambda_1, \lambda_2, \ldots, \lambda_n].$$

The left multiplication of $\mathbf{P}^{-1}$ on both sides of equation yields

$$\mathbf{P}^{-1}\mathbf{A}\mathbf{P} = \mathbf{P}^{-1}\mathbf{P}diag[\lambda_1, \lambda_2, \ldots, \lambda_n] = diag[\lambda_1, \lambda_2, \ldots, \lambda_n].$$

Consider a new variable of $\mathbf{y}_k = \mathbf{P}^{-1}\mathbf{x}_k$. Thus, application of $\mathbf{x}_k = \mathbf{P}\mathbf{y}_k$ to Eq. (B.2) yields

$$\mathbf{y}_{k+1} = \mathbf{P}^{-1}\mathbf{x}_{k+1} = \mathbf{P}^{-1}\mathbf{A}\mathbf{x}_k = \mathbf{P}^{-1}\mathbf{A}\mathbf{P}\mathbf{y}_k = diag[\lambda_1, \lambda_2, \ldots, \lambda_n]\mathbf{y}_k.$$

Using $\mathbf{x}_k = \mathbf{P}\mathbf{y}_k$ and $\mathbf{y}_k = \mathbf{P}^{-1}\mathbf{x}_k$, we have

$$\mathbf{x}_{k+1} = \mathbf{P}diag[\lambda_1, \lambda_2, \ldots, \lambda_n]\mathbf{P}^{-1}\mathbf{x}_k.$$

where

$$\mathbf{E} = diag[\lambda_1, \lambda_2, \ldots, \lambda_n].$$

Further, one obtains

$$\mathbf{x}_k = (\mathbf{P}\mathbf{E}\mathbf{P}^{-1})^k\mathbf{x}_0 = (\mathbf{P}\mathbf{E}^k\mathbf{P}^{-1})\mathbf{x}_0 = \mathbf{P}diag[\lambda_1^k, \lambda_2^k, \ldots, \lambda_n^k]\mathbf{P}^{-1}\mathbf{x}_0.$$

This theorem is proved. ∎

The eigenvector of $\mathbf{v}_i$ is assumed as

$$\mathbf{v}_i = \begin{Bmatrix} 1 \\ \mathbf{r}_i \end{Bmatrix} v_i. \tag{B.17}$$

From Eq. (B.12), we have

$$\begin{bmatrix} a_{11} - \lambda_i & \mathbf{b}_{1\times(n-1)} \\ \mathbf{c}_{(n-1)\times 1} & \mathbf{A}_{11} - \lambda_i \mathbf{I}_{(n-1)\times(n-1)} \end{bmatrix} \begin{Bmatrix} 1 \\ \mathbf{r}_i \end{Bmatrix} v_i = \mathbf{0}, \tag{B.18}$$

where the minor of matrix $\mathbf{A}$ is $\mathbf{A}_{11}$, and other vectors are defined by

$$\begin{aligned} \mathbf{c}_{(n-1)\times 1} &= (a_{i1})_{(n-1)\times 1}\ (i = 2, 3, \ldots, n), \\ \mathbf{b}_{1\times(n-1)} &= (a_{1j})_{1\times(n-1)}\ (j = 2, 3, \ldots, n), \\ \mathbf{A}_{11} &= (a_{ij})_{(n-1)\times(n-1)}\ (i, j = 2, 3, \ldots, n). \end{aligned} \tag{B.19}$$

Thus,

$$\mathbf{r}_i = (\mathbf{A}_{11} - \lambda_i \mathbf{I}_{(n-1)\times(n-1)})^{-1}\mathbf{c}_{n\times 1}. \tag{B.20}$$

The solution of discrete linear dynamical system in Eq. (B.2) is

$$\begin{aligned} \mathbf{x}_{k+1} &= \sum\nolimits_{i=1}^{n} C_i \lambda_i \mathbf{v}_k^{(i)} = [\mathbf{v}_k^{(1)}, \mathbf{v}_k^{(2)}, \ldots, \mathbf{v}_k^{(n)}]diag[\lambda_1, \lambda_2, \ldots, \lambda_n]\mathbf{C} \\ &= \mathbf{P}diag[\lambda_1, \lambda_2, \ldots, , \lambda_n]\mathbf{C} \end{aligned} \tag{B.21}$$

where

$$\mathbf{C} = (C_1, C_2, \ldots, C_n)^{\mathrm{T}}. \tag{B.22}$$

If $\mathbf{x}_{k+1} = \mathbf{x}_k$, $diag[\lambda_1, \lambda_2, \ldots, \lambda_n] = \mathbf{I}$. Thus

$$\mathbf{P}^{-1}\mathbf{x}_k = \mathbf{C}. \tag{B.23}$$

Therefore, the solution is expressed by

$$\mathbf{x}_{k+1} = \mathbf{P}diag[\lambda_1, \lambda_2, \ldots, \lambda_n]\mathbf{P}^{-1}\mathbf{x}_k. \tag{B.24}$$

The two methods give the same expression.

Theorem B.2 *Consider a linear dynamical system $\mathbf{x}_{k+1} = \mathbf{A}\mathbf{x}_k$ in Eq. (B.2) with the initial state of $\mathbf{x}_k$. If the distinct complex eigenvalues of the $2n \times 2n$ matrix $\mathbf{A}$ are $\lambda_i = \alpha_i + \mathbf{i}\beta_i$ and $\bar{\lambda}_i = \alpha_i - \mathbf{i}\beta_i$ ($i = 1, 2, \ldots, n$ and $\mathbf{i} = \sqrt{-1}$) with corresponding eigenvectors $\mathbf{w}_k^{(i)} = \mathbf{u}_k^{(i)} + \mathbf{i}\mathbf{v}_k^{(i)}$ and $\bar{\mathbf{w}}_k^{(i)} = \mathbf{u}_k^{(i)} - \mathbf{i}\mathbf{v}_k^{(i)}$, then the corresponding eigenvectors $\mathbf{u}_k^{(i)}$* and *$\mathbf{v}_k^{(i)}$ ($i = 1, 2, \ldots, n$) are determined by*

$$\begin{aligned} &(\mathbf{A} - (\alpha_i + \mathbf{i}\beta_i)\mathbf{I})(\mathbf{u}_k^{(i)} + \mathbf{i}\mathbf{v}_k^{(i)}) = \mathbf{0}, \text{ or} \\ &(\mathbf{A} - (\alpha_i - \mathbf{i}\beta_i)\mathbf{I})(\mathbf{u}_k^{(i)} - \mathbf{i}\mathbf{v}_k^{(i)}) = \mathbf{0} \end{aligned} \tag{B.25}$$

which forms a basis in $\Omega \subseteq \mathscr{R}^{2n}$. The corresponding eigenvector matrix of $\mathbf{P} = [\mathbf{u}_k^{(1)}, \mathbf{v}_k^{(1)}, \mathbf{u}_k^{(2)}, \mathbf{v}_k^{(2)}, \ldots, \mathbf{u}_k^{(n)}, \mathbf{v}_k^{(n)}]$ is invertible and

$$\mathbf{P}^{-1}\mathbf{A}\mathbf{P} = diag(\mathbf{B}_1, \mathbf{B}_2, \ldots, \mathbf{B}_n), \tag{B.26}$$

where

$$\mathbf{B}_i = \begin{bmatrix} \alpha_i & \beta_i \\ -\beta_i & \alpha_i \end{bmatrix} (i = 1, 2, \ldots, n). \tag{B.27}$$

Thus, with the initial state of $\mathbf{x}_k$, the solution of the linear dynamical system in Eq. (B.2) is

$$\mathbf{x}_{k+1} = \mathbf{P}\mathbf{E}\mathbf{P}^{-1}\mathbf{x}_k \tag{B.28}$$

where the diagonal matrix $\mathbf{E}$ is given by

$$\begin{aligned} &\mathbf{E} = diag[\mathbf{E}_1, \mathbf{E}_2, \ldots, \mathbf{E}_n], \\ &\mathbf{E}_i = r_i \begin{bmatrix} \cos\theta_i & \sin\theta_i \\ -\sin\theta_i & \cos\theta_i \end{bmatrix}, \\ &\text{with } r_i = \sqrt{\alpha_i^2 + \beta_i^2} \text{ and } \theta_i = \arctan\frac{\beta_i}{\alpha_i}, \\ &\alpha_i = r_i\cos\theta_i \text{ and } \beta_i = r_i\sin\theta_i. \end{aligned} \tag{B.29}$$

The iteration solution of the linear dynamical system in Eq. (B.2) is

$$\mathbf{x}_k = \mathbf{P}\mathbf{E}^k\mathbf{P}^{-1}\mathbf{x}_0 = \mathbf{P}\mathbf{E}(k)\mathbf{P}^{-1}\mathbf{x}_0 \tag{B.30}$$

where

$$\mathbf{E}(k) = \mathbf{E}^k = diag[\mathbf{E}_1(k), \mathbf{E}_2(k), \ldots, \mathbf{E}_n(k)],$$
$$\mathbf{E}_i(k) = r_i^k \begin{bmatrix} \cos k\theta_i & \sin k\theta_i \\ -\sin k\theta_i & \cos k\theta_i \end{bmatrix}. \tag{B.31}$$

Proof Assuming $\mathbf{x}_{k+1} = \lambda \mathbf{x}_k$ and $\mathbf{x}_k = c\mathbf{v}_k$, equation (B.2) gives $c(\mathbf{A} - \lambda\mathbf{I})\mathbf{v}_k = 0$. Since $\det(\mathbf{A} - \lambda\mathbf{I}) = 0$ gives $\lambda_i = \alpha_i + \mathbf{i}\beta_i$ and $\bar{\lambda}_i = \alpha_i - \mathbf{i}\beta_i$. From Eq. (B.25), we have

$$(\mathbf{A} - \alpha_i\mathbf{I})\mathbf{u}_k^{(i)} + \beta_i\mathbf{I}\mathbf{v}_k^{(i)} = \mathbf{0},$$
$$-\beta_i\mathbf{I}u_k^{(i)} + (\mathbf{A} - \alpha_i\mathbf{I})\mathbf{v}_k^{(i)} = \mathbf{0}.$$

$$\mathbf{A}u_k^{(i)} = (\mathbf{u}_k^{(i)}, \mathbf{v}_k^{(i)})\begin{Bmatrix} \alpha_i \\ -\beta_i \end{Bmatrix} \text{and } \mathbf{A}\mathbf{v}_k^{(i)} = (\mathbf{u}_k^{(i)}, \mathbf{v}_k^{(i)})\begin{Bmatrix} \beta_i \\ \alpha_i \end{Bmatrix}.$$

$$\mathbf{A}(\mathbf{u}_k^{(i)}, \mathbf{v}_k^{(i)}) = (\mathbf{u}_k^{(i)}, \mathbf{v}_k^{(i)})\begin{bmatrix} \alpha_i & \beta_i \\ -\beta_i & \alpha_i \end{bmatrix}.$$

Assembling $\mathbf{A}(\mathbf{u}_k^{(i)}, \mathbf{v}_k^{(i)})$ for $(i = 1, 2, \ldots, n)$ gives

$$\begin{aligned}\mathbf{AP} &= \mathbf{P}diag(\begin{bmatrix} \alpha_1 & \beta_1 \\ -\beta_1 & \alpha_1 \end{bmatrix}, \begin{bmatrix} \alpha_2 & \beta_2 \\ -\beta_2 & \alpha_2 \end{bmatrix}, \ldots, \begin{bmatrix} \alpha_n & \beta_n \\ -\beta_n & \alpha_n \end{bmatrix}) \\ &= \mathbf{P}diag(\mathbf{E}_1, \mathbf{E}_2, \ldots, \mathbf{E}_n)\end{aligned}$$

where

$$\mathbf{E}_i = r_i\begin{bmatrix} \cos\theta_i & \sin\theta_i \\ -\sin\theta_i & \cos\theta_i \end{bmatrix},$$
$$\alpha_i = r_i\cos\theta_i \text{ and } \beta_i = r_i\sin\theta_i;$$

$$r_i = \sqrt{\alpha_i^2 + \beta_i^2} \text{ and } \theta_i = \arctan\frac{\beta_i}{\alpha_i},$$
$$\mathbf{P} = (\mathbf{u}_k^{(1)}, \mathbf{v}_k^{(1)}, \mathbf{u}_k^{(2)}, \mathbf{v}_k^{(2)}, \ldots, \mathbf{u}_k^{(n)}, \mathbf{v}_k^{(n)}).$$

The left multiplication of $\mathbf{P}^{-1}$ on both sides of equation yields

$$\begin{aligned}\mathbf{P}^{-1}\mathbf{AP} &= \mathbf{P}^{-1}\mathbf{P}diag(\mathbf{E}_1, \mathbf{E}_2, \ldots, \mathbf{E}_n) \\ &= diag(\mathbf{E}_1, \mathbf{E}_2, \ldots, \mathbf{E}_n).\end{aligned}$$

Consider a new variable $\mathbf{y}_k = \mathbf{P}^{-1}\mathbf{x}_k$. Thus, application of $\mathbf{x}_k = \mathbf{P}y_k$ to Eq. (B.2) yields

$$\begin{aligned}\mathbf{y}_{k+1} &= \mathbf{P}^{-1}\mathbf{x}_{k+1} = \mathbf{P}^{-1}\mathbf{A}\mathbf{x}_k = \mathbf{P}^{-1}\mathbf{AP}\mathbf{y}_k \\ &= diag(\mathbf{E}_1, \mathbf{E}_2, \ldots, \mathbf{E}_n)\mathbf{y}_k.\end{aligned}$$

Using $\mathbf{x}_k = \mathbf{P}\mathbf{y}_k$ and $\mathbf{y}_k = \mathbf{P}^{-1}\mathbf{x}_k$, we have

$$\mathbf{x}_{k+1} = \mathbf{P}diag[\mathbf{E}_1, \mathbf{E}_2, \ldots, \mathbf{E}_n]\mathbf{P}^{-1}\mathbf{x}_k = \mathbf{P}\mathbf{E}\mathbf{P}^{-1}\mathbf{x}_k.$$

Therefore,

$$\mathbf{x}_k = (\mathbf{P}\mathbf{E}\mathbf{P}^{-1})^k\mathbf{x}_0 = \mathbf{P}\mathbf{E}^k\mathbf{P}^{-1}\mathbf{x}_0 = \mathbf{P}\mathbf{E}(k)\mathbf{P}^{-1}\mathbf{x}_0.$$

where

$$\mathbf{E}(k) = \mathbf{E}^k = diag[\mathbf{E}_1(k), \mathbf{E}_2(k), \ldots, \mathbf{E}_n(k)],$$
$$\mathbf{E}_i(k) = r_i^k \begin{bmatrix} \cos k\theta_i & \sin k\theta_i \\ -\sin k\theta_i & \cos k\theta_i \end{bmatrix}$$

This theorem is proved. ■

The conjugate complex eigenvectors are assumed as

$$\mathbf{u}_k^{(i)} + \mathbf{i}\mathbf{v}_k^{(i)} = C_k^{(i)}\left\{ \begin{matrix} 1 \\ \mathbf{r}_k^{(i)} \end{matrix} \right\} \text{ and } \mathbf{u}_k^{(i)} - \mathbf{i}\mathbf{v}_k^{(i)} = \bar{C}_k^{(i)}\left\{ \begin{matrix} 1 \\ \bar{\mathbf{r}}_k^{(i)} \end{matrix} \right\} \tag{B.32}$$

where the conjugate complex constants are assumed as

$$\begin{aligned} C_k^{(i)} &= \frac{1}{2}(M_k^{(i)} - \mathbf{i}N_k^{(i)}) \text{ and } \bar{C}_k^{(i)} = \frac{1}{2}(M_k^{(i)} + \mathbf{i}N_k^{(i)}). \\ \mathbf{r}_k^{(i)} &= \mathbf{U}_k^{(i)} + \mathbf{i}V_k^{(i)} \text{ and } \bar{\mathbf{r}}_k^{(i)} = \mathbf{U}_k^{(i)} - \mathbf{i}V_k^{(i)}. \end{aligned} \tag{B.33}$$

From Eq. (B.25), we have

$$\begin{bmatrix} a_{11} - \alpha_i - \mathbf{i}\beta_i & \mathbf{b}_{1\times(n-1)} \\ \mathbf{c}_{(n-1)\times 1} & \mathbf{A}_{11} - (\alpha_i + \mathbf{i}\beta_i)\mathbf{I}_{(n-1)\times(n-1)} \end{bmatrix} \left\{ \begin{matrix} 1 \\ \mathbf{r}_k^{(i)} \end{matrix} \right\} C_k^{(i)} = \mathbf{0}. \tag{B.34}$$

Thus, the foregoing equation gives

$$\mathbf{c} + [(\mathbf{A}_{11} - \alpha_i\mathbf{I}) - \mathbf{i}\beta_i\mathbf{I}]\mathbf{r}_k^{(i)} = \mathbf{0}, \tag{B.35}$$

$$\mathbf{r}_k^{(i)} = \left[(\mathbf{A}_{11} - \alpha_i\mathbf{I})^2 + \beta_i^2\mathbf{I}\right]^{-1}[(\mathbf{A}_{11} - \alpha_i\mathbf{I}) + \mathbf{i}\beta_i\mathbf{I}]\mathbf{c}_k^{(i)} = \mathbf{U}_k^{(i)} + \mathbf{i}\mathbf{V}_k^{(i)} \tag{B.36}$$

where

$$\begin{aligned} \mathbf{U}_k^{(i)} &= \left[(\mathbf{A}_{11} - \alpha_i\mathbf{I})^2 + \beta_i^2\mathbf{I}\right]^{-1}(\mathbf{A}_{11} - \alpha_i\mathbf{I})\mathbf{c}_k^{(i)}, \\ \mathbf{V}_k^{(i)} &= \left[(\mathbf{A}_{11} - \alpha_i\mathbf{I})^2 + \beta_i^2\mathbf{I}\right]^{-1}\beta_i\mathbf{c}_k^{(i)}. \end{aligned} \tag{B.37}$$

The solution of the linear dynamical system in Eq. (B.2) is

$$\begin{aligned}
\mathbf{x}_{k+1} &= \sum\nolimits_{i=1}^{n} C_k^{(i)}(\mathbf{u}_k^{(i)} + \mathbf{i}\mathbf{v}_k^{(i)})(\alpha_i + \mathbf{i}\beta_i) + \bar{C}_k^{(i)}(\mathbf{u}_k^{(i)} - \mathbf{i}\mathbf{v}_k^{(i)})(\alpha_i - \mathbf{i}\beta_i) \\
&= \sum\nolimits_{i=1}^{n} r_i \left[\left\{ \begin{array}{c} M_k^{(i)} \\ M_k^{(i)}\mathbf{U}_k^{(i)} + N_k^{(i)}\mathbf{V}_k^{(i)} \end{array} \right\} \cos\theta_i + \left\{ \begin{array}{c} N_k^{(i)} \\ N_k^{(i)}\mathbf{U}_k^{(i)} - M_k^{(i)}\mathbf{V}_k^{(i)} \end{array} \right\} \sin\theta_i \right] \\
&= \sum\nolimits_{i=1}^{n} (\mathbf{u}_i, \mathbf{v}_i) r_i \begin{bmatrix} \cos\theta_i & \sin\theta_i \\ -\sin\theta_i & \cos\theta_i \end{bmatrix} \left\{ \begin{array}{c} M_k^{(i)} \\ N_k^{(i)} \end{array} \right\} \\
&= \mathbf{PEC}
\end{aligned} \tag{B.38}$$

where

$$\begin{aligned}
&\mathbf{P} = [\mathbf{u}_1, \mathbf{v}_1, \ldots, \mathbf{u}_n, \mathbf{v}_n], \mathbf{E} = diag[\mathbf{E}_1, \mathbf{E}_2, \ldots, \mathbf{E}_n], \\
&\mathbf{C} = (M_1, N_1, \ldots, M_n, N_n)^{\mathrm{T}}, \\
&\mathbf{E}_i = r^i \begin{bmatrix} \cos\theta_i & \sin\theta_i \\ -\sin\theta_i & \cos\theta_i \end{bmatrix}, \mathbf{u}_k^{(i)} = \left\{ \begin{array}{c} 1 \\ \mathbf{U}_k^{(i)} \end{array} \right\} \text{ and } \mathbf{v}_k^{(i)} = \left\{ \begin{array}{c} 0 \\ \mathbf{V}_k^{(i)} \end{array} \right\}.
\end{aligned} \tag{B.39}$$

With the initial state of $\mathbf{x}_{k+1} = \mathbf{x}_k$, one obtains $\mathbf{E} = \mathbf{I}$. Thus,

$$\mathbf{C} = \mathbf{P}^{-1}\mathbf{x}_k. \tag{B.40}$$

Therefore, the solution is expressed by

$$\mathbf{x}_{k+1} = \mathbf{P}diag[\mathbf{E}_1, \mathbf{E}_2, \ldots, \mathbf{E}_n]\mathbf{P}^{-1}\mathbf{x}_k = \mathbf{PEP}^{-1}\mathbf{x}_k. \tag{B.41}$$

The two methods give the same expression.

Theorem B.3 *Consider a linear dynamical system* $\mathbf{x}_{k+1} = \mathbf{A}\mathbf{x}_k$ *in Eq.* (B.2) *with the initial state of* $\mathbf{x}_k$. *If the distinct complex eigenvalues of the* $n \times n$ *matrix* $\mathbf{A}$ *are* $\lambda_i = \alpha_i + \mathbf{i}\beta_i$ *and* $\bar{\lambda}_i = \alpha_i - \mathbf{i}\beta_i$ $(i = 1, 2, \ldots, p$ *and* $\mathbf{i} = \sqrt{-1})$ *with corresponding eigenvectors* $\mathbf{w}_k^{(i)} = \mathbf{u}_k^{(i)} + \mathbf{i}\mathbf{v}_k^{(i)}$ *and* $\bar{\mathbf{w}}_k^{(i)} = \mathbf{u}_k^{(i)} - \mathbf{i}\mathbf{v}_k^{(i)}$, *and* $(n-2p)$ *distinct real eigenvalues of* $\lambda_{2p+1}, \lambda_{2p+2}, \ldots, \lambda_n$, *then the corresponding eigenvectors* $\mathbf{u}_k^{(i)}$ *and* $\mathbf{v}_k^{(i)}$ *for complex eigenvalues* $(\lambda_i, \bar{\lambda}_i)$ $(i = 1, 2, \ldots, p)$ *are determined by*

$$\begin{aligned}
&(\mathbf{A} - (\alpha_i + \mathbf{i}\beta_i)\mathbf{I})(\mathbf{u}_k^{(i)} + \mathbf{i}\mathbf{v}_k^{(i)}) = \mathbf{0}, \text{ or} \\
&(\mathbf{A} - (\alpha_i - \mathbf{i}\beta_i)\mathbf{I})(\mathbf{u}_k^{(i)} - \mathbf{i}\mathbf{v}_k^{(i)}) = \mathbf{0}
\end{aligned} \tag{B.42}$$

and a set of corresponding eigenvectors $\{\mathbf{v}_k^{(2p+1)}, \mathbf{v}_k^{(2p+2)}, \ldots, \mathbf{v}_k^{(n)}\}$ *is determined by*

$$(\mathbf{A} - \lambda_i\mathbf{I})\mathbf{v}_k^{(i)} = \mathbf{0} \tag{B.43}$$

which forms a basis in $\Omega \subseteq \mathscr{R}^n$*. The eigenvector matrix of*

$$\mathbf{P} = [\mathbf{u}_k^{(1)}, \mathbf{v}_k^{(1)}, \mathbf{u}_k^{(2)}, \mathbf{v}_k^{(2)}, \ldots, \mathbf{u}_k^{(p)}, \mathbf{v}_k^{(p)}, \mathbf{v}_k^{(2p+1)}, \mathbf{v}_k^{(2p+2)}, \ldots, \mathbf{v}_k^{(n)}] \tag{B.44}$$

is invertible and

$$\mathbf{P}^{-1}\mathbf{A}\mathbf{P} = diag(\mathbf{E}_1, \mathbf{E}_2, \ldots, \mathbf{E}_p, \lambda_{2p+1}, \lambda_{2p+2}, \ldots, \lambda_n). \tag{B.45}$$

where

$$\mathbf{E}_i = r_i \begin{bmatrix} \cos\theta_i & \sin\theta_i \\ -\sin\theta_i & \cos\theta_i \end{bmatrix} (i = 1, 2, \ldots, n). \tag{B.46}$$

Thus, with an initial state of $\mathbf{x}_k$*, the solution of linear dynamical system in Eq.* (B.2) *is*

$$\begin{aligned} \mathbf{x}_{k+1} &= \mathbf{P} diag[\mathbf{E}_1, \mathbf{E}_2, \ldots, \mathbf{E}_p, \lambda_{2p+1}, \lambda_{2p+2}, \ldots, \lambda_n]\mathbf{P}^{-1}\mathbf{x}_k \\ &= \mathbf{P}\mathbf{E}\mathbf{P}^{-1}\mathbf{x}_k. \end{aligned} \tag{B.47}$$

The iterative solution of linear dynamical system in Eq. (B.2) *is*

$$\begin{aligned} \mathbf{x}_k &= \mathbf{P} diag[\mathbf{E}_1(k), \mathbf{E}_2(k), \ldots, \mathbf{E}_p(k), \lambda_{2p+1}^k, \lambda_{2p+2}^k, \ldots, \lambda_n^k]\mathbf{P}^{-1}\mathbf{x}_0 \\ &= \mathbf{P}\mathbf{E}(k)\mathbf{P}^{-1}\mathbf{x}_0 \end{aligned} \tag{B.48}$$

where

$$\mathbf{E}_i(k) = r_i^k \begin{bmatrix} \cos k\theta_i & \sin k\theta_i \\ -\sin k\theta_i & \cos k\theta_i \end{bmatrix} (i = 1, 2, \ldots, n). \tag{B.49}$$

Proof The proof of the theorem is from the proof of Theorems B.1 and B.2. ■

B.3 Linear Discrete Systems with Repeated Eigenvalues

In this section, the solution for a discrete dynamical system possessing repeated eigenvalues will be discussed. The case of repeated real eigenvalues will be discussed first, and then the case of repeated complex eigenvalues will be presented. Finally, the solutions for nonhomogeneous discrete dynamical systems will be presented.

Theorem B.4 *Consider a linear dynamical system* $\mathbf{x}_{k+1} = \mathbf{A}\mathbf{x}_k$ *in Eq.* (B.2) *with the initial state of* $\mathbf{x}_k$*. There is a repeated eigenvalue* λ_i *with m-times among the real eigenvalues* $\lambda_1, \lambda_2, \ldots, \lambda_n$ *of the* $n \times n$ *matrix* $\mathbf{A}$*. If a set of generalized eigenvectors* $\{\mathbf{v}_k^{(1)}, \mathbf{v}_k^{(2)}, \ldots, \mathbf{v}_k^{(n)}\}$ *forms a basis in* $\Omega \subseteq \mathscr{R}^n$*. The eigenvector matrix of* $\mathbf{P} = [\mathbf{v}_k^{(1)}, \mathbf{v}_k^{(2)}, \ldots, \mathbf{v}_k^{(n)}]$ *is invertible. For the repeated eigenvalue* λ_i*, the matrix* $\mathbf{A}$ *can be decomposed by*

$$\mathbf{A} = \mathbf{S} + \mathbf{N} \tag{B.50}$$

where

$$\mathbf{P}^{-1}\mathbf{S}\mathbf{P} = diag[\lambda_i]_{n\times n}, \tag{B.51}$$

the matrix $\mathbf{N} = \mathbf{A} - \mathbf{S}$ *is nilpotent of order* $m \le n$ $(\mathbf{N}^m = \mathbf{0})$ *with* $\mathbf{SN} = \mathbf{NS}$.

$$\mathbf{P}^{-1}\mathbf{A}\mathbf{P} = diag[\lambda_1, \ldots, \lambda_{i-1}, \underbrace{\lambda_i, \ldots, \lambda_i}_{m}, \lambda_{i+m}, \ldots, \lambda_n]. \tag{B.52}$$

Thus, with an initial state of $\mathbf{x}_{k+1} = \mathbf{x}_k$, *the solution of the discrete linear dynamical system in Eq.* (B.2) *is*

$$\mathbf{x}_{k+1} = \mathbf{P}[\mathbf{E} + (\mathbf{P}^{-1}\mathbf{N}\mathbf{P})]\mathbf{P}^{-1}\mathbf{x}_k \tag{B.53}$$

where

$$\mathbf{E} = diag[\lambda_1, \ldots, \lambda_{i-1}, \underbrace{\lambda_i, \ldots, \lambda_i}_{m}, \lambda_{i+m}, \ldots, \lambda_n]. \tag{B.54}$$

The iterative solution of linear dynamical system in Eq. (B.2) *is*

$$\mathbf{x}_k = \mathbf{P}\left[\mathbf{E} + (\mathbf{P}^{-1}\mathbf{N}\mathbf{P})\right]^k\mathbf{P}^{-1}\mathbf{x}_0. \tag{B.55}$$

Proof For the repeated real eigenvalue λ_i of the matrix $\mathbf{A}$, consider the corresponding solution as

$$\begin{aligned}
\mathbf{x}_{k+1}^{(i+j)} &= \bar{C}_{k+1}^{(i+j)}\mathbf{v}_k^{(i+j)} + \lambda_i C_k^{(i+j)}\mathbf{v}_k^{(i+j)},\\
\mathbf{x}_k^{(i+j)} &= C_k^{(i+j)}\mathbf{v}_k^{(i+j)},
\end{aligned}$$

$$\bar{C}_{k+1}^{(i+j)}\mathbf{v}_k^{(i+j)} + \lambda_i C_k^{(i+j)}\mathbf{v}_k^{(i+j)} = \mathbf{A}C_k^{(i+j)}\mathbf{v}_k^{(i+j)}.$$

Therefore,

$$\bar{C}_{k+1}^{(i+j)}\mathbf{v}_k^{(i+j)} = (\mathbf{A} - \lambda_i\mathbf{I})C_k^{(i+j)}\mathbf{v}_k^{(i+j)}.$$

Consider the constant vector and eigenvector matrix as

$$\mathbf{C}_k^{(i+j)} = (\underbrace{0, \ldots, 0}_{i+j-1}, C_k^{(i+j)}, \underbrace{0, \ldots, 0}_{n-i-j})^{\mathrm{T}}, \bar{\mathbf{C}}_{k+1}^{(i+j)} = (\underbrace{0, \ldots, 0}_{i+j-1}, \bar{C}_{k+1}^{(i+j)}, \underbrace{0, \ldots, 0}_{n-i-j})^{\mathrm{T}};$$

$$\mathbf{P} = (\mathbf{v}_k^{(1)}, \ldots, \mathbf{v}_k^{(i-1)}, \mathbf{v}_k^{(i)}, \ldots, \mathbf{v}_k^{(i+m-1)}, \mathbf{v}_k^{(i+m)}, \ldots, \mathbf{v}_k^{(n)}).$$

Thus

$$\mathbf{P}\bar{\mathbf{C}}_{k+1}^{(i+j)} = (\mathbf{A} - \lambda_i\mathbf{I})\mathbf{P}C_k^{(i+j)} \Rightarrow \bar{\mathbf{C}}_{k+1}^{(i+j)} = \mathbf{P}^{-1}(\mathbf{A} - \lambda_i\mathbf{I})\mathbf{P}C_k^{(i+j)}.$$

Let $\mathbf{A} = \mathbf{S} + \mathbf{N}$, thus

$$\begin{aligned}\bar{\mathbf{C}}_{k+1}^{(i+j)} &= \mathbf{P}^{-1}(\mathbf{A} - \lambda_i\mathbf{I})\mathbf{P}C_k^{(i+j)}\\ &= \mathbf{P}^{-1}(\mathbf{S} + \mathbf{N} - \lambda_i\mathbf{I})\mathbf{P}C_k^{(i+j)}\\ &= (\mathbf{P}^{-1}\mathbf{SP} - \lambda_i\mathbf{I} + \mathbf{P}^{-1}\mathbf{NP})\mathbf{C}_k^{(i+j)}.\end{aligned}$$

Because of $\mathbf{P}^{-1}\mathbf{SP} = diag[\lambda_i]$, the solution of the foregoing equation is

$$\bar{\mathbf{C}}_{k+1}^{(i+j)} = (\mathbf{P}^{-1}\mathbf{NP})\mathbf{C}_k^{(i+j)}.$$

Thus,

$$\begin{aligned}\mathbf{x}_{k+1}^{(i)} &= \sum\nolimits_{j=0}^{m-1}(\bar{C}_{k+1}^{(i+j)} + \lambda_i C_k^{(i+j)})\mathbf{v}_k^{(i+j)}\\ &= \lambda_i\mathbf{P}C_k^{(i)} + \mathbf{P}\bar{\mathbf{C}}_{k+1}^{(i+j)} = \mathbf{P}[\lambda_i\mathbf{I} + (\mathbf{P}^{-1}\mathbf{NP})]\mathbf{C}_k^{(i)}.\end{aligned}$$

Let

$$\mathbf{C} = (C_1, \ldots, C_{i-1}, \underbrace{C_i, \ldots, C_i}_{m}, C_{i+m}, \ldots, C_n)^{\mathrm{T}},$$

$$\mathbf{P} = (\mathbf{v}_1\ , \ldots, \mathbf{v}_{i-1}, \underbrace{\mathbf{v}_i, \ldots, \mathbf{v}_{i+m-1}}_{m}; \mathbf{v}_{j+m}, \ldots, \mathbf{v}_n).$$

Thus, there is a relation

$$\mathbf{P}^{-1}\mathbf{AP} = diag[\lambda_1, \ldots, \lambda_{i-1}, \underbrace{\lambda_i, \ldots, \lambda_i}_{m}, \lambda_{i+m}, \ldots, \lambda_n],$$

and the resultant solution is

$$\mathbf{x}_{k+1} = \mathbf{x}_k^{(1)} + \ldots + \mathbf{x}_k^{(i-1)} + \underbrace{\mathbf{x}_k^{(i)} + \mathbf{x}_k^{(i+1)} + \ldots + \mathbf{x}_k^{(i+m-1)}}_{m} + \mathbf{x}_k^{(i+m)} + \ldots + \mathbf{x}_k^{(n)}$$

$$\begin{aligned}\mathbf{x}_{k+1} &= \mathbf{P}[diag(\lambda_1, \ldots, \lambda_{i-1}, \underbrace{\lambda_i, \ldots, \lambda_i}_{m}, \lambda_{i+m}, \ldots, \lambda_n) + (\mathbf{P}^{-1}\mathbf{NP})]\mathbf{P}^{-1}\mathbf{x}_k\\ &= \mathbf{P}\big[\mathbf{E} + (\mathbf{P}^{-1}\mathbf{NP})\big]\mathbf{P}^{-1}\mathbf{x}_k\end{aligned}$$

where

$$\mathbf{E} = diag(\lambda_1, \ldots, \lambda_{i-1}, \underbrace{\lambda_i, \ldots, \lambda_i}_{m}, \lambda_{i+m}, \ldots, \lambda_n).$$

Therefore, the iterative solution of linear dynamical system in Eq. (B.2) is

$$\mathbf{x}_k = \mathbf{P}\big[\mathbf{E} + (\mathbf{P}^{-1}\mathbf{NP})\big]^k\mathbf{P}^{-1}\mathbf{x}_0.$$

This theorem is proved. ■

Consider the solution for repeated eigenvalues of a linear dynamical system as

$$\mathbf{x}_{k+1}^{(i+j)} = \lambda_i C_k^{(i+j)} \mathbf{v}_k^{(i+j)} + C_k^{(i+j+1)} \mathbf{v}_k^{(i+j+1)} \quad (j = 0, 1, \ldots, m-2). \tag{B.56}$$

Submission of (B.56) into $\mathbf{x}_{k+1} = \mathbf{A}\mathbf{x}_k$ in Eq. (B.2) gives

$$\sum_{j=0}^{m-1} (\mathbf{A} - \lambda_i \mathbf{I})(C_k^{(i+j)} \mathbf{v}_k^{(i+j)}) + (C_k^{(i+j+1)} \mathbf{v}_k^{(i+j+1)}) = \mathbf{0}. \tag{B.57}$$

Thus

$$(\mathbf{A} - \lambda_i \mathbf{I})(C_k^{(i+j)} \mathbf{v}_k^{(i+j)}) - (C_k^{(i+j+1)} \mathbf{v}_k^{(i+j+1)}) = \mathbf{0} \quad (j = 0, 1, 2, \ldots, m-2). \tag{B.58}$$

With $(\mathbf{A} - \lambda_i \mathbf{I})\mathbf{v}_k^{(i+m-1)} = \mathbf{0}$, once eigenvectors are determined, the constants $C_k^{(i+j)}$ are obtained. On the other hand, let

$$C_k^{(i+j)} = C_k^{(i+j+1)}. \tag{B.59}$$

Thus, one obtains

$$\begin{aligned} &(\mathbf{A} - \lambda_i \mathbf{I})\mathbf{v}_k^{(i+m-1)} = \mathbf{0}, \\ &(\mathbf{A} - \lambda_i \mathbf{I})\mathbf{v}_k^{(i+j)} = \mathbf{v}_k^{(i+j+1)} (j = 0, 1, 2, \ldots, m-2). \end{aligned} \tag{B.60}$$

Deformation of Eq. (B.60) gives

$$\begin{aligned} &\mathbf{A}\mathbf{v}_k^{(i+m-1)} = \lambda_i \mathbf{v}_k^{(i+m-1)}, \\ &\mathbf{A}\mathbf{v}_k^{(i+j)} = \lambda_i \mathbf{v}_k^{(i+j)} + \mathbf{v}_k^{(i+j+1)} (j = 0, 1, 2, \ldots, m-2), \\ &\mathbf{A}(\mathbf{0}, \ldots, \mathbf{0}, \mathbf{v}_k^{(i)}, \mathbf{v}_k^{(i+1)}, \ldots, \mathbf{v}_k^{(i+m-1)}, \mathbf{0}, \ldots, \mathbf{0}) \\ &= (\mathbf{0}, \ldots, \mathbf{0}, \mathbf{v}_k^{(i)}, \mathbf{v}_k^{(i+1)}, \ldots, \mathbf{v}_k^{(i+m-1)}, \mathbf{0}, \ldots, \mathbf{0})\mathbf{B}^{(i)}; \end{aligned} \tag{B.61}$$

where the Jordan matrix is

$$\begin{aligned} &\mathbf{B}^{(i)} = diag(0_{(i-1)\times(i-1)}, \mathbf{B}_k^{(i)}, \mathbf{0}_{(n-m-i+1)\times(n-m-i+1)}), \\ &\mathbf{B}_k^{(i)} = \begin{bmatrix} \lambda_i & 0 & 0 & \ldots & 0 & 0 \\ 1 & \lambda_i & 0 & \ldots & 0 & 0 \\ 0 & 1 & \lambda_i & \ldots & 0 & 0 \\ \vdots & \vdots & \vdots & \ldots & \vdots & \vdots \\ 0 & 0 & 0 & \ldots & \lambda_i & 0 \\ 0 & 0 & 0 & \ldots & 1 & \lambda_i \end{bmatrix}_{m\times m}. \end{aligned} \tag{B.62}$$

Thus

$$\begin{aligned}\mathbf{AP} &= \mathbf{P}diag(\lambda_1, \ldots, \lambda_{i-1}, \mathbf{B}^{(i)}|_{m\times m}, \lambda_{i+m}, \ldots, \lambda_n), \\ \mathbf{P}^{-1}\mathbf{AP} &= diag(\lambda_1, \ldots, \lambda_{i-1}, \mathbf{B}^{(i)}|_{m\times m}, \lambda_{i+m}, \ldots \lambda_n),\end{aligned} \tag{B.63}$$

where

$$\begin{aligned}\mathbf{P} &= (\mathbf{v}_k^{(1)}, \ldots, \mathbf{v}_k^{(i-1)}, \mathbf{v}_k^{(i)}, \mathbf{v}_k^{(i+1)}, \ldots, \mathbf{v}_k^{(i+m-1)}, \mathbf{v}_k^{(i+m)}, \ldots, \mathbf{v}_k^{(n)}) \\ &= (\mathbf{v}_k^{(1)}, \mathbf{v}_k^{(2)}, \ldots, \mathbf{v}_k^{(n)}).\end{aligned} \tag{B.64}$$

With Eqs. (B.59), Eq. (B.56) becomes

$$\begin{aligned}\mathbf{x}_{k+1}^{(i)} &= \sum\nolimits_{j=0}^{m-1} \lambda_i C_k^{(i+j)} \mathbf{v}_k^{(i+j)} + (C_k^{(i+j+1)} \mathbf{v}_k^{(i+j+1)}) \\ &= (\mathbf{v}_k^{(1)}, \ldots, \mathbf{v}_k^{(i-1)}, \mathbf{v}_k^{(i)}, \mathbf{v}_k^{(i+1)}, \ldots, \mathbf{v}_k^{(i+m-1)}, \mathbf{v}_k^{(i+m)}, \ldots, \mathbf{v}_k^{(n)})\mathbf{B}_k\mathbf{C}_k \\ &= \mathbf{PB}_k\mathbf{C}_k.\end{aligned} \tag{B.65}$$

$$\begin{aligned}\mathbf{C}_k &= (C_k^{(1)}, \ldots, C_k^{(i-1)}, \underbrace{C_k^{(i)}, \ldots, C_k^{(i+m-1)}}_{m}, C_k^{(i+m)}, \ldots, C_k^{(n)})^{\mathrm{T}}, \\ \mathbf{B}_k &= diag(\mathbf{0}_{(i-1)\times(i-1)}, \mathbf{B}_k^{(i)}, \mathbf{0}_{(n-i-m+1)\times(n-i-m+1)}).\end{aligned} \tag{B.66}$$

Therefore,

$$\begin{aligned}\mathbf{x}_{k+1} &= \sum\nolimits_{j=1}^{i-1} \mathbf{x}_{k+1}^{(j)} + \mathbf{x}_{k+1}^{(i)} + \sum\nolimits_{j=i+m-1}^{n} \mathbf{x}_{k+1}^{(j)} \\ &= \mathbf{P}diag(\lambda_1, \ldots, \lambda_{i-1}, \underbrace{0, \ldots, 0}_{m}, \lambda_{i+m}, \ldots, \lambda_n)\mathbf{C}_k \\ &\quad + \mathbf{P}diag(\mathbf{0}_{(i-1)\times(i-1)}, \mathbf{B}_k^{(i)}, \mathbf{0}_{(n-i-m+1)\times(n-i-m+1)})\mathbf{C}_k\end{aligned} \tag{B.67}$$

where

$$\bar{\mathbf{E}} = diag(\lambda_1, \ldots, \lambda_{i-1}, \mathbf{B}_k^{(i)}, \lambda_{i+m}, \ldots, \lambda_n). \tag{B.68}$$

If $\mathbf{x}_{k+1} = \mathbf{x}_k$, $diag[\lambda_1, \lambda_2, \ldots, \lambda_n] = \mathbf{I}$. Thus,

$$\mathbf{P}^{-1}\mathbf{x}_k = \mathbf{C}_k. \tag{B.69}$$

One obtains

$$\mathbf{x}_{k+1} = \mathbf{P}\bar{\mathbf{E}}\mathbf{P}^{-1}\mathbf{x}_k = \mathbf{P}\bar{\mathbf{E}}\mathbf{P}^{-1}\mathbf{x}_k. \tag{B.70}$$

Deformation of $\bar{\mathbf{E}}$ gives

$$\begin{aligned}\bar{\mathbf{E}} &= diag(\lambda_1, \ldots, \lambda_{i-1}, \underbrace{0, \ldots, 0}_{m}, \lambda_{i+m}, \ldots, \lambda_n) \\ &\quad + diag(0, \ldots, 0, \mathbf{B}^{(i)}|_{m\times m}, 0, \ldots, 0)\end{aligned} \tag{B.71}$$

and

$$\mathbf{B}^{(i)} = (\lambda_i\mathbf{I} + \mathbf{PNP}^{-1}). \tag{B.72}$$

where

$$\mathbf{N} = diag(\mathbf{0}_{(i-1)\times(i-1)}, \bar{\mathbf{N}}|_{m\times m}, \mathbf{0}_{(n-i-m+1)\times(n-i-m+1)}). \tag{B.73}$$

The $m \times m$ nilpotent matrix of order m is

$$\bar{\mathbf{N}} = \begin{bmatrix} 0 & 0 & 0 & \dots & 0 & 0 \\ 1 & 0 & 0 & \dots & 0 & 0 \\ 0 & 1 & 0 & \dots & 0 & 0 \\ \vdots & \vdots & \vdots & \dots & \vdots & \vdots \\ 0 & 0 & 0 & \dots & 0 & 0 \\ 0 & 0 & 0 & \dots & 1 & 0 \end{bmatrix}_{m\times m}. \tag{B.74}$$

Finally,

$$\mathbf{x}_{k+1} = \mathbf{P}\left[\mathbf{E} + \mathbf{P}^{-1}\mathbf{N}\mathbf{P}\right]\mathbf{P}^{-1}\mathbf{x}_k \tag{B.75}$$

where

$$\mathbf{E} = diag(\lambda_1, \ldots, \lambda_{i-1}, \underbrace{\lambda_i, \ldots, \lambda_i}_{m}, \lambda_{i+m}, \ldots, \lambda_n). \tag{B.76}$$

Therefore, the iterative solution is

$$\mathbf{x}_k = \mathbf{P}\left[\mathbf{E} + \mathbf{P}^{-1}\mathbf{N}\mathbf{P}\right]^k\mathbf{P}^{-1}\mathbf{x}_0 \tag{B.77}$$

where

$$\begin{aligned} \mathbf{N} &= diag(\mathbf{0}_{(i-1)\times(i-1)}, \bar{\mathbf{N}}|_{m\times m}, \mathbf{0}_{(n-i-m+1)\times(n-i-m+1)}), \\ \mathbf{N}^2 &= diag(\mathbf{0}_{(i-1)\times(i-1)}, \bar{\mathbf{N}}^2|_{m\times m}, \mathbf{0}_{(n-i-m+1)\times(n-i-m+1)}), \\ &\vdots \\ \mathbf{N}^{m-1} &= diag(\mathbf{0}_{(i-1)\times(i-1)}, \bar{\mathbf{N}}^{m-1}|_{m\times m}, \mathbf{0}_{(n-i-m+1)\times(n-i-m+1)}). \end{aligned} \tag{B.78}$$

The $m \times m$ nilpotent matrix of order m (i.e.,$\bar{\mathbf{N}}$) has

$$\bar{\mathbf{N}}^2 = \begin{bmatrix} 0 & 0 & 0 & \dots & 0 & 0 \\ 0 & 0 & 0 & \dots & 0 & 0 \\ 1 & 0 & 0 & \dots & 0 & 0 \\ 0 & 1 & 0 & \dots & 0 & 0 \\ \vdots & \vdots & \vdots & \vdots & \vdots & \vdots \\ 0 & 0 & 0 & \dots & 0 & 0 \end{bmatrix}_{m\times m}, \cdots, \bar{\mathbf{N}}^{m-1} = \begin{bmatrix} 0 & 0 & 0 & \dots & 0 & 0 \\ 0 & 0 & 0 & \dots & 0 & 0 \\ 0 & 0 & 0 & \dots & 0 & 0 \\ \vdots & \vdots & \vdots & \dots & \vdots & \vdots \\ 0 & 0 & 0 & \dots & 0 & 0 \\ 1 & 0 & 0 & \dots & 0 & 0 \end{bmatrix}_{m\times m}. \tag{B.79}$$

Theorem B.5 *Consider a discrete linear dynamical system* $\mathbf{x}_{k+1} = \mathbf{A}\mathbf{x}_k$ *in Eq.* (B.2) *with the initial state of* $\mathbf{x}_k$. *A pair of repeated complex eigenvalue with m-times among the n-pairs of complex eigenvalues of the* $2n \times 2n$ *matrix* $\mathbf{A}$ *is*

$\lambda_j = \alpha_j + \mathbf{i}\beta_j$ and $\bar{\lambda}_j = \alpha_j - \mathbf{i}\beta_j$ ($j = 1, 2, \ldots, n$ and $\mathbf{i} = \sqrt{-1}$). The corresponding eigenvectors are $\mathbf{w}_j = \mathbf{u}_j + \mathbf{i}v_j$ and $\bar{\mathbf{w}}_j = \mathbf{u}_j - \mathbf{i}v_j$. If the corresponding eigenvector matrix of $\mathbf{P} = [\mathbf{u}_1, \mathbf{v}_1, \mathbf{u}_2, \mathbf{v}_2, \ldots, \mathbf{u}_n, \mathbf{v}_n]$ is invertible as a basis in $\Omega \subseteq R^{2n}$. For the repeated complex eigenvalue λ_j, the matrix $\mathbf{A}$ can be decomposed by

$$\mathbf{A} = \mathbf{S} + \mathbf{N} \tag{B.80}$$

where

$$\mathbf{P}^{-1}\mathbf{S}\mathbf{P} = diag(\begin{bmatrix} \alpha_j & \beta_j \\ -\beta_j & \alpha_j \end{bmatrix})_{n\times n}, \tag{B.81}$$

the matrix $\mathbf{N} = \mathbf{A} - \mathbf{S}$ is nilpotent of order $m \leq n$ (i.e., $\mathbf{N}^m = \mathbf{0}$) with $\mathbf{SN} = \mathbf{NS}$.

$$\mathbf{P}^{-1}\mathbf{A}\mathbf{P} = diag[\mathbf{B}_1, \ldots, \mathbf{B}_{i-1}, \underbrace{\mathbf{B}_i, \ldots, \mathbf{B}_i}_{m}, \mathbf{B}_{i+m}, \ldots, \mathbf{B}_n] \tag{B.82}$$

where

$$\mathbf{B}_k = \begin{bmatrix} \alpha_k & \beta_k \\ -\beta_k & \alpha_k \end{bmatrix} (k = 1, 2, \ldots, n). \tag{B.83}$$

Thus, with an initial state of $\mathbf{x}_0$, the solution of linear dynamical system in Eq. (B.2) is

$$\mathbf{x}_{k+1} = \mathbf{P}\big[\mathbf{E} + (\mathbf{P}^{-1}\mathbf{N}\mathbf{P})\big]\mathbf{P}^{-1}\mathbf{x}_k \tag{B.84}$$

where the diagonal matrix $\mathbf{E}$ is given by

$$\left.\begin{aligned} &\mathbf{E} = diag[\mathbf{E}_1, \ldots, \mathbf{E}_{i-1}, \underbrace{\mathbf{E}_i, \ldots, \mathbf{E}_i}_{m}, \mathbf{E}_{i+m}, \ldots, \mathbf{E}_n], \\ &\mathbf{E}_j = r_j \begin{bmatrix} \cos\theta_j & \sin\theta_j \\ -\sin\theta_j & \cos\theta_j \end{bmatrix} (j = 1, 2, \ldots, n). \end{aligned}\right\} \tag{B.85}$$

The iterative solution of linear dynamical system in Eq. (B.2) is

$$\mathbf{x}_k = \mathbf{P}\big[\mathbf{E} + (\mathbf{P}^{-1}\mathbf{N}\mathbf{P})\big]^k\mathbf{P}^{-1}\mathbf{x}_0. \tag{B.86}$$

Proof Consider a pair of repeated complex eigenvalues with $\lambda_j = \alpha_i + \mathbf{i}\beta_j$ and $\bar{\lambda}_j = \alpha_i - \mathbf{i}\beta_j$ of the matrix $\mathbf{A}$, the method of coefficient variation should be adopted. Thus a pair of solutions relative to the two conjugate complex eigenvalue is given by

$$\mathbf{x}^{(i+j)}_{(k+1)+} = \mathbf{C}^{(i+j)}_k(\alpha_i + \mathbf{i}\beta_i) + \mathbf{B}^{(i+j)}_{k+1} \text{ and } \mathbf{x}^{(i+j)}_{(k+1)-} = \bar{\mathbf{C}}^{(i+j)}_k(\alpha_i - \mathbf{i}\beta_i) + \bar{\mathbf{B}}^{(i+j)}_{k+1}.$$

Assume the coefficient vectors for complex eigenvalues as

$$\mathbf{C}_k^{(i+j)} = \mathbf{U}_k^{(i+j)} + \mathbf{i}V_k^{(i+j)} = \frac{1}{2}(U_k^{(i+j)} - \mathbf{i}V_k^{(i+j)})(\mathbf{u}_k^{(i+j)} + \mathbf{i}v_k^{(i+j)})$$
$$= \frac{1}{2}(U_k^{(i+j)}\mathbf{u}_k^{(i+j)} + V_k^{(i+j)}\mathbf{v}_k^{(i+j)}) - \frac{1}{2}\mathbf{i}(V_k^{(i+j)}\mathbf{u}_k^{(i+j)} - U_k^{(i+j)}\mathbf{v}_k^{(i+j)}),$$
$$\mathbf{B}_{k+1}^{(i+j)} = \tilde{\mathbf{U}}_{k+1}^{(i+j)} + \mathbf{i}\tilde{V}_{k+1}^{(i+j)} = \frac{1}{2}(\tilde{U}_{k+1}^{(i+j)} - \mathbf{i}\tilde{V}_{k+1}^{(i+j)})(\mathbf{u}_k^{(i+j)} + \mathbf{i}v_k^{(i+j)})$$
$$= \frac{1}{2}(\tilde{U}_{k+1}^{(i+j)}\mathbf{u}_k^{(i+j)} + \tilde{V}_{k+1}^{(i+j)}\mathbf{v}_k^{(i+j)}) - \frac{1}{2}\mathbf{i}(\tilde{V}_{k+1}^{(i+j)}\mathbf{u}_k^{(i+j)} - \tilde{U}_{k+1}^{(i+j)}\mathbf{v}_k^{(i+j)}),$$

and

$$\bar{\mathbf{C}}_k^{(i+j)} = \mathbf{U}_k^{(i+j)} - \mathbf{i}V_k^{(i+j)} = \frac{1}{2}(U_k^{(i+j)} + \mathbf{i}V_k^{(i+j)})(\mathbf{u}_k^{(i+j)} - \mathbf{i}v_k^{(i+j)})$$
$$= \frac{1}{2}(U_k^{(i+j)}\mathbf{u}_k^{(i+j)} + V_k^{(i+j)}\mathbf{v}_k^{(i+j)}) + \frac{1}{2}\mathbf{i}(V_k^{(i+j)}\mathbf{u}_k^{(i+j)} - U_k^{(i+j)}\mathbf{v}_k^{(i+j)}),$$
$$\bar{\mathbf{B}}_{k+1}^{(i+j)} = \tilde{\mathbf{U}}_{k+1}^{(i+j)} - \mathbf{i}\tilde{V}_{k+1}^{(i+j)} = \frac{1}{2}(\tilde{U}_{k+1}^{(i+j)} + \mathbf{i}\tilde{V}_{k+1}^{(i+j)})(\mathbf{u}_k^{(i+j)} - \mathbf{i}v_k^{(i+j)})$$
$$= \frac{1}{2}(\tilde{U}_{k+1}^{(i+j)}\mathbf{u}_k^{(i+j)} + \tilde{V}_{k+1}^{(i+j)}\mathbf{v}_k^{(i+j)}) + \frac{1}{2}\mathbf{i}(\tilde{V}_{k+1}^{(i+j)}\mathbf{u}_k^{(i+j)} - \tilde{U}_{k+1}^{(i+j)}\mathbf{v}_k^{(i+j)}).$$

Thus,

$$\mathbf{x}_{(k+1)+}^{(i+j)} + \mathbf{x}_{(k+1)-}^{(i+j)} = \mathbf{C}_k^{(i+j)}(\alpha_i + \mathbf{i}\beta_i) + \bar{\mathbf{C}}_k^{(i+j)}(\alpha_i - \mathbf{i}\beta_i) + \mathbf{B}_k^{(i+j)} + \bar{\mathbf{B}}_k^{(i+j)}$$
$$= (\mathbf{U}_k^{(i+j)} + \mathbf{i}\mathbf{V}_k^{(i+j)})(\alpha_i + \mathbf{i}\beta_i) + (\mathbf{U}_k^{(i+j)} - \mathbf{i}\mathbf{V}_k^{(i+j)})(\alpha_i - \mathbf{i}\beta_i)$$
$$+ (\tilde{\mathbf{U}}_{k+1}^{(i+j)} + \mathbf{i}\tilde{\mathbf{V}}_{k+1}^{(i+j)}) + (\tilde{\mathbf{U}}_{k+1}^{(i+j)} - \mathbf{i}\tilde{\mathbf{V}}_{k+1}^{(i+j)})$$
$$= (\mathbf{u}_k^{(i+j)}, \mathbf{v}_k^{(i+j)}) \begin{bmatrix} \alpha_i & \beta_i \\ -\beta_i & \alpha_i \end{bmatrix} \begin{Bmatrix} U_k^{(i+j)} \\ V_k^{(i+j)} \end{Bmatrix} + (\mathbf{u}_k^{(i+j)}, \mathbf{v}_k^{(i+j)}) \begin{Bmatrix} \tilde{U}_{k+1}^{(i+j)} \\ \tilde{V}_{k+1}^{(i+j)} \end{Bmatrix}.$$

Further,

$$\mathbf{A}(\mathbf{x}_{k+}^{(i+j)} + \mathbf{x}_{k-}^{(i+j)}) = \mathbf{A}(\mathbf{u}_k^{(i+j)}, \mathbf{v}_k^{(i+j)}) \begin{bmatrix} \alpha_i & \beta_i \\ -\beta_i & \alpha_i \end{bmatrix} \begin{Bmatrix} U_k^{(i+j)} \\ V_k^{(i+j)} \end{Bmatrix}.$$

The equation of $\mathbf{x}_{(k+1)+}^{(i)} + \mathbf{x}_{(k+1)-}^{(i)} = \mathbf{A}(\mathbf{x}_{k+}^{(i)} + \mathbf{x}_{k-}^{(i)})$ gives

$$(\mathbf{u}_k^{(i+j)}, \mathbf{v}_k^{(i+j)})(\begin{Bmatrix} \tilde{U}_{k+1}^{(i+j)} \\ \tilde{V}_{k+1}^{(i+j)} \end{Bmatrix}) = \left\{ \mathbf{A} - \begin{bmatrix} \alpha_j & \beta_j \\ \beta_j & \alpha_j \end{bmatrix} \mathbf{I}_{2n\times 2n} \right\} (\mathbf{u}_k^{(i+j)}, \mathbf{v}_k^{(i+j)}) \begin{Bmatrix} U_{k+1}^{(i+j)} \\ V_{k+1}^{(i+j)} \end{Bmatrix}.$$

Consider the constant vector and eigenvector matrix as

$$\mathbf{D}_k^{(i+j)} = (0,0,\ldots,0,0,U_k^{(i+j)},V_k^{(i+j)},0,0,\ldots,0,0)^{\mathbf{T}}_{1\times 2n},$$
$$\tilde{\mathbf{D}}_{k+1}^{(i+j)} = (0,0,\ldots,0,0,\tilde{U}_{k+1}^{(i+j)},\tilde{V}_{k+1}^{(i+j)},0,0,\ldots,0,0)^{\mathbf{T}}_{1\times 2n},$$
$$\mathbf{P} = (\mathbf{u}_1,\mathbf{v}_1\ ,\ldots,\mathbf{u}_i,\mathbf{v}_i,\ldots,\mathbf{u}_n,\mathbf{v}_n).$$

Thus

$$\mathbf{P}\tilde{\mathbf{D}}_{k+1}^{(i+j)} = (\mathbf{A} - \begin{bmatrix} \alpha_i & \beta_i \\ -\beta_i & \alpha_i \end{bmatrix}\mathbf{I}_{2n\times 2n})\mathbf{P}\mathbf{D}_k^{(i+j)}.$$

Let $\mathbf{A} = \mathbf{S} + \mathbf{N}$, thus

$$\tilde{\mathbf{D}}_{k+1}^{(i+j)} = (\mathbf{P}^{-1}\mathbf{S}\mathbf{P} - \mathbf{B}_i\mathbf{I}_{2n\times 2n} + \mathbf{P}^{-1}\mathbf{N}\mathbf{P})\mathbf{D}_k^{(i+j)}$$

where

$$\mathbf{B}_i = \begin{bmatrix} \alpha_i & \beta_i \\ -\beta_i & \alpha_i \end{bmatrix}, \mathbf{I}_{2\times 2} = \begin{bmatrix} 1 & 0 \\ 0 & 1 \end{bmatrix} \text{ and}$$
$$\mathbf{I}_{2n\times 2n} = diag(\mathbf{I}_{2\times 2},\mathbf{I}_{2\times 2},\ldots,\mathbf{I}_{2\times 2})_{n\times n}.$$

Because of

$$\mathbf{P}^{-1}\mathbf{S}\mathbf{P} = diag(\begin{bmatrix} \alpha_i & -\beta_i \\ \beta_i & \alpha_i \end{bmatrix})_{n\times n} = diag(\mathbf{B}_i)_{n\times n} = \mathbf{B}_i\mathbf{I}_{2n\times 2n},$$

the solution of the foregoing equation is

$$\tilde{\mathbf{D}}_{k+1}^{(i+j)} = (\mathbf{P}^{-1}\mathbf{N}\mathbf{P})\mathbf{D}_k^{(i+j)}.$$

Further

$$\mathbf{x}_{(k+1)+}^{(i)} + \mathbf{x}_{(k+1)-}^{(i)} = \mathbf{P}(\begin{bmatrix} \cos\theta_j & \sin\theta_j \\ -\sin\theta_j & \cos\theta_j \end{bmatrix}\mathbf{I}_{2m\times 2m} + \mathbf{P}^{-1}\mathbf{N}\mathbf{P})\mathbf{D}_k^{(j)}.$$

Consider the total solution of the complex eigenvalues. Let

$$\mathbf{D}_k = (U_k^{(1)},V_k^{(1)},\ldots,U_k^{(i-1)},V_k^{(i-1)},\underbrace{U_k^{(i)},V_k^{(i)},\ldots,U_k^{(i+m-1)},V_k^{(i+m-1)}}_{m},$$
$$U_k^{(i+m)},V_k^{(i+m)},\ldots U_k^{(n)},V_k^{(n)})^{\mathrm{T}}.$$

Thus, there is a relation

$$\mathbf{P}^{-1}\mathbf{A}\mathbf{P} = diag[\mathbf{B}_1,\ldots,\mathbf{B}_{i-1},\underbrace{\mathbf{B}_i,\ldots,\mathbf{B}_i}_{m},\mathbf{B}_{j+m},\ldots,\mathbf{B}_n]$$

and, the resultant solution for the repeated eigenvalues is

$$\begin{aligned}\mathbf{x}_{k+1} &= \sum\nolimits_{l=0}^{n}(\mathbf{x}^{(l)}_{(k+1)+}+\mathbf{x}^{(l)}_{(k+1)-})(1-\delta^{j}_{l})+\sum\nolimits_{j=0}^{m-1}(\mathbf{x}^{(j)}_{(k+1)+}+\mathbf{x}^{(j)}_{(k+1)-})\\ &= \mathbf{P}\big[\mathbf{E}+(\mathbf{P}^{-1}\mathbf{N}\mathbf{P})\big]\mathbf{D}_{k}.\end{aligned}$$

where the diagonal matrix $\mathbf{E}$ is given by

$$\mathbf{E} = diag[\mathbf{E}_1,\ldots,\mathbf{E}_{i-1},\underbrace{\mathbf{E}_i,\ldots,\mathbf{E}_i}_{m},\mathbf{E}_{i+m},\ldots,\mathbf{E}_n],$$

$$\mathbf{E}_j = r_j\begin{bmatrix}\cos\theta_j & \sin\theta_j\\ -\sin\theta_j & \cos\theta_j\end{bmatrix}(j=1,2,\ldots,n).$$

For $k=0$, using $\mathbf{x}_{k+1}=\mathbf{x}_k$ gives $\mathbf{E} = \mathbf{I}$ and one obtains $\mathbf{D}_k = \mathbf{P}^{-1}\mathbf{x}_k$. Thus, the solution is

$$\mathbf{x}_{k+1} = \mathbf{P}\big[\mathbf{E}+(\mathbf{P}^{-1}\mathbf{N}\mathbf{P})\big]\mathbf{P}^{-1}\mathbf{x}_k.$$

The iterative solution is

$$\mathbf{x}_{k} = \mathbf{P}\big[\mathbf{E}+(\mathbf{P}^{-1}\mathbf{N}\mathbf{P})\big]^{k}\mathbf{P}^{-1}\mathbf{x}_0.$$

This theorem is proved. ∎

Consider the solution for repeated eigenvalues of a linear dynamical system as

$$\begin{aligned}\mathbf{x}^{(i)}_{(k+1)+} &= \sum\nolimits_{j=0}^{m-1} C^{(i+j)}_{k}\mathbf{c}^{(i+j)}_{k}(\alpha_j+\mathbf{i}\beta_j)+C^{(i+j+1)}_{k}\mathbf{c}^{(i+j+1)}_{k},\\ \mathbf{x}^{(i)}_{(k+1)-} &= \sum\nolimits_{j=0}^{m-1} \bar{C}^{(i+j)}_{k}\bar{\mathbf{c}}^{(i+j)}_{k}(\alpha_j-\mathbf{i}\beta_j)+\bar{C}^{(i+j+1)}_{k}\bar{\mathbf{c}}^{(i+j+1)}_{k}.\end{aligned} \tag{B.87}$$

Submission of Eq. (B.87) into $\mathbf{x}^{(i+j+1)}_{(k+1)+} = \mathbf{A}\mathbf{x}^{(i+j)}_{k+}$ and $\mathbf{x}^{(i+j+1)}_{(k+1)-} = \mathbf{A}\mathbf{x}^{(i+j)}_{k-}$ gives

$$\begin{aligned}&\sum\nolimits_{j=0}^{m-1}[\mathbf{A}-(\alpha_j+\mathbf{i}\beta_j)\mathbf{I}_{2n\times 2n}]C^{(i+j)}_{k}\mathbf{c}^{(i+j)}_{k}-C^{(i+j+1)}_{k}\mathbf{c}^{(i+j+1)}_{k}=\mathbf{0},\\ &\sum\nolimits_{j=0}^{m-1}[\mathbf{A}-(\alpha_j-\mathbf{i}\beta_j)\mathbf{I}_{2n\times 2n}]\bar{C}^{(i+j)}_{k}\bar{\mathbf{c}}^{(i+j)}_{k}-\bar{C}^{(i+j+1)}_{k}\bar{\mathbf{c}}^{(i+j+1)}_{k}=\mathbf{0}.\end{aligned} \tag{B.88}$$

Thus

$$\begin{aligned}&[\mathbf{A}-(\alpha_j+\mathbf{i}\beta_j)\mathbf{I}_{2n\times 2n}]C^{(i+j)}_{k}\mathbf{c}^{(i+j)}_{k}-C^{(i+j+1)}_{k}\mathbf{c}^{(i+j+1)}_{k}=\mathbf{0},\\ &[\mathbf{A}-(\alpha_j+\mathbf{i}\beta_j)\mathbf{I}_{2n\times 2n}]\bar{C}^{(i+j)}_{k}\bar{\mathbf{c}}^{(i+j)}_{k}-\bar{C}^{(i+j+1)}_{k}\bar{\mathbf{c}}^{(i+j+1)}_{k}=\mathbf{0}\\ &(i=0,1,2,\ldots m-2).\end{aligned} \tag{B.89}$$

With $[\mathbf{A}-(\alpha_j+\mathbf{i}\beta_j)\mathbf{I}_{2n\times 2n}]\mathbf{c}^{(i+m-1)}_{k}=0$ and $[\mathbf{A}-(\alpha_j-\mathbf{i}\beta_j)\mathbf{I}_{2n\times 2n}]\bar{\mathbf{c}}^{(i+m-1)}_{k}=0$, once eigenvectors are determined, the constants $C^{(j)}_{k}$ are obtained. On the other hand, let

$$C^{(i+j+1)}_{k} = C^{(i+j)}_{k} \text{ and } \bar{C}^{(i+j)}_{k} = \bar{C}^{(i+j+1)}_{k}. \tag{B.90}$$

Thus, one obtains

$$\begin{aligned}
&[\mathbf{A}-(\alpha_j+\mathbf{i}\beta_j)\mathbf{I}_{2n\times 2n}]\mathbf{c}_k^{(i+m-1)}=\mathbf{0},\\
&[\mathbf{A}-(\alpha_j+\mathbf{i}\beta_j)\mathbf{I}_{2n\times 2n}]\mathbf{c}_k^{(i+j)}-\mathbf{c}_k^{(i+j+1)}=\mathbf{0},\\
&[\mathbf{A}-(\alpha_j-\mathbf{i}\beta_j)\mathbf{I}_{2n\times 2n}]\bar{\mathbf{c}}_k^{(i+m-1)}=\mathbf{0},\\
&[\mathbf{A}-(\alpha_j-\mathbf{i}\beta_j)\mathbf{I}_{2n\times 2n}]\bar{\mathbf{c}}_k^{(i+j)}-\bar{\mathbf{c}}_k^{(i+j+1)}=\mathbf{0}\\
&(i=0,1,2,\ldots,m-2).
\end{aligned} \tag{B.91}$$

Assuming

$$\mathbf{c}_k^{(i+j)}=\mathbf{u}_k^{(i+j)}+\mathbf{i}\mathbf{v}_k^{(i+j)} \text{ and } \bar{\mathbf{c}}_k^{(i+j)}=\mathbf{u}_k^{(i+j)}-\mathbf{i}\mathbf{v}_k^{(i+j)}, \tag{B.92}$$

deformation of Eq. (B.91) gives

$$\begin{aligned}
&\mathbf{A}(\mathbf{u}_k^{(i+m-1)},\mathbf{v}_k^{(i+m-1)})=(\mathbf{u}_k^{(i+m-1)},\mathbf{v}_k^{(i+m-1)})\begin{bmatrix}\alpha_i & \beta_i\\ -\beta_i & \alpha_i\end{bmatrix},\\
&\mathbf{A}(\mathbf{u}_k^{(i+j)},\mathbf{v}_k^{(i+j)})=(\mathbf{u}_k^{(i+j)},\mathbf{v}_k^{(i+j)})\begin{bmatrix}\alpha_i & \beta_i\\ -\beta_i & \alpha_i\end{bmatrix}+(\mathbf{u}_k^{(i+j+1)},\mathbf{v}_{k+1}^{(i+j+1)})\\
&(j=0,1,2,\ldots,m-2);
\end{aligned} \tag{B.93}$$

$$\begin{aligned}
&\mathbf{A}(\underbrace{\mathbf{0},\ldots,\mathbf{0}}_{i-1},\underbrace{\mathbf{u}_k^{(i)},\mathbf{v}_k^{(i)},\mathbf{u}_k^{(i+1)},\mathbf{v}_k^{(i+1)},\ldots,\mathbf{u}_k^{(i+m-1)},\mathbf{v}_k^{(i+m-1)}}_{m},\underbrace{\mathbf{0},\ldots,\mathbf{0}}_{n-(i+m)+1})\\
&=(\underbrace{\mathbf{0},\ldots,\mathbf{0}}_{i-1+1},\underbrace{\mathbf{u}_k^{(i)},\mathbf{v}_k^{(i)},\mathbf{u}_k^{(i+1)},\mathbf{v}_k^{(i+1)},\ldots,\mathbf{u}_{m-1}^{(i+m-1)},\mathbf{v}_{m-1}^{(i+m-1)}}_{m},\underbrace{\mathbf{0},\ldots,\mathbf{0}}_{n-(i+m)+1})\mathbf{B}^{(i)}\mathbf{I}_{n\times n},
\end{aligned} \tag{B.94}$$

where the Jordan matrix is

$$\begin{aligned}
&\mathbf{B}^{(i)}=\begin{bmatrix}
\mathbf{D}^{(i)} & \mathbf{0} & \mathbf{0} & \ldots & \mathbf{0} & \mathbf{0}\\
\mathbf{I}_{2\times 2} & \mathbf{D}^{(i)} & \mathbf{0} & \ldots & \mathbf{0} & \mathbf{0}\\
\mathbf{0} & \mathbf{I}_{2\times 2} & \mathbf{D}^{(i)} & \ldots & \mathbf{0} & \mathbf{0}\\
\vdots & \vdots & \vdots & \vdots & \vdots & \vdots\\
\mathbf{0} & \mathbf{0} & \mathbf{0} & \ldots & \mathbf{D}^{(i)} & \mathbf{0}\\
\mathbf{0} & \mathbf{0} & \mathbf{0} & \ldots & \mathbf{I}_{2\times 2} & \mathbf{D}^{(i)}
\end{bmatrix}_{2m\times 2m},\\
&\mathbf{D}^{(i)}=\begin{bmatrix}\alpha_i & \beta_i\\ -\beta_i & \alpha_i\end{bmatrix},\\
&\mathbf{I}_{n\times n}=diag(\mathbf{I}_{(i-1)\times(i-1)},\mathbf{I}_{m\times m},\mathbf{I}_{p\times p}) \text{ with } p=n-i-m+1.
\end{aligned} \tag{B.95}$$

Thus

$$\begin{aligned}
\mathbf{AP}&=\mathbf{P}diag(\mathbf{D}^{(1)},\ldots,\mathbf{D}^{(i-1)},\mathbf{B}^{(i)},\mathbf{D}^{(i+m)},\ldots,\mathbf{D}^{(n)}),\\
\mathbf{P}^{-1}\mathbf{AP}&=diag(\mathbf{D}^{(1)},\ldots,\mathbf{D}^{(i-1)},\mathbf{B}^{(i)},\mathbf{D}^{(i+m)},\ldots,\mathbf{D}^{(n)})
\end{aligned} \tag{B.96}$$

where

$$\begin{aligned}\mathbf{P} &= (\mathbf{u}_k^{(1)}, \mathbf{v}_k^{(1)}, \ldots, \mathbf{u}_k^{(i-1)}, \mathbf{v}_k^{(i-1)}, \mathbf{u}_k^{(i)}, \mathbf{v}_k^{(i)}, \mathbf{u}_k^{(i+1)}, \mathbf{v}_k^{(i+1)}, \\ &\quad \ldots, \mathbf{u}_k^{(i+m-1)}, \mathbf{v}_k^{(i+m-1)}, \mathbf{u}_k^{(i+m)}, \mathbf{v}_k^{(i+m)}, \ldots, \mathbf{u}_k^{(n)}, \mathbf{v}_k^{(n)}) \\ &= (\mathbf{u}_k^{(1)}, \mathbf{v}_k^{(1)}, \mathbf{u}_k^{(2)}, \mathbf{v}_k^{(2)}, \ldots, \mathbf{u}_k^{(n)}, \mathbf{v}_k^{(n)}).\end{aligned} \tag{B.97}$$

Suppose two conjugate constants are

$$C_k^{(i+j)} = \frac{1}{2}(U_k^{(i+j)} - \mathbf{i}V_k^{(i+j)}) \text{ and } \bar{C}_k^{(i+j)} = \frac{1}{2}(U_k^{(i+j)} + \mathbf{i}V_k^{(i+j)}). \tag{B.98}$$

With Eqs. (B.90), equation (B.87) becomes

$$\begin{aligned}\mathbf{x}_{k+1}^{(i)} &= \mathbf{x}_{(k+1)+}^{(i)} + \mathbf{x}_{(k+1)-}^{(i)} \\ &= \sum\nolimits_{j=0}^{m-1} C_k^{(i+j)} \mathbf{c}_k^{(i+j)} (\alpha_j + \mathbf{i}\beta_j) + C_k^{(i+j+1)} \mathbf{c}_k^{(i+j+1)} \\ &+ \sum\nolimits_{j=0}^{m-1} \bar{C}_k^{(i+j)} \bar{\mathbf{c}}_k^{(i+j)} (\alpha_j - \mathbf{i}\beta_j) + \bar{C}_k^{(i+j+1)} \bar{\mathbf{c}}_k^{(i+j+1)} \\ &= (\mathbf{u}_k^{(i)}, \mathbf{v}_k^{(i)}, \mathbf{u}_k^{(i+1)}, \mathbf{v}_k^{(i+1)}, \ldots, \mathbf{u}_k^{(i+m-1)}, \mathbf{v}_k^{(i+m-1)}) \mathbf{B}_i \mathbf{C}_k^{(i)}\end{aligned} \tag{B.99}$$

where

$$\mathbf{B}_i = \mathbf{D}^{(i)} \mathbf{I}_{n \times n} + (\mathbf{P}^{-1}\mathbf{N}\mathbf{P}), \tag{B.100}$$

$$\mathbf{C}_k^{(i)} = (\underbrace{U_k^{(i)}, V_k^{(i)}, \ldots, U_k^{(i+m-1)}, V_k^{(i+m-1)}}_{m})^{\mathrm{T}}. \tag{B.101}$$

Therefore,

$$\begin{aligned}\mathbf{x}_{k+1} &= \mathbf{P} diag(\mathbf{E}_1, \ldots, \mathbf{E}_{i-1}, \mathbf{B}_i, \mathbf{E}_{i+m}, \ldots, \mathbf{E}_n) \mathbf{C}_k \\ &= \mathbf{P}\mathbf{E}\mathbf{C}_k\end{aligned} \tag{B.102}$$

where

$$\begin{aligned}\mathbf{C}_k &= (U_k^{(1)}, V_k^{(1)}, \ldots, U_k^{(i-1)}, V_k^{(i-1)}, \underbrace{U_k^{(i)}, V_k^{(i)}, \ldots, U_k^{(i+m-1)}, V_k^{(i+m-1)}}_{m}, \\ &\quad U_k^{(i+m)}, V_k^{(i+m)}, \ldots, U_k^{(n)}, V_k^{(n)})^{\mathrm{T}}. \\ \mathbf{E} &= \mathbf{P} diag(\mathbf{E}_1, \ldots, \mathbf{E}_{i-1}, \underbrace{\mathbf{E}_i, \ldots, \mathbf{E}_i}_{m}, \mathbf{E}_{i+m}, \ldots, \mathbf{E}_n).\end{aligned} \tag{B.103}$$

For initial conditions, we have $\mathbf{x}_k = \mathbf{P}\mathbf{C}_k$. So $\mathbf{C}_k = \mathbf{P}^{-1}\mathbf{x}_k$. Further,

$$\mathbf{x}_{k+1} = \mathbf{P}\big[\mathbf{E} + (\mathbf{P}^{-1}\mathbf{N}\mathbf{P})\big]\mathbf{P}^{-1}\mathbf{x}_k. \tag{B.104}$$

The $2m \times 2m$ nilpotent matrix of order m is

$$\overline{\mathbf{N}} = \begin{bmatrix} \mathbf{0} & \mathbf{0} & \mathbf{0} & \dots & \mathbf{0} & \mathbf{0} \\ \mathbf{I}_{2\times 2} & \mathbf{0} & \mathbf{0} & \dots & \mathbf{0} & \mathbf{0} \\ \mathbf{0} & \mathbf{I}_{2\times 2} & \mathbf{0} & \dots & \mathbf{0} & \mathbf{0} \\ \vdots & \vdots & \vdots & \vdots & \vdots & \vdots \\ \mathbf{0} & \mathbf{0} & \mathbf{0} & \dots & \mathbf{0} & \mathbf{0} \\ \mathbf{0} & \mathbf{0} & \mathbf{0} & \dots & \mathbf{I}_{2\times 2} & \mathbf{0} \end{bmatrix}_{2m\times 2m}, \tag{B.105}$$

where

$$\mathbf{N} = diag(\mathbf{0}_{2(i-1)\times 2(i-1)}, \overline{\mathbf{N}}|_{2m\times 2m}, \mathbf{0}_{2(n-i-m+1)\times 2(n-i-m+1)}). \tag{B.106}$$

Finally,

$$\mathbf{x}_k = \mathbf{P}\big[\mathbf{E} + (\mathbf{P}^{-1}\mathbf{N}\mathbf{P})\big]^k \mathbf{P}^{-1}\mathbf{x}_0. \tag{B.107}$$

where the $2m \times 2m$ nilpotent matrix of order m $\overline{\mathbf{N}}$ has the following property:

$$\begin{aligned} \overline{\mathbf{N}}^2 &= \begin{bmatrix} \mathbf{0} & \mathbf{0} & \mathbf{0} & \dots & \mathbf{0} & \mathbf{0} \\ \mathbf{0} & \mathbf{0} & \mathbf{0} & \dots & \mathbf{0} & \mathbf{0} \\ \mathbf{I}_{2\times 2} & \mathbf{0} & \mathbf{0} & \dots & \mathbf{0} & \mathbf{0} \\ \mathbf{0} & \mathbf{I}_{2\times 2} & \mathbf{0} & \dots & \mathbf{0} & \mathbf{0} \\ \vdots & \vdots & \vdots & \vdots & \vdots & \vdots \\ \mathbf{0} & \mathbf{0} & \mathbf{0} & \dots & \mathbf{0} & \mathbf{0} \end{bmatrix}_{2m\times 2m}, \\ &\vdots \\ \overline{\mathbf{N}}^{m-1} &= \begin{bmatrix} \mathbf{0} & \mathbf{0} & \mathbf{0} & \dots & \mathbf{0} & \mathbf{0} \\ \mathbf{0} & \mathbf{0} & \mathbf{0} & \dots & \mathbf{0} & \mathbf{0} \\ \mathbf{0} & \mathbf{0} & \mathbf{0} & \dots & \mathbf{0} & \mathbf{0} \\ \vdots & \vdots & \vdots & \vdots & \vdots & \vdots \\ \mathbf{0} & \mathbf{0} & \mathbf{0} & \dots & \mathbf{0} & \mathbf{0} \\ \mathbf{I}_{2\times 2} & \mathbf{0} & \mathbf{0} & \dots & \mathbf{0} & \mathbf{0} \end{bmatrix}_{2m\times 2m} \end{aligned} \tag{B.108}$$

where

$$\begin{aligned} \mathbf{N} &= diag(\mathbf{0}_{2(i-1)\times 2(i-1)}, \overline{\mathbf{N}}|_{2m\times 2m}, \mathbf{0}_{2(n-i-m+1)\times 2(n-i-m+1)}), \\ \mathbf{N}^2 &= diag(\mathbf{0}_{2(i-1)\times 2(i-1)}, \overline{\mathbf{N}}^2|_{2m\times 2m}, \mathbf{0}_{2(n-i-m+1)\times 2(n-i-m+1)}), \\ &\vdots \\ \mathbf{N}^{m-1} &= diag(\mathbf{0}_{2(i-1)\times 2(i-1)}, \overline{\mathbf{N}}^{m-1}|_{2m\times 2m}, \mathbf{0}_{2(n-i-m+1)\times 2(n-i-m+1)}). \end{aligned} \tag{B.109}$$

From the previous discussion, the solutions for homogenous discrete dynamical systems were presented for distinct and repeated eigenvalues. For a nonhomogenous discrete dynamical system, the corresponding solution is presented through the following theorem.

Theorem B.6 *For a linear dynamical system* $\mathbf{x}_{k+1} = \mathbf{A}\mathbf{x}_k + \mathbf{B}$ *in Eq.* (B.1) *with the initial state of* $\mathbf{x}_k$, *the solution of Eq.* (B.1) *is given by*

$$\mathbf{x}_k = \mathbf{P}\mathbf{E}^k\mathbf{P}^{-1}(\mathbf{x}_0 - \mathbf{x}^*) + \mathbf{x}^*, \tag{B.110}$$

where

$$\mathbf{x}_k^{\cdot} = (\mathbf{I} - \mathbf{A})^{-1}\mathbf{B}. \tag{B.111}$$

Proof Letting $\mathbf{x}_{k+1} = \mathbf{x}_k = \mathbf{x}^*$, then $\mathbf{x}_k^{\cdot} = (\mathbf{I} - \mathbf{A})^{-1}\mathbf{B}$.

$$\mathbf{x}_{k+1} - \mathbf{x}_k^{\cdot} = \mathbf{A}(\mathbf{x}_k - \mathbf{x}_k^{\cdot}) \Rightarrow \mathbf{y}_{k+1} = \mathbf{A}\mathbf{y}_k.$$

Since

$$\mathbf{A} = \mathbf{P}\mathbf{E}\mathbf{P}^{-1},$$

one obtains

$$\mathbf{y}_{k+1} = \mathbf{P}\mathbf{E}\mathbf{P}^{-1}\mathbf{y}_k \Rightarrow \mathbf{y}_k = \mathbf{P}\mathbf{E}^k\mathbf{P}^{-1}\mathbf{y}_0.$$

So,

$$\mathbf{x}_k = \mathbf{P}\mathbf{E}\mathbf{P}^{-1}(\mathbf{x}_0 - \mathbf{x}^*) + \mathbf{x}^*.$$

This theorem is proved. ■

B.4 Stability and Boundary

In this section, the stability of discrete dynamical systems will be presented. Compared to continuous dynamical systems, discrete dynamical systems possess much richer stability characteristics.

Definition B.4 For a discrete linear dynamical system $\mathbf{x}_{k+1} = \mathbf{A}\mathbf{x}_k$ in Eq. (B.2), consider a real eigenvalue λ_i of matrix $\mathbf{A}(i \in N = \{1, 2, \ldots, n\})$ and there is a corresponding eigenvector $\mathbf{v}_i$. On the invariant eigenvector $\mathbf{v}_k^{(i)} = \mathbf{v}_i$, consider $\mathbf{x}_k^{(i)} = c_k^{(i)}\mathbf{v}_i$ and $\mathbf{x}_{k+1}^{(i)} = c_{k+1}^{(i)}\mathbf{v}_i = \lambda_i c_k^{(i)}\mathbf{v}_i$, thus, $c_{k+1}^{(i)} = \lambda_i c_k^{(i)}$.

(i) $\mathbf{x}_k^{(i)}$ on the direction $\mathbf{v}_i$ is stable if

$$\lim_{k\to\infty} |c_k^{(i)}| = \lim_{k\to\infty} |(\lambda_i)^k| \times |c_0^{(i)}| = 0 \text{ for } |\lambda_i| < 1. \tag{B.112}$$

(ii) $\mathbf{x}_k^{(i)}$ on the direction $\mathbf{v}_i$ is unstable if

$$\lim_{k\to\infty} |c_k^{(i)}| = \lim_{k\to\infty} |(\lambda_i)^k| \times |c_0^{(i)}| = \infty \text{ for } |\lambda_i| > 1. \quad \text{(B.113)}$$

(iii) $\mathbf{x}_k^{(i)}$ on the direction $\mathbf{v}_i$ is invariant if

$$\lim_{k\to\infty} c_k^{(i)} = \lim_{k\to\infty} (\lambda_i)^k c_0^{(i)} = c_0^{(i)} \text{ for } \lambda_i = 1. \quad \text{(B.114)}$$

(iv) $\mathbf{x}_k^{(i)}$ on the direction $\mathbf{v}_i$ is flipped if

$$\left.\begin{aligned} \lim_{2k\to\infty} c_k^{(i)} &= \lim_{2k\to\infty} (\lambda_i)^{2k} \times c_0^{(i)} = c_0^{(i)} \\ \lim_{2k+1\to\infty} c_k^{(i)} &= \lim_{2k+1\to\infty} (\lambda_i)^{2k+1} \times c_0^{(i)} = -c_0^{(i)} \end{aligned}\right\} \text{for } \lambda_i = -1. \quad \text{(B.115)}$$

(v) $\mathbf{x}_k^{(i)}$ on the direction $\mathbf{v}_i$ is degenerate if

$$c_k^{(i)} = (\lambda_i)^k c_0^{(i)} = 0 \text{ for } \lambda_i = 0. \quad \text{(B.116)}$$

Definition B.5 For a discrete linear dynamical system $\mathbf{x}_{k+1} = \mathbf{A}\mathbf{x}_k$ in Eq. (B.2), consider a pair of complex eigenvalue $\alpha_i \pm \mathbf{i}\beta_i$ of matrix $\mathbf{A}(i \in N = \{1, 2, \ldots, n\}$, $\mathbf{i} = \sqrt{-1})$ and there is a corresponding eigenvector $\mathbf{u}_i \pm \mathbf{i}v_i$. On the invariant plane of $(\mathbf{u}_k^{(i)}, \mathbf{v}_k^{(i)}) = (\mathbf{u}_i, \mathbf{v}_i)$, consider $\mathbf{x}_k^{(i)} = \mathbf{x}_{k+}^{(i)} + \mathbf{x}_{k-}^{(i)}$ with

$$\mathbf{x}_k^{(i)} = c_k^{(i)}\mathbf{u}_i + d_k^{(i)}\mathbf{v}_i,\ \mathbf{x}_{k+1}^{(i)} = c_{k+1}^{(i)}\mathbf{u}_i + d_{k+1}^{(i)}\mathbf{v}_i. \quad \text{(B.117)}$$

Thus, $\mathbf{c}_k^{(i)} = (c_k^{(i)}, d_k^{(i)})^{\mathrm{T}}$ with

$$\mathbf{c}_{k+1}^{(i)} = \mathbf{E}_i\mathbf{c}_k^{(i)} = r_i\mathbf{R}_i\mathbf{c}_k^{(i)} \quad \text{(B.118)}$$

where

$$\mathbf{E}_i = \begin{bmatrix} \alpha_i & \beta_i \\ -\beta_i & \alpha_i \end{bmatrix} \text{ and } \mathbf{R}_i = \begin{bmatrix} \cos\theta_i & \sin\theta_i \\ -\sin\theta_i & \cos\theta_i \end{bmatrix}, \quad \text{(B.119)}$$
$$r_i = \sqrt{\alpha_i^2 + \beta_i^2}, \cos\theta_i = \alpha_i/r_i \text{ and } \sin\theta_i = \beta_i/r;$$

and

$$\mathbf{E}_i^k = \begin{bmatrix} \alpha_i & \beta_i \\ -\beta_i & \alpha_i \end{bmatrix}^k \text{ and } \mathbf{R}_i^k = \begin{bmatrix} \cos k\theta_i & \sin k\theta_i \\ -\sin k\theta_i & \cos k\theta_i \end{bmatrix}. \quad \text{(B.120)}$$

(i) $\mathbf{x}_k^{(i)}$ on the plane of $(\mathbf{u}_i, \mathbf{v}_i)$ is spirally stable if

$$\lim_{k\to\infty}||\mathbf{c}_k^{(i)}|| = \lim_{k\to\infty} r_i^k||\mathbf{R}_i^k|| \times ||\mathbf{c}_0^{(i)}|| = 0 \text{ for } r_i = |\lambda_i|<1. \tag{B.121}$$

(ii) $\mathbf{x}_k^{(i)}$ on the plane of $(\mathbf{u}_i, \mathbf{v}_i)$ is spirally unstable if

$$\lim_{k\to\infty}||\mathbf{c}_k^{(i)}|| = \lim_{k\to\infty} r_i^k||\mathbf{R}_i^k|| \times ||\mathbf{c}_0^{(i)}|| = \infty \text{ for } r_i = |\lambda_i| > 1. \tag{B.122}$$

(iii) $\mathbf{x}_k^{(i)}$ on the plane of $(\mathbf{u}_i, \mathbf{v}_i)$ is on the invariant circles if,

$$||\mathbf{c}_k^{(i)}|| = r_i^k||\mathbf{R}_i^k|| \times ||\mathbf{c}_0^{(i)}|| = ||\mathbf{c}_0^{(i)}|| \text{ for } r_i = |\lambda_i| = 1. \tag{B.123}$$

(iv) $\mathbf{x}_k^{(i)}$ on the plane of $(\mathbf{u}_i, \mathbf{v}_i)$ is degenerate in the direction of $\mathbf{u}_i$ if $\beta_i = 0$.

Definition B.6 For a discrete linear dynamical system of $\mathbf{x}_{k+1} = \mathbf{A}\mathbf{x}_k$ in Eq. (B.2), the matrix $\mathbf{A}$ has n_1 real eigenvalues $|\lambda_j|<1$ $(j \in N_1)$, n_2 real eigenvalues $|\lambda_j| > 1$ $(j \in N_2)$, n_3 real eigenvalues $\lambda_j = 1$ $(j \in N_3)$, and n_4 real eigenvalues $\lambda_j = -1$ $(j \in N_4)$. Set $N = \{1, 2, \ldots, n\}$ and $N_i = \{i_1, i_2, \ldots, i_{n_i}\} \cup \varnothing$ $(i = 1, 2, 3, 4)$ with $i_m \in N$ $(m = 1, 2, \ldots, n_i)$. $N_i \subseteq N \cup \varnothing$, $\cup_{i=1}^3 N_i = N$, $N_i \cap N_p = \varnothing$ $(p \neq i)$ and $\Sigma_{i=1}^3 n_i = n$. $N_i = \varnothing$ if $n_i = 0$. The corresponding eigenvectors for contraction, expansion, invariance and flip oscillation are $\{\mathbf{v}_j\}$ $(j \in N_i)$ $(i = 1, 2, 3, 4)$, respectively. The stable, unstable, invariant and flip subspaces of $\mathbf{x}_{k+1} = \mathbf{A}\mathbf{x}_k$ in Eq. (B.2) are linear subspace spanned by $\{\mathbf{v}_j\}$ $(j \in N_i)$ $(i = 1, 2, 3, 4)$, respectively,

$$\begin{aligned}
\mathscr{E}^{\mathrm{s}} &= span\{\mathbf{v}_j|(\mathbf{A} - \lambda_j\mathbf{I})\mathbf{v}_j = \mathbf{0}, |\lambda_j|<1, j \in N_1 \subseteq N \cup \varnothing\};\\
\mathscr{E}^{\mathrm{u}} &= span\{\mathbf{v}_j|(\mathbf{A} - \lambda_j\mathbf{I})\mathbf{v}_j = \mathbf{0}, |\lambda_j| > 1, j \in N_2 \subseteq N \cup \varnothing\};\\
\mathscr{E}^{\mathrm{i}} &= span\{\mathbf{v}_j|(\mathbf{A} - \lambda_j\mathbf{I})\mathbf{v}_j = \mathbf{0}, \lambda_j = 1, j \in N_3 \subseteq N \cup \varnothing\};\\
\mathscr{E}^{\mathrm{f}} &= span\{\mathbf{v}_j|(\mathbf{A} - \lambda_j\mathbf{I})\mathbf{v}_j = \mathbf{0}, \lambda_j = -1, j \in N_4 \subseteq N \cup \varnothing\};
\end{aligned} \tag{B.124}$$

where

$$\begin{aligned}
&\mathscr{E}^{\mathrm{s}} = \mathscr{E}^{\mathrm{s}}_{\mathrm{m}} \cup \mathscr{E}^{\mathrm{s}}_{\mathrm{o}} \cup \mathscr{E}^{\mathrm{s}}_{\mathrm{z}} \text{ with}\\
&\mathscr{E}^{\mathrm{s}}_{\mathrm{m}} = span\{\mathbf{v}_j|(\mathbf{A} - \lambda_j\mathbf{I})\mathbf{v}_j = \mathbf{0}, 0<\lambda_j<1, j \in N_1^{\mathrm{m}} \subseteq N \cup \varnothing\};\\
&\mathscr{E}^{\mathrm{s}}_{\mathrm{o}} = span\{\mathbf{v}_j|(\mathbf{A} - \lambda_j\mathbf{I})\mathbf{v}_j = \mathbf{0}, -1<\lambda_j<0, j \in N_1^{\mathrm{o}} \subseteq N \cup \varnothing\};\\
&\mathscr{E}^{\mathrm{s}}_{\mathrm{z}} = span\{\mathbf{v}_j|(\mathbf{A} - \lambda_j\mathbf{I})\mathbf{v}_j = \mathbf{0}, \lambda_j = 0, j \in N_1^{\mathrm{z}} \subseteq N \cup \varnothing\};
\end{aligned} \tag{B.125}$$

$$\begin{aligned}
&\mathscr{E}^{\mathrm{u}} = \mathscr{E}^{\mathrm{u}}_{\mathrm{m}} \cup \mathscr{E}^{\mathrm{u}}_{\mathrm{o}} \text{ with}\\
&\mathscr{E}^{\mathrm{u}}_{\mathrm{m}} = span\{\mathbf{v}_j|(\mathbf{A} - \lambda_j\mathbf{I})\mathbf{v}_j = \mathbf{0}, \lambda_j > 1, j \in N_2^{\mathrm{m}} \subseteq N \cup \varnothing\};\\
&\mathscr{E}^{\mathrm{u}}_{\mathrm{o}} = span\{\mathbf{v}_j|(\mathbf{A} - \lambda_j\mathbf{I})\mathbf{v}_j = \mathbf{0}, \lambda_j< -1, j \in N_2^{\mathrm{o}} \subseteq N \cup \varnothing\};
\end{aligned} \tag{B.126}$$

where subscripts "m" and "o" represent the monotonic and oscillatory evolutions.

Definition B.7 For a discrete linear dynamical system $\mathbf{x}_{k+1} = \mathbf{A}\mathbf{x}_k$ in Eq. (B.2), the matrix $\mathbf{A}$ has complex eigenvalues $\alpha_j \pm \mathbf{i}\beta_j$ with eigenvectors $\mathbf{u}_j \pm \mathbf{i}\mathbf{v}_j$ $(j \in \{1, 2, \ldots, n\})$ and the base of vector is

$$\mathbf{B} = \{\mathbf{u}_1, \mathbf{v}_1, \ldots, \mathbf{u}_j, \mathbf{v}_j, \ldots, \mathbf{u}_n, \mathbf{v}_n\}. \tag{B.127}$$

The stable, unstable, center subspaces of $\mathbf{x}_{k+1} = \mathbf{A}\mathbf{x}_k$ in Eq. (B.2) are linear subspaces spanned by $\{\mathbf{u}_j, \mathbf{v}_j\}$ $(j \in N_i,\ i = 1, 2, 3)$, respectively. $N = \{1, 2, \ldots, n\}$ and $N_i = \{i_1, i_2, \ldots, i_{n_i}\} \cup \varnothing \subseteq N \cup \varnothing$ with $i_m \in N$ $(m = 1, 2, \ldots, n_i, i = 1, 2, 3)$. $\cup_{i=1}^{3} N_i = N$ with $N_i \cap N_p = \varnothing (p \neq i)$ and $\Sigma_{i=1}^{3} n_i = n$. $N_i = \varnothing$ if $n_i = 0$. The stable, unstable, center subspaces of $\mathbf{x}_{k+1} = \mathbf{A}\mathbf{x}_k$ in Eq. (B.2) are defined by

$$\begin{aligned}
\mathscr{E}^{\mathrm{s}} &= span\left\{ (\mathbf{u}_j, \mathbf{v}_j) \left| \begin{array}{c} r_j = \sqrt{\alpha_j^2 + \beta_j^2} < 1, \\ (\mathbf{A} - (\alpha_j \pm \mathbf{i}\beta_j)\mathbf{I})(\mathbf{u}_j \pm \mathbf{i}\mathbf{v}_j) = \mathbf{0}, \\ j \in N_1 \subseteq \{1, 2, \ldots, n\} \cup \varnothing \end{array} \right. \right\}; \\
\mathscr{E}^{\mathrm{u}} &= span\left\{ (\mathbf{u}_j, \mathbf{v}_j) \left| \begin{array}{c} r_j = \sqrt{\alpha_j^2 + \beta_j^2} > 1, \\ (\mathbf{A} - (\alpha_j \pm \mathbf{i}\beta_j)\mathbf{I})(\mathbf{u}_j \pm \mathbf{i}\mathbf{v}_j) = \mathbf{0}, \\ j \in N_2 \subseteq \{1, 2, \ldots, n\} \cup \varnothing \end{array} \right. \right\}; \\
\mathscr{E}^{\mathrm{c}} &= span\left\{ (\mathbf{u}_j, \mathbf{v}_j) \left| \begin{array}{c} r_j = \sqrt{\alpha_j^2 + \beta_j^2} = 1, \\ (\mathbf{A} - (\alpha_j \pm \mathbf{i}\beta_j)\mathbf{I})(\mathbf{u}_j \pm \mathbf{i}\mathbf{v}_j) = \mathbf{0}, \\ j \in N_3 \subseteq \{1, 2, \ldots, n\} \cup \varnothing \end{array} \right. \right\}.
\end{aligned} \tag{B.128}$$

Definition B.8 For a discrete linear dynamical system $\mathbf{x}_{k+1} = \mathbf{A}\mathbf{x}_k$ in Eq. (B.2),

(i) the linear discrete system is stable if

$$\lim_{k\to\infty} \|\mathbf{x}_k\| = \lim_{k\to\infty} \|\mathbf{A}^k \mathbf{x}_0\| = 0 \text{ for } \mathbf{x}_0 \in \Omega \subseteq \mathscr{R}^n, \tag{B.129}$$

(ii) the linear discrete system is unstable if

$$\lim_{k\to\infty} \|\mathbf{x}_k\| = \lim_{k\to\infty} \|\mathbf{A}^k \mathbf{x}_0\| = \infty \text{ for } \mathbf{x}_0 \in \Omega \subseteq \mathscr{R}^n, \tag{B.130}$$

(iii) the discrete origin of the linear system is a center if,

$$\lim_{k\to\infty} \|\mathbf{x}_k\| = \lim_{k\to\infty} \|\mathbf{A}^k \mathbf{x}_0\| = C \text{ for } \mathbf{x}_0 \in \Omega \subseteq \mathscr{R}^n. \tag{B.131}$$

Theorem B.7 *Consider a discrete linear dynamical system* $\mathbf{x}_{k+1} = \mathbf{A}\mathbf{x}_k$ *in Eq.* (B.2) *and the matrix* $\mathbf{A}$ *possesses n eigenvalues* λ_i $(i = 1, 2, \ldots n)$. *Let* $N = \{1, 2, \ldots, n\}$, *and* $N = \cup_{j=3}^{3} N_j$ *with* $N_j \cap N_p = \varnothing$ $(j, p = 1, 2, 3,\ j \neq p)$.

(i) *If* $|\lambda_i| > 1$ *for* $i \in N$, *the linear discrete system is unstable.*

(ii) *If* $|\lambda_i|<1$ *for all* $i \in N$ *with distinct eigenvalues, the linear discrete system is stable.*

(iii) If$|\lambda_i|<1$ *for* all $i \in N_1 \neq N$*and* $|\lambda_j|=1$ *for* all $j \in N_3 \subseteq N$ and $N_2=\varnothing$ *with distinct eigenvalues, the linear discrete system is stable.*

(iv) If $|\lambda_i|<1$ *for* all $i \in N_1 \neq N$*and* $|\lambda_j|=1$ *for* all$j \in N_3 \subseteq N$ and $N_2=\varnothing$ *with repeated eigenvalues with the* mth *-order nilpotent matrix* $\mathbf{N}^m=\mathbf{0}$ $(1<m\leq n)$, *the linear discrete system is unstable.*

(v) *If* $|\lambda_i|<1$ *for* all $i \in N_1 \neq N$*and* $|\lambda_j|=1$ *for* all $j \in N_3 \subseteq N$ and $N_2=\varnothing$ *with repeated eigenvalues with* $\mathbf{N}=\mathbf{0}$,*the linear discrete system is stable.*

Proof For the different real eigenvalues, from Eq. (B.15),

$$\mathbf{x}_k=\mathbf{P}diag[\lambda_1^k,\lambda_2^k,\ldots,\lambda_n^k]\mathbf{P}^{-1}\mathbf{x}_0=\mathbf{P}\mathbf{E}^k\mathbf{P}^{-1}\mathbf{x}_0.$$

For non-degenerate matrix $\mathbf{P}$ ($\det\mathbf{P}\neq 0$), and

$$\lim_{k\to\infty}|\lambda_i^k|=\begin{cases}0 \text{ if } |\lambda_i|<1,\\ \infty \text{ if } |\lambda_i|>1,\\ 1 \text{ if } |\lambda_i|=1\end{cases} \text{ for } i=1,2,\ldots,n.$$

Thus for an arbitrary $\mathbf{x}_0$, the following relations exist

$$\lim_{k\to\infty}||\mathbf{x}_k||=0, \text{ for } |\lambda_i|<1\ (i=1,2,\ldots,n),$$

$$\lim_{k\to\infty}||\mathbf{x}_k||=\infty, \text{ for } |\lambda_i|>1\ (i\in\{1,2,\ldots,n\}),$$

$$\lim_{k\to\infty}||\mathbf{x}_k||=||\mathbf{x}_0||, \text{ for } |\lambda_i|=1\ (i=1,2,\ldots,n),$$

$$\lim_{k\to\infty}||\mathbf{x}_k||=\text{C}, \text{ for } |\lambda_i|=1 \text{ and } |\lambda_j|<1\ (\text{all } j\in\{1,2,\ldots,n\}, j\neq i),$$

$$\lim_{k\to\infty}||\mathbf{x}_k||=\infty, \text{ for } |\ \lambda_i|=1 \text{ and } |\ \lambda_j|>1\ (\text{all } j\in\{1,2,\ldots,n\}, j\neq i).$$

For different complex eigenvalues, from Eq. (B.48) with $n=2m$

$$\mathbf{x}_k=\mathbf{P}\mathbf{E}(k)\mathbf{P}^{-1}\mathbf{x}_0=\mathbf{P}diag(\mathbf{E}_1(k),\mathbf{E}_2(k),\ldots,\mathbf{E}_m(k))\mathbf{P}^{-1}\mathbf{x}_0$$

where

$$||\mathbf{E}_i(k)||=2|r_i|^k(|\cos k\theta_i|+|\sin k\theta_i|)(i=1,2,\ldots,m);$$

$$\text{with } |\cos k\theta_i|\leq 1 \text{ and } |\sin k\theta_i|\leq 1;$$

$$|r_i|=\sqrt{\alpha_i^2+\beta_i^2}, \cos\theta_i=\alpha_i/\sqrt{\alpha_i^2+\beta_i^2} \text{ and } \sin\theta_i=\beta_i/\sqrt{\alpha_i^2+\beta_i^2}.$$

$$\lim_{k\to\infty} ||\mathbf{E}_i(k)|| = \begin{cases} 0 \text{ if } |\lambda_i| \equiv r_i < 1, \\ \infty \text{ if } |\lambda_i| \equiv r_i > 1, \\ 1 \text{ if } |\lambda_i| \equiv r_i = 1 \end{cases} \text{ for } i = 1, 2, \ldots, m.$$

Since non-degenerate matrix $\mathbf{P}$ ($\det \mathbf{P} \neq 0$), as $k \to \infty$,

$$\lim_{k\to\infty} ||\mathbf{x}_k|| = \begin{cases} 0 \text{ for } |\lambda_i| < 1 \ (i = 1, 2, \ldots, n), \\ \infty \text{ for } |\lambda_i| > 1 \ (i \in \{1, 2, \ldots, n\}), \\ ||\mathbf{x}_0|| \text{ for } |\lambda_i| = 1 \ (i = 1, 2, \ldots, n), \\ \text{C for } |\lambda_i| = 1 \text{ and } |\lambda_j| < 1 \ (\text{all } j \in \{1, 2, \ldots, n\}, j \neq i), \\ \infty \text{ for } |\lambda_i| = 1 \text{ and } |\lambda_j| > 1 \ (\text{all } j \in \{1, 2, \ldots, n\}, j \neq i). \end{cases}$$

For the repeated eigenvalues λ_i with the mth-order nilpotent matrix of $\mathbf{N}$, one obtains $\mathbf{N}^m = \mathbf{0}$. Consider

$$\left[\mathbf{E} + (\mathbf{P}^{-1}\mathbf{N}\mathbf{P})\right]^k = \sum_{j=0}^{m} C_k^j \mathbf{E}^{k-j}(\mathbf{P}^{-1}\mathbf{N}^j\mathbf{P}) \text{ with } C_k^j = \frac{k!}{(k-j)!j!}.$$

For the repeated real eigenvalues λ_i , with Eq. (B.53), one obtains

$$\mathbf{E}^k = diag[\lambda_1^k, \ldots \lambda_{i-1}^k, \underbrace{\lambda_i^k, \ldots, \lambda_i^k}_{m}, \lambda_{i+m}^k, \ldots, \lambda_n^k],$$

Thus,

$$\lim_{k\to\infty} ||\mathbf{N}^j|| = K \text{ and } ||\mathbf{E}^{k-j}|| = \sum_{i=0}^{n} |\lambda_i^{k-j}|,$$

$$\lim_{k\to\infty} C_k^j |\lambda_i^{k-j}| = \lim_{k\to\infty} \frac{k!}{(k-j)!j!} |\lambda_i^k| = \begin{cases} 0 \text{ if } |\lambda_i| < 1, \\ \infty \text{ if } |\lambda_i| > 1, \\ \infty \text{ if } |\lambda_i| = 1 \end{cases}$$

for $i = 1, 2, \ldots, n$ and $j = 1, 2, \ldots, m - 1$.

From Eq. (B.54),

$$\begin{aligned} \mathbf{x}_k &= \mathbf{P}\left[\mathbf{E} + (\mathbf{P}^{-1}\mathbf{N}\mathbf{P})\right]^k \mathbf{P}^{-1}\mathbf{x}_0 \\ &= \mathbf{P}\left[\sum_{j=0}^{m} C_k^j \mathbf{E}^{k-j}(\mathbf{P}^{-1}\mathbf{N}^j\mathbf{P})\right]\mathbf{P}^{-1}\mathbf{x}_0. \end{aligned}$$

As $k \to \infty$,

$$\lim_{k\to\infty} ||\mathbf{x}_k|| = \begin{cases} 0 \text{ for } |\lambda_i| < 1 \ (i = 1, 2, \ldots, n), \\ \infty \text{ for } |\lambda_i| > 1 \ (i \in \{1, 2, \ldots, n\}). \end{cases}$$

If $|\lambda_i| = 1 (i \in \{1, 2, \ldots, n\})$ is relative to repeated eigenvalues,

$$\lim_{k\to\infty} ||\mathbf{x}_k|| = \infty \text{ for } |\lambda_j| < 1 \text{ (all } j \in \{1, 2, \ldots, n\}, j \neq i),$$

$$\lim_{k\to\infty} ||\mathbf{x}_k|| = \infty \text{ for } |\lambda_j| > 1 \text{ (all } j \in \{1, 2, \ldots, n\}, j \neq i).$$

If $|\lambda_i| = 1 (i \in \{1, 2, \ldots, n\})$ is relative to non-repeated eigenvalues,

$$\lim_{k\to\infty} ||\mathbf{x}_k|| = C \text{ for } |\lambda_j| < 1 \text{ (all } j \in \{1, 2, \ldots, n\}, j \neq i),$$

$$\lim_{k\to\infty} ||\mathbf{x}_k|| = \infty \text{ for } |\lambda_j| > 1 \text{ (all } j \in \{1, 2, \ldots, n\}, j \neq i).$$

For the repeated complex eigenvalues, with Eq. (B.83), one obtains

$$\mathbf{E}^k = diag[\mathbf{E}_1, \ldots, \mathbf{E}_{i-1}, \underbrace{\mathbf{E}_i, \ldots, \mathbf{E}_i}_{m}, \mathbf{E}_{i+m-1}, \ldots, \mathbf{E}_n],$$

$$\mathbf{E}_i^k = r_i^k \begin{bmatrix} \cos k\theta_i & \sin k\theta_i \\ -\sin k\theta_i & \cos k\theta_i \end{bmatrix} (i = 1, 2, \ldots, n).$$

Thus,

$$\lim_{k\to\infty} ||\mathbf{N}^l|| = K \text{ and } ||\mathbf{E}^{k-l}|| = \sum\nolimits_{i=0}^{n} ||\mathbf{E}_i^{k-l}||;$$

$$||\mathbf{E}_i^{k-l}|| = 2|r_i|^{k-l}(|\cos(k-l)\theta_i| + |\sin(k-l)\theta_i|) \ (i = 1, 2, \ldots, n);$$

with $|\cos(k-l)\theta_i| \leq 1$ and $|\sin(k-l)\theta_i| \leq 1$;

$$|r_i| = \sqrt{\alpha_i^2 + \beta_i^2}, \cos\theta_i = \alpha_j / \sqrt{\alpha_i^2 + \beta_i^2} \text{ and } \sin\theta_i = \beta_i / \sqrt{\alpha_i^2 + \beta_i^2}.$$

$$\lim_{k\to\infty} ||C_k^j \mathbf{E}_i^{k-l}|| = \lim_{k\to\infty} \frac{k!}{(k-l)!l!} ||\mathbf{E}_i^{k-l}|| = \begin{cases} 0 \text{ if } |\lambda_i| < 1, \\ \infty \text{ if } |\lambda_i| > 1, \\ \infty \text{ if } |\lambda_i| = 1 \end{cases}$$

for $i = 1, 2, \ldots, n$ and $l = 1, 2, \ldots, m-1$.

From Eq. (B.84) for repeated complex eigenvalues, one achieves

$$\begin{aligned} \mathbf{x}_k &= \left[\mathbf{E} + (\mathbf{P}^{-1}\mathbf{N}\mathbf{P})\right]^k \mathbf{x}_0 \\ &= \mathbf{P}\left[\sum\nolimits_{j=0}^{m} C_k^j \mathbf{E}^{k-j} (\mathbf{P}^{-1}\mathbf{N}^j\mathbf{P})\right] \mathbf{P}^{-1}\mathbf{x}_0. \end{aligned}$$

As $k \to \infty$,

$$\lim_{k\to\infty} ||\mathbf{x}_k|| = \begin{cases} 0 \text{ for } |\lambda_i| < 1 \ (i = 1, 2, \ldots, n), \\ \infty \text{ for } |\lambda_i| > 1 \ (i \in \{1, 2, \ldots, n\}). \end{cases}$$

If $|\lambda_i| = 1 (i \in \{1,2,\ldots,n\})$ is relative to repeated complex eigenvalue,

$$\lim_{k\to\infty} ||\mathbf{x}_k|| = \infty \text{ for } |\lambda_j| < 1 \ (j \in \{1,2,\ldots,n\}, j \neq i),$$
$$\lim_{k\to\infty} ||\mathbf{x}_k|| = \infty \text{ for } |\lambda_j| > 1 \ (j \in \{1,2,\ldots,n\}, j \neq i).$$

If $|\lambda_i| = 1 (i \in \{1,2,\ldots,n\})$ is relative to non-repeated complex eigenvalue,

$$\lim_{k\to\infty} ||\mathbf{x}_k|| = C \text{ for } |\lambda_j| < 1 \ (j \in \{1,2,\ldots,n\}, j \neq i),$$
$$\lim_{k\to\infty} ||\mathbf{x}_k|| = \infty \text{ for } |\lambda_j| > 1 \ (j \in \{1,2,\ldots,n\}, j \neq i).$$

For repeated eigenvalue $\lambda_i = \lambda (i = 1,2,\ldots,n)$ with the zero-order nilpotent matrix of $\mathbf{N}$, one obtains

$$\lim_{k\to\infty} ||\mathbf{x}_k|| = \begin{cases} 0 & \text{for } |\lambda_i| = \lambda < 1 \ (i = 1,2,\ldots,n), \\ \infty & \text{for } |\lambda_i| = \lambda > 1 \ (i = 1,2,\ldots,n), \\ ||\mathbf{x}_0|| & \text{for } |\lambda_i| = \lambda = 1 \ (i = 1,2,\ldots,n). \end{cases}$$

From the above facts and Definition B.6, all the cases in this theorem exist, vice versa. This theorem is proved. ■

Definition B.9 Consider a discrete linear dynamical system $\mathbf{x}_{k+1} = \mathbf{A}\mathbf{x}_k$ in Eq. (B.2), the matrix $\mathbf{A}$ possesses n eigenvalues λ_i $(i = 1, 2, \ldots n)$.

(i) The origin is called a hyperbolic fixed point for the linear discrete system if $|\lambda_i| \neq 1$ $(i = 1, 2, \ldots, n)$.
(ii) The origin is called a sink for the linear discrete system if $|\lambda_i| < 1$ $(i = 1, 2, \ldots, n)$.
(iii) The origin is called a source for the linear discrete system if $|\lambda_i| > 1$ $(i = 1, 2, \ldots, n)$.
(iv) The origin is called a center for the linear discrete system if $|\lambda_i| = 1$ $(i = 1, 2, \ldots, n)$ with distinct eigenvalues.
(v) The origin is called a source for the linear discrete system if $|\lambda_i| = 1$ $(i \in \{1, 2, \ldots, n\})$ with at least one repeated eigenvalues with the mth-order nilpotent matrix $\mathbf{N}^m = \mathbf{0}$ $(1 < m < n)$.

Definition B.10 For an n-dimensional, linear discrete system $\mathbf{x}_{k+1} = \mathbf{A}\mathbf{x}_k$ in Eq. (B.2), the matrix $\mathbf{A}$ possesses n real eigenvalues λ_i $(i = 1, 2, \ldots, n)$.

(i) The origin is called a stable node for the linear discrete system if $|\lambda_i| < 1$ $(i = 1, 2, \ldots, n)$.
(ii) The origin is called an unstable node for the linear discrete system if $|\lambda_i| > 1$ $(i = 1, 2, \ldots, n)$.
(iii) The origin is called an $(l_1 : l_2)$ -saddle for the linear discrete system if at least one $|\lambda_i| > 1$ $(i \in L_1 \subset \{1, 2, \ldots, n\})$ and the other $|\lambda_j| < 1$ $(j \in L_2 \subset \{1, 2, \ldots, n\})$ with $L_1 \cup L_2 = \{1, 2, \ldots, n\}$ and $L_1 \cap L_2 = \varnothing$.
(iv) The origin is called an lth -order degenerate case for the linear discrete system if $\lambda_i = 0$ $(i \in L \subseteq \{1, 2, \ldots, n\})$.

Definition B.11 For a discrete linear dynamical system $\mathbf{x}_{k+1} = \mathbf{A}\mathbf{x}_k$ in Eq. (B.2), the matrix $\mathbf{A}$ possesses n-pairs of complex eigenvalues λ_i $(i = 1, 2, \ldots, n)$.

(i) The origin is called a spiral sink for the linear discrete system if $|\lambda_i| < 1$ $(i = 1, 2, \ldots, n)$ and $\mathrm{Im}\lambda_j \neq 0$ $(j \in \{1, 2, \ldots, n\})$.
(ii) The origin is called a spiral source for the linear discrete system if $|\lambda_i| > 1$ $(i = 1, 2, \ldots, n)$ with $\mathrm{Im}\lambda_j \neq 0$ $(j \in \{1, 2, \ldots, n\})$.
(iii) The origin is called a center for the linear discrete system if $|\lambda_i| = 1$ with distinct $\mathrm{Im}\lambda_i \neq 0$ $(i \in \{1, 2, \ldots, n\})$.

The generalized structures of stability characteristics for iterative solutions of linear dynamical systems in Eq. (B.2) will be given as follows.

Definition B.12 Consider a discrete linear dynamical system $\mathbf{x}_{k+1} = \mathbf{A}\mathbf{x}_k$ in Eq. (B.2), the matrix $\mathbf{A}$ possesses n eigenvalues λ_i $(i = 1, 2, \ldots, n)$. Set $N = \{1, 2, \ldots, m, m+1, \ldots, (n-m)/2\}$, $N_p = \{p_1, p_2, \ldots, p_{n_p}\} \cup \varnothing$ with $p_q \in N$ $(q = 1, 2, \ldots, n_p,\ p = 1, 2, \ldots, 7)$, $\Sigma_{p=1}^{4} n_p = m$ and $2\Sigma_{p=5}^{7} n_p = n - m$. $\cup_{p=1}^{7} N_p = N$ and $N_p \cap N_1 = \varnothing$ $(l \neq p)$. $N_p = \varnothing$ if $n_p = 0$. $N_\alpha = N_\alpha^{\mathrm{m}} \cup N_\alpha^{\mathrm{o}}$ $(\alpha = 1, 2)$ and $N_\alpha^{\mathrm{m}} \cap N_\alpha^{\mathrm{o}} = \varnothing$ with $n_\alpha^{\mathrm{m}} + n_\alpha^{\mathrm{o}} = n_\alpha$ where superscripts "m" and "o" represent monotonic and oscillatory evolutions. The matrix $\mathbf{A}$ possesses n_1 -stable, n_2 -unstable, n_3 -invariant and n_4 -flip real eigenvectors plus n_5 -stable, n_6 -unstable and n_7 -center pairs of complex eigenvectors. Without repeated complex eigenvalues of $|\lambda_k| = 1 (k \in N_3 \cup N_4 \cup N_7)$, an iterative response of $\mathbf{x}_{k+1} = \mathbf{A}\mathbf{x}_k$ is an $([n_1^{\mathrm{m}}, n_1^{\mathrm{o}}] : [n_2^{\mathrm{m}}, n_2^{\mathrm{o}}] : [n_3; \kappa_3] : [n_4; \kappa_4] | n_5 : n_6 : n_7)$ flow. With repeated complex eigenvalues of $|\lambda_k| = 1$ $(k \in N_3 \cup N_4 \cup N_7)$, an iterative response of $\mathbf{x}_{k+1} = \mathbf{A}\mathbf{x}_k$ is an $([n_1^{\mathrm{m}}, n_1^{\mathrm{o}}] : [n_2^{\mathrm{m}}, n_2^{\mathrm{o}}] : [n_3; \kappa_3] : [n_4; \kappa_4] | n_5 : n_6 : [n_7, l; \boldsymbol{\kappa}_7])$ flow, where $\kappa_p \in \{\varnothing, m_p\} (p = 3, 4)$, $\boldsymbol{\kappa}_7 = (\kappa_{71}, \kappa_{72}, \ldots, \kappa_{7l})^{\mathrm{T}}$ with $\kappa_{7s} \in \{\varnothing, m_{7s}\} (s = 1, 2, \ldots, l)$. The meanings of notations in the aforementioned structures are defined as follows:

(i) $[n_1^{\mathrm{m}}, n_1^{\mathrm{o}}]$ represents n_1 -sinks with n_1^{m} -monotonic convergence and n_1^{o} -oscillatory convergence among n_1 -directions of $\mathbf{v}_i (i \in N_1)$ if $|\lambda_i| < 1$ $(k \in N_1$ and $1 \leq n_1 \leq m)$ with distinct or repeated eigenvalues.
(ii) $[n_2^{\mathrm{m}}, n_2^{\mathrm{o}}]$ represents n_2 -sources with n_2^{m} -monotonic divergence and n_2^{o} -oscillatory divergence among n_2 -directions of $\mathbf{v}_i (i \in N_2)$ if $|\lambda_i| > 1$ $(k \in N_2$ and $1 \leq n_2 \leq m)$ with distinct or repeated eigenvalues.
(iii) $n_3 = 1$ represents an invariant center on 1-direction of $\mathbf{v}_i (i \in N_3)$ if $\lambda_k = 1$ $(i \in N_3$ and $n_3 = 1)$.
(iv) $n_4 = 1$ represents a flip center on 1-direction of $\mathbf{v}_i$ $(i \in N_4)$ if $\lambda_i = -1$ $(i \in N_4$ and $n_4 = 1)$.
(v) n_5 represents n_5 -spiral sinks on n_5 -pairs of $(\mathbf{u}_i, \mathbf{v}_i)(i \in N_5)$ if $|\lambda_i| < 1$ and $\mathrm{Im}\lambda_i \neq 0$ $(i \in N_5$ and $1 \leq n_5 \leq (n-m)/2)$ with distinct or repeated eigenvalues.
(vi) n_6 represents n_6 -spiral sources on n_6 -directions of $(\mathbf{u}_i, \mathbf{v}_i)$ $(i \in N_6)$ if $|\lambda_i| > 1$ and $\mathrm{Im}\lambda_i \neq 0$ $(i \in N_6$ and $1 \leq n_6 \leq (n-m)/2))$ with distinct or repeated eigenvalues.

(vii) n_7 represents n_7 -invariant centers on n_7 -pairs of $(\mathbf{u}_k, \mathbf{v}_k)(k \in N_7)$ if $|\lambda_i| = 1$ and $\mathrm{Im}\lambda_i \neq 0$ $(k \in N_7$ and $1 \leq n_7 \leq (n-m)/2))$ with distinct eigenvalues.

(viii) $\varnothing$ represents empty or none if $n_j = 0$ $(j \in \{1, 2, \ldots, 7\})$.

(ix) $[n_3; \kappa_3]$ represents $(n_3 - \kappa_3)$ invariant centers on $(n_3 - \kappa_3)$ directions of $\mathbf{v}_{i_3}$ $(i_3 \in N_3)$ and κ_3 -sources in κ_3 -directions of $\mathbf{v}_{j_3} (j_3 \in N_3$ and $j_3 \neq i_3)$ if $\lambda_i = 1$ $(i \in N_3$ and $n_3 \leq m)$ with the $(\kappa_3 + 1)$th -order nilpotent matrix $\mathbf{N}_3^{\kappa_3+1} = \mathbf{0}$ $(0 < \kappa_3 \leq n_3 - 1)$.

(x) $[n_3; \varnothing]$ represents n_3 invariant centers on n_3 -directions of $\mathbf{v}_i$ $(i \in N_3)$ if $\lambda_i = 1$ $(i \in N_3$ and $1 < n_3 \leq m)$ with a nilpotent matrix $\mathbf{N}_3 = \mathbf{0}$.

(xi) $[n_4; \kappa_4]$ represents $(n_4 - \kappa_4)$ flip oscillatory centers on $(n_4 - \kappa_4)$ directions of $\mathbf{v}_{i_4}$ $(i_4 \in N_4)$ and κ_4 -sources in κ_4 -directions of $\mathbf{v}_{j_4}$ $(j_4 \in N_4$ and $j_4 \neq i_4)$ if $\lambda_i = -1$ $(i \in N_4$ and $n_4 \leq m)$ with the $(\kappa_4 + 1)$th -order nilpotent matrix $\mathbf{N}_4^{\kappa_4+1} = \mathbf{0}$ $(0 < \kappa_4 \leq n_4 - 1)$.

(xii) $[n_4; \varnothing]$ represents n_4 flip oscillatory centers on n_4 -directions of $\mathbf{v}_i$ $(i \in N_3)$ if $\lambda_i = -1$ $(i \in N_4$ and $1 < n_4 \leq m)$ with a nilpotent matrix $\mathbf{N}_4 = \mathbf{0}$.

(xiii) $[n_7, l; \boldsymbol{\kappa}_7]$ represents $(n_7 \sum_{s=1}^{l} - \kappa_{7s})$ invariant centers on $(n_7 \sum_{s=1}^{l} - \kappa_{7s})$ pairs of $(\mathbf{u}_{i_7}, \mathbf{v}_{i_7})(i_7 \in N_7)$ and $\sum_{s=1}^{l} \kappa_{7s}$ sources on $\sum_{s=1}^{l} \kappa_{7s}$ pairs of $(\mathbf{u}_{j_7}, \mathbf{v}_{j_7})$ $(j_7 \in N_7$ and $j_7 \neq i_7)$ if $|\lambda_i| = 1$ and $\mathrm{Im}\lambda_i \neq 0$ $(i \in N_7$ and $n_7 \leq (n-m)/2)$ for $\sum_{s=1}^{l} \kappa_{7s}$ pairs of repeated eigenvalues with the $(\kappa_{7s} + 1)$th -order nilpotent matrix $\mathbf{N}_7^{\kappa_{7s}+1} = \mathbf{0}$ $(0 < \kappa_{7s} \leq l,\ s = 1, 2, \ldots, l)$.

(xiv) $[n_7, l; \varnothing]$ represents n_7 -invariant centers on n_7 -pairs of $(\mathbf{u}_i, \mathbf{v}_i)$ $(i \in N_6)$ if $|\lambda_i| = \mathbf{1}$ and $\mathrm{Im}\lambda_i \neq 0$ $(i \in N_7$ and $1 \leq n_7 \leq (n-m)/2)$ for $\sum_{s=1}^{l} \kappa_{7s}$ pairs of repeated eigenvalues with a nilpotent matrix $\mathbf{N}_7 = \mathbf{0}$.

Definition B.13 Consider a discrete linear dynamical system $\mathbf{x}_{k+1} = \mathbf{A}\mathbf{x}_k$ in Eq. (B.2), the matrix $\mathbf{A}$ possesses n eigenvalues λ_i $(i = 1, 2, \ldots, n)$. Set $N = \{1, 2, \ldots, m, m+1, \ldots, (n-m)/2\}$, $N_p = \{p_1, p_2, \ldots, p_{n_p}\} \cup \varnothing$ with $p_q \in N$ $(q = 1, 2, \ldots, n_p,\ p = 1, 2, \ldots, 7)$, $\Sigma_{p=1}^{4} n_p = m$ and $2\Sigma_{p=5}^{7} n_p = n - m$. $\cup_{p=1}^{7} N_p = N$ and $N_p \cap N_l = \varnothing (l \neq p)$. $N_p = \varnothing$ if $n_p = 0$. $N_\alpha = N_\alpha^{\mathrm{m}} \cup N_\alpha^{\mathrm{o}}$ $(\alpha = 1, 2)$ and $N_\alpha^{\mathrm{m}} \cap N_\alpha^{\mathrm{o}} = \varnothing$ with $n_\alpha^{\mathrm{m}} + n_\alpha^{\mathrm{o}} = n_\alpha$ where superscripts "m" and "o" represent monotonic and oscillatory evolutions. The matrix $\mathbf{A}$ possesses n_1 -stable, n_2 -unstable, n_3 -invariant and n_4 -flip real eigenvectors plus n_5 -stable, n_6 -unstable and n_7 -center pairs of complex eigenvectors. Without repeated complex eigenvalues of $|\lambda_k| = 1 (k \in N_3 \cup N_4 \cup N_7)$, an iterative response of $\mathbf{x}_{k+1} = \mathbf{A}\mathbf{x}_k$ is an $([n_1^{\mathrm{m}}, n_1^{\mathrm{o}}] : [n_2^{\mathrm{m}}, n_2^{\mathrm{o}}] : [n_3; \kappa_3] : [n_4; \kappa_4] | n_5 : n_6 : n_7)$ flow. With repeated complex eigenvalues of $|\lambda_k| = 1$ $(k \in N_3 \cup N_4 \cup N_7)$, an iterative response of $\mathbf{x}_{k+1} = \mathbf{A}\mathbf{x}_k$ is an $([n_1^{\mathrm{m}}, n_1^{\mathrm{o}}] : [n_2^{\mathrm{m}}, n_2^{\mathrm{o}}] : [n_3; \kappa_3] : [n_4; \kappa_4] | n_5 : n_6 : [n_7, l; \boldsymbol{\kappa}_7])$ flow, where $\kappa_p \in \{\varnothing, m_p\} (p = 3, 4)$, $\boldsymbol{\kappa}_7 = (\kappa_{71}, \kappa_{72}, \ldots, \kappa_{7l})^{\mathrm{T}}$ with $\kappa_{7s} \in \{\varnothing, m_{7s}\} (s = 1, 2, \ldots, l)$.

I. *Non-degenerate cases*

(i) The origin is an $([n_1^{\mathrm{m}}, n_1^{\mathrm{o}}] : [n_2^{\mathrm{m}}, n_2^{\mathrm{o}}] : \varnothing : \varnothing | n_5 : n_6 : \varnothing)$ hyperbolic point for the linear discrete system.

(ii) The origin is an $([n_1^{\mathrm{m}}, n_1^{\mathrm{o}}] : [\varnothing,\varnothing] : \varnothing : \varnothing | n_5 : \varnothing : \varnothing)$ -sink for the linear discrete system.
(iii) The origin is an $([\varnothing,\varnothing] : [n_2^{\mathrm{m}}, n_2^{\mathrm{o}}] : \varnothing : \varnothing | \varnothing : n_6 : \varnothing)$ -source for the linear discrete system.
(iv) The origin is an $([\varnothing,\varnothing] : [\varnothing,\varnothing] : \varnothing : \varnothing | \varnothing : \varnothing : n/2)$ -circular center for the linear discrete system.
(v) The origin is an $([\varnothing,\varnothing] : [\varnothing,\varnothing] : \varnothing : \varnothing | \varnothing : \varnothing : [n/2, l; \varnothing])$ -circular center for the linear discrete system.
(vi) The origin is an $([\varnothing,\varnothing] : [\varnothing,\varnothing] : \varnothing : \varnothing | \varnothing : \varnothing : [n/2, l; \boldsymbol{\kappa}_7])$ -point for the linear discrete system.
(vii) The origin is an $([n_1^{\mathrm{m}}, n_1^{\mathrm{o}}] : [\varnothing,\varnothing] : \varnothing : \varnothing | n_5 : \varnothing : n_7)$ -point for the linear discrete system.
(viii) The origin is an $([\varnothing,\varnothing] : [n_2^{\mathrm{m}}, n_2^{\mathrm{o}}] : \varnothing : \varnothing | \varnothing : n_6 : n_7)$ -point for the linear discrete system.
(ix) The origin is an $([n_1^{\mathrm{m}}, n_1^{\mathrm{o}}] : [n_2^{\mathrm{m}}, n_2^{\mathrm{o}}] : \varnothing : \varnothing | n_5 : n_6 : n_7)$ -point for the linear discrete system.

II. *Simple special cases*

(i) The origin is an $([\varnothing,\varnothing] : [\varnothing,\varnothing] : [n; \varnothing] : \varnothing | \varnothing : \varnothing : \varnothing)$ -invariant center (or static center) the linear discrete system.
(ii) The origin is an $([\varnothing,\varnothing] : [\varnothing,\varnothing] : [n; m_3] : \varnothing | \varnothing : \varnothing : \varnothing)$ -point for the linear system.
(iii) The origin is an $([\varnothing,\varnothing] : [\varnothing,\varnothing] : \varnothing : [n; \varnothing] | \varnothing : \varnothing : \varnothing)$ -flip center for the linear discrete system
(iv) The origin is an $([\varnothing,\varnothing] : [\varnothing,\varnothing] : \varnothing : [n; m_4] | \varnothing : \varnothing : \varnothing)$ -point for the linear discrete system.
(v) The origin is an $([\varnothing,\varnothing] : [\varnothing,\varnothing] : [n_3; \kappa_3] : [n_4; \kappa_4] | \varnothing : \varnothing : \varnothing)$ -point for the linear discrete system.
(vi) The origin is an $([\varnothing,\varnothing] : [\varnothing,\varnothing] : [1; \varnothing] : [n_4; \kappa_4] | \varnothing : \varnothing : \varnothing)$ -point for the linear discrete system.
(vii) The origin is an $([\varnothing,\varnothing] : [\varnothing,\varnothing] : [n_3; \kappa_3] : [1; \varnothing] | \varnothing : \varnothing : \varnothing)$ -point for the linear discrete system.
(viii) The origin is an $([\varnothing,\varnothing] : [\varnothing,\varnothing] : [n_3; \kappa_3] : [\varnothing; \varnothing] | \varnothing : \varnothing : n_7)$ -point for the linear discrete system.
(ix) The origin is an $([\varnothing,\varnothing] : [\varnothing,\varnothing] : [1; \varnothing] : [\varnothing; \varnothing] | \varnothing : \varnothing : n_7)$ -point for the linear discrete system.
(x) The origin is an $([\varnothing,\varnothing] : [\varnothing,\varnothing] : [n_3; \kappa_3] : [\varnothing; \varnothing] | \varnothing : \varnothing : [n_7, l; \boldsymbol{\kappa}_7])$ -point for the linear discrete system.
(xi) The origin is an $([\varnothing,\varnothing] : [\varnothing,\varnothing] : [\varnothing; \varnothing] : [n_4; \kappa_4] | \varnothing : \varnothing : n_7)$ -point for the linear discrete system.
(xii) The origin is an $([\varnothing,\varnothing] : [\varnothing,\varnothing] : [\varnothing; \varnothing] : [n_4; \kappa_4] | \varnothing : \varnothing : [n_7, l; \boldsymbol{\kappa}_7])$ -point for the linear discrete system.
(xiii) The origin is an $([\varnothing,\varnothing] : [\varnothing,\varnothing] : [n_3; \kappa_3] : [n_4; \kappa_4] | \varnothing : \varnothing : n_7)$ -point for the linear discrete system.

(xiv) The origin is an $([\varnothing,\varnothing]:[\varnothing,\varnothing]:[n_3;\kappa_3]:[n_4;\kappa_4]|\varnothing:\varnothing:[n_7,l;\boldsymbol{\kappa}_7])$ -point for the linear discrete system.

III. *Complex special cases*

(i) The origin is an $([n_1^{\mathrm{m}},n_1^{\mathrm{o}}]:[n_2^{\mathrm{m}},n_2^{\mathrm{o}}]:[1;\varnothing]:[\varnothing;\varnothing]|n_5:n_6:n_7)$ -point for the linear discrete system.
(ii) The origin is an $([n_1^{\mathrm{m}},n_1^{\mathrm{o}}]:[n_2^{\mathrm{m}},n_2^{\mathrm{o}}]:[1;\varnothing]:[\varnothing;\varnothing]|n_5:n_6:[n_7,l;\boldsymbol{\kappa}_7])$ -point for the linear discrete system.
(iii) The origin is an $([n_1^{\mathrm{m}},n_1^{\mathrm{o}}]:[n_2^{\mathrm{m}},n_2^{\mathrm{o}}]:[\varnothing;\varnothing]:[1;\varnothing]|n_5:n_6:n_7)$ -point for the linear discrete system.
(iv) The origin is an $([n_1^{\mathrm{m}},n_1^{\mathrm{o}}]:[n_2^{\mathrm{m}},n_2^{\mathrm{o}}]:[\varnothing;\varnothing]:[1;\varnothing]|n_5:n_6:[n_7,l;\boldsymbol{\kappa}_7])$ -point for the linear discrete system.
(v) The origin is an $([n_1^{\mathrm{m}},n_1^{\mathrm{o}}]:[n_2^{\mathrm{m}},n_2^{\mathrm{o}}]:[n_3;\kappa_3]:[n_4;\kappa_4]|n_5:n_6:n_7)$ -point for the linear discrete system.
(vi) The origin is an $([n_1^{\mathrm{m}},n_1^{\mathrm{o}}]:[n_2^{\mathrm{m}},n_2^{\mathrm{o}}]:[n_3;\kappa_3]:[n_4;\kappa_4]|n_5:n_6:[n_7,l;\boldsymbol{\kappa}_7])$ -point for the linear discrete system.

IV. *Simple critical cases*

(i) An $([n_1^{\mathrm{m}},n_1^{\mathrm{o}}]:[n_2^{\mathrm{m}},n_2^{\mathrm{o}}]:1:\varnothing|n_5:n_6:\varnothing)$ state of the origin for the linear discrete system is a boundary of its $([n_1^{\mathrm{m}}+1,n_1^{\mathrm{o}}]:[n_2^{\mathrm{m}},n_2^{\mathrm{o}}]:\varnothing:\varnothing|n_5:n_6:\varnothing)$ spiral saddle and $([n_1^{\mathrm{m}},n_1^{\mathrm{o}}]:[n_2^{\mathrm{m}}+1,n_2^{\mathrm{o}}]:\varnothing:\varnothing|n_5:n_6:\varnothing)$ spiral saddle.
(ii) An $([n_1^{\mathrm{m}},n_1^{\mathrm{o}}]:[n_2^{\mathrm{m}},n_2^{\mathrm{o}}]:\varnothing:1|n_5:n_6:\varnothing)$ state of the origin for the linear discrete system is a boundary of its $([n_1^{\mathrm{m}},n_1^{\mathrm{o}}+1]:[n_2^{\mathrm{m}},n_2^{\mathrm{o}}]:\varnothing:\varnothing|n_5:n_6:\varnothing)$ spiral saddle and $([n_1^{\mathrm{m}},n_1^{\mathrm{o}}]:[n_2^{\mathrm{m}},n_2^{\mathrm{o}}+1]:\varnothing:\varnothing|n_5:n_6:\varnothing)$ spiral saddle.
(iii) An $([n_1^{\mathrm{m}},n_1^{\mathrm{o}}]:[\varnothing,\varnothing]:1:\varnothing|n_5:\varnothing:\varnothing)$ state of the origin for the linear discrete system is a boundary of its $([n_1^{\mathrm{m}}+1,n_1^{\mathrm{o}}]:[\varnothing,\varnothing]:\varnothing:\varnothing|n_5:\varnothing:\varnothing)$ spiral sink and $([n_1^{\mathrm{m}},n_1^{\mathrm{o}}]:[1,\varnothing]:\varnothing:\varnothing|n_5:\varnothing:\varnothing)$ spiral saddle.
(iv) An $([n_1^{\mathrm{m}},n_1^{\mathrm{o}}]:[\varnothing,\varnothing]:\varnothing:1|n_5:\varnothing:\varnothing)$ state of the origin for the linear discrete system is a boundary of its $([n_1^{\mathrm{m}},n_1^{\mathrm{o}}+1]:[\varnothing,\varnothing]:\varnothing:\varnothing|n_5:\varnothing:\varnothing)$ sink and $([n_1^{\mathrm{m}},n_1^{\mathrm{o}}]:[\varnothing,1]:\varnothing:\varnothing|n_5:\varnothing:\varnothing)$ spiral saddle.
(v) An $([\varnothing,\varnothing]:[n_2^{\mathrm{m}},n_2^{\mathrm{o}}]:1:\varnothing|\varnothing:n_6:\varnothing)$ state of the origin for the linear discrete system is a boundary of its $([\varnothing,\varnothing]:[n_2^{\mathrm{m}}+1,n_2^{\mathrm{o}}]:\varnothing:\varnothing|\varnothing:n_6:\varnothing)$ spiral source and $([1,\varnothing]:[n_2^{\mathrm{m}},n_2^{\mathrm{o}}]:\varnothing:\varnothing|\varnothing:n_6:\varnothing)$ spiral saddle.
(vi) An $([\varnothing,\varnothing]:[n_2^{\mathrm{m}},n_2^{\mathrm{o}}]:\varnothing:1|\varnothing:n_6:\varnothing)$ state of the origin for the linear discrete system is a boundary of its $([\varnothing,\varnothing]:[n_2^{\mathrm{m}},n_2^{\mathrm{o}}+1]:\varnothing:\varnothing|\varnothing:n_6:\varnothing)$ spiral source and $([\varnothing,1]:[n_2^{\mathrm{m}},n_2^{\mathrm{o}}]:\varnothing:\varnothing|\varnothing:n_6:\varnothing)$ spiral saddle.

(vii) An $([n_1^{\mathrm{m}}, n_1^{\mathrm{o}}] : [n_2^{\mathrm{m}}, n_2^{\mathrm{o}}] : \varnothing : \varnothing | n_5 : n_6 : 1)$ state of the origin for the linear discrete system is a boundary of its $([n_1^{\mathrm{m}}, n_1^{\mathrm{o}}] : [n_2^{\mathrm{m}}, n_2^{\mathrm{o}}] : \varnothing : \varnothing | n_5 + 1 : n_6 : \varnothing)$ spiral saddle and $([n_1^{\mathrm{m}}, n_1^{\mathrm{o}}] : [n_2^{\mathrm{m}}, n_2^{\mathrm{o}}] : \varnothing : \varnothing | n_5 : n_6 + 1 : \varnothing)$ saddle.

(viii) An $([n_1^{\mathrm{m}}, n_1^{\mathrm{o}}] : [\varnothing, \varnothing] : \varnothing : \varnothing | n_5 : \varnothing : 1)$ state of the origin for the linear discrete system is a boundary of its $([n_1^{\mathrm{m}}, n_1^{\mathrm{o}}] : [\varnothing, \varnothing] : \varnothing : \varnothing | n_5 + 1 : \varnothing : \varnothing)$ spiral sink and $([n_1^{\mathrm{m}}, n_1^{\mathrm{o}}] : [\varnothing, \varnothing] : \varnothing : \varnothing | n_5 : 1 : \varnothing)$ spiral saddle.

(ix) An $([\varnothing, \varnothing] : [n_2^{\mathrm{m}}, n_2^{\mathrm{o}}] : \varnothing : \varnothing | \varnothing : n_6 : 1)$ state of the origin for the linear discrete system is a boundary of its $([\varnothing, \varnothing] : [n_2^{\mathrm{m}}, n_2^{\mathrm{o}}] : \varnothing : \varnothing | \varnothing : n_6 + 1 : \varnothing)$ spiral source and $([\varnothing, \varnothing] : [n_2^{\mathrm{m}}, n_2^{\mathrm{o}}] : \varnothing : \varnothing | 1 : n_6 : \varnothing)$ spiral saddle.

(x) An $([n_1^{\mathrm{m}}, n_1^{\mathrm{o}}] : [n_2^{\mathrm{m}}, n_2^{\mathrm{o}}] : 1 : \varnothing | n_5 : n_6 : n_7)$ state of the origin for the linear discrete system is a boundary of its $([n_1^{\mathrm{m}} + 1, n_1^{\mathrm{o}}] : [n_2^{\mathrm{m}}, n_2^{\mathrm{o}}] : \varnothing : \varnothing | n_5 : n_6 : n_7)$ state and $([n_1^{\mathrm{m}}, n_1^{\mathrm{o}}] : [n_2^{\mathrm{m}} + 1, n_2^{\mathrm{o}}] : \varnothing : \varnothing | n_5 : n_6 : n_7)$ state.

(xi) An $([n_1^{\mathrm{m}}, n_1^{\mathrm{o}}] : [n_2^{\mathrm{m}}, n_2^{\mathrm{o}}] : \varnothing : \varnothing : 1 | n_5 : n_6 : n_7)$ state of the origin for the linear discrete system is a boundary of its $([n_1^{\mathrm{m}}, n_1^{\mathrm{o}} + 1] : [n_2^{\mathrm{m}}, n_2^{\mathrm{o}}] : \varnothing : \varnothing | n_5 : n_6 : n_7)$ state and $([n_1^{\mathrm{m}}, n_1^{\mathrm{o}}] : [n_2^{\mathrm{m}}, n_2^{\mathrm{o}} + 1] : \varnothing : \varnothing | n_5 : n_6 : n_7)$ state.

(xii) An $([n_1^{\mathrm{m}}, n_1^{\mathrm{o}}] : [n_2^{\mathrm{m}}, n_2^{\mathrm{o}}] : 1 : \varnothing | n_5 : n_6 : [n_7, l; \varnothing])$ state of the origin for the linear discrete system is a boundary of its $([n_1^{\mathrm{m}} + 1, n_1^{\mathrm{o}}] : [n_2^{\mathrm{m}}, n_2^{\mathrm{o}}] : \varnothing : \varnothing | n_5 : n_6 : [n_7, l; \boldsymbol{\kappa}_7])$ state and $([n_1^{\mathrm{m}}, n_1^{\mathrm{o}}] : [n_2^{\mathrm{m}} + 1, n_2^{\mathrm{o}}] : \varnothing : \varnothing | n_5 : n_6 : [n_7, l; \boldsymbol{\kappa}_7])$ state.

(xiii) An $([n_1^{\mathrm{m}}, n_1^{\mathrm{o}}] : [n_2^{\mathrm{m}}, n_2^{\mathrm{o}}] : \varnothing : 1 | n_5 : n_6 : [n_7, l; \boldsymbol{\kappa}_7])$ state of the origin for the linear discrete system is a boundary of its $([n_1^{\mathrm{m}}, n_1^{\mathrm{o}} + 1] : [n_2^{\mathrm{m}}, n_2^{\mathrm{o}}] : \varnothing : \varnothing | n_5 : n_6 : [n_7, l; \boldsymbol{\kappa}_7])$ state and $([n_1^{\mathrm{m}}, n_1^{\mathrm{o}}] : [n_2^{\mathrm{m}}, n_2^{\mathrm{o}} + 1] : \varnothing : \varnothing | n_5 : n_6 : [n_7, l; \boldsymbol{\kappa}_7]))$ state.

(xiv) An $([n_1^{\mathrm{m}}, n_1^{\mathrm{o}}] : [n_2^{\mathrm{m}}, n_2^{\mathrm{o}}] : \varnothing : \varnothing | n_5 : n_6 : n_7 + 1)$ state of the origin for the linear discrete system is a boundary of its $([n_1^{\mathrm{m}}, n_1^{\mathrm{o}}] : [n_2^{\mathrm{m}}, n_2^{\mathrm{o}}] : \varnothing : \varnothing | n_5 + 1 : n_6 : n_7)$ state and $([n_1^{\mathrm{m}}, n_1^{\mathrm{o}}] : [n_2^{\mathrm{m}}, n_2^{\mathrm{o}}] : \varnothing : \varnothing | n_5 : n_6 + 1 : n_7)$ state.

(xv) An $([n_1^{\mathrm{m}}, n_1^{\mathrm{o}}] : [n_2^{\mathrm{m}}, n_2^{\mathrm{o}}] : [n_3; \kappa_3] : [n_4; \kappa_4] | n_5 : n_6 : n_7 + 1)$ state of the origin for the linear discrete system is a boundary of its $([n_1^{\mathrm{m}}, n_1^{\mathrm{o}}] : [n_2^{\mathrm{m}}, n_2^{\mathrm{o}}] : [n_3; \kappa_3] : [n_4; \kappa_4] | n_5 + 1 : n_6 : n_7)$ state and $([n_1^{\mathrm{m}}, n_1^{\mathrm{o}}] : [n_2^{\mathrm{m}}, n_2^{\mathrm{o}}] : [n_3; \kappa_3] : [n_4; \kappa_4] | n_5 : n_6 + 1 : n_7)$ state.

V. *Complex critical cases*

(i) An $([n_1^{\mathrm{m}}, n_1^{\mathrm{o}}] : [n_2^{\mathrm{m}}, n_2^{\mathrm{o}}] : [n_3; \kappa_3] : \varnothing | n_5 : n_6 : n_7)$ state of the origin for the linear discrete system is a boundary of its $([n_1^{\mathrm{m}} + n_3, n_1^{\mathrm{o}}] : [n_2^{\mathrm{m}}, n_2^{\mathrm{o}}] : \varnothing : \varnothing | n_5 : n_6 : n_7)$ state and $([n_1^{\mathrm{m}}, n_1^{\mathrm{o}}] : [n_2^{\mathrm{m}} + n_3, n_2^{\mathrm{o}}] : \varnothing : \varnothing | n_5 : n_6 : n_7)$ state.

(ii) An $([n_1^{\mathrm{m}}, n_1^{\mathrm{o}}] : [n_2^{\mathrm{m}}, n_2^{\mathrm{o}}] : \varnothing : [n_4; \kappa_4] | n_5 : n_6 : n_7)$ state of the origin for the linear discrete system is a boundary of its $([n_1^{\mathrm{m}}, n_1^{\mathrm{o}} + n_4] : [n_2^{\mathrm{m}}, n_2^{\mathrm{o}}] : \varnothing : \varnothing | n_5 : n_6 : n_7)$ state and $([n_1^{\mathrm{m}}, n_1^{\mathrm{o}}] : [n_2^{\mathrm{m}}, n_2^{\mathrm{o}} + n_4] : \varnothing : \varnothing | n_5 : n_6 : n_7)$ state.

(iii) An $([n_1^m, n_1^o] : [n_2^m, n_2^o] : [n_3 + k_3; \kappa_3] : \varnothing | n_5 : n_6 : n_7)$ state of the origin for the linear discrete system is a boundary of its $([n_1^m + k_3, n_1^o] : [n_2^m, n_2^o] : [n_3; \kappa_3'] : \varnothing | n_5 : n_6 : n_7)$ state and $([n_1^m, n_1^o] : [n_2^m + k_3, n_2^o] : [n_3; \kappa_3'] : \varnothing | n_5 : n_6 : n_7)$ state.

(iv) An $([n_1^m, n_1^o] : [n_2^m, n_2^o] : \varnothing : [n_4 + k_4; \kappa_4] | n_5 : n_6 : n_7)$ state of the origin for the linear discrete system is a boundary of its $([n_1^m, n_1^o + k_4] : [n_2^m, n_2^o] : \varnothing : [n_4; \kappa_4'] | dn_5 : n_6 : n_7)$ state and $([n_1^m, n_1^o] : [n_2^m, n_2^o + k_4] : \varnothing : [n_4; \kappa_4'] | n_5 : n_6 : n_7)$ state.

(v) An $([n_1^m, n_1^o] : [n_2^m, n_2^o] : [n_3 + k_3; \kappa_3] : [n_4 + k_4; \kappa_4] | n_5 : n_6 : n_7)$ state of the origin for the linear discrete system is a boundary of its $([n_1^m + k_3, n_1^o + k_4] : [n_2^m, n_2^o] : [n_3; \kappa_3'] : [n_4; \kappa_4'] | n_5 : n_6 : n_7)$ state and $([n_1^m, n_1^o] : [n_2^m + k_3, n_2^o + k_4] : [n_3; \kappa_3''] : [n_4; \kappa_4''] | n_5 : n_6 : n_7)$ state.

(vi) An $([n_1^m, n_1^o] : [n_2^m, n_2^o] : [n_3 + k_3; \kappa_3] : \varnothing | n_5 : n_6 : [n_7, l; \boldsymbol{\kappa}_7])$ state of the origin for the linear discrete system is a boundary of its $([n_1^m + k_3, n_1^o] : [n_2^m, n_2^o] : [n_3; \kappa_3'] : \varnothing | n_5 : n_6 : [n_7, l; \boldsymbol{\kappa}_7])$ state and $([n_1^m, n_1^o] : [n_2^m + k_3, n_2^o] : [n_3; \kappa_3'] : \varnothing | \; n_5 : n_6 : [n_7, l; \boldsymbol{\kappa}_7])$ state.

(vii) An $([n_1^m, n_1^o] : [n_2^m, n_2^o] : \varnothing : [n_4 + k_4; \kappa_4] | n_5 : n_6 : [n_7, l; \boldsymbol{\kappa}_7])$ state of the origin for the linear discrete system is a boundary of its $([n_1^m, n_1^o + k_4] : [n_2^m, n_2^o] : \varnothing : [n_4; \kappa_4'] | n_5 : n_6 : [n_7, l; \boldsymbol{\kappa}_7])$ state and $([n_1^m, n_1^o] : [n_2^m, n_2^o + k_4] : \varnothing : [n_4; \kappa_4'] | \; n_5 : n_6 : [n_7, l; \boldsymbol{\kappa}_7])$ state.

(viii) An $([n_1^m, n_1^o] : [n_2^m, n_2^o] : [n_3 + k_3; \kappa_3] : [n_4 + k_4; \kappa_4] | n_5 : n_6 : [n_7, l; \boldsymbol{\kappa}_7])$ state of the origin for the linear discrete system is a boundary of its $([n_1^m + k_3, n_1^o + k_4] : [n_2^m, n_2^o] : [n_3; \kappa_3'] : [n_4; \kappa_4'] | n_5 : n_6 : [n_7, l; \boldsymbol{\kappa}_7])$ state and $([n_1^m, n_1^o] : [n_2^m + k_3, n_2^o + k_4] : [n_3; \kappa_3'] : [n_4; \kappa_4'] | n_5 : n_6 : [n_7, l; \boldsymbol{\kappa}_7])$ state.

(ix) An $([n_1^m, n_1^o] : [n_2^m, n_2^o] : [n_3 + k_3; \kappa_3] : [n_4 + k_4; \kappa_4] | n_5 : n_6 : [n_7 + k_7, l; \boldsymbol{\kappa}_7])$ state of the origin for the linear discrete system is a boundary of its $([n_1^m + k_3, n_1^o + k_4] : [n_2^m, n_2^o] : [n_3; \kappa_3] : [n_4; \kappa_4] | n_5 + k_7 : n_6 : [n_7, l; \boldsymbol{\kappa}_7'])$ state and $([n_1^m, n_1^o] : [n_2^m + k_3, n_2^o + k_4] : [n_3; \kappa_3] : [n_4; \kappa_4] | n_5 : n_6 + k_7 : [n_7, l; \boldsymbol{\kappa}_7'])$ state.

Definition B.14 Consider a discrete linear dynamical system $\mathbf{x}_{k+1} = \mathbf{A}\mathbf{x}_k$ in Eq. (B.2), the matrix $\mathbf{A}$ possesses n eigenvalues λ_i $(i = 1, 2, \ldots, n)$. Set $N = \{1, 2, \ldots, n\}$, $N_p = \{p_1, p_2, \ldots, p_{n_p}\} \cup \varnothing$ with $p_q \in N$ $(q = 1, 2, \ldots, n_p,\ p = 1, 2, 3, 4)$ and $\Sigma_{p=1}^4 n_p = n$. $\cup_{p=1}^4 N_p = N$ and $N_p \cap N_l = \varnothing$ $(l \neq p)$. $N_p = \varnothing$ if $n_p = 0$. $N_\alpha = N_\alpha^m \cup N_\alpha^o$ $(\alpha = 1, 2)$ and $N_\alpha^m \cap N_\alpha^o = \varnothing$ with $n_\alpha^m + n_\alpha^o = n_\alpha$ where superscripts "m" and "o" represent monotonic and oscillatory evolutions. The matrix $\mathbf{A}$ possesses n_1 -stable, n_2 -unstable, n_3 -invariant, and n_4 -flip real eigenvectors. An iterative response of $\mathbf{x}_{k+1} = \mathbf{A}\mathbf{x}_k$ is an $([n_1^m, n_1^m] : [n_2^m, n_2^o] : [n_3; \kappa_3] : [n_4; \kappa_4]|$ flow. $\kappa_p \in \{\varnothing, m_p\}$ $(p = 3, 4)$.

I. *Non-degenerate cases*

(i) The origin is an $([n_1^m, n_1^o] : [n_2^m, n_2^o] : \varnothing : \varnothing|$ saddle for the linear discrete system.

(ii) The origin is an $([n_1^{\rm m}, n_1^{\rm o}] : [\varnothing, \varnothing] : \varnothing : \varnothing|$ -sink for the linear discrete system.
(iii) The origin is an $([\varnothing, \varnothing] : [n_2^{\rm m}, n_2^{\rm o}] : \varnothing : \varnothing|$ -source for the linear discrete system.

II. *Simple special cases*

(i) The origin is an $([\varnothing, \varnothing] : [\varnothing, \varnothing] : [n; \varnothing] : \varnothing|$ -invariant center (or static center) the linear discrete system.
(ii) The origin is an $([\varnothing, \varnothing] : [\varnothing, \varnothing] : [n; m_3] : \varnothing|$ -point for the linear system.
(iii) The origin is an $([\varnothing, \varnothing] : [\varnothing, \varnothing] : \varnothing : [n; \varnothing]|$ -flip center for the linear discrete system
(iv) The origin is an $([\varnothing, \varnothing] : [\varnothing, \varnothing] : \varnothing : [n; m_4]|$ -point for the linear discrete system.
(v) The origin is an $([\varnothing, \varnothing] : [\varnothing, \varnothing] : [n_3; \kappa_3] : [n_4; \kappa_4]|$ -point for the linear discrete system.
(vi) The origin is an $([\varnothing, \varnothing] : [\varnothing, \varnothing] : [1; \varnothing] : [n; \kappa_4]|$ -point for the linear discrete system.
(vii) The origin is an $([\varnothing, \varnothing] : [\varnothing, \varnothing] : [n_3; \kappa_3] : [1; \varnothing]|$ -point for the linear discrete system.
(viii) The origin is an $([\varnothing, \varnothing] : [\varnothing, \varnothing] : [n_3; \kappa_3] : [\varnothing; \varnothing]|$ -point for the linear discrete system.
(ix) The origin is an $([\varnothing, \varnothing] : [\varnothing, \varnothing] : [\varnothing; \varnothing] : [n_4; \kappa_4]|$ -point for the linear discrete system.

III. *Complex special cases*

(i) The origin is an $([n_1^{\rm m}, n_1^{\rm o}] : [n_2^{\rm m}, n_2^{\rm o}] : [1; \varnothing] : [\varnothing; \varnothing]|$ -point for the linear discrete system.
(ii) The origin is an $([n_1^{\rm m}, n_1^{\rm o}] : [n_2^{\rm m}, n_2^{\rm o}] : [\varnothing; \varnothing] : [1; \varnothing]|$ -point for the linear discrete system.
(iii) The origin is an $([n_1^{\rm m}, n_1^{\rm o}] : [n_2^{\rm m}, n_2^{\rm o}] : [n_3; \kappa_3] : [n_4; \kappa_4]|$ -point for the linear discrete system.

IV. *Simple critical cases*

(i) An $([n_1^{\rm m}, n_1^{\rm o}] : [n_2^{\rm m}, n_2^{\rm o}] : 1 : \varnothing|$ state of the origin for the linear discrete system is a boundary of its $([n_1^{\rm m}+1, n_1^{\rm o}] : [n_2^{\rm m}, n_2^{\rm o}] : \varnothing : \varnothing|$ saddle and $([n_1^{\rm m}, n_1^{\rm o}] : [n_2^{\rm m}+1, n_2^{\rm o}] : \varnothing : \varnothing|$ saddle.
(ii) An $([n_1^{\rm m}, n_1^{\rm o}] : [n_2^{\rm m}, n_2^{\rm o}] : \varnothing : 1|$ state of the origin for the linear discrete system is a boundary of its $([n_1^{\rm m}, n_1^{\rm o}+1] : [n_2^{\rm m}, n_2^{\rm o}] : \varnothing : \varnothing|$ saddle and $([n_1^{\rm m}, n_1^{\rm o}] : [n_2^{\rm m}, n_2^{\rm o}+1] : \varnothing : \varnothing|$ saddle.
(iii) An $([n_1^{\rm m}, n_1^{\rm o}] : [\varnothing, \varnothing] : 1 : \varnothing|$ state of the origin for the linear discrete system is a boundary of its $([n_1^{\rm m}+1, n_1^{\rm o}] : [\varnothing, \varnothing] : \varnothing : \varnothing|$ sink and $([n_1^{\rm m}, n_1^{\rm o}] : [1, \varnothing] : \varnothing : \varnothing|$ saddle.

(iv) An $([n_1^{\rm m}, n_1^{\rm o}] : [\varnothing, \varnothing] : \varnothing : 1|$ state of the origin for the linear discrete system is a boundary of its $([n_1^{\rm m}, n_1^{\rm o}+1] : [\varnothing, \varnothing] : \varnothing : \varnothing|$ sink and $([n_1^{\rm m}, n_1^{\rm o}] : [\varnothing, 1] : \varnothing : \varnothing|$ saddle.

(v) An $([\varnothing, \varnothing] : [n_2^{\rm m}, n_2^{\rm o}] : 1 : \varnothing|$ state of the origin for the linear discrete system is a boundary of its $([\varnothing, \varnothing] : [n_2^{\rm m}+1, n_2^{\rm o}] : \varnothing : \varnothing|$ source and $([1, \varnothing] : [n_2^{\rm m}, n_2^{\rm o}] : \varnothing : \varnothing|$ saddle.

(vi) An $([\varnothing, \varnothing] : [n_2^{\rm m}, n_2^{\rm o}] : \varnothing : 1|$ state of the origin for the linear discrete system is a boundary of its $([\varnothing, \varnothing] : [n_2^{\rm m}, n_2^{\rm o}+1] : \varnothing : \varnothing|$ source and $([\varnothing, 1] : [n_2^{\rm m}, n_2^{\rm o}] : \varnothing : \varnothing|$ saddle.

(vii) An $([n_1^{\rm m}, n_1^{\rm o}] : [n_2^{\rm m}, n_2^{\rm o}] : 1 : \varnothing|$ state of the origin for the linear discrete system is a boundary of its $([n_1^{\rm m}+1, n_1^{\rm o}] : [n_2^{\rm m}, n_2^{\rm o}] : \varnothing : \varnothing|$ saddle and $([n_1^{\rm m}, n_1^{\rm o}] : [n_2^{\rm m}+1, n_2^{\rm o}] : \varnothing : \varnothing|$ saddle.

(viii) An $([n_1^{\rm m}, n_1^{\rm o}] : [n_2^{\rm m}, n_2^{\rm o}] : \varnothing : 1|$ state of the origin for the linear discrete system is a boundary of its $([n_1^{\rm m}, n_1^{\rm o}+1] : [n_2^{\rm m}, n_2^{\rm o}] : \varnothing : \varnothing|$ state and $([n_1^{\rm m}, n_1^{\rm o}] : [n_2^{\rm m}, n_2^{\rm o}+1] : \varnothing : \varnothing|$ saddle.

V. *Complex critical cases*

(i) An $([n_1^{\rm m}, n_1^{\rm o}] : [n_2^{\rm m}, n_2^{\rm o}] : [n_3; \kappa_3] : \varnothing|$ state of the origin for the linear discrete system is a boundary of its $([n_1^{\rm m}+n_3, n_1^{\rm o}] : [n_2^{\rm m}, n_2^{\rm o}] : \varnothing : \varnothing|$ saddle and $([n_1^{\rm m}, n_1^{\rm o}] : [n_2^{\rm m}+n_3, n_2^{\rm o}] : \varnothing : \varnothing|$ saddle.

(ii) An $([n_1^{\rm m}, n_1^{\rm o}] : [n_2^{\rm m}, n_2^{\rm o}] : \varnothing : [n_4; \kappa_4]|$ state of the origin for the linear discrete system is a boundary of its $([n_1^{\rm m}, n_1^{\rm o}+n_4] : [n_2^{\rm m}, n_2^{\rm o}] : \varnothing : \varnothing|$ saddle and $([n_1^{\rm m}, n_1^{\rm o}] : [n_2^{\rm m}, n_2^{\rm o}+n_4] : \varnothing : \varnothing|$ saddle.

(iii) An $([n_1^{\rm m}, n_1^{\rm o}] : [n_2^{\rm m}, n_2^{\rm o}] : [n_3 \mid k_3, \kappa_3] : \varnothing|$ state of the origin for the linear discrete system is a boundary of its $([n_1^{\rm m}+k_3, n_1^{\rm o}] : [n_2^{\rm m}, n_2^{\rm o}] : [n_3; \kappa_3'] : \varnothing|$ state and $([n_1^{\rm m}, n_1^{\rm o}] : [n_2^{\rm m}+k_3, n_2^{\rm o}] : [n_3; \kappa_3'] : \varnothing|$ state.

(iv) An $([n_1^{\rm m}, n_1^{\rm o}] : [n_2^{\rm m}, n_2^{\rm o}] : \varnothing : [n_4+k_4; \kappa_4]|$ state of the origin for the linear discrete system is a boundary of its $([n_1^{\rm m}, n_1^{\rm o}+k_4] : [n_2^{\rm m}, n_2^{\rm o}] : \varnothing : [n_4; \kappa_4']|$ state and $([n_1^{\rm m}, n_1^{\rm o}] : [n_2^{\rm m}, n_2^{\rm o}+k_4] : \varnothing : [n_4; \kappa_4']|$ state.

(v) An $([n_1^{\rm m}, n_1^{\rm o}] : [n_2^{\rm m}, n_2^{\rm o}] : [n_3+k_3; \kappa_3] : [n_4+k_4; \kappa_4]|$ state of the origin for the linear discrete system is a boundary of its $([n_1^{\rm m}+k_3, n_1^{\rm o}+k_4] : [n_2^{\rm m}, n_2^{\rm o}] : [n_3; \kappa_3'] : [n_4; \kappa_4']|$ state and $([n_1^{\rm m}, n_1^{\rm om}] : [n_2^{\rm m}+k_3, n_2^{\rm o}+k_4] : [n_3; \kappa_3'] : [n_4; \kappa_4']|$ state.

Definition B.15 Consider a $2n$-dimensional, linear discrete system $\mathbf{x}_{k+1} = \mathbf{A}\mathbf{x}_k$ in Eq. (B.2), the matrix $\mathbf{A}$ possesses n-pairs of eigenvalues λ_i $(i = 1, 2, \ldots, n)$. Set $N = \{1, 2, \ldots, n\}$, $N_p = \{p_1, p_2, \ldots, p_{n_p}\} \cup \varnothing$ with $p_q \in N$ $(q = 1, 2, \ldots, n_p,\ p = 5, 6, 7)$ and $\Sigma_{p=5}^{7} n_p = n$. $\cup_{p=5}^{7} N_p = N$ and $N_p \cap N_l = \varnothing$ $(l \neq p)$. $N_p = \varnothing$ if $n_p = 0$. The matrix $\mathbf{A}$ possesses n_5 -stable, n_6 -unstable and n_7 -center pairs of complex eigenvectors. Without repeated complex eigenvalues of $|\lambda_k| = 1 (k \in N_7)$, an iterative response of $\mathbf{x}_{k+1} = \mathbf{A}\mathbf{x}_k$ is an $|n_5 : n_6 : n_7)$ flow. With repeated complex eigenvalues of $|\lambda_k| = 1$ $(k \in N_7)$, an iterative response of $\mathbf{x}_{k+1} = \mathbf{A}\mathbf{x}_k$ is an $|n_5 : n_6 : [n_7, l; \boldsymbol{\kappa}_7])$ flow, where $\boldsymbol{\kappa}_7 = (\kappa_{71}, \kappa_{72}, \ldots, \kappa_{7l})^{\rm T}$ with $\kappa_{7s} \in \{\varnothing, m_{7s}\}$ $(s = 1, 2, \ldots, l)$.

I. *Non-degenerate cases*

(i) The origin is an $|n_5 : n_6 : \varnothing)$ spiral hyperbolic point for the linear discrete system.
(ii) The origin is an $|n : \varnothing : \varnothing)$ spiral sink for the linear discrete system.
(iii) The origin is an $|\varnothing : n : \varnothing)$ spiral source for the linear discrete system.
(iv) The origin is an $|\varnothing : \varnothing : n)$ -circular center for the linear discrete system.
(v) The origin is an $|n_5 : \varnothing : n_7)$ -point for the linear discrete system.
(vi) The origin is an $|\varnothing : n_6 : n_7)$ -point for the linear discrete system.
(vii) The origin is an $|n_5 : n_6 : n_7)$ -point for the linear discrete system.

II. *Special cases*

(i) The origin is an $|\varnothing : \varnothing : [n, l; \varnothing])$-circular center for the linear discrete system.
(ii) The origin is an $|\varnothing : \varnothing : [n, l; \boldsymbol{\kappa}_7])$ -point for the linear discrete system.
(iii) The origin is an $|n_5 : \varnothing : [n_7, l; \boldsymbol{\kappa}_7])$ -point for the linear discrete system.
(iv) The origin is an $|\varnothing : n_6 : [n_7, l; \boldsymbol{\kappa}_7])$-point for the linear discrete system.
(v) The origin is an $|n_5 : n_6 : [n_7, l; \boldsymbol{\kappa}_7])$ -point for the linear discrete system.

III. *Simple critical cases*

(i) An $|n_5 : n_6 : 1)$ state of the origin for the linear discrete system is a boundary of its $|n_5 + 1 : n_6 : \varnothing)$ spiral saddle and $|n_5 : n_6 + 1 : \varnothing)$ saddle.
(ii) An $|n_5 : \varnothing : 1)$ state of the origin for the linear discrete system is a boundary of its $|n_5 + 1 : \varnothing : \varnothing)$ spiral sink and $|n_5 : 1 : \varnothing)$ spiral saddle.
(iii) An $|\varnothing : n_6 : 1)$ state of the origin for the linear discrete system is a boundary of its $|\varnothing : n_6 + 1 : \varnothing)$ spiral source and $|1 : n_6 : \varnothing)$ spiral saddle.
(iv) An $|n_5 : n_6 : n_7 + 1)$ state of the origin for the linear discrete system is a boundary of its $|n_5 + 1 : n_6 : n_7)$ state and $|n_5 : n_6 + 1 : n_7)$ state.
(v) An $|\varnothing : n_6 : n_7 + 1)$ state of the origin for the linear discrete system is a boundary of its $|1 : n_6 : n_7)$ state and $|n_5 : n_6 + 1 : n_7)$ state.
(vi) An $|n_5 : \varnothing : n_7 + 1)$ state of the origin for the linear discrete system is a boundary of its $|n_5 + 1 : \varnothing : n_7)$ state and $|n_5 : 1 : n_7)$ state.

IV. *Complex critical cases*

(i) An $|n_5 : n_6 : [n_7, l; \boldsymbol{\kappa}_7])$ state of the origin for the linear discrete system is a boundary of its $|n_5 + n_7 : n_6 : \varnothing)$ and $|n_5 : n_6 + n_7 : \varnothing)$ spiral saddles.

(ii) An $|n_5 : n_6 : [n_7 + k_7, l; \boldsymbol{\kappa}_7])$ state of the origin for the linear discrete system is a boundary of its $|n_5 + k_7 : n_6 : [n_7, l; \boldsymbol{\kappa}'_7])$ and $|n_5 : n_6 + k_7 : [n_7, l; \boldsymbol{\kappa}'_7])$ states.
(iii) An $|n_5 : n_6 : [n_7 + k_5 - k_6, l_2; \boldsymbol{\kappa}_7])$ state of the origin for the linear discrete system is a boundary of its $|n_5 + k_5 : n_6 : [n_7, l_1; \boldsymbol{\kappa}'_7])$ and $|n_5 : n_6 + k_6 : [n_7, l_3; \boldsymbol{\kappa}''_7])$ states.

B.5 Lower-Dimensional Discrete Systems

Consider a one-dimensional linear system as

$$x_{k+1} = \lambda x_k \tag{B.132}$$

with initial condition $x_k = x_0$. The iterative solution is

$$x_k = \lambda^k x_0. \tag{B.133}$$

The following properties of the solution exist.

(i) $\lim_{k\to\infty} |x_k| = 0$, and the system to the origin is stable if $|\lambda| < 1$;
(ii) $\lim_{k\to\infty} |x_k| = \infty$, and the system to the origin is unstable if $|\lambda| > 1$;
(iii) $x_k = x_0$ $(k = 1, 2, \cdots, \infty)$, and the system is invariant if $\lambda = 1$;
(iv) $x_k \in \{x_0, -x_0\}$ with $x_{2m} \neq x_{2m+1} (k, m = 1, 2, \cdots, \infty)$, and the system is symmetrically flipped if $\lambda = -1$.

The above solutions are illustrated in Fig. B.1. The solutions and phase lines for the unstable, stable and invariant linear systems are presented in Fig. B.1a–f, respectively. The gray points are values of λ. In Fig. B.1a, the $([\varnothing, \varnothing] : [1, \varnothing] : \varnothing : \varnothing|$ monotonical source is an unstable node of the first kind with $\lambda > 1$. In Fig. B.1b, the $([\varnothing, \varnothing] : [\varnothing, 1] : \varnothing : \varnothing|$ oscillatory source is an unstable node of the second kind with $-1 < \lambda$. In Fig. B.1c, the $([1, \varnothing] : [\varnothing, \varnothing] : \varnothing : \varnothing|$ monotonical sink is a stable node of the first kind with $0 < \lambda < 1$. In Fig. B.1d, the $([\varnothing, 1] : ([\varnothing, 1] : [\varnothing, \varnothing] : \varnothing : \varnothing|$ -oscillatory sink is a stable node of the second with $-1 < \lambda < 0$. In Fig. B.1e, the$([\varnothing, \varnothing] : [\varnothing, \varnothing] : 1 : \varnothing|$ - state with $\lambda = 1$ is the boundary of source and sink, which is also called the stability boundary of the first kind. In Fig. B.1f, the $([\varnothing, \varnothing] : [\varnothing, \varnothing] : \varnothing : 1|$ state with $\lambda = -1$ is an flip boundary of flip source and flip sink, which is also called the stability boundary of the second kind. In addition, the $([0, 0] : [\varnothing, \varnothing] : \varnothing : \varnothing|$ direct sink is a direct stable node of the first kind with $\lambda = 0$, as shown in Fig. B.1g. This sink is the boundary for the monotonical and oscillatory sinks. This special case will not affect the stability but it changes trajectory of the descrete system.

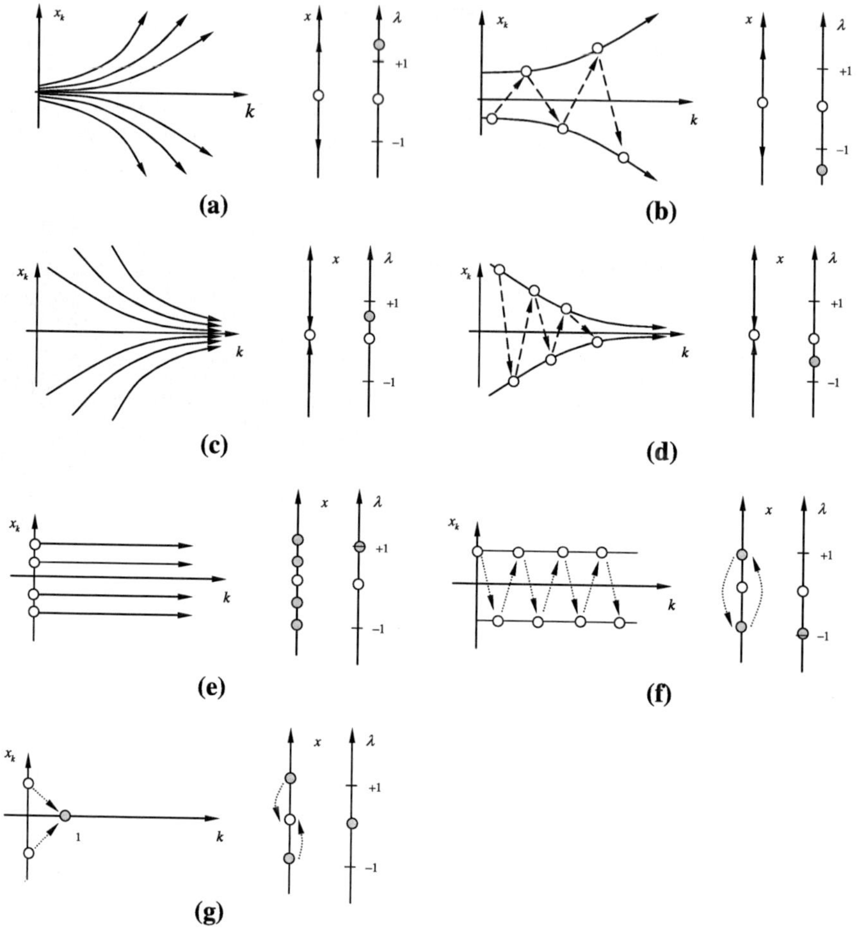

Fig. B.1 Solution and phase line of $x_{k+1} = \lambda x_k$: **a** $([\varnothing,\varnothing] : [1,\varnothing] : \varnothing : \varnothing|$ -monotonic source $(\lambda > 1)$, **b** $([\varnothing,\varnothing] : [\varnothing,1] : \varnothing : \varnothing|$ -oscillatory source $(-1<\lambda)$, **c** $([1,\varnothing] : [\varnothing,\varnothing] : \varnothing : \varnothing|$ - monotonic sink with $0<\lambda<1)$, **d** $([\varnothing,1] : [\varnothing,\varnothing] : \varnothing : \varnothing|$ - oscillaotory sink $(-1<\lambda<0)$, **e** $([\varnothing,\varnothing] : [\varnothing,\varnothing] : 1 : \varnothing|$ -boundary of source and sink $(\lambda = 1)$, **f** $([\varnothing,\varnothing] : [\varnothing,\varnothing] : \varnothing : 1|$ - flip boundary of source and sink $(\lambda = -1)$ and **g** $([0,0] : [\varnothing,\varnothing] : \varnothing : \varnothing|$ - direct sink $(\lambda = 0)$

Consider a one-dimensional linear system with external excitation

$$x_{k+1} = \lambda x_k + b \tag{B.134}$$

with initial condition x_0. With the fixed point x^*, the solution is

$$x_{k+1} = \lambda(x_k - x^*) + x^* \Rightarrow x_k = \lambda^k(x_0 - x^*) + x^*. \tag{B.135}$$

B.5.1 Planar Discrete Linear Systems

Consider a two-dimensional linear system as

$$\mathbf{x}_{k+1} = \mathbf{A}\mathbf{x}_k \tag{B.136}$$

with initial condition $\mathbf{x}_k$, and

$$\mathbf{A} = \begin{bmatrix} a_{11} & a_{12} \\ a_{21} & a_{22} \end{bmatrix}. \tag{B.137}$$

If $\det \mathbf{A} \neq 0$, $\mathbf{x}_k^* = \mathbf{0}$ is an unique fixed point. With a nonsingular transform matrix $\mathbf{P}$, $\mathbf{B} = \mathbf{P}^{-1}\mathbf{A}\mathbf{P}$. With $\mathbf{x}_k = \mathbf{P}\mathbf{y}_k$,

$$\mathbf{y}_{k+1} = \mathbf{B}\mathbf{y}_k \tag{B.138}$$

where

$$\mathbf{B} = \begin{bmatrix} \lambda_1 & 0 \\ 0 & \lambda_2 \end{bmatrix}. \tag{B.139}$$

There are four cases:

(A) For two real distinct eigenvalues ($\lambda_1 \neq \lambda_2$), the solution is expressed by

$$\mathbf{B} = \begin{bmatrix} \lambda_1 & 0 \\ 0 & \lambda_2 \end{bmatrix} \text{ and } \mathbf{y}_k = \begin{bmatrix} \lambda_1^k & 0 \\ 0 & \lambda_1^k \end{bmatrix}\mathbf{y}_0. \tag{B.140}$$

If $|\lambda_k| < 1$ $(k = 1, 2)$, the origin is a stable node. If $|\lambda_k| > 1$ $(k = 1, 2)$, the origin is an unstable node. The corresponding phase portraits and eigenvalue diagrams for the stable and unstable nodes of the linear systems are sketched in Figs. B.2–B.4.

The origin is called a saddle of the linear system if $|\lambda_i| > 1 (i \in \{1, 2\})$ and $|\lambda_j| < 1$ $(j \in \{1, 2\}$ and $j \neq i)$. The linear system is unstable. The corresponding phase portraits and eigenvalue diagram are presented in Figs. B.5 and B.6. On the eigenvector direction, the discrete states will come to or leave the origin.

The linear discrete system possess a saddle-stable node boundary of the *first* kind to the origin if $\lambda_i = 1 (i \in \{1, 2\})$ and $|\lambda_j| < 1$ $(j \in \{1, 2\}$ and $j \neq i)$.

The linear discrete system possess a saddle-stable node boundary of the *second* kind to the origin if $\lambda_i = -1$ $(i \in \{1, 2\})$ and $|\lambda_j| < 1 (j \in \{1, 2\}$ and $j \neq i)$. The phase portraits and eigenvalue diagram are presented in Figs. B.7 and B.8. The linear system is critically stable. The origin is called the center of the linear system. The linear discrete system possess a saddle-*unstable* node boundary of the *first* kind to the origin if $\lambda_i = 1 (i \in \{1, 2\})$ and $|\lambda_j| > 1$ $(j \in \{1, 2\}$ and $j \neq i)$. The linear discrete system possess a saddle-*unstable* node boundary of the *second* kind to the origin if $\lambda_i = -1$ $(i \in \{1, 2\})$ and $|\lambda_j| > 1$ $(j \in \{1, 2\}$ and $j \neq i)$. The

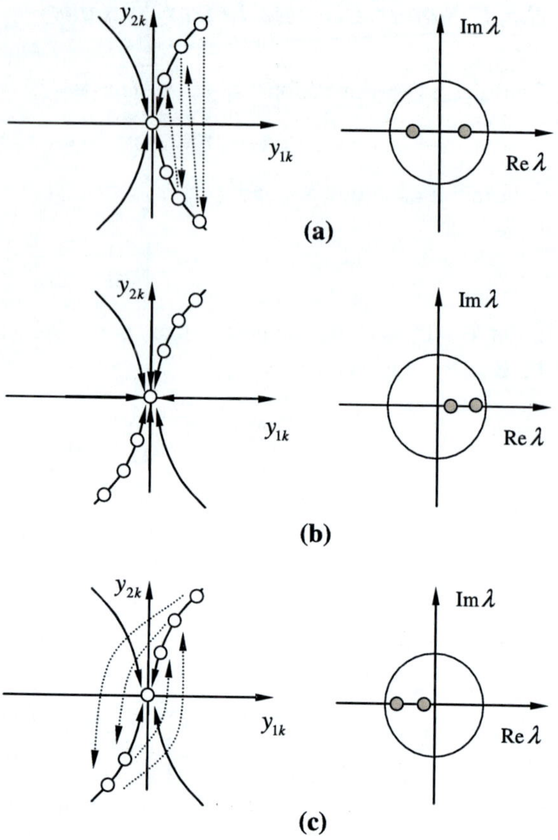

Fig. B.2 Sink or stable node at the origin for $\mathbf{y}_{k+1} = \mathbf{B}\mathbf{y}_k$:**a** $([1,1] : [\varnothing,\varnothing] : \varnothing : \varnothing|$ -sink (or stable node of the third kind) $(-1<\lambda_1<0<\lambda_2<1)$, **b** $([2,\varnothing] : [\varnothing,\varnothing] : \varnothing : \varnothing|$ -sink (or stable node of the first kind) $(0<\lambda_1<\lambda_2<1)$ and **c** $([\varnothing,2] : [\varnothing,\varnothing] : \varnothing : \varnothing|$ -sink (or stable node of the second) $(-1<\lambda_1<\lambda_2<0)$

linear system is unstable. This is the boundary for unstable nodes and saddles. If $\lambda_1 = -1$ and $\lambda_2 = 1$, this case is a critical case for saddle-node boundary of the first or second kind. The phase portraits and eigenvalue diagram are presented in Figs. B.9–B.11.

(B) For two real repeated eigenvalues $(\lambda_1 = \lambda_2 = \lambda)$, the solution is given by

$$\begin{aligned} \mathbf{B} &= \begin{bmatrix} \lambda & 1 \\ 0 & \lambda \end{bmatrix} \text{ and } \mathbf{y}_k = \begin{bmatrix} \lambda^k & k\lambda^{k-1} \\ 0 & \lambda^k \end{bmatrix} \mathbf{y}_0. \\ \mathbf{B} &= \begin{bmatrix} \lambda & 0 \\ 0 & \lambda \end{bmatrix} \text{ and } \mathbf{y}_k = \begin{bmatrix} \lambda^k & 0 \\ 0 & \lambda^k \end{bmatrix} \mathbf{y}_0. \end{aligned} \tag{B.141}$$

For repeated eigenvalues $|\lambda_k| = |\lambda| < 1$ $(k = 1, 2)$, the origin is a stable node. If repeated eigenvalues $|\lambda_k| = |\lambda| > 1$ $(k = 1, 2)$, the origin is also an unstable node. The phase portraits and eigenvalue diagram for the stable and unstable nodes are

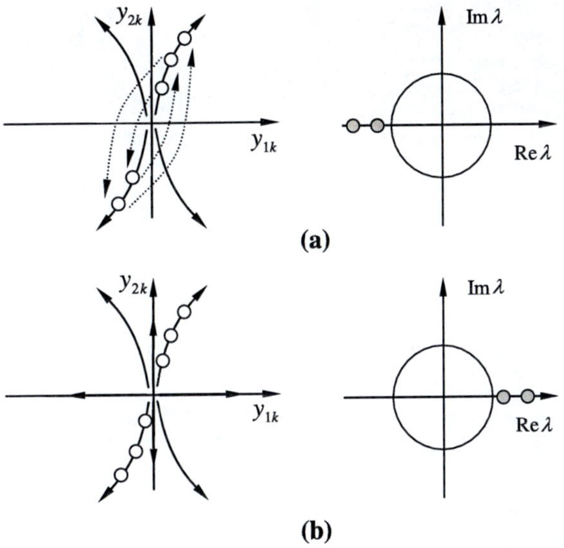

Fig. B.3 Source or unstable node of $\mathbf{y}_{k+1} = \mathbf{B}\mathbf{y}_k$: **a** $([\varnothing,\varnothing] : [\varnothing,2] : \varnothing : \varnothing|$ - source (or unstable node of the second kind) $(\lambda_1 < \lambda_2 < -1)$, **b** $([\varnothing,\varnothing] : [2,\varnothing] : \varnothing : \varnothing|$ - source (or unstable node of the first kind) $(\lambda_1 > \lambda_2 > 1)$

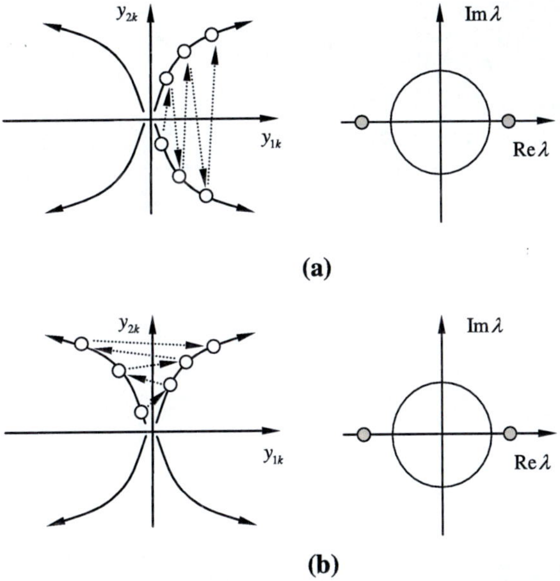

Fig. B.4 $([\varnothing,\varnothing] : [1,1] : \varnothing : \varnothing|$ -source or unstable node of the third kind for $\mathbf{y}_{k+1} = \mathbf{B}\mathbf{y}_k$: **a** $\lambda_2 < -1$ and $\lambda_1 > 1$, **b** $\lambda_1 < -1$ and $\lambda_2 > 1$

shown in Figs. B.12 and B.13. For the second equation of Eq. (B.141), the points for stable and unstable nodes exist on the line in phase plane.

From the first equation of Eq. (B.141), for $\lambda = 1$, then one obtains $y_{2k} = y_{20}$ and $y_{1k} = y_{10} + ky_{20}$. If $y_{20} \neq 0$, this is the unstable case as $k \to \infty$. If $y_{20} = 0$, the linear discrete system has a fixed point. For the second equation of Eq. (B.141), the

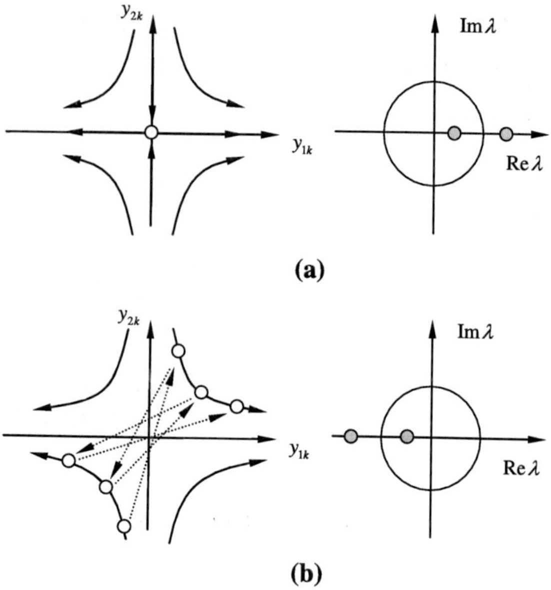

Fig. B.5 Saddle of $\mathbf{y}_{k+1} = \mathbf{B}\mathbf{y}_k$: **a** $([1,\varnothing]:[1,\varnothing]:\varnothing:\varnothing|$ -saddle $(0<\lambda_1<1<\lambda_2)$ and **b** $([\varnothing,1]:[\varnothing,1]:\varnothing:\varnothing|$ -saddle $(\lambda_1<-1<\lambda_2<0)$

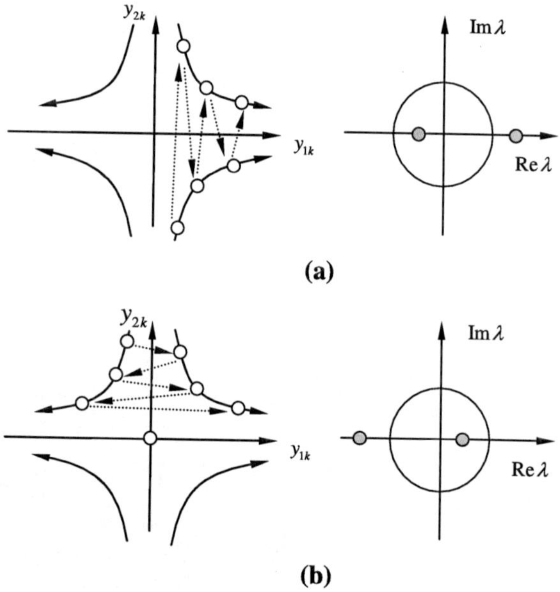

Fig. B.6 Saddle of $\mathbf{y}_{k+1} = \mathbf{B}\mathbf{y}_k$: **a** $([\varnothing,1]:[1,\varnothing]:\varnothing:\varnothing|$ -saddle $(-1<\lambda_1<0$ and $\lambda_2>1)$ and **b** $([1,\varnothing]:[\varnothing,1]:\varnothing:\varnothing|$ -saddle $(\lambda_1<-1$ and $0<\lambda_2<1)$

discrete system possesses a fixed point, which is a critical point for saddle-node boundary and flutter (Neimark) boundary. If $\lambda=-1$, then $y_{2k}=(-1)^k y_{20}$ and $y_{1k}=(-1)^k y_{10}+(-1)^{k-1} k y_{20}$. If $y_{20}\neq 0$, this is an unstable case. If $y_{20}=0$, the

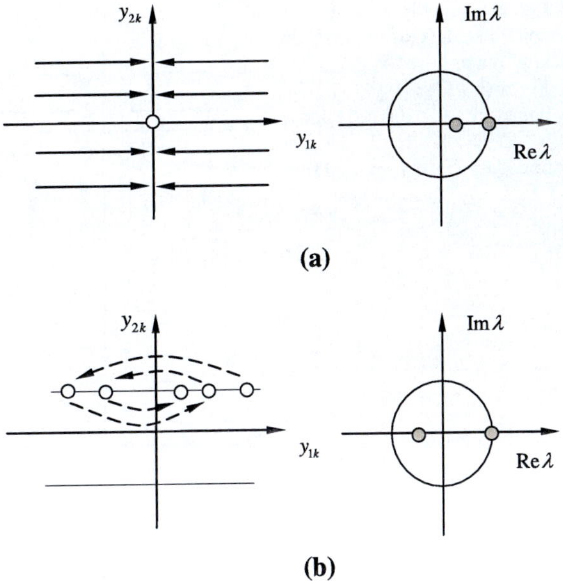

Fig. B.7 Saddle-stable node boundary for the first kind for $\mathbf{y}_{k+1} = \mathbf{B}\mathbf{y}_k$: **a** $([1,\varnothing]:[\varnothing,\varnothing]:1:\varnothing|$ boundary ($\lambda_1 \in (0,1)$ and $\lambda_2 = 1$), **b** $([\varnothing,1]:[\varnothing,\varnothing]:1:\varnothing|$ boundary ($\lambda_1 \in (-1,0)$ and $\lambda_2 = 1$)

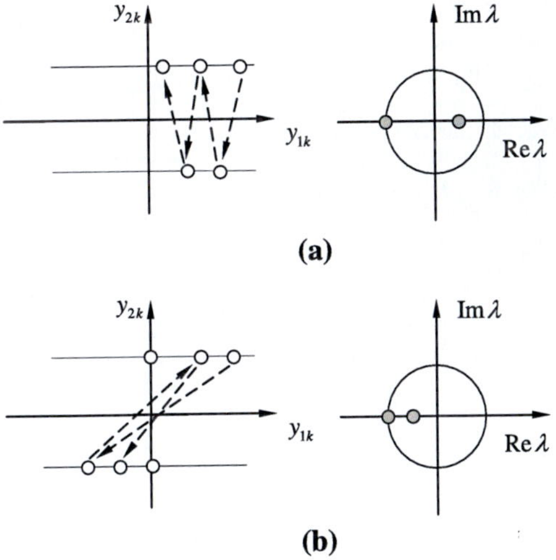

Fig. B.8 Saddle-stable node boundary for the second kind (or flip boundary) for $\mathbf{y}_{k+1} = \mathbf{B}\mathbf{y}_k$: **a** $([1,\varnothing]:[\varnothing,\varnothing]:\varnothing:1|$ boundary ($\lambda_1 \in (0,1)$ and $\lambda_2 = -1$), **b** $([\varnothing,1]:[\varnothing,\varnothing]:\varnothing:1|$ boundary ($\lambda_1 \in (-1,0)$ and $\lambda_2 = -1$)

discrete system has a pair of flip points in direction of y_1. For the second equation of Eq. (B.141) with $\lambda = -1$, the discrete system possesses a pair of flipped points $y_{2k} = (-1)^k y_{20}$ and $y_{1k} = (-1)^k y_{10}$. This is the critical point for saddle-node

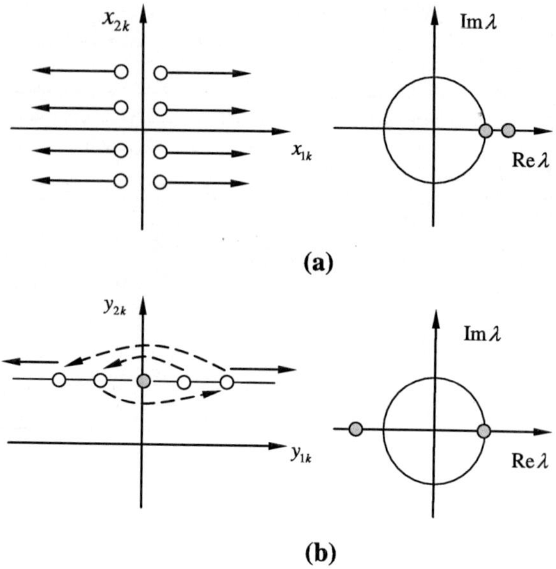

Fig. B.9 Saddle-unstable node boundary of the first kind for $\mathbf{y}_{k+1} = \mathbf{B}\mathbf{y}_k$ ($\lambda_2 = 1$) : **a** ($[\varnothing,\varnothing] : [1,\varnothing] : 1 : \varnothing|$ boundary ($\lambda_1 \in (1,\infty)$) and **b** ($[\varnothing,\varnothing] : [\varnothing,1] : 1 : \varnothing|$ boundary ($\lambda_1 \in (-\infty,-1)$)

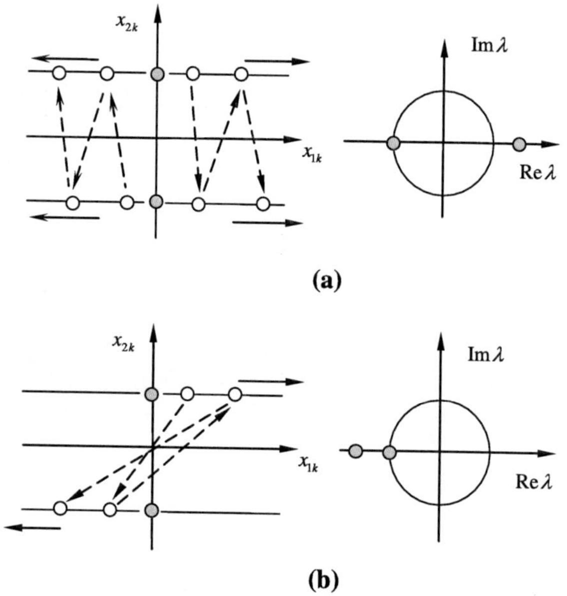

Fig. B.10 Saddle-unstable node boundary of the second kind (flip boundary) for $\mathbf{y}_{k+1} = \mathbf{B}\mathbf{y}_k$: **a** ($[\varnothing,\varnothing] : [1,\varnothing] : \varnothing : 1|$ -boundary ($\lambda_1 \in (1,\infty)$) and **b** ($[\varnothing,\varnothing] : [\varnothing,1] : \varnothing : 1|$ boundary ($\lambda_1 \in (-\infty,-1)$)

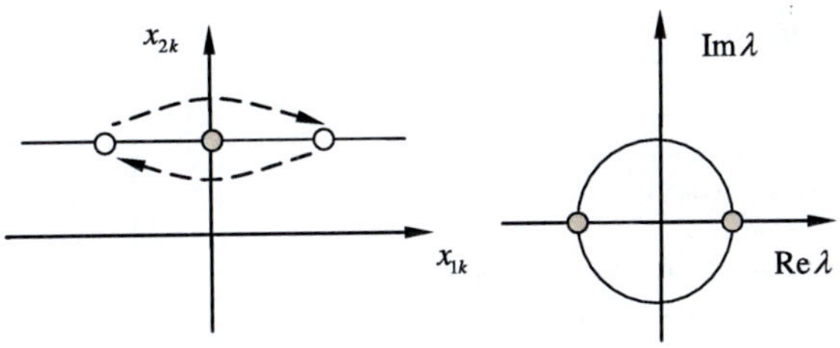

Fig. B.11 Critical case of saddle-node boundary for $\mathbf{y}_{k+1} = \mathbf{B}\mathbf{y}_k$ and eigenvalue diagram ($[\varnothing, \varnothing] : [\varnothing, \varnothing] : 1 : 1|$ -critical point ($\lambda_1 = -1$ and $\lambda_2 = 1$)

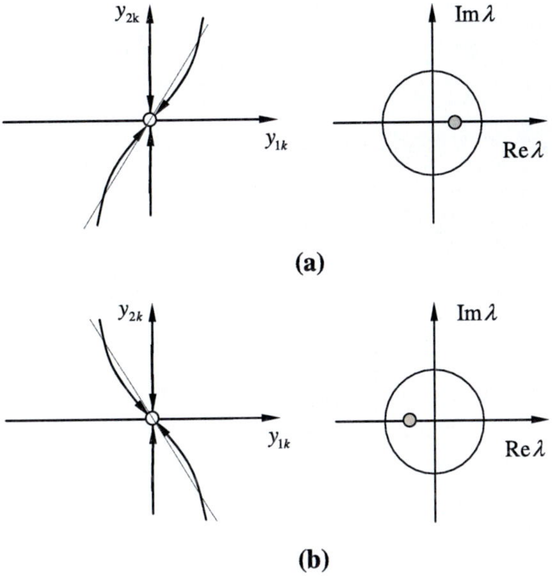

Fig. B.12 Stable nodes of $\mathbf{y}_{k+1} = \mathbf{B}\mathbf{y}_k$: **a** ($[\varnothing, 2] : [\varnothing, \varnothing] : \varnothing : \varnothing|$ sink with $\lambda_i = \lambda \in (0, 1)$ $(i = 1, 2)$, **b** ($[2, \varnothing] : [\varnothing, \varnothing] : \varnothing : \varnothing|$ sink with $\lambda_i = \lambda \in (-1, 0)$ $(i = 1, 2)$

boundary (flip) and flutter boundary. The illustrations are given in Figs. B.14 and B.15.

(C) For $\lambda_1 = \alpha + \mathbf{i}\beta$ and $\lambda_2 = \alpha - \mathbf{i}\beta$, the solution is given by

$$\mathbf{B} = \begin{bmatrix} \alpha & \beta \\ -\beta & \alpha \end{bmatrix} \text{ and } \mathbf{y}_k = r^k \begin{bmatrix} \cos k\theta & \sin sk\theta \\ -\sin k\theta & \cos k\theta \end{bmatrix} \mathbf{y}_0. \qquad (B.142)$$
$$r = \sqrt{\alpha^2 + \beta^2}, \alpha = r\cos\theta, \beta = r\sin\theta.$$

The origin is called a focus of the linear system if the imaginary part of two complex eigenvalues are nonzero ($\mathrm{Im}\,\lambda_k = \beta \neq 0$ for $k = 1, 2$). The origin is called a stable focus if the magnitude of two complex eigenvalues is less than one ($r<1$). The origin is called an unstable focus if the magnitude of two complex eigenvalues is greater than one ($r > 1$). If the magnitude of two complex eigenvalues equals to one ($r = 1$), the origin is a center for the discrete system. The discrete system possesses a flutter boundary (or Neimark boundary). From the

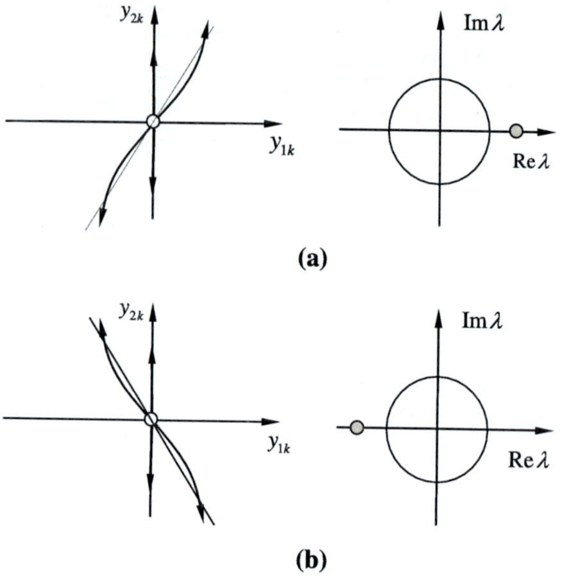

Fig. B.13 Unstable nodes of $\mathbf{y}_{k+1} = \mathbf{B}\mathbf{y}_k$: **a** $([\varnothing,\varnothing] : [2,\varnothing] : \varnothing : \varnothing|$ monotonic source $\lambda_i = \lambda \in (1,\infty)$ $(i = 1,2)$ and **b** $([\varnothing,\varnothing] : [\varnothing,2] : \varnothing : \varnothing|$ oscillatory source $\lambda_i = \lambda \in (-\infty,-1)(i = 1,2)$

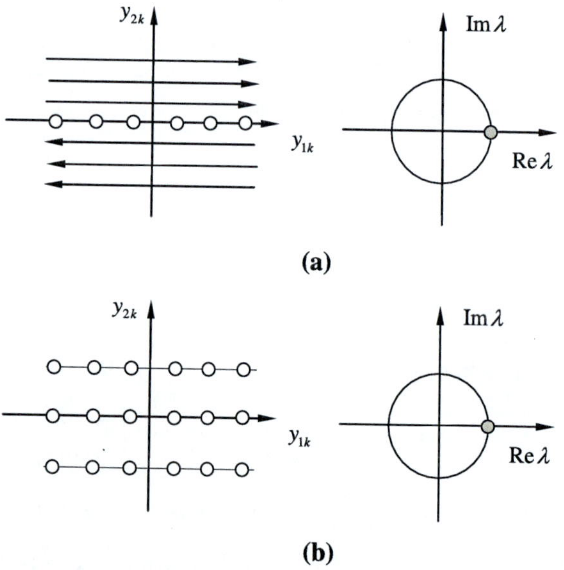

Fig. B.14 Phase portrait and eigenvalue diagram $\mathbf{y}_{k+1} = \mathbf{B}\mathbf{y}_k$: **a** $(\varnothing : \varnothing : [2;1] : \varnothing|$ -critical case (unstable source) with the second-order nilpotent matrix ($\lambda_i = \lambda = 1$ and $b_{12} = 1,\ i = 1,2$), **b** $(\varnothing : \varnothing : [2;\varnothing] : \varnothing|$ -critical boundary with the first-order nilpotent matrix for saddle-node boundary of the first kind or Neimark boundary ($\lambda_i = \lambda = 1$ and $b_{12} = 0,\ i = 1,2$)

solutions, the phase portraits and eigenvalue diagram for stable and unstable focuses are presented in Figs. B.16 and B.17. The eigenvalues are a pair of complex eigenvalues in or out the unit circle. The initial point for the unstable focus cannot be selected at the origin. For the stable focus, the solution of the linear system will approach the origin as $k \to \infty$. On the flutter boundary $(r = 1)$

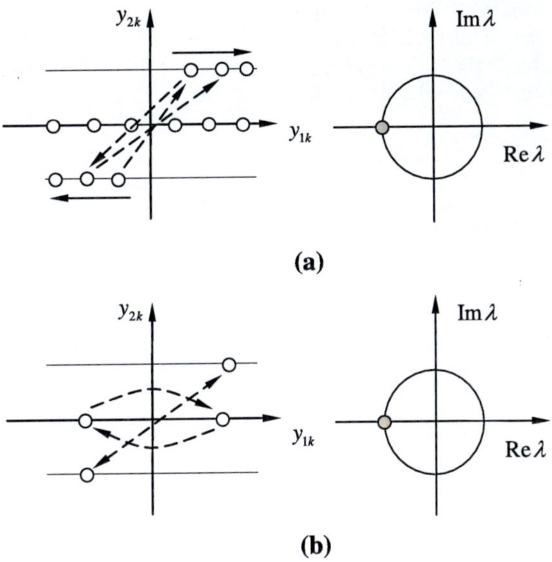

Fig. B.15 Phase portrait and eigenvalue diagram of $\mathbf{y}_{k+1} = \mathbf{B}\mathbf{y}_k$: **a** $(\varnothing : \varnothing : \varnothing : [2;1]|$ -flip critical case (or unstable node) with the second-order nilpotent matrix ($\lambda_i = \lambda = -1$ and $b_{12} = 1,\ i = 1,2$), **b** $(\varnothing : \varnothing] : \varnothing : [2;\varnothing|$ -critical flip case with the first-order nilpotent matrix for saddle-node boundary of the *second* kind (flip boundary) or Neimark boundary ($\lambda_i = \lambda = -1$ and $b_{12} = 0,\ i = 1,2$)

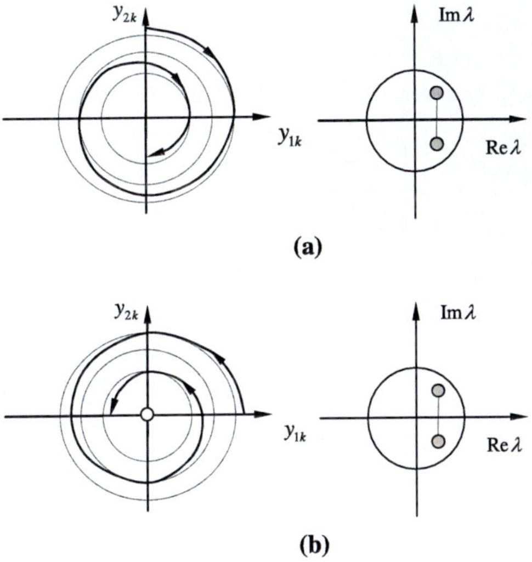

Fig. B.16 $|1 : \varnothing : \varnothing)$ -spiral sink (or stable focus) and eigenvalue diagram of $\mathbf{y}_{k+1} = \mathbf{B}\mathbf{y}_k$: **a** $\beta > 0$ and **b** $\beta < 0$

of discrete systems, the iterative points will oscillate on the circular curves, as shown in Fig. B.18.

(D) The origin is called the sink of the linear system in Eq. (B.136) if the magnitudes of all eigenvalues are less than zero ($|\lambda_i| < 1$ for $i = 1,2$). The origin is called the source of the linear system in Eq. (B.136) if the magnitudes of all eigenvalues are greater than zero ($|\lambda_i| > 1$ for $k = 1,2$) . Compared to the node and saddle-nodes, the stable and unstable focuses make discrete states spirally come to the origin or spirally leave for infinity, respectively.

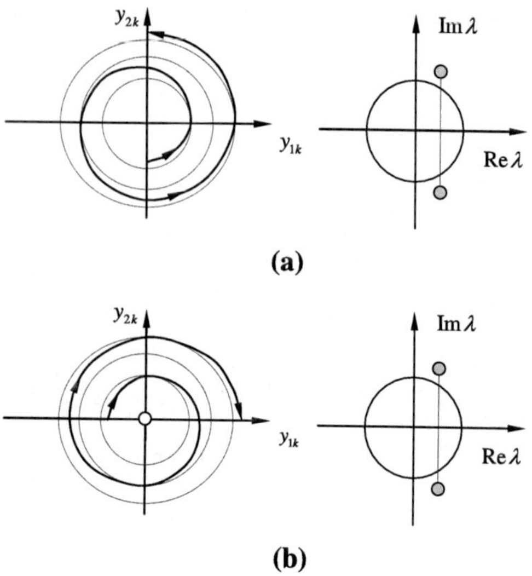

Fig. B.17 $|\varnothing : 1 : \varnothing)$ -spiral source (unstable focus) and eigenvalue diagram of $\mathbf{y}_{k+1} = \mathbf{B}\mathbf{y}_k$: **a** $\beta > 0$ and **b** $\beta < 0$

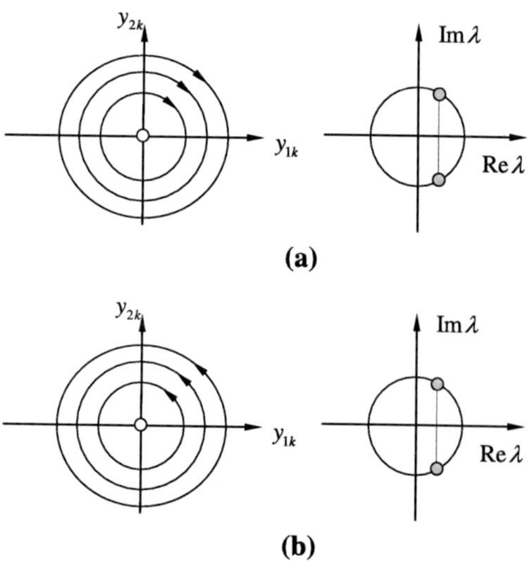

Fig. B.18 $|\varnothing : \varnothing : 1)$ -chatter (Neimark) boundary and eigenvalue diagram of $\mathbf{y}_{k+1} = \mathbf{B}\mathbf{y}_k$: **a** $\beta > 0$ and **b** $\beta < 0$

The eigenvalues of $\mathbf{A}$ are determined by $\det(\mathbf{A} - \lambda\mathbf{I}) = 0$, i.e.,

$$\lambda^2 - \mathrm{tr}(\mathbf{A})\lambda + \det(\mathbf{A}) = 0, \tag{B.143}$$

where

$$\mathrm{tr}(\mathbf{A}) = a_{11} + a_{22} \text{ and } \det(\mathbf{A}) = \begin{vmatrix} a_{11} & a_{12} \\ a_{21} & a_{22} \end{vmatrix}. \tag{B.144}$$

The corresponding eigenvalues are

$$\lambda_{1,2} = \frac{\mathrm{tr}(\mathbf{A}) \pm \sqrt{\Delta}}{2} \text{ and } \Delta = (\mathrm{tr}(\mathbf{A}))^2 - 4\det(\mathbf{A}). \tag{B.145}$$

The linear system in Eq. (B.136) possesses

(i) a saddle at the origin for real eigenvalues of $|\lambda_i| < 1$ ($i \in \{1,2\}$) and $|\lambda_j| > 1$ ($j \in \{1,2\}$ and $j \neq i$);
(ii) a stable node at the origin for real eigenvalues of $|\lambda_i| < 1$ ($i = 1,2$);
(iii) an unstable node at the origin for real eigenvalues of $|\lambda_i| > 1$ ($i = 1,2$);
(iv) a stable focus at the origin for complex eigenvalues of $|\lambda_i| < 1$ ($i = 1,2$);
(v) an unstable focus at the origin for complex eigenvalues of $|\lambda_i| > 1$ ($i = 1,2$);
(vi) a flutter phenomena (Neimark boundary) at the origin for complex eigenvalues of $|\lambda_i| = 1$ ($i = 1,2$), i.e.,

$$\det(\mathbf{A}) = 1; \tag{B.146}$$

(vii) saddle-stable node boundary of the first kind at the origin for real eigenvalues of $\lambda_i = 1$ and $|\lambda_j| < 1$ ($i,j \in \{1,2\}$ and $j \neq i$), i.e.,

$$\begin{gathered} \mathrm{tr}(\mathbf{A}) = 1 + \det(\mathbf{A}) \text{ for } i \in \{1,2\} \\ |\lambda_j| < 1 \text{ for } j \in \{1,2\} \text{ and } j \neq i; \end{gathered} \tag{B.147}$$

(viii) saddle-unstable node boundary of the first kind at the origin for real eigenvalues of $\lambda_i = 1$ and $|\lambda_j| > 1$ ($i,j \in \{1,2\}$ and $j \neq i$), i.e.,

$$\begin{gathered} \mathrm{tr}(\mathbf{A}) = 1 + \det(\mathbf{A}) \text{ for } i \in \{1,2\} \\ |\lambda_j| > 1 \text{ for } j \in \{1,2\} \text{ and } j \neq i; \end{gathered} \tag{B.148}$$

(ix) saddle-stable node boundary of the second kind (flip boundary) at the origin for real eigenvalues of $\lambda_i = -1$ and $|\lambda_j| < 1$ ($i,j \in \{1,2\}$ and $j \neq i$), i.e.,

$$\begin{gathered} \mathrm{tr}(\mathbf{A}) + \det(\mathbf{A}) + 1 = 0 \text{ for } i \in \{1,2\} \\ |\lambda_j| < 1 \text{ for } j \in \{1,2\} \text{ and } j \neq i; \end{gathered} \tag{B.149}$$

(x) saddle-unstable node boundary of the second kind (flip boundary) at the origin for real eigenvalues of $\lambda_i = -1$ and $|\lambda_j| < 1$ ($j \in \{1,2\}$ and$j \neq i$), i.e.,

$$\begin{gathered} \mathrm{tr}(\mathbf{A}) + \det(\mathbf{A}) + 1 = 0 \text{ for } i \in \{1,2\} \\ |\lambda_j| > 1 \text{ for } j \in \{1,2\} \text{ and } j \neq i; \end{gathered} \tag{B.150}$$

(xi) saddle-node boundary of the third kind at the origin for real eigenvalues of $\lambda_i = -1$ and $\lambda_j = 1$ ($i,j \in \{1,2\}$ and $j \neq i$), i.e.,

$$\mathrm{tr}(\mathbf{A}) = 0 \text{ and } \det(\mathbf{A}) = -1 \tag{B.151}$$

from which there are eight possibilities;
(xii) a degenerate fixed point at the origin for $\det(\mathbf{A}) = 0$, which is reduced to the one-dimensional case.

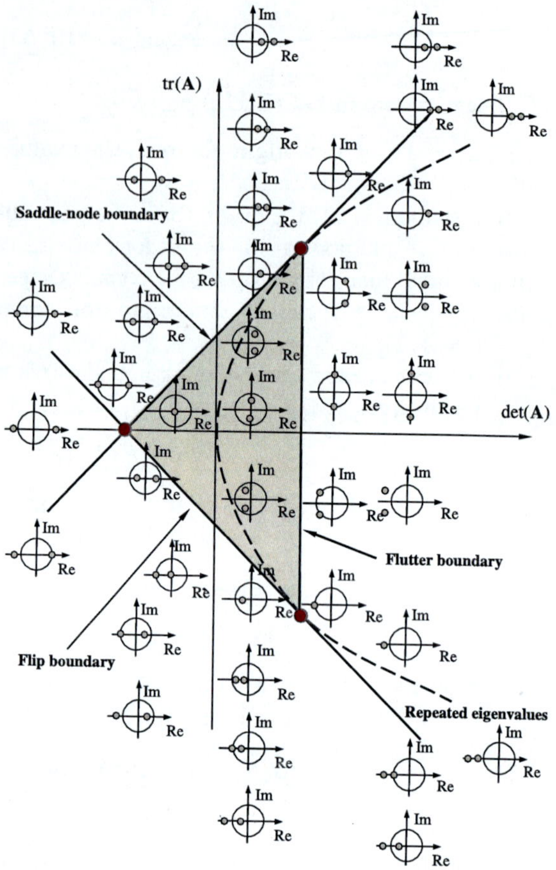

Fig. B.19 Stability and its boundary diagram through trace tr(**A**) and determinant det(**A**)

The summarization of stability and its boundary for the linear discrete system in Eq. (B.116) are intuitively illustrated in Fig. B.19 through the complex plane of eigenvalue. The shaded area is for stable nodes and stable focus. The area above the shaded area is for unstable node, and the area below the shaded area is for stable node. The left area of the axis of tr(**A**) outside of the shaded area is for saddle. The vertical line is for center with $\det(\mathbf{A}) = 1$ and $|\mathrm{tr}(\mathbf{A})| < 2$, which is also called the flutter boundary (Neimark boundary). For $\det(\mathbf{A}) > 1$, the area between the dashed curves are for unstable focus. The dashed parabolic curve is a boundary of complex and real eigenvalues. The upper line is the saddle-node boundary of the first kind (saddle-node boundary). The lower line is the saddle-node boundary of the second kind (flip boundary). The left point of the shaded triangle is the saddle-node of the third kind. The phase portrait is based on the transformed system in Eq. (B.138). The solutions of $\mathbf{x}_{k+1} = \mathbf{A}\mathbf{x}_k$ in Eq. (B.136) is given by $\mathbf{x}_k = \mathbf{P}y_k$.

B.5.2 Three-Dimensional Discrete Systems

Consider a three-dimensional discrete, linear system as

$$\mathbf{x}_{k+1} = \mathbf{A}\mathbf{x}_k \quad \text{(B.152)}$$

with initial condition $\mathbf{x}_0$, and

$$\mathbf{A} = \begin{bmatrix} a_{11} & a_{12} & a_{13} \\ a_{21} & a_{22} & a_{23} \\ a_{31} & a_{32} & a_{33} \end{bmatrix}. \quad \text{(B.153)}$$

If $\det \mathbf{A} \neq 0$, $\mathbf{x}_k^* = \mathbf{0}$ is an unique fixed point. With a nonsingular transform matrix $\mathbf{P}$, $\mathbf{B} = \mathbf{P}^{-1}\mathbf{A}\mathbf{P}$. With

$$\mathbf{x}_k = \mathbf{P}\mathbf{y}_k, \; \mathbf{y}_{k+1} = \mathbf{B}\mathbf{y}_k \quad \text{(B.154)}$$

where

$$\mathbf{B} = \begin{bmatrix} \lambda_1 & 0 & 0 \\ 0 & \lambda_2 & 0 \\ 0 & 0 & \lambda_3 \end{bmatrix}. \quad \text{(B.155)}$$

(A) If three real eigenvalues are different $(\lambda_1 \neq \lambda_2 \neq \lambda_3)$, the solution is

$$\mathbf{B} = diag(\lambda_1, \lambda_2, \lambda_3) \text{ and } \mathbf{y}_k = diag(\lambda_1^k, \lambda_2^k, \lambda_3^k)\mathbf{y}_0. \quad \text{(B.156)}$$

The origin is called a node of the discrete system if three real eigenvalues are inside or outside the unit circle. If $|\lambda_i| < 1$ $(i = 1, 2, 3)$, the origin is a stable node. If $|\lambda_i| > 1$ $(i = 1, 2, 3)$, the origin is an unstable node. The phase portraits and eigenvalue diagrams for the linear discrete system with stable and unstable nodes at the origin are sketched in Fig. B.20a and b with one eighth view. For two phase portraits, $\lambda_i > 0$ $(i = 1, 2, 3)$ is adopted. Thus, the iterative points of evolution solutions in phase portrait decrease (or increase) monotonically for the stable or unstable node in such an eighth view. All flows will monotonically come to the origin as the stable node. However, flows in a linear discrete system with an unstable node at the origin will monotonically leave away from the origin. If $\lambda_i < 0$ $(i \in \{1, 2, 3\})$, the phase portrait will be oscillatory. The iterative points of the evolution solutions in phase portraits will be oscillatory in the direction of $\mathbf{v}_i$ $(i \in \{1, 2, 3\})$. If three eigenvalues are negative $(\lambda_i < 0,\ i = 1, 2, 3)$, the three directions will be oscillatory. Therefore, the eigenvalue diagrams for the other sinks and sources are presented in Figs. B.21 and B.22 , repsectively.

The origin is called a saddle-node of the linear system if three real eigenvalues distribute inside and outside eigenvalues. If $|\lambda_i| < 1$ $(i = 1, 2)$ with $|\lambda_3| > 1$, the origin is a saddle-node with two-directional attraction and one-directional

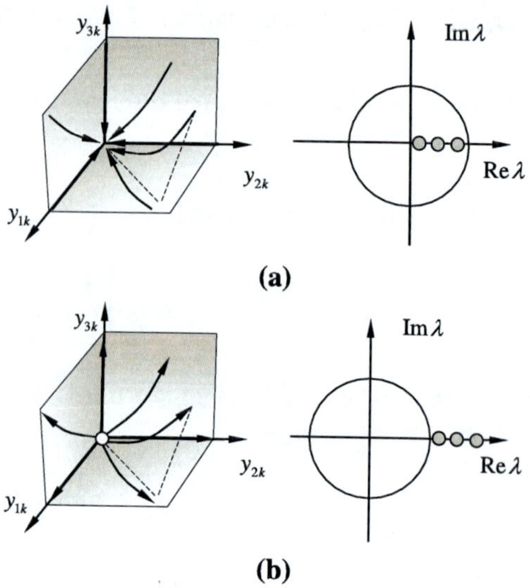

Fig. B.20 Phase portrait and eigenvalue diagrams: **a** $([3,\varnothing]:\varnothing:\varnothing:\varnothing|$ sink ($\lambda_i \in (0,1)$, $i=1,2,\ 3$) and **b** $(\varnothing:[3,\varnothing]:\varnothing:\varnothing|$ source ($\lambda_i \in (1,\infty)$, $i=1,2,3$)

expansion. If $|\lambda_i|>1$ $(i=1,2)$ with $|\lambda_3|<1$, the origin is a saddle-node with one-directional attraction and two-directional expansion. The phase portraits and eigenvalue diagrams for the linear system with two saddle-nodes at the origin are sketched in Fig. B.23a and b with one-eighth view for $\lambda_i>0$ $(i=1,2,3)$. The flows in the linear discrete systems with saddle-nodes shrink in the attraction direction(s) and stretch in the expansion direction(s). Once again, if $\lambda_i<0$ $(i\in\{1,2,3\})$, the phase portrait will be oscillatory. The iterative points of the evolution solutions in phase portraits will be oscillatory in the direction of $\mathbf{v}_i$ $(i\in\{1,2,3\})$. If three eigenvalues are negative $(\lambda_i<0,\ i=1,2,3)$, the three directions will be oscillatory. For all possible cases, the corresponding eigenvalue diagrams are presented in Figs. B.24 and B.25.

If $|\lambda_i|<1$ $(i=1,2)$ with $\lambda_3=1$, the linear discrete system has a saddle–stable node boundary of the first kind to the origin. If $|\lambda_i|>1$ $(i=1,2)$ with $\lambda_3=1$, the linear discrete system possesses a saddle-unstable node boundary of the first kind to the origin. Eigenvalue diagrams for the six critical states are presented in Fig. B.26.

If $|\lambda_i|<1$ $(i=2,3)$ with $\lambda_1=-1$, the linear discrete system has a saddle-stable node boundary of the second kind to the origin. If $|\lambda_i|>1$ $(i=2,3)$ with $\lambda_1=-1$, the linear discrete system possesses a saddle-unstable node boundary of the second kind to the origin. For the saddle-stable node boundary, there are three cases. For this boundary is often called the saddle-node boundary. For the saddle-unstable stable node boundary, three are also three cases. For this boundary is often called the saddle-node boundary, as shown in Fig. B.27 through eigenvalue diagrams.

If $|\lambda_1|<1$ and $|\lambda_2|>1$ with $\lambda_3=1$, the linear discrete system has the saddle–saddle boundary of the first kind to the origin. For the saddle–saddle boundary of the first kind, there are four cases, as shown in Fig.B.28 through eigenvalue diagrams. If $|\lambda_1|<1$ and $|\lambda_2|>1$ with $\lambda_3=-1$, the linear discrete system

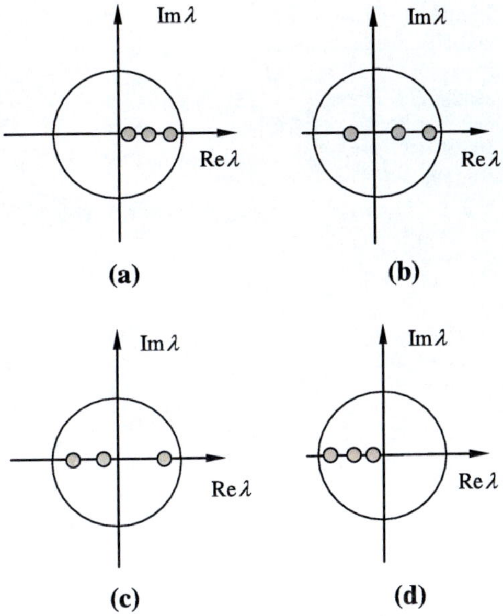

Fig. B.21 Eigenvalue diagrams for four stable nodes ($\lambda_i \in (-1,1)$, $i = 1,2,3$) : **a** ($[3,\varnothing] : \varnothing : \varnothing : \varnothing|$ sink, **b** ($[2,1] : \varnothing : \varnothing : \varnothing|$ sink, **c** ($[1,2] : \varnothing : \varnothing : \varnothing|$ -sink and **d** ($[\varnothing,3] : \varnothing : \varnothing : \varnothing|$ sink

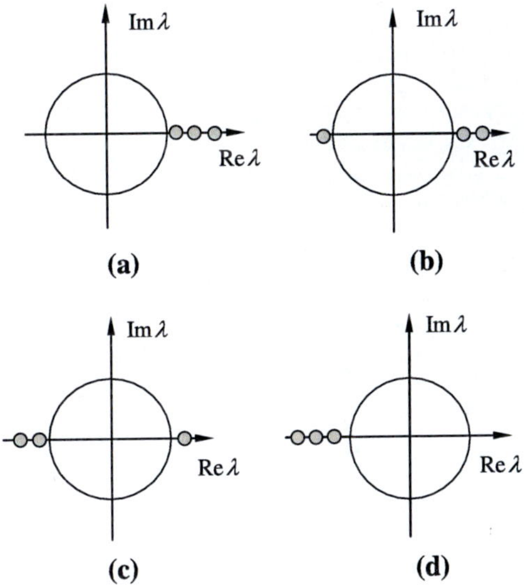

Fig. B.22 Eigenvalue diagrams for four sources (unstable node) with $|\lambda_i| > 1 (i = 1,2,3)$: **a** ($\varnothing : [3,\varnothing] : \varnothing : \varnothing|$ -source, **b** ($\varnothing : [2,1] : \varnothing : \varnothing|$ -source, **c** ($\varnothing : [1,2] : \varnothing : \varnothing|$ -source and **d** ($\varnothing : [\varnothing,3] : \varnothing : \varnothing|$ -source

possesses a saddle–saddle boundary of the second kind to the origin. For the saddle–saddle boundary of the second kind, three are also four cases, as shown in Fig. B.29 through eigenvalue diagrams. If $\lambda_1 = -1$ and $|\lambda_2| < 1$ with $\lambda_3 = 1$, the linear discrete system has the saddle-node boundary of the third kind to the origin. There are two critical states ($[1,\varnothing] : [\varnothing,\varnothing] : 1 : 1|$ and ($[\varnothing,1] : [\varnothing,\varnothing] : 1 : 1|$ states. The ($[0,0] : [\varnothing,\varnothing] : 1 : 1|$ state is a special state.

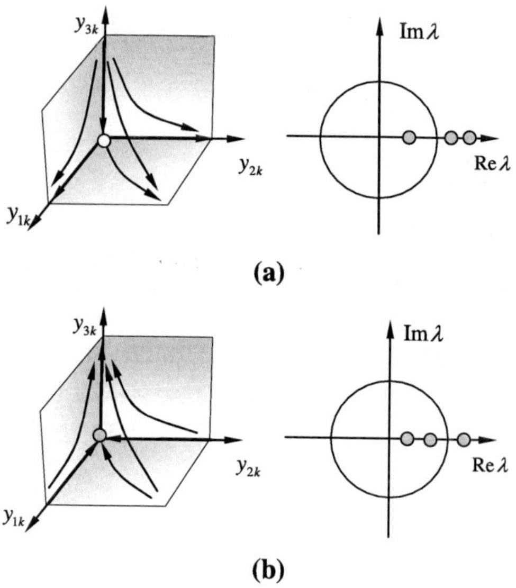

Fig. B.23 Phase portraits and eigenvalue diagrams: **a** $([1,\varnothing]:[2,\varnothing]:\varnothing:\varnothing|$ -saddle $\lambda_i \in (1,\infty)$ $(i=1,2)$ and $\lambda_3 \in (0,1)$, **b** $([2,\varnothing]:[1,\varnothing]:\varnothing:\varnothing|$ -saddle $\lambda_i \in (0,1)$ $(i=1,2)$ and$\lambda_3 \in (1,\infty)$

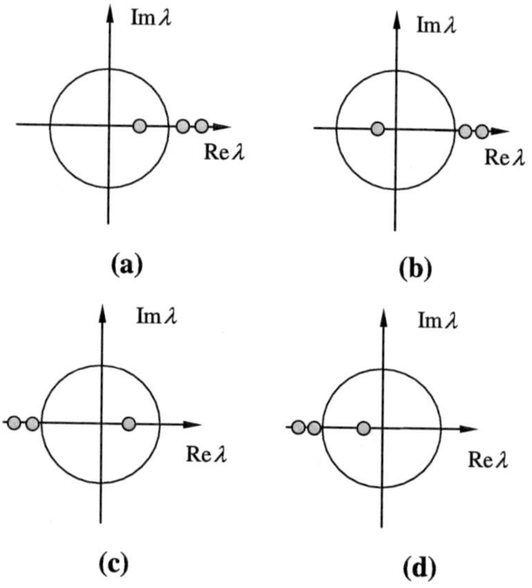

Fig. B.24 Eigenvalue diagrams for four saddles with $|\lambda_1|<1$ and $|\lambda_{2,3}|>1$: **a** $([1,\varnothing]:[2,\varnothing]:\varnothing:\varnothing|$ -saddle, **b** $([\varnothing,1]:[2,\varnothing]:\varnothing:\varnothing|$ -saddle, **c** $([1,\varnothing]:[\varnothing,2]:\varnothing:\varnothing|$ -saddle and **d** $([\varnothing,1]:[\varnothing,2]:\varnothing:\varnothing|$ -saddle

If $\lambda_1=-1$ and $|\lambda_2|>1$ with $\lambda_3=1$, the linear discrete system possesses a saddle-unstable node boundary of the fourth kind to the origin. There are two critical states $([\varnothing,\varnothing]:[1,\varnothing]:1:1|$ and $([\varnothing,\varnothing]:[\varnothing,1]:1:1|$ states. The above mentioned cases will not be illustrated herein.

(B) For two repeated real eigenvalues ($\lambda_1=\lambda_2=\lambda$ and λ_3), the solutions are

$$\mathbf{B}=daig(\lambda,\lambda,\lambda_3) \text{ and } \mathbf{y}_k=daig(\lambda^k,\lambda^k,\lambda_3^k)\mathbf{y}_0. \tag{B.157}$$

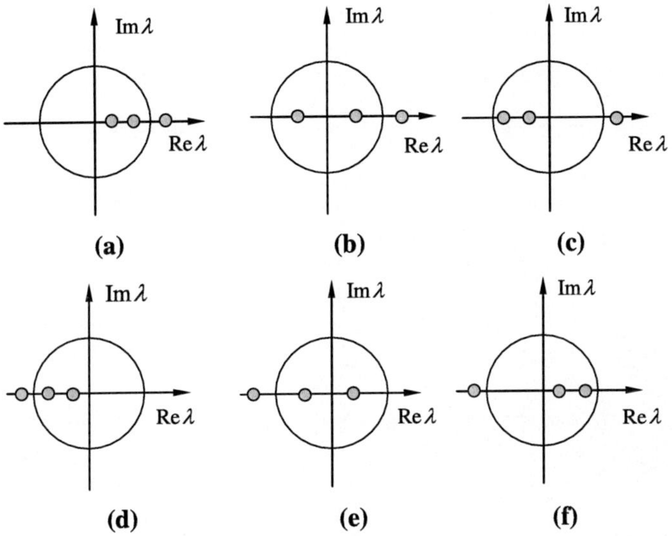

Fig. B.25 Eigenvalue diagrams for six saddles with $|\lambda_1|<1$ and $|\lambda_{2,3}|>1$: **a** $([2,\varnothing]:[1,\varnothing]:\varnothing:\varnothing|$ -saddle, **b** $([1,1]:[1,\varnothing]:\varnothing:\varnothing|$ -saddle and **c** $([\varnothing,2]:[1,\varnothing]:\varnothing:\varnothing|$ -saddle, **d** $([\varnothing,2]:[\varnothing,1]:\varnothing:\varnothing|$ -saddle, **e** $([1,1]:[\varnothing,1]:\varnothing:\varnothing|$ -saddle and **f** $([2,\varnothing]:[\varnothing,1]:\varnothing:\varnothing|$ -saddle

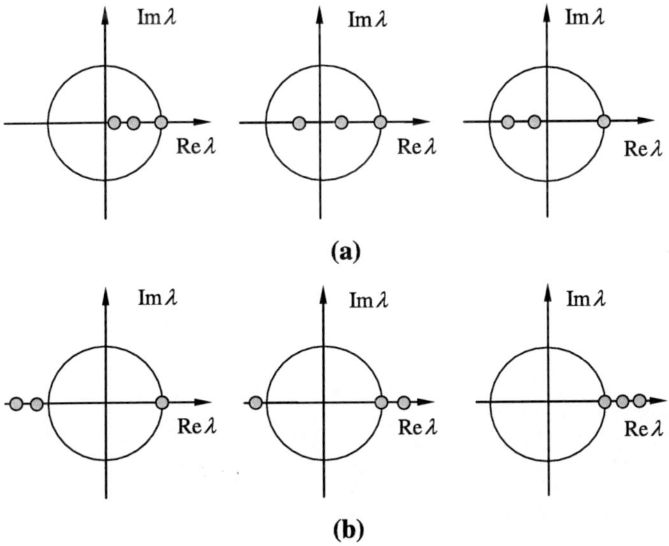

Fig. B.26 Eigenvalue diagrams $(\lambda_1=1)$: **a** saddle-stable node boundary of the first kind $|\lambda_{2,3}|<1$ with $([2,\varnothing]:\varnothing:1:\varnothing|$, $([1,1]:\varnothing:1:\varnothing|$, $([\varnothing,2]:\varnothing:1:\varnothing|$ states, **b** saddle-unstable node boundary of the first kind $|\lambda_{2,3}|>1$ with $(\varnothing:[\varnothing,2]:1:\varnothing|$, $(\varnothing:[1,1]:1:\varnothing|$, $(\varnothing:[2,\varnothing]:1:\varnothing|$ states

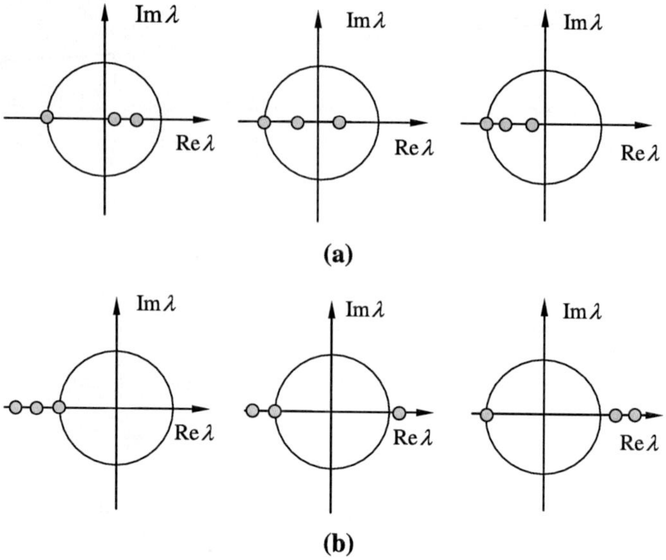

Fig. B.27 Eigenvalue diagrams ($\lambda_1 = -1$) : **a** saddle-stable node boundary of the second kind ($|\lambda_{2,3}|<1$) with $([2,\varnothing]:\varnothing:\varnothing:1|$, $([1,1]:\varnothing:\varnothing:1|$, $([\varnothing,2]:\varnothing:\varnothing:1|$ states, **b** saddle-unstable node boundary of the second kind ($|\lambda_{2,3}|>1$) with $(\varnothing:[\varnothing,2]:\varnothing:1|$, $(\varnothing:[1,1]:\varnothing:1|$, $(\varnothing:[2,\varnothing]:\varnothing:1|$- states

Fig. B.28 Eigenvalue diagrams ($\lambda_1 = 1$) for saddle–saddle boundary of the first kind: **a** $|\lambda_2|<1$ for$([1,\varnothing]:[1,\varnothing]:1:\varnothing|$ -state and $([\varnothing,1]:[1,\varnothing]:1:\varnothing|$ -state, and **b** $|\lambda_3|>1$ for $([1,\varnothing]:[\varnothing,1]:1:\varnothing|$ -state and $([\varnothing,1]:[\varnothing,1]:1:\varnothing|$ -state

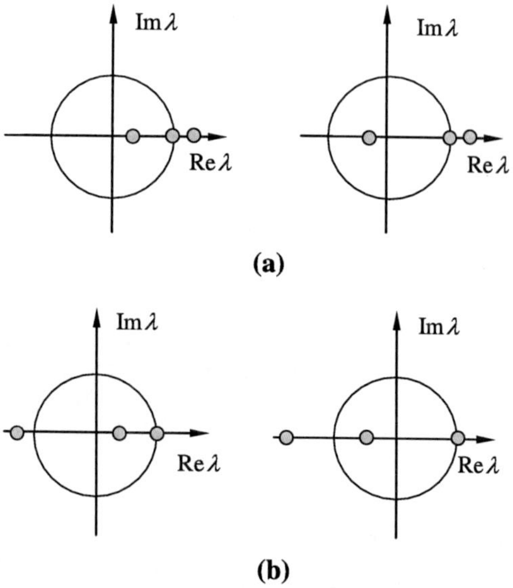

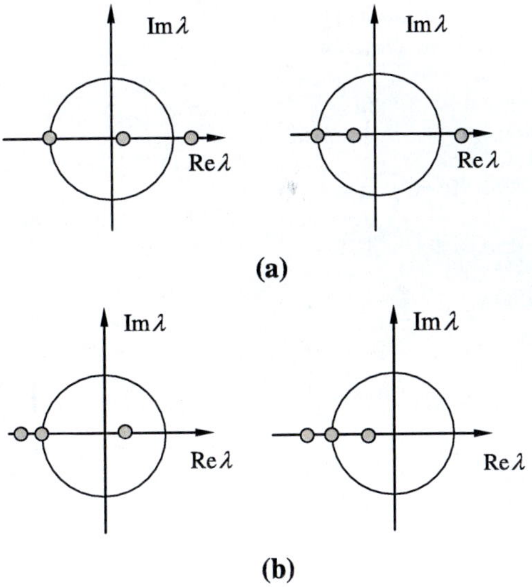

Fig. B.29 Eigenvalue diagrams ($\lambda_1 = -1$) for saddle–saddle boundary of the second kind: **a** $|\lambda_2| < 1$ and $\lambda_3 > 1$ for $([1, \varnothing] : [1, \varnothing] : \varnothing : 1|$ -state and $([\varnothing, 1] : [1, \varnothing] : \varnothing : 1|$ -state, **b** $|\lambda_2| < 1$ and $\lambda_3 < 1$ for $([1, \varnothing] : [\varnothing, 1] : \varnothing : 1|$ -state and $([\varnothing, 1] : [\varnothing, 1] : \varnothing : 1|$ -state

$$\mathbf{B} = \begin{bmatrix} \lambda & 1 & 0 \\ 0 & \lambda & 0 \\ 0 & 0 & \lambda_3 \end{bmatrix} \text{ and } \mathbf{y}_k = \begin{bmatrix} \lambda^k & k\lambda^{k-1} & 0 \\ 0 & \lambda^k & 0 \\ 0 & 0 & \lambda_3^k \end{bmatrix} \mathbf{y}_0. \tag{B.158}$$

The stability characteristics of Eq. (B.158) with two repeated real eigenvalues ($\lambda_3 \neq \lambda$) are similar to the case of three real distinct eigenvalues. The origin is (a) a stable node (sink) with ($|\lambda| < 1$ and $|\lambda_3| < 1$), (b) an unstable node (source) with ($|\lambda| > 1$ and $|\lambda_3| > 1$), and (c) a saddle-node ($|\lambda| < 1$ and $|\lambda_3| > 1$ or $|\lambda| \geq 1$ and $|\lambda_3| < 1$) for the linear system. For $\lambda_3 = 1 (\lambda_3 = -1)$, the linear system to the origin possesses (a) a saddle-stable node boundary of the first (second) kind with $|\lambda| < 1$, and (b) a saddle-unstable node boundary of the first (second) kind with $|\lambda| \geq 1$. However, for Eq. (B.157), the origin is a stable node (sink) with ($|\lambda| < 1$ and $|\lambda_3| < 1$), an unstable node (source) with ($|\lambda| > 1$ and $|\lambda_3| > 1$), and a saddle-node ($|\lambda| < 1$ and $|\lambda_3| > 1$ or $|\lambda| > 1$ and $|\lambda_3| < 1$) for the linear system. For $\lambda_i = 1$ or ($\lambda_i = -1$) with ($i \in \{1, 2, 3\}$), the linear system possesses (a) a saddle-stable node of the first (second) kind to the origin with $|\lambda_j| < 1 (j \in \{1, 2, 3\}$ and $j \neq i$) and (b) a saddle-unstable node of the first (second) kind to the origin with $|\lambda_j| > 1 (j \in \{1, 2, 3\}$ and $j \neq i$). Notice that $\lambda_1 = \lambda_2 = \lambda$. The phase portraits and eigenvalue diagram will not be presented, which can be illustrated as 2-D linear discrete systems.

(C) For three repeated real eigenvalues ($\lambda_1 = \lambda_2 = \lambda_3 = \lambda$), the solutions are

$$\mathbf{B} = diag(\lambda, \lambda, \lambda) \text{ and } \mathbf{y}_k = diag(\lambda^k, \lambda^k, \lambda^k)\mathbf{y}_0. \tag{B.159}$$

$$\mathbf{B} = \begin{bmatrix} \lambda & 1 & 0 \\ 0 & \lambda & 0 \\ 0 & 0 & \lambda \end{bmatrix} \text{ and } \mathbf{y}_k = \begin{bmatrix} \lambda^k & k\lambda^{k-1} & 0 \\ 0 & \lambda^k & 0 \\ 0 & 0 & \lambda^k \end{bmatrix} \mathbf{y}_0. \tag{B.160}$$

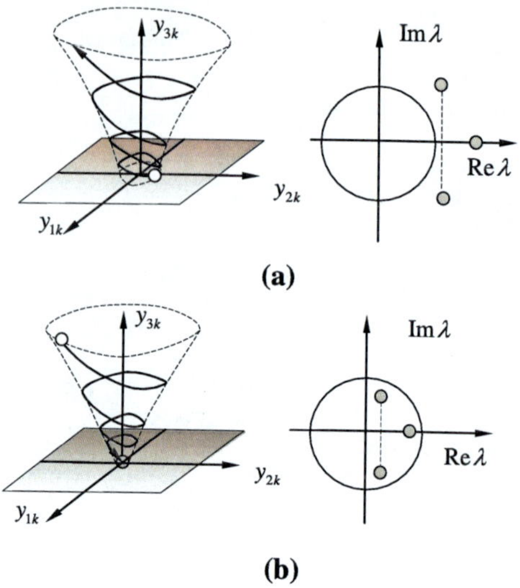

Fig. B.30 Spiral flows and eigenvalue diagram: **a** $(\varnothing:[1,\varnothing]:\varnothing:\varnothing|\varnothing:1:\varnothing)$ - spiral source $|\lambda_i|>1$ $(i=1,2,3)$ and $\mathrm{Im}\lambda_j\neq 0$ $(j=1,2)$; and **b** $([1,\varnothing]:\varnothing:\varnothing:\varnothing|1:\varnothing:\varnothing)$ -spiral sink $|\lambda_i|<1$ $(i=1,2,3)$ and $\mathrm{Im}\lambda_j\neq 0$ $(j=1,2)$

$$\mathbf{B}=\begin{bmatrix}\lambda & 1 & 0\\ 0 & \lambda & 1\\ 0 & 0 & \lambda\end{bmatrix} \text{ and } \mathbf{y}(t)=\mathbf{y}_k=\begin{bmatrix}\lambda^k & k\lambda^{k-1} & \frac{1}{2}k(k-1)\lambda^{k-2}\\ 0 & \lambda^k & k\lambda^{k-1}\\ 0 & 0 & \lambda^k\end{bmatrix}\mathbf{y}_0. \quad \text{(B.161)}$$

The stability characteristics of Eq. (B.152) with three repeated real eigenvalues are similar to the case of three real distinct eigenvalues. The origin is a stable node (sink) with $|\lambda|<1$, an unstable node (source) with $|\lambda|>1$ for the linear system. The phase portraits and eigenvalue diagrams will not be presented.

(D) For $(\lambda_{1,2}=\alpha\pm\mathbf{i}\beta)$ and $\mathrm{Im}\,\lambda_3=0$, the solution is

$$\mathbf{B}=\begin{bmatrix}\alpha & \beta & 0\\ -\beta & \alpha & 0\\ 0 & 0 & \lambda_3\end{bmatrix}, \text{ and } \mathbf{y}(t)=\begin{bmatrix}r^k\cos k\theta & r^k\sin k\theta & 0\\ -r^k\sin k\theta & r^k\cos k\theta & 0\\ 0 & 0 & \lambda_3^k\end{bmatrix}\mathbf{y}_0. \quad \text{(B.162)}$$

The origin is called a spiral focus of the discrete linear system if the real parts of three eigenvalues are inside or outside unit circle. If $|\lambda_i|<1$ $(i=1,2,3)$, the origin is a spiral sink for this linear discrete system, and there is a pair of complex eigenvalue and a real eigenvalue. If $|\lambda_i|>1$ $(i=1,2,3)$, the origin is a spiral source (Tornado) for the linear discrete system. The linear discrete system with stable and unstable spiral focuses at the origin are sketched in Fig.B.30a and b with a half space. The spiral flows and eigenvalue diagrams are presented. Eigenvalues $\lambda_{1,2}$ are complex and $\lambda_3>0$ is real. Thus the spiral sink (or source) flow increases (or deceases) monotonically. All flows spirally come to the origin as the stable spiral sink. However, flows in the linear system with an unstable spiral source at the origin will spirally leave away from the origin like a tornado. For a pair of complex eigenvalues with a real eigenvalue of $\lambda_3<0$, the spiral sink and source flows will be oscillatory in the eigenvector of $\mathbf{v}_3$, which will not be sketched. For

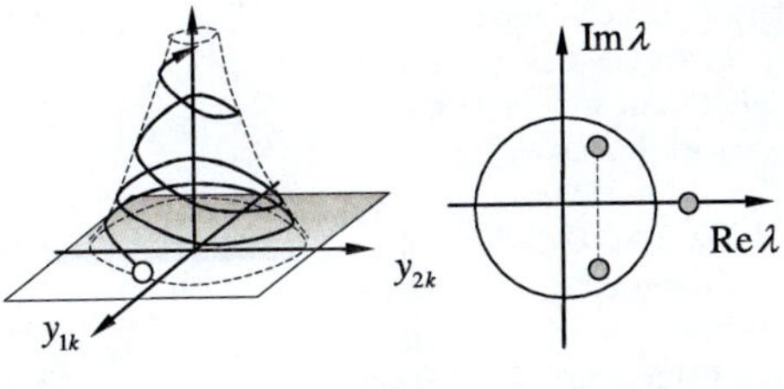

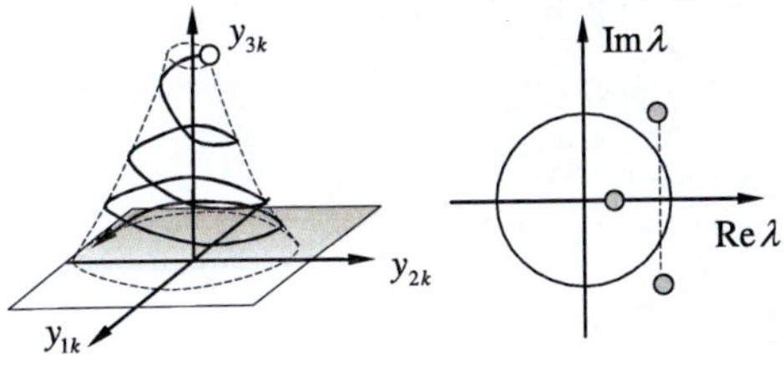

Fig. B.31 Spiral saddle flows and eigenvalue diagrams: **a** $(\varnothing : [1,\varnothing] : \varnothing : \varnothing|1 : \varnothing : \varnothing)$ spiral saddle ($|\lambda_i|<1$ with $\text{Im}\lambda_i \neq 0$ $(i = 1,2)$ and $\lambda_3 \in (1,\infty)$) and **b** $([1,\varnothing] : \varnothing : \varnothing : \varnothing|\varnothing : 1 : \varnothing)$ spiral saddle ($|\lambda_i| > 1$ with $\text{Im}\lambda_i \neq 0$ $(i = 1,2)$ and $\lambda_3 \in (0,1)$)

this case, the iterative points of the evolution solution for the linear discrete system will jump from the positive to negative space to the plane of (y_{1k}, y_{2k}).

The origin is called a spiral attraction and expansion of the linear system if the three eigenvalues are outside and inside the unit circle. For a pair of complex eigenvalues $|\lambda_i|<1$ $(i = 1,2)$ with $\lambda_3 \in (1,\infty)$, the origin is a spiral attraction with an expansion. For $|\lambda_i| > 1$ $(i = 1,2)$ with $\lambda_3 \in (0,1)$, the origin is a spiral expansion with an attraction. Flows and eigenvalue diagrams for the two cases of the linear system are sketched in Fig. B.31a and b, respectively. The origin is called a cylindrical spiral saddle-focus of the linear discrete system for a pair of complex eigenvalues $|\lambda_i| = 1$ $(i = 1,2)$ with a real eigenvalue $|\lambda_3| \neq 1$. For a pair of complex eigenvalues $|\lambda_i| = 1$ $(i = 1,2)$ with $|\lambda_3| > 1$, the origin is a cylindrical spiral saddle-source boundary. For a pair of complex eigenvalues $|\lambda_i| = 1$ $(i = 1,2)$ with $|\lambda_3|<1$, the origin is a cylindrical spiral saddle-sink boundary. Cylindrically spiral saddle-sink and spiral saddle-source boundaries (or Niemark boundary) and eigenvalue diagrams are sketched in Fig. B.32a and b with $\lambda_3 \in (0,1)$ and $\lambda_3 \in (1,\infty)$.

The origin is called a circular saddle-node boundary for the linear discrete system for a pair of complex eigenvalues $|\lambda_i| = 1$ $(i = 1,2)$ with a real eigenvalue $|\lambda_3| = 1$. For a pair of complex eigenvalues $|\lambda_i| = 1$ $(i = 1,2)$ with $\lambda_3 = 1$, the origin is circular saddle. For a pair of complex eigenvalues $|\lambda_i| = 1$ $(i = 1,2)$ with $\lambda_3 = -1$, the origin is a center of a stable cylindrical spiral. Circular saddle-node boundary (or circular Niemark-node boundary) and eigenvalue diagrams are sketched in Fig. B.33a and b with $\lambda_3 \in (0,1)$ and $\lambda_3 \in (1,\infty)$. In Fig. B.34a and b, the eigenvalue distribution for the first and second kind spiral sources are $\lambda_3 \in (-\infty,1)$ and $\lambda_3 \in (1,\infty)$ with $|\lambda_i| > 1$ $(i = 1,2)$. Three kinds of spiral saddles for 3-dimensional system are for $\lambda_3 \in$

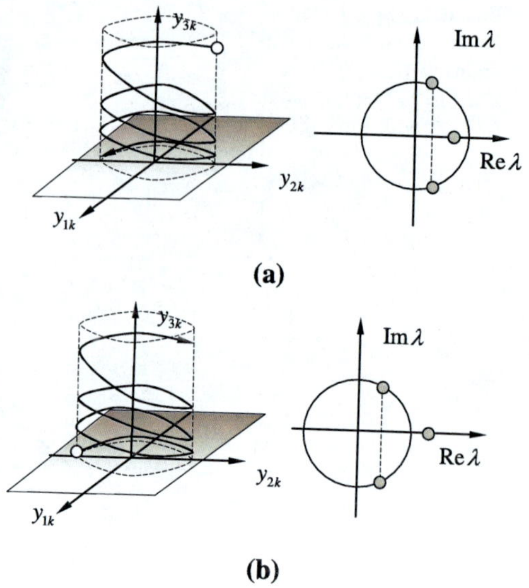

Fig. B.32 Cylindrically spiral saddle-sink and spiral saddle-source boundaries (or Niemark boundary) and eigenvalue diagram: **a** $([1,\varnothing]:\varnothing:\varnothing:\varnothing|\varnothing:\varnothing:1)$ -cylindrically spiral saddle-sink flow (or Neimark-sink boundary) ($|\lambda_i| = 1$ with $\text{Im}\lambda_i \neq 0$ $(i = 1,2)$ and $\lambda_3 \in (0,1)$), and **b** $(\varnothing:[1,\varnothing]:\varnothing:\varnothing|\varnothing:\varnothing:1)$ -cylindrically spiral saddle-source flow (or Neimark-source boundary) ($|\lambda_i| = 1$ with $\text{Im}\lambda_i \neq 0$ $(i = 1,2)$ and $\lambda_3 \in (1,\infty)$)

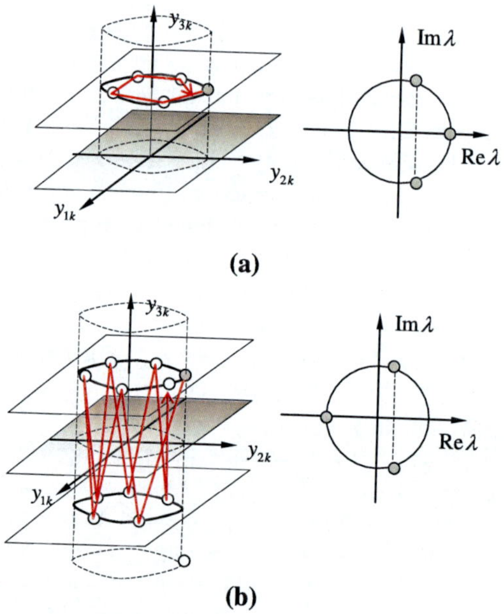

Fig. B.33 Circular saddle-node boundary and eigenvalue diagrams: **a** $(\varnothing:\varnothing:1:\varnothing|\varnothing:\varnothing:1)$ -spiral saddle-node boundary of the first kind ($|\lambda_i| = 1$ with $\text{Im}\lambda_i \neq 0$ $(i = 1,2)$ and $\lambda_3 = 1$), **b** $(\varnothing:\varnothing:\varnothing:1|\varnothing:\varnothing:1)$ -spiral saddle-node boundary of the second kind ($|\lambda_i| = 1$ with $\text{Im}\lambda_i \neq 0$ and $\lambda_3 = -1$)

$(-1,1)$ with $|\lambda_i| > 1$ $(i = 1,2)$, as in shown Fig. B.35a–c. Two boundaries for spiral saddle and spiral source are also presented in Fig. B.36a and b. In addition, there are two kinds of spiral saddles with $\lambda_3 > 1$ and $\lambda_3 < -1$ with $|\lambda_i| < 1$ $(i = 1,2)$ in Fig. B.37a and b. The corresponding spiral sinks are presented in Fig. B.38a–c. Such spiral saddles with the spiral sinks have boundaries as in Fig. B.39a and b.

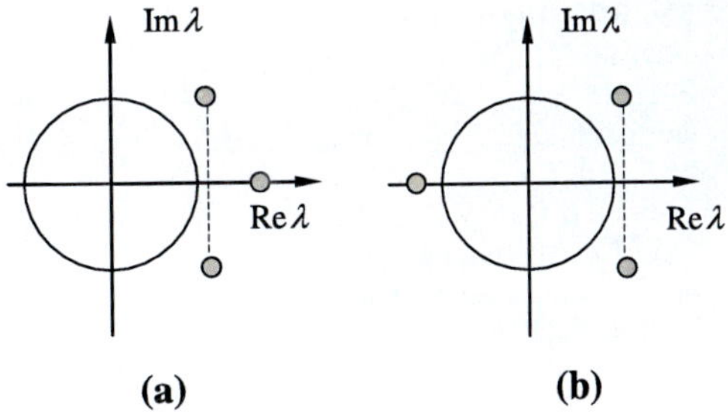

Fig. B.34 Eigenvalue diagrams: **a** $(\varnothing:[1,\varnothing]:\varnothing:\varnothing|\varnothing:1:\varnothing)$ -spiral source of the first kind ($|\lambda_i|>1$ with $\mathrm{Im}\lambda_i \neq 0$ $(i=1,2)$ and $\lambda_3>1$),**b** $(\varnothing:[\varnothing,1]:\varnothing:\varnothing|\varnothing:1:\varnothing)$ -spiral source of the second kind ($|\lambda_i|>1$ with $\mathrm{Im}\lambda_i \neq 0$ and $\lambda_3<-1$)

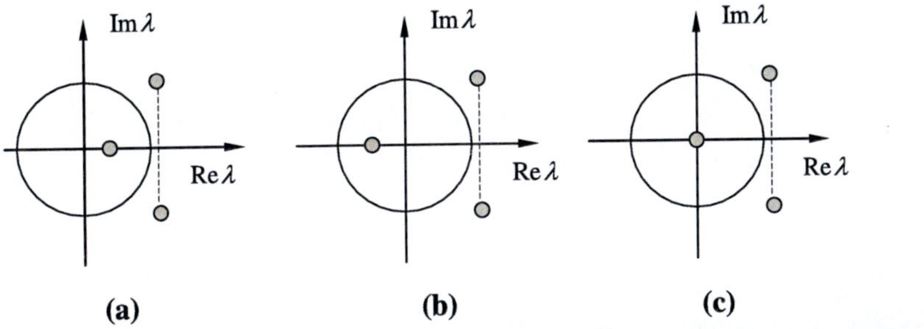

Fig. B.35 Eigenvalue diagrams: **a** $([1,\varnothing]:\varnothing:\varnothing:\varnothing|\varnothing:1:\varnothing)$ -spiral saddle of the first kind ($|\lambda_i|>1$ with $\mathrm{Im}\lambda_i \neq 0$ $(i=1,2)$ and $\lambda_3\in(0,1)$),**b** $([\varnothing,1]:\varnothing:\varnothing:\varnothing|\varnothing:1:\varnothing)$ spiral saddle of the second kind ($|\lambda_i|>1$ with $\mathrm{Im}\lambda_i \neq 0$ and $\lambda_3\in(-1,0)$) and **c** $([0,0]:\varnothing:\varnothing:\varnothing|\varnothing:1:\varnothing)$ spiral saddle of the third kind ($|\lambda_i|>1$ with $\mathrm{Im}\lambda_i \neq 0$ and $\lambda_3=0$)

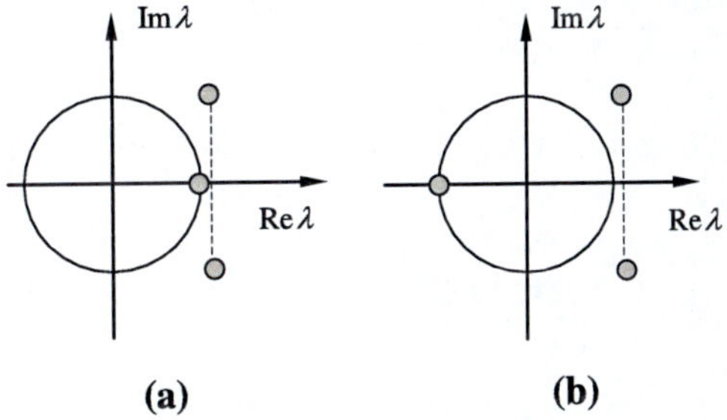

Fig. B.36 Eigenvalue diagrams: **a** $(\varnothing:\varnothing:1:\varnothing|\varnothing:1:\varnothing)$ spiral saddle-source boundary of the first kind ($|\lambda_i|>1$ with $\mathrm{Im}\lambda_i \neq 0$ $(i=1,2)$ and $\lambda_3=1$), and **b** $(\varnothing:\varnothing:\varnothing:1|\varnothing:1:\varnothing)$ spiral saddle-source boundary of the second kind ($|\lambda_i|>1$ with $\mathrm{Im}\lambda_i \neq 0$ and $\lambda_3=-1$)

The eigenvalues of $\mathbf{A}$ in Eq. (B.152) are determined by $\det(\mathbf{A}-\lambda\mathbf{I})=0$, i.e.,

$$\lambda^3+I_1\lambda^2+I_2\lambda+I_3=0 \tag{B.163}$$

where

$$\begin{aligned} I_1 &= \mathrm{tr}(\mathbf{A}) = a_{11}+a_{22}+a_{33},\\ I_2 &= a_{11}a_{22}+a_{22}a_{33}+a_{33}a_{11}-a_{12}a_{21}-a_{23}a_{32}-a_{13}a_{31},\\ I_3 &= \det(\mathbf{A}) = a_{11}a_{22}a_{33}+a_{12}a_{23}a_{31}+a_{13}a_{32}a_{21}\\ &\quad -a_{11}a_{32}a_{23}-a_{22}a_{13}a_{31}-a_{33}a_{12}a_{21}. \end{aligned} \tag{B.164}$$

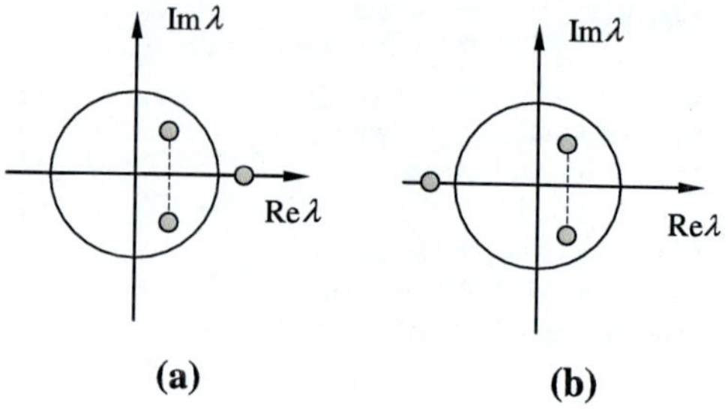

Fig. B.37 Eigenvalue diagrams: **a** $(\varnothing:[1,\varnothing]:\varnothing:\varnothing|1:\varnothing:\varnothing)$ spiral saddle of the fourth kind ($|\lambda_i|<1$ with $\mathrm{Im}\lambda_i\neq 0$ $(i=1,2)$ and $\lambda_3>1$), **b** $(\varnothing:[\varnothing,1]:\varnothing:\varnothing|1:\varnothing:\varnothing)$ spiral saddle of the fifth kind ($|\lambda_i|<1$ with $\mathrm{Im}\lambda_i\neq 0$ and $\lambda_3<-1$)

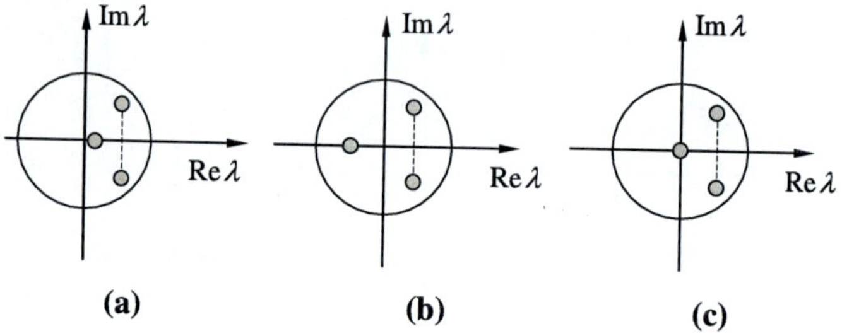

Fig. B.38 Eigenvalue diagrams: **a** $([1,\varnothing]:\varnothing:\varnothing:\varnothing|1:\varnothing:\varnothing)$ -spiral sink of the first kind ($|\lambda_i|<1$ with $\mathrm{Im}\lambda_i\neq 0$ $(i=1,2)$ and $\lambda_3\in(0,1)$),**b** $([\varnothing,1]:\varnothing:\varnothing:\varnothing|1:\varnothing:\varnothing)$ spiral sink of the second kind ($|\lambda_i|<1$ with $\mathrm{Im}\lambda_i\neq 0$ and $\lambda_3\in(-1,0)$) and **c** $([0,0]:\varnothing:\varnothing:\varnothing|1:\varnothing:\varnothing)$ spiral sink of the third kind ($|\lambda_i|<1$ with $\mathrm{Im}\lambda_i\neq 0$ and $\lambda_3=0$)

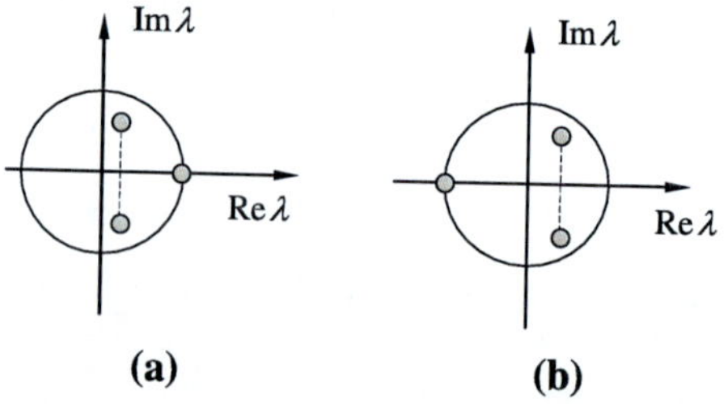

Fig. B.39 Eigenvalue diagrams: **a** $(\varnothing:\varnothing:1:\varnothing|1:\varnothing:\varnothing)$ spiral saddle-sink boundary of the first kind ($|\lambda_i|>1$ with $\mathrm{Im}\lambda_i\neq 0$ $(i=1,2)$ and $\lambda_3=1$), and **b** $(\varnothing:\varnothing:\varnothing:1|1:\varnothing:\varnothing)$ -spiral saddle-sink boundary of the second kind ($|\lambda_i|>1$ with $\mathrm{Im}\lambda_i\neq 0$ and $\lambda_3=-1$)

The corresponding eigenvalues are

$$\begin{aligned}
\lambda_1 &= \sqrt[3]{\Delta_1}+\sqrt[3]{\Delta_2},\\
\lambda_2 &= \omega_1\sqrt[3]{\Delta_1}+\omega_2\sqrt[3]{\Delta_2},\\
\lambda_3 &= \omega_2\sqrt[3]{\Delta_1}+\omega_1\sqrt[3]{\Delta_2}
\end{aligned} \tag{B.165}$$

where

$$
\begin{aligned}
\omega_1 &= \frac{-1 + \mathbf{i}\sqrt{3}}{2} \text{ and } \omega_2 = \frac{-1 - \mathbf{i}\sqrt{3}}{2}; \\
\Delta_{1,2} &= -\frac{q}{2} \pm \sqrt{\Delta} \text{ and } \Delta = (\frac{q}{2})^2 + (\frac{p}{3})^3; \\
p &= I_2 - \frac{1}{3}I_1^2 \text{ and } q = I_3 - \frac{1}{3}I_1 I_2 + \frac{2}{27}I_1^3.
\end{aligned}
\tag{B.166}
$$

The linear system in Eq. (B.152) possesses the following characteristics

(i) For $\Delta > 0$, the matrix $\mathbf{A}$ has one real eigenvalue and a pair of complex eigenvalues. The spiral sink, spiral source and a spiral-exponential attraction and expansion exist at the origin.
(ii) For $\Delta = 0$ and $p = q = 0$, the matrix $\mathbf{A}$ has three repeated eigenvalues. Stable and unstable nodes exist at the origin.
(iii) For $\Delta = 0$ and $q^2 = -4p^3/27 \neq 0$, the matrix $\mathbf{A}$ has two repeated eigenvalues. Sink, sources and saddle-node exist at the origin.
(iv) For $\Delta < 0$, the matrix $\mathbf{A}$ has three different eigenvales. Sink, sources and saddle-node exist at the origin.
(v) A degenerate fixed point at the origin for $\det(\mathbf{A}) = 0$. For the degenerate case, the readers can discuss the solutions as in 2-dimensional linear systems.

Index

A. C. J. Luo, *Regularity and Complexity in Dynamical Systems*,
DOI: 10.1007/978-1-4614-1524-4,